EVOLUTIONARY OPTIMIZATION ALGORITHMS

EVOLUTIONARY OPTIMIZATION ALGORITHMS

Biologically-Inspired and Population-Based Approaches to Computer Intelligence

Dan Simon
Cleveland State University

WILEY

Published by John Wiley & Sons, Inc., Hoboken, New Jersey

Published simultaneously in Canada

Library of Congress Cataloging-in-Publication Data is available.

ISBN 978-0-470-93741-9

10 9 8 7 6 5 4 3 2 1

SHORT TABLE OF CONTENTS

DETAILED TABLE OF CONTENTS

PART III MORE RECENT EVOLUTIONARY ALGORITHMS

PART V APPENDICES

ACKNOWLEDGMENTS

I would like to thank everyone who has financially supported my research in evolutionary optimization during the two decades that I have been involved in this fascinating area of study: Hossny El-Sherief at the TRW Systems Integration Group, Dorin Dragotoniu at TRW Vehicle Safety Systems, Sanjay Garg and Donald Simon at the NASA Glenn Controls and Dynamics Branch, Dimitar Filev at the Ford Motor Company, Brian Davis and William Smith at the Cleveland Clinic, the National Science Foundation, and Cleveland State University. I would also like to thank the students and colleagues who have worked with me and published papers in the area of evolutionary algorithms: Jeff Abell, Dawei Du, Mehmet Ergezer, Brent Gardner, Boris Igelnik, Paul Lozovyy, Haiping Ma, Berney Montavon, Mirela Ovreiu, Rick Rarick, Hanz Richter, David Sadey, Sergey Samorezov, Nina Scheidegger, Arpit Shah, Steve Szatmary, George Thomas, Oliver Tiber, Ton van den Bogert, Arun Venkatesan, and Tim Wilmot. Finally, I would like to thank those who read preliminary drafts of this material and made many helpful suggestions to improve it: Emile Aarts, Dan Ashlock, Forrest Bennett, Hans-Georg Beyer, Maurice Clerc, Carlos Coello Coello, Kalyanmoy Deb, Gusz Eiben, Jin-Kao Hao, Yaochu Jin, Pedro Larrañaga, Mahamed Omran, Kenneth V. Price, Hans-Paul Schwefel, Thomas Stützle, Hamid Tizhoosh, Darrell Whitley, and three anonymous reviewers of the original book proposal. These reviewers do not necessarily endorse this book, but it is much better because of their suggestions and comments.

Dan Simon

ACRONYMS

ABC	artificial bee colony
ACM	adaptive cultural model
ACO	ant colony optimization
ACR	acronym
ACS	ant colony system
ADF	automatically defined function
AFSA	artificial fish swarm algorithm
ANTS	approximated non-deterministic tree search
AS	ant system
ASCHEA	adaptive segregational constraint handling evolutionary algorithm
BBO	biogeography-based optimization
BFOA	bacterial foraging optimization algorithm
BMDA	bivariate marginal distribution algorithm
BOA	Bayesian optimization algorithm
CA	cultural algorithm
CAEP	cultural algorithm-influenced evolutionary programming
CE	cross entropy
CEC	Congress on Evolutionary Computation

CDF	cumulative distribution function
cGA	compact genetic algorithm
CMA-ES	covariance matrix adaptation evolution strategy
CMSA-ES	covariance matrix self-adaptive evolution strategy
COMIT	combining optimizers with mutual information trees
CX	cycle crossover
DACE	design and analysis of computer experiments
DAFHEA	dynamic approximate fitness based hybrid evolutionary algorithm
DE	differential evolution
DEMO	diversity evolutionary multi-objective optimizer
ϵ-MOEA	ϵ-based multi-objective evolutionary algorithm
EA	evolutionary algorithm
EBNA	estimation of Bayesian networks algorithm
ECGA	extended compact genetic algorithm
EDA	estimation of distribution algorithm
EGNA	estimation of Gaussian network algorithm
EMNA	estimation of multivariate normal algorithm
EP	evolutionary programming
ES	evolution strategy
FDA	factorized distribution algorithm
FIPS	fully informed particle swarm
FSM	finite state machine
GA	genetic algorithm
GENOCOP	genetic algorithm for numerical optimization of constrained problems
GOM	generalized other model
GP	genetic programming, or genetic program
GSA	gravitational search algorithm
GSO	group search optimizer
GSO	Glowworm search optimization
HBA	honey bee algorithm
hBOA	hierarchical Bayesian optimization algorithm
HCwL	hill climbing with learning
HLGA	Hajela-Lin genetic algorithm
HS	harmony search
HSI	habitat suitability index
IDEA	iterated density estimation algorithm

IDEA	infeasibility driven evolutionary algorithm
IUMDA	incremental univariate marginal distribution algorithm
MMAS	max-min ant system
MMES	multimembered evolution strategy
MIMIC	mutual information maximization for input clustering
MOBBO	multi-objective biogeography-based optimization
MOEA	multi-objective evolutionary algorithm
MOGA	multi-objective genetic algorithm
MOP	multi-objective optimization problem
MPM	marginal product model
$N(\mu, \sigma^2)$	normal PDF with a mean μ and a variance σ^2
$N(\mu, \Sigma)$	multidimensional normal PDF with mean μ and covariance Σ
NFL	no free lunch
NPBBO	niched Pareto biogeography-based optimization
NPGA	niched Pareto genetic algorithm
NPSO	negative reinforcement particle swarm optimization
NSBBO	nondominated sorting biogeography-based optimization
NSGA	nondominated sorting genetic algorithm
OBBO	oppositional biogeography-based optimization
OBL	opposition-based learning
OBX	order-based crossover
OX	order crossover
PAES	Pareto archived evolution strategy
PBIL	population based incremental learning
PDF	probability density function
PID	proportional integral derivative
PMBGA	probabilistic model-building genetic algorithm
PMX	partially matched crossover
PSO	particle swarm optimization
QED	quod erat demonstrandum: "that which was to be demonstrated"
RMS	root mean square
RV	random variable
SA	simulated annealing
SBX	simulated binary crossover
SAPF	self-adaptive penalty function
SCE	shuffled complex evolution
SEMO	simple evolutionary multi-objective optimizer

SFLA	shuffled frog leaping algorithm
SGA	stochastic gradient ascent
SHCLVND	stochastic hill climbing with learning by vectors of normal distributions
SIV	suitability index variable
SPBBO	strength Pareto biogeography-based optimization
SPEA	strength Pareto evolutionary algorithm
TLBO	teaching-learning-based optimization
TS	tabu search
TSP	traveling salesman problem
$U[a, b]$	uniform PDF that is nonzero on the domain $[a, b]$. This may refer to a continuous PDF or a discrete PDF depending on the context.
UMDA	univariate marginal distribution algorithm
UMDA_c^G	continuous Gaussian univariate marginal distribution algorithm
VEBBO	vector evaluated biogeography-based optimization
VEGA	vector evaluated genetic algorithm

LIST OF ALGORITHMS

Chapter 6: Evolution Strategies

Chapter 7: Genetic Programming

Chapter 8: Evolutionary Algorithm Variations

Chapter 9: Simulated Annealing

Chapter 10: Ant Colony Optimization

Chapter 11: Particle Swarm Optimization

Chapter 12: Differential Evolution

Chapter 18: Combinatorial Optimization

Chapter 19: Constrained Optimization

Chapter 20: Multi-Objective Optimization

Chapter 21: Expensive, Noisy, and Dynamic Fitness Functions

Appendix C: Benchmark Optimization Functions

PART I

INTRODUCTION TO EVOLUTIONARY OPTIMIZATION

CHAPTER 1

Introduction

> But ask the animals, and they will teach you, or the birds of the air, and they will tell you; or speak to the earth, and it will teach you, or let the fish of the sea inform you.
>
> —Job 12:7–9

This book discusses approaches to the solution of optimization problems. In particular, we[1] discuss evolutionary algorithms (EAs) for optimization. Although the book includes some mathematical theory, it should not be considered a mathematics text. It is more of an engineering or applied computer science text. The optimization approaches in this book are all given with the goal of eventual implementation in software. The aim of this book is to present evolutionary optimization algorithms in the most clear yet rigorous way possible, while also providing enough advanced material and references so that the reader is prepared to contribute new material to the state of the art.

[1]This book uses the common practice of referring to a generic third person with the word *we*. Sometimes, the book uses *we* to refer to the reader and the author. Other times, the book uses *we* to indicate that it is speaking on behalf of the general population of teachers and researchers in the areas of evolutionary algorithms and optimization. The distinction should be clear from the context. Do not read too much into the use of the word *we*; it is a matter of writing style rather than a claim to authority.

Overview of the Chapter

This chapter begins in Section 1.1 with an overview of the mathematical notation that we use in this book. The list of acronyms starting on page xxiii might also be useful to the reader. Section 1.2 gives some reasons why I decided to write this book about EAs, what I hope to accomplish with it, and why I think that it is distinctive in view of all of the other excellent EA books that are available. Section 1.3 discusses the prerequisites the are expected from a reader of this book. Section 1.4 discusses the philosophy of the homework assignments in this book, and the availability of the solution manual. Section 1.5 summarizes the mathematical notation that we use in this book. The reader is encouraged to regularly remember that section when encountering unfamiliar notation, and also to begin using it himself in homework assignments and in his own research. Section 1.6 gives a descriptive outline of the book. This leads into Section 1.7, which gives some important pointers to the instructor regarding some ways that he could teach a course from this book. That section also gives the instructor some advice about which chapters are more important than others.

1.1 TERMINOLOGY

Some authors use the term *evolutionary computing* to refer to EAs. This emphasizes the point that EAs are implemented in computers. However, evolutionary computing could refer to algorithms that are not used for optimization; for example, the first genetic algorithms (GAs) were not used for optimization per se, but were intended to study the process of natural selection (see Chapter 3). This book is geared towards evolutionary optimization algorithms, which are more specific than evolutionary computing.

Others use the term *population-based optimization* to refer to EAs. This emphasizes the point that EAs generally consist of a population of candidate solutions to some problem, and as time passes, the population evolves to a better solution to the problem. However, many EAs can consist of only a single candidate solution at each iteration (for example, hill climbing and evolution strategies). EAs are more general than population-based optimization because EAs include single-individual algorithms.

Some authors use the term *computer intelligence* or *computational intelligence* to refer to EAs. This is often done to distinguish EAs from expert systems, which have traditionally been referred to as *artificial intelligence.* Expert systems model deductive reasoning, while evolutionary algorithms model inductive reasoning. However, sometimes EAs are considered a type of artificial intelligence. Computer intelligence is a more general term than evolutionary algorithm, and includes technologies like neural networks, fuzzy systems, and artificial life. These technologies can be used for applications other than optimization. Therefore, depending on one's perspective, EAs might be more general or more specific than computer intelligence.

Soft computing is another term that is related to EAs. Soft computing is a contrast to hard computing. Hard computing refers to exact, precise, numerically rigorous calculations. Soft computing refers to less exact calculations, such as those that humans perform during their daily routines. Soft computing algorithms calculate generally good (but inexact) solutions to problems that are difficult, noisy, multimodal, and multi-objective. Therefore, EAs are a subset of soft computing.

Other authors use terms like *nature-inspired computing* or *bio-inspired computing* to refer to EAs. However, some EAs, like differential evolution and estimation of distribution algorithms, might not be motivated by nature. Other EAs, like evolution strategies and opposition-based learning, have a very weak connection with natural processes. EAs are more general than nature-inspired algorithms because EAs include non-biologically motivated algorithms.

Another oft-used term for EAs is *machine learning.* Machine learning is the study of computer algorithms that learn from experience. However, this field often includes many algorithms other than EAs. Machine learning is generally considered to be more broad than EAs, and includes fields such as reinforcement learning, neural networks, clustering, support vector machines, and others.

Some authors like to use the term *heuristic algorithms* to refer to EAs. *Heuristic* comes from the Greek word $\eta\upsilon\rho\iota\sigma\kappa\omega$, which is transliterated as *eurisko* in English. The word means *find* or *discover.* It is also the source of the English exclamation *eureka,* which we use to express triumph when we discover something or solve a problem. Heuristic algorithms are methods that use rules of thumb or common sense approaches to solve a problem. Heuristic algorithms usually are not expected to find the best answer to a problem, but are only expected to find solutions that are "close enough" to the best. The term *metaheuristic* is used to describe a family of heuristic algorithms. Most, if not all, of the EAs that we discuss in this book can be implemented in many different ways and with many different options and parameters. Therefore, they can all be called metaheuristics. For example, the family of all ant colony optimization algorithms can be called the ant colony metaheuristic.

Most authors separate EAs from swarm intelligence. A swarm intelligence algorithm is one that is based on swarms that occur in nature (for example, swarms of ants or birds). Ant colony optimization (Chapter 10) and particle swarm optimization (Chapter 11) are two prominent swarm algorithms, and many researchers insist that they should not be classified as EAs. However, some authors consider swarm intelligence as a subset of EAs. For example, one of the inventors of particle swarm optimization refers to it as an EA [Shi and Eberhart, 1999]. Since swarm intelligence algorithms execute in the same general way as EAs, that is, by evolving a population of candidate problem solutions which improve with each iteration, we consider swarm intelligence to be an EA.

Terminology is imprecise and context-dependent, but in this book we settle on the term *evolutionary algorithm* to refer to an algorithm that evolves a problem solution over many iterations. Typically, one iteration of an EA is called a *generation* in keeping with its biological foundation. However, this simple definition of an EA is not perfect because, for example, it implies that gradient descent is an EA, and no one is prepared to admit that. So the terminology in the EA field is not uniform and can be confusing. We use the tongue-in-cheek definition that an algorithm is an EA if it is generally considered to be an EA. This circularity is bothersome at first, but those of us who work in the field get used to it after a while. After all, natural selection is defined as the survival of the fittest, and fitness is defined as those who are most likely to survive.

1.2 WHY ANOTHER BOOK ON EVOLUTIONARY ALGORITHMS?

There are many fine books on EAs, which raises the question: Why yet another textbook on the topic of EAs? The reason that this book has been written is to offer a pedagogical approach, perspective, and material, that is not available in any other single book. In particular, the hope is that this book will offer the following:

- A straightforward, bottom-up approach that assists the reader in obtaining a clear but theoretically rigorous understanding of EAs is given in the book. Many books discuss a variety of EAs as cookbook algorithms without any theoretical support. Other books read more like research monographs than textbooks, and are not entirely accessible to the average engineering student. This book tries to strike a balance by presenting easy-to-implement algorithms, along with some rigorous theory and discussion of trade-offs.

- Simple examples that provide the reader with an intuitive understanding of EA math, equations, and theory, are given in the book. Many books present EA theory, and then give examples or problems that are not amenable to an intuitive understanding. However, it is possible to present simple examples and problems that require only paper and pencil to solve. These simple problems allow the student to more directly see how the theory works itself out in practice.

- MATLAB®-based source code for all of the examples in the book is available at the author's web site.[2] A number of other texts supply source code, but it is often incomplete or outdated, which is frustrating for the reader. The author's email address is also available on the web site, and I enthusiastically welcome feedback, comments, suggestions for improvements, and corrections. Of course, web addresses are subject to obsolescence, but this book contains algorithmic, high-level pseudocode listings that are more permanent than any specific software listings. Note that the examples and the MATLAB code are not intended as efficient or competitive optimization algorithms; they are instead intended only to allow the reader to gain a basic understanding of the underlying concepts. Any serious research or application should rely on the sample code only as a preliminary starting point.

- This book includes theory and recently-developed EAs that are not available in most other textbooks. These topics include Markov theory models of EAs, dynamic system models of EAs, artificial bee colony algorithms, biogeography-based optimization, opposition-based learning, artificial fish swarm algorithms, shuffled frog leaping, bacterial foraging optimization, and many others. These topics are recent additions to the state of the art, and their coverage in this book is not matched in any other books. However, this book is not intended to survey the state-of-the-art in any particular area of EA research. This book is instead intended to provide a high-level overview of many areas of EA research so that the reader can gain a broad understanding of EAs, and so that the reader can be well-positioned to pursue additional studies in the state-of-the-art.

[2]See `http://academic.csuohio.edu/simond/EvolutionaryOptimization` – if the address changes, it should be easy to find with an internet search.

1.3 PREREQUISITES

In general, a student will not gain anything from a course like this without writing his own EA software. Therefore, competent programming skills could be listed as a prerequisite. At the university where I teach this course to electrical and computer engineering students, there are no specific course prerequisites; the prerequisite for undergraduates is senior standing, and there are no prerequisites for graduate students. However, I assume that undergraduates at the senior level, and graduate students, are good programmers.

The notation used in the book assumes that the reader is familiar with the standard mathematical notations that are used in algebra, geometry, set theory, and calculus. Therefore, another prerequisite for understanding this book is a level of mathematical maturity that is typical of an advanced senior undergraduate student. The mathematical notation is described in Section 1.5. If the reader can understand the notation described in that section, then there is a good chance that he will also be able to follow the discussion in the rest of the book.

The mathematics in the theoretical sections of this book (Chapter 4, Section 7.6, much of Chapter 13, and a few other scattered sections) require an understanding of probability and linear systems theory. It will be difficult for a student to follow that material unless he has had a graduate course in those two subjects. A course geared towards undergraduates should probably skip that material.

1.4 HOMEWORK PROBLEMS

The problems at the end of each chapter have been written to give flexibility to the instructor and student. The problems include written exercises and computer exercises. The written exercises are intended to strengthen the student's grasp of the theory, deepen the student's intuitive understanding of the concepts, and develop the student's analytical skills. The computer exercises are intended to help the student develop research skills, and learn how to apply the theory to the types of problems that are typically encountered in industry. Both types of problems are important for gaining proficiency with EAs. The distinction between written exercises and computer exercises is not strict but is more of a fuzzy division. That is, some of the written exercises might require some computer work, and the computer exercises require some analysis. The instructor might have EA-related assignments in mind based on his own interests. Semester-length, project-based assignments are often instructive for topics such as this. For example, students could be assigned to solve some practical optimization problem using the EAs discussed in this book, applying one EA per chapter, and then comparing the performance of the EAs and their variations at the end of the semester.

A solution manual to all of the problems in the text (both written exercises and computer exercises) is available from the publisher for instructors. Course instructors are encouraged to contact the publisher for further information about how to obtain the solution manual. In order to protect the integrity of the homework assignments, the solution manual will be provided only to course instructors.

1.5 NOTATION

Unfortunately, the English language does not have a gender-neutral, singular, third-person pronoun. Therefore, we use the term *he* or *him* to refer to a generic third person, whether male or female. This convention can feel awkward to both writers and readers, but it seems to be the most satisfactory resolution to a difficult solution.

The list below describes some of the mathematical notation in this book.

- $x \leftarrow y$ is a computational notation that indicates that y is assigned to the variable x. For example, consider the following algorithm:

$$\begin{aligned} a &= \text{coefficient of } x^2 \\ b &= \text{coefficient of } x^1 \\ c &= \text{coefficient of } x^0 \\ x^* &\leftarrow (-b + \sqrt{b^2 - 4ac})/(2a) \end{aligned}$$

  The first three lines are not assignment statements in the algorithm; they simply describe or define the values of a, b, and c. These three parameters could have been set by the user, or by some other algorithm or process. The last line, however, is an assignment statement that indicates the value on the right side of the arrow is written to x^*.

- $df(\cdot)/dx$ is the total derivative of $f(\cdot)$ with respect to x. For example, suppose that $y = 2x$ and $f(x, y) = 2x + 3y$. Then $f(x, y) = 8x$ and $df(\cdot)/dx = 8$.

- $f_x(\cdot)$, also denoted as $\partial f(\cdot)/\partial x$, is the partial derivative of $f(\cdot)$ with respect to x. For example, suppose again that $y = 2x$ and $f(x, y) = 2x + 3y$. Then $f_x(x, y) = 2$.

- $\{x : x \in S\}$ is the set of all x such that x belongs to the set S. A similar notation is used to denote those values of x that satisfy any other particular condition. For example, $\{x : x^2 = 4\}$ is the same as $\{x : x \in \{-2, +2\}\}$, which is the same as $\{-2, +2\}$.

- $[a, b]$ is the closed interval between a and b, which means $\{x : a \leq x \leq b\}$. This might be a set of integers or a set of real numbers, depending on the context.

- (a, b) is the open interval between a and b, which is $\{x : a < x < b\}$. This might be a set of integers or a set of real numbers, depending on the context.

- If is it understood from the context that $i \in S$, the $\{x_i\}$ is shorthand for $\{x_i : i \in S\}$. For example, if $i \in [1, N]$, then $\{x_i\} = \{x_1, x_2, \cdots, x_N\}$.

- $S_1 \cup S_2$ is the set of all x such that x belongs to either set S_1 or set S_2. For example, if $S_1 = \{1, 2, 3\}$ and $S_2 = \{7, 8\}$, then $S_1 \cup S_2 = \{1, 2, 3, 7, 8\}$.

- $|S|$ is the number of elements in the set S. For example, if $S = \{i : i \in [4, 8]\}$, then $|S| = 5$. If $S = \{3, 19, \pi, \sqrt{2}\}$, then $|S| = 4$. If $S = \{\alpha : 1 < \alpha < 3\}$, then $|S| = \infty$.

- $\emptyset$ is the empty set. $|\emptyset| = 0$.

- $x \bmod y$ is the remainder after x is divided by y. For example, $8 \bmod 3 = 2$.

- $\lceil x \rceil$ is the ceiling of x; that is, the smallest integer that is greater than or equal to x. For example, $\lceil 3.9 \rceil = 4$, and $\lceil 5 \rceil = 5$.

- $\lfloor x \rfloor$ is the floor of x; that is, the largest integer that is less than or equal to x For example, $\lfloor 3.9 \rfloor = 3$, and $\lfloor 5 \rfloor = 5$.

- $\min_x f(x)$ indicates the problem of finding the value of x that gives the smallest value of $f(x)$. Also, it can indicate the smallest value of $f(x)$. For example, suppose that $f(x) = (x-1)^2$. Then we can solve the problem $\min_x f(x)$ using calculus or by graphing the function $f(x)$ and visually noting the smallest value of $f(x)$. We find for this example that $\min_x f(x) = 0$. A similar definition holds for $\max_x f(x)$.

- $\arg\min_x f(x)$ is the value of x that results in the smallest value of $f(x)$. For example, suppose again that $f(x) = (x-1)^2$. The smallest value of $f(x)$ is 0, which occurs when $x = 1$, so for this example $\arg\min_x f(x) = 1$. A similar definition holds for $\arg\max_x f(x)$.

- R^s is the set of all real s-element vectors. It may indicate either column vectors or row vectors, depending on the context.

- $R^{s \times p}$ is the set of all real $s \times p$ matrices.

- $\{y_k\}_{k=L}^{U}$ is the set of all y_k, where the integer k ranges from L to U. For example, $\{y_k\}_{k=2}^{5} = \{y_2, y_3, y_4, y_5\}$.

- $\{y_k\}$ is the set of all y_k, where the integer k ranges from a context-dependent lower limit to a context-dependent upper limit. For example, suppose the context indicates that there are three values: y_1, y_2, and y_3. Then $\{y_k\} = \{y_1, y_2, y_3\}$.

- $\exists$ means "there exists," and $\nexists$ means "there does not exist." For example, if $Y = \{6, 1, 9\}$, then $\exists y < 2 : y \in Y$. However, $\nexists y > 10 : y \in Y$.

- $A \implies B$ means that A implies B. For example, $(x > 10) \implies (x > 5)$.

- I is the identity matrix. Its dimensions depend on the context.

See the list of acronyms on page xxiii for more notation.

1.6 OUTLINE OF THE BOOK

This book is divided into six parts.

1. Part I consists of this introduction, and one more chapter that covers introductory material related to optimization. It introduces different types of optimization problems, the simple-but-effective hill climbing algorithm, and concludes with a discussion about what makes an algorithm intelligent.

2. Part II discusses the four EAs that are commonly considered to be the classics:

    - Genetic algorithms;
    - Evolutionary programming;
    - Evolution strategies;
    - Genetic programming.

    Part II also includes a chapter that discusses approaches for the mathematical analysis of GAs. Part II concludes with a chapter that discusses some of the many algorithmic variations that can be used in these classic algorithms. These same variations can also be used in the more recent EAs, which are covered in the next part.

3. Part III discusses some of the more recent EAs. Some of these are not really that recent, dating back to the 1980s, but others date back only to the first decade of the 21st century.

4. Part IV discusses special types of optimization problems, and shows how the EAs of the earlier chapters can be modified to solve them. These special types of problems include:

    - Combinatorial problems, whose domain consists of integers;
    - Constrained problems, whose domain is restricted to a known set;
    - Multi-objective problems, in which it is desired to minimize more than one objective simultaneously; and
    - Problems with noisy or expensive fitness functions for which it is difficult to precisely obtain the performance of a candidate solution, or for which it is computationally expensive to evaluate the performance of a candidate solution.

5. Part V includes several appendices that discuss topics that are important or interesting.

    - Appendix A offers some miscellaneous, practical advice for the EA student and researcher.
    - Appendix B discusses the no-free-lunch theorem, which tells us that, on average, all optimization algorithms perform the same. It also discusses how statistics should be used to evaluate the differences between EAs.
    - Appendix C gives some standard benchmark functions that can be used to compare the performance of different EAs.

1.7 A COURSE BASED ON THIS BOOK

Any course based on this book should start with Chapters 1 and 2, which give an overview of optimization problems. From that point on, the remaining chapters can be studied in almost any order, depending on the preference and interests of the instructor. The obvious exceptions are that the study of genetic algorithms (Chapter 3) should precede the study of their mathematical models (Chapter 4).

Also, at least one chapter in Parts II or III (that is, at least one specific EA) needs to be covered in detail before any of the chapters in Part IV.

Most courses will, at a minimum, cover Chapters 3 and 5–7 to give the student a background in the classic EAs. If the students have sufficient mathematical sophistication, and if there is time, then the course should also include Chapter 4 somewhere along the line. Chapter 4 is important for graduate students because it helps them see that EAs are not only a qualitative subject, but there can and should be some theoretical basis for them also. Too much EA research today is based on minor algorithmic adjustments without any mathematical support. Many EA practitioners only care about getting results, which is fine, but academic researchers need to be involved in theory as well as practice.

The chapters in Parts III and IV can be covered on the basis of the instructor's or the students' specific interests.

The appendices are not included in the main part of the book because they are not about EAs per se, but the importance of the appendices should not be underestimated. In particular, the material in Appendices B and C are of critical importance and should be included in every EA course. I recommend that these two appendices be discussed in some detail immediately after the first chapter in Parts II or III.

Putting the above advice together, here is a proposed outline for a one-semester graduate course.

- Chapters 1 and 2.
- Chapter 3.
- Appendices B and C.
- Chapters 4–8. I recommend skipping Chapter 4 for most undergraduate students and for short courses.
- A few chapters in Part III, based on the instructor's preference. At the risk of starting an "EA war" with my readers, I will go out on a limb and claim that ACO, PSO, and DE are among the most important "other" EAs, and so the instructor should cover Chapters 10–12 at a minimum.
- A few chapters in Part IV, based on the instructor's preference and the available time.

CHAPTER 2

Optimization

Optimization saturates what we do and drives almost every aspect of engineering.
—Dennis Bernstein [Bernstein, 2006]

As indicated by the above quote, optimization is a part of almost everything that we do. Personnel schedules need to be optimized, teaching styles need to be optimized, economic systems need to be optimized, game strategies need to be optimized, biological systems need to be optimized, and health care systems need to be optimized. Optimization is a fascinating area of study not only because of its algorithmic and theoretical content, but also because of its universal applicability.

Overview of the Chapter

This chapter gives a brief overview of optimization (Section 2.1), including optimization subject to constraints (Section 2.2), optimization problems that have multiple goals (Section 2.3), and optimization problems that have multiple solutions (Section 2.4). Most of our work in this book is focused on continuous optimization problems, that is, problems where the independent variable can vary continuously. However, problems where the independent variable is restricted to a finite set, which are called combinatorial problems, are also of great interest, and we introduce them in Section 2.5. We present a simple, general-purpose optimization algorithm called hill climbing in Section 2.6, and we also discuss some of its variations. Finally, we discuss a few concepts related to the nature of intelligence in Section 2.7, and show how they relate to the evolutionary optimization algorithms that we present in the later chapters.

2.1 UNCONSTRAINED OPTIMIZATION

Optimization is applicable to virtually all areas of life. Optimization algorithms can be applied to everything from aardvark breeding to zygote research. The possible applications of EAs are limited only by the engineer's imagination, which is why EAs have become so widely researched and applied in the past few decades.

For example, in engineering, EAs are used to find the best robot trajectories for a certain task. Suppose that you have a robot in your manufacturing plant, and you want it to perform its task in such a way that it finishes as quickly as possible, or uses the least power possible. How can you figure out the best possible path for the robot? There are so many possible paths that finding the best solution is a daunting task. But EAs can make the task manageable (if not exactly easy), and at least find a good solution (if not exactly the best). Robots are very nonlinear, so the search space for robotic optimization problems ends up with lots of peaks and valleys, like what we saw in the simple example presented earlier in this chapter. But with robotics problems the situation is even worse, because the peaks and valleys lie in a multidimensional space (instead of the simple three-dimensional space that we saw earlier). The complexity of robotics optimization problems makes them a natural target for EAs. This idea has been applied for both stationary robots (i.e., robot arms) and mobile robots.

EAs have also been used to train neural networks and fuzzy logic systems. In neural networks we have to figure out the network architecture and the neuron weights in order to get the best possible network performance. Again, there are so many possibilities that the task is formidable. But EAs can be used to find the best configuration and the best weights. We have the same issue with fuzzy logic systems. What rule base should we use? How many membership functions should we use? What membership function shapes should we use? GAs can (and have) helped solve these difficult optimization problems.

EAs have also been used for medical diagnosis. For example, after a biopsy, how do medical professionals recognize which cells are cancerous and which are not? What features should they look for in order to diagnose cancer? Which features are the most important, and which ones are irrelevant? Which features are important only if the patient belongs to a certain demographic? A EA can help make these types of decisions. The EA always needs a professional to get it started and to train it, but after that the EA can actually outperform its teacher. Not only does the EA never get tired or fatigued, but it can extract patterns from data that may be too subtle for humans to recognize. EAs have been used for the diagnosis of several different types of cancer.

After a disease has been diagnosed, the next difficult question involves the management of the disease. For example, after cancer has been detected, what is the best treatment for the patient? How often should radiation be administered, what kind of radiation should it be, and in what doses? How should side effects be treated? This is another complex optimization problem. The wrong kind of treatment can do more harm than good. The determination of the right kind of treatment is a complicated function of things like cancer type, cancer location, patient demographics, general health, and other factors. So GAs are used to not only diagnose illness, but also to plan treatment.

EAs should be considered any time that you want to solve a difficult problem. That does not mean that EAs are always the best choice for the job. Calculators don't use EA software to add numbers, because there are much simpler and more effective algorithms available. But EAs should at least be considered for any complex problem. If you want to design a housing project or a transportation system, an EA might be the answer. If you want to design a complex electrical circuit or a computer program, an EA might be able to do the job.

An optimization problem can be written as a minimization problem or as a maximization problem. Sometimes we try to minimize a function and sometimes we try to maximize a function. These two problems are easily converted to the other form:

$$\begin{aligned} \min_x f(x) &\iff \max_x[-f(x)] \\ \max_x f(x) &\iff \min_x[-f(x)]. \end{aligned} \tag{2.1}$$

The function $f(x)$ is called the objective function, and the vector x is called the independent variable, or decision variable. Note that the terms independent variable and decision variable sometimes refer to the entire vector x, and sometimes refer to specific elements in x, depending on the context. Elements of x are also called solution features. The number of elements in x is called the dimension of the problem. As we see from Equation (2.1), any algorithm that is designed to minimize a function can easily be used to maximize a function, and any algorithm that is designed to maximize a function can easily be used to minimize a function. When we try to minimize a function, we call the function value the cost function. When we try to maximize a function, we call the function value the fitness.

$$\begin{aligned} \min_x f(x) &\Rightarrow f(x) \text{ is called "cost" or "objective"} \\ \max_x f(x) &\Rightarrow f(x) \text{ is called "fitness" or "objective."} \end{aligned} \tag{2.2}$$

■ EXAMPLE 2.1

This example illustrates the terminology that we use in this book. Suppose that we want to minimize the function

$$f(x, y, z) = (x-1)^2 + (y+2)^2 + (z-5)^2 + 3. \tag{2.3}$$

The variables x, y, and z are called the independent variables, the decision variables, or the solution features; all three terms are equivalent. This is a three-dimensional problem. $f(x, y, z)$ is called the objective function or the cost function. We can change the problem to a maximization problem by defining $g(x, y, z) = -f(x, y, z)$ and trying to maximize $g(x, y, z)$. The function $g(x, y, z)$ is called the objective function or the fitness function. The solution to the problem $\min f(x, y, z)$ is the same as the solution to the problem $\max g(x, y, z)$, and is $x = 1$, $y = -2$, and $z = 5$. However, the optimal value of $f(x, y, z)$ is the negative of the optimal value of $g(x, y, z)$.

□

Sometimes optimization is easy and can be accomplished using analytical methods, as we see in the following example.

■ EXAMPLE 2.2

Consider the problem

$$\min_x f(x), \text{ where } f(x) = x^4 + 5x^3 + 4x^2 - 4x + 1. \tag{2.4}$$

A plot of $f(x)$ is shown in Figure 2.1. Since $f(x)$ is a quartic polynomial (also called fourth order or fourth degree), we know that it has at most three stationary points, that is, three values of x at which its derivative $f'(x) = 0$. These points are seen from Figure 2.1 to occur at $x = -2.96$, $x = -1.10$, and $x = 0.31$. We can confirm that $f'(x)$, which is equal to $4x^3 + 15x^2 + 8x - 4$, is zero at these three values of x. We can further find that the second derivative of $f(x)$ at these three points is

$$f''(x) = 12x^2 + 30x + 8 = \begin{cases} 24.33, & x = -2.96 \\ -10.48, & x = -1.10 \\ 18.45, & x = 0.31 \end{cases} \tag{2.5}$$

Recall that the second derivative of a function at a local minimum is positive, and the second derivative of a function at a local maximum is negative. The values of $f''(x)$ at the stationary points therefore confirm that $x = -2.96$ is a local minimum, $x = -1.10$ is a local maximum, and $x = 0.31$ is another local minimum.

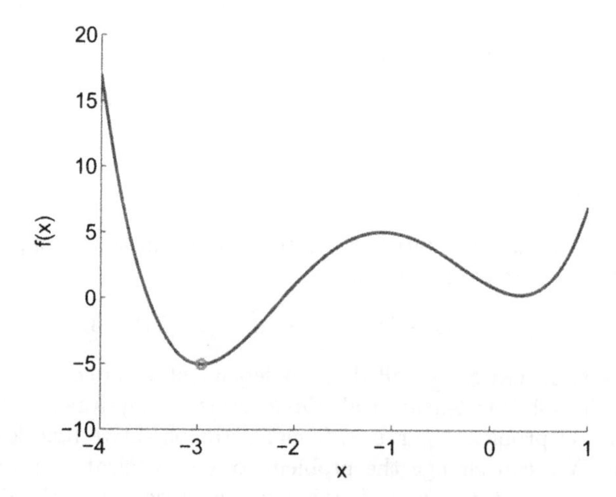

Figure 2.1 Example 2.2: A simple minimization problem. $f(x)$ has two local minima and one global minimum, which occurs at $x = -2.96$.

□

The function of Example 2.2 has two local minima and one global minimum. Note that the global minimum is also a local minimum. For some functions $\min_x f(x)$ occurs at more than one value of x; if that occurs then $f(x)$ has multiple global minima. A local minimum x^* can be defined as

$$f(x^*) < f(x) \text{ for all } x \text{ such that } ||x - x^*|| < \epsilon \tag{2.6}$$

where $|| \cdot ||$ is some distance metric, and $\epsilon > 0$ is some user-defined neighborhood size. In Figure 2.1 we see that $x = 0.31$ is a local optimum if, for example, the neighborhood size $\epsilon = 1$, but it is not a local optimum if $\epsilon = 4$. A global minimum x^* can be defined as

$$f(x^*) \leq f(x) \text{ for all } x. \tag{2.7}$$

2.2 CONSTRAINED OPTIMIZATION

Many times an optimization problem is constrained. That is, we are presented with the problem of minimizing some function $f(x)$ with restrictions on the allowable values of x, as in the following example.

■ EXAMPLE 2.3

Consider the problem

$$\begin{aligned} \min_x f(x) \quad & \text{where} \quad f(x) = x^4 + 5x^3 + 4x^2 - 4x + 1 \\ & \text{and} \quad x \geq -1.5. \end{aligned} \tag{2.8}$$

This is the same problem as that in Example 2.2 except that x is constrained. A plot of $f(x)$ and the allowable values of x are shown in Figure 2.2, and an examination of the plot reveals the constrained minimum. To solve this problem analytically, we find the three stationary points of $f(x)$ as in Example 2.2 while ignoring the constraint. We find that the two local minima occur at $x = -2.96$ and $x = 0.31$, as in Example 2.2. We see that only one of these values, $x = 0.31$, satisfies the constraint. Next we must evaluate $f(x)$ on the constraint boundary to see if it is smaller than at the local minimum $x = 0.31$. We find that

$$f(x) = \begin{cases} 4.19 & \text{for } x = \ -1.50 \\ 0.30 & \text{for } x = \ \ 0.31 \end{cases}. \tag{2.9}$$

We see that $x = 0.31$ is the minimizing value of x for the constrained minimization problem.

If the constraint boundary were farther to the left, then the minimizing value of x would occur at the constraint boundary rather than at the local minimum $x = 0.31$. If the constraint boundary were left of $x = -2.96$, then the minimizing value of x for the constrained minimization problem would be the same as that for the unconstrained minimization problem.

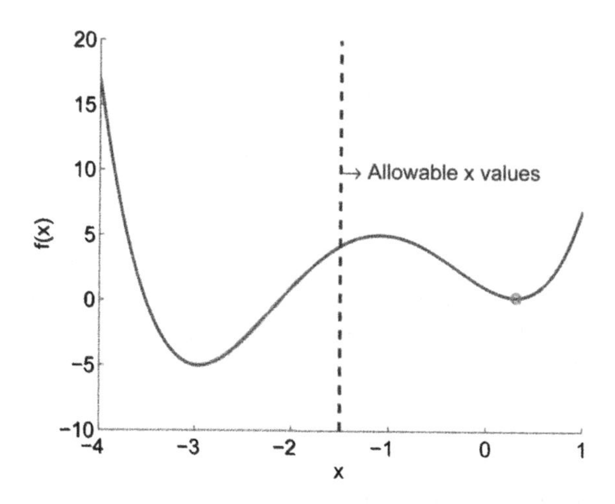

Figure 2.2 Example 2.3: A simple constrained minimization problem. The constrained minimum occurs at $x = 0.31$.

□

Real-world optimization problems almost always have constraints. Also, the optimizing value of the independent variable almost always occurs on the constraint boundary in real-world optimization problems. This is not surprising because we normally expect to obtain the best engineering design, allocation of resources, or other optimization goal, by using all of the available energy, or force, or some other resource [Bernstein, 2006]. Constraints are therefore important in almost all real-world optimization problems. Chapter 19 presents a more detailed discussion of constrained evolutionary optimization.

2.3 MULTI-OBJECTIVE OPTIMIZATION

Not only are real-world optimization problems constrained, but they are also multi-objective. This means that we are interested in minimizing more than one measure simultaneously. For example, in a motor control problem we might be interested in minimizing tracking error while also minimizing power consumption. We could get a very small tracking error at the expense of high power consumption, or we could allow a large tracking error while using very little power. In the extreme case, we could turn off the motor to achieve zero power consumption, but then our tracking error would not be very good.

■ EXAMPLE 2.4

Consider the problem

$$\min_x [f(x) \text{ and } g(x)], \quad \text{where} \quad f(x) = x^4 + 5x^3 + 4x^2 - 4x + 1$$
$$\text{and} \quad g(x) = 2(x+1)^2. \tag{2.10}$$

The first minimization objective, $f(x)$, is the same as that in Example 2.2. But now we also want to minimize $g(x)$. A plot of $f(x)$, $g(x)$, and their minima, are shown in Figure 2.3. An examination of the plot reveals that $x = -2.96$ minimizes $f(x)$, while $x = -1$ minimizes $g(x)$. It is not clear what the most preferable value of x would be for this problem because we have two conflicting objectives. However, it should be obvious from Figure 2.3 that we would never want $x < -2.96$ or $x > 0.31$. If x decreases from -2.96, or increases from 0.31, objectives $f(x)$ and $g(x)$ both increase, which is clearly undesirable.

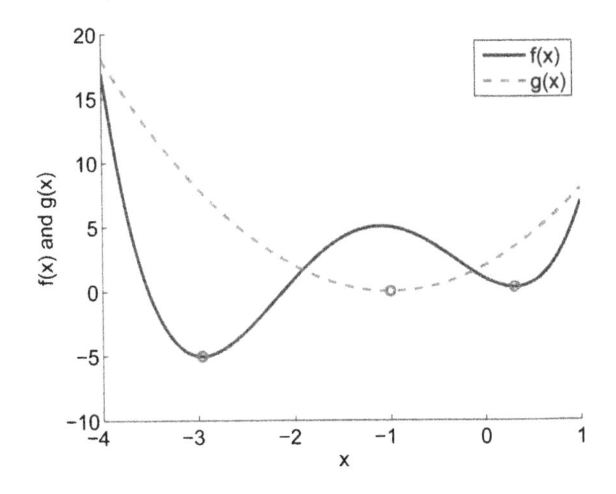

Figure 2.3 Example 2.4: A simple multi-objective minimization problem. $f(x)$ has two minima and $g(x)$ has one minimum. The two objectives conflict.

One way to evaluate this problem is to plot $g(x)$ as a function of $f(x)$. This is shown in Figure 2.4, where we have varied x from -3.4 to 0.8. Below we discuss each section of the plot.

- $x \in [-3.4, -2.96]$: As x increases from -3.4 to -2.96, both $f(x)$ and $g(x)$ decrease. Therefore, we will never choose $x < -2.96$.

- $x \in [-2.96, -1]$: As x increases from -2.96 to -1, $g(x)$ decreases while $f(x)$ increases.

- $x \in [-1, 0]$: As x increases from -1 to 0, $g(x)$ increases while $f(x)$ decreases. However, on this part of the plot, even though $g(x)$ is increasing, it is still less than $g(x)$ for $x \in [-2, -1]$. Therefore, $x \in [-1, 0]$ is preferable to $x \in [-2, -1]$.

- $x \in [0, 0.31]$: As x increases from 0 to 0.31, $g(x)$ increases while $f(x)$ decreases. We see from the plot that for $x \in [0, 0.31]$, $g(x)$ is greater than it is on the $x \in [-2.96, -2]$ part of the plot. Therefore, we will never want to choose $x \in [0, 0.31]$.

- $x \in [0.31, 0.8]$: Finally, as x increases from 0.31 to 0.8, both $f(x)$ and $g(x)$ increase. Therefore, we will never choose $x > 0.31$.

Summarizing the above results, we graph potentially desirable values of $f(x)$ and $g(x)$ with the solid line in Figure 2.4. For values of x on the solid line, we cannot find any other x values that will simultaneously decrease both $f(x)$ and $g(x)$. The solid line is called the Pareto front, and the corresponding set of x values is called the Pareto set.

$$\begin{aligned} \text{Pareto set:} \quad & x^* = \{x : x \in [-2.96, -2] \text{ or } x \in [-1, 0]\} \\ \text{Pareto front:} \quad & \{(f(x), g(x)) : x \in x^*\}. \end{aligned} \tag{2.11}$$

After we obtain the Pareto set, we cannot say anything more about the optimal value of x. It is a question of engineering judgment to select a point along the Pareto front as the ultimate solution. The Pareto front gives a set of reasonable choices, but any choice of x from the Pareto set still entails a tradeoff between the two objectives.

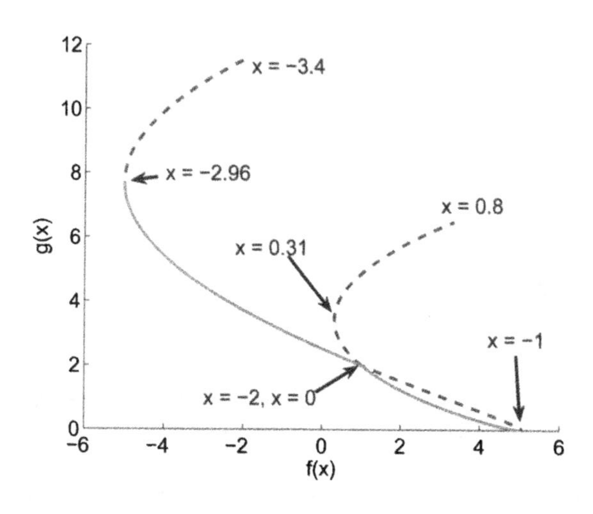

Figure 2.4 Example 2.4: This figure shows $g(x)$ as a function of $f(x)$ for a simple multi-objective minimization problem as x varies from -3.4 to 0.8. The solid line is the Pareto front.

□

Example 2.4 is a fairly simple multi-objective optimization problem with only two objectives. A typical real-world optimization problem involves many more than just two objectives, and so its Pareto front is difficult to obtain. Even if we could obtain the Pareto front, we would not be able to visualize it because of its high dimensionality. Chapter 20 presents a more detailed discussion of evolutionary multi-objective optimization.

2.4 MULTIMODAL OPTIMIZATION

A multimodal optimization problem is a problem that has more than one local minimum. We saw an example of a multimodal problem in Figure 2.1, but there were only two local minimum in that problem so it was fairly easy to handle. Some problems, however, have many local minima and it can be challenging to discover which minimum is the global minimum.

■ **EXAMPLE 2.5**

Consider the problem

$$\min_{x,y} f(x,y), \text{ where} \tag{2.12}$$

$$f(x,y) = e - 20\exp\left(-0.2\sqrt{\frac{x^2+y^2}{2}}\right) - \exp\left(\frac{\cos(2\pi x)+\cos(2\pi y)}{2}\right).$$

This is the two-dimensional Ackley function. It is plotted in Figure 2.5 and is defined in Appendix C.1.2. A plot like Figure 2.5 is often called a fitness landscape because it graphically illustrates how a fitness or cost function varies with independent variables. We cannot illustrate a plot like Figure 2.5 in more than two dimensions, but even if we have a problem with more than two dimensions, fitness or cost as a function of independent variables is still called a fitness landscape. Figure 2.5 shows that even in two dimensions, the Ackley function has many local minima. Imagine how many minima it has in 20 or 30 dimensions. We could attack this problem by taking the derivative of $f(x,y)$ with respect to x and y, and then solving for $f_x(x,y) = f_y(x,y) = 0$ to find the local minima. However, solving these simultaneous equations could be difficult.

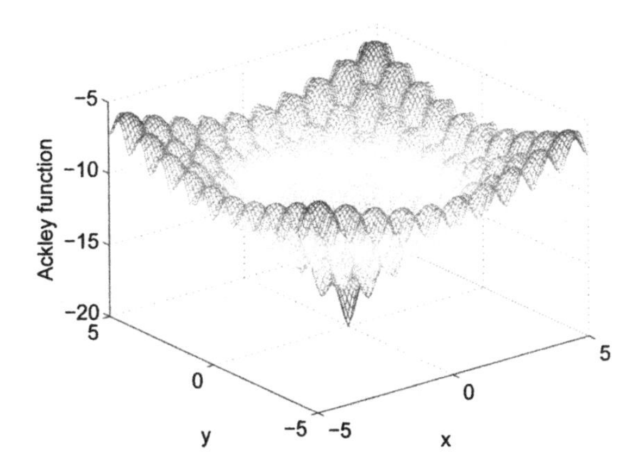

Figure 2.5 Example 2.5: The two-dimension Ackley function.

□

The previous example shows why EAs are useful. We were able to solve Examples 2.2, 2.3, and 2.4 using graphical methods or calculus, but many real-world problems are more like Example 2.5 except with more independent variables, with multiple objectives, and with constraints. With these types of problems, methods based on calculus or graphics fall short, and EAs can give better results.

2.5 COMBINATORIAL OPTIMIZATION

Up until now we have considered continuous optimization problems; that is, the independent variables have been allowed to vary continuously. However, there are many optimization problems for which the independent variables are restricted to a set of discrete values. These types of problems are called combinatorial optimization problems.

■ EXAMPLE 2.6

Suppose a business person wants to visit four branch offices, starting and ending at his home office. The home office is in city A, and the branch offices are in cities B, C, and D. The business person wants to visit the branch offices in the order that minimizes his total travel distance. There are six possible solutions S_i to this problem:

$$\begin{aligned}
S_1 &: \quad A \to B \to C \to D \to A \\
S_2 &: \quad A \to B \to D \to C \to A \\
S_3 &: \quad A \to C \to B \to D \to A \\
S_4 &: \quad A \to C \to D \to B \to A \\
S_5 &: \quad A \to D \to B \to C \to A \\
S_6 &: \quad A \to D \to C \to B \to A.
\end{aligned} \tag{2.13}$$

It is easy for the business person to solve this problem by calculating the total distance for each of the six possible solutions.[1]

□

The problem of Example 2.6 is called the closed traveling salesman problem (TSP).[2] We can easily enumerate all of the possible solutions for a four-city TSP. Searching through all possible solutions of a combinatorial problem is called a brute-force search, or exhaustive search. If you have time to do that, then it is often the best way to solve a combinatorial problem because it guarantees a solution.

But how many possible solutions exist for a general $n-$city TSP? A little thought shows that there are $(n-1)!$ possible solutions. This number grows very rapidly, and for modest values of n it is not possible to calculate all possible solutions. Suppose the business person needs to visit one city in each of the 50 states in the USA. The number of possible solutions is $49! = 6.1 \times 10^{62}$. Could modern computers

[1] Actually, the problem is simpler than it seems at first. S_1 is the reverse of S_6, and so S_1 and S_6 have the same total distance. The same can be said for S_2 and S_4, and S_3 and S_5.

[2] The open TSP is the problem of visiting all cities exactly once without returning to the starting city.

calculate this number of possible solutions? The universe is about 15 billion years old, which is 4.7×10^{17} seconds. Suppose a trillion computers were running since the beginning of the universe, and suppose each of those trillion computers could calculate the distance for a trillion possible solutions every second. Then we would have calculated the distance for a total of 4.7×10^{41} possible solutions. We would not even have scratched the surface of solving the 50-city TSP.

Here is another way of looking at the complexity of the TSP. There are somewhere between 10^{20} and 10^{24} grains of sand on earth [Welland, 2009]. If each grain of sand on earth were an earth-like planet with the same amount of sand that earth has, then the number of possible 50-city TSP routes would still be much greater than the total number of mini-grains of sand. Obviously it is impossible to do a brute force search for the solution of such a large problem.

We see that some problems are so large that a brute-force approach is simply not feasible. Also, combinatorial problems like the TSP don't have continuous independent variables and so they cannot be solved with derivatives. Although we can never be sure that we have the best solution to a combinatorial problem unless we try every possible solution, EAs provide a powerful way to find good solutions. EAs are not magic, but they can help find at least a good solution (if not the best solution) to these types of large, multidimensional problems. Potential solutions in an EA share information with each other and eventually come to a "consensus" of the best solution. We cannot prove that it is the best solution; we would have to look every possible solution to prove that we have found the best. But when compare EA solutions with other types of solutions, we see that EAs work pretty well. Chapter 18 discusses evolutionary combinatorial optimization problems in more detail.

2.6 HILL CLIMBING

This section presents a simple optimization algorithm called hill climbing. Actually, hill climbing is a family of algorithms with many variations. Some researchers consider hill climbing to be a simple EA, while others consider it a non-evolutionary algorithm. When we encounter a new optimization problem, hill climbing is often a good first choice for solving it because it is simple, surprisingly effective, has many simple variations, and provides a good benchmark with which to compare more complicated algorithms such as EAs. The idea of hill climbing is so straightforward that it must have been invented many times, and long ago, so it is difficult to determine its origin.

If you want to get to the highest point in a landscape, one reasonable strategy is to simply take a step in the direction of steepest ascent. After that step, you re-evaluate the slope of the hill, and again take a step in the direction of steepest ascent. This process is continued until there are no directions which lead you higher, at which point you have reached the top of a hill. This is a local search strategy, and is called hill climbing.

A better strategy would be to look around, estimate where the highest point is, and then estimate the best way to get there. This would eliminate the problem of zigzagging your way to the top, or getting stuck at the top of a small hill that is lower than the globally highest point. But if visibility is low, a local search strategy may be your best course of action.

Hill climbing may or may not work well, depending on the shape of the hill, the number of local maxima, and your initial position. Hill climbing can be used by itself as an optimization algorithm. It can also be combined with an EA, which would combine the global search ability of an EA with the local search ability of hill climbing. There are several varieties of hill climbing strategies [Mitchell, 1998], a few of which we discuss in this section.

Figure 2.6 shows the steepest ascent hill climbing algorithm. This algorithm proceeds conservatively, changing only one solution feature at a time, and replacing the current best solution with the best one-feature change.

$x_0 \leftarrow$ randomly generated individual
While not(termination criterion)
 Compute the fitness $f(x_0)$ of x_0
 For each solution feature $q = 1, \cdots, n$
  $x_q \leftarrow x_0$
  Replace the q-th solution feature of x_q with a random mutation
  Compute the fitness $f(x_q)$ of x_q
 Next solution feature
 $x' \leftarrow \arg\max_{x_q}(f(x_q) : q \in [0, n])$
 If $x_0 = x'$ then
  $x_0 \leftarrow$ randomly generated individual
 else
  $x_0 \leftarrow x'$
 End if
Next generation

Figure 2.6 The above pseudo-code outlines the steepest ascent hill climbing algorithm for the maximization of the n-dimensional function $f(x)$. Note that x_q is equal to x_0 but with its q-th feature mutated.

Figure 2.7 shows the next ascent hill climbing algorithm, also called simple hill climbing. This algorithm, like the steepest ascent hill climbing algorithm, changes only one solution feature at a time. But the next ascent hill climbing algorithm is more greedy because as soon as a better solution is found, the current solution is replaced.

The next two hill climbing algorithms randomly select which solution feature to mutate, and therefore fall under the general classification of stochastic hill climbing. Figure 2.8 shows the random mutation hill climbing algorithm. This algorithm is very similar to the next ascent hill climbing algorithm, except that the mutated solution feature is chosen randomly.

Figure 2.9 shows the adaptive hill climbing algorithm. This algorithm is similar to the random mutation hill climbing algorithm, except that every solution feature is mutated with some probability before the mutated solution is compared with the current best solution.

The results of a hill climbing algorithm can strongly depend on the initial condition x_0 in Figures 2.6–2.9. It therefore makes sense to try hill climbing with several different randomly generated initial conditions. This approach of putting a hill climbing algorithm inside an initial condition loop is called random restart hill climbing.

```
x_0 ← randomly generated individual
While not(termination criterion)
    Compute the fitness f(x_0) of x_0
    ReplaceFlag ← false
    For each solution feature q = 1, ···, n
        x_q ← x_0
        Replace the q-th solution feature of x_q with a random mutation
        Compute the fitness f(x_q) of x_q
        If f(x_q) > f(x_0) then
            x_0 ← x_q
            ReplaceFlag ← true
        End if
    Next solution feature
    If not(ReplaceFlag)
        x_0 ← randomly generated individual
    End if
Next generation
```

Figure 2.7 The above pseudo-code outlines the next ascent hill climbing algorithm for the maximization of the n-dimensional function $f(x)$. Note that x_q is equal to x_0 but with its q-th feature mutated.

```
x_0 ← randomly generated individual
While not(termination criterion)
    Compute the fitness f(x_0) of x_0
    q ← randomly chosen solution feature index ∈ [1, n]
    x_1 ← x_0
    Replace the q-th solution feature of x_1 with a random mutation
    Compute the fitness f(x_1) of x_1
    If f(x_1) > f(x_0) then
        x_0 ← x_1
    End if
Next generation
```

Figure 2.8 The above pseudo-code outlines the random mutation hill climbing algorithm for the maximization of the n-dimensional function $f(x)$. Note that x_1 is equal to x_0 but with a random feature mutated.

```
Initialize p_m ∈ [0, 1] as the probability of mutation
x_0 ← randomly generated individual
While not(termination criterion)
    Compute the fitness f(x_0) of x_0
    x_1 ← x_0
    For each solution feature q = 1, ···, n
        Generate a uniformly distributed random number r ∈ [0, 1]
        If r < p_m then
            Replace the q-th solution feature of x_1 with a random mutation
        End if
    Next solution feature
    Compute the fitness f(x_1) of x_1
    If f(x_1) > f(x_0) then
        x_0 ← x_1
    End if
Next generation
```

Figure 2.9 The above pseudo-code outlines the adaptive hill climbing algorithm for the maximization of the n-dimensional function $f(x)$. Note that x_q is equal to x_0 but with its q-th feature mutated.

■ EXAMPLE 2.7

We simulate the four hill climbing algorithms on a set of 20-dimensional benchmark problems (see Appendix C.1). Note that the benchmark problems in Appendix C.1 are minimization problems, and so we adapt the hill climbing algorithms in a straightforward way to obtain hill descending algorithms. We run each algorithm on each benchmark 50 times, each time with a different initial condition. For adaptive hill climbing, we use $p_m = 0.1$. We terminate each hill climbing algorithm after 1,000 fitness function evaluations.

Table 2.1 shows the results. Note that steepest ascent hill climbing (Figure 2.6) requires at least n fitness function evaluations every generation ($n = 20$ in our examples), while random mutation hill climbing (Figure 2.8) requires only one fitness function evaluation every generation. The vast majority of the computational effort of a heuristic algorithm is typically consumed by fitness function evaluations (see Chapter 21). Therefore, for a fair comparison of different optimization algorithms, the number of fitness function evaluations should be the same for the algorithms that are being compared. If different optimization algorithms have the same number of fitness function evaluations per generation (for example, steepest ascent hill climbing in Figure 2.6 and next ascent hill climbing in Figure 2.7), then it would be fair to compare the algorithms on the basis of the number of generations.

Table 2.1 shows that random mutation hill climbing performs the best on 12 of the 14 benchmarks, and on average it far outperforms the other hill climbing methods. Adaptive hill climbing and steepest ascent hill climbing each perform slightly better than random mutation hill climbing on one benchmark. However, the performance of adaptive hill climbing can be strongly dependent on the mutation rate (see Problem 2.11), and we did not make any effort to find a good mutation rate in this example.

Benchmark	Steepest Ascent	Next Ascent	Random Mutation	Adaptive
Ackley	2.27	1.82	1.00	1.70
Fletcher	2.62	1.87	1.00	1.68
Griewank	9.58	4.41	1.00	3.81
Penalty #1	26624	2160	1.00	281
Penalty #2	99347	5690	1.00	4178
Quartic	133.94	29.99	1.00	25.61
Rastrigin	3.76	2.52	1.00	2.10
Rosenbrock	2.68	1.50	1.00	1.72
Schwefel 1.2	1.63	1.37	1.24	1.00
Schwefel 2.21	1.00	1.75	1.02	1.12
Schwefel 2.22	3.65	2.73	1.00	2.30
Schwefel 2.26	5.05	3.63	1.00	2.91
Sphere	17.97	7.32	1.00	6.09
Step	16.58	6.78	1.00	6.52
Average	9012	565	1.02	323

Table 2.1 Example 2.7: Relative performance of hill climbing algorithms. The table shows the normalized minimum found by four hill climbing algorithms, averaged over 50 Monte Carlo simulations. Since hill climbing is stochastic, your results may vary.

□

2.6.1 Biased Optimization Algorithms

We now mention one important caveat related to benchmark functions that we must keep in mind during our optimization studies. This caveat includes two related statements: first, many benchmark cost functions have their minima near the middle of the search domain; and second, many optimization algorithms are biased toward the middle of the search domain.[3] We discuss this biasing phenomenon in more detail in Appendix C.7, which we encourage the reader to carefully study before performing any serious research. Many of the simulation results in this book are based on benchmarks with minima at the center of the search domain. Those results are not intended to accurately portray optimization algorithm performance, but are only intended to illustrate the application of a particular optimization algorithm. We need to implement the unbiasing approach of Appendix C.7 before we can safely make conclusions about optimization algorithm performance.

[3]This is not necessarily the case for the hill climbing algorithms in this section, but it is the case for many of the evolutionary algorithms that we discuss later in this book.

2.6.2 The Importance of Monte Carlo Simulations

Note that in Example 2.7 we averaged 50 simulation results to plot the performance of the binary GA and the continuous GA. Showing the results of a single simulation does not prove anything, because the results depend on a random number generator. We can obtain very few valid conclusions from a single simulation, or from a single experiment. We discuss this in more detail in Appendix B, but we also mention it here since this is the first place in this book that we have averaged the results of multiple simulations.

Multiple simulations that are used for performance analysis are often called Monte Carlo simulations. The name came from John von Neumann, Stanislaw Ulam, and Nicholas Metropolis, during their work in the 1940s on nuclear weapons. Much of their work involved the analysis of the results of multiple experiments, and there was an obvious connection between the statistical analysis of their experiments and the statistical properties of gambling. This connection, in conjunction with the fact that Ulam's uncle was a notorious gambler at the casinos of Monte Carlo in Monaco, led to the name "Monte Carlo simulations" [Metropolis, 1987].

2.7 INTELLIGENCE

In the early days of computing researchers realized that computers were very good at things that humans did poorly, like calculating the trajectory of a ballistic missile. But computers were (and still are) not effective at tasks that humans can do well, like recognizing a face. This led to attempts to mimic biological behavior in an effort to make computers better at such tasks. These efforts resulted in technologies like fuzzy systems, neural networks, genetic algorithms, and other EAs. EAs are therefore considered to be a part of the general category of computer intelligence.

In our development of EAs, we try to create algorithms that are *intelligent.* But what does it mean to be intelligent? Does it mean that our EAs can score high on an IQ test? This section discusses the meaning of intelligence and some of its characteristics: adaptation, randomness, communication, feedback, exploration, and exploitation. These are the characteristics that we implement in EAs in our search for intelligent algorithms.

2.7.1 Adaptation

We usually consider adaptation to changing environments as a feature of intelligence. Suppose you learn how to assemble a widget, and then your supervisor asks you to assemble a doohickey, which you've never seen before. If you are intelligent, you will be able to generalize what you know about widgets, and you will be able to assemble the doohickey. However, if you are not so intelligent, then you will have to be taught the specific details of doohickey assembly.

However, adaptive controllers [Åström and Wittenmark, 2008] are not considered intelligent. A virus that can survive extreme environments is not considered intelligent. We thus conclude that adaptation is a necessary but not sufficient condition for intelligence. We try to design EAs that can be adapted to a wide class of problems. Adaptability in an EA is only one of many criteria for a successful EA.

2.7.2 Randomness

We usually think of randomness in negative terms. We don't like unpredictability in our lives and so we try to avoid it, and we try to control our environment. However, some degree of randomness is a necessary component of intelligence. Think of a zebra that is running from a lion. If the zebra runs in a straight line and at a constant speed, it will be easy to catch. But an intelligent zebra will zigzag and move unpredictably to avoid its predator. Conversely, think of a lion that is trying to catch a zebra. The lion stalks the herd of zebras each day. If the lion waits at the same place and at the same time every day, it will be easy to avoid. But an intelligent lion will strike at different places and different times and in an unpredictable way. Randomness is a characteristic of intelligence.

Too much randomness will be counterproductive. If the zebra randomly decides to lie down while being chased, we would be right to question its intelligence. If a lion randomly decides to dig a hole in its search for a zebra, we would be right to question its intelligence. So randomness is a feature of intelligence, but only within limitations.

Our EA designs will include some component of randomness. If we exclude randomness, our EAs will not work well. But if we use too much randomness, they will not work well either. We will need to use the right amount of randomness in our EA designs. Of course, as discussed earlier, EAs are adaptable. Therefore, good EAs will perform well over a range of randomness measures. We cannot expect the EAs to be so adaptable that we can use any level of randomness, but they will be adaptable enough so that the exact randomness measure will not be critical.

2.7.3 Communication

Communication is a feature of intelligence. Consider a genius who takes an IQ test, except the genius has no way of communicating. He will fail the IQ test even though he is a genius. Many deaf, dumb, and autistic individuals fail IQ tests even though they are quite intelligent. Children who are raised without human interaction are not creative, intelligent, happy, or well adjusted [Newton, 2004]. Their lack of communication with others during their formative years prevents them from developing any intellectual capacity beyond a young child. Their years of isolation are irrecoverable, and they cannot learn to communicate or adapt to society.

Intelligence not only involves communication, but it is also *emergent*. That is, intelligence arises from a population of individuals. A single individual cannot be intelligent. It can be argued that there are many intelligent individuals in the world, and even if such an individual were isolated he would still be intelligent. However, such individuals gained their intelligence only through interaction with others. A single ant wanders aimlessly and accomplishes nothing, but a colony of ants can find the shortest path to food, build elaborate networks of tunnels, and organize themselves as a self-sustaining community. Likewise, a single individual will never accomplish anything if he never has any interaction with a community. A community, however, can send a man to the moon, connect billions of people through the Internet, and build food and water supply systems in the desert.

We see that intelligence and communication form a positive feedback loop. Communication is required to develop intelligence, and intelligence is required to communicate. But the main point here is that communication is a feature of intelligence. This is why most EAs involve a population of candidate solutions to some problem. Those candidate solutions, which we call *individuals*, communicate with each other and learn from each other's successes and failures. Over time, the population of individuals evolves a good solution to the optimization problem.

2.7.4 Feedback

Feedback is a fundamental characteristic of intelligence. This involves adaptation, which was discussed above. A system cannot adapt if it cannot sense and react to its environment. However, feedback involves more than adaptation; it also involves learning. When we make mistakes, we change so that we don't repeat those mistakes.[4] However, even more importantly, when *others* make mistakes, we adjust our behavior so that we don't repeat those mistakes. Failure provides negative feedback. Conversely, success (our's and others') provides positive feedback and influences us to adopt those behaviors to which we attribute success. We often see others who don't seem to learn from mistakes, and who don't adopt behaviors that are proven to lead to success; we don't consider such people to be very intelligent.

Feedback is also the basis for many natural phenomena. The water cycle consists of an endless succession of rain and evaporation. More rain leads to more evaporation, and more evaporation leads to more rain. Since this includes a fixed amount of water, the water cycle leads to a stable amount of moisture on the surface of the earth and in the sky. If this feedback mechanism were somehow disturbed, there would be a lot of difficulties for life, including floods and drought.

The sugar/insulin balance in the human body is another feedback mechanism. The more sugar we eat, the more insulin our pancreas produces; the more insulin our pancreas produces, the more sugar is absorbed from the blood. Too much sugar in the blood leads to hyperglycemia, and too little sugar in the blood leads to hypoglycemia. Diabetes is the disturbance of the sugar/insulin feedback mechanism, and can lead to serious and long-term health problems.

This characterization of feedback as a hallmark of intelligence is often recognized in intelligent control theory. Feedback is not a sufficient condition for intelligence. No one would call a proportional controller intelligent, and no one would call a mechanical thermostat intelligent. Feedback is a necessary, but not sufficient, condition for intelligence.

Our EA designs will incorporate positive and negative feedback. An EA without feedback will not be very effective, but an EA with feedback has satisfied this necessary condition for intelligence.

2.7.5 Exploration and Exploitation

Exploration is the search for new ideas or new strategies. *Exploitation* is the use of existing ideas and strategies that have proven successful in the past. Exploration is high-risk; a lot of new ideas waste time and lead to dead ends. However, exploration can also be high-return; a lot of new ideas pay off in ways that we could not have imagined. Exploitation is closely related to the feedback strategies discussed

[4]Albert Einstein is reputed to have defined insanity as doing the same thing over and over again and expecting different results.

previously. Someone who is intelligent uses what they know and what they have instead of constantly reinventing the wheel. But someone who is intelligent is also open to new ideas, and is willing to take calculated risks. Intelligence includes the proper balance of exploration and exploitation. The proper balance of exploration and exploitation depends on how regular our environment is. If our environment is rapidly changing, then our knowledge quickly becomes obsolete and we cannot rely as much on exploitation. However, if our environment is highly consistent, then our knowledge is dependable and it may not make sense to try very many new ideas.

Our EA designs will need a proper balance of exploration and exploitation to be successful. Too much exploration is similar to too much randomness, which we discussed earlier, and will probably not give good optimization results. But too much exploitation is related to too little randomness. The proper balance of exploration and exploitation in EAs was called "the optimal allocation of trials" by John Holland, one of the pioneers of genetic algorithms [Holland, 1975].

2.8 CONCLUSION

The key point of this chapter is that optimization is a fundamental aspect of engineering and problem solving. When we try to optimize a function, we refer to the function as an objective function. When we try to minimize a function, we refer to it as a cost function. When we try to maximize a function, we refer to it as a fitness function. Any optimization problem can easily be converted back and form between a minimization problem and a maximization problem. Some special types of problems that we introduced in this chapter are constrained problems, multi-objective problems, and multimodal problems. Almost all real-world optimization problems are constrained, multi-objective, and multimodal. Another special class of problems is combinatorial problems, in which the independent variables belong to a finite set.

We introduced hill climbing in this chapter, which is a simple but effective optimization algorithm. There are many different types of hill climbing algorithms. Although they are usually quite simple, they provide a nice benchmark with which to compare more complicated optimization algorithms. Finally, we mentioned some features of natural intelligence, and discussed how we can implement those features in our evolutionary algorithms to justify the label *intelligent*.

PROBLEMS

Written Exercises

2.1 Consider the problem $\min f(x)$, where

$$f(x) = 40 + \sum_{i=1}^{4} x_i^2 - 10\cos(2\pi x_i).$$

Note that $f(x)$ is the Rastrigin function – see Section C.1.11.

a) What are the independent variables of $f(x)$? What are the decision variables of $f(x)$? What are the solution features of $f(x)$?

b) What is the dimension of this problem?

c) What is the solution to this problem?

d) Rewrite this problem as a maximization problem.

2.2 Consider the function $f(x) = \sin x$.

a) How many local minima does $f(x)$ have? What are the function values at the local minima, and what are the locally minimizing values of x?

b) How many global minima does $f(x)$ have? What are the function values at the global minima, and what are the globally minimizing values of x?

2.3 Consider the function $f(x) = x^3 + 4x^2 - 4x + 1$.

a) How many local minima does $f(x)$ have? What are the function values at the local minima, and what are the locally minimizing values of x?

b) How many local maxima does $f(x)$ have? What are the function values at the local maxima, and what are the locally maximizing values of x?

c) How many global minima does $f(x)$ have?

d) How many global maxima does $f(x)$ have?

2.4 Consider the same function as in Problem 2.3, $f(x) = x^3 + 4x^2 - 4x + 1$, but with the constraint $x \in [-5, 3]$.

a) How many local minima does $f(x)$ have? What are the function values at the local minima, and what are the locally minimizing values of x?

b) How many local maxima does $f(x)$ have? What are the function values at the local maxima, and what are the locally maximizing values of x?

c) How many global minima does $f(x)$ have? What is the function value at the global minimum, and what is the globally minimizing values of x?

d) How many global maxima does $f(x)$ have? What is the function value at the global maximum, and what is the globally maximizing values of x?

2.5 Recall that Figure 2.4 shows the Pareto front for a two-objective problem in which the goal is to minimize both objectives.

a) Sketch a possible set of points in the (f, g)-plane and the Pareto front for a problem in which the goal is to maximize $f(x)$ and minimize $g(x)$.

b) Sketch a possible set of points in the (f, g)-plane and the Pareto front for a problem in which the goal is to minimize $f(x)$ and maximize $g(x)$.

c) Sketch a possible set of points in the (f, g)-plane and the Pareto front for a problem in which the goal is to maximize both $f(x)$ and $g(x)$.

2.6 How many unique closed paths exist through N cities? By *unique* we mean that the starting city does not matter, and the direction of travel does not matter. For example, in a four-city problem with cities A, B, C, and D, we consider route $A \to B \to C \to D \to A$ equivalent to routes $D \to C \to B \to A \to D$ and $B \to C \to D \to A \to B$.

2.7 Consider the closed TSP with the cities in Table 2.2.

City	x	y
A	5	9
B	9	8
C	−6	−8
D	9	−2
E	−5	9
F	4	−7
G	−9	1

Table 2.2 TSP coordinates of cities for Problem 2.7.

a) How many closed routes exist through these seven cities?
b) Is it easy to see the solution by looking at the coordinates in Table 2.2?
c) Plot the coordinates. Is it easy to see the solution from the plot? What is the optimal solution? This problem shows that looking at a problem in a different way might help us find a solution.

2.8 Given an arbitrary maximization problem $f(x)$ and a random initial candidate solution x_0, what is the probability that the steepest ascent hill climbing algorithm will find a value x' such that $f(x') > f(x_0)$ after the first generation?

Computer Exercises

2.9 Plot the function of Problem 2.4 with the local and global optima clearly indicated.

2.10 Consider the multi-objective optimization problem $\min\{f_1, f_2\}$, where

$$f_1(x_1, x_2) = x_1^2 + x_2, \text{ and } f_2(x_1, x_2) = x_1 + x_2^2$$

and x_1 and x_2 are both constrained to $[-10, 10]$.

a) Calculate $f_1(x_1, x_2)$ and $f_2(x_1, x_2)$ for all allowable integer values of x_1 and x_2, and plot the points in (f_1, f_2) space (a total of $21^2 = 441$ points). Clearly indicate the Pareto front on the plot.
b) Given the resolution that you used in part (a), give a mathematical description of the Pareto set. Plot the Pareto set in (x_1, x_2) space.

2.11 Adaptive hill climbing.

a) Run 20 Monte Carlo simulations of the adaptive hill climbing algorithm, with 1,000 generations per Monte Carlo simulation, for the two-dimensional Ackley function. Record the minimum value achieved by each Monte Carlo simulation, and compute the average. Do this for 10 different mutation rates, $p_m = k/10$ for $k \in [1, 10]$, and record your results in Table 2.3. What is the best mutation rate?

b) Repeat part (a). Do you get the same, or similar, results? What do you conclude about the number of Monte Carlo simulations that you need to get reproducible results for this problem?

c) Repeat part (a) for the 10-dimensional Ackley function. What do you conclude about the relationship between the optimal mutation rate and the problem dimension?

p_m	Average Result
0.1	
0.2	
0.3	
0.4	
0.5	
0.6	
0.7	
0.8	
0.9	
1.0	

Table 2.3 Complete this table for Problem 2.11.

PART II

CLASSIC EVOLUTIONARY ALGORITHMS

CHAPTER 3

Genetic Algorithms

Genetic algorithms are NOT function optimizers.

—Kenneth De Jong [De Jong, 1992]

Genetic algorithms (GAs) are the earliest, most well-known, and most widely-used EAs. GAs are simulations of natural selection that can solve optimization problems. In spite of the above quote by Kenneth De Jong, GAs often serve as effective optimization tools. De Jong's quote emphasizes the point that GAs were originally developed to study adaptive systems rather than to optimize functions. GAs comprise a much more broad class of systems than function optimizers. We can use GAs to study the dynamics of adaptive systems [Mitchell, 1998, Chapter 4], to provide advice to fashion designers [Kim and Cho, 2000], to provide design tradeoffs to bridge designers [Furuta et al., 1995], and for many other non-optimization applications. Sometimes the dividing line between an optimization algorithm and a non-optimization algorithm is fuzzy because all algorithms attempt to function as well as possible. In any case, our main interest in GAs in this book is their specific application as optimization algorithms. To begin our study of GAs, we observe some basic features of natural selection.

1. A biological system includes a population of individuals, many of which have the ability to reproduce.

2. The individuals have a finite life span.
3. There is variation in the population.
4. The ability to survive is positively correlated with the ability to reproduce.

Genetic algorithms simulate each of these features of natural selection. Given an optimization problem, we create a population of candidate solutions, which we call individuals. Some solutions are good, and some are not so good. The good individuals have a relatively high chance of reproducing, while the poor individuals have a relatively low chance of reproducing. Parents beget children, and then the parents drop out of the population to make way for their offspring. As generations come and go, the population becomes more fit. Sometimes one or more "supermen" evolve to become highly fit individuals that can provide near-optimal solutions to our engineering problem.

Overview of the Chapter

This chapter gives an overview of natural genetics, and also of artificial genetic algorithms for optimization problems. Since we are just getting started with EAs in this chapter, we spend more time in this chapter on history and biological underpinnings than we do in most of the later chapters. The reader who wants to jump right in to the study of GAs can safely skip the first three sections without seriously jeopardizing their understanding of GAs. Section 3.1 briefly discusses the history of the science of genetics, focusing on the work of Charles Darwin and Gregor Mendel in the 19th century. Section 3.2 reviews the science of genetics, which forms the foundation of GAs. Section 3.3 provides a history of computer simulations of genetics, beginning in the 1940s with biologists who were interested in studying natural selection, and ending with the explosion of GA research in the 1970s and 1980s.

Section 3.4 develops a simple binary GA in a methodical, step-by-step manner. The GA is based on natural genetics, and so it represents solutions to optimization problems as chromosomes with binary alleles. The binary GA is naturally suited to optimization problems whose domain is comprised of an n-dimensional binary search space, or at least to problems whose domains are discrete.

We can use bit strings to represent candidate solutions to optimization problems with continuous domains, provided that we use enough bits to give us the required resolution. But it is more natural to represent candidate solutions to continuous-domain problems as vectors of real numbers. Section 3.5 therefore extends GAs to continuous-domain problems.

3.1 THE HISTORY OF GENETICS

Genetics is the study of heredity and variation in living organisms. This section gives a brief history of the development of modern genetics, focusing on the work of Charles Darwin, the father of evolution, and Gregor Mendel, the father of genetics.

3.1.1 Charles Darwin

Charles Darwin was born in England in 1809, the son of the wealthy doctor Robert Darwin. Charles's privileged position in life allowed him to wander from one interest to another as a young man, apparently destined to waste his life in lazy meanderings. His father was a hard-working man, but as often happens, hard work by the father resulted in laziness in the son. "You care for nothing but shooting, dogs, and rat-catching; and you will be a disgrace to yourself and all your family," Robert told his son [Darwin et al., 2002, page 10]. Robert tried to involve his son in his medical practice, but Charles was not interested, and besides, he hated the sight of blood. So Robert sent his son to Cambridge University to study for the ministry.

Charles wasn't really interested in his studies at Cambridge; the only thing he was interested in was the outdoors. He spent all of his time exploring and studying nature, reading the books of great naturalists, and collecting beetles. His life began to take on some focus as he became more proficient as a naturalist. Charles began meeting professors and other students who shared his interest in nature. He began making plans to leave his ministerial studies and pursue his true passion in life. He was finally becoming ambitious.

In 1831, at the age of 22, Charles applied for a position on the Beagle, a ship that was commissioned to survey the southern tip of South America for the English government. Charles was accepted for the position under the condition that he pay his own way.

What would you do if you were the father of Charles Darwin? You've sent your son to school and have paid for three years of ministerial training, and now he comes to you asking for funds to pay for a five-year sea voyage as a naturalist. You would say no, of course. And that is just what Robert told his son – at first. Fortunately for Charles, Robert recognized that his son was becoming a man and had found a passion in life. Robert eventually allowed himself to become convinced, and he agreed to fund the excursion.

During the five-year voyage of the Beagle, Charles lived on the 90-foot long, 25-foot wide ship with over 70 other sailors. The ship also contained surveying equipment and enough supplies to last for several months at sea. For Charles it must have been a difficult transition from his life of ease, but he made the most of it. He spent his time at sea reading and studying. When the ship stopped at islands or at the South American mainland, he collected animals and sent them back to England on the next available ship. He collected a huge variety of species during his travels. Similar species on neighboring islands were different enough from one another that each seemed to have adapted to its own particular environment.

Charles returned home to England in 1836 at the age of 27. Almost immediately after his return to England, he began working on his book Origin of Species [Darwin, 1859], which would end up being a decades-long project. He also was active in writing journal papers and speaking at conferences. He continued studying and learning as he began putting together a coherent theory of natural selection. Natural selection says that the most fit individuals survive and pass on their characteristics to their offspring. This is how adaptation takes place – through the "survival of the fittest."

As Charles continued working on his book, he was hesitant to publicize his theory of evolution. Having studied the ministry for three years, he knew that his theory could generate a storm of controversy because of its possible contradiction with the Bible. He wanted to build an air-tight case and a true magnum opus before publishing his results. However, in 1858 he received a paper from Alfred Wallace, a naturalist who was traveling in the South Pacific. Wallace had independently arrived at many of the same ideas as Darwin, and sent a paper to Darwin asking for his help in publishing it.

Darwin was in somewhat of a predicament. He had a choice to make. He could "lose" Wallace's paper,[1] publish his own results, and claim precedence for his theory of evolution. Or he could submit Wallace's paper for publication and allow Wallace to have precedence. To his credit, Darwin decided to try to strike a balance. He quickly wrote his own paper, and then presented both his and Wallace's paper at the next available conference. He then put the finishing touches on his book, which ended up being much shorter than he originally intended[2] due to his haste to stake a claim to the credit that he deserved. The Origin of Species was published in 1859, and the first printing of 1,250 copies sold out in one day. Darwin was on the fast track to becoming the most famous and controversial scientist of his generation.

Although Darwin's theory of evolution quickly gained scientific credibility, like all new theories, it was not without detractors. First, it seemed to go against the Bible's teaching of the special creation of all species and thus was susceptible to attacks by religious leaders. Second, Darwin did not have any idea how traits were passed by parents to their offspring. In some ways it is surprising that his theory gained acceptance as quickly as it did in spite of his lack of explanation for heredity. He observed that it happened and so he postulated natural selection, but he could not say how it happened.

Darwin, along with other scientists of his time, had two fundamental misconceptions about heredity. First, he believed that the traits of parents could be blended in their offspring; for example, the child of a black mouse and a white mouse might be gray. Second, he believed that acquired traits could be passed to offspring. For example, a man who lifts weights and becomes strong will tend to have strong children because of his weight lifting.

Darwin, a child of privilege, developed the theory of natural selection, but it would be left to a child of poverty, Gregor Mendel, to prove it.

3.1.2 Gregor Mendel

Gregor Mendel was the first to understand and explain how heredity occurs. He was born as Johann Mendel in 1822 to a poor farmer in Czechoslovakia [Bankston, 2005]. His father needed him to help on the farm, but young Johann was much more suited to academics than physical labor. His parents could barely afford it, but they sent him to school in order to help him gain the opportunities in life that they lacked. In spite of his parents' support and his own part-time work, he could barely survive financially as a student. His financial situation was much the same as many graduate students today, except that there were fewer opportunities for financial aid.

At the age of 21, Mendel heard about a nearby monastery where he could continue his education without any financial worries. He would have to take a vow of poverty and celibacy, but the financial benefits were too good to turn down. Mendel

[1] Who would have doubted Darwin's claim that a letter mailed from the South Pacific to England in 1858 had never arrived?

[2] This all worked out for the best in the long run. Darwin's book was about 500 pages. If he had finished it the way he wanted to, it would have been several hundred pages longer and would have been more daunting to potential readers. By making the book shorter he increased his readership and subsequent success.

was not particularly religious, but he jumped at the opportunity and joined the Augustinian order of the St. Thomas Monastery.

Augustine, who lived in the Roman empire around 400 AD, was one of the greatest intellectual leaders in Christian history. His theology emphasized the ability of God to communicate to man through secular knowledge. The monastery that Mendel joined, in keeping with the philosophy of its namesake, encouraged learning in all areas of life and was thus a perfect fit for Mendel. It was "worldly" compared with many other monasteries. The monks did not have to punish themselves, or spend all day praying, or take vows of silence. They only had to study, believing that God spoke to them through their learning. In accordance with tradition, Johann Mendel gave himself a new name when he joined St. Thomas; his name was now Gregor Mendel.

As a monk, Mendel continued to take classes at universities, taught science at nearby schools, read, and conducted his own research at the monastery. The research that he pursued involved the breeding of plants, and in particular, peas. This was a fitting avenue for his creative talents because of his background on his father's farm.

As Mendel experimented with peas, he noticed that they had various traits. Some were smooth, while others were rough; some were more green, while others were more yellow; some had buds in one location, others had buds in another location. As Mendel experimented, he realized that the traits were controlled by some unseen unit of heredity, which he called elements. Some of the elements were strong and tended to have more control over the peas' traits. Other elements were weak and had less control over the peas' traits. Today, we use the word gene instead of element, and we say that genes are either dominant or recessive, rather than strong or weak. But it was Mendel who first understood genetics, heredity, and dominance. Mendel's work was the missing link in Darwin's theory, and explained how natural selection worked.

Mendel presented his findings at a conference in 1865. This was only six years after Darwin's publication of Origin of Species, but for some reason Mendel's audience did not realize the magnitude of what he had discovered. The reception of Mendel's scientific breakthrough could not have been more different than that of Darwin's. Whereas Darwin became immediately famous, Mendel's work was ignored. He continued to work in obscurity at St. Thomas, publishing a few papers here and there, all of which were essentially overlooked by the scientific community.

Mendel became the administrative leader of St. Thomas in 1868 and didn't have much time for science after that.[3] Mendel died in 1884. His work on genetics was finally rediscovered by the Dutch biologist Hugo de Vries, the German botanist Carl Correns, and the Austrian agronomist Erich von Tschermak, all around 1900.

3.2 THE SCIENCE OF GENETICS

Each of our individual characteristics, or traits, is controlled by a pair of genes. Human genetics are therefore called diploid, meaning that the genes for each trait occur in pairs. Some plants and animals are haploid, meaning that each trait is

[3]As often happens in academia today, a thriving career in research was tragically cut short by a promotion to administration.

determined by a single set of genes. Other organisms are polyploid, meaning that each trait is determined by more than two sets of genes.

In diploid genetics, some genetic values are dominant while others are recessive. If a dominant and recessive gene both appear in an individual, then the dominant gene will determine the trait that appears in the individual. The only way that a recessive gene can determine the trait is if both genes are the same.

■ EXAMPLE 3.1

Consider three individuals: Chris has two brown-eye genes, Kim has two green-eye genes, and Terry has a brown-eye gene and a green-eye gene. Since Chris has two brown-eye genes, Chris has brown eyes. Since Kim has two green-eye genes, Kim has green eyes. Since Terry has one brown-eye gene and one green-eye gene, and the brown-eye gene is dominant, Terry has brown eyes.

- Chris: brown/brown → brown eyes
- Kim: green/green → green eyes
- Terry: brown/green → brown eyes

If Chris and Terry mate, they will each contribute one eye-color gene to their offspring. Their offspring could therefore have either two brown-eye genes, or else one brown-eye gene and one green-eye gene. All of their offspring will have brown eyes since brown is dominant.

If Chris and Kim mate, their offspring will all have one brown-eye gene from Chris and one green-eye gene from Kim. All of their offspring will have brown eyes since brown is dominant.

If Terry and Kim mate, their offspring will have either one brown-eye gene and one green-eye gene, or two green-eye genes. Their offspring could either have brown eyes or green eyes.

□

Now suppose that there is some evolutionary benefit in having green eyes. For example, females may be highly attracted to males with green eyes. Or perhaps green eyes are more receptive to certain frequencies of light, which enables green-eyed individuals to be more successful in hunting. In that case the green-eyed individuals will be more likely to survive than the brown-eyed individuals. Furthermore, the green-eyes will be more likely to be strong and successful than the brown-eyes, which will give them more opportunities to reproduce. This will affect the gene pool of the human race by increasing the number of green-eye genes while reducing the number of brown-eye genes. This is called natural selection, or survival of the fittest, and this is what Darwin deduced during his journey on the Beagle during the 1830s.

Sometimes a mutation affects the parents' offspring. In this case the genes are not passed intact from parent to offspring, but the genes are instead changed. Mutations are caused by the fundamental imperfection of life and biological processes, including radiation and illness.

The vast majority of mutations are neutral; they have little or no effect on the offspring because of the remarkable resilience and redundancy of biology. These neutral mutations are important in biology's search for improved fitness, and we will see later in this book that mutations are also usually important in an EA's search for improved fitness. However, the majority of mutations that measurably affect biological offspring are harmful. In fact, we can say that almost all such mutations are harmful. There are more than 6,000 commonly-occurring single-gene mutations that cause disease, and these occur in one of every 200 births. Some genetic disorders appear at birth while others may show up only later in life. For example, many types of cancer have a genetic component.

However, once in a great while a mutation appears that is actually beneficial. For example, suppose that in Example 3.1, one of Terry and Kim's offspring has a mutation that results in a purple-eye gene. Suppose that this offspring with purple eyes has sharper vision than average because of the correlation of purple irises with higher corneal adaptability. This allows the purple-eyed mutant to be more successful in hunting. He becomes strong and successful because of his purple eyes, and this allows him more opportunities to mate. If his purple-eye gene is dominant, then all of his offspring will have purple eyes, and the mutation will spread throughout the species. If his purple-eye gene is recessive, he could have offspring with purple eyes if he finds another purple-eye mutation to mate with. Mutation, accompanied by natural selection, helps improve the survivability of the species. Without mutation the species would become stagnant. Mutation is generally harmful to individuals, but ironically it is beneficial to the species as a whole.

3.3 THE HISTORY OF GENETIC ALGORITHMS

1903 was a good year for technology. The Marconi Company began the first regular trans-Atlantic radio broadcast, the Wright brothers successfully completed their first airplane flight, and Neumann Janos was born in Budapest, Hungary. Neumann's genius displayed itself at a young age in his voracious reading and his mathematical aptitude. His parents, both of whom were from educated upper-class families, recognized early on that he was a prodigy, but they were careful not to push him too hard. By the time he was 23 years old he had an undergraduate degree in chemical engineering and a PhD in mathematics. He continued to be productive as an academic professional, and in 1929 accepted a faculty position at Princeton University in New Jersey. His name was now John von Neumann, and in 1933 he became one of the original members of Princeton's Institute of Advanced Studies.

During his wide-ranging career at Princeton, von Neumann made fundamental contributions to mathematics, physics, and economics. He was one of the leaders of the atomic bomb effort during World War II, and he was also one of the pioneers of the invention of the digital computer. There were others who were just as influential (or perhaps even more influential) in the development of digital computing – for example, Alan Turing (who worked with von Neumann at Princeton), and John Mauchly and John Eckert (who led the construction of ENIAC, the first computer, in the 1940s). But it was von Neumann who first realized that program instructions should be stored in the same way in a computer as program data. To this day, such machines are called "von Neumann machines."

After the war, von Neumann became interested in artificial intelligence. In 1953 he invited Italian-Norwegian mathematician Nils Barricelli to Princeton to study artificial life. Barricelli used the new digital computers to write simulations of evolutionary processes. He was not interested in biological evolution, and he was not interested in solving optimization problems. He wanted to create artificial life inside a computer by using processes that are found in nature (e.g., reproduction and mutation). In 1953 he wrote, "A series of numerical experiments are being made with the aim of verifying the possibility of an evolution similar to that of living organisms, taking place in an artificially created universe" [Dyson, 1998, page 111]. Barricelli became the first person to write genetic algorithm software. His first work on the subject was published in Italian in 1954 with the title "Esempi numerici di processi di evoluzione" (Numerical models of evolutionary processes) [Barricelli, 1954].

Alexander Fraser, born in London in 1923, followed shortly after Barricelli and used computer programs to simulate evolution. His education and career took him to Hong Kong, New Zealand, Scotland, and finally, in the 1950s, to the Commonwealth Scientific and Industrial Research Organisation in Sydney, Australia. Fraser was not an engineer; he was a biologist, and he was interested in evolution. He couldn't observe evolution happening in the world around him because it was too slow, requiring time periods on the order of millions of years. So Fraser decided that he would study evolution by creating his own universe inside of a digital computer. That way he could speed up the process and observe how evolution really worked. In 1957 Fraser wrote a paper titled "Simulation of genetic systems by automatic digital computers" [Fraser, 1957] becoming the first to use computer simulations for the express purpose of studying biological evolution. He published many papers about his work, mostly in biology journals. In the late 1950s and 1960s, many other biologists followed in his steps and began using computers to simulate biological evolution.

Hans-Joachim Bremermann, a mathematician and physicist, also performed early computer simulations of biological evolution. His first work on the subject was published as a technical report in 1958 while he was a professor at the University of Washington, and was titled "The evolution of intelligence" [Fogel and Anderson, 2000]. Bremermann worked for most of his career at the University of California, Berkeley, where in the 1960s he used computer simulations to study the operation of complex systems, especially evolution. But his computer programs didn't just model evolution – they also simulated parasite/host interactions, pattern recognition by the human brain, and immune system response.

George Box, born in 1919 in England, was also interested in artificial evolution, but unlike his predecessors, he wasn't interested in artificial life or evolution for its own sake. He wanted to solve real-world problems. Box used statistics to analyze the design and results of experiments, then he became an industrial engineer and used statistics to optimize manufacturing processes. What's the best way to lay out the machines on the plant floor to maximize the production of widgets? What's the best way to schedule the flow of material through the plant? During the 1950s, Box developed a technique that he called "evolutionary operation" as a way of optimizing an industrial process while it was operating. His work was not a GA per se, but it did use the idea of evolution via an accumulation of many incremental changes to optimize an engineering design. His first paper on the subject was published in 1957 with the title "Evolutionary operation: A method for increasing industrial productivity" [Box, 1957].

George Friedman, like George Box, was also a practical man. For his 1956 Master's thesis at UCLA, he designed a robot that could learn how to build electric circuits to control its own behavior. The title of his thesis was "Selective Feedback Computers for Engineering Synthesis and Nervous System Analogy" [Friedman, 1998], [Fogel, 2006]. His work was similar to today's GAs, although he used the term "selective feedback computer" to describe his approach. The last paragraph of his conclusion states, "The concepts and schematic illustrations in this paper, while not conclusively demonstrating the usefulness of [GAs] ... did at least indicate a possible area for further investigation." Indeed! Now, more than a half century after Friedman's thesis, thousands of technical articles are published every year on the topic of genetic algorithms.

Another pioneer in the area of genetic algorithms was Lawrence Fogel, who began working on GAs in 1962. In 1966, along with Alvin (Al) Owens and Michael (Jack) Walsh, he wrote the first book about GAs: Artificial Intelligence through Simulated Evolution [Fogel et al., 1966]. Fogel's early work in genetic algorithms was motivated by engineering problems such as the prediction of signals, modeling combat, and controlling engineering systems. Lawrence Fogel's son, David Fogel, edited an important volume that contains 31 foundational papers about GAs and related topics [Fogel, 1998].

After the seminal work of Barricelli, Fraser, Bremermann, Box, and Friedman in the 1950s, others began using genetic algorithms to study biological evolution and to solve engineering problems. Some important advances in genetic algorithms were made in the 1960s by John Holland, a professor of psychology, electrical engineering, and computer science at the University of Michigan. In the 1960s Holland was interested in adaptive systems. He wasn't necessarily interested in evolution or optimization, but rather in how systems adapt to their surroundings. He began teaching and conducting research in these areas, and in 1975 he wrote his famous book *Adaptation in Natural and Artificial Systems* [Holland, 1975]. The book became a classic because of its presentation of the mathematics of evolution. Also in 1975, Holland's student Kenneth De Jong finished his doctoral dissertation, titled "An analysis of the behavior of a class of genetic adaptive systems." De Jong's dissertation was the first systematic and thorough investigation of the use of GAs for optimization. De Jong used a set of sample problems to explore the effects of various GA parameters on optimization performance. His work was so thorough that for a long time any optimization paper that did not include De Jong's benchmark problems was considered inadequate.

It was in the 1970s and 1980s that GA research increased exponentially. This was probably due to several factors. One factor was the increased computing power that became available with the popularization and commercialization of the transistor in the 1950s, which exponentially increased computing capabilities. Another factor was the increased interest in biologically-motivated algorithms as researchers saw the limitations of conventional computing. Fuzzy logic and neural network research, two other biologically-motivated computing algorithms, also increased exponentially in the 1970s and 1980s, even though those paradigms do not require much computing power.

3.4 A SIMPLE BINARY GENETIC ALGORITHM

Suppose you have a problem that you want to solve. If you can represent each possible solution to the problem as a bit string, then a GA might be able to solve the problem. Each potential solution is called a "candidate solution" or "individual." A group of individuals is called the "population" of the GA. This means that we need to encode each problem parameter as a bit string. This section introduces GAs by giving a couple of simple examples. We do not present these examples as realistic problems, but we present them as straightforward problems that nicely illustrate the essential features of GAs.

3.4.1 A Genetic Algorithm for Robot Design

Suppose that our problem involves the design of a low-weight mobile robot that has enough power to navigate rough terrain, and enough range that it does not need to return to its home base too often. The parameters that we need to specify in our robot design include the motor type and size, and the power source type and size. The motor type and size might be encoded as follows:

$$\begin{aligned}
000 &= \text{5-volt step motor} \\
001 &= \text{9-volt step motor} \\
010 &= \text{12-volt step motor} \\
011 &= \text{24-volt step motor} \\
100 &= \text{5-volt servo motor} \\
101 &= \text{9-volt servo motor} \\
110 &= \text{12-volt servo motor} \\
111 &= \text{24-volt servo motor.}
\end{aligned} \tag{3.1}$$

The power source type and size might be encoded as follows:

$$\begin{aligned}
000 &= \text{12-volt nickel-cadmium battery} \\
001 &= \text{24-volt nickel-cadmium battery} \\
010 &= \text{12-volt lithium-ion battery} \\
011 &= \text{24-volt lithium-ion battery} \\
100 &= \text{12-volt solar panel} \\
101 &= \text{24-volt solar panel} \\
110 &= \text{12-volt fusion reactor} \\
111 &= \text{24-volt fusion reactor.}
\end{aligned} \tag{3.2}$$

The encoding of the system parameters is a crucial aspect of the GA and will have a significant influence on whether the GA really works or not.

After we have decided on an encoding scheme, we need to decide how to evaluate the "fitness" of each potential problem solution. In our robot example, we might have a formula that relates robot weight to motor type/size and to power source type/size. We might have other formulas that relate the robot power to the motor and to the power source, and that relates robot range to the motor and to the power source. We can't simulate evolution if we don't have a good definition for fitness. Alternatively, we might have a computer simulation in which we could input motor type/size and power source type/size, and output a measure of how well our design works. This is where the GA designer's understanding of the problem is critical. There are no hard-and-fast rules for defining the fitness function of the GA problem. It is up to the GA designer to understand the problem well enough so that he can define a fitness function that makes sense. In our example, we might have a fitness function that looks like the following:

$$\text{Fitness} = \text{Range (hours)} + \text{Power (Watts)} - \text{Weight (kilograms)}. \tag{3.3}$$

The range, power, and weight might each be complicated functions involving motor type and power source type, or they might be determined by the output of simulation software or hardware experiments.[4]

We begin the GA by randomly generating a set of individuals. Consider two of the individuals in a GA population. The first individual is a robot design with a 12-volt step motor and a 24-volt solar panel, and the second individual is a robot design with a 9-volt servo motor and a 24-volt nickel-cadmium battery. These two individuals are specified as follows:

$$\begin{aligned} \text{Individual 1} &= \underbrace{\text{12-volt step motor}}_{010}, \underbrace{\text{24-volt solar panel}}_{101} \\ \text{Individual 2} &= \underbrace{\text{9-volt servo motor}}_{101}, \underbrace{\text{24-volt NiCad battery}}_{001}. \end{aligned} \tag{3.4}$$

Individual 1 is encoded with the bit string 010101, and individual 2 is encoded with the bit string 101001. Each bit is called an allele. A sequence of bits in an individual that contains information about some trait of that individual is called a gene. Specific genes are called genotypes, and the problem-specific parameter that a genotype represents is called a phenotype. In our robot example, each individual has two genes: one for the motor size/type, and one for the battery size/type. Individual 1 has a motor genotype of 010, which corresponds to a phenotype of "12-volt step motor," and a battery genotype of 101, which corresponds to a phenotype of "24-volt solar battery." The collection of all genes in an individual is called a chromosome. Individual 1 has a chromosome of 010101.

3.4.2 Selection and Crossover

A GA might have many individuals. Typical GAs have dozens or hundreds of individuals. The two individuals above could mate, just as individuals in biological populations mate. In order to mate, we cause them to "cross over," which means that each individual shares some of its genetic information with its offspring. To find the crossover point, we choose a random number between 1 and 5. Suppose that we choose the random number 2. That means the two individuals swap all of their alleles after the second bit position in each chromosome, as shown in Figure 3.1.

[4]Because of unit differences, we can't add hours and watts and kilograms together, so we would need some scaling parameters to weight the relative importance of each term, and to convert the quantities to units that can be added together. Nevertheless, this equation gives a general idea of how to formulate a fitness equation.

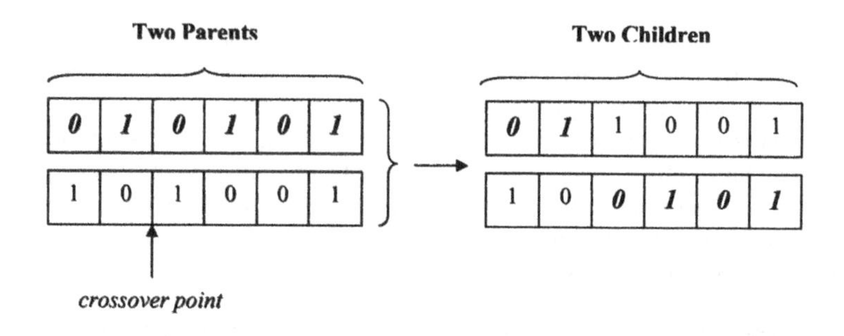

Figure 3.1 Illustration of crossover in a binary GA. The crossover point is randomly chosen. The two parents produce two children.

The two parents have mated (i.e., crossed over) to produce two children. Each child receives some genetic information from one parent, and the other genetic information from the other parent. The parents die and the children survive to continue the evolutionary process. This event is called one generation of the GA.

Just as in biology, some of the children will have high fitness while others will have low fitness. Low-fitness individuals have a high probability of dying in their generation; that is, they are removed from the GA simulation. High-fitness individuals survive to cross over with other high-fitness individuals, and thereby produce a new generation of individuals. This process is continued until the GA finds an acceptable solution to the optimization problem.

At some point in our GA software we will have to decide which individuals mate to produce children. This decision is based on the fitness of the individuals in the population. The most fit individuals are likely to mate to produce children, while the least fit individuals are unlikely to find mates and therefore are likely to die without producing any offspring.

One common way to select parents is roulette-wheel selection, which is also called fitness-proportional selection, or fitness-proportionate selection. Suppose we have four individuals in our population. (A real GA would have many more than four individuals, but this example is just for the sake of illustration.) Suppose that the individual fitnesses are evaluated as follows:

$$\begin{aligned} \text{Individual 1: Fitness} &= 10 \\ \text{Individual 2: Fitness} &= 20 \\ \text{Individual 3: Fitness} &= 30 \\ \text{Individual 4: Fitness} &= 40 \end{aligned} \tag{3.5}$$

Individual 4 is the most fit and Individual 1 is the least fit. We create a roulette wheel with each slot area corresponding to the fitness of one of the individuals. In our example, the roulette wheel is shown in Figure 3.2.

To keep the population size constant from one generation to the next, we pick two pairs of mates. This will produce four children total. To pick the first pair of mates, we spin an imaginary (computer simulated) spinner, and wherever it ends up in the roulette wheel decides who the first parent is. We see from the roulette wheel in Figure 3.2 that Individual 1 has a 10% chance of being selected, Individual 2 has a 20% chance of being selection, Individual 3 has a 30% chance

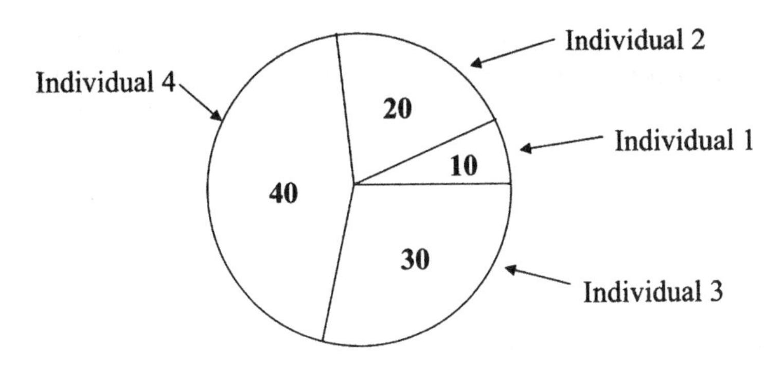

Figure 3.2 The above pie chart illustrates roulette-wheel selection in a GA for a four-member population. Each individual is assigned a slice that is proportional to its fitness. Each individual's selection as a parent is proportional to its slice in the roulette wheel.

of being selected, and Individual 4 has a 40% chance of being selected. In other words, each individual has a probability of being selected that is proportional to its fitness. Next we spin the roulette wheel a second time to select a second parent. If the roulette wheel stops at the same individual as the first spin, then we spin again – a parent can't mate with itself. After we have two parents, they cross over to produce two children. We then repeat the process to obtain two more parents, mate them, and obtain two more children. This process continues to repeat until the population of children is the same size as the population of parents. This idea is illustrated in Figure 3.3.

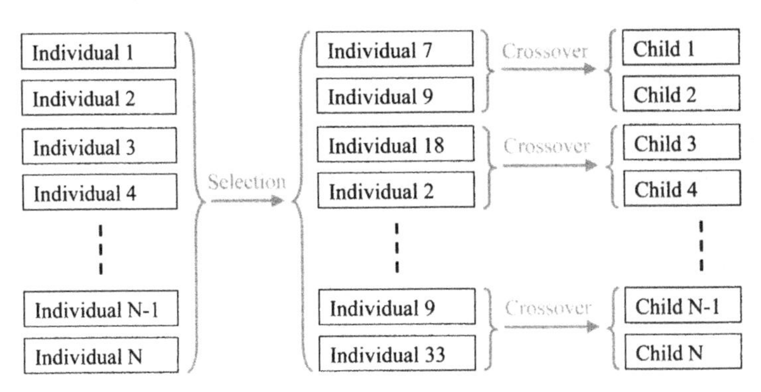

Figure 3.3 Illustration of the crossover of population of parents to create a population of children. The initial population of N individuals on the left undergoes a selection process, perhaps roulette-wheel selection, to create a set of N parents. Some individuals may be selected more than once, while other individuals may not be selected at all. Then each pair of parents in the middle crosses over to create a pair of children. Adapted from [Whitley, 2001].

Given the fitness values shown in Figure 3.2, the process of spinning the roulette wheel to select a parent can be accomplished as shown in Figure 3.4. We repeat the process of Figure 3.4 four times to select four parents, which create four children for the next generation.

In general, given a population of N individuals, Figure 3.5 shows the pseudo-code to select a parent using roulette-wheel selection. We repeat the process of Figure 3.5 as many times as necessary to select parents for the creation of children for the next generation.

Generate a uniformly distributed random number $r \in [0, 1]$
If $r < 0.1$ then
    Parent = Individual 1
else if $r < 0.3$ then
    Parent = Individual 2
else if $r < 0.6$ then
    Parent = Individual 3
else
    Parent = Individual 4
End if

Figure 3.4 The above pseudo-code shows how to select one parent based on the roulette wheel of Figure 3.2.

$x_i = i$-th individual in population, $i \in [1, N]$
$f_i \leftarrow \text{fitness}(x_i)$ for $i \in [1, N]$
$f_{\text{sum}} = \sum_{i=1}^{N} f_i$
Generate a uniformly distributed random number $r \in [0, f_{\text{sum}}]$
$F \leftarrow f_1$
$k \leftarrow 1$
While $F < r$
    $k \leftarrow k + 1$
    $F \leftarrow F + f_k$
End while
Parent $\leftarrow x_k$

Figure 3.5 The above pseudo-code shows how to select one parent from N individuals using roulette-wheel selection. This code assumes that the fitness value $f_i \geq 0$ for all $i \in [1, N]$.

3.4.3 Mutation

The final step in the GA is called mutation. Mutation in biology is relatively rare, at least in so far as it noticeably affects offspring. In most GA implementations, mutation is also rare (on the order of 2%). But we cannot say, in general, what the correct setting should be for a GA mutation rate. The best mutation rate depends on the problem, population size, encoding, and other factors. Regardless of its frequency, mutation is important because it allows the evolutionary process to explore new potential solutions to the problem. If some genetic information is missing from the population, mutation provides the possibility of injecting that information into the population. This is important in biological evolution, but even more important in GAs. This is because GAs typically have such small population sizes that inbreeding can easily become a problem, and evolutionary dead ends are more common in GAs than in biological evolution. In biological evolution we typically talk about populations of millions, while in GAs we talk about populations of dozens or hundreds.

To implement mutation, we select a mutation probability, say 1%. This means that after the crossover process produces offspring, each bit in each child has a 1% probability of flipping to the opposite value (a 1 changes to a 0, or a 0 changes to a 1). Mutation is simple, but it's important to select a reasonable mutation probability. Too high of a mutation probability makes the GA behave like a random search, which is not usually a great way to solve a problem. Too low of a mutation probability results in problems with inbreeding and evolutionary dead ends, which also prevents the GA from finding a good solution.

If we have a population of N individuals x_i, where each individual has n bits, and our mutation rate is ρ, then at the end of each generation, we flip each bit in each individual with a probability of ρ:

$$
\begin{aligned}
r &\leftarrow U[0,1] \\
x_i(k) &\leftarrow \begin{cases} x_i(k) & \text{if } r \geq \rho \\ 0 & \text{if } r < \rho \text{ and } x_i(k) = 1 \\ 1 & \text{if } r < \rho \text{ and } x_i(k) = 0 \end{cases}
\end{aligned} \tag{3.6}
$$

for $i \in [1, N]$ and $k \in [1, n]$, where $U[0,1]$ is a random number that is uniformly distributed on $[0,1]$.

3.4.4 GA Summary

This section has given a simple example of a GA for robot design, and a discussion of selection, crossover, and mutation in GAs. Now we bring all of this material together to summarize the outline of a GA in Figure 3.6.

3.4.5 GA Tuning Parameters and Examples

Figure 3.6 outlines a simple GA, but we can see a lot of flexibility in the implementation of the algorithm of Figure 3.6. For instance, the stopping criteria for a GA can include a few different options – the same options as with any other iterative optimization algorithm. One possibility is that the GA can run for a predetermined number of generations. Another possibility is for the GA to run until the fitness

```
Parents ← {randomly generated population}
While not(termination criterion)
    Calculate the fitness of each parent in the population
    Children ← ∅
    While |Children| < |Parents|
        Use fitnesses to probabilistically select a pair of parents for mating
        Mate the parents to create children c1 and c2
        Children ← Children ∪ {c1, c2}
    Loop
    Randomly mutate some of the children
    Parents ← Children
Next generation
```

Figure 3.6 The above pseudo-code illustrates a simple genetic algorithm.

of the best individual is better than some user-defined threshold. If our problem is to find a solution that is "good enough," then this might be a reasonable stopping criterion. Another possibility is for the GA to run until the fitness of the best individual stops improving from one generation to the next. This indicates that the evolutionary process has reached a plateau and cannot improve any further.

There are a number of parameters that the GA designer needs to specify to obtain good results. The selection of these parameters can often spell the difference between success and failure. Some of these parameters include the following:

1. An encoding scheme that maps problem solutions to bit strings. Some examples below illustrate binary encoding of real numbers, Section 3.5 discusses GAs with real-valued parameters, and Section 8.3 discusses gray coding in binary GAs.

2. A fitness function that maps problem solutions to fitness values.

3. Population size.

4. Selection method. Above we talked about roulette-wheel selection, but other types of selection are also possible, including tournament selection, rank selection, and many other variations. Section 8.7 discusses some of these options.

5. Mutation rate. A GA that uses a mutation rate that is too high will degenerate into a random search. But a GA that uses a mutation rate that is too low will not be able to sufficiently explore the search space.

6. Fitness scaling. This defines how the fitness function is implemented. Sometimes a fitness function is poorly defined so that all of the individuals have fitness values that are very close to each other. If the fitness values are clumped together, the selection process cannot distinguish well between high-fit and low-fit individuals. This prevents the more fit individuals from propagating to the next generation. The opposite problem also occurs sometimes; the fitness values are spread apart too much, so that low-fitness individuals don't have

any chance of being selection for reproduction. Section 8.7 discusses fitness scaling.

7. Crossover type. Above we talked about crossover at one point in each chromosome pair, but we could cross over at multiple points also. Section 8.8 discusses different types of crossover.

8. Speciation/incest. Some GA researchers allow individuals to mate only if they are similar enough to each other; that is, only if they belong to the same "species." Other GA researchers allow individuals to mate only if they are different enough from each other; that is, only if they belong to different "families." Section 8.6 discusses some of these ideas.

These issues also apply to EAs other than GAs, and so Chapter 8 discusses these issues and several others.

■ EXAMPLE 3.2

Consider the minimization problem of Example 2.2:

$$\min_x f(x) \text{ where } f(x) = x^4 + 5x^3 + 4x^2 - 4x + 1. \tag{3.7}$$

Suppose we know ahead of time, somehow, that the minimum of $f(x)$ occurs in the domain $x \in [-4, -1]$. We choose to encode x with four bits:

$$\begin{aligned}
0000 &= -4.0, & 0001 &= -3.8, \\
0010 &= -3.6, & 0011 &= -3.4, \\
0100 &= -3.2, & 0101 &= -3.0, \\
0110 &= -2.8, & 0111 &= -2.6, \\
1000 &= -2.4, & 1001 &= -2.2, \\
1010 &= -2.0, & 1011 &= -1.8, \\
1100 &= -1.6, & 1101 &= -1.4, \\
1110 &= -1.2, & 1111 &= -1.0.
\end{aligned} \tag{3.8}$$

The encoding scheme is a tradeoff between accuracy and complexity. More bits will give us more resolution, but will also make the GA more complicated. Consider a randomly generated initial population of four individuals:

$$\begin{aligned}
x_1 &= 1100, \\
x_2 &= 1011, \\
x_3 &= 0010, \\
x_4 &= 1001.
\end{aligned} \tag{3.9}$$

We want to minimize $f(x)$, but GAs are designed to maximize fitness.[5] We therefore need to convert the minimization problem to a maximization problem so that it fits in the GA framework. We can do this by maximizing the

[5]This statement assumes that we use roulette wheel selection. Some of the selection methods that we discuss in Section 8.7 do not assume that the underlying problem is a maximization problem.

negative of $f(x)$. The fitness values are thus obtained by decoding the individuals using the genotype/phenotype combinations shown in Equation (3.8) and then evaluating $-f(x)$:

$$\begin{aligned}
\text{fitness}(x_1) &= -f(-1.6) = -3.71 \\
\text{fitness}(x_2) &= -f(-1.8) = -2.50 \\
\text{fitness}(x_3) &= -f(-3.6) = -1.92 \\
\text{fitness}(x_4) &= -f(-2.2) = +0.65.
\end{aligned} \tag{3.10}$$

Now we arbitrarily add some offset to each fitness value so that they are all greater than 0. This is necessary so that we can later assign percentage fitness values to each individual:

$$\begin{aligned}
f_1 &= -3.71 + 10 = 6.29 \\
f_2 &= -2.50 + 10 = 7.50 \\
f_3 &= -1.92 + 10 = 8.08 \\
f_4 &= +0.65 + 10 = 10.65.
\end{aligned} \tag{3.11}$$

Now we compute the relative fitness values of each individual. The relative fitness value of each individual is its probability of selection when the roulette wheel is spun:

$$\begin{aligned}
p_1 &= f_1/(f_1 + f_2 + f_3 + f_4) = 0.19 \\
p_2 &= f_2/(f_1 + f_2 + f_3 + f_4) = 0.23 \\
p_3 &= f_3/(f_1 + f_2 + f_3 + f_4) = 0.25 \\
p_4 &= f_4/(f_1 + f_2 + f_3 + f_4) = 0.33.
\end{aligned} \tag{3.12}$$

The initial population is summarized in Table 3.1. Table 3.1 shows us that x_2 and x_3 both have about a 25% chance of selection on each roulette wheel spin, while x_4 has almost twice the probability of selection as x_1. To begin the first GA generation, we generate four uniformly distributed random numbers in the domain $[0, 1]$ and use them to select four parents. Suppose that this process results in the selection of x_3, x_4, x_4, and x_1. This means that we want to cross x_3 and x_4 to get two children, and x_4 and x_1 to get two more children. Remember that the crossover point is randomly selected for each pair of parents. This is shown in Table 3.2.

Individual Number	Genotype	Phenotype	Fitness	Selection Probability
x_1	1100	−1.4	−4.56	0.19
x_2	1011	−1.8	−2.50	0.23
x_3	0010	−3.6	−1.92	0.25
x_4	1001	−2.2	+0.65	0.33

Table 3.1 Example 3.2: Initial population for a simple GA.

Parents		Children	
Individual	Genotype	Genotype	Fitness
x_3	***0*** 010	0001	−8.11
x_4	1***001***	1010	−1.00
x_4	***10*** 01	1000	+2.30
x_1	11***00***	1101	−4.56

Table 3.2 Example 3.2: Crossover for a simple GA. The randomly-chosen crossover points are indicated with bold italics. The crossover point is between the first two bits for the first set of parents, and in the middle of the chromosome for the second set of parents.

We see from Table 3.2 that the best child has a fitness of 2.30, which is better than the best individual of the initial generation (0.65). The GA has taken a significant step toward optimizing $f(x)$. There are no guarantees that the children will be better than the parents, but this simple example illustrates how a GA can home in on the solution to an optimization problem.

■ **EXAMPLE 3.3**

Consider the minimization problem of Example 2.5:

$$\min_{x,y} f(x,y), \text{ where} \tag{3.13}$$

$$f(x,y) = e - 20\exp\left(-0.2\sqrt{\frac{x^2+y^2}{2}}\right) - \exp\left(\frac{\cos(2\pi x)+\cos(2\pi y)}{2}\right).$$

Suppose, as in Example 2.5, that x and y can both range from -5 to $+5$. We need to decide on the resolution that we want to use for x and y in our GA. If we want a resolution of 0.25 or better for each independent variable x and y, then we need to use six bits for both x and y:

$$\begin{aligned} x \text{ genotype} &= x_g \in [0,63] \\ x \text{ phenotype} &= -5 + \frac{10x_g}{63} \in [-5,5] \\ y \text{ genotype} &= y_g \in [0,63] \\ y \text{ phenotype} &= -5 + \frac{10y_g}{63} \in [-5,5]. \end{aligned} \tag{3.14}$$

This gives us a resolution of $10/63 = 0.159$ for each bit of x_g and y_g. Let us run a GA for 10 generations to try to minimize $f(x,y)$. We need to decide on a population size and a mutation rate. Let us use a population size of 20 and a mutation rate of 2% per bit. A typical GA run gives the results shown in Figure 3.7. Each run will be different because of the random numbers used in the GA, but Figure 3.7 shows typical results. As the number of generations increases, we see a decrease in both the minimum cost (that is, the cost of the best individual) and the average cost of the population.

Figure 3.8 shows a contour plot of $f(x, y)$, along with the location on the plot of the GA individuals, at the first, fourth, seventh, and tenth generations. We see from Figure 3.8 that the population is initially scattered throughout the domain because of our random initialization. As the GA progresses, the population begins to cluster, and the individuals tend to move closer to the minimum, which is at the center of the plot.

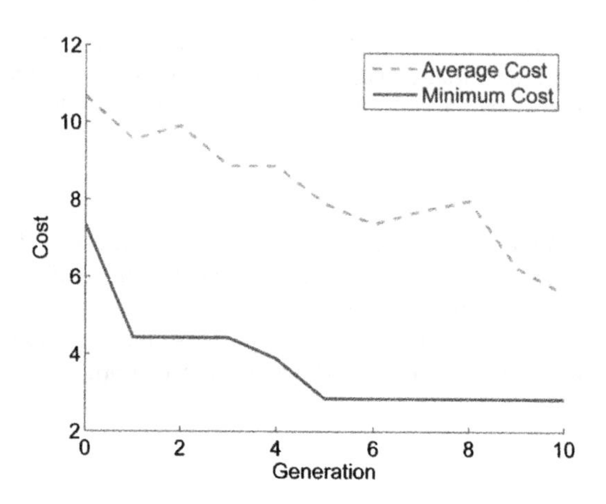

Figure 3.7 Example 3.3: Typical GA simulation results for the minimization of the two-dimensional Ackley function.

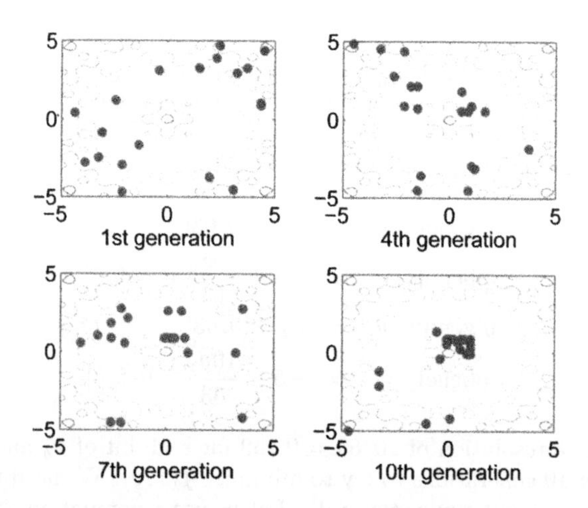

Figure 3.8 Example 3.3: Typical GA simulation results for the minimization of the two-dimensional Ackley function. As the GA progresses, the individuals in the population gradually begin to cluster together and move toward the minimum, which is at the center of the contour plot.

The contour plot of Figure 3.8 shows how difficult it might be to minimize a multimodal, high dimensional function. You can imagine that you are standing on a landscape with hills and valleys of Figure 3.8, and you want to find the lowest point of the landscape. It would be difficult to do because there are so many peaks and valleys. However, a group of individuals scattered throughout the landscape would have a better chance of finding the lowest point. The individuals could learn from each other: "This valley looks pretty low; let's explore it," says one. "No, this one looks lower; come on over here!" The individuals cooperate with each other and together find the lowest point in the valley. That is similar to how a GA and other EAs work. The individuals in the population work together to find a good solution to the problem.

3.5 A SIMPLE CONTINUOUS GENETIC ALGORITHM

Figure 3.6 outlines a simple binary GA, and that is exactly the algorithm that we used in Examples 3.2 and 3.3. However, the problems in those examples are defined on a continuous domain, and so we had to discretize the domain to apply the binary GA. It would be simpler and more natural if we could apply a GA directly to the continuous domain of the problems. We use the term continuous GAs, or real-coded GAs, to refer to GAs that operate directly on continuous variables.

The extension of GAs from binary domains to continuous domains is pretty straightforward. In fact, we can still use the algorithm of Figure 3.6 – we just need to modify some of the steps in that algorithm. Look at the operations in Figure 3.6 and consider how they might work on an optimization problem with a continuous domain.

1. In Figure 3.6, we first generate a random initial population. We can easily do this on a continuous domain. Suppose that we want to generate N individuals in our GA. Then we denote the i-th individual as x_i for $i \in [1, N]$. Also suppose that we want to minimize an n-dimensional function on a continuous domain. Then we use $x_i(k)$ to denote the k-th element of x_i:

$$x_i = \begin{bmatrix} x_i(1) & x_i(2) & \cdots & x_i(n) \end{bmatrix}. \tag{3.15}$$

Suppose that the search domain of the k-th dimension is $[x_{\min}(k),\, x_{\max}(k)]$:

$$x_i(k) \in [x_{\min}(k),\, x_{\max}(k)] \tag{3.16}$$

for $i \in [1, N]$ and $k \in [1, n]$. We can generate a random initial population, as in the first line of Figure 3.6, as follows:

For $i = 1$ to N
    For $k = 1$ to n
        $x_i(k) \leftarrow U[x_{\min}(k),\, x_{\max}(k)]$
    Next k
Next i

That is, we simply set each $x_i(k)$ equal to a realization of a random variable that is uniformly distributed between $x_{\min}(k)$ and $x_{\max}(k)$.

2. Next, we begin the "while not(termination criterion) loop" in Figure 3.6. The first step in that loop is to calculate the fitness of each individual. If we are trying to maximize $f(x)$, then we calculate the fitness of each x_i by computing $f(x_i)$. If we are trying to minimize $f(x)$, then we calculate the fitness of each x_i by computing the negative of $f(x_i)$.

3. Next, we begin the "while |Children| < |Parents|" loop in Figure 3.6. The first step in that loop is to "use fitnesses to probabilistically select a pair of parents for mating." We perform this step using roulette-wheel selection, as we discussed in Section 3.4.2. We discuss other options for this step in Section 8.7, but for now we simply use roulette-wheel selection.

4. Next, we perform the "Mate the parents" step in Figure 3.6 to create two children. We perform this step using single-point crossover as illustrated in Figure 3.1. The only difference is that we combine continuous-domain individuals rather than binary-domain individuals. We illustrate single-point crossover for continuous-domain individuals in Figure 3.9. We discuss other types of crossover for continuous GAs in Section 8.8.

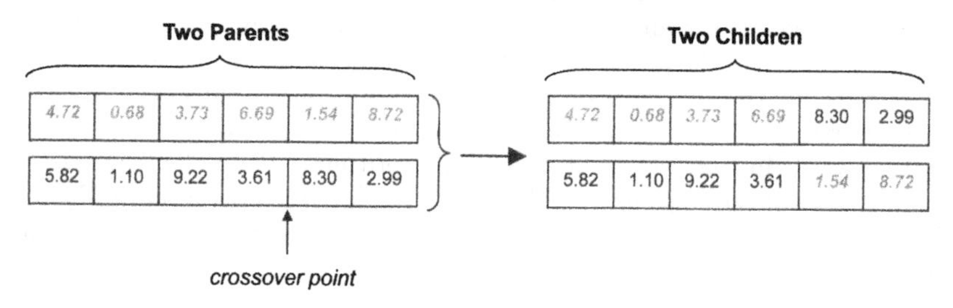

Figure 3.9 Illustration of crossover in a continuous-domain GA. The crossover point is randomly chosen. The two parents produce two children.

5. Next, we perform the "Randomly mutate" step in Figure 3.6. In binary EAs, mutation is a straightforward operation, as shown in Equation (3.6). In a continuous GA, we mutate $x_i(k)$ by assigning it a random number that is generated from a uniform distribution on the search domain:

$$\begin{aligned} r &\leftarrow U[0,1] \\ x_i(k) &\leftarrow \begin{cases} x_i(k) & \text{if } r \geq \rho \\ U[x_{\min}(k), x_{\max}(k)] & \text{if } r < \rho \end{cases} \end{aligned} \tag{3.17}$$

for $i \in [1, N]$ and $k \in [1, n]$, where ρ is the mutation rate. We discuss other possibilities for mutation in continuous-domain GAs in Section 8.9.

Mutation in Continuous GAs

Note that a given mutation rate has a different effect in a binary GA than in a continuous GA. If we have a continuous-domain problem with n dimensions and a mutation rate of ρ_c, then each solution feature of each child has a probability of ρ_c of being mutated. For example, in Figure 3.9, each of the six components of both children has a mutation probability of ρ_c. Also, mutation in a continuous GA results in the solution feature being taken from a uniform distribution between its minimum and maximum possibility values, as shown in Equation (3.17).[6]

However, in a binary GA, we discretize each dimension of each individual. If we discretize a continuous dimension into m bits and use a mutation rate of ρ_b, then each bit has a probability of ρ_b of being mutated. That means that each bit has a probability of $1 - \rho_b$ of not being mutated. Therefore, the probability of each dimension not being mutated is equal to the probability that all m of its bits are not mutated, which is equal to $(1 - \rho_b)^m$. Therefore, the probability of the m-th dimension being mutated is $1 - (1 - \rho_b)^m$. Furthermore, if mutation does occur, then the mutated dimension is *not* uniformly distributed between its minimum and maximum values; it's distribution instead depends on which bit is mutated.

We can obtain the mutation rate ρ_c for a continuous-domain problem that has an effect that is approximately equal to the mutation rate ρ_b for a discrete problem. As we discussed above, if a binary GA for a discrete problem with m bits per dimension has a mutation rate of ρ_b, then the probability that any given dimension is not mutated is equal to $(1 - \rho_b)^m$. This can be approximated with a first-order Taylor series:

$$\begin{aligned} \Pr(\text{no mutation in a binary GA}) &= (1 - \rho_b)^m \\ &\approx 1 - m\rho_b \end{aligned} \tag{3.18}$$

where the approximation is valid for small ρ_b. If a GA for a continuous problem has a mutation rate of ρ_c, then the probability that any given dimension is not mutated is equal to $1 - \rho_c$. Equating this probability with Equation (3.18) gives

$$\begin{aligned} 1 - \rho_c &= 1 - m\rho_b \\ \rho_c &= m\rho_b. \end{aligned} \tag{3.19}$$

Therefore, the mutation process in a binary GA with m bits per dimension and a mutation rate of ρ_b, is *approximately* equivalent to the mutation process in a continuous GA with a mutation rate of $m\rho_b$. We stress the word *approximately* in the previous sentence because it is not clear that equivalent mutation rates in binary and continuous GAs give equivalent results. This is because he distribution of the magnitude of a binary GA mutation is different than that of a continuous GA mutation. An interesting topic for further work would be a thorough study of the equivalence of binary and continuous GA mutations.

■ **EXAMPLE 3.4**

Consider the minimization problem of Example 3.3:

$$\min_{x,y} f(x,y), \text{ where} \tag{3.20}$$

$$f(x,y) = e - 20\exp\left(-0.2\sqrt{\frac{x^2+y^2}{2}}\right) - \exp\left(\frac{\cos(2\pi x) + \cos(2\pi y)}{2}\right).$$

Suppose that x and y can both range from -1 to $+1$. In Example 3.3, we discretized the search domain so that we could apply a binary GA. However,

[6]Uniform mutation is probably the most classic type of mutation in continuous GAs. However, we can also choose from many other types of mutation as described in Section 8.9.

since the problem is defined on a continuous domain, it is more natural to use a continuous GA. In this example we run both the binary GA and the continuous GA for 20 generations with a population size of 10. For the binary GA, we use four bits per dimension and a mutation rate of 2% per bit, as in Example 3.3. To keep the effect of mutation approximately the same for the continuous GA as for the binary GA, we use a mutation rate of 8% in the continuous GA. We also use an elitism factor of 1, which means that we keep the best individual in the population from one generation to the next (see Section 8.4).

Figure 3.10 shows the best individual found at each generation, averaged over 50 simulations. We see that the continuous GA is significantly better than the binary GA. For continuous-domain problems, we generally (but not always) get better performance with a binary GA as we use more bits, and we get the best performance if we use a continuous GA.

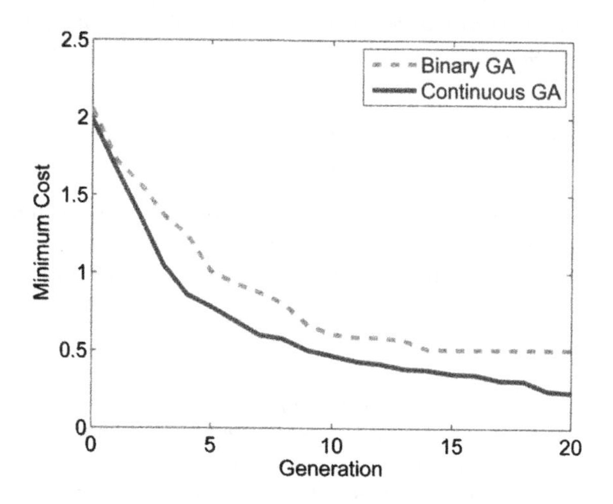

Figure 3.10 Example 3.4: Binary GA vs. continuous GA performance for the two-dimensional Ackley function. The plot shows the cost of the best individual at each generation, averaged over 50 simulations.

□

It is interesting to note that continuous GAs have a somewhat controversial history. Since GAs were originally developed for binary representations, and since all of the early GA theory was geared towards binary GAs, researchers were skeptical about the rise of continuous GAs in the 1980s [Goldberg, 1991]. However, it is difficult to argue with the success of continuous GAs, their ease of use, and their relatively recent theoretical support.

3.6 CONCLUSION

The genetic algorithm was one of the first evolutionary algorithms, and today it is probably the most popular. Recent years have seen the introduction of many competing EAs, but GAs remain popular because of their familiarity, their ease of implementation, their intuitive appeal, and their good performance on a variety of problems.

Many books and survey papers have been written about GAs over the years. David Goldberg's book [Goldberg, 1989a] was one of the first books about GAs, but like early books in many other subjects, it has aged well and is still popular because of its clear exposition. There are many other good books about GAs, including [Mitchell, 1998], [Michalewicz, 1996], [Haupt and Haupt, 2004], and [Reeves and Rowe, 2003], which is notable because of its strong emphasis on theory. Some popular tutorial papers include [Bäck and Schwefel, 1993], [Whitley, 1994], and [Whitley, 2001].

In view of the huge number of books and papers about GAs, this chapter is a necessarily brief introduction to the topic. We have neglected many GA-related issues in this chapter – not because we believe that they are unimportant, but simply because our perspective is limited. Some of these issues include messy GAs, which have variable-length chromosomes [Goldberg, 1989b], [Mitchell, 1998]; gender-based GAs, which simulate multiple genders in the GA population and are often used for multi-objective optimization (Chapter 20) [Lis and Eiben, 1997]; island GAs, which includes subpopulations [Whitley et al., 1998]; cellular GAs, which impose a specific spatial relationship among the individuals in the population [Whitley, 1994]; and covariance matrix adaptation, which is a local search strategy that can augment any EA [Hansen et al., 2003].

There are also many variations on the basic GA that we presented in this chapter. Some of those variations are extremely important, and can make the difference between success and failure in a GA application. Chapter 8 discusses many variations that apply to GAs and other EAs.

PROBLEMS

Written Exercises

3.1 Section 3.4.1 gave a simple example for how we could represent robot design parameters in a GA. Suppose that we have a GA individual given by the bit string 110010.

a) What is the chromosome for this GA individual?

b) What are the genotypes and phenotypes for this individual?

3.2 We want to use a binary GA to find x to a resolution of 0.1 to minimize the two-dimensional Rastrigin function (see Section C.1.11) on the domain $[-5, 5]$.

a) How many genes do we need for each chromosome?

b) How many bits do we need in each gene?

c) Given your answer to part (b), what is the resolution of each element of x?

3.3 We have a GA with 10 individuals $\{x_i\}$, and the fitness of x_i is $f(x_i) = i$ for $i \in [1, 10]$. We use roulette wheel selection to select 10 parents for crossover. The first two parents mate to create two children, and the next two mate to create two more children, and so forth.

a) What is the probability that the most fit individual will mate with itself at least once to create two cloned children?

b) Repeat part (a) for the least fit individual.

3.4 We have a GA with 10 individuals $\{x_i\}$, and the fitness of x_i is $f(x_i) = i$ for $i \in [1, 10]$. We use roulette wheel selection to select 10 parents for crossover.

a) What is the probability that x_{10} is not selected at all after 10 spins of the roulette wheel?

b) What is the probability that x_{10} is selected exactly once after 10 spins of the roulette wheel?

c) What is the probability that x_{10} is selected more than once after 10 spins of the roulette wheel?

3.5 Roulette wheel selection assumes that the fitness values of the population satisfy $f(x_i) \geq 0$ for $i \in [1, N]$. Suppose you have a population with fitness values $\{-10, -5, 0, 2, 3\}$. How would you propose modifying those fitness values so that you could use roulette wheel selection?

3.6 Roulette wheel selection assumes that the population is characterized by fitness values $\{f(x_i)\}$, where higher fitness values are better than lower fitness values. Suppose we have a problem whose population is characterized by cost values $\{c(x_i)\}$, where lower cost values are better than higher cost values, and $c(x_i) > 0$ for all i. How could you modify the cost values to use roulette wheel selection?

3.7 We have two parents in a binary GA, each with n bits. The i-th bit in parent 1 is different than the i-th bit in parent 2 for $i \in [1, n]$. We randomly select a crossover point $c \in [1, n]$. What is the probability that the children are clones of (that is, identical to) the parents?

3.8 Suppose we have N randomly initialized individuals in a binary GA, where each individual is comprised of n bits.

a) What is the probability that the i-th bit of each individual is the same for a given i?

b) What is the probability that the i-th bit of each individual is *not* the same for all $i \in [1, n]$?

c) Recall that $\exp(-am) \approx 1 - am$ for small am, and $(1 - a)^m \approx 1 - am$ for small values of a. Use these facts to approximate your answer to part (b) as an exponential.

d) Use your answer to part (c) to find the population size N that is required to obtain a probability p that both alleles occur at each bit position of a randomly initialized population.

e) Suppose we want to randomly initialize a population of individuals, each with 100 bits, so that there is a 99.9% or greater chance that both alleles occur at each bit position. Use your answer to part (d) to obtain the minimum population size.

3.9 We have a binary GA with a population size of N and a mutation rate of ρ, and each individual is comprised of n bits.

a) What is the probability that we will not mutate any bits in the entire population for one generation?

b) Use your answer to part (a) to find the minimum mutation rate ρ for a given population size N and bit length n such that the probability of no mutations during each generation is no greater than P_{none}.

c) Use your answer to part (b) to find the minimum mutation rate ρ such that the probability of not mutating any bits is 0.01% when $N = 100$ and $n = 100$.

Computer Exercises

3.10 Write a computer simulation to confirm your answers to Problem 3.3.

3.11 Write a computer simulation to confirm your answer to Problem 3.8.

3.12 The one-max problem is the search for a string of n bits with as many 1's as possible. The fitness of a bit string is the number of 1's. Of course, we can easily solve this by simply writing n consecutive 1's, but in this problem we are interested in seeing if a GA can solve the one-max problem. Write a GA to solve the one-max problem. Use $n = 30$, generation limit $= 100$, population size $= 20$, and mutation rate $= 1\%$.

a) Plot the fitness of the best individual, and the average fitness of the population, as a function of generation number.

b) Run 50 Monte Carlo simulations of your GA. This will result in 50 plots of the fitness of the best individual as a function of generation number. Plot the average of those 50 plots. Denote the average of the 50 best fitness values at the 100th generation as $\bar{f}(x^*)$. What is $\bar{f}(x^*)$?

c) Repeat part (b) with a population size of 40. How does $\bar{f}(x^*)$ change compared to your answer from part (b)? Why?

d) Set the population size back to 20 and change the mutation rate to 5%. How does $\bar{f}(x^*)$ change compared to your answer from part (b)? Why?

e) Set the mutation rate to 0%. How does $\bar{f}(x^*)$ change compared to your answer from part (b)? Why?

f) Instead of setting the fitness equal to the number of 1's, set the fitness equal to the number of 1's plus 50. Now repeat part (b). How does $\bar{f}(x^*)$ change compared to your answer from part (b)? Why?

g) As in part (b), set fitness equal to the number of 1's; but then for all individuals with fitness less than average, set fitness to 0. How does $\bar{f}(x^*)$ change compared to your answer from part (b)? Why?

3.13 Write a continuous GA to minimize the sphere function (see Section C.1.1). Set the search domain in each dimension to $[-5, +5]$, the problem dimension to 20, the generation limit to 100, the population size to 20, and the mutation rate to 1%. For roulette wheel selection, we need to map the cost values $c(x_i)$ to fitness values $f(x_i)$. Do this as follows: $f(x_i) = 1/c(x_i)$.

a) Plot the cost of the best individual, and the average cost of the population, as a function of generation number.

b) Run 50 Monte Carlo simulations of your GA. This will result in 50 plots of the cost of the best individual as a function of generation number. Plot the average of those 50 plots. Denote the average of the 50 best cost values at the 100th generation as $\bar{c}(x^*)$. What is $\bar{c}(x^*)$?

c) Repeat part (b) with a mutation rate of 2%. How does $\bar{c}(x^*)$ change compared to your answer from part (b)? Repeat with a mutation rate of 5%.

CHAPTER 4

Mathematical Models of Genetic Algorithms

But program source code is not necessarily the most perspicuous description possible.
—Michael Vose [unpublished course notes, 2010]

The study of evolutionary algorithms (EAs) has often been ad-hoc, simulation-based, heuristic, and non-analytic. Historically, engineers have been more concerned with the question of *whether* EAs work, rather than *how* or *why* they work. However, with the maturing of EA research in the last couple of decades of the 20th century, engineers began focusing more on the *how* and *why* questions. In this chapter, we discuss some ways to answer these questions for GAs. This chapter is the most technical and mathematical one in this book. The student who wants only a working knowledge of EAs can skip this chapter. However, it is important for the student who wants to become well-informed and well-rounded in the area of EA research to understand the ideas in this chapter. The student who takes the time and effort to understand this material might find unexpected and brand new avenues of research.

Overview of the Chapter

One of the early answers to the *how* and *why* questions was schema theory, which analyzes the growth and decay over time of various bit combinations in GAs, and so we discuss this in Section 4.1. Some of the more recent mathematical analyses of GAs have relied on Markov models and dynamic system models, and we also explore those approaches in this chapter. These models have their own shortcomings, but their shortcomings are in the area of implementation and computing resources rather than theory. Section 4.2 gives an overview of Markov theory, which was developed by the Russian mathematician Andrey Markov[1] in 1906 [Seneta, 1966]. Markov theory has become a fundamental area of mathematics, with applications in physics, chemistry, computer science, social science, engineering, biology, music, athletics, and other surprising areas. We will see in this chapter that Markov theory can also provide insight into GA behavior. Section 4.3 presents some notation and preliminary results that we will use in later sections as we develop Markov models and dynamic system models.

Section 4.4 develops a Markov chain model for a GA that uses fitness-based selection, followed by mutation, followed by single-point crossover. Unfortunately, the dimension of the Markov model grows factorially (i.e., faster than exponentially) with the population size and the search space size. This limits its application to very small problems. However, Markov models are still useful for giving exact results without the need to rely on the random nature of stochastic simulations.

Section 4.5 develops a dynamic system model for a GA. The dynamic system model is based on the Markov model, but the application is quite different. The Markov model gives the steady-state probability as the generation count approaches infinity of each possible population. The dynamic system model gives the time-varying proportion of each individual in the search space as the population size approaches infinity.

4.1 SCHEMA THEORY

Consider the simple problem $\max_x f(x)$, where $f(x) = x^2$. Suppose we encode x as a 5-bit integer, where the bit string 00000 represents decimal 0, and the bit string 11111 represents decimal 31. The maximum of $f(x)$ occurs when $x = 11111$. Not only that, but any bit string that begins with a 1 is better than every bit string that begins with a 0. This leads to the concept of a schema. A schema is a bit pattern that describes a set of individuals, where an $*$ is used to represent a "don't care" bit. For example, the bit strings 11000 and 10011 both belong to the schema $1****$. This schema is a very high-fitness schema for the function x^2. Any bit string that belongs to this schema is better than every bit string that does not belong to it. GAs combine schemata in a way that results in a highly fit individuals.

Consider bit strings of length two. The *schemata* (plural of schema) with length two are $**$, $0*$, $1*$, $*0$, $*1$, 00, 01, 10, and 11. There are a total of nine unique schemata of length two. In general, there are a total of 3^l schemata of length l.

Now consider the number of schemata to which a bit string belongs. As an example, notice that 01 belongs to four schemata: 01, $*1$, $0*$, and $**$. In general, a bit string of length l belongs to 2^l schemata.

Now consider a population of N bit strings, each of length l. Each bit string in the population belongs to a certain set of schemata. We say that the union of these N sets of schemata is the set to which the entire population belongs. If all the bit

[1] Markov's son, Andrey Markov Jr., was also an accomplished mathematician.

strings are identical, then each bit string belongs to the same 2^l schemata, and the entire population belongs to 2^l schemata. At the other extreme, all the bit strings may be unique and not belong to any of the same schemata except for the universal schema $** \cdots **$. In this case the entire population belongs to $N2^l - (N-1)$ schemata. We see that a population of N bit strings, each of length l, belongs to somewhere between 2^l and $(N(2^l - 1) + 1)$ schemata.

The number of defined bits (that is, non-asterisks) in a schema is called the order o of the schema. For example, $o(1***0) = 2$, and $o(0*11*) = 3$.

The number of bits from the left-most defined bit to the right-most defined bit in a schema is called is defining length δ. For example, $\delta(1***0) = 4$, $\delta(0*11*) = 3$, and $\delta(1****) = 0$.

A bit string that belongs to a schema is called an instance of the schema. For example, the schema $0*11*$ has four instances: 00110, 00111, 01110, and 01111. In general, the number of instances that a schema has is equal to 2^A, where A is the number of asterisks in the schema. Note that $A = l - o$.

We use the notation $m(h,t)$ to represent the number of instances of schema h at generation t in a GA. We use $f(x)$ to denote the fitness of the bit string x. We use $f(h,t)$ to denote the average fitness of the instances of schema h in the population at generation t:

$$f(h,t) = \frac{\sum_{x \in h} f(x)}{m(h,t)}. \tag{4.1}$$

We use $\bar{f}(t)$ to denote the average fitness of the entire population at generation t. If we use roulette-wheel selection to choose the parents of the next generation, we see that the expected number of instances of h after selection is

$$\begin{aligned} E\left[m(h,t+1)\right] &= \frac{\sum_{x \in h} f(x)}{\bar{f}(t)} \\ &= \frac{f(h,t)m(h,t)}{\bar{f}(t)}. \end{aligned} \tag{4.2}$$

Next we perform crossover with probability p_c. We assume that the crossover point is between bits, and never at the end of a bit string. We obtain two children from each pair of parents. What is the probability that crossover will destroy a schema? Let us look at a few examples.

- Consider the schema $h = 1****$. Crossover will never destroy this schema. If an instance of this schema crosses with another bit string, at least one child will be an instance of h.

- Consider the schema $h = 11***$. If an instance of this schema crosses with bit string x, the crossover point could be one of four places. If the crossover point is between the two most significant bits, then the schema might be destroyed, depending on the value of x. However, if the crossover point is to the right of that point (three other possible crossover points), then the schema will never be destroyed; at least one child will be an instance of h. We see that the probability of destroying the schema h is less than or equal to $1/4$, depending on where crossover occurs.

- Consider the schema $h = 1*1**$. If an instance of this schema crosses with bit string x, the crossover point could be one of four places. If the crossover point is between the two 1 bits (two possible crossover points), then the schema might be destroyed, depending on the value of x. However, if the crossover point is to the right of the rightmost 1 bit (two other possible crossover points), then the schema will never be destroyed; at least one child will be an instance of h. We see that the probability of destroying the schema h is less than or equal to 1/2, depending on where crossover occurs.

Generalizing the above, we see that the probability that crossover will destroy a schema, if it occurs, is less than or equal to $\delta/(l-1)$. The probability that crossover occurs at all is p_c, so the total probability that crossover destroys a schema is less than or equal to $p_c\delta/(l-1)$. Therefore, the probability that a schema will survive crossover is

$$p_s \geq 1 - p_c\left(\frac{\delta}{l-1}\right). \tag{4.3}$$

Next we perform mutation with a probability of p_m per bit. The number of defined (non-asterisk) bits in h is the order of h and is denoted as $o(h)$. The probability that a defined bit mutates is p_m, and the probability that it does not mutate is $1-p_m$. Therefore, the probability that none of the defined bits mutate is $(1-p_m)^{o(h)}$.

This probability is of the form $g(x) = (1-x)^y$. The Taylor series expansion of $g(x)$ around x_0 is

$$g(x) = \sum_{n=0}^{\infty} g^{(n)}(x_0)\frac{(x-x_0)^n}{n!}. \tag{4.4}$$

Setting $x_0 = 0$ gives

$$\begin{aligned} g(x) &= \sum_{n=0}^{\infty} g^{(n)}(0)\frac{x^n}{n!} \\ &= 1 - xy + \frac{x^2y(y-1)}{2!} - \frac{x^3y(y-1)(y-2)}{3!} + \cdots \\ &\approx 1 - xy \quad \text{for } xy \ll 1. \end{aligned} \tag{4.5}$$

So if $p_m o(h) \ll 1$, then $(1-p_m)^{o(h)} \approx 1 - p_m o(h)$. Combining this with Equations (4.2) and (4.3) gives

$$\begin{aligned} E\left[m(h,t+1)\right] &\geq \frac{f(h,t)m(h,t)}{\bar{f}(t)}\left(1 - p_c\left(\frac{\delta}{l-1}\right)\right)(1 - p_m o(h)) \\ &\approx \frac{f(h,t)m(h,t)}{\bar{f}(t)}\left(1 - p_c\left(\frac{\delta}{l-1}\right) - p_m o(h)\right). \end{aligned} \tag{4.6}$$

Suppose that a schema is short; that is, its defining length δ is small. Then $\delta/(l-1) \ll 1$. Suppose that we use a low mutation rate, and a schema is of low order; that is, there are not many defined bits. Then $p_m o(h) \ll 1$. Suppose that a schema has above-average fitness; that is, $f(h)/\bar{f}(t) = k > 1$, where k is some constant. Finally, suppose that we have a large population so that $E\left[m(h,t+1)\right] \approx m(h,t+1)$. Then we can approximately write

$$m(h,t+1) \geq km(h,t) = k^t m(h,0). \tag{4.7}$$

This results in the following theorem, which is called the schema theorem.

Theorem 4.1 *Short, low-order schemata with above-average fitness values receive an exponentially increasing number of representatives in a GA population.*

The schema theorem is often written as Equation (4.6) or (4.7).

Schema theory originated with John Holland in the 1970s [Holland, 1975] and quickly gained a foothold in GA research. Schema theory was so dominant in the 1980s that GA implementations were suspect if they violated its assumptions (for example, if they used rank-based rather than fitness-based selection [Whitley, 1989]). A description of how schema theory works on a simple example is provided in [Goldberg, 1989a, Chapter 2].

However, some counterexamples to schema theory are provided in [Reeves and Rowe, 2003, Section 3.2]. That is, the schema theorem is not always useful. This is because of the following.

- Schema theory applies to arbitrary subsets of the search space. Consider Table 3.2 in Example 3.2. We see that x_1 and x_4 both belong to schema $h = 1{*}0{*}$. But these are the least fit and most fit individuals in the population, and so these two individuals do not really have anything to do with each other, besides the fact that they are both members of h. There is nothing special about h, so the schema theorem does not give useful information about h.

- Schema theory does not recognize that similar bit strings might not belong to the same schema. In Example 3.2, we see that 0111 and 1000 are neighbors in the search space, but do not belong to any common schema except the universal schema $****$. This problem can be alleviated with gray coding, but even then, depending on the search space, neighbors in the search space may not have a close relationship in fitness space.

- Schema theory tells us the number of schema instances that survive from one generation to the next, but it is more important *which* schema instances survive. This is closely related to item 1 above. Again looking at Example 3.2, we see that x_1 and x_4 both belong to schema $h = 1{*}0{*}$. Schema theory tells us if h survives from one generation to the next, but we are much more interested in the survival of x_4 than of x_1.

- Schema theory gives us the *expected* number of schema instances. But the stochastic nature of GAs results in different behavior each time the GA runs. The expected number of schema instances is equal to the actual number of schema instances only as the population size approaches infinity.

- No schema can both increase exponentially and have above-average fitness. If a schema increases exponentially, then it will soon dominate the population, at which time the average fitness of the population will approximately equal the fitness of the schema. The approximation $f(h)/\bar{f}(t) = k$, in the paragraph before Theorem 4.1 above, where k is a constant, is therefore incorrect. Related to this idea is the fact that most GAs operate with a population size of 100 or fewer. Such small population sizes cannot support exponential growth of any schema for more than a couple of generations.

By the 1990s, an overemphasis on the shortcomings of schema theory resulted in extreme statements like the following: "I will say – since it is no longer controversial

– that the 'schema theorem' explains virtually nothing about SGA [simple genetic algorithm] behavior" [Vose, 1999, page xi]. The pendulum has swung from one side (over-reliance on schema theory) to the other (complete dismissal of schema theory). This high variance is typical of many new theories. Schema theory is true, but it has limitations. A balanced view of the benefits and shortcomings of schema theory is given in [Reeves and Rowe, 2003, Chapter 3].

4.2 MARKOV CHAINS

Suppose that we have a discrete-time system that can be described by a set of discrete states $S = \{S_1, \cdots, S_n\}$. For instance, the weather might be described by the set of states $S = \{\text{rainy}, \text{nice}, \text{snowy}\}$. We use the notation $S(t)$ to denote the state at time step t. The initial state is $S(0)$, the state at the next time step is $S(1)$, and so on. The system state might change from one time step to the next, or it might remain in the same state from one time step to the next. The transition from one state to another is entirely probabilistic. In a first-order Markov process, also called a first-order Markov chain, the probability that the system transitions to any given state at the next time step depends only on the current state; that is, the probability is independent of all previous states. The probability that the system transitions from state i to state j from one time step to the next is denoted by p_{ij}. Therefore,

$$\sum_{j=1}^{n} p_{ij} = 1 \tag{4.8}$$

for all i. We form the $n \times n$ matrix P, where p_{ij} is the element in the i-th row and j-th column. P is called the transition matrix, probability matrix, or stochastic matrix, of the Markov process.[2] The sum of the elements of each row of P is 1.

■ **EXAMPLE 4.1**

The land of Oz never has two nice days in a row [Kemeny et al., 1974]. If it is a nice day, then the next day has a 50% chance of rain and a 50% chance of snow. If it rains, then the next day has a 50% chance of rain again, a 25% chance of snow, and a 25% chance of nice weather. If it snows, then the next day has a 50% chance of snow again, a 25% chance of rain, and a 25% chance of nice weather. We see that the weather forecast for a given day depends solely on the weather of the previous day. If we assign states R, N, and S, to rain, nice weather, and snow respectively, then we can form a Markov matrix that represents the probability of various weather transitions:

$$P = \begin{array}{c} \begin{array}{ccc} R & N & S \end{array} \\ \left[\begin{array}{ccc} 1/2 & 1/4 & 1/4 \\ 1/2 & 0 & 1/2 \\ 1/4 & 1/4 & 1/2 \end{array}\right] \end{array} \begin{array}{c} \\ R \\ N \\ S \end{array} \tag{4.9}$$

[2]More precisely, the matrix that we have defined is called a *right* transition matrix. Some books and papers denote the transition probability as p_{ji}, and define the Markov transition matrix as the transpose of the one that we have defined. Their matrix is called a *left* transition matrix, and the sum of the elements of each column is 1.

□

Suppose a Markov process begins in state i at time 0. We know from our previous discussion that the probability that the process is in state j at time 1, given that the process was in state i at time 0, is given by $\Pr(S(1) = S_j | S(0) = S_i) = p_{ij}$. Next we consider the following time step. We can use the total probability theorem [Mitzenmacher and Upfal, 2005] to find the probability that the process is in state 1 at time 2 as

$$\begin{aligned}
\Pr(S(2) = S_1 | S(0) = S_i) &= \Pr(S(1) = S_1 | S(0) = S_i)p_{11} + \\
&\quad \Pr(S(1) = S_2 | S(0) = S_i)p_{21} + \cdots + \\
&\quad \Pr(S(1) = S_n | S(0) = S_i)p_{n1} \\
&= \sum_{k=1}^{n} \Pr(S(1) = S_k | S(0) = S_i)p_{k1} \\
&= \sum_{k=1}^{n} p_{ik}p_{k1}. \qquad (4.10)
\end{aligned}$$

Generalizing the above development, we find that the probability that the process is in state j at time 2 is given by

$$\Pr(S(2) = S_j | S(0) = S_i) = \sum_{k=1}^{n} p_{ik}p_{kj}. \qquad (4.11)$$

But this is equal to the element in the i-th row and j-th column of the square of P; that is,

$$\Pr(S(2) = S_j | S(0) = S_i) = \left[P^2\right]_{ij}. \qquad (4.12)$$

Continuing this line of reasoning in an inductive manner, we find that

$$\Pr(S(t) = S_j | S(0) = S_i) = \left[P^t\right]_{ij}. \qquad (4.13)$$

That is, the probability that the Markov process transitions from state i to state j after t time steps is equal to the element in the i-th row and j-th column of P^t.

In Example 4.1, we can compute P^t for various values of t to obtain

$$\begin{aligned}
P &= \begin{bmatrix} 0.5000 & 0.2500 & 0.2500 \\ 0.5000 & 0.0000 & 0.5000 \\ 0.2500 & 0.2500 & 0.5000 \end{bmatrix} \\
P^2 &= \begin{bmatrix} 0.4375 & 0.1875 & 0.3750 \\ 0.3750 & 0.2500 & 0.3750 \\ 0.3750 & 0.1875 & 0.4375 \end{bmatrix} \\
P^4 &= \begin{bmatrix} 0.4023 & 0.1992 & 0.3984 \\ 0.3984 & 0.2031 & 0.3984 \\ 0.3984 & 0.1992 & 0.4023 \end{bmatrix} \\
P^8 &= \begin{bmatrix} 0.4000 & 0.2000 & 0.4000 \\ 0.4000 & 0.2000 & 0.4000 \\ 0.4000 & 0.2000 & 0.4000 \end{bmatrix}. \qquad (4.14)
\end{aligned}$$

Interestingly, P^t converges as $t \to \infty$ to a matrix with identical rows. This does not happen for all transition matrices, but it happens for a certain subset as specificed in the following theorem.

Theorem 4.2 *A regular $n \times n$ transition matrix P, also called a primitive transition matrix, is one for which all elements of P^t are nonzero for some t. If P is a regular transition matrix, then*

1. $\lim_{t\to\infty} P^t = P_\infty$;
2. *All rows of P_∞ are identical and are denoted as p_{ss};*
3. *Each element of p_{ss} is positive;*
4. *The probability that the Markov process is in the i-th state after an infinite number of transitions is equal to the i-th element of p_{ss};*
5. *p_{ss}^T is the eigenvector of P^T corresponding to the eigenvalue 1, normalized so that its elements sum to 1;*
6. *If we form the matrices P_i, $i \in [1, n]$, by replacing the i-th column of P with zeros, then the i-th element of p_{ss} is given as*

$$p_{ss,i} = \frac{|P_i - I|}{\sum_{j=1}^{n} |P_j - I|} \tag{4.15}$$

where I is the $n \times n$ identity matrix, and $|\cdot|$ is the determinant operator.

Proof: The first five properties above comprise the fundamental limit theorem for regular Markov chains and are proven in [Grinstead and Snell, 1997, Chapter 11] and other books on Markov chains. For more information on concepts like determinants, eigenvalues, and eigenvectors, read any linear systems text [Simon, 2006, Chapter 1]. The last property of Theorem 4.2 is proven in [Davis and Principe, 1993].

□

■ EXAMPLE 4.2

Using Equation (4.14) and applying Theorem 4.2 to Example 4.1, we see that any given day in the distant future has a 40% probability of rain, a 20% probability of sun, and a 40% probability of snow. Therefore, 40% of the days in Oz are rainy, 20% are sunny, and 40% are snowy. Furthermore, we can find the eigenvalues of P^T as 1, -0.25, and 0.25. The eigenvector corresponding to the eigenvalue 1 is $[\ 0.4 \quad 0.2 \quad 0.4\ ]^T$.

□

Now suppose that we don't know the initial state of the Markov process, but we do know the probabilities for each state; the probability that the initial state $S(0)$

is equal to S_k is given by $p_k(0)$, $k \in [1, n]$. Then we can use the total probability theorem [Mitzenmacher and Upfal, 2005] to obtain

$$\begin{aligned}
\Pr(S(1) = S_i) &= \Pr(S(0) = S_1)p_{1i} + \Pr(S(0) = S_2)p_{2i} + \cdots + \\
&\quad \Pr(S(0) = S_n)p_{ni} \\
&= \sum_{k=1}^{n} \Pr(S(0) = S_k)p_{ki} \\
&= \sum_{k=1}^{n} p_{ki}p_k(0).
\end{aligned} \tag{4.16}$$

Generalizing the above equation, we obtain

$$\begin{bmatrix} \Pr(S(1) = S_1) \\ \cdots \\ \Pr(S(1) = S_n) \end{bmatrix}^T = p^T(0)P \tag{4.17}$$

where $p(0)$ is the column vector comprised of $p_k(0)$, $k \in [1, n]$. Generalizing this development for multiple time steps, we obtain

$$p^T(t) = \begin{bmatrix} \Pr(S(t) = S_1) \\ \cdots \\ \Pr(S(t) = S_n) \end{bmatrix}^T = p^T(0)P^t. \tag{4.18}$$

■ EXAMPLE 4.3

Today's weather forecast in Oz is 80% sun and 20% snow. What is the weather forecast for two days from now?

From Equation (4.18), $p^T(2) = p^T(0)P^2$, where P is given in Example 4.1 and $p(0) = [\ 0.0 \quad 0.8 \quad 0.2\ ]^T$. This gives $p(2) = [\ 0.3750 \quad 0.2375 \quad 0.3875\ ]^T$. That is, two days from now, there is a 37.5% chance of rain, a 23.75% chance of sun, and a 38.75% chance of snow.

□

■ EXAMPLE 4.4

Consider a simple hill-climbing EA comprised of a single individual [Reeves and Rowe, 2003, page 112]. The goal of the EA is to minimize $f(x)$. We use x_i to denote the candidate solution at the i-th generation. Each generation we randomly mutate x_i to obtain x_i'. If $f(x_i') < f(x_i)$, then we set $x_{i+1} = x_i'$.

If $f(x_i') > f(x_i)$, then we use the following logic to determine x_{i+1}. If we had set $x_{k+1} = x_k'$ at the previous generation k at which $f(x_k') > f(x_k)$, then we set $x_{i+1} = x_i'$ with a 10% probability, and $x_{i+1} = x_i$ with a 90% probability. If, however, we had set $x_{k+1} = x_k$ at the previous generation k at which $f(x_k') > f(x_k)$, then we set $x_{i+1} = x_i'$ with a 50% probability, and we set $x_{i+1} = x_i$ with a 50% probability. This EA is greedy in that it always accepts a beneficial mutation. However, it also includes some exploration in that it sometimes accepts a detrimental mutation. The probability of accepting a detrimental mutation varies depending on whether or not the previous detrimental mutation was accepted. The algorithm for this hill-climbing EA is shown in Figure 4.1.

Initialize x_1 to a random candidate solution
Intialize AcceptFlag to false
For $i = 1, 2, \cdots$
  Mutate x_i to get x_i'
  If $f(x_i') < f(x_i)$
    $x_{i+1} \leftarrow x_i'$
  else
    If AcceptFlag
      $\Pr(x_{i+1} \leftarrow x_i') = 0.1$, and $\Pr(x_{i+1} \leftarrow x_i) = 0.9$
    else
      $\Pr(x_{i+1} \leftarrow x_i') = 0.5$, and $\Pr(x_{i+1} \leftarrow x_i) = 0.5$
    end if
    AcceptFlag $\leftarrow (x_{i+1} = x_i')$
  end if
Next i

Figure 4.1 The above pseudo-code outlines the single-individual hill-climbing EA of Example 4.4. *AcceptFlag* indicates if the previous detrimental mutation replaced the candidate solution.

We can analyze this EA by considering what happens if x_i' is worse than x_i. We use Z_k to denote the state the k-th time that $f(x_i') > f(x_i)$. We define Y_1 as the "accept" state; that is, $x_{i+1} \leftarrow x_i'$. We define Y_2 as the "reject" state; that is, $x_{i+1} \leftarrow x_i$. Then, by examining the algorithm of Figure 4.1, we can write

$$\begin{aligned}
\Pr(Z_k = Y_1 | Z_{k-1} = Y_1) &= 0.1 \\
\Pr(Z_k = Y_2 | Z_{k-1} = Y_1) &= 0.9 \\
\Pr(Z_k = Y_1 | Z_{k-1} = Y_2) &= 0.5 \\
\Pr(Z_k = Y_2 | Z_{k-1} = Y_2) &= 0.5.
\end{aligned} \tag{4.19}$$

This equation shows that the transition matrix is

$$P = \begin{bmatrix} 0.1 & 0.9 \\ 0.5 & 0.5 \end{bmatrix}. \tag{4.20}$$

Notice that the rows of P sum to 1. We also see that all the elements of P^t are nonzero for some t (actually, for all t in this case) so P is a regular transition matrix. Theorem 4.2 assures us that: (1) P^t converges as $t \to \infty$; (2) All the rows of P^∞ are identical; (3) Each element of P^∞ is positive; (4) The probability that the Markov process is in state Y_i after an infinite number of transitions is equal to the i-th element of each row of P^∞; and (5) Each row of P^∞ is equal to the transpose of the eigenvector corresponding to the eigenvalue 1 of P^T.

We use a numerical calculation to find

$$P^\infty = \frac{1}{14}\begin{bmatrix} 5 & 9 \\ 5 & 9 \end{bmatrix}. \tag{4.21}$$

We also find that the eigenvalues of P^T are equal to -0.4 and 1, and the eigenvector corresponding to the 1 eigenvalue is $[\ 5/14 \quad 9/14\ ]^T$. These results tell us that in the long run, the ratio of acceptances to rejections of detrimental mutations is 5/9.

□

4.3 MARKOV MODEL NOTATION FOR EVOLUTIONARY ALGORITHMS

In this section we define the notation that we will use later to derive a Markov model and dynamic system model for EAs. Markov models can be valuable tools for analyzing EAs because they give us exact results. We can run simulations to investigate the performance of EAs, but simulations can be misleading. For instance, a set of Monte Carlo simulations might happen to give misleading results due to the particular sequence of random numbers generated during the simulation. Also, the random number generator using in the EA simulation may be incorrect, which happens more often than we would like to think, and which would give misleading results [Savicky and Robnik-Sikonja, 2008]. Finally, the number of Monte Carlo simulations to estimate highly improbable outcomes might be so high as to not be attainable in a reasonable amount of computational time. The Markov model results that we derive avoid all of these pitfalls and give exact results. The drawback of the Markov models is the high amount of computational effort that is required for their implementation.

We will focus on EAs with a population size N operating in a discrete search space of cardinality n. We will assume that the search space consists of all q-bit binary strings, so that $n = 2^q$. We use x_i to denote the i-th bit string in the search space. We use v to denote the population vector; that is, v_i is the number of x_i individuals in the population. We see that

$$\sum_{i=1}^{n} v_i = N. \tag{4.22}$$

This equation simply means that the total number of individuals in the population is equal to N. We use y_k to denote the k-th individual in the population. The population Y of the EA can be represented as

$$\begin{aligned} Y &= \{y_1, \cdots, y_N\} \\ &= \{\underbrace{x_1, x_1, \cdots, x_1}_{v_1 \text{ copies}}, \underbrace{x_2, x_2, \cdots, x_2}_{v_2 \text{ copies}}, \cdots \underbrace{x_n, x_n, \cdots, x_n}_{v_n \text{ copies}}\} \end{aligned} \tag{4.23}$$

where the y_i's have been ordered to group identical individuals. We use T to denote the total number of possible populations Y. That is, T is the number of $n \times 1$ integer vectors v such that $\sum_{i=1}^{n} v_i = N$ and $v_i \in [0, N]$.

■ EXAMPLE 4.5

Suppose that $N = 2$ and $n = 4$; that is, the search space consists of the bit strings $\{00, 01, 10, 11\}$, and there are two individuals in the EA. The search space individuals are

$$\begin{aligned} x_1 &= 00, & x_2 &= 01, \\ x_3 &= 10, & x_4 &= 11. \end{aligned} \tag{4.24}$$

The possible populations include the following:

$$\begin{aligned} &\{00, 00\}, && \{00, 01\}, \\ &\{00, 10\}, && \{00, 11\}, \\ &\{01, 01\}, && \{01, 10\}, \\ &\{01, 11\}, && \{10, 10\}, \\ &\{10, 11\}, && \{11, 11\}. \end{aligned} \tag{4.25}$$

We see that $T = 10$ for this example.

□

How many possible EA populations exist for a population size of N in a search space of cardinality n? It can be shown [Nix and Vose, 1992] that T is given by the following binomial coefficient, also called the choose function:

$$T = \binom{n + N - 1}{N}. \tag{4.26}$$

We can also use the multinomial theorem [Chuan-Chong and Khee-Meng, 1992], [Simon et al., 2011a] to find T. The multinomial theorem can be stated in several ways, including the following: given K classes of objects, the number of different ways that N objects can be selected, independent of order, while choosing from each class no more than M times, is the coefficient q_N in the polynomial

$$\begin{aligned} q(x) &= (1 + x + x^2 + \cdots + x^M)^K \\ &= 1 + q_1 x + q_2 x^2 + \cdots + q_N x^N + \cdots + x^{MK}. \end{aligned} \tag{4.27}$$

Our EA population vector v is an n-element vector where each element is an integer between 0 and N inclusive, and whose elements sum to N. T is the number of unique population vectors v. So T is the number of ways that N objects can be selected, independent of order, from n classes of objects while choosing from each class no more than N times. Applying the multinomial theorem (4.27) to this problem gives

$$\begin{aligned} T &= q_N \\ \text{where } q(x) &= (1 + x + x^2 + \cdots + x^N)^n \\ &= 1 + q_1 x + q_2 x^2 + \cdots + q_N x^N + \cdots + x^{Nn}. \end{aligned} \tag{4.28}$$

We can also use a different form of the multinomial theorem to find T [Chuan-Chong and Khee-Meng, 1992], [Simon et al., 2011a]. The multinomial theorem can be stated as follows:

$$
\begin{aligned}
(x_1 + x_2 + \cdots + x_N)^n &= \sum_{S(k)} \frac{n!}{\prod_{j=0}^{N} k_j!} \prod_{j=0}^{N} x_j^{k_j} \\
&= \sum_{S(k)} \prod_{i=0}^{N} \binom{\sum_{j=0}^{i} k_j}{k_i} \prod_{j=0}^{N} x_j^{k_j} \qquad (4.29) \\
\text{where } S(k) &= \{k \in \mathbf{R}^N : k_j \in \{0, 1, \cdots, n\}, \sum_{j=0}^{N} k_j = n\}.
\end{aligned}
$$

Now consider the polynomial $(x^0 + x^1 + x^2 + \cdots + x^N)^n$. From the multinomial theorem of Equation (4.29) we see that the coefficient of $[(x^0)^{k_0}(x^1)^{k_1} \cdots (x^N)^{k_N}]$ is given by

$$\prod_{i=0}^{N} \binom{\sum_{j=0}^{i} k_j}{k_i}. \qquad (4.30)$$

If we add up these terms for all k_j such that

$$\sum_{j=0}^{N} j k_j = N \qquad (4.31)$$

then we obtain the coefficient of x^N. But Equation (4.28) shows that T is equal to the coefficient of x^N. Therefore

$$
\begin{aligned}
T &= \sum_{S'(k)} \prod_{i=0}^{N} \binom{\sum_{j=0}^{i} k_j}{k_i} \qquad (4.32) \\
\text{where } S'(k) &= \{k \in \mathbf{R}^{N+1} : k_j \in \{0, 1, \cdots, n\}, \sum_{j=0}^{N} k_j = n, \sum_{j=0}^{N} j k_j = n\}.
\end{aligned}
$$

Equations (4.26), (4.28), and (4.32) give equivalent expressions for T.

■ EXAMPLE 4.6

This example is taken from [Simon et al., 2011a]. Suppose that we have a two-bit search space ($q = 2$, $n = 4$) and an EA population size $N = 4$. Equation (4.26) gives

$$T = \binom{7}{4} = 35. \qquad (4.33)$$

Equation (4.28) gives

$$
\begin{aligned}
q(x) &= (1 + x + x^2 + x^3 + x^4)^4 \\
&= 1 + \cdots + 35x^4 + \cdots + x^{16} \qquad (4.34)
\end{aligned}
$$

which means that $T = 35$. Equation (4.32) gives the following:

$$\begin{aligned}
T &= \sum_{S'(k)} \prod_{i=0}^{4} \begin{pmatrix} \sum_{j=0}^{i} k_j \\ k_i \end{pmatrix} \\
\text{where } S'(k) &= \left\{ k \in \mathbf{R}^5 : k_j \in \{0, 1, \cdots, 4\}, \sum_{j=0}^{4} k_j = 4, \sum_{j=0}^{4} j k_j = 4 \right\} \\
&= \{ (\ 3\ \ 0\ \ 0\ \ 0\ \ 1\ ), (\ 2\ \ 1\ \ 0\ \ 1\ \ 0\ ), (\ 2\ \ 0\ \ 2\ \ 0\ \ 0\ ), \\
&\quad\ (\ 1\ \ 2\ \ 1\ \ 0\ \ 0\ ), (\ 0\ \ 4\ \ 0\ \ 0\ \ 0\ ) \}. \qquad (4.35)
\end{aligned}$$

This means that $T = 4 + 12 + 6 + 12 + 1 = 35$. We see that all three methods for the calculation of T give the same result.

□

4.4 MARKOV MODELS OF GENETIC ALGORITHMS

Markov models were first used to model GAs in [Nix and Vose, 1992], [Davis and Principe, 1991], and citeDavis93, and are further explained in [Reeves and Rowe, 2003] and [Vose, 1999]. As we saw in Chapter 3, a GA consists of selection, crossover, and mutation. For the purposes of Markov modeling, we will switch the order of crossover and mutation, so we will consider a GA which consists of selection, mutation, and crossover, in that order.

4.4.1 Selection

First we consider fitness-proportional (that is, roulette-wheel) selection. The probability of selecting an x_i individual with one spin of the roulette wheel is proportional to the fitness of the x_i individual, multiplied by the number of x_i individuals in the population. This probability is normalized so that all probabilities sum to 1. As defined in the previous section, v_i is the number of x_i individuals in the population. Therefore, the probability of selecting an x_i individual with one spin of the roulette wheel is

$$P_s(x_i|v) = \frac{v_i f_i}{\sum_{j=1}^{n} v_j f_j} \qquad (4.36)$$

for $i \in [1, n]$, where n is the cardinality of the search space, and f_j is the fitness of x_j. We use the notation $P_s(x_i|v)$ to show that the probability of selection an x_i individual depends on the population vector v. Given a population of N individuals, suppose that we spin the roulette wheel N times to select N parents. Each spin of the roulette wheel has n possible outcomes $\{x_1, \cdots, x_n\}$. The probability of obtaining outcome x_i at each spin is equal to $P_s(x_i|v)$. Let $U = [\ U_1\ \ \cdots\ \ U_n\ ]$ be a vector of random variables where U_i denotes the total number of times that x_i occurs in N spins of the roulette wheel, and let $u = [\ u_1\ \ \cdots\ \ u_n\ ]$ be a realization of U. Multinomial distribution theory [Evans et al., 2000] tells us that

$$\Pr_s(u|v) = N! \prod_{i=1}^{n} \frac{[P_s(x_i|v)]^{u_i}}{u_i!}. \qquad (4.37)$$

This gives us the probability of obtaining the population vector u after N roulette-wheel spins if we start with the population vector v. We use the subscript s on $\Pr_s(u|v)$ to denote that we consider only selection (not mutation or crossover).

Now recall that a Markov transition matrix contains all of the probabilities of transitioning from one state to another. Equation (4.37) gives us the probability of transitioning from one population vector v to another population vector u. There are T possible population vectors, as discussed in the previous section. Therefore, if we calculate Equation (4.37) for each possible u and each possible v, we will obtain a $T \times T$ Markov transition matrix which gives an exact probabilistic model of a selection-only GA. Each entry of the transition matrix contains the probability of transitioning from some particular population vector to some other population vector.

4.4.2 Mutation

Now suppose that after selection, we implement mutation on the selected individuals. Define M_{ji} as the probability that x_j mutates to x_i. Then the probability of obtaining an x_i individual after a single spin of the roulette wheel, followed by a single chance of mutation, is

$$P_{sm}(x_i|v) = \sum_{j=1}^{n} M_{ji} P_s(x_j|v) \tag{4.38}$$

for $i \in [1, n]$. This means that we can write the n-element vector whose i-th element is equal to $P_{sm}(x_i|v)$ as follows:

$$P_{sm}(x|v) = M^T P_s(x|v) \tag{4.39}$$

where M is the matrix containing M_{ji} in the j-th row and i-th column, and $P_s(x|v)$ is the n-element vector whose j-th element is $P_s(x_j|v)$. Now we use multinomial distribution theory again to find that

$$\Pr_{sm}(u|v) = N! \prod_{i=1}^{n} \frac{[P_{sm}(x_i|v)]^{u_i}}{u_i!}. \tag{4.40}$$

This gives us the probability of obtaining the population vector u if we start with the population vector v, after both selection and mutation take place. If we calculate Equation (4.40) for each of the T possible u and v population vectors, we will have a $T \times T$ Markov transition matrix which gives an exact probabilistic model of a GA which consists of both selection and mutation.

If mutation is defined so that $M_{ji} > 0$ for all i and j, then $\Pr_{sm}(u|v) > 0$ for all u and v. This means that the Markov transition matrix will contain all positive entries, which means that the transition matrix will be regular. Theorem 4.2 tells us that there will be a unique nonzero probability for obtaining each possible population distribution. This means that in the long run, each possible population distribution will occur for a nonzero percent of time. These percentages can be calculated using Theorem 4.2 and the transition matrix obtained from Equation (4.40). The GA will not converge to any specific population, but will endlessly wander throughout the search space, hitting each possible population for the percent of time given in Theorem 4.2.

■ **EXAMPLE 4.7**

Suppose we have a four-element search space with individuals $x = \{00, 01, 10, 11\}$. Suppose that each bit in each individual has a 10% chance of mutation. The probability that 00 remains equal to 00 after a mutation chance is equal to the probability that that first 0 bit remains unchanged (90%), multiplied by the probability that the second 0 bit remains unchanges (90%), which gives a probability of 0.81. This gives M_{11}, which is the probability that x_1 remains unchanged after a mutation chance. The probability that 00 will change to 01 is equal to the probability that that first 0 bit remains unchanged (90%), multiplied by the probability that the second 0 bit changes to a 1 (10%), which gives a probability of $M_{12} = 0.09$. Continuing along these lines, we find that

$$M = \begin{bmatrix} 0.81 & 0.09 & 0.09 & 0.01 \\ 0.09 & 0.81 & 0.01 & 0.09 \\ 0.09 & 0.01 & 0.81 & 0.09 \\ 0.01 & 0.09 & 0.09 & 0.81 \end{bmatrix}. \tag{4.41}$$

Note that M is symmetric (that is, M is equal to its transpose M^T). This is typically (but not always) the case, which means that it is equally likely for x_i to mutate to form x_j, as it is for x_j to mutate to form x_i.

□

4.4.3 Crossover

Now suppose that after selection and mutation, we implement crossover. We let r_{jki} denote the probability that x_j and x_k cross to form x_i. Then the probability of obtaining an x_i individual after two spins of the roulette wheel, followed by a single chance of mutation for each selected individual, followed by crossover, is

$$P_{smc}(x_i|v) = \sum_{j=1}^{n}\sum_{k=1}^{n} r_{jki} P_{sm}(x_j|v) P_{sm}(x_k|v). \tag{4.42}$$

Now we use multinomial distribution theory again to find that

$$\Pr_{smc}(u|v) = N! \prod_{i=1}^{n} \frac{[P_{smc}(x_i|v)]^{u_i}}{u_i!}. \tag{4.43}$$

This gives us the probability of obtaining the population vector u if we start with the population vector v, after selection, mutation, and crossover take place.

■ **EXAMPLE 4.8**

Suppose we have a four-element search space with individuals $\{x_1, x_2, x_3, x_4\} = \{00, 01, 10, 11\}$. Suppose that we implement crossover by randomly setting $b = 1$ or $b = 2$ with equal probability, and then concatenating bits $1 \to b$ from the first parent with bits $(b+1) \to 2$ from the second parent. Some of the crossover possibilities can be written as follows:

$$\begin{aligned} 00 \times 00 &\rightarrow 00 \\ 00 \times 01 &\rightarrow 01 \text{ or } 00 \\ 00 \times 10 &\rightarrow 00 \\ 00 \times 11 &\rightarrow 01 \text{ or } 00. \end{aligned} \tag{4.44}$$

This gives crossover probabilities

$$\begin{array}{llll} r_{111} = 1.0, & r_{112} = 0.0, & r_{113} = 0.0, & r_{114} = 0.0 \\ r_{121} = 0.5, & r_{122} = 0.5, & r_{123} = 0.0, & r_{124} = 0.0 \\ r_{131} = 1.0, & r_{132} = 0.0, & r_{133} = 0.0, & r_{134} = 0.0 \\ r_{141} = 0.5, & r_{142} = 0.5, & r_{143} = 0.0, & r_{144} = 0.0. \end{array} \tag{4.45}$$

The other r_{jki} values can be calculated similarly.

□

EXAMPLE 4.9

In this example we consider the three-bit one-max problem. Each individual's fitness value is proportional to the number of ones in the individual:

$$\begin{array}{llll} f(000) = 1, & f(001) = 2, & f(010) = 2, & f(011) = 3, \\ f(100) = 2, & f(101) = 3, & f(110) = 3, & f(111) = 4. \end{array} \tag{4.46}$$

Suppose each bit has a 10% probability of mutation, which gives the mutation matrix derived in Example 4.7. After selection and mutation, we perform crossover with a probability of 90%. If crossover is selected, then crossover is performed by selecting a random bit position $b \in [1, q-1]$, where q is the number of bits in each individual. We then concatenate bits $1 \rightarrow b$ from the first parent with bits $(b+1) \rightarrow q$ from the second parent.

Let's use a population size $N = 3$. There are $(n + N - 1)$-choose-N = 10-choose-3 = 120 possible population distributions. We can use Equation (4.43) to calculate the probability of transitioning between each of the 120 population distributions, which gives us a 120×120 transition matrix P. We can then calculate the probability of each possible population distribution in three different ways:

1. We can use the Davis-Principe result of Equation (4.15);
2. From Theorem 4.2, we can numerically raise P to ever-increasing higher powers until it converges, and then use any of the rows of P^∞ to observe the probability of each possible population;
3. We can calculate the eigendata of P^T and find the eigenvector corresponding to the 1 eigenvalue.

Each of these approaches give us the same set of 120 probabilities for the 120 population distributions. We find that the probability that the population contains all optimal individuals, that is, each individual is equal to the bit string 111, is 6.1%. The probability that the population contains no optimal individuals is 51.1%. Figure 4.2 shows the results of a simulation of 20,000 generations, and shows that the simulation results closely match the Markov results. The simulation results are approximate, will vary from one run to the next, and will equal the Markov results only as the number of generations approaches infinity.

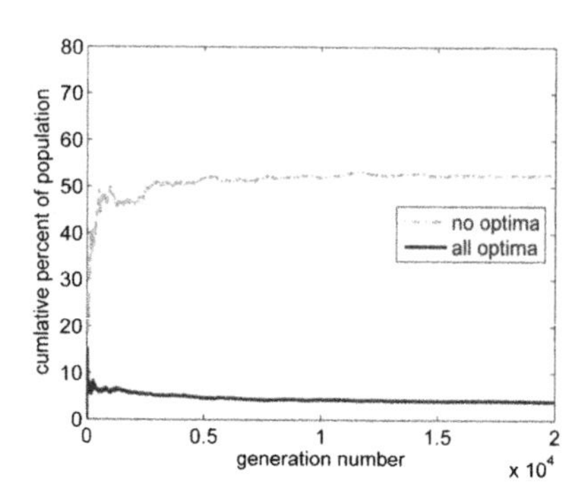

Figure 4.2 Example 4.9: Three-bit one-max simulation results. Markov theory predicts that the percentage of no optima is 51.1% and the percentage of all optima is 6.1%.

□

■ EXAMPLE 4.10

Here we repeat Example 4.9 except we use the following fitness values:

$$\begin{array}{llll} f(000) = 5, & f(001) = 2, & f(010) = 2, & f(011) = 3, \\ f(100) = 2, & f(101) = 3, & f(110) = 3, & f(111) = 4. \end{array} \tag{4.47}$$

These fitness values are the same as those in Equation (4.46), except that we made the 000 bit string the most fit individual. This is called a deceptive problem because usually when we add a 1 bit to one of the above individuals its fitness increases. The exception is that 111 is not the most fit individual, but rather 000 is the most fit individual.

As in Example 4.9, we calculate a set of 120 probabilities for the 120 population distributions. We find that the probability that the population contains all optimal individuals, that is, each individual is equal to the bit string 000, is 5.9%. This is smaller than the probability of all optima in Example 4.9, which was 6.1%. The probability that the population contains no optimal individuals is 65.2%. This is larger than the probability of no optima in Example 4.9, which was 51.1%. This example illustrates that deceptive problems are more difficult to solve than problems with a more regular structure. Figure 4.3 shows the results of a simulation of 20,000 generations, and shows that the simulation results closely match the Markov results.

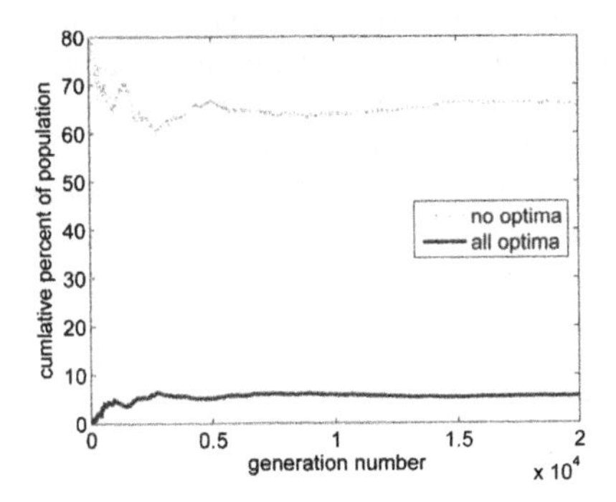

Figure 4.3 Example 4.10: Three-bit deceptive problem simulation results. Markov theory predicts that the percentage of no optima is 65.2% and the percentage of all optima is 5.9%.

□

The Curse of Dimensionality: *The curse of dimensionality* is a phrase which was originally used in the context of dynamic programming [Bellman, 1961]. However, it applies even more appropriately to Markov models for GAs. The size of the transition matrix of a Markov model of an EA is $T \times T$, where $T = (N + n - 1)$-choose-N. The transition matrix dimensions for some combinations of population size N, and search space cardinality n, which is equal to 2^q for q-bit search spaces, are shown in Table 4.1. We see that the transition matrix dimension grows ridiculously large for problems of even modest dimension. This seems to indicate that Markov modeling is interesting only from a theoretical viewpoint, and does not have any practical applications. However, there are a couple of reasons that such a response may be premature.

# bits q	$n = 2^q$	N	T
10	2^{10}	10	10^{23}
10	2^{10}	20	10^{42}
20	2^{20}	20	10^{102}
50	2^{50}	50	10^{688}

Table 4.1 Markov transition matrix dimensions for various search space cardinalities n and population sizes N. Adapted from [Reeves and Rowe, 2003, page 131].

First, although we cannot apply Markov models to realistically-sized problems, Markov models still give us exact probabilities for small problems. This allows us to look at the advantages and disadvantages of different EAs for small problems, assuming that we have Markov models for EAs other than GAs. This is exactly what we do in [Simon et al., 2011b] when we compare GAs with BBO. A lot of research in EAs today is focused on simulations. The problem with simulations is that their outcomes depend so strongly on implementation details and on the specific random number generator that is used. Also, if some event has a very small probability of occurring, then it would take many simulations to discover that probability. Simulation results are useful and necessary, but they must always be taken with a dash of skepticism and a grain of salt.

Second, the dimension of the Markov transition matrices can be reduced. Our Markov models include T states, but many of these states are very similar to each other. For example, consider a GA with a search space cardinality of 10 and a population size of 10. Table 4.1 shows us that the Markov model has 10^{23} states, but these include the states

$$\begin{aligned} v(1) &= \{5,5,0,0,0,0,0,0,0,0\} \\ v(2) &= \{4,6,0,0,0,0,0,0,0,0\} \\ v(3) &= \{6,4,0,0,0,0,0,0,0,0\}. \end{aligned} \tag{4.48}$$

These three states are so similar that it makes sense to group them together and consider them as a single state. We can do this with many other states to get a new Markov model with a reduced state space. Each state in the reduced-order model consists of a group of the original states. The transition matrix would then specify the probability of transitioning from one group of original states to another group of original states. This idea was proposed in [Spears and De Jong, 1997] and is further discussed in [Reeves and Rowe, 2003]. It is hard to imagine how to group states to reduce a $10^{23} \times 10^{23}$ matrix to a manageable size, but at least this idea allows us to handle larger problems than we would be able to otherwise.

4.5 DYNAMIC SYSTEM MODELS OF GENETIC ALGORITHMS

In this section we use the Markov model of the previous section to derive a dynamic system model of GAs. The Markov model gives us the probability of occurrence of each population distribution as the number of generations approaches infinity. The dynamic system model that we derive here is quite different; it will give us the percentage of each individual in the population as a function of time as the population size approaches infinity. The view of a GA as a dynamic system was originally published in [Nix and Vose, 1992], [Vose, 1990], [Vose and Liepins, 1991], and is explained further in [Reeves and Rowe, 2003], [Vose, 1999].

Recall from Equation (4.22) that $v = [\ v_1 \quad \cdots \quad v_n\ ]^T$ is the population vector, v_i is the number of x_i individuals in the population, and the elements of v sum to N, which is the population size. We define the proportionality vector as

$$p = v/N \tag{4.49}$$

which means that the elements of p sum to 1.

4.5.1 Selection

To find a dynamic system model of a GA with selection only (i.e., no mutation or crossover), we can divide the numerator and denominator of Equation (4.36) by N to write the probability of selecting individual x_i from a population described by population vector v as follows:

$$\begin{aligned} P_s(x_i|v) &= \frac{p_i f_i}{\sum_{j=1}^{n} p_j f_j} \\ &= \frac{p_i f_i}{f^T p} \end{aligned} \tag{4.50}$$

where f is the column vector of fitness values. Writing Equation (4.50) for $i \in [1, n]$ and combining all n equations gives

$$P_s(x|v) = \begin{bmatrix} P_s(x_1|v) \\ \cdots \\ P_s(x_n|v) \end{bmatrix} = \frac{\text{diag}(f)p}{f^T p} \tag{4.51}$$

where diag(f) is the $n \times n$ diagonal matrix whose diagonal entries are comprised of the elements of f.

The law of large numbers tells us that the average of the results obtained from a large number of trials should be close to the expected value of a single trial [Grinstead and Snell, 1997]. This means that as the population size becomes large, the proportion of selections of each individual x_i will be close to $P_s(x_i|v)$. But the number of selections of x_i is simply equal to v_i at the next generation. Therefore, for large population sizes, Equation (4.50) can be written as

$$p_i(t) = \frac{p_i(t-1) f_i}{\sum_{j=1}^{n} p_j(t-1) f_j}. \tag{4.52}$$

where t is the generation number.

Now suppose that

$$p_i(t) = \frac{p_i(0) f_i^t}{\sum_{j=1}^{n} p_j(0) f_j^t}. \tag{4.53}$$

This is clearly true for $t = 1$, as can be seen from Equation (4.52). Supposing that Equation (4.53) holds for $t - 1$, the numerator of Equation (4.52) can be written as

$$\begin{aligned} f_i p_i(t-1) &= f_i \frac{p_i(0) f_i^{t-1}}{\sum_{j=1}^{n} p_j(0) f_j^{t-1}} \\ &= \frac{p_i(0) f_i^t}{\sum_{j=1}^{n} p_j(0) f_j^{t-1}} \end{aligned} \tag{4.54}$$

and the denominator of Equation (4.52) can be written as

$$\begin{aligned} \sum_{j=1}^{n} p_j(t-1) f_j &= \sum_{j=1}^{n} f_j \frac{p_j(0) f_j^{t-1}}{\sum_{k=1}^{n} p_j(0) f_k^{t-1}} \\ &= \sum_{j=1}^{n} \frac{f_j^t p_j(0)}{\sum_{k=1}^{n} f_k^{t-1} p_k(0)}. \end{aligned} \tag{4.55}$$

Substituting Equations (4.54) and (4.55) into Equation (4.52) gives

$$p_i(t) = \frac{p_i(0) f_i^t}{\sum_{k=1}^{n} f_k^{t-1} p_k(0)}. \tag{4.56}$$

This equation gives the proportionality vector as a function of time, as a function of the fitness values, and as a function of the initial proportionality vector, when only selection (no mutation or crossover) is implemented in a GA.

■ EXAMPLE 4.11

As in Example 4.9, we consider the three-bit one-max problem with fitness values

$$\begin{aligned} &f(000)=1, \quad f(001)=2, \quad f(010)=2, \quad f(011)=3, \\ &f(100)=2, \quad f(101)=3, \quad f(110)=3, \quad f(111)=4. \end{aligned} \tag{4.57}$$

Suppose the initial proportionality vector is

$$p(0) = \left[\begin{array}{cccccccc} 0.93 & 0.01 & 0.01 & 0.01 & 0.01 & 0.01 & 0.01 & 0.01 \end{array} \right]^T. \tag{4.58}$$

93% of the initial population is comprised of the least fit individual, and only 1% of the population is comprised of the most fit individual. Figure 4.4 shows a plot of Equation (4.56). We see that as the GA population evolves, x_4, x_6, and x_7, which are the second best individuals, initially gain much of the population that originally belonged to p_1. The least fit individual, x_1, quickly is removed from the population by the selection process. p_2, p_3, and p_5 are not shown in the figure. It does not take very many generations before the entire population converges to x_8, the optimal individual.

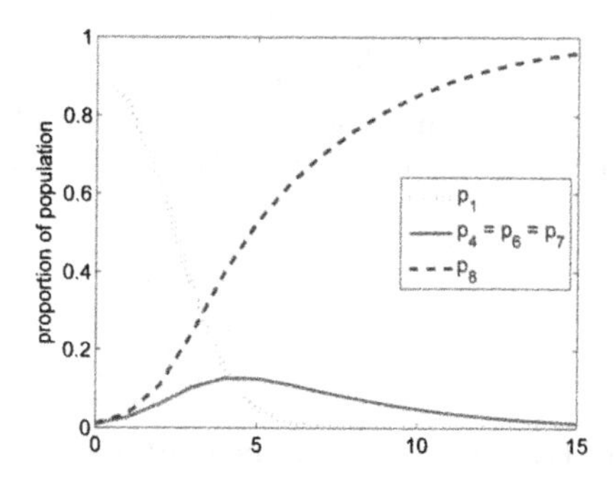

Figure 4.4 Population proportionality vector evolution for Example 4.11. Even though the best individual, x_8, starts with only 1% of the population, it quickly converges to 100%. The least fit individual, x_1, starts with 93% of the population but quickly decreases to 0%.

□

We have discussed the dynamic system model for fitness-proportional selection, but other types of selection, such as tournament selection and rank selection, can also be modeled as a dynamic system [Reeves and Rowe, 2003], [Vose, 1999].

4.5.2 Mutation

Equation (4.51), along with the law of large numbers, tells us that

$$p(t) = \frac{\text{diag}(f)p(t-1)}{f^T p(t-1)} \text{ (selection only).} \tag{4.59}$$

If selection is followed by mutation, and M_{ji} is the probability that x_j mutatates to x_i, then we can use a derivation similar to Equation (4.38) to obtain

$$p(t) = \frac{M^T \text{diag}(f)p(t-1)}{f^T p(t-1)} \text{ (selection and mutation).} \tag{4.60}$$

If $p(t)$ reaches a steady state value, then we can write $p_{ss} = p(t-1) = p(t)$ to write Equation (4.60) as

$$\begin{aligned} p_{ss} &= \frac{M^T \text{diag}(f)p_{ss}}{f^T p_{ss}} \\ M^T \text{diag}(f)p_{ss} &= \left(f^T p_{ss}\right) p_{ss}. \end{aligned} \tag{4.61}$$

This equation is of the form $Ap = \lambda p$, where λ is an eigenvalue of A, and p is an eigenvector of A. We see that the steady-state proportionality vector of a selection-mutation GA (i.e., no crossover) is an eigenvector of $M^T \text{diag}(f)$.

■ **EXAMPLE 4.12**

As in Example 4.10, we consider the three-bit deceptive problem with fitness values

$$\begin{array}{llll} f(000) = 5, & f(001) = 2, & f(010) = 2, & f(011) = 3, \\ f(100) = 2, & f(101) = 3, & f(110) = 3, & f(111) = 4. \end{array} \tag{4.62}$$

We use a mutation rate of 2% per bit in this example. For this problem, we obtain

$$M^T \text{diag}(f) = \tag{4.63}$$
$$\begin{bmatrix} 4.706 & 0.038 & 0.038 & 0.001 & 0.038 & 0.001 & 0.001 & 0.000 \\ 0.096 & 1.882 & 0.001 & 0.058 & 0.001 & 0.058 & 0.000 & 0.002 \\ 0.096 & 0.001 & 1.882 & 0.058 & 0.001 & 0.000 & 0.058 & 0.002 \\ 0.002 & 0.038 & 0.038 & 2.824 & 0.000 & 0.001 & 0.001 & 0.077 \\ 0.096 & 0.001 & 0.001 & 0.000 & 1.882 & 0.058 & 0.058 & 0.002 \\ 0.002 & 0.038 & 0.000 & 0.001 & 0.038 & 2.824 & 0.001 & 0.077 \\ 0.002 & 0.000 & 0.038 & 0.001 & 0.038 & 0.001 & 2.824 & 0.077 \\ 0.000 & 0.001 & 0.001 & 0.058 & 0.001 & 0.058 & 0.058 & 3.765 \end{bmatrix}.$$

We calculate the eigenvectors of $M^T \text{diag}(f)$ as indicated by Equation (4.61) and scale each eigenvector so that its elements sum to 1. Recall that eigenvectors of a matrix are invariant up to a scaling value; that is, if p is an eigenvector, then cp is also an eigenvector for any nonzero constant c. Since the each eigenvector represents a proportionality vector, its elements must sum to 1 as indicated by Equation (4.49). We obtain eight eigenvectors, but only one of them is comprised entirely of positive elements, and so there is only one steady-state proportionality vector:

$$\begin{aligned} p_{ss}(1) &= \begin{bmatrix} 0.90074 & 0.03070 & 0.03070 & 0.00221 \\ 0.03070 & 0.00221 & 0.00221 & 0.0005 \end{bmatrix}^T . \end{aligned} \tag{4.64}$$

This indicates that the GA will converge to a population consisting of 90.074% of x_1 individuals, 3.07% each of x_2, x_3, and x_5 individuals, and so on. Over 90% of the GA population will consist of optimal individuals. However, there is also an eigenvector of $M^T\text{diag}(f)$ that contains only one negative element:

$$\begin{aligned} p_{ss}(2) &= \begin{bmatrix} -0.0008 & 0.0045 & 0.0045 & 0.0644 \\ 0.0045 & 0.0644 & 0.0644 & 0.7941 \end{bmatrix}^T . \end{aligned} \tag{4.65}$$

This is called a metastable point [Reeves and Rowe, 2003], and it includes a high percentage (79.41%) of x_8 individuals, which is the second most fit individual in the population. Any proportionality vector close to $p_{ss}(2)$ will tend to stay there since $p_{ss}(2)$ is a fixed point of Equation (4.61). However, $p_{ss}(2)$ is not a valid proportionality vector since it has a negative element, and so even though the GA population is attracted to $p_{ss}(2)$, eventually the population will drift away from it and will converge to $p_{ss}(1)$. Figure 4.5 shows the results of a simulation of the selection-mutation GA. We used a population size $N = 500$, and an initial proportionality vector of

$$p(0) = \begin{bmatrix} 0.0 & 0.0 & 0.0 & 0.1 & 0.0 & 0.1 & 0.1 & 0.7 \end{bmatrix}^T \tag{4.66}$$

which is close to the metastable point $p_{ss}(2)$. We see from Figure 4.5 that for about 30 generations the population stays close to its original distribution, which is comprised of 70% of x_8 individuals, and which is close to the metastable point $p_{ss}(2)$. After about 30 generations, the population quickly converges to the stable point $p_{ss}(1)$, which is comprised of about 90% of x_1 individuals. Note that if the simulation is run again it will give different results because of the random number generator that is used for selection and mutation.

Figure 4.6 shows p_1 and p_8 from Equation (4.60) for 100 generations. This gives an exact proportion of x_1 and x_8 individuals as the population size approaches infinity. We can see that Figures 4.5 and 4.6 are similar, but Figure 4.5 is the result of a finite population size simulation and will change each time the simulation is run due to the random number generator that is used. Figure 4.6, on the other hand, is exact.

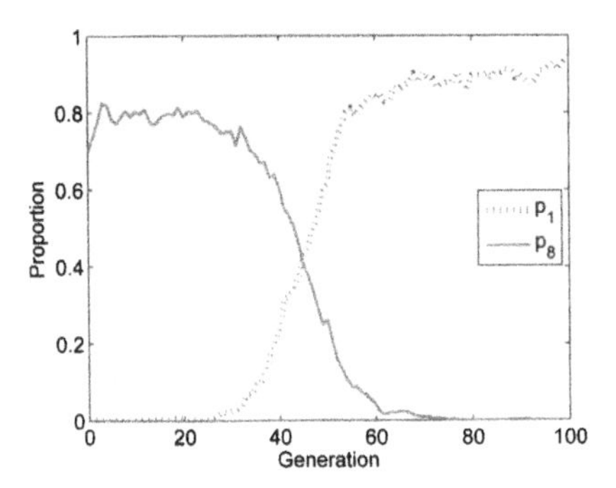

Figure 4.5 Simulation results for Example 4.12. The population hovers around the metastable point, which is comprised of 70% of x_1 individuals, before eventually converging to the stable point of 90% of x_8 individuals. Results will change from one simulation to the next due to the stochastic nature of the simulation.

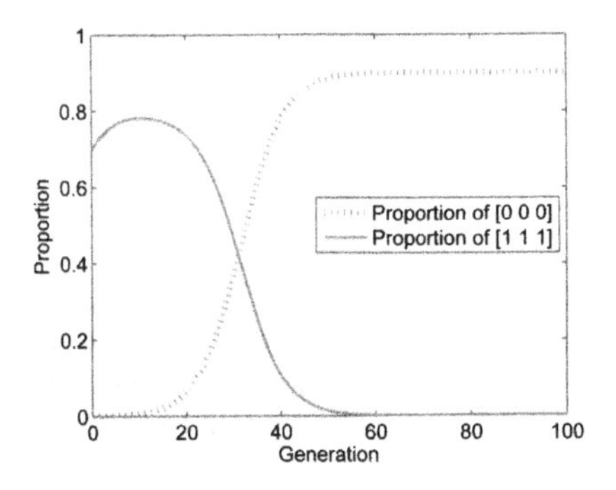

Figure 4.6 Analytical results for Example 4.12. Compare with Figure 4.5. Analytical results do not depend on random number generators.

□

4.5.3 Crossover

As in Section 4.4.3, we use r_{jki} to denote the probability that x_j and x_k cross to form x_i. If the population is specified by the proportionality vector p in an infinite population, then the probability that x_i is obtained from a random crossover is derived as follows:

$$
\begin{aligned}
P_c(x_i|p) &= \sum_{j=1}^{n}\sum_{k=1}^{n} p_j p_k r_{jki} = \sum_{k=1}^{n} p_k \sum_{j=1}^{n} p_j r_{jki} \\
&= \sum_{k=1}^{n} p_k \begin{bmatrix} p_1 & \cdots & p_n \end{bmatrix} \begin{bmatrix} r_{1ki} \\ \vdots \\ r_{nki} \end{bmatrix} \\
&= \begin{bmatrix} p_1 & \cdots & p_n \end{bmatrix} \sum_{k=1}^{n} p_k \begin{bmatrix} r_{1ki} \\ \vdots \\ r_{nki} \end{bmatrix} \\
&= p^T \begin{bmatrix} \sum_{k=1}^{n} r_{1ki} p_k \\ \cdots \\ \sum_{k=1}^{n} r_{nki} p_k \end{bmatrix} \\
&= p^T \begin{bmatrix} \begin{bmatrix} r_{11i} & \vdots & r_{1ni} \end{bmatrix} p \\ \vdots \\ \begin{bmatrix} r_{n1i} & \cdots & r_{nni} \end{bmatrix} p \end{bmatrix} \\
&= p^T \begin{bmatrix} r_{11i} & \cdots & r_{1ni} \\ \vdots & \ddots & \vdots \\ r_{n1i} & \cdots & r_{nni} \end{bmatrix} p \\
&= p^T R_i p
\end{aligned} \tag{4.67}
$$

where the element in the j-th row and k-th column of R_i is r_{jki}, the probability that x_j and x_k cross to form x_i. We again using the law of large numbers [Grinstead and Snell, 1997] to find that in the limit as the population size N approaches infinity, crossover changes the proportion of x_i individuals as follows:

$$\hat{p}_i = P_c(x_i|p) = p^T R_i p. \tag{4.68}$$

Although R_i is often nonsymmetric, the quadratic $P_c(x_i|p)$ can always be written using a symmetric matrix as follows.

$$
\begin{aligned}
P_c(x_i|p) &= p^T R_i p \\
&= \frac{1}{2} p^T R_i p + \frac{1}{2}\left(p^T R_i p\right)^T
\end{aligned} \tag{4.69}
$$

where the second line follows because $p^T R_i p$ is a scalar, and the transpose of a scalar is equal to the scalar. Therefore, recalling that $(ABC)^T = C^T B^T A^T$,

$$
\begin{aligned}
P_c(x_i|p) &= \frac{1}{2} p^T R_i p + \frac{1}{2} p^T R_i^T p \\
&= \frac{1}{2} p^T (R_i + R_i^T) p \\
&= p^T \hat{R}_i p
\end{aligned} \tag{4.70}
$$

where the symmetric matrix $\hat{R}_i$ is given by

$$\hat{R}_i = \frac{1}{2}(R_i + R_i^T). \tag{4.71}$$

EXAMPLE 4.13

As in Example 4.8, suppose we have a four-element search space with individuals $x = \{x_1, x_2, x_3, x_4\} = \{00, 01, 10, 11\}$. We implement crossover by randomly setting $b = 1$ or $b = 2$ with equal probability, and then concatenating bits $1 \to b$ from the first parent with bits $(b+1) \to 2$ from the second parent. The crossover possibilities can be written as

$$\begin{aligned}
00 \times 00 &\to 00 \\
00 \times 01 &\to 01 \text{ or } 00 \\
00 \times 10 &\to 00 \\
00 \times 11 &\to 01 \text{ or } 00 \\
01 \times 00 &\to 00 \text{ or } 01 \\
01 \times 01 &\to 01 \\
01 \times 10 &\to 00 \text{ or } 01 \\
01 \times 11 &\to 01 \\
10 \times 00 &\to 10 \\
10 \times 01 &\to 11 \text{ or } 10 \\
10 \times 10 &\to 10 \\
10 \times 11 &\to 11 \text{ or } 10 \\
11 \times 00 &\to 10 \text{ or } 11 \\
11 \times 01 &\to 11 \\
11 \times 10 &\to 10 \text{ or } 11 \\
11 \times 11 &\to 11.
\end{aligned} \tag{4.72}$$

This gives r_{jk1} crossover probabilities, which are the probabilities that x_j and x_k cross to give $x_1 = 00$, as follows:

$$\begin{array}{llll}
r_{111} = 1.0, & r_{121} = 0.5, & r_{131} = 1.0, & r_{141} = 0.5 \\
r_{211} = 0.5, & r_{221} = 0.0, & r_{231} = 0.5, & r_{241} = 0.0 \\
r_{311} = 0.0, & r_{321} = 0.0, & r_{331} = 0.0, & r_{341} = 0.0 \\
r_{411} = 0.0, & r_{421} = 0.0, & r_{431} = 0.0, & r_{441} = 0.0
\end{array} \tag{4.73}$$

which results in the crossover matrix

$$R_1 = \begin{bmatrix} 1.0 & 0.5 & 1.0 & 0.5 \\ 0.5 & 0.0 & 0.5 & 0.0 \\ 0.0 & 0.0 & 0.0 & 0.0 \\ 0.0 & 0.0 & 0.0 & 0.0 \end{bmatrix}. \tag{4.74}$$

R_1 is clearly nonsymmetric, but $P_c(x_i|p)$ can still be written using the symmetric matrix

$$\begin{aligned}
P_c(x_1|p) &= p^T \hat{R}_1 p \\
\text{where } \hat{R}_1 = \frac{1}{2}(R_1 + R_1^T) &= \begin{bmatrix} 1.0 & 0.5 & 0.5 & 0.25 \\ 0.5 & 0.0 & 0.25 & 0.0 \\ 0.5 & 0.25 & 0.0 & 0.0 \\ 0.25 & 0.0 & 0.0 & 0.0 \end{bmatrix}.
\end{aligned} \tag{4.75}$$

The other $\hat{R}_i$ matrices can be found similarly.

□

Now suppose we have a GA with selection, mutation, and crossover, in that order. We have a proportionality vector p at generation $t-1$. Selection and mutation modify p as shown in Equation (4.60):

$$p(t) = \frac{M^T \text{diag}(f) p(t-1)}{f^T p(t-1)}. \tag{4.76}$$

Crossover modifies p_i as shown in Equation (4.68). However, p on the right side of Equation (4.68) has already been modified by selection and mutation to result in the p shown in Equation (4.76). Therefore, the sequence of selection, mutation, and crossover, results in $\hat{p}_i$ as shown in Equation (4.68), but with the p on the right side of Equation (4.68) replaced by the p resulting from the selection and mutation of Equation (4.76):

$$\begin{aligned} p_i(t) &= \left[\frac{M^T \text{diag}(f) p(t-1)}{f^T p(t-1)}\right]^T R_i \left[\frac{M^T \text{diag}(f) p(t-1)}{f^T p(t-1)}\right] \\ &= \frac{p^T(t-1) \text{diag}(f) M R_i M^T \text{diag}(f) p(t-1)}{\left(f^T p(t-1)\right)^2}. \end{aligned} \tag{4.77}$$

R_i can be replaced with $\hat{R}_i$ in Equation (4.77) to give an equivalent expression. Equation (4.77) gives an exact, analytic expression for the dynamics of the proportion of x_i individuals in an infinite population.

We see that we need to calculate the dynamic system model of Equation (4.77) for $i \in [1, n]$) at each generation, where n is the search space size. The matrices in Equation (4.77) are $n \times n$, and the computational effort of matrix multiplication is proportional to n^3 if implemented with standard algorithms. Therefore, the dynamic system model requires computation on the order of n^4. This is much less computational effort than the Markov model requires, but it still grows very rapidly as the search space size n increases, and it is still requires unattainable computational resources for even moderately-sized problems.

■ EXAMPLE 4.14

Once again we consider the three-bit one-max problem (see Example 4.9) in which each individual's fitness is proportional to the number of ones. We use a crossover probability of 90%, a mutation probability of 1% per bit, a population size of 1,000, and an initial population proportionality vector of

$$p(0) = \begin{bmatrix} 0.8 & 0.1 & 0.1 & 0.0 & 0.0 & 0.0 & 0.0 & 0.0 \end{bmatrix}^T. \tag{4.78}$$

Figure 4.7 shows the percent of optimal individuals in the population from a single simulation, along with the exact theoretical results of Equation (4.77). The simulation results match the theory nicely, but the simulation results are approximate and will vary from one run to the next, while the theory is exact.

Now suppose that we change the initial population proportionality vector to

$$p(0) = \begin{bmatrix} 0.0 & 0.1 & 0.1 & 0.0 & 0.0 & 0.0 & 0.0 & 0.8 \end{bmatrix}^T. \tag{4.79}$$

Figure 4.8 shows the percent of *least fit* individuals from a single simulation, along with the exact theoretical results. Since the probability of obtaining a least-fit individual is so low, the simulation results show a couple of spikes in the graph due to random mutations. The spikes look large given the graph scale, but they are actually quite small, peaking at only 0.2%. The theoretical results, however, are exact. They show that the proportion of least-fit individuals initially increases for a few generations due to mutation, and then quickly decreases to the steady-state value of precisely 0.00502%. It would take many, many simulations to arrive at this conclusion. Even after thousands of simulations, the wrong conclusion may be reached, depending on the integrity of the random number generator that is used.

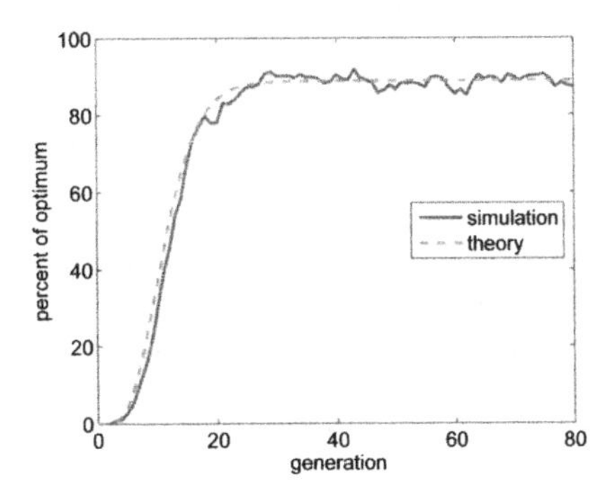

Figure 4.7 Proportion of most-fit individuals for Example 4.14.

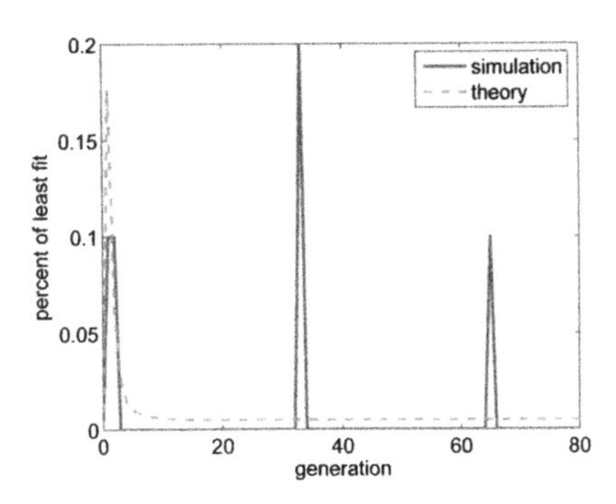

Figure 4.8 Proportion of least-fit individuals for Example 4.14.

□

4.6 CONCLUSION

In this chapter we outlined Markov models and dynamic system models for GAs. These models, which were first developed in the 1990s, give theoretically exact results, whereas simulations change from one run to the next due to the random number generator that is used for selection, crossover, and mutation. The size of the Markov model increases factorially with the population size and with the search space cardinality. The dynamic system model increases with n^4, where n is the search space cardinality. These computational requirements restrict the application of the Markov models and dynamic system models to very small problems. However, the models are still useful for comparing different implementations of GAs and for comparing different EAs, as we see in [Simon et al., 2011b]. Some additional ideas and developments along these directions can be found in [Reeves and Rowe, 2003], [Vose, 1999].

Markov modeling and dynamic system modeling are very mature fields and many general results have been obtained. There is a lot of room for the additional application of these subjects to GAs and other EAs.

Other methods can also be used to model or analyze the behavior of GAs. For example, the field of statistical mechanics involves averaging many molecular particles to model the behavior of a group of molecules, and we can use this idea to model GA behavior with large populations [Reeves and Rowe, 2003, Chapter 7]. We can also use the Fourier and Walsh transforms can to analyze GA behavior [Vose and Wright, 1998a], [Vose and Wright, 1998b]. Finally, we can use Price's selection and covariance theorem to mathematically model GAs [Poli et al., 2008, Chapter 3].

The ideas presented in this chapter could be applied to many other EAs besides GAs. We do this in [Simon et al., 2011a] and [Simon, 2011a] for biogeography-based optimization, and other researchers apply these ideas to other EAs, but there is still a lot of room for the application of Markov models and dynamic system models to EAs. This would allow for comparisons and contrasts between various EAs on an analytical level, at least for small problems, rather than reliance on simulation. Simuation is necessary in our study of EAs, but it should be used to support theory.

PROBLEMS

Written Exercises

4.1 How many schemata of length 2 exist? How many of them are order 0, how many are order 1, and how many are order 2?

4.2 How many schemata of length 3 exist? How many of them are order 0, how many are order 1, how many are order 2, and how many are order 3?

4.3 How many schemata of length l exist?

a) How many of them are order 0?
b) How many of them are order 1?
c) How many of them are order 2?
d) How many of them are order 3?
e) How many of them are order p?

4.4 Suppose that instances of schema h have fitness values that are 25% greater than the average fitness of a GA population. Suppose that the destruction probability of h under mutation and crossover are neglible. Suppose that the GA is initialized with a single instance of h. Determine the generation number when h will overtake the population for population sizes of 20, 50, 100, and 200 [Goldberg, 1989a].

4.5 Suppose that we have a GA with 2-bit individuals such that the probability of mutating from any bit string x_i to any other bit string x_j is p_m for all $j \neq i$. What is the mutation matrix? Verify that each row sums to 1.

4.6 Suppose that we have a GA with 2-bit individuals such that the probability of mutating a 0 bit is p_0, and the probability of mutating a 1 bit is p_1. What is the mutation matrix? Verify that each row sums to 1.

4.7 Calculate r_{2ij} in Example 4.8 for $i \in [1, 4]$ and $j \in [1, 4]$.

4.8 Find R_2 and $\hat{R}_2$ for Example 4.13.

4.9 Suppose that the population at the t-th generation consists entirely of optimal individuals. Suppose that mutation is implemented so that the probabilty of an optimal individual mutating to a different individual is 0. Use Equation (4.77) to show that the population at the $(t + 1)$-st generation consists entirely of optimal individuals.

Computer Exercises

4.10 Consider a GA in which each individual consists of a single bit. Let m_1 denote the number of instances of schema $h_1 = 1$, and let f_1 denote its fitness. Let m_0 denote the number of instances of schema $h_0 = 0$, and let f_0 denote its fitness.

Suppose that the GA has an infinitely large population and uses reproduction and mutation (no crossover). Derive a recursive equation for $p(t)$, which is the ratio m_1/m_0 at the t-th generation [Goldberg, 1989a].

a) Suppose that the GA is initialized with $p(0) = 1$. Plot $p(t)$ for the first 100 generations when the mutation rate is 10% for fitness ratios $f_1/f_0 = 10$, 2, and 1.1.

b) Repeat for a mutation rate of 1%.

c) Repeat for a mutation rate of 0.1%.

d) Give an intuitive explanation of your results.

4.11 Verify Property 6 of Theorem 4.2 for the transition matrix of Example 4.1.

4.12 A certain professor gives difficult exams. 70% of students who currently have an A in the course will drop to a B or worse after each exam. 20% of students who currently have a B or worse will raise their grade to an A after each exam. Given an infinite number of exams, how many students will earn an A in the course?

4.13 Use Equations (4.26) and (4.28) to calculate the number of possible populations in a GA with a population size of 10 in which each individual is comprised of 6 bits.

4.14 Repeat Example 4.10 with the following fitness values:

$$\begin{array}{llll} f(000) = 7, & f(001) = 2, & f(010) = 2, & f(011) = 4, \\ f(100) = 2, & f(101) = 4, & f(110) = 4, & f(111) = 6. \end{array}$$

What do you get for the probability of no optimal solutions? How does this probability compare with that obtained in Example 4.10? How can you explain the difference?

4.15 Repeat Example 4.10 with a mutation rate of 1%. What do you get for the probability of no optimal solutions? How does this probability compare with that obtained in Example 4.10? How can you explain the difference?

4.16 Repeat Example 4.9 with the change that if the optimal solution is obtained, then it is never mutated. How does this change the mutation matrix? What do you get for the probability of all optimal solutions? How can you explain your results?

CHAPTER 5

Evolutionary Programming

Success in predicting an environment is a prerequisite to intelligent behavior.
—Lawrence Fogel [Fogel, 1999, page 3]

Evolutionary programming (EP) was invented by Lawrence Fogel, along with his coworkers Al Owens and Jack Walsh, in the 1960s [Fogel et al., 1966], [Fogel, 1999]. An EP evolves a population of individuals but does not involve recombination. New individuals are created only by mutation.

EP was originally invented to evolve finite state machines (FSMs). An FSM is a virtual machine that generates a sequence of outputs from a sequence of inputs. The output sequence generation is determined not only by the inputs, but also by a set of states and state transition rules. Lawrence Fogel considered prediction to be a key component of intelligence. Therefore, he considered the development of FSMs that could predict the next output of some process as a key step toward the development of computational intelligence.

Overview of the Chapter

Section 5.1 gives an overview of EP for continuous problem domains. Although EP was originally described as operating on a discrete domain, today it is often (perhaps usually) implemented on continuous domains, and so that is how we describe it in general terms. Section 5.2 describes finite state machines (FSMs) and shows how EPs can optimize them. FSMs are interesting because they can be used to model many different types of systems, including computer programs, digital electronics, control systems, and classifier systems.

Section 5.3 discusses Fogel's original method of EP for discrete problem domains. Section 5.4 discusses the prisoner's dilemma, which is a classic game theory problem. Prisoner's dilemma solutions can be represented as FSMs, and so EPs can find optimal solutions to the prisoner's dilemma. Section 5.5 discusses the artificial ant problem, which uses EP to evolve an FSM with which an ant can navigate a grid to efficiently find food.

5.1 CONTINUOUS EVOLUTIONARY PROGRAMMING

Suppose we want to minimize $f(x)$, where x is an n-dimensional vector. Assume that $f(x) \geq 0$ for all x. An EP begins with a randomly-generated population of individuals $\{x_i\}$, $i \in [0, N]$. We create children $\{x_i'\}$ as follows:

$$x_i' = x_i + r_i\sqrt{\beta f(x_i) + \gamma}, \quad i \in [1, N] \tag{5.1}$$

where r_i is a random n-element vector, each of whose elements is taken from a Gaussian distribution with a mean of 0 and a variance of 1, and β and γ are EP tuning parameters. The variance of the mutation of x_i is $(\beta f(x_i) + \gamma)$. If $\beta = 0$, then all individuals have the same average mutation magnitude. If $\beta > 0$, then an individual with a low cost does not mutate as much as an individual with a high cost. Often $\beta = 1$ and $\gamma = 0$ are used as default, standard EP parameter values. We will see in Chapter 6 that an EP with a population size of one is equivalent to a two-membered ES.

An examination of Equation (5.1) reveals some of the implementation issues associated with EP [Bäck, 1996, Section 2.2].

- First, cost values $f(x)$ should be shifted so that they are always non-negative. This is not difficult in practice but it is still something that needs to be done.

- Second, β and γ need to be tuned. The default values are $\beta = 1$ and $\gamma = 0$, but there is no reason to suppose that these values will be effective. For example, suppose the x_i values have a large domain, and we use the default values $\beta = 1$ and $\gamma = 0$. Then the mutation in Equation (5.1) will be very small relative to the value of x_i, which will result in very slow convergence, or no convergence at all. Conversely, if the x_i values have a very small domain, then the default values of β and γ will result in mutations that are unreasonably large; that is, the mutations will result in x_i values that are outside of the domain.

- Third, if $\beta > 0$ (the typical case) and all of the cost values are high, then $(\beta f(x_i) + \gamma)$ will be approximately constant for all x_i, which will result in an approximately constant expected mutation for all individuals, regardless of their cost value. Even if an individual improves its cost by a beneficial mutation, that improvement is likely to be reversed by a detrimental mutation.

For example, suppose that cost values range from a minimum of $f(x_1) = 1000$ to a maximum of $f(x_N) = 1100$. Individual x_1 is relatively much better than x_N, but the cost values are scaled in such a way that both x_1 and x_N are mutated by approximately the same magnitude. However, this is not a problem that is exclusive to EP. This issue of cost function value scaling applies to other EAs also, and we will discuss it further in Section 8.7.

After Equation (5.1) generates N children, we have $2N$ individuals: $\{x_i\}$ and $\{x'_i\}$. We select the best N from these $2N$ individuals to form the population at the next generation. A basic EP algorithm is summarized in Figure 5.1.

Select non-negative EP parameters β and γ. Nominally $\beta = 1$ and $\gamma = 0$.
$\{x_i\} \leftarrow$ {randomly generated population}, $i \in [1, N]$
While not(termination criterion)
    Calculate the cost $f(x_i)$ of each individual in the population
    For each individual x_i, $i \in [1, N]$
        Generate a random vector r_i with each element $\sim N(0, 1)$
        $x'_i \leftarrow x_i + r_i\sqrt{\beta f(x_i) + \gamma}$
    Next individual
    $\{x_i\} \leftarrow$ best N individuals from $\{x_i, x'_i\}$
Next generation

Figure 5.1 The above pseudo-code outlines a basic evolutionary program (EP) for minimizing $f(x)$.

Different options can be used in EP to select individuals for the next generation from $\{x_i, x'_i\}$. Figure 5.1 shows that this is done deterministically; the best N individuals are selected from $\{x_i, x'_i\}$. However, selection could also be done probabilistically. For example, roulette-wheel selection could be used N times to select N individuals from $\{x_i, x'_i\}$, or tournament selection could be used, or various other selection methods could be used (see Section 8.7).

An EP is often written so that not only the candidate solutions evolve, but also their mutation variances evolve. This is often called a meta EP and is summarized in Figure 5.2. In a meta EP, each individual x_i is associated with a mutation variance v_i. The mutation variances themselves mutate in the search for an optimal mutation variance. We constrain the mutation variances in Figure 5.2 to a minimum of ϵ, which is a user-defined tuning parameter. A meta EP can speed convergence by automatically adapting mutation variances, but it can also slow down convergence, depending on the specific problem.

Select non-negative EP parameters ϵ and c. Nominally $\epsilon \ll 1$ and $c = 1$.
$\{x_i\} \leftarrow$ {randomly generated population}, $i \in [1, N]$
$\{v_i\} \leftarrow$ {randomly generated variances}, $i \in [1, N]$
While not(termination criterion)
    Calculate the cost $f(x_i)$ of each individual in the population
    For each individual x_i, $i \in [1, N]$
        Generate random vectors r_{xi} and r_{vi} with each element $\sim N(0, 1)$
        $x_i' \leftarrow x_i + r_{xi}\sqrt{v_i}$
        $v_i' \leftarrow v_i + r_{vi}\sqrt{cv_i}$
        $v_i' \leftarrow \max(v_i', \epsilon)$
    Next individual
    $\{x_i\} \leftarrow$ best N individuals from $\{x_i, x_i'\}$
    $\{v_i\} \leftarrow$ variances that correspond to $\{x_i\}$
Next generation

Figure 5.2 The above pseudo-code outlines a meta evolutionary program (EP) for minimizing $f(x)$. Note that v_i is associated with the individual x_i for all $i \in [1, N]$.

■ EXAMPLE 5.1

In this example we use an EP to optimize the Griewank and Ackley test functions (see Appendix C for the definitions of these benchmarks). We use 20 dimensions in each benchmark. We run the standard EP of Figure 5.1 with $\beta = (x_{\max} - x_{\min})/10$ and $\gamma = 0$. We use a population size of 50 and we normalize the cost of each individual so that $f(x_i) \in [1, 2]$ at every generation.

We also use the meta EP of Figure 5.2 with $c = 1$ and $\epsilon = \beta/10$. Figures 5.3 and 5.4 show the minimum cost of the population as a function of generation number, averaged over 20 Monte Carlo simulations. We see that the meta EP converges much better than the standard EP on the Griewank function, but much worse than the standard EP on the Ackley function. This is no doubt due to the different domains of the two functions. The domain of each independent variable is ± 600 in the Griewank function, but only ± 30 in the Ackley function. A comparison of the numbers on the vertical axes of the two figures shows that the ranges of the two functions are also very different.

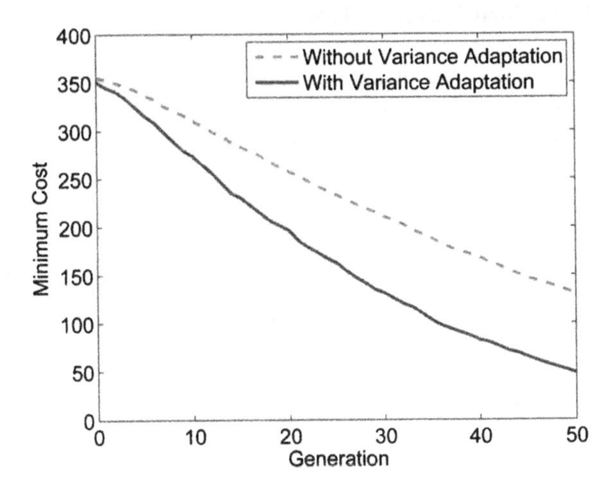

Figure 5.3 Example 5.1: EP convergence on the 20-dimensional Griewank function, averaged over 20 Monte Carlo simulations. The meta EP converges much faster than the standard EP.

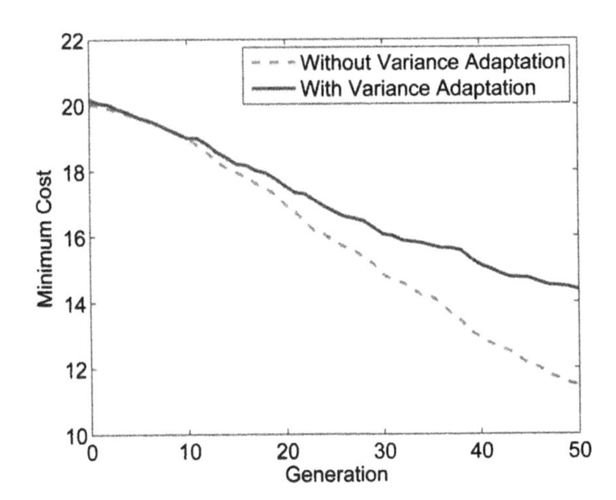

Figure 5.4 Example 5.1: EP convergence on the 20-dimensional Ackley function, averaged over 20 Monte Carlo simulations. The standard EP converges much faster than the meta EP.

□

5.2 FINITE STATE MACHINE OPTIMIZATION

EP was originally invented to evolve finite state machines (FSMs). An FSM generates a sequence of outputs as a function of an internal state and a sequence of inputs. Figure 5.5 shows an example of an FSM. It has four states, A, B, C, and D; it has two possible inputs, 0 and 1, which are shown on the left of each forward-slash in the figure; and it has three possible outputs, a, b, and c, which are shown on the right of each forward-slash. The arrow at the top right of the figure shows that the FSM begins in state C. The arrows show how the state transitions after a particular input. The labels on the lines show the input/output combinations. Figure 5.5 can also be depicted in tabular form as shown in Table 5.1.

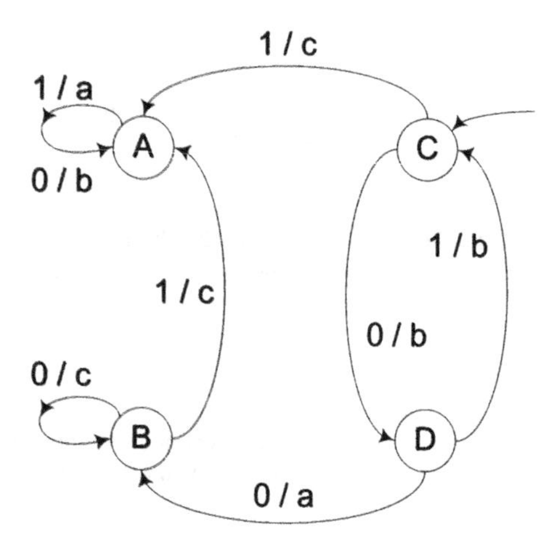

Figure 5.5 Finite state machine of Table 5.1 shown in diagram form. This FSM has four states. The pair beside each arrow shows the input and the corresponding output if the FSM is in the state at the tail of the arrow. The arrow at the top right shows that the FSM begins in state C.

Current State	A	A	B	B	C	C	D	D
Input	0	1	0	1	0	1	0	1
Next State	A	A	B	A	D	A	B	C
Output	b	a	c	c	b	c	a	b

Table 5.1 Finite state machine of Figure 5.5 in tabular form.

Suppose that we want to create an FSM that replicates a certain output sequence from a certain input sequence. For example, we might know that the input sequence

$$\text{Input} = \{1, 0, 1, 0, 1, 0, 0, 1, 1, 0, 1, 0\} \tag{5.2}$$

should result in the output sequence

$$\text{Output} = \{0, 0, 1, 1, 1, 1, 0, 1, 1, 0, 0, 1\}. \tag{5.3}$$

Can we create a state machine that will produce the desired behavior? This is an optimization problem: we want to evolve an FSM that minimizes the difference between the FSM behavior and the goal behavior. We can represent a state machine in the format

$$S = \left[\; (\; \text{output}_0 \quad \text{next state}_0 \;) \quad (\; \text{output}_1 \quad \text{next state}_1 \;) \quad \cdots \; \right]^T. \tag{5.4}$$

We assume, without loss of generality, that the FSM begins in state 1. The elements of S are arranged as

$$\begin{aligned} S(1) &= \text{output if the FSM is in state 1 and the input is 0} \\ S(2) &= \text{next state if the FSM is in state 1 and the input is 0} \\ S(3) &= \text{output if the FSM is in state 1 and the input is 1} \\ S(4) &= \text{next state if the FSM is in state 1 and the input is 1} \\ S(5) &= \text{output if the FSM is in state 2 and the input is 0} \\ &\;\;\vdots \\ S(4n) &= \text{next state if the FSM is in state } n \text{ and the input is 1} \end{aligned} \tag{5.5}$$

where we have assumed that the input is binary. We can easily extend this structure to the case of non-binary inputs. We see that S is a $4n$-element column vector that describes the FSM, where n is the number of states. We can apply the input of Equation (5.2) to an FSM, and define the error cost of the FSM as

$$\text{Cost} = \sum_{i=1}^{12} \left| (\text{Desired Output})_i - (\text{FSM Output})_i \right| \tag{5.6}$$

where $(\text{Desired Output})_i$ is the i-th output of Equation (5.3), and there is a sequence of 12 outputs. We can then use the EP algorithm of Figure 5.1 or 5.2 to evolve an FSM to minimize Equation (5.6).

One implementation detail that we need to consider is that x_i is a continuous variable in Figure 5.1; but in FSM evolution, the elements of each individual are integers that are constrained to specific domains. Equation (5.5) shows that for an FSM with binary outputs, $S(i) \in [0, 1]$ for odd values of i, and $S(i) \in [1, n]$ for even values of i, where n is the number of states. This can be handled by simply performing the mutations shown in Figure 5.1, then constraining the elements of x'_i to the appropriate domain, and finally rounding the elements of x'_i to the nearest integers. Again, other approaches are possible and are left to the research and creativity of the reader.

Other implementation details include tuning of β and γ, and scaling the cost values $f(x_i)$ appropriately before using them to obtain mutation variances.

■ EXAMPLE 5.2

In this example we use EP to evolve an FSM to minimize the cost function of Equation (5.6), where the inputs and outputs are given in Equations (5.2) and (5.3). We use four states in each EP individual, $\beta = 1$, $\gamma = 0$, and a population size of 5. We use the EP algorithm of Figure 5.1 to evolve the population of FSMs. We scale the cost of each individual so that at each generation the costs $\in [1, 2]$. Figure 5.6 shows the convergence of the cost function for one EP simulation. An S vector that gives zero cost was found as follows:

$$S = \begin{bmatrix} 1 & 3 & 1 & 2, & 1 & 1 & 0 & 4, & 0 & 1 & 0 & 2, & 1 & 4 & 0 & 2 \end{bmatrix} \tag{5.7}$$

which corresponds to the FSM shown in Figure 5.7.

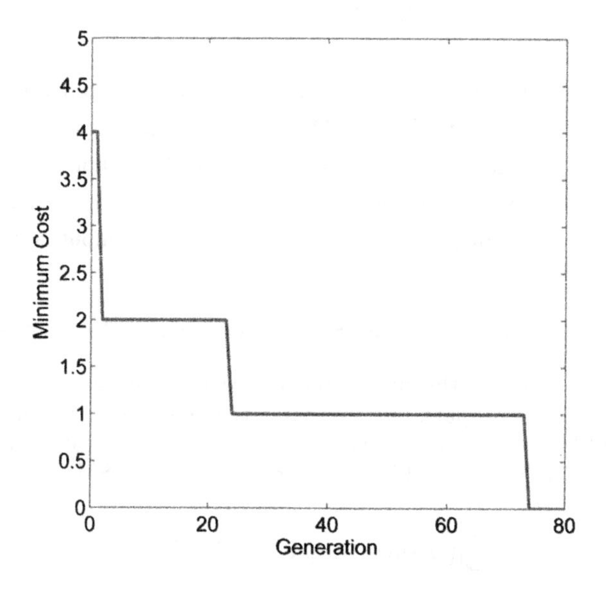

Figure 5.6 Example 5.2: Convergence of the finite state machine.

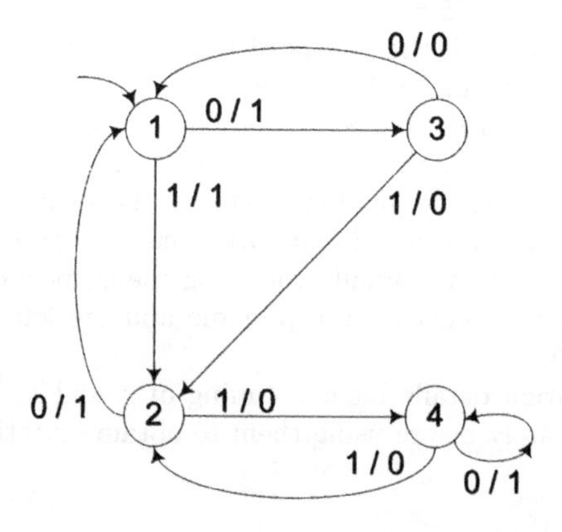

Figure 5.7 The finite state machine evolved by the EP of Example 5.2. If the input to this FSM is given by Equation (5.2), it creates the output of Equation (5.3).

□

5.3 DISCRETE EVOLUTIONARY PROGRAMMING

Fogel's original EP for FSM generation was implemented differently than what we described in the previous section [Fogel et al., 1966], [Fogel, 1999]. His implementation was directly applicable to integer domains. His approach can be used not only for FSM optimization, but for any problem that is defined on a discrete domain. We give a summary of his approach in Figure 5.8.

$\{x_i\} \leftarrow$ {randomly generated population}, $i \in [1, N]$
While not(termination criterion)
    Calculate the cost $f(x_i)$ of each individual in the population
    For each individual x_i, $i \in [1, N]$
        $x'_i \leftarrow$ random mutation of x_i
    Next individual
    $\{x_i\} \leftarrow$ best N individuals from $\{x_i, x'_i\}$
Next generation

Figure 5.8 The above pseudo-code outlines Fogel's evolutionary program for discrete optimization problems. Note that this algorithm is a generalization of Figure 5.1.

The "random mutation" in Figure 5.8 depends entirely on the specific problem that we are trying to solve. As an example, the random mutation that Fogel used for FSM optimization was selected randomly as one of the following.

- Add a state with random input/output and input/transition pairs.
- Delete a state. Any state transitions to the deleted state are redirected to another randomly-selected state.
- Randomly change an input/output pair for a randomly-selected state.
- Randomly change an input/transition pair for a randomly-selected state.
- Randomly change the initial state.

Fogel also suggested adding a penalty to the cost function that was proportional to the complexity of the state machine. This biases the selection of the best N individuals at the end of each generation to simpler state machines. This idea allows us to not only find FSMs to generate a desired pattern, but also to find *simple* FSMs to generate a desired pattern.

■ EXAMPLE 5.3

In this example we try to find a state machine that can generate prime numbers. We use 0 to indicate non-prime numbers, and 1 to indicate prime numbers. The input to the state machine at each time step is the prime indicator (0 for false, 1 for true) of the previous time step. This gives the input and output sequences

$$\begin{aligned} \text{Input} &= \{0,1,1,0,1,0,1,0,0,0,1,0,\cdots\} \\ \text{Desired Output} &= \{\ \ 1,1,0,1,0,1,0,0,0,1,0,\cdots\}. \end{aligned} \tag{5.8}$$

The input sequence corresponds to the fact that 1 is nonprime; 2 and 3 are prime; 4 is nonprime; 5 is prime; 6 is nonprime; 7 is prime; 8, 9, and 10 are nonprime; 11 is prime; 12 is nonprime; and so on. The output sequence is equal to the input sequence delayed by one time step. We use the first 100 positive integers to evaluate the performance of an FSM, so the input and output sequences are each 99 bits long; the input sequence corresponds to integers 1–99, and the desired output sequence corresponds to integers 2–100. We use a population size $N = 20$. For each individual at each EP generation, we randomly select one of the five mutations described earlier in this section. We simulate the EP with a cost penalty for the number of FSM states, and also without a penalty. Figure 5.9 shows the best FSM for each generation, averaged over 100 Monte Carlo simulations. The average number of states is 4.7 if there is no penalty for the number of states, and 2.8 if there is a cost penalty of $n/2$ for the number of states, where n is the number of states. Figure 5.9 shows that the best cost that we can achieve is between 18 and 19, which is not that impressive considering that there are only 25 prime numbers in the first 100 positive integers. A state machine that always generated a 0 would have a cost of 25.

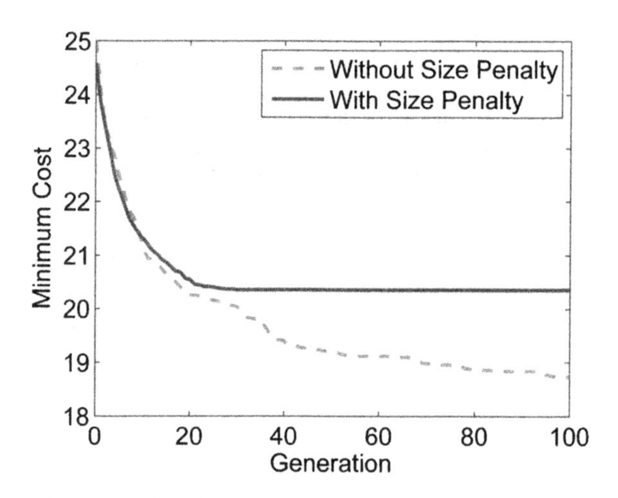

Figure 5.9 Example 5.3: Convergence of the finite state machine for prime number prediction, averaged over 100 Monte Carlo simulations.

□

5.4 THE PRISONER'S DILEMMA

The prisoner's dilemma is a classic game theory problem. Suppose that two crime suspects are arrested by the police. The police independently question the suspects. The police offer each suspect immunity from prosecution if he betrays his accomplice. A confession by either suspect will give the police enough evidence to imprison his accomplice for a long period of time. However, if both suspects remain silent, then the police will not have enough evidence to imprison the suspects for very long. Herein lies the quandry for the prisoners, who are not allowed to communicate with each other. If each suspect remains silent (cooperate with each other), then both suspects receive a suspended sentence. If each suspect talks (defect against each other), then both suspects receive a medium sentence. However, if one suspect cooperates and one defects, then the suspect who defects goes free, while the suspect who cooperates receives a long sentence. The prisoner's dilemma is summarized in Table 5.2.

	Prisoner B Cooperates	Prisoner B Defects
Prisoner A Cooperates	Prisoner A: 1 Year Prisoner B: 1 Year	Prisoner A: 10 Years Prisoner B: Free
Prisoner A Defects	Prisoner A: Free Prisoner B: 10 Years	Prisoner A: 5 Years Prisoner B: 5 Years

Table 5.2 Prisoner's dilemma cost matrix.

Suppose that you are Prisoner A. If your accomplice cooperates, then you can go free if you defect. If your accomplice defects, then you can get 5 years instead of 10 years by defecting. Therefore, it seems that no matter what your accomplice does, you should defect. However, if both prisoners use this strategy, then both prisoners will defect and receive 5 year sentences. If both prisoners cooperate with each other, then they will both receive only 1 year sentences. Selfish decisions result in both prisoners being worse off than if they act in the interest of their accomplice. This is why the problem is called a dilemma.

In the iterated prisoner's dilemma, the prisoner's dilemma game is played several times, and each player's goal is to maximize his total benefit (that is, minimize his total prison time) over all of the games [Axelrod, 2006]. As you play the game by making a choice to cooperate or defect, you remember your accomplice's previous decisions. Therefore, if your accomplice repeatedly defects, you can choose to defect to maximize your benefit. If your accomplice cooperates, then you can choose to cooperate to maintain mutual cooperation and mutual benefit. The word "iterated" is often omitted from the term "iterated prisoner's dilemma." So the term "prisoner's dilemma" can refer to one round of Table 5.2, or multiple rounds.

There are several strategies that have been proposed for the prisoner's dilemma, each of which can be conveniently represented as an FSM [Ashlock, 2009], [Rubinstein, 1986]. One strategy is to cooperate all the time. This optimistic strategy is depicted with the one-state FSM shown in Figure 5.10. We begin by cooperating on our first turn and moving to state 1. If our opponent's previous decision was C, then we output C and remain in state 1. If our opponent's previous decision was D, we do the same – output C and remain in state 1.

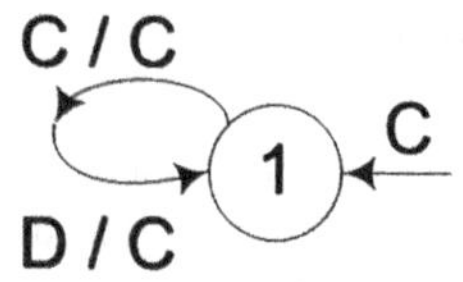

Figure 5.10 Finite state machine for the always-cooperate strategy in the prisoner's dilemma.

Another strategy, the tit-for-tat strategy, follows the philosophy "do unto others as they have done unto you." We begin by cooperating on our first turn and moving to state 1. We cooperate if our opponent cooperated on his previous move, and defect if our opponent defected on his previous move. This strategy is shown in Figure 5.11.

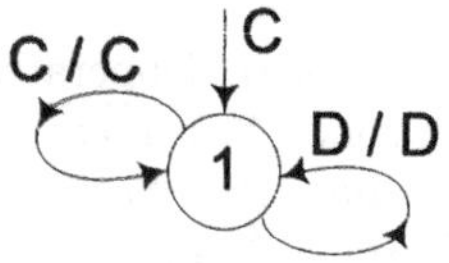

Figure 5.11 Finite state machine for the tit-for-tat strategy in the prisoner's dilemma.

Another strategy, tit-for-two-tats, is a little more hopeful and forgiving than tit-for-tat. We cooperate unless our opponent defects for two consecutive turns. This strategy is shown in Figure 5.12.

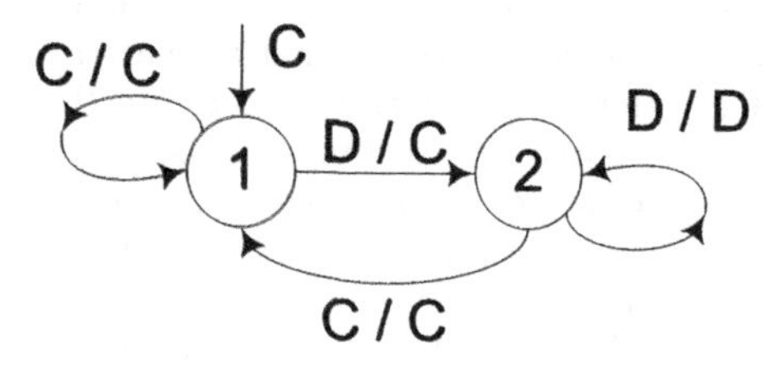

Figure 5.12 Finite state machine for the tit-for-two-tats strategy in the prisoner's dilemma.

The grim strategy is very unforgiving. We begin optimistically by cooperating on our first turn and moving to state 1, and we continue to cooperate as long as our opponent cooperates. However, if our opponent defects, then we never again cooperate. This strategy is shown in Figure 5.13.

In the punish strategy, we take some revenge on our opponent for defecting, but eventually we forgive. If our opponent defects, then we defect, and we continue to defect until our opponent cooperates for three consecutive turns. Only after our opponent cooperates for three consecutive turns do we once again cooperate. This strategy is shown in Figure 5.14.

The prisoner's dilemma has been studied a lot because it has many applications, including peer-to-peer file sharing [Ellis and Yao, 2007], advertising strate-

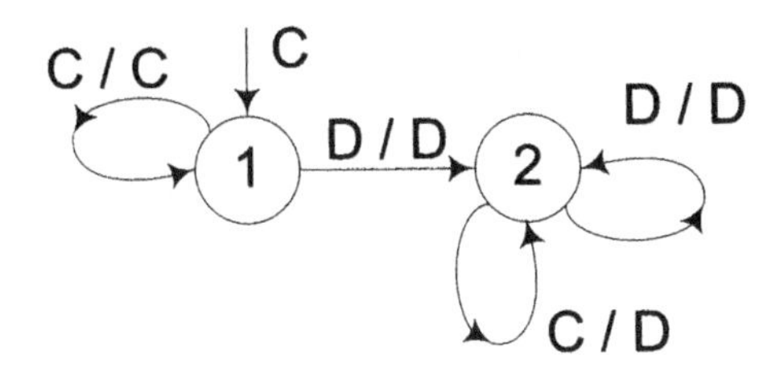

Figure 5.13 Finite state machine for the grim strategy in the prisoner's dilemma.

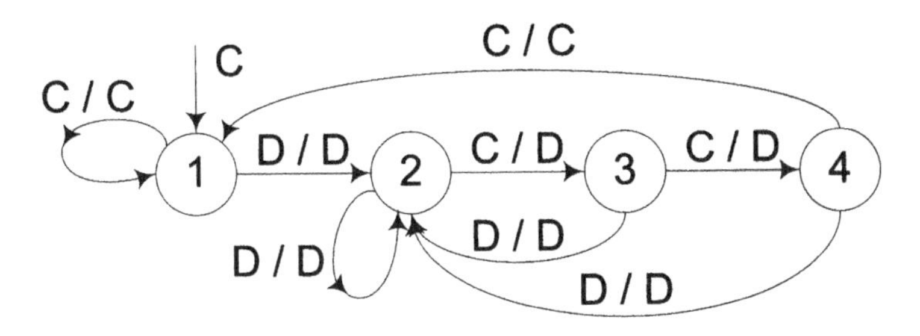

Figure 5.14 Finite state machine for the punish strategy in the prisoner's dilemma.

gies among competing companies [Corfman and Lehmann, 1994], politics [Grieco, 1988], cheating in sports and other areas [Ehrnborg and Rosén, 2009], and many others [Poundstone, 1993].

■ EXAMPLE 5.4

In this example we evolve an FSM to minimize the cost in a prisoner's dilemma game. The FSM for a prisoner's dilemma can be represented as shown in Equation (5.5), except we add one more integer at the beginning of the vector to indicate the first move. We use 0 to indicate cooperate, and 1 to indicate defect. We create four random but constant opponents, each with a four-state FSM strategy. We run an EP with $\beta = 1$ and $\gamma = 0$. We randomly initialize an EP population size of 5, with each individual containing four states. Each EP individual plays 10 games against each of the four random but constant opponents to evaluate its performance. In this example, the state machines for the four random but constant opponents are

$$\begin{aligned} S_1 &= \left[\, 0, \;\; 0\;\;3\;\;0\;\;3, \;\; 0\;\;3\;\;1\;\;2, \;\; 1\;\;4\;\;0\;\;2, \;\; 1\;\;3\;\;1\;\;3 \,\right] \\ S_2 &= \left[\, 0, \;\; 0\;\;1\;\;0\;\;4, \;\; 1\;\;2\;\;0\;\;2, \;\; 0\;\;4\;\;0\;\;3, \;\; 0\;\;2\;\;0\;\;4 \,\right] \\ S_3 &= \left[\, 0, \;\; 1\;\;4\;\;0\;\;4, \;\; 0\;\;1\;\;0\;\;1, \;\; 1\;\;4\;\;0\;\;2, \;\; 0\;\;1\;\;1\;\;4 \,\right] \\ S_4 &= \left[\, 1, \;\; 0\;\;4\;\;1\;\;4, \;\; 0\;\;4\;\;0\;\;1, \;\; 0\;\;3\;\;0\;\;3, \;\; 0\;\;3\;\;0\;\;2 \,\right]. \end{aligned} \quad (5.9)$$

We use the EP algorithm of Figure 5.1 to evolve a population of FSMs. Figure 5.15 shows the convergence of the EP cost function for one simulation.

The S vector that gives minimum cost was found to be

$$S = \left[\begin{array}{ccccccccccccccccc} 1, & 1 & 2 & 1 & 2, & 0 & 1 & 1 & 3, & 1 & 4 & 0 & 1, & 1 & 1 & 0 & 1 \end{array}\right] \tag{5.10}$$

which corresponds to the FSM shown in Figure 5.16.

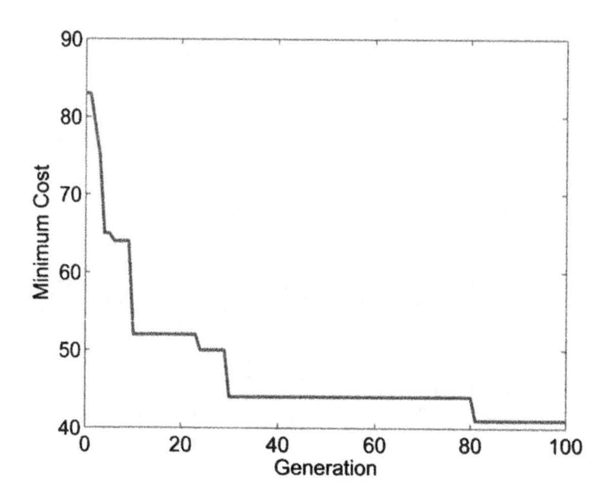

Figure 5.15 Example 5.4: Convergence of the cost of the prisoner's dilemma finite state machine.

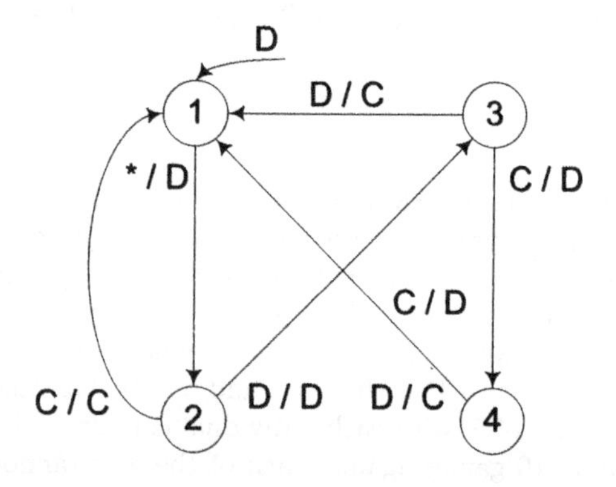

Figure 5.16 Example 5.4: The best finite state machine evolved by the EP. The asterisk coming out of state 1 means either C or D; that is, if the FSM is in state 1, then regardless of the opponent's previous move (C or D), the FSM's output will be D and the next state will be state 2.

□

We can also do interesting experiments in which an EP population evolves by playing against itself. That is, each individual in the EP plays every other individual in the EP to evaluate individual costs.

There are many variations of the prisoner's dilemma. For example, we have assumed that each individual can choose one of two possible moves at each turn. However, we could also assume levels of cooperation so that each turn could involve a continuum of moves, along with a continuum of associated costs. 0 could represent total cooperation, 1 could represent total defection, and anything in between 0 and 1 could represent varying degrees of cooperation and defection [Harrald and Fogel, 1996]. Another variation is to allow each individual to voluntarily terminate the game at any desired time [Delahaye and Mathieu, 1995]. Another variation is to allow more than two players to play simultaneously. In this case the cost for a given player is often a function of the player's move and the number of opponents who cooperate [Bonacich et al., 1976]. Another complication could be introduced if the cost matrix of Table 5.2 changes with time [Worden and Levin, 2007].

5.5 THE ARTIFICIAL ANT PROBLEM

In this section we discuss the artificial ant problem (not to be confused with ant colony optimization), which is another famous problem that can be solved with an FSM. The artificial ant problem was introduced in 1990 [Jefferson et al., 2003] and is nicely described in [Koza, 1992, Section 3.3.2]. An artificial ant is placed on a 32×32 toroidal grid that has food in 90 of the 1024 squares. The ant's sensory capabilities are extremely limited; he can only sense whether or not there is food in the square directly in front of him. At each square, he can make one of three moves: he can move one square forward in the direction in which he is facing, in which case he will eat the food in that square, if it is present; or he can turn right while remaining in his present square; or he can turn left while remaining in his present square. The trail is referred to as the Sante Fe trail, and is shown in Figure 5.17.

The ant begins in the (1,1) square, which is the bottom left corner of the grid (although technically speaking there are not any "corners" since the grid is toroidal), and the initial orientation of the ant is facing to the right. He senses food ahead of him in Figure 5.17, and so he should move forward to the next square at the (2,1) coordinate to eat that food. While in the (2,1) square, he will again sense food in front of him, and so he should move forward to the (3,1) square to eat that food. While in the (3,1) square, he will sense food in front of him, and so he should move forward to the (4,1) square to eat that food. But now he encounters a snag in his heretofore predictable and satisfying journey. While in the (4,1) square, he will sense that there is no food in front of him. Should he move forward anyway, hoping to find food in the next square? Or should he turn left or right, hoping to find food beside his current position? If he turns left, then he will sense food in the (4,2) square and remain on the optimal trail. But if he turns right, then he will sense food in the (4,32) square (remember that the grid is toroidal). This will provide him with some short-term gratification, but will eventually lead him astray.

The artificial ant problem consists of finding an FSM to guide the ant through the Sante Fe trail so that the ant eats all of the food in the fewest moves possible. A step forward, a turn to the right, and a turn to the left, are each considered one move. The optimal path through the grid, which is achieved by moving through the black and gray squares in Figure 5.17, consists of 167 moves.

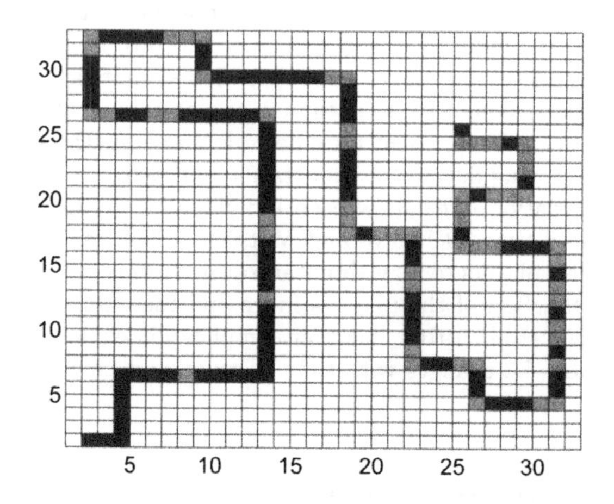

Figure 5.17 The 32 × 32 Sante Fe trail. An ant is placed in the lower left corner facing the right. The white squares are empty, and the black squares have food. The gray squares are also empty, but are shown as gray to better illustrate the ant's optimal path through the grid.

We can use EP to evolve a solution to the artificial ant problem. First we decide how many states we want to use. Suppose that we decide we want to use five states. Then we encode an FSM with the following sequence of integers:

$$\begin{aligned} &1_{0,m}, 1_{0,s}, 1_{1,m}, 1_{1,s}, \\ &2_{0,m}, 2_{0,s}, 2_{1,m}, 2_{1,s}, \\ &\vdots \\ &5_{0,m}, 5_{0,s}, 5_{1,m}, 5_{1,s} \quad . \end{aligned} \tag{5.11}$$

The notation used in the above FSM representation is as follows:

- $n_{0,m}$ is the move that the ant makes if he is currently in state n and does not sense food directly in front of him. We set $n_{0,m}$ = 0, 1, or 2 to respectively indicate a move forward, a turn to the right, or a turn to the left.
- $n_{0,s}$ is the state to which the ant transitions if he is currently in state n and does not sense food directly in front of him.
- $n_{1,m}$ is the move that the ant makes if he is currently in state n and senses food directly in front of him.
- $n_{1,s}$ is the state to which the ant transitions if he is currently in state n and senses food directly in front of him.

We thus encode an FSM with $4N$ integers, where N is the number of states. We assume that the ant always begins in state 1.

We can evaluate each FSM in an EP population and see how well it navigates through the grid. We initialize the ant by placing it in the lower left corner facing right. We put an upper limit of 500 on the number of moves that the ant can make with each FSM. The cost of an FSM is measured by the number of moves that the ant requires to eat all of the food in the grid. If the ant has not eaten all of the food after 500 moves, then the cost is equal to 500 plus the number of food squares that the ant has not reached. Figure 5.18 shows the progress of one EP simulation for five-state FSMs with a population size of 100.

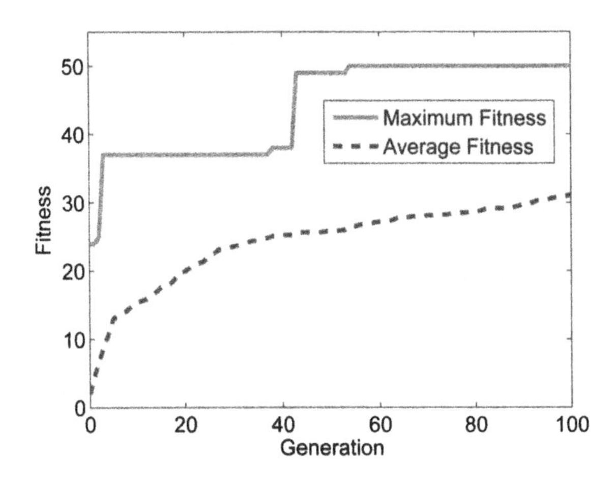

Figure 5.18 The progress of one EP simulation for FSM evolution for solving the artificial ant problem. Each FSM has five states, and the number of moves is limited to 500. The best FSM at initialization enables the ant to eat 24 of 90 food pellets. After 100 generations, the best FSM enables the ant to eat 50 food pellets.

The average number of food pellets that the ants eat depends on how many states we use in the FSMs. If we use too few states, then we do not have enough flexibility to find a good solution. If we use too many states, then EP performance improves, but the improvement may not be worth the increased computer run time. The average amount of food eaten by each ant in the population after 100 generations in an EP with a population size of 100 is given as follows:

$$\begin{aligned}
\text{FSM dimension} &= 4 &:& \text{ 50.1 food pellets} \\
\text{FSM dimension} &= 6 &:& \text{ 60.5 food pellets} \\
\text{FSM dimension} &= 8 &:& \text{ 62.5 food pellets} \\
\text{FSM dimension} &= 10 &:& \text{ 63.1 food pellets} \\
\text{FSM dimension} &= 12 &:& \text{ 63.8 food pellets.}
\end{aligned} \tag{5.12}$$

We see that there is a big jump in performance if we increase the number of states from four to six, but after that, increases in the number of states result in smaller improvements.

After several Monte Carlo runs, we found that the best FSM that the EP evolved had 12 states. However, five of the states were never reached, so the FSM actually included only seven operational states. This FSM is depicted in the format of Equation (5.11) as the 28-element array shown in Table 5.3.

Figure 5.19 shows the FSM in graphical format. An ant navigating with this FSM was able to eat all 90 food pellets in 349 moves, which is slightly more than twice the minimum number of moves required. A close look at Figure 5.19 shows us some inefficiencies in the FSM. For example, when in state 4, if the ant senses food, then it turns right and proceeds to state 7. However, any time the ant senses food, we intuitively expect that he should move forward to eat the food. It seems that the "1,R" label coming from state 4 is wasteful. We could correct this problem by forcing all FSMs to move forward whenever they sense food. This would be a way of incorporating problem-specific information into our EP, which could possibly improve EP performance. In general, we should always try to incorporate problem-specific information into our EAs to improve performance.

	Food Not Sensed		Food Sensed	
	Move	Next State	Move	Next State
State 1	2	2	0	5
State 2	1	3	0	3
State 3	1	4	0	1
State 4	2	5	1	7
State 5	0	1	0	6
State 6	0	5	0	1
State 7	2	6	0	7

Table 5.3 The best finite state machine evolved by EP for the 32×32 Sante Fe trail. Moves are labeled as follows: 0 = move forward, 1 = turn right, and 2 = turn left. This FSM is shown in graphical form in Figure 5.19.

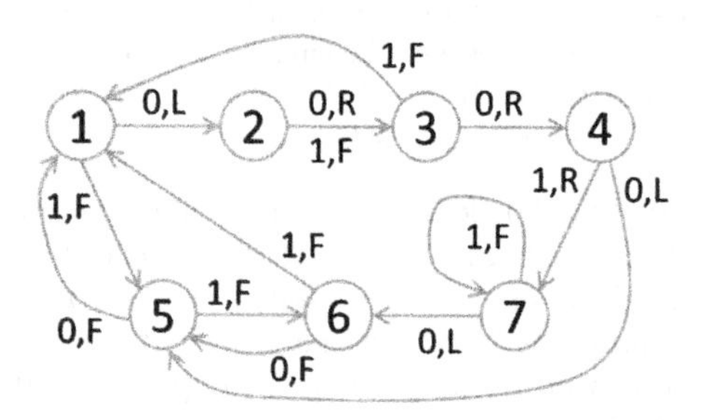

Figure 5.19 The best finite state machine evolved by EP for the 32×32 Sante Fe trail. The outputs of each state are labeled (f, s), where $f = 0$ indicates that food is not sensed, and $f = 1$ indicates that food is sensed; and $s = F$ indicates a move forward, $s = L$ indicates a turn to the left, and $s = R$ indicates a turn to the right. This FSM is shown in tabular form in Table 5.3.

Other versions of the artificial ant problem include the Los Altos Hills trail [Koza, 1992, Section 7.2], the San Mateo trail [Koza, 1994, Chapter 12], and the John Muir trail [Jefferson et al., 2003].

5.6 CONCLUSION

Historically, EP has often been used to find optimal FSMs. However, we emphasize two important points to conclude this chapter. First, we can use EP as a general-purpose algorithm to solve any optimization problem, and EP is in fact a popular algorithm for general-purpose optimization. Second, we can solve FSM problems not only with EP, but also with any of the other EAs discussed in this book. The prisoner's dilemma and its variations are general optimization problems that can be solved with many types of optimization algorithms. The reason that we devote so much of this chapter to the prisoner's dilemma is because EP was originally developed to solve FSMs. In conclusion, we note that several books and papers discuss EP from other perspectives, and sometimes in more detail than this chapter, including [Bäck and Schwefel, 1993], [Bäck, 1996], and [Yao et al., 1999].

PROBLEMS

Written Exercises

5.1 The EP mutation variance

$$\sigma_i^2 = \beta f(x_i) + \gamma$$

is often referred to as a linear relationship between $f(x_i)$ and σ_i^2. Actually, it is not linear, but is affine. How should the above equation be written to obtain a linear relationship between $f(x_i)$ and σ_i^2?

5.2 An elevator can be in one of two states: on the first floor, or on the second floor. It can take one of two inputs: the user can press the first floor button, or the second floor button. Write an FSM for this system in both graphical and tabular form.

5.3 Expand the system of Problem 5.2 so that the elevator has four states (on the first floor, on the second floor, traveling from the first to the second floor, and traveling from the second to the first floor), and so that it has three inputs (the user pressed the first floor button, the second floor button, or nothing). Write an FSM for this system in both graphical and tabular form.

5.4 Write the always-cooperate prisoner's dilemma strategy in the vector form of Equation (5.9). Do the same for the tit-for-tat strategy, the tit-for-two-tats strategy, and the grim strategy. Based on your vector forms, which strategies are more similar: the always-cooperate and tit-for-tat strategies, or the tit-for-two-tats and grim strategies?

5.5 Suppose the always-cooperate, tit-for-tat, tit-for-two-tats, and grim strategies compete against each other in a prisoner's dilemma competition. Which one will win?

5.6 The FSM of Figure 5.20 has been suggested for an artificial ant in the Sante Fe trail, where input 0 means that no food is sensed; input 1 means that food is sensed; and outputs L, R, and F mean move left, right, and forward, respectively [Meuleau et al., 1999], [Kim, 2006]. Write this FSM in the format of Equation (5.11).

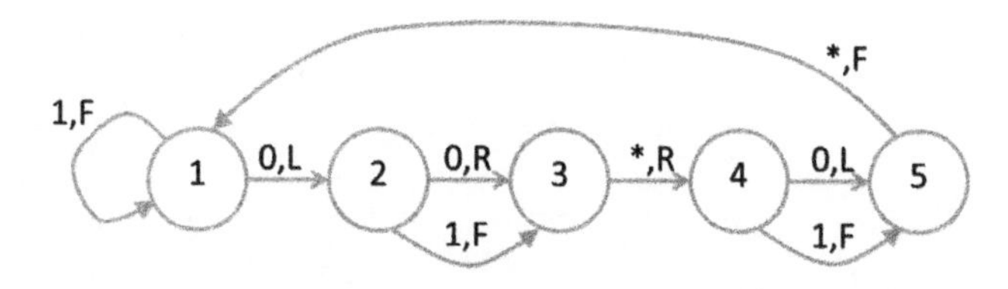

Figure 5.20 Problem 5.6: FSM for the artificial ant problem.

5.7 What is the minimum number of moves required for an artificial ant to visit every square of the Sante Fe trail?

5.8 Suppose an artificial ant visits β unique random squares of the 32×32 Sante Fe trail. What is the probability that the ant will find all of the food?

Computer Exercises

5.9 Simulate the EP of Figure 5.1 to minimize the 10-dimensional sphere function. Use the search domain $[-5.12, +5.12]$, $\beta = (x_{\max} - x_{\min})/10 = 1.024$, $\gamma = 0$, population size $= 50$, and generation limit $= 50$. When calculating the mutation variance, normalize the cost function values so that $f(x_i) \in [1, 2]$ for all i.

- **a)** What is the best solution obtained by the EP, averaged over 20 Monte Carlo simulations?
- **b)** Replace the variance with the function $\beta f^2(x_i)$ and repeat.
- **c)** Replace the variance with the function $\beta\sqrt{f(x_i)}$ and repeat.
- **d)** Use β as the variance for all i and repeat.

5.10 Rerun Example 5.3 with an FSM size penalty of n, where n is the number of states. What is the number of states and the cost of the best FSM, averaged over 100 Monte Carlo simulations? How do your results compare to Example 5.3, which used an FSM size penalty of $n/2$?

5.11 Given a prisoner's dilemma opponent that defects every turn, we know that our best strategy is to also defect every turn. Use an EP to evolve an FSM that performs as well as possible against an always-defect opponent.

5.12 How many moves does it take for an ant using the FSM of Problem 5.6 to eat all of the food on the Sante Fe trail?

5.13 Use your answer to Problem 5.8 to plot the probability, for $\beta \in [1, 1024]$, that an ant will find all of the food in the Sante Fe trail after visiting β unique random squares. Use a log scale for the probability axis. How many squares must the ant visit to have at least a 50% chance of finding all of the food?

CHAPTER 6

Evolution Strategies

> The first ES version operated with just one offspring per 'generation' because we did not have a population of objects to operate with.
>
> —Hans-Paul Schwefel [Schwefel and Mendes, 2010]

An early European foray into EAs occurred at the Technical University of Berlin in the 1960s by three students who were trying to find optimal body shapes in a wind tunnel to minimize air resistance. The students, Ingo Rechenberg, Hans-Paul Schwefel, and Peter Bienert, had difficulty solving their problem analytically. So they came up with the idea of trying random changes to the body shapes, selecting those that worked best, and repeating the process until they found good solutions to their problem.

Rechenberg's first publication on evolution strategy (ES), which is also called evolutionary strategy, was in 1964 [Rechenberg, 1998]. Interestingly, the first ES implementations were experimental. Computational resources were not sufficient for high-fidelity simulations, so fitness functions were obtained experimentally, and mutations were implemented on physical hardware. Rechenberg received his doctorate for his efforts in 1970 and later published his work in book form [Rechenberg, 1973]. Although the book is written in German, it is still interesting to non-German readers because of its graphical depictions of optimization processes. The book shows the evolution of wing shapes to minimize drag in an air-flow field, rocket

Evolutionary Optimization Algorithms, First Edition. By Dan J. Simon

nozzle shapes to minimize the drag of fuel as it passes through the nozzle, and pipe shapes to minimize the drag of fluid as it passes through the pipe. These early algorithms were called *cybernetic solution paths.*

Schwefel received his doctorate in 1975, and later wrote several books about ES [Schwefel, 1977], [Schwefel, 1981], [Schwefel, 1995]. Since 1985 he has been with the University of Dortmund. Bienert received his doctorate in 1967. An interesting account of the early years of ES is given in an interview with Schwefel in [Schwefel and Mendes, 2010].

Overview of the Chapter

Section 6.1 discusses the (1+1)-ES. This was the first ES that was used, and it is the most simple. It consists solely of mutation, and does not involve recombination. Section 6.2 derives the 1/5 rule for ES, which tells us how to adjust the mutation rate to obtain the best performance, and which can be skipped by readers who are not interested in mathematical proofs. Section 6.3 generalizes the (1+1)-ES to obtain an algorithm with μ parents at each generation, where μ is a user-defined constant. The parents combine to form a single child, which might become a part of the next generation if it is fit enough. Several options are available for recombination. Section 6.4 is a further generalization which results in λ children at each generation. Section 6.5 discusses how we can adapt the mutation rate to dramatically improve ES performance. These adaptation options include the state-of-the-art algorithms CMA-ES and CMSA-ES.

6.1 THE (1+1) EVOLUTION STRATEGY

Suppose that $f(x)$ is a function of a real random vector x, and that we want to maximize the fitness $f(x)$. The original ES algorithm operated by initializing a single candidate solution and evaluating its fitness. The candidate solution was then mutated, and the mutated individual's fitness was evaluated. The best of the two candidate solutions (parent and child) formed the starting point for the next generation. The original ES was designed for discrete problems, used small mutations in a discrete search space, and thus tended to get trapped in a local optimum. The original ES was therefore modified to use continuous mutations in continuous search spaces [Beyer and Schwefel, 2002]. This algorithm is summarized in Figure 6.1.

Figure 6.1 is called a (1+1)-ES because each generation consists of 1 parent and 1 child, and the best individual is chosen from the parent and child as the individual in the next generation. The (1+1)-ES, also called the two-membered ES is very similar to the hill climbing strategies of Section 2.6. It is also the same as an EP with a population size of 1 (see Section 5.1). The following theorem guarantees that the (1+1)-ES eventually finds the global maximum of $f(x)$.

Theorem 6.1 *If $f(x)$ is a continuous function defined on a closed domain with a global optimum $f^*(x)$, then*

$$\lim_{t\to\infty} f(x) = f^*(x) \tag{6.1}$$

where t is the generation number.

```
Initialize the non-negative mutation variance σ²
x₀ ← randomly generated individual
While not(termination criterion)
    Generate a random vector r with rᵢ ~ N(0, σ²) for i ∈ [1, n]
    x₁ ← x₀ + r
    If x₁ is better than x₀ then
        x₀ ← x₁
    End if
Next generation
```

Figure 6.1 The above pseudo-code outlines the (1+1) evolution strategy, where n is the problem dimension, and x_1 is equal to x_0 but with each element mutated.

Theorem 6.1 is proven in [Devroye, 1978], [Rudolph, 1992], [Bäck, 1996, Theorem 7], and [Michalewicz, 1996]. It also agrees with our intuition. Since we use random mutations to explore the search space, given enough time, we will eventually explore the entire search space (to within computer precision) and find the global optimum.

The σ^2 variance in the (1+1)-ES of Figure 6.1 is a tuning parameter. The value of σ is a tradeoff.

- σ should be large enough so that mutations can reach all areas of the search space in a reasonable period of time.
- σ should be small enough so that the search can find the optimal solution within the user's desired resolution.

It may be appropriate to decrease σ as the ES progresses. At the beginning of the ES, large values of σ will allow the ES to conduct a coarse-grained search and get close to the optimal solution. Toward the end of the ES, smaller values of σ will allow the ES to fine-tune its candidate solution and converge to the optimal solution with better resolution.

The mutation in Figure 6.1 is called *isotropic* because the mutation of each element of x_0 has the same variance. In practice we might want to implement non-isotropic mutations as follows:

$$x_1 \leftarrow x_0 + N(0, \Sigma) \tag{6.2}$$

where Σ is an $n \times n$ diagonal matrix with diagonal elements σ_i for $i \in [1, n]$. This means that each element of x_0 is mutated with a different variance. We would assign each σ_i independently, depending on the domain of the i-th element of x and the shape of the objective function in that dimension.

Rechenberg analyzed the (1+1)-ES for some simple optimization problems and concluded that 20% of mutations should result in improvements in the fitness function $f(x)$ [Rechenberg, 1973], [Bäck, 1996, Section 2.1.7]. We reproduce some of his analysis in Section 6.2. If the mutation sucess rate is higher than 20%, then the mutations are too small, which leads to small improvements, which results in long convergence times. If the mutation success rate is lower than 20%, then the mutations are too large, which leads to large but infrequent improvements, and this also results in long convergence times. Rechenberg's work led to the 1/5 rule:

> *In the (1+1)-ES, if the ratio of successful mutations to total mutations is less than 1/5, then the standard deviation σ should be decreased. If the ratio is more than 1/5, then the standard deviation should be increased.*

This rule really applies only to a couple of specific objective functions as we will see in Section 6.2, but it has proven to be a useful guideline for general ES implementations. But the 1/5 rule raises the question, By how much should the standard deviation be decreased or increased? Schwefel theoretically derived the factor by which to decrease or increase σ:

$$\begin{aligned} \text{standard deviation decrease:} \quad & \sigma \leftarrow c\sigma \\ \text{standard deviation increase:} \quad & \sigma \leftarrow \sigma/c \\ \text{where} \quad & c = 0.817. \end{aligned} \tag{6.3}$$

These results lead to the adaptive (1+1)-ES shown in Figure 6.2. The adaptive (1+1)-ES requires that we define a moving window length G. We want G to be large enough to get a good idea of the success rate of the ES mutations, but not so large that the adaptation of σ is sluggish. The recommendation in [Beyer and Schwefel, 2002] is

$$G = \min(n, 30) \tag{6.4}$$

where n is the problem dimension.

Initialize the non-negative mutation variance σ^2
$x_0 \leftarrow$ randomly generated individual
While not(termination criterion)
    Generate a random vector r with $r_i \sim N(0, \sigma^2)$ for $i \in [1, n]$
    $x_1 \leftarrow x_0 + r$
    If x_1 is better than x_0 then
        $x_0 \leftarrow x_1$
    End if
    $\phi \leftarrow$ proportion of successful mutations during the past G generations
    If $\phi < 1/5$
        $\sigma \leftarrow c^2\sigma$
    else if $\phi > 1/5$
        $\sigma \leftarrow \sigma/c^2$
    End if
Next generation

Figure 6.2 The above pseudo-code outlines the adaptive (1+1) evolution strategy where n is the problem dimension. x_1 is equal to x_0 but with each feature mutated. ϕ is the proportion of mutations during the past G generations that result in x_1 being better than x_0. The mutation variance is automatically adjusted to increase the rate of convergence. The nominal value of c is 0.817.

EXAMPLE 6.1

In this example, we use the (1+1)-ES to optimize the 20-dimensional Ackley benchmark function (see Appendix C.1.2). We compare the standard (1+1)-ES shown in Figure 6.1 with the adaptive (1+1)-ES shown in Figure 6.2. For the adaptive ES, we keep track of the number of successful mutations and the total number of mutations. The total number of mutations is equal to the number of generations. Every 20 generations, we examine the mutation success rate, and adjust the standard deviations as shown in Equation (6.3). Figure 6.3 compares the average convergence rate of the standard (1+1)-ES and the adaptive (1+1)-ES. We see that the adaptive ES converges much faster than the standard ES. Figure 6.4 shows a typical profile of the mutation success rate and the mutation standard deviation. We see that when the success rate has been greater than 20% over the previous 20-generation time span, the mutation standard deviation is automatically increased; when the success rate has been less than 20%, the mutation standard deviation is automatically decreased.

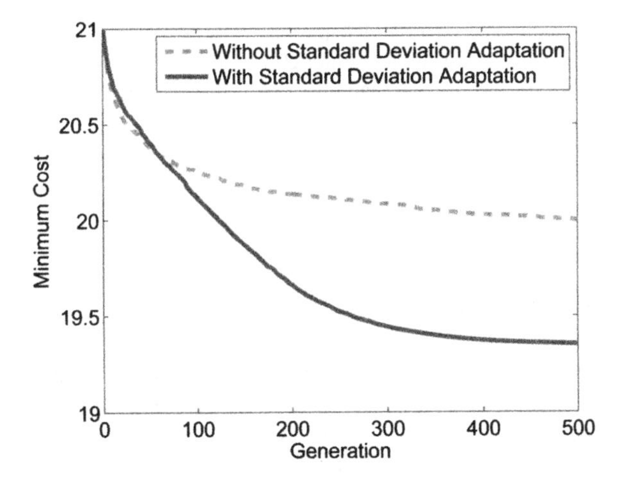

Figure 6.3 Example 6.1: The convergence of the (1+1)-ES algorithms, averaged over 100 simulations. The adaptive ES, which automatically adjusts the mutation standard deviations, converges much faster than the standard ES.

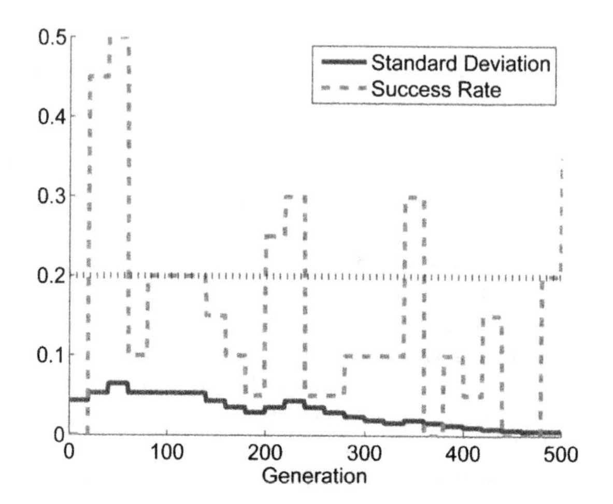

Figure 6.4 Example 6.1: The mutation success rate and mutation standard deviation of the adaptive (1+1)-ES. The adaptive ES automatically increases the mutation standard deviation when the success rate is greater than 20%, and decreases the mutation standard deviation when the success rate is less than 20%. This illustrates the 1/5 rule.

□

6.2 THE 1/5 RULE: A DERIVATION

This section derives the 1/5 rule, which states that approximately 20% of all mutations should lead to an improvement in the mutated ES individual. This section is motivated by [Rechenberg, 1973, Chapters 14–15], and can be skipped by readers who are not interested in the details of mathematical proofs.

Suppose we have an n-dimensional minimization problem with the cost function $f(x)$, where $x = \begin{bmatrix} x_1 & \cdots & x_n \end{bmatrix}^T$. We focus in this section on the corridor problem, which has the domain

$$\begin{aligned} x_1 &\in [0, \infty) \\ x_j &\in (-\infty, \infty), \quad j \in [2, n]. \end{aligned} \tag{6.5}$$

The corridor problem has the cost function

$$f(x) = \begin{cases} c_0 + c_1 x_1, & \text{if } x_j \in [-b, b] \text{ for all } j \in [2, n] \\ \infty, & \text{otherwise} \end{cases} \tag{6.6}$$

where c_0, c_1, and b are positive constants. This is called the corridor problem because the cost improves as x_1 decreases, but only if x is in the corridor $x_j \in [-b, b]$ for all $j \in [2, n]$.

Recall from Figures 6.1 and 6.2 that a given ES individual x_0 is mutated according to the equation $x_1 \leftarrow x_0 + r$, where r is a random n-element vector.[1] We use x_{0j} to refer to the j-th element of the x_0 vector, and x_{1j} to refer to the j-th element of the x_1 vector, for $j \in [1, n]$.

The mutation of each element of x_0 is selected from a Gaussian distribution with zero mean and variance σ^2. The PDF of x_{1j} can therefore be written as

$$\text{PDF}(x_{1j}) = \frac{1}{\sigma\sqrt{2\pi}} \exp[-(x_{1j} - x_{0j})^2/(2\sigma^2)], \quad j \in [1, n]. \tag{6.7}$$

A mutation that improves x requires the occurrence of n independent events:

$$\begin{aligned} r_1 &< 0 \\ x_{1j} &\in [-b, b], \quad j \in [2, n] \end{aligned} \tag{6.8}$$

where r_1 is the mutation of the first element of x_0 (that is, $x_{11} \leftarrow x_{01} + r_1$). Therefore, ϕ', the expected magnitude of a useful mutation, is

$$\begin{aligned} \phi' &= |E(r_1 | r_1 < 0)| \prod_{j=2}^{n} \text{Prob}(x_{1j} \in [-b, b]) \\ &= \left| \int_{-\infty}^{0} \frac{r_1}{\sigma\sqrt{2\pi}} \exp(-r_1^2/(2\sigma^2))\, dr_1 \right| \prod_{j=2}^{n} \int_{-b}^{b} \frac{1}{\sigma\sqrt{2\pi}} \exp(-(x_{1j} - x_{0j})^2/(2\sigma^2))\, dx_{1j} \\ &= \frac{\sigma}{\sqrt{2\pi}} \prod_{j=2}^{n} \frac{1}{2} \left[\text{erf}\left(\frac{b - x_{0j}}{\sigma\sqrt{2}} \right) + \text{erf}\left(\frac{b + x_{0j}}{\sigma\sqrt{2}} \right) \right] \end{aligned} \tag{6.9}$$

(see Problem 6.2) where $\text{erf}(\cdot)$ is the error function:

$$\text{erf}(x) = \frac{2}{\sqrt{\pi}} \int_0^x \exp(-t^2)\, dt, \quad x \geq 0. \tag{6.10}$$

The expected magnitude of a useful mutation, given that x_0 was in the $[-b, b]$ corridor before the mutation, can be written as

$$\begin{aligned} \phi &= E(\phi' \,|\, x_{0j} \in [-b, b] \text{ for all } j \in [2, n]) \\ &= \int_{-b}^{b} \cdots \int_{-b}^{b} \phi' \,\text{PDF}(x_{02}) \cdots \text{PDF}(x_{0n})\, dx_{02} \cdots dx_{0n}. \end{aligned} \tag{6.11}$$

Given that $x_{0j} \in [-b, b]$, we assume that x_{0j} is uniformly distributed in $[-b, b]$ (an admittedly unproven assumption), which gives

$$\phi = \frac{\sigma}{\sqrt{2\pi}} \prod_{j=2}^{n} \int_{-b}^{b} \frac{1}{2} \left[\text{erf}\left(\frac{b - x_{0j}}{\sigma\sqrt{2}} \right) + \text{erf}\left(\frac{b + x_{0j}}{\sigma\sqrt{2}} \right) \right] \left(\frac{1}{2b} \right) dx_{0j}. \tag{6.12}$$

Now recall that

$$\int \text{erf}(z)\, dz = z\, \text{erf}(z) + \frac{\exp(-z^2)}{\sqrt{\pi}}. \tag{6.13}$$

[1]There is a temporary inconsistency in notation here. x_0 and x_1 in Figures 6.1 and 6.2 refer to an ES candidate solution before and after mutation, while x_1 in Equations (6.5) and (6.6) refer to the first element of the x vector.

Using this in Equation (6.12) gives

$$\begin{aligned}\phi &= \frac{\sigma}{\sqrt{2\pi}(4b)^{n-1}} \prod_{j=2}^{n} \left[4b \operatorname{erf}\left(\frac{2b}{\sigma\sqrt{2}}\right) + \frac{2\sigma\sqrt{2}(\exp(-2b^2/\sigma^2) - 1)}{\sqrt{\pi}} \right] \\ &= \frac{\sigma}{\sqrt{2\pi}} \left[\operatorname{erf}\left(\frac{2b}{\sigma\sqrt{2}}\right) - \frac{\sigma}{b\sqrt{2\pi}}(1 - \exp(-2b^2/\sigma^2)) \right]^{n-1} . \end{aligned} \tag{6.14}$$

Recall that $\lim_{x\to\infty} \operatorname{erf}(x) = 1$, and $\lim_{x\to\infty} \exp(-x) = 0$. Therefore,

$$\phi \approx \frac{\sigma}{\sqrt{2\pi}} \left(1 - \frac{\sigma}{b\sqrt{2\pi}}\right)^{n-1} \quad \text{for large } b/\sigma. \tag{6.15}$$

ϕ is the expected value of a useful mutation, which we would like to be large. We can find the values of σ that result in extreme values of ϕ by taking the derivative of ϕ with respect to σ, and setting the result to 0. We find that

$$\frac{d\phi}{d\sigma} = \frac{1}{\sqrt{2\pi}} \left(1 - \frac{\sigma}{b\sqrt{2\pi}}\right)^{n-1} - \frac{\sigma(n-1)}{b2\pi} \left(1 - \frac{\sigma}{b\sqrt{2\pi}}\right)^{n-2} . \tag{6.16}$$

Setting this derivative to 0 and solving for σ results in

$$\sigma^* = b\sqrt{2\pi}/n \tag{6.17}$$

which gives the largest possible value of ϕ and hence the largest expected magnitude of a useful mutation.

Now consider the probability w' that a mutation is useful. A mutation is useful if the n independent events shown in Equation (6.8) occur. The probability that this occurs can be written as

$$w' = \text{Prob}(r_1 < 0) \prod_{j=2}^{n} \text{Prob}(x_{1j} \in [-b, b]). \tag{6.18}$$

Since r_1 is zero-mean, the probability that $r_1 < 0$ is one-half, so the above equation can be written as

$$w' = \frac{1}{2} \prod_{j=2}^{n} \text{Prob}(x_{1j} \in [-b, b]). \tag{6.19}$$

Comparing the above equation with Equation (6.9), we see that

$$w' = \frac{\sqrt{2\pi}}{2\sigma} \phi'. \tag{6.20}$$

Now consider the expected value w of the probability that a mutation is useful, given that x_0 was in the $[-b, b]$ corridor before the mutation. This can be written as follows:

$$\begin{aligned} w &= E(w' | x_{0j} \in [-b, b] \text{ for all } j \in [2, n]) \\ &= \int_{-b}^{b} \cdots \int_{-b}^{b} w' \, \text{PDF}(x_{02}) \cdots \text{PDF}(x_{0n}) \, dx_{02} \cdots dx_{0n}. \end{aligned} \tag{6.21}$$

Comparing this equation with Equations (6.11) and (6.20), we see that

$$w = \frac{\sqrt{2\pi}}{2\sigma}\phi. \tag{6.22}$$

Now we substitute for ϕ from Equation (6.15) to obtain

$$w = \left(\frac{\sqrt{2\pi}}{2\sigma}\right)\left(\frac{\sigma}{\sqrt{2\pi}}\right)\left(1 - \frac{\sigma}{b\sqrt{2\pi}}\right)^{n-1}. \tag{6.23}$$

Next we substitute for the optimal value of σ from Equation (6.17) to obtain the optimal value of w:

$$w^* = \frac{1}{2}\left(1 - 1/n\right)^{n-1}. \tag{6.24}$$

Recall that $\exp(-x) \approx 1 - x$ for small x. Therefore, $\exp(-1/n) \approx 1 - 1/n$ for large n. This gives

$$w^* = \frac{1}{2}\left(\exp(-1/n)\right)^{n-1} = \frac{1}{2e} \approx 0.18. \tag{6.25}$$

The optimal standard deviation σ^* results in a mutation magnitude that gives improvements 18% of the time.

Rechenberg also analyzed a sphere function, where the objective was to minimize

$$f(x) = \sum_{j=1}^{n} x_j^2. \tag{6.26}$$

He found that the optimal mutation success rate for the sphere function was 27%.

These results apply only to specific functions, and they were derived under simplifying approximations, but they resulted in a rule of thumb called the 1/5 rule that has proven useful in many problems. To maximize convergence rate, the standard deviations in an ES should be adjusted to give a 1/5 ratio of successful to total mutations.

6.3 THE (μ+1) EVOLUTION STRATEGY

The first generalization of the (1+1)-ES is the (μ+1)-ES. In the (μ+1)-ES, μ parents are used at each generation, where μ is a user-defined parameter. Each parent also has an associated σ vector that controls the magnitude of mutations. The parents combine with each other to form a single child, and then the child is mutated. The best μ individuals are chosen from among the μ parents and the child, and they become the μ parents of the next generation. This algorithm is summarized in Figure 6.5. Since the $(\mu + 1)$-ES retains the best individuals each generation, it is elitist; that is, its best individual never gets worse from one generation to the next (see Section 8.4). The $(\mu + 1)$-ES is also called the steady-state ES. Since only one individual is removed from the overall population at the end of each generation, this strategy could be called *extinction of the worst*, which is the flip side of *survival of the fittest*.

The parents in Figure 6.5 combine both their solution features and their mutation variances. However, Figure 6.5 does not include the type of mutation variance adaptation that we saw in Figure 6.2. In fact, after a certain number of generations that is proportional to μ, the σ values in Figure 6.5 will collapse to a single value, which may or may not reflect an appropriate mutation strength [Beyer, 1998]. Figure 6.5 should probably not be implemented as shown, but is a stepping stone to the more effective self-adaptive ES of Section 6.5.

$\{(x_k, \sigma_k)\} \leftarrow$ randomly generated individuals, $k \in [1, \mu]$.
Each x_k is a candidate solution, and each σ_k is a standard deviation vector.
Note that $x_k \in R^n$, and $\sigma_k \in R^n$ with each element positive.
While not(termination criterion)
  Randomly select two parents from the population $\{(x_k, \sigma_k)\}$
  Use a recombination method to combine the two parents and obtain a child,
    which is denoted as $(x_{\mu+1}, \sigma_{\mu+1})$
  $\Sigma_{\mu+1} \leftarrow \text{diag}(\sigma^2_{\mu+1,1}, \cdots, \sigma^2_{\mu+1,n}) \in R^{n \times n}$
  Generate a random vector r from $N(0, \Sigma_{\mu+1})$
  $x_{\mu+1} \leftarrow x_{\mu+1} + r$
  Remove the worst individual from the population: that is,
    $\{(x_k, \sigma_k)\} \leftarrow$ the best μ individuals from $\{(x_1, \sigma_1), \cdots, (x_{\mu+1}, \sigma_{\mu+1})\}$
Next generation

Figure 6.5 The above pseudo-code outlines the $(\mu+1)$ evolution strategy, where n is the problem dimension.

The recombination step in Figure 6.5 says, "Use a recombination method to combine the two parents ..." There are various ways to perform recombination.[2] Discrete sexual crossover works by randomly selecting each child element from either x_p or x_q, and randomly selecting each child standard deviation from either σ_p or σ_q, where each selection is independent of the others. This type of crossover is described with the word *discrete* because each child feature comes from a single parent, and it is described with the word *sexual* because each child feature comes from one of two parents. Discrete sexual crossover is illustrated in Figure 6.6.

Another recombination option is intermediate sexual crossover. In this option, child features are set to the midpoint of their parent features; hence the designation *intermediate*. Intermediate sexual crossover is illustrated in Figure 6.7.

Another recombination option is global crossover, or panmictic crossover. A panmictic population is one in which each individual is a potential mate for every other individual. In discrete global crossover, each child feature comes from a parent selected randomly from the entire population. This is illustrated in Figure 6.8. The discrete crossover options of Figures 6.6 and 6.8 are also called dominant crossover.

Global crossover can be combined with intermediate crossover to obtain intermediate global crossover. In this option, each child feature is a linear combination of a randomly-selected pair of parents. This is illustrated in Figure 6.9. Intermediate global crossover is the type that is usually used in practice, and it is also recommended for well-understood theoretical reasons [Beyer and Schwefel, 2002].

[2]The ES community generally prefers the term *recombination* or *mixing* rather than *crossover*. Throughout this book we use these terms synonymously.

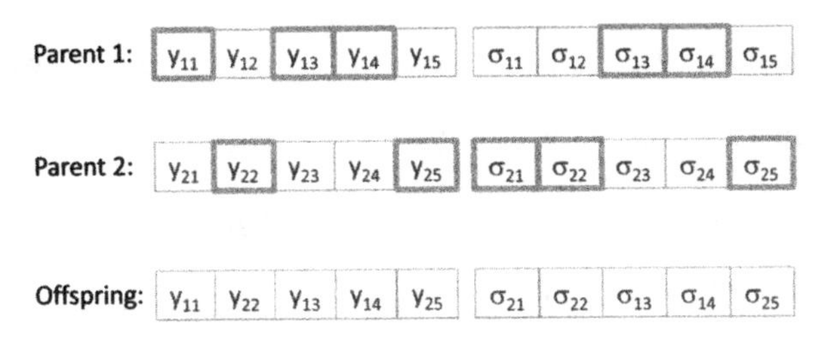

Figure 6.6 Discrete sexual crossover in an ES, where the problem dimension $n = 5$. Each solution feature and standard deviation in the child is randomly selected from one of two parents.

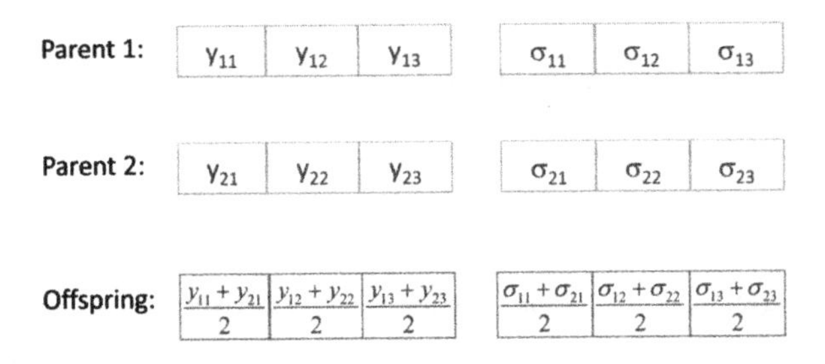

Figure 6.7 Intermediate sexual crossover in an ES, where the problem dimension $n = 3$. Each solution feature and standard deviation in the child is half way between the two parents.

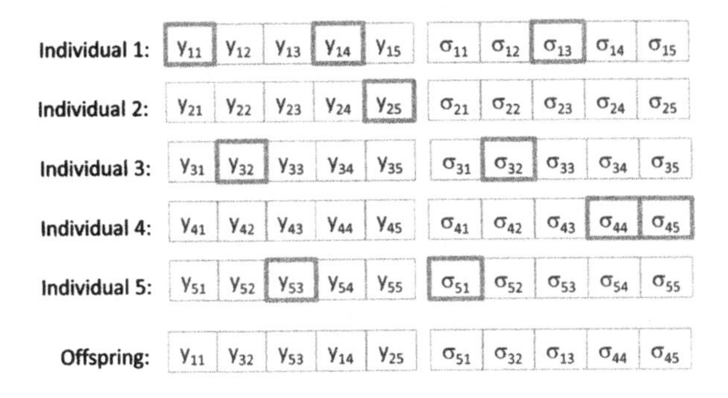

Figure 6.8 Discrete global crossover in a five-member ES ($\mu = 5$), where the problem dimension $n = 5$. Each solution feature and standard deviation in the child is randomly selected from the entire population.

Other types of crossover can also be used. For example, one parent $x_{p(0)}$ could be selected for the child, and n other parents could be selected, $\{x_{p(k)}\}$ for $k \in [1, n]$, one for each solution feature. Then the k-th child solution feature $x_{\mu+1,k}$ could be generated by crossing $x_{p(0)}$ with $x_{p(k)}$ for $k \in [1, n]$. Likewise, the k-th child standard deviation $\sigma_{\mu+1,k}$ could be generated by crossing $\sigma_{p(0)}$ with $\sigma_{p(k)}$. See Section 8.8 for additional types of crossover operators.

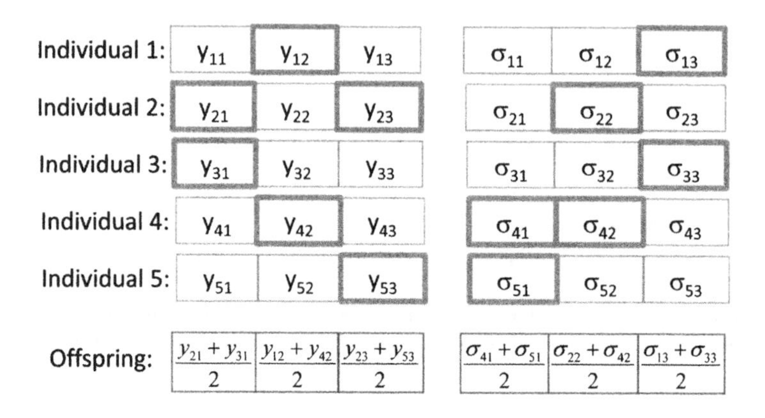

Figure 6.9 Illustration of intermediate global crossover in a five-member ES ($\mu = 5$), where the problem dimension $n = 3$. Each solution feature and standard deviation in the child is half way between two randomly-selected parents.

6.4 $(\mu + \lambda)$ AND (μ, λ) EVOLUTION STRATEGIES

The next evolution strategy generalization is the $(\mu + \lambda)$-ES. In the $(\mu + \lambda)$-ES, we have a population size of μ, and we generate λ children each generation. After the generation of the children, we have a total of $(\mu + \lambda)$ individuals, which includes both parents and children. We select the best μ of those individuals as the parents of the next generation.

Another commonly-used evolution strategy is the (μ, λ)-ES. In the (μ, λ)-ES, the parents of the next generation are selected as the best μ individuals from among the λ children. In other words, none of the μ parents survive to the next generation; instead, a subset of μ of the λ children are chosen to become the parents of the next generation. In the (μ, λ)-ES, we need to make sure that we choose $\lambda \geq \mu$. Parents of the previous generation never survive to the next generation. The life of each individual is limited to one generation.

If $\mu > 1$ in the $(\mu+\lambda)$-ES or (μ, λ)-ES, then the ES is called multi-membered. In spite of the success of these generalizations, there were initially strong objections to setting μ and λ greater than 1. The argument against $\lambda > 1$ was that the exploitation of information would be delayed. The argument against $\mu > 1$ was that survival of inferior individuals would delay the progress of the ES [De Jong et al., 1997].

The (μ, λ)-ES often works better than the $(\mu + \lambda)$-ES when the fitness function is noisy or time-varying (Chapter 21). In the $(\mu + \lambda)$-ES, a given individual (x, σ) may have a good fitness but be unlikely to improve due to an inappropriate σ. So the (x, σ) individual may remain in the population for many generations without improving, which wastes a place in the population. The (μ, λ)-ES solves this problem by forcing all individuals out of the population after one generation, and allowing only the best children to survive. This helps restrict survival in the next generation to those children with a good σ, which is a σ that results in a mutation vector that allows improvement in x. [Beyer and Schwefel, 2002] recommends the (μ, λ)-ES for problems with unbounded search spaces, and the $(\mu + \lambda)$-ES for problems with discrete search spaces.

Figure 6.10 summarizes the $(\mu+\lambda)$-ES and the (μ, λ)-ES. Note that if σ_k = constant for all k, then the σ_k values will remain unchanged from one generation to the next. Figure 6.10, like Figure 6.5, does not include mutation variance adaptation. Therefore, Figure 6.10 should not be implemented as shown, but is a stepping stone to the self-adaptive ES. We will generalize Figure 6.10 to obtain the self-adaptive $(\mu + \lambda)$-ES and (μ, λ)-ES in the next section.

$\{(x_k, \sigma_k)\} \leftarrow$ randomly generated individuals, $k \in [1, \mu]$.
Each x_k is a candidate solution, and each σ_k is a standard deviation vector.
Note that $x_k \in R^n$, and $\sigma_k \in R^n$ with each element positive.
While not(termination criterion)
    For $k = 1, \cdots, \lambda$
        Randomly select two parents from $\{(x_k, \sigma_k)\}$
        Use a recombination method to combine the two parents and obtain a
            child, which is denoted as (x'_k, σ'_k)
        $\Sigma'_k \leftarrow \text{diag}((\sigma'_{k1})^2, \cdots, (\sigma'_{kn})^2) \in R^{n\times n}$
        Generate a random vector r from $N(0, \Sigma'_k)$
        $x'_k \leftarrow x'_k + r$
    Next k
    If this is a $(\mu + \lambda)$-ES then
        $\{(x_k, \sigma_k)\} \leftarrow$ the best μ individuals from $\{(x_k, \sigma_k)\} \cup \{(x'_k, \sigma'_k)\}$
    else if this is a (μ, λ)-ES then
        $\{(x_k, \sigma_k)\} \leftarrow$ the best μ individuals from $\{(x'_k, \sigma'_k)\}$
    End if
Next generation

Figure 6.10 The above pseudo-code outlines the $(\mu + \lambda)$ and (μ, λ) evolution strategies, where n is the problem dimension.

■ EXAMPLE 6.2

In this example we compare the $(\mu + \lambda)$ and (μ, λ) evolution strategies. We run both strategies with $\mu = 10$, $\lambda = 20$, discrete sexual crossover, and a problem dimension of 20. Figure 6.11 shows the average performance of both algorithms on the Schwefel 2.26 benchmark. We see that the $(\mu + \lambda)$-ES outperforms the (μ, λ)-ES. In general, this is expected because the (μ, λ)-ES might throw away a good solution due to its limitation of one generation for the lifetime of each individual. However, the performance comparison between the two ES variations depends on the problem. Figure 6.12 shows the average performance of the two algorithms on the Ackley benchmark. The $(\mu + \lambda)$-ES initially outperforms the (μ, λ)-ES, but eventually the (μ, λ)-ES catches up and performs better than the $(\mu + \lambda)$-ES. Sometimes the (μ, λ)-ES is advantageous. This is not only because of its greater adaptability to noisy and time-varying fitnesses; for some functions its greater emphasis on exploration results in better performance than the $(\mu + \lambda)$-ES.

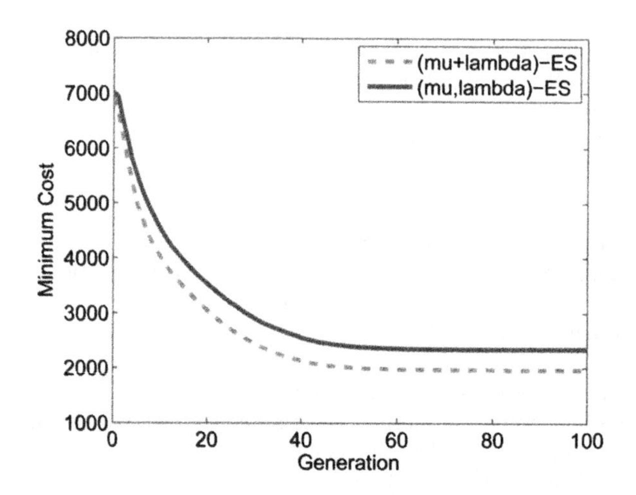

Figure 6.11 Example 6.2: The convergence rate of the $(\mu + \lambda)$-ES and the (μ, λ)-ES on the Schwefel 2.26 benchmark, averaged over 100 simulations. The $(\mu + \lambda)$-ES clearly outperforms the (μ, λ)-ES.

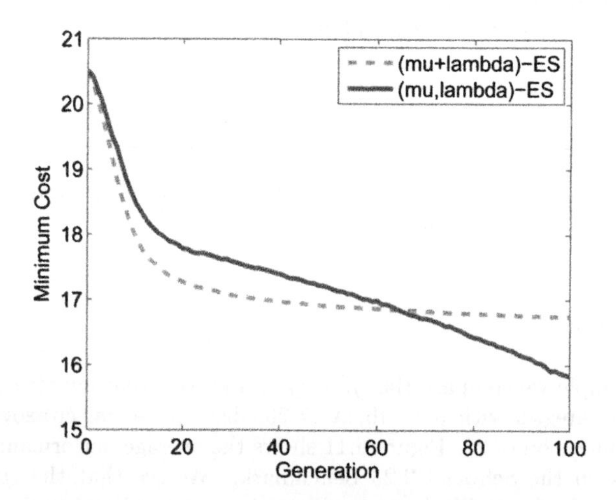

Figure 6.12 Example 6.2: The convergence rate of the $(\mu + \lambda)$-ES and the (μ, λ)-ES on the Ackley benchmark, averaged over 100 simulations. The $(\mu + \lambda)$-ES initially outperforms the (μ, λ)-ES, but eventually the (μ, λ)-ES does much better than the (μ, λ)-ES.

□

The $(\mu, \kappa, \lambda, \rho)$ Evolution Strategy

Recall in Figure 6.10 that each child has two parents. But there is no reason to restrict the number of parents to two. We could instead combine more than two parents, and we use ρ to indicate the number of parents that contribute to each child. We discussed some multi-parent crossover operators at the end of Section 6.3, and we discuss additional possibilities in Section 8.8.

We could also set a maximum lifetime for each individual in the population, which we denote with κ. If the maximum lifetime is one generation, then $\kappa = 1$ and we have a (μ, λ)-ES because parents are not allowed to survive to the next generation. If the maximum lifetime is unlimited, then $\kappa = \infty$ and we have a $(\mu + \lambda)$-ES because there is no restriction on allowing parents to survive to the next generation; as long as a parent is one of the μ most fit individuals in the combined child/parent population, then it survives to the next generation, regardless of how long it has been in the population. In general we may want to limit the lifetime of ES individuals to prevent stagnation, especially for time-varying problems (see Section 21.2).

Combining these two generalizations results in the $(\mu, \kappa, \lambda, \rho)$-ES [Schwefel, 1995]. The population of the $(\mu, \kappa, \lambda, \rho)$-ES has μ parents, each individual has a maximum lifetime of κ generations, and each generation produces λ children, each of whom has ρ parents.

6.5 SELF-ADAPTIVE EVOLUTION STRATEGIES

The ES algorithms that we have studied do not give us a lot of options for adjusting the standard deviations σ_{kj} of the mutations. Our only option so far is the adaptive (1+1)-ES of Figure 6.2, which adjusts standard deviations based on the mutation success ratio. This can be generalized to $(1 + \lambda)$ ES algorithms by examining all λ of the mutations at each generation, and keeping track of how many of them result in improvements. However, there is no clear way to generalize this idea to $(\mu + \lambda)$ or (μ, λ) ES algorithms when $\mu > 1$. The children in this case are comprised not only of mutations, but also of combinations of their parents. Therefore, it may not be meaningful to determine the appropriate mutation rate by comparing the fitness of the child to that of its parents.

However, just like we mutate the solution features $\{x_i\}$ for $i \in [1, n]$ to search for an optimum x, we can also mutate the elements $\{\sigma_i\}$ of the standard deviation vector to search for an optimum σ. After a child (x', σ') is created, we mutate the child as follows [Schwefel, 1977], [Bäck, 1996, Section 2.1.2]:

$$\begin{aligned} \sigma_i' &\leftarrow \sigma_i' \exp(\tau' \rho_0 + \tau \rho_i) \\ x_i' &\leftarrow x_i' + \sigma_i' r_i \end{aligned} \tag{6.27}$$

for $i \in [1, n]$, where ρ_0, ρ_i, and r_i, are scalar random variables taken from $N(0, 1)$; and τ and τ' are tuning parameters. The factor $\tau' \rho_0$ allows for a general change in the mutation rate of x', and the factors $\tau \rho_i$ allow for changes in the mutation rates of specific elements of x'. The form of the σ' mutation guarantees that σ' remains positive.

Note that ρ_0 and ρ_i are equally likely to be positive as they are to be negative. This means that the exponential in Equation (6.27) is equally likely to be greater than one as it is to be less than one. This in turn means that σ_i' is just as likely to increase as it is to decrease. Schwefel concludes that this mutation approach is robust to changes in τ and τ', but he suggests setting them as follows [Schwefel, 1977], [Bäck, 1996, Section 2.1.2]:

$$\begin{aligned} \tau &= P_1 \left(\sqrt{2\sqrt{n}}\right)^{-1} \\ \tau' &= P_2 \left(\sqrt{2n}\right)^{-1} \end{aligned} \qquad (6.28)$$

where n is the problem dimension, and P_1 and P_2 are proportionality constants which are typically equal to 1.

It is important to implement the mutation in the order indicated by Equation (6.27); that is, σ' needs to be mutated before x' is mutated. This is because σ' needs to be used to mutate x' so that the fitness of x' indicates, as accurately as possible, the appropriateness of σ'. These ideas lead to the self-adaptive $(\mu + \lambda)$ and (μ, λ) evolution strategies shown in Figure 6.13. Note that Figure 6.13 is called the *self-adaptive* ES, in contrast to the simpler *adaptive* ES idea of Figure 6.2. The self-adaptive ES, introduced in [Rechenberg, 1973], is perhaps the most important contribution of the ES to evolutionary algorithm research and practice. Today virtually all EAs use some type of self-adaptation to adjust algorithmic tuning parameters. In addition, [Beyer and Deb, 2001] has shown that even EAs without explicit self-adaptation can exhibit self-adaptive behavior. The interpretation of EAs as self-adaptive algorithms, and the effect of self-adaptive behavior on EA performance, remain as important tasks for future research.

The algorithm of Figure 6.13 that the mutation covariance matrix Σ_k' is diagonal. In general, we could use a non-diagonal covariance to generate the mutation vector r. We could then try to optimize the entire covariance matrix rather than just the diagonal elements [Bäck, 1996, Section 2.1].

■ EXAMPLE 6.3

In this example, we use the $(\mu + \lambda)$-ES to optimize the Ackley benchmark function. We compare the standard $(\mu + \lambda)$-ES shown in Figure 6.10, and the self-adaptive $(\mu + \lambda)$-ES shown in Figure 6.13. We use both algorithms with $\mu = 10$, $\lambda = 20$, discrete sexual crossover, and a problem dimension $n = 20$. We use the standard values for τ and τ' shown in Equation (6.28), and the standard values $P_1 = P_2 = 1$. Figure 6.14 compares the average convergence rates of the standard $(\mu + \lambda)$-ES and the self-adaptive $(\mu + \lambda)$-ES. We see that the self-adaptive ES converges much faster than the standard ES. Figure 6.15 shows the standard deviation values σ_{ki}, $i \in [1, 20]$, for the best individual (x_k, σ_k) in the population at the last generation. Figure 6.15 shows that the standard deviations have evolved differently for different feature numbers. They have evolved in a way that attempts to optimize the effectiveness of the mutations.

□

Initialize constants τ and τ' as shown in Equation (6.28).
$\{(x_k, \sigma_k)\} \leftarrow$ randomly generated individuals, $k \in [1, \mu]$.
Each x_k is a candidate solution, and each σ_k is a standard deviation vector.
Note that $x_k \in R^n$ and $\sigma_k \in R^n$.
While not(termination criterion)
    For $k = 1, \cdots, \lambda$
        Randomly select two parents from $\{(x_k, \sigma_k)\}$
        Use a recombination method to combine the two parents and obtain
            a child, which is denoted as (x'_k, σ'_k)
        Generate a random scalar ρ_0 from $N(0, 1)$
        Generate a random vector $\begin{bmatrix} \rho_1 & \cdots & \rho_n \end{bmatrix}$ from $N(0, I)$
        $\sigma'_{ki} \leftarrow \sigma'_{ki} \exp(\tau' \rho_0 + \tau \rho_i)$ for $i \in [1, n]$
        $\Sigma'_k \leftarrow \text{diag}((\sigma'_{k1})^2, \cdots, (\sigma'_{kn})^2) \in R^{n \times n}$
        Generate a random vector r from $N(0, \Sigma'_k)$
        $x'_k \leftarrow x'_k + r$
    Next k
    If this is a $(\mu + \lambda)$-ES then
        $\{(x_k, \sigma_k)\} \leftarrow$ the best μ individuals from $\{(x_k, \sigma_k)\} \cup \{(x'_k, \sigma'_k)\}$
    else if this is a (μ, λ)-ES then
        $\{(x_k, \sigma_k)\} \leftarrow$ the best μ individuals from $\{(x'_k, \sigma'_k)\}$
    End if
Next generation

Figure 6.13 The above pseudo-code outlines the self-adaptive $(\mu+\lambda)$ and (μ, λ) evolution strategies, where n is the problem dimension.

■ EXAMPLE 6.4

This example is the same as Example 6.3 except we use the (μ, λ)-ES and the Griewank benchmark function. Figure 6.16 compares the average convergence rates of the standard (μ, λ)-ES and the self-adaptive (μ, λ)-ES. We see that the self-adaptive ES performs very poorly. The reason is that although the σ' mutation in Equation (6.27) has a median of 1, which means that σ' is equally likely to increase as it is to decrease, the σ' mutation has a mean that is greater than 1. The argument of the exponential function in Equation (6.27) is the sum of two zero-mean Gaussian random variables. Suppose for simplicity sake that the argument x of an exponential function $\exp(x)$ is a zero-mean Gaussian random variable with a variance of 1. Then the exponential has a median of 1, but it has a mean of

$$\begin{aligned} E[\exp(x)] &= \int_{-\infty}^{\infty} \text{PDF}(x) \exp(x)\, dx \\ &= \int_{-\infty}^{\infty} \frac{1}{\sqrt{2\pi}} \exp(-x^2/2) \exp(x)\, dx \\ &= \exp(1/2) \approx 1.65. \end{aligned} \tag{6.29}$$

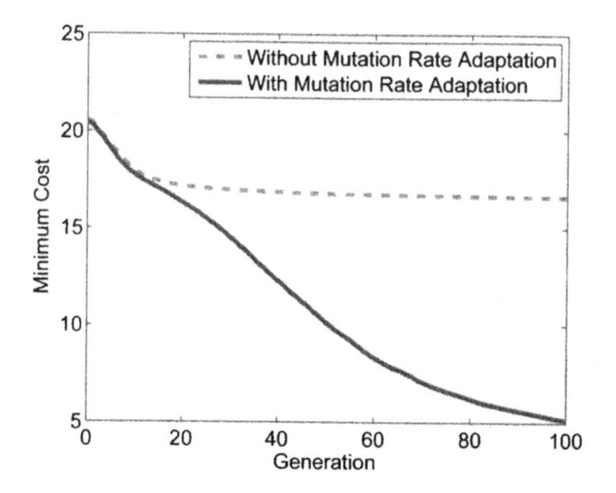

Figure 6.14 Example 6.3: The convergence of the standard and self-adaptive $(\mu + \lambda)$-ES algorithms on the 20-dimensional Ackley function, averaged over 100 simulations. The self-adaptive ES, which automatically adjusts the mutation standard deviations, converges much faster than the standard ES.

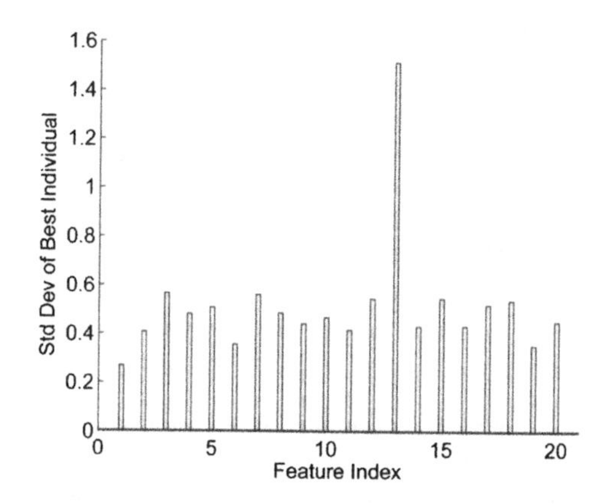

Figure 6.15 Example 6.3: The standard deviation values of the best individual at the final generation for the self-adaptive ES as applied to the 20-dimensional Ackley function. There are 20 standard deviations, corresponding to the 20-dimensional optimization problem. The self-adaptive ES seeks to adjust the mutation standard deviations in a way that maximizes the effectiveness of the mutations.

We see that the σ' mutation tends to increase σ' more than it tends to decrease σ'. This can result in a preponderance of large σ' values in the offspring. If all the parents are discarded at the end of each generation, as they are in the (μ, λ)-ES, unacceptably large σ' values can perpetuate in the population and lead to poor performance.

□

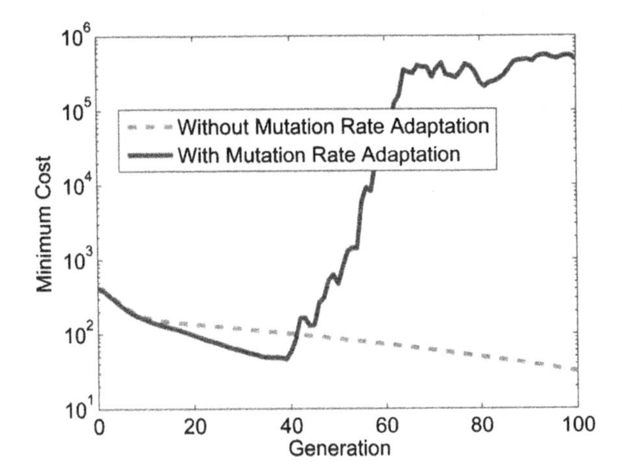

Figure 6.16 Example 6.4: The convergence rate of the standard and self-adaptive (μ, λ)-ES approaches on the 20-dimensional Griewank function, averaged over 100 simulations. The self-adaptive ES, which automatically adjusts the mutation standard deviations, performs very poorly.

Covariance Matrix Adaptation

One successful ES variation is CMA-ES, where CMA is an acronym for covariance matrix adaptation [Hansen, 2010]. The goal of CMA-ES, which has shown a lot of success on benchmark functions, is to fit (as well as possible) the distribution of the ES mutations to the contour of the objective function. This attempted fit can perfectly succeed only for quadratic objective functions, but many objective functions can be approximated with a quadratic near their optimum. The drawbacks of CMA-ES are its complicated adaptation strategy, and its complicated tuning parameter settings. Here we present a simplified version of CMA-ES which is called covariance matrix self-adaptive ES (CMSA-ES) [Beyer and Sendhoff, 2008], The idea of CMSA-ES is to learn the shape of the search space during evolution, and adapt the mutation variance. Figure 6.17 outlines CMSA-ES.

In Figure 6.17, τ is a learning parameter that determines the adaptation speed of the σ_k values. The σ_k values govern the mutation strength. Note that each σ_k is a scalar, which is in contrast to the self-adaptive ES of Figure 6.13. The time constant τ_c determines the adaptation speed of the covariance matrix C, which governs the relative magnitudes and correlations of the mutations along each dimension. Recommended values for τ and τ_c are as follows [Beyer and Sendhoff, 2008]:

$$\begin{aligned} \tau &\leftarrow 1/\sqrt{2n} \\ \tau_c &\leftarrow 1 + n(n+1)/(2\mu) \end{aligned} \tag{6.30}$$

although more efficient performance might be obtained with time-varying values of τ and τ_c.

The square root $\sqrt{C}$ in Figure 6.17 could be calculated a couple of different ways. The original CMA-ES used spectral decomposition, or eigenvalue decomposition

[Hansen and Ostermeier, 2001], and CMSA-ES uses Cholesky decomposition [Beyer and Sendhoff, 2008].

Figure 6.17 shows the averages $\bar{\sigma}$, $\bar{x}$, and $\hat{S}$ as simple averages. However, we could also compute them as weighted averages, which means that we could give more weight to more fit individuals. In view of its nice tradeoff between simplicity and effectiveness on benchmark problems, CMSA-ES seems to hold much promise for future research, including hybridization with other EAs.

Initialize constants τ and τ_c
$C \leftarrow I = n \times n$ identity matrix
$\{(x_k, \sigma_k)\} \leftarrow$ randomly generated individuals, $k \in [1, \mu]$.
Each x_k is a candidate solution, and each σ_k is a standard deviation.
Note that $x_k \in R^n$ and $\sigma_k \in R$.
While not(termination criterion)
    $\bar{\sigma} \leftarrow \sum_{k=1}^{\mu} \sigma_k / \mu$
    $\bar{x} \leftarrow \sum_{k=1}^{\mu} x_k / \mu$
    For $k = 1, \cdots, \lambda$
        $r \leftarrow N(0, 1) =$ Gaussian random scalar
        $\sigma_k \leftarrow \bar{\sigma} \exp(r\tau)$
        $R \leftarrow N(0, I) = n$-dimensional Gaussian random vector
        $s_k \leftarrow \sqrt{C} R$
        $z_k \leftarrow \sigma_k s_k$
        $x_k \leftarrow \bar{x} + z_k$
    Next k
    $\hat{S} \leftarrow \sum_{k=1}^{\lambda} s_k s_k^T / \lambda$
    $C \leftarrow (1 - 1/\tau_c) C + \hat{S} / \tau_c$
Next generation

Figure 6.17 The above pseudo-code outlines the covariance matrix self-adaptive evolution strategy (CMSA-ES), where n is the problem dimension. See the text for details.

6.6 CONCLUSION

We have discussed the original (1+1)-ES, the more general $(\mu + 1)$-ES, and the even more general $(\mu + \lambda)$-ES. We have also discussed the (μ, λ)-ES. We see that an ES is similar to a GA, but GAs were originally developed by encoding candidate solutions as bit strings, while ES has always operated on continuous parameters. Although GAs are often developed to operate on continuous parameters, this remains a philosophical difference between the two algorithms: ES tends to operate on representations that are closer to the problem statement, while GAs tend to operate on representations that are farther removed from the original problem statement. Another difference between the two algorithms is that GAs emphasize recombination, while ES emphasizes mutation. This can guide our choice of algorithms when we are faced with a particular optimization problem. If exploration is more important than exploitation for a particular problem, then we might want to use an ES. However, if exploitation is more important, then we might want to use a GA.[3]

In all of the ES variations that we have discussed thus far, the selection mechanism is deterministic. That is, the best μ individuals are chosen for the next generation. As a variation on this approach, we could instead perform a probabilistic selection of individuals for the next generation. For example, in the $(\mu + \lambda)$-ES, we could use roulette-wheel selection to probabilistically select the parents of the next generation, where each section of the roulette wheel is proportional to the fitness of a corresponding individual. Work in this direction remains as future research.

Additional material on ES is presented in [Bäck and Schwefel, 1993] and [Beyer, 2010]. A Markov model of ES is presented in [Francqis, 1998]. Evolution strategies for multi-objective problems (see Chapter 20) are discussed in [Rudolph and Schwefel, 2008].

[3]See Section 2.7.5 for a discussion of exploration and exploitation.

PROBLEMS

Written Exercises

6.1 The conclusion to this chapter says that an ES might be more appropriate for a problem where exploration is needed, but a GA might be more appropriate for a problem where exploitation is needed. In what type of problem would exploration be more desirable, and in what type of problem would exploitation be more desirable?

6.2 Show that the second and third lines of Equation (6.9) are equal.

6.3 Use Equation (6.16) to derive Equation (6.17).

6.4 Suppose we use the (1+1)-ES to minimize the one-dimensional sphere function. What is the probability that a mutation with a standard deviation of σ will improve a candidate solution x_0?

6.5 Equation (6.27) shows that the standard deviation of the mutation of the self-adaptive ES changes by an exponential factor. Equation (6.29) shows that the mean of that factor is greater than 1, which may lead to poor ES performance. How could you change Equation (6.27) so that the mean of the factor is 1?

Computer Exercises

6.6 Simulate the adaptive (1+1)-ES of Figure 6.2 to minimize the 10-dimensional sphere function (see Section C.1.1) on a domain of $[-5.12, +5.12]$. Initialize the standard deviation of the mutation of each dimension to $0.1/(2\sqrt{3})$. Simulate for 500 generations, and record the cost at each generation. Run 50 simulations like this, and average the 50 cost values at each generation. Plot the average cost values as a function of generation number. Do this for $c = 0.6, 0.8$, and 1.0. Which value of c gives the best performance?

6.7 Repeat Problem 6.6, but instead of simulating for three different values of c, use the default value of c and simulate for three different mutation success ratio thresholds ϕ_{thresh}. That is, instead of using the default value $\phi_{\text{thresh}} = 1/5$, use $\phi_{\text{thresh}} = 0.01, 0.2$, and 0.4. Which value of ϕ_{thresh} gives the best performance?

6.8 Plot the two-dimensional corridor function on the domain $[-50, +50]$ with the constants $c_0 = 0$, $c_1 = 1$, and $b = 10$.

6.9 Use the $(\mu + \lambda)$-ES to minimize the 10-dimensional sphere function on the domain $[-5.12, +5.12]$ with $\mu = 10$ and $\lambda = 20$. Set the standard deviation of the mutation of each dimension to $0.1/(2\sqrt{3})$. Simulate for 100 generations, and record the minimum cost at each generation. Run 50 simulations like this, and average the 50 minimum cost values at each generation. Plot the average minimum cost values as a function of generation number. Do this for discrete sexual, discrete global, intermediate sexual, and intermediate global crossover. Which type of crossover gives the best performance?

6.10 Recall that Equation (6.29) showed that if $x \sim N(0,1)$, then $\exp(x)$ has a median of 1 and a mean of about 1.65. Equation (6.27) shows that the standard deviation of the mutation of the self-adaptive ES changes by an exponential factor. Numerically approximate the median and mean value of the exponential factor if $n = 10$. How does an increase in n affect the median and mean of the exponential factor?

6.11 P_1 and P_2 in Equation (6.28) have default values of 1, but perhaps other values would give better performance. Use the self-adaptive $(\mu+\lambda)$-ES to minimize the 10-dimensional sphere function on the domain $[-5.12, +5.12]$ with $\mu = 10$ and $\lambda = 20$. Initialize the standard deviation of the mutation of each dimension to $0.1/(2\sqrt{3})$. Simulate for 100 generations, and record the minimum cost at each generation. Run 50 simulations like this, and average the 50 minimum cost values at each generation. Plot the average minimum cost values as a function of generation number. Do this for $P_1 = P_2 = 0.1$, $P_1 = P_2 = 1$, and $P_1 = P_2 = 10$. Which value of P_1 and P_2 gives the best performance?

CHAPTER 7

Genetic Programming

Machines would be more useful if they could learn to perform tasks for which they were not given precise methods.

—Richard Friedberg [Friedberg, 1958]

Genetic algorithms and similar EAs are powerful optimization techniques, but they have an inherent limitation: they incorporate the assumed solution structure in the representation of their candidate solutions. For instance, if we want to use a GA to solve a continuous optimization problem with 10 variables, then the GA chromosome is typically represented as $(x_1, x_2, \cdots, x_{10})$. This is both an advantage and a disadvantage for GAs. It is an advantage because it allows the engineer to encode problem-specific information into the solution representation. If we know that our target solution can be nicely represented with 10 real parameters, then defining the chromosome as $(x_1, x_2, \cdots, x_{10})$ makes a lot of sense. However, we may not know which parameters need to be optimized in a given problem. Also, we may not know the structure of the parameters that need to be optimized. Are the parameters real numbers, or state space machines, or computer programs, or complex arrays, or time schedules, or something else?

Genetic programming (GP) is an attempt to generalize EAs to an algorithm that can learn not only the best solution to a problem given a specific structure, but that can also learn the optimal structure. GP evolves computer programs to solve

Evolutionary Optimization Algorithms, First Edition. By Dan J. Simon

optimization problems. This is the distinctive feature of GP compared with other EAs; other EAs evolve solutions, while GP evolves programs that can compute solutions. In fact, this was one of the original goals of the artificial intelligence community. Arthur Samuel, one of the early American pioneers of artificial intelligence, wrote in 1959, "Programming computers to learn from experience should eventually eliminate the need for much of this detailed programming effort."

The fundamental features of GP can be summarized in three basic principles [Koza, 1992, Chapter 2]. First, GP, which evolves computer programs, gives us the flexibility to obtain solution methods to a wide variety of problems. Many engineering problems can be solved with structures that are organized like a computer program, decision tree, or network architecture. Second, GP does not constrain its solutions nearly as much as other EAs; the evolved programs have the freedom to assume the size, shape, and structure that is best suited to the problem at hand. Third, GP evolves computer programs using induction. This is both a strength and a weakness. GP does not evolve programs by building them, as a human would, in a deductive and logical manner. However, some problems are not amenable to deduction. If we want to write a computer program on the basis of a set of training samples, it would be difficult to do so with standard computer programming techniques. But this is the way that GP, like other EAs, operates. GP inductively constructs optimal computer programs.

Early Results in Genetic Programming

Alan Turing, one of the fathers of computer science and artificial intelligence, envisioned something like GP when he wrote in a famous paper in 1950, "We cannot expect to find a good child-machine at the first attempt. One must experiment with teaching one such machine and see how well it learns. One can then try another and see if it is better or worse" [Turing, 1950, page 456]. Richard Friedberg, who studied computer intelligence in the 1950s and later went on to a career in medicine, was one of the first to work on problems that could be classified as genetic programming. Friedberg wrote computer programs that could evolve other computer programs, which could then themselves solve problems. His work was published in the late 1950s with the title "A Learning Machine" [Friedberg, 1958], [Friedberg et al., 1958]. He took a number of shortcuts in his work because of the limited computing capability of the time. For instance, he grouped similar programs together and assumed that their fitness was correlated so that he could reduce the number of fitness calculations. This was a remarkably prescient forerunner to the many methods for fitness computation reduction that we see today in EAs (see Chapter 21). Computing power (speed and memory) increased by a factor of about one million during the 50 years from 1960 to 2010, but we are still just as concerned today about processing power as Friedberg was in the 1950s.

A precursor to modern GP was the variable-length GA developed by Stephen Smith in his 1980 doctoral dissertation [Smith, 1980], in which each individual in the GA population represented a set of decision rules. Another early work which foreshadowed today's GP was Richard Forsyth's paper in 1981, which evolved pattern classification rules [Forsyth, 1981]. Nichael Cramer wrote perhaps the first explicit GP paper in 1985 [Cramer, 1985], which was partly based on Smith's dissertation. In 1990, Hugo de Garis used the term "genetic programming" to refer to the optimization of neural networks using genetic algorithms [de Garis, 1990], but his use of the term has since been superceded by today's definition, which defines GP as the evolution of computer programs. John Koza's 1992 book, which is still an excellent introduction to the topic, was instrumental in popularizing GP research [Koza, 1992]. Koza has since written three additional GP books [Koza, 1994], [Koza et al., 1999], [Koza et al., 2005] that discuss practical applications and more advanced aspects of GP. More discussion of the early history of GP can be found in [Koza, 2010].

Overview of the Chapter

We begin this chapter with a preliminary discussion of the computer programming language Lisp in Section 7.1. Lisp is often used for GP because the structure of Lisp is amenable to EA operations like crossover and mutation. Section 7.2 gives a basic overview of GP, including some of the design choices that we need to make. Section 7.3 discusses an example of GP for minimum time control. Section 7.4 discusses GP bloat, which is the tendency of GP solutions to uncontrollably increase in size. Section 7.5 discusses the use of GP for evolving solutions other than computer programs, including electric circuits and other engineering designs. Section 7.6 discusses some ways that GP performance can be modeled mathematically, especially using schema theory; this section can be skipped by the reader who wants only a working knowledge of GP practice. Section 7.7 summarizes this chapter and provides suggestions for future GP research.

7.1 LISP: THE LANGUAGE OF GENETIC PROGRAMMING

Genetic programming is often implemented in the Lisp computer language because the structure of Lisp ties in so nicely with computer program crossover and mutation. This section provides an overview of Lisp, and provides a conceptual description of how Lisp programs can be combined to create new programs.

The evolution of computer programs is challenging because programs are not usually represented in a way that makes mutation and crossover feasible. This is the primary obstacle that we need to overcome to evolve computer programs. For example, consider two standard MATLAB programs:

Program 1	Program 2
if $x < 1$	for $i = 1 : 10$
$z = [1, 2, 3, 4, 5]$;	if $x > 5$
else	$z(i) = x^i$;
for $i = 1 : 5$	else
$z(i) = i/x$;	$z(i) = x/i$;
end	end
end	end

How could we perform crossover on these programs? A crossover operation that does not carefully consider syntax will result in an illegal program (that is, a program that does not run, or even compile). For example, if we replace the first two lines in the left program above with the first two lines from the right program, we obtain the following:

```
Child Program
for i = 1 : 10
      if x > 5
else
      for i = 1 : 5
            z(i) = i/x;
      end
end
```

The child program above is clearly not a legal program. Most of the crossover operations and mutation operations that we perform on MATLAB programs will result in illegal programs. The same can be said for programs written in most other popular languages (C, Java, Fortran, Basic, Perl, Python, and so on). All of these languages have similar structures, and so none of them can be easily mutated or crossed over with other programs.

However, there is one language that is suitable for crossover and mutation. That language is called Lisp [Winston and Horn, 1989]. Lisp, invented in 1958, is probably the second oldest programming language, only one year newer than Fortran. Lisp is an acronym for "list processing." Linked lists are one of the major structures in Lisp, and this made it a likely choice for artificial intelligence applications (that is, expert systems and their chains of inference rules) in its early days. Lisp is not particularly popular any more because it is different from other languages. However, it has become popular among GP researchers and practitioners because of its suitability for crossover and mutation.

Lisp program code is written with parentheses, with the function name followed by its arguments. For example, the following code adds x and 3:

$$(+\ x\ 3)\ .$$

This is an example of prefix notation because the mathematical operator precedes its inputs. A parenthetical expression in Lisp is also called an s-expression, which is short for symbolic expression. All s-expressions can be viewed as functions that return the value that they compute. S-expressions that compute multiple values return the last value that they compute. Not only does $(+\ x\ 3)$ add x and 3, but it returns $x + 3$ to the next higher level of function execution.

We present a few more examples. The following code computes the cosine of $(x + 3)$:

$$(\cos\ (+\ x\ 3))\ .$$

The following code computes the minimum of $\cos(x + 3)$ and $z/14$:

$$(\min\ (\cos\ (+\ x\ 3))\ (/\ z\ 14))\ .$$

The following code copies the value of y to x if $z > 4$:

$$(\text{if}\ (>\ z\ 4)\ (\text{setf}\ x\ y))\ .$$

Note that s-expressions are like sets in that s-expressions can contain other s-expressions. In the above s-expression, $(>\ z\ 4)$ and $(\text{setf}\ x\ y)$ are both s-expressions that are part of the higher-level s-expression $(\text{if}\ (>\ z\ 4)\ (\text{setf}\ x\ y))$.

The reason Lisp code is evolvable is that s-expressions directly correspond to tree structures, also called syntax trees. For example, a Lisp s-expression for the calculation of $xy + |z|$ can be written as follows:

$$(+\ (*\ x\ y)\ (\text{abs}\ z))\ .$$

This s-expression can be represented with the syntax tree shown in Figure 7.1. We interpret a syntax tree like the one in Figure 7.1 by working our way up from the bottom. Figure 7.1 shows that x and y are at the bottom of the tree, and are connected to each other with a multiplication operator. This gives us the expression xy, or the s-expression $(*\ x\ y)$. We see that this sub-s-expression corresponds to a subtree in Figure 7.1. The symbols that appear at the bottom of a syntax tree (for example, x, y, and z in Figure 7.1) are called *leaves*.

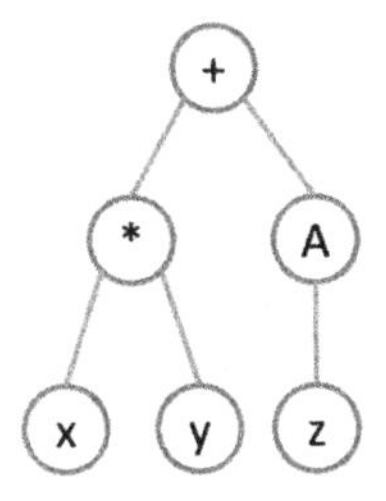

Figure 7.1 Syntax tree for the function $xy+|z|$, which is represented with the s-expression $(+\ (*\ x\ y)\ (\text{abs}\ z))$. The "A" node represents the absolute value operator.

Figure 7.1 also shows that z is at the bottom of the tree, and is operated on by the absolute value function. This gives us the expression $|z|$, or the s-expression (abs z). We see again that the sub-s-expression corresponds to a subtree in Figure 7.1.

Finally, Figure 7.1 shows that xy and $|z|$ meet at the addition node at the top of the tree. This gives us the expression $xy + |z|$, or the s-expression $(+\ (*\ x\ y)\ (\text{abs}\ z))$. We see that this high-level s-expression corresponds to the entire tree structure in Figure 7.1.

For another example, consider a function that returns $(x + y)$ if $t > 5$, and $(x + 2 + z)$ otherwise:

```
If t > 5
      return (x + y)
else
      return (x + 2 + z)
End
```

This function can be written in Lisp notation as:

$$(\text{if}\ (>\ t\ 5)\ (+\ x\ y)\ (+\ x\ 2\ z))\ .$$

Figure 7.2 shows the syntax tree for this function. Note that many Lisp functions, like the addition function in the above example, can take a variable number of arguments.

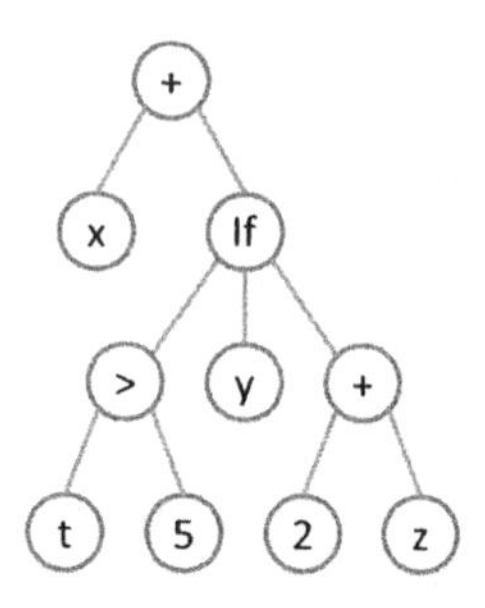

Figure 7.2 Syntax tree for the function, "If $t > 5$ then return $(x + y)$, else return $(x + 2 + z)$."

Crossover with Lisp Programs

The correspondence between s-expressions and subtrees makes it conceptually straightforward to perform operations like crossover and mutation on Lisp computer programs. For example, consider the following functions:

$$\begin{array}{llcl} \text{Parent 1:} & xy + |z| & \Longrightarrow & (+\ (*\ x\ y))\ \text{abs}\ z) \\ \text{Parent 2:} & (x+z)x - (z + y/x) & \Longrightarrow & (-\ (*\ (+\ x\ z)\ x)\ (+\ z\ (/\ y\ x))). \end{array} \tag{7.1}$$

These two parent functions are shown in Figure 7.3. We can create two child functions by randomly choosing a crossover point in each parent, and swapping the subtrees that lie below those points. For example, suppose that we choose the multiplication node in Parent 1, and the second addition node in Parent 2, as crossover points. Figure 7.3 shows how the subtrees in the parents that lie below those points can be swapped to create child functions. Child functions that are created in this way are always valid syntax trees.

Now consider crossover between the s-expressions of Equation (7.1). The following equation emphasizes the parenthetical pairs that correspond to the subtrees of Figure 7.3, and shows how swapping parenthetical pairs in the two original s-expressions (parents) creates new s-expressions (children):

$$\left.\begin{array}{r} (+\ [\ \boldsymbol{*\ x\ y}\ ])\ \text{abs}\ z) \\ (-\ (*\ (+\ x\ z)\ x)\ \{\ \boldsymbol{+\ z\ (\ /\ y\ x\ )}\ \}\ ) \end{array}\right\} \Longrightarrow \left\{\begin{array}{l} (+\ \{\ \boldsymbol{+\ z\ (\ /\ y\ x\ )}\ \}\ )\ \text{abs}\ z) \\ (-\ (*\ (+\ x\ z)\ x)\ [\ \boldsymbol{*\ x\ y}\ ]\ ) \end{array}\right. \tag{7.2}$$

where the crossed-over s-expressions are shown in bold font. This is called tree-based crossover. Any s-expression in a syntax tree can be replaced by any other s-expression, and the syntax tree will remain valid. This is what makes Lisp a perfect language for GP. If we want to perform crossover between two Lisp programs, we simply find a random left parenthesis in Parent 1, then find the matching right parenthesis; the text between those two parentheses forms a valid s-expression. Similarly, we find a random left parenthesis in Parent 2, then find the matching right parenthesis, to obtain another s-expression. After we swap the two s-expressions, we have two children. We can perform mutation in a similar way by replacing a randomly selected s-expression with a randomly-generated s-expression, which is called tree-based mutation.

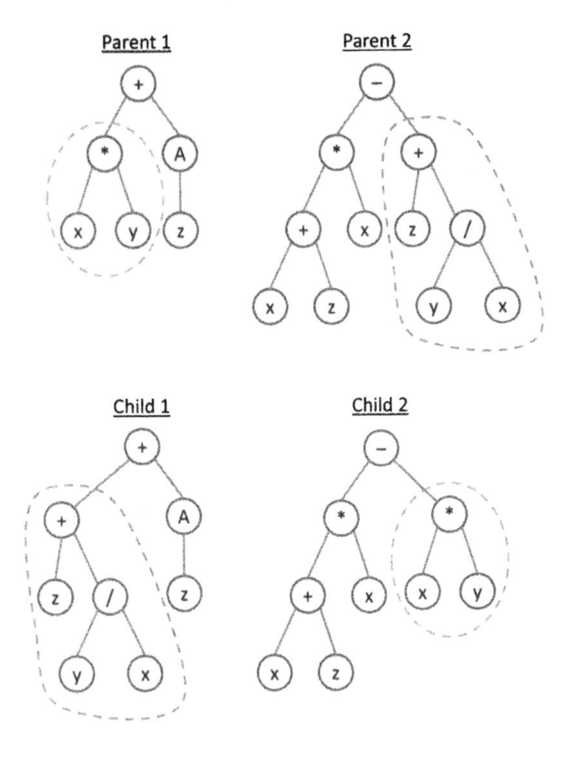

Figure 7.3 Two syntax trees (parents) cross over to produce two new syntax trees (children). Crossover performed in this way always results in valid child syntax trees.

Figure 7.4 shows an additional crossover example. We have the same two parents as in Figure 7.3, but we randomly select the z node as the crossover point in Parent 1, and the division node as the crossover point in Parent 2. The crossover operation of Figure 7.4 is represented in s-expression notation as follows (where, as before, the crossed-over s-expressions are shown in bold font):

$$\left.\begin{array}{r} (+\ (\ *\ x\ y\ ))\ \text{abs}\ \boldsymbol{z}) \\ (-\ (*\ (+\ x\ z)\ x)\ (\ +\ z\ [\ /\ \boldsymbol{y}\ \boldsymbol{x}\ ]\ )\ ) \end{array}\right\} \Longrightarrow \left\{\begin{array}{l} (+\ (\ *\ x\ y\ ))\ \text{abs}\ [\ /\ \boldsymbol{y}\ \boldsymbol{x}\ ]\ ) \\ (-\ (*\ (+\ x\ z)\ x)\ (\ +\ z\ \boldsymbol{z}\ )\ )\ . \end{array}\right. \tag{7.3}$$

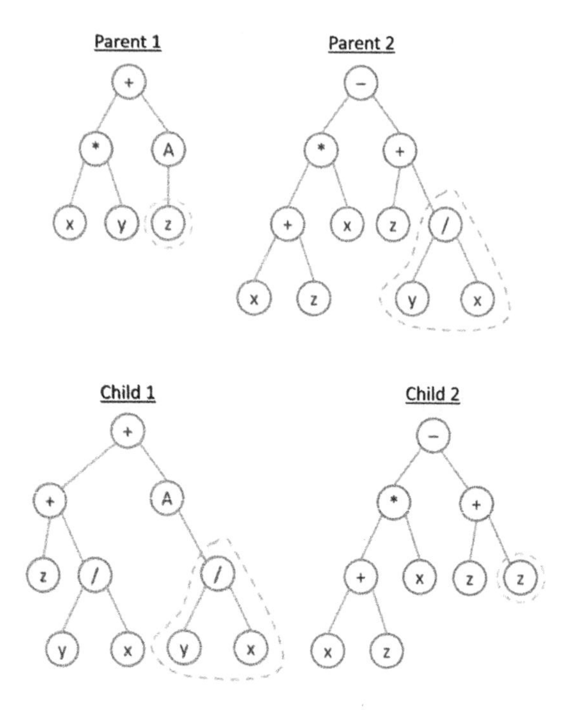

Figure 7.4 Two syntax trees (parents) cross over to produces two new syntax trees (children). The parents shown here are the same as those in Figure 7.3, but the crossover points are chosen differently.

7.2 THE FUNDAMENTALS OF GENETIC PROGRAMMING

Now that we know how to combine Lisp programs, we have the tools to generalize EAs to the evolution of computer programs. Figure 7.5 shows a simple outline of GP. We see that it is similar to a genetic algorithm, but the GA evolves solutions to an optimization problem, while the GP evolves computer programs which can themselves solve an optimization problem.

We need to make some basic decisions before we can implement a GP.

1. What is the fitness measure in Figure 7.5?
2. What is the termination criterion in Figure 7.5?
3. What is the terminal set for the evolving computer programs? That is, what symbols can appear at the leaves of the syntax trees?
4. What is the function set for the evolving computer programs? That is, what functions can appear at the non-terminal nodes of the syntax trees?
5. How should we generate the initial population of computer programs?
6. What other parameters do we need to determine to control GP execution?

Some of these decisions are also required for other EAs, but some of them are specific to GP. The following sections discuss each of these issues in turn.

```
Parents ← {randomly generated computer programs}
While not(termination criterion)
    Calculate the fitness of each parent in the population
    Children ← ∅
    While |Children| < |Parents|
        Use fitnesses to probabilistically select parents p1 and p2
        Mate p1 and p2 to create children c1 and c2
        Children ← Children ∪ {c1, c2}
    Loop
    Randomly mutate some of the children
    Parents ← Children
Next generation
```

Figure 7.5 A conceptual overview of a simple genetic program.

7.2.1 Fitness Measure

What is the fitness measure in Figure 7.5? This decision must be made for all EAs, but the decision is more complicated with GP. A computer program needs to work well for a wide variety of inputs, a variety of initial conditions, and a variety of environments. For example, a program to find a fuel-efficient satellite trajectory from one orbit to another should work well for various satellite parameters and various orbits. Therefore, many different conditions must be used when determining the fitness of a computer program. For a given computer program, each computer input set and operating condition returns its own "subfitness." How should we combine these subfitnesses to obtain a single fitness measure for the computer program? Should we use average performance? Should we try to maximize worst-case performance? Should we use some combination of the two? These questions naturally lead to multi-objective optimization (Chapter 20), although it not necessary to use multi-objective optimization in GP.

7.2.2 Termination Criteria

What is the termination criterion in Figure 7.5? This question needs to be answered for all EAs (see Section 8.2), but it may be especially important for GP. This is because the fitness measure is usually more computationally demanding in GP than in other EAs. The choice of the termination criterion could determine whether or not the GP is successful. As with other EAs, the termination criterion for GP could include factors such as number of iterations, number of fitness evaluations, run time, best fitness value, change in best fitness over several generations, or standard deviation of fitness values over the entire population.

7.2.3 Terminal Set

What is the terminal set for the evolving computer programs? This set describes the symbols that can appear at the leaves of the syntax trees. The terminal set is the set of all possible inputs to the evolving computer programs. This set includes variables that are input to the computer program, along with constants that we think might be important. The constants could include basic integers like 0 and 1, and also constants that may be important for the particular optimization problem (π, e, and so on). The syntax trees in Figure 7.3 have three terminals: x, y, and z. Some constants can be obtained implicitly; for example, $x - x = 0$, and $x/x = 1$. So as long as we have a subtraction and division function, we do not really need the 0 and 1 constants. However, most GP implementations should include constants in their terminal sets.

We can also use random numbers in the terminal set, but usually we do not want a random number to change after it is generated. These type of random numbers are called ephemeral random constants [Koza, 1992, Chapter 26]. Ephemeral random constants are obtained by specifying a quantity denoted as $\mathcal{R}$ in the terminal set. If $\mathcal{R}$ is chosen as a terminal during population initialization, we generate a random number r_1 between given limits, and insert r_1 into the GP individual. From that point on, that particular value r_1 does not change. However, if $\mathcal{R}$ is chosen again for initialization of another individual, or for mutation, then we generate a new random constant r_2 for that realization. The choice of the limits within which to generate ephemeral random constants is another GP design decision.

Defining the terminal set for a GP application is a balancing act. If we use a set that is too small, then the GP will not be able to effectively solve our problem. However, if we use a terminal set that is too large, then it may be too difficult for the GP to find a good solution in a reasonable time. Koza studies this issue in [Koza, 1992, Chapter 24] for the simple problem of discovering the program $x^3 + x^2 + x$ on the basis of 20 test cases. For this problem, the only terminal that the GP needs is x. When the terminal set is the minimal set $\{x\}$, the GP finds the correct program within 50 generations 99.8% of the time. Table 7.1 shows how the probability of success decreases when extra members (random floating point numbers) are added to the terminal set of the GP. For this simple problem, the probability of success decreases linearly with the number of extraneous variables in the terminal set. The good news is that even when 32 of the 33 members in the terminal set are extraneous, GP is still able to solve the problem 35% of the time.

7.2.4 Function Set

What is the function set for the evolving computer programs? This set describes the functions that can appear at the non-terminal nodes of the syntax trees, such as the following.

- Standard mathematical operators can be included in the function set (for example, addition, subtraction, multiplication, division, absolute value).
- Problem-specific functions that we think are important for our particular optimization problem can be included in the function set (for example, exponential functions, logarithmic functions, trigonometric functions, filters, integrators, differentiators).

Number of extra variables	Probability of success (percent)
0	99.8
1	96.6
4	84.0
8	67.0
16	57.0
32	35.0

Table 7.1 GP probability of success after 50 generations for discovering the program $x^3 + x^2 + x$. The population size was 1,000. The data is obtained from [Koza, 1992, Chapter 24].

- Conditional tests can be included in the function set (for example, greater than, less than, equal to).
- Logic functions can be included in the function set, if we think that they could be applicable to the solution of our particular optimization problem (for example, and, nand, or, xor, nor, not).
- Variable assignment functions can be included in the function set.
- Loop statements can be included in the function set (for example, while loops, for loops).
- Subroutine calls can be included in the function set, if we have a set of predefined functions that we have created for our problem.

The syntax trees in Figure 7.3 include five functions: addition, subtraction, multiplication, division, and absolute value. We need to find the right balance in our definition of the function set and the terminal set. The sets need to be large enough to be able to represent a solution to our problem, but if they are too large, then the search space will be so large that the GP will have a hard time finding a good solution.

Some functions need to be modified for GP because the syntax trees evolve might not have legal function arguments. For example, GP could evolve the s-expression (/ x 0), which is division by zero. This would result in a Lisp error, which would cause the GP to terminate. Therefore, instead of using the standard division operator in Lisp, we can define a division operator DIV that protects against division by zero, and that also protects against overflow due to division by a very small number:

```
(defun DIV (x y)                          ; define a protected division function
(if (< (abs y) ε) (return-from DIV 1))   ; return 1 if the divisor is very small
return-from DIV (/ x y))                  ; else return x/y
```
(7.4)

where ϵ is a very small positive constant, like 10^{-20}. Equation (7.4) shows the Lisp syntax for defining a protected division routine.[1] The DIV function returns 1 if the

[1]Note that any text following a semicolon in a Lisp function is interpreted as a comment.

divisor has a very small magnitude. We may need to redefine other functions in a similar way (logarithm functions, inverse trigonometric functions, and so on) to make sure that the functions in our function set can handle all possible inputs.

7.2.5 Initialization

How should we generate the initial population of computer programs? We have two basic options for initialization, which are referred to as the full method and the grow method. We can also combine these options to get a third option, which is referred to as the ramped half-and-half method [Koza, 1992].

The *full method* creates programs such that the number of nodes from each terminal node to the top-level node is D_c, a user-specified constant. D_c is called the depth of the syntax tree. As an example, Parent 1 in Figure 7.3 has a depth of three, while Parent 2 has a depth of four. Parent 1 in Figure 7.3 is a full syntax tree because there are three nodes from each terminal node to the top-level addition node. However, Parent 2 is not a full syntax tree because some of the program branches have a depth of four while others only have a depth of three.

We can use recursion to generate random syntax trees. For example, if we want to generate a syntax tree with a structure like Parent 2 in Figure 7.3, we first generate the subtraction node at the top level and note that it requires two arguments. For the first argument, we generate the multiplication node and note that it requires two arguments. This process continues for each node and each argument until we have generated enough levels to reach the desired depth. When we reach the desired depth, we generate a random terminal node to complete that branch of the syntax tree. Figure 7.6 illustrates the concept for a recursive algorithm that generates random computer programs. We can generate a random syntax tree by calling routine GrowProgramFull(D_c, 1), where D_c is our desired syntax tree depth. GrowProgramFull calls itself each time it needs to add another layer in its growing syntax tree.

The *grow method* of initialization creates programs such that the number of nodes from each terminal node to the top-level node is less than or equal to D_c. If the parents in Figure 7.3 were created by random initialization, then Parent 1 might have been generated with either the full method or the grow method, while Parent 2 was definitely generated with the grow method since it is not a full syntax tree. The grow method can be implemented the same way as the full method, except that when we generate a random node at depths less than D_c, either a function or terminal node can be generated. If a function node is generated, the syntax tree continues to grow. As with the full method, when we reach the maximum depth D_c, we generate a random terminal to complete that branch of the syntax tree. Figure 7.7 illustrates the concept for a recursive algorithm that generates random computer programs with the grow method.

The *ramped half-and-half method* generates half of the initial population with the full method, and half with the grow method. Also, it generates an equal number of syntax trees for each value of depth between 2 and D_c, which is the maximum allowable depth specified by the user. Figure 7.8 illustrates the concept of ramped half-and-half syntax tree initialization.

```
function [SyntaxTree] = GrowProgramFull(Depth, NumArgs)
SyntaxTree ← ∅
For i = 1 to NumArgs
    If Depth = 1
        SyntaxTree ← Random terminal
    else
        NewFunction ← Randomly chosen function
        NewNumArgs ← Number of arguments required by NewFunction
        SyntaxTree ← (NewFunction + GrowProgramFull(Depth−1, NewNumArgs))
    End
Next i
```

Figure 7.6 A conceptual view of a recursive algorithm to grow random syntax trees in s-expression form with the full method. This routine is initially called with the syntax GrowProgramFull(D_c, 1), where D_c is the desired depth of the random syntax tree. The plus operator indicates string concatenation. Note that this algorithm is conceptual; it does not include all of the details required for valid syntax tree generation, such as correct parenthesis placement.

```
function [SyntaxTree] = GrowProgramGrow(Depth, NumArgs)
SyntaxTree ← ∅
For i = 1 to NumArgs
    If Depth = 1
        SyntaxTree ← Random terminal
    else
        NewNode ← Randomly chosen function or terminal
        If NewNode is a terminal
            SyntaxTree ← (SyntaxTree + NewNode)
        else
            NewNumArgs ← Number of arguments required by NewNode
            SyntaxTree ← (NewNode + GrowProgramGrow(Depth−1, NewNumArgs))
        End
    End
Next i
```

Figure 7.7 A conceptual view of a recursive algorithm to grow random syntax trees in s-expression form with the grow method. This routine is initially called with the syntax GrowProgramGrow(D_c, 1), where D_c is the desired depth of the random syntax tree. As with Figure 7.6, the plus operator indicates string concatenation, but this algorithm does not include all of the details required for implementation.

D_c = maximum syntax tree depth
N = population size
For $i = 1$ to N
    Depth $\leftarrow U[2, D_c]$
    $r \leftarrow U[0, 1]$
    If $r < 0.5$
        SyntaxTree(i) $\leftarrow$ GrowProgramGrow(Depth, 1)
    else
        SyntraxTree(i) $\leftarrow$ GrowProgramFull(Depth, 1)
    End
Next i

Figure 7.8 Algorithm to create an initial GP population with the ramped half-and-half method. $U[2, D_c]$ is a random integer uniformly distributed on $[2, D_c]$, and $U[0, 1]$ is a random real number uniformly distributed on $[0, 1]$. This algorithm calls the routines of Figures 7.6 and 7.7.

Koza experimented with the three different types of initializations described above for some simple GP problems [Koza, 1992, Chapter 25]. He found a difference in the probability of GP success depending on which initialization method was used, as shown in Table 7.2. The table shows that the ramped half-and-half initialization method is generally much better than the other two initialization methods.

Problem	Full	Grow	Ramped Half-and-Half
Symbolic Regression	3%	17%	23%
Boolean Logic	42%	53%	66%
Artificial Ant	14%	50%	46%
Linear Equation	6%	37%	53%

Table 7.2 GP probability of success for various problems and various initialization methods. This data is obtained from [Koza, 1992, Chapter 25].

To conclude our discussion of initialization, we note that it is often advantageous to seed the initial population of an EA with some known good individuals. These good individuals may be user-generated individuals, or they may come from some other optimization algorithm or other source. However, seeding does not necessarily improve EA performance. If there are only a few good individuals in the initial population, and the rest of the individuals are relatively poor randomly-generated individuals, then the few good individuals could dominate the selection process, and the poor individuals might quickly die out. This could result in an evolutionary dead end and premature convergence, otherwise known as "survival of the mediocre" [Koza, 1992, page 104]. However, the chances that this negative event occurs depends on the type of selection that we use (see Section 8.7). If we use roulette-wheel selection, then selection pressure is high and a few fit individuals are likely to quickly dominate the population. If we use tournament selection, then selection pressure is much lower and the probability of a few fit individuals dominating the population is correspondingly lower.

7.2.6 Genetic Programming Parameters

What are the parameters that control GP execution? These parameters include those that are used for other EAs, but also include GP-specific parameters.

1. We need to specify the selection method by which parents are chosen to participate in crossover. We could use fitness-proportional selection, tournament selection, or some other method. In fact, we could use any of the selection methods discussed in Section 8.7.

   This is also a good place to mention that we could implement tree-based crossover more intelligently than simply selecting random crossover points. There are some subtrees that are more useful than others, and we may not want to break up those subtrees. We could quantify the fitness of subtrees by obtaining correlations between crossover points and the fitness of child programs, and then using those correlations to bias the selection of future crossover points [Iba and de Garis, 1996].

2. We need to specify the population size. Since there are so many degrees of freedom in computer programs, GP usually has larger populations than other EAs. GP usually has a population size of at least 500, and often has a population size of several thousand.

3. We need to specify the mutation method. Various GP mutation methods have been used over the years, some of which are described as follows.

   (a) We can select a random node, and replace everything below that node with a randomly-generated syntax subtree. This is called subtree mutation [Koza, 1992, page 106]. This is equivalent to crossing a program with a randomly generated program, and is also called headless chicken crossover [Angeline, 1997].

   (b) Expansion mutation replaces a terminal with a randomly-generated subtree. This is equivalent to subtree mutation if the replaced node in subtree mutation is a terminal.

   (c) We can replace a randomly selected node or terminal with a new randomly generated node or terminal. This is called point mutation or node replacement mutation, and requires that the arity of the replaced node be equal to the arity of the replacement node.[2] For example, we could replace an addition operation with a multiplication operation, or we could replace an absolute value operation with a sine operation.

   (d) Hoist mutation creates a new program that is a randomly selected subtree of the parent program [Kinnear, 1994].

   (e) Shrink mutation replaces a randomly chosen syntax subtree with a randomly selected terminal [Angeline, 1996a]; this is also called collapse

[2]The arity of a function is equal to the number of its arguments. For example, a constant has an arity of 0, the absolute value function has an arity of 1, and an addition function can have an arity of two or more.

subtree mutation. Hoist mutation and shrink mutation were originally introduced to reduce code bloat (see Section 7.4).

(f) Permutation mutation randomly permutes the arguments of a randomly selected function [Koza, 1992]. For example, we could replace the x and y arguments of a division function. Of course, this type of mutation does not have any affect on commutative functions.

(g) We can randomly mutate constants in a program [Schoenauer et al., 1996].

We often implement mutation in such a way that the mutated program replaces the original program only if it is more fit. This idea of replace-only-if-more-fit can be applied to mutation in any EA.

4. We need to specify the mutation probability p_m. This is similar to other EAs. Mutation in a GP with N individuals is often implemented with a method similar to the following:

```
For each candidate computer program x_i, where i ∈ [1, N]
    Generate a random number r uniformly distributed in [0, 1]
    If r < p_m
        Randomly select a node k in computer program x_i
        Replace the selected subtree starting at node k with a
            randomly-generated subtree
    End
Next computer program
```

The large population size that is used in GP, along with the large number of possible nodes at which crossover can occur, usually means that good GP results do not depend on mutation [Koza, 1992, Chapter 25]. Often we can get good results with $p_m = 0$. However, mutation may still be desirable just in case an important terminal or function is lost from the population. If that occurs, mutation is the only way that it could re-enter the population.

5. We need to specify the crossover probability p_c. This is similar to GAs. After selecting two parents in Figure 7.5, we can either use crossover to combine them, or we can instead clone them for the next generation. The line:

```
Mate p_1 and p_2 to create children c_1 and c_2
```

in Figure 7.5 would then be replaced with something like the following:

```
Generate a random number r uniformly distributed on [0, 1]
If r < p_c
    Mate p_1 and p_2 to create children c_1 and c_2
else
    c_1 ← p_1
    c_2 ← p_2
End
```

Most experience suggests that crossover is an important aspect of GP and should be used with a probability $p_c \geq 0.9$ [Koza, 1992, Chapter 25].

6. We need to decide whether or not to use elitism. As with any other EA, we can save the best m computer programs in GP from one generation to the next to make sure they are not lost in the following generation (see Section 8.4). The parameter m is called the elitism parameter. Elitism can be implemented in several different ways. For example, we could archive the best m individuals at the end of a generation, create the children for the next generation as usual, and then replace the worst m children with the elites from the previous generation. Alternatively, we could copy the m elites to the first m children each generation, and then create only $(N - m)$ additional children each generation (where N is the population size).

7. We need to specify D_i, the maximum program size of the initial population. A program's size can be quantified by its depth, which measures the maximum number of nodes between the highest level and the lowest level (inclusive). For example, Parent 1 in Figure 7.3 has a depth of three, while Parent 2 has a depth of four.

8. We also need to specify D_c, the maximum depth of child programs. During GP operation, child programs can grow larger and larger with each succeeding generation. If a maximum depth is not enforced, then child programs can become unreasonably long, wasting space and execution time; this is called GP bloat (Section 7.4). The maximum depth D_c can be enforced in several ways. One way is to replace a child with one of its parents if the child's depth exceeds D_c. Another way is to redo the crossover operation if the child's depth exceeds D_c. Yet another way is to examine the parent syntax trees before choosing their crossover points, and constrain the randomly selected crossover points so that D_c will not be exceeded by the children's depths.

9. We need to decide whether or not we want to allow a terminal node in a syntax tree to be replaced with a subtree during crossover. Figure 7.4 shows that the z terminal in Parent 1 is selected for crossover, and is replaced with a subtree in Child 1. We use p_i to denote the probability of crossover at an internal node. When selecting a crossover point, we generate a random number r uniformly distributed on $[0, 1]$. If r is less than p_i, then we select a terminal node for crossover; that is, we select a symbol in the syntax tree that is not immediately preceded by a left parenthesis. However, if r is greater than p_i, then we select an s-expression for crossover; that is, we select a subtree that is surrounded by matching left and right parentheses for crossover.

10. We need to decide whether or not to worry about duplicate individuals in the population. Duplicate individuals are a waste of computer resources. In EAs with relatively small search spaces or small populations, duplicates can arise quite often, and dealing with duplicates can be an important aspect of the EA (see Section 8.6.1). However, in GP, the search space is so large that duplicates rarely occur. Therefore, we usually do not need to worry about duplicate individuals in GP.

7.3 GENETIC PROGRAMMING FOR MINIMUM TIME CONTROL

In this section, which is motivated by [Koza, 1992, Section 7.1], we demonstrate the use of GP for minimum time control of a second-order Newtonian system. A second-order Newtonian system is a simple position-velocity-acceleration system that satisfies the equations

$$\begin{aligned} \dot{x} &= v \\ \dot{v} &= u \end{aligned} \tag{7.5}$$

where x is position, v is velocity, and u is the commanded acceleration. That is, the derivative of position is velocity, and the derivative of velocity is acceleration. We consider motion only in one dimension. The problem is to find the acceleration profile $u(t)$ to drive the system from some initial position $x(0)$ and velocity $v(0)$, to $x(t_f) = 0$ and $v(t_f) = 0$, in the minimum time t_f. Intuition tells us approximately how to accomplish this: we accelerate as fast as we can in one direction until we reach a certain position, and then we accelerate as fast as we can in the opposite direction until we reach $x(0) = v(0) = 0$.[3]

We assume for the sake of simplicity, and without loss of generality, that the maximum acceleration magnitude is 1, and that we can acceleration in either direction. The minimum time control problem is illustrated at the top of Figure 7.9. We accelerate in the positive direction (toward the right) until reaching the strategic point labeled "Switch." Then we accelerate in the negative direction (toward the left) until reaching the goal. Note that the vehicle's velocity is toward the right for the entire time period. If the timing is right, we will reach the goal with zero velocity.

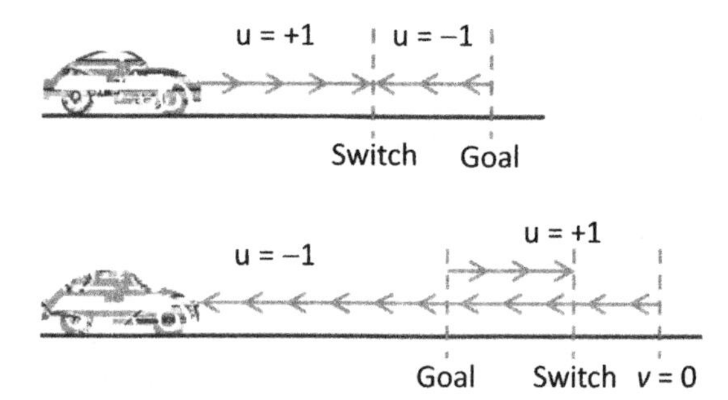

Figure 7.9 Illustration of minimum time control. In the top figure, we accelerate to the right to the switching point, then accelerate to the left, and reach the goal with zero velocity. In the bottom figure, our initial velocity is so high tha we twill inevitably overshoot the goal. In this case we accelerate to the left, overshoot the goal, switch the acceleration to the right at the switching point, and reach the goal with zero velocity.

It may be that we have such a high initial velocity that we cannot stop before the goal. In this case we will inevitably overshoot the goal, and so we must return back

[3]In fact, this is what we observe in teenage male drivers at every stoplight: accelerate as fast as possible when the light turns green, and then at a carefully chosen point before the next stoplight, slam on the brakes. If the teenager's timing is right, the car will stop precisely at the next red light, and the travel time between stoplights will be minimized.

to the goal, reaching it with zero velocity. This situation is a little less intuitive, and is shown at the bottom of Figure 7.9. The minimum time solution is to first accelerate as much as possible in the negative direction (toward the left). The vehicle overshoots the goal. Eventually the vehicle will reach zero velocity, at which point the vehicle begins moving toward the left. We continue accelerating in the negative direction until reaching the strategic point labeled "Switch," at which time we begin accelerating in the positive direction (toward the right) until returning to the goal. Again, if the timing is right, we will reach the goal with zero velocity.

The minimum-time control problem is a classic optimal control problem with many aerospace applications, and is studied in detail in many optimal control books [Kirk, 2004]. The solution is called bang-bang control, because for any initial condition $x(0)$ and $v(0)$, the solution consists of one time period of maximum acceleration in one direction, followed by a time period of maximum acceleration in the other direction. The minimum time control problem can be represented in graphical form with a phase plane diagram as shown in Figure 7.10. We assume for simplicity that the vehicle mass is 2. In this case, the curve drawn in Figure 7.10, which is called the switching curve, is given by

$$x = -v|v|/2. \tag{7.6}$$

The goal is to reach the origin $x = 0$ and $v = 0$ in minimum time from any initial point in the phase plane. If the position and velocity is above the switching curve, then we should apply maximum acceleration in the negative direction. If the position and velocity is below the switching curve, then we should apply maximum acceleration in the positive direction. This will take us on a trajectory that reaches the switching curve, at which point we will reverse the direction of the acceleration. Then we will follow the switching curve to the origin of the phase plane.

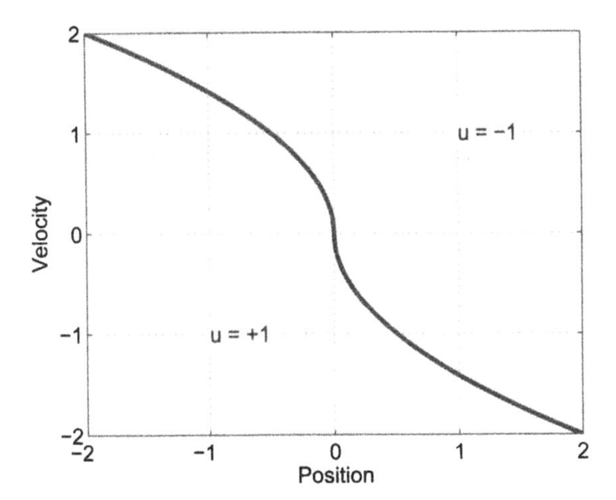

Figure 7.10 Switching curve for minimum time control. If the position and velocity lie above the switching curve, then the acceleration should be maximum in the negative direction. If the position and velocity lie below the curve, then the acceleration should be maximum in the positive direction.

Figure 7.11 illustrates the optimal trajectory for the initial condition $x(0) = -0.5$ and $v(0) = 1.5$. This corresponds to the bottom picture in Figure 7.9. The vehicle is moving too fast to stop before reaching $x = 0$. Therefore, we apply the maximum accleration in the negative direction until reaching the switching curve; note that the vehicle passes through $v = 0$ during the time of maximum negative acceleration. When the vehicle reaches the switching curve, we apply the maximum acceleration in the forward direction. The trajectory reaches the origin of the phase plane ($x = 0$ and $v = 0$) in the minimum possible time.

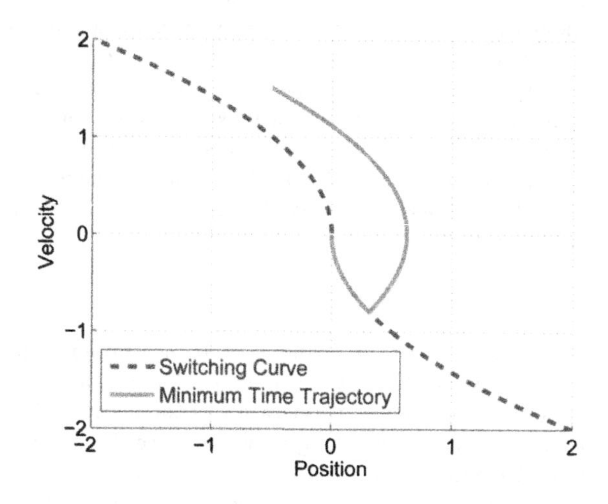

Figure 7.11 Minimum time trajectory for initial condition $x(0) = -0.5$ and $v(0) = 1.5$. The acceleration is -1 above the switching curve, and $+1$ after the trajectory reaches the switching curve.

Now we use GP to try to evolve a minimum time control program for this problem. We define two special Lisp functions for this problem. The first is the protected division operator shown in Equation (7.4), and the second is the greater-than operator:

$$\begin{aligned}&\text{(defun GT } (x\ y)\\&\text{(if } (>\ x\ y) \text{ (return-from GT 1) (return-from GT } -1\text{))).}\end{aligned} \tag{7.7}$$

The GT function returns 1 if $x > y$, and returns -1 otherwise.

To evaluate the cost of a program, we take 20 random initial points in the (x, v) phase plane, with $|x| < 0.75$ and $|v| < 0.75$, and see if the program can bring each of the (x, v) pairs to the origin within 10 seconds. If the program is successful for an initial condition, then the cost contribution of that simulation is the time required to bring (x, v) to the origin. If the program is not successful within 10 seconds, then the cost contribution of that simulation is 10. The total cost of a computer program is the average of all 20 cost contributions. Table 7.3 summarizes the GP parameters for this problem, which are mainly based on [Koza, 1992, Section 7.1].

GP Option	Setting
Objective	Find the minimum time vehicle control program
Terminal set	x (position), v (velocity), -1
Function set	$+$, $-$, $*$, DIV, GT, ABS
Cost	Time to bring the vehicle to the phase plane origin, averaged over 20 random initial conditions
Generation limit	50
Population size	500
Maximum initial tree depth	6
Initialization method	Ramped half-and-half
Maximum tree depth	17
Probability of crossover	0.9
Probability of mutation	0
Number of elites	2
Selection method	Tournament (see Section 8.7.6)

Table 7.3 GP parameters for the minimum time vehicle control problem.

Figure 7.12 shows the cost of the best GP solution as a function of generation number. The best computer program is found after less than 10 generations for this particular run, but the average performance of the entire population continues to decrease during the entire 50 generations. For most GP problems, it takes much longer than 10 generations to find the best solution. The reason this particular run was quicker than the average GP run might be because the problem is relatively easy, or it might simply be a statistical fluke. The best solution obtained by the GP is

$$\begin{aligned} u \;=\; & (\, *\, (\, \mathrm{GT}\, (\, -\, (\, \mathrm{DIV}\; x\; v)\, (\, -\; -1\; v)\,)\, (\, \mathrm{GT}\, (\, +\; v\; x)\, (\, \mathrm{DIV}\; x\; v)\,)\,) \\ & (\, \mathrm{DIV}\, (\, \mathrm{GT}\, (\, +\; x\; v)\, (\, +\; v\; x)\,)\, (\, \mathrm{GT}\, (\, +\; v\; x)\; x\,)\,)\,)\,)\, . \end{aligned} \tag{7.8}$$

The switching curve for this control is plotted in Figure 7.13, along with the theoretically optimal switching curve. For $v < 0$ the two curves are very similar. For $v > 0$ there is more of a difference between the curves, but the general shape is still similar.

The time that it takes the vehicle to reach the origin of the phase plane, averaged over 10,000 random initial conditions in the state space $x \in [-0.75, +0.75]$ and $v \in [-0.75, +0.75]$, is about 1.53 seconds for the optimal switching curve and 1.50 seconds for the GP switching curve. Interestingly, the GP switching actually performs slightly better than the optimal switching curve! This is not possible theoretically, but practice and theory do not always match.[4] In practice, there are implementation issues that make it possible to perform better than the theoretically optimal strategy. For example, we terminated our simulation when $|x| < 0.01$ and $|v| < 0.01$, and considered such small values a complete success. In theory, we can reach the origin with exactly zero error, but in practice, we cannot. Also, we used a step size of $\tau = 0.02$ seconds to simulate the dynamic system. Rather than

[4]In theory, practice and theory should match. In practice, they do not.

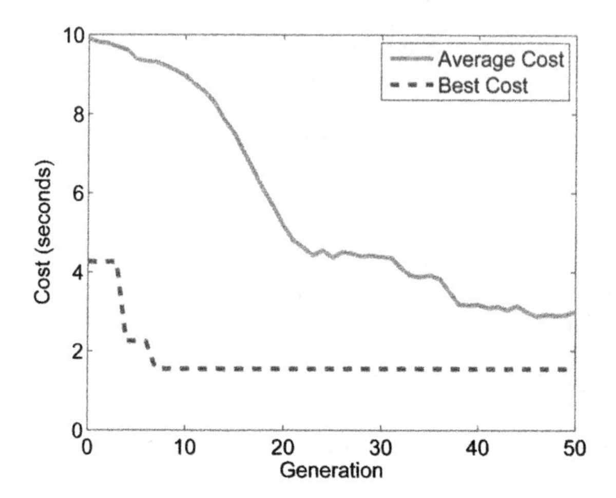

Figure 7.12 GP performance for the minimum time control problem. The best solution is found after less than 10 generations for this particular run.

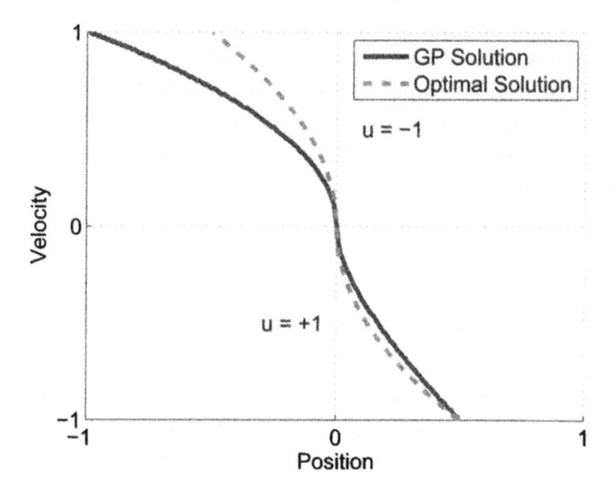

Figure 7.13 The best switching curve obtained by the GP for the minimum time control problem, along with the theoretically optimal switching curve.

computing the exact continuous time solution to

$$\begin{aligned} \dot{v} &= u \\ \dot{x} &= v \end{aligned} \tag{7.9}$$

we instead approximated the solution as

$$\begin{aligned} v_{k+1} &= v_k + \tau u_k \\ x_{k+1} &= x_k + \tau(v_k + v_{k+1})/2 \end{aligned} \tag{7.10}$$

where k is the time index, which ranges from 0 to 500 (that is, 0 to 10 seconds). That is, we used rectangular integration for the solution of velocity, and trapezoidal integration for the solution of position [Simon, 2006, Chapter 1]. These differences between theory and practice may result in more control chattering along the optimal switching curve than is present along the GP-generated switching curve.[5] The reader can replicate the results in this section by following the steps described in Problem 7.13.

Theory versus Practice

The superiority of the GP switching curve over the theoretically optimal solution raises an important point regarding the difference between theory and practice. Engineering solutions are often generated on the basis of theory, but as any practicing engineer knows, theoretical results need to be modified to take real-world considerations into account. This example shows that a GP may be able to take real-world considerations into account to find a solution that is better than the theoretically optimal solution to a problem.

It may be easier to learn optimal control theory and solve the minimum time control problem in a more traditional way, rather than learning how to use a GP. But it may not. This example shows us that GP might be able to find solutions to problems that we lack the expertise to solve on our own. It further shows the possibility of finding "better-than-optimal" solutions when practical considerations are taken into account.

7.4 GENETIC PROGRAMMING BLOAT

Genetic programming can result in a programs that become unreasonably long, and that require high levels of computational effort. The extra code that evolves during GP goes by several different names, including introns, junk code, fluff, ineffective code, hitchhiker code, and invisible code [Langdon and Poli, 2002, Chapter 11]. Some examples of introns include the following:

$$\begin{aligned} &(\text{not } (\text{not } x)) \\ &(+\ x\ 0) \\ &(\text{if } (>1\ 2)\ x\ y). \end{aligned} \tag{7.11}$$

Any serious GP implementation needs to protect against bloat to prevent uncontrolled increases in code length. There are several ways to protect against bloat.

The first way of combatting code bloat is to use a maximum depth parameter D_c, as discussed in Section 7.2.6. However, this is a balancing act. If D_c is too small, then we limit the search space of the GP, and we may reduce the fitness of the best program that it can find.

The second way to combat code bloat is to adjust our implementations of crossover and mutation to combat bloat. For example, size fair crossover chooses crossover points that balance the size of the parent code fragments, and that therefore results in children that are no larger than the parents [Langdon, 2000]. If

[5] Koza reports an average time of 2.13 seconds for the theoretically optimal switching curve [Koza, 1992, Section 7.1], again pointing to differences in the implementation of the dynamic system equations.

subtree mutation is used, it can be adjusted to guarantee that the size of the mutated program is limited [Kinnear, 1993]. Hoist mutation removes code to decrease the length of a program [Kinnear, 1994]. One-point crossover is a method that we have not discussed in this chapter, but it automatically limits the depth of children to the depth of the largest parent [Poli et al., 2008, Chapter 5].

The third way to combat bloat is to penalize long programs in the selection, reproduction, and crossover operations. This idea can be implemented in several different ways. For example, we can add a cost penalty to large programs:

$$\text{Penalized Cost} \leftarrow \text{Cost} + \text{Program Size}. \tag{7.12}$$

In general, it will require more fitness evaluations to find a good solution if large programs are penalized [Koza, 1992, Chapter 25]. On the other hand, those fitness evaluations will be faster since the programs will be smaller. This approach essentially biases selection toward those programs that are shorter. This idea has been called parsimony pressure, Occam's razor, and minimum description length [Langdon and Poli, 2002, Chapter 11]. Another approach to penalize long programs is the Tarpeian method [Poli, 2003], which sets the selection probability of randomly selected longer-than-average programs to zero. As we change the frequency with which this is done, we adjust the anti-bloat capability of the method. This has the additional benefit of reducing execution time, since programs with zero selection probability do not need to be evaluated.

There are also other ways to fight code bloat, such as using multi-objective optimization with the two objectives of program fitness and program length (see Chapter 20, automatically removing excess code, and using automatically defined functions (ADFs). These methods are more complex, and do not guarantee the prevention of code bloat [Langdon and Poli, 2002, Chapter 11].

Finally, we mention that code bloat may be beneficial in some circumstances. Bloat has a biological analogy, and it may help computer programs protect their children against the effects of harmful crossover [Angeline, 1996b], [Nordin et al., 1996]. This gives rise to the term *effective fitness*, which indicates not only how fit a computer program is, but how fit its children are likely to be [Banzhaf et al., 1998, Chapter 7]. A fit parent computer program may not be likely to produce fit children if the parent program is fragile, but a fit parent with a lot of bloat may be more likely to produce fit children. Consider the crossover operation between two computer programs. A good program with a lot of unused code is likely to result in fit children after crossover, because the crossover point is likely to occur at an unused portion of the parent.

We can also talk about *effective complexity*. The absolute complexity of a computer program is a function of its length and structure, while its effective complexity is a function of the length and structure of the non-bloat portion (that is, the active portion) of the program.

7.5 EVOLVING ENTITIES OTHER THAN COMPUTER PROGRAMS

The accomplishments of GP are impressive. [Koza, 1992] gives examples of GP for discovering trigonometric identities, discovering scientific laws, solving mathematical equations in symbolic form, inducing the symbolic form for a sequence, and finding programs for image compression. He also gives control problem examples, including balancing an inverted pendulum on a moving cart, and backing up a tractor-trailer.

But GP still has not seen widespread use compared to other EAs. There are various reasons for this lag [Koza, 1992, Chapter 1].

1. Engineers have been trained to find correct solutions to problems. In school, there is often only one correct answer to a homework problem. GP finds solutions that are only approximately correct, and so this deters its use in practice. However, in the real engineering world, all of our solutions are approximation, if for no other reason than the assumptions that we make (explicitly or implicitly) while deriving our solutions. The lack of correctness is a theoretical roadblock to the use of GP, but it need not be a practical roadblock.

2. Engineers have been trained to find solutions by incremental improvements. GP follows this approach, but it also encourages searches in blind alleys. Poor programs need to evolve before good programs are achieved. In the real world, failure is a stigma, but the most successful engineers recognize that failure is a prerequisite to success [Petroski, 1992]. This is true not only for humans, but also for GP.

3. Engineers have been trained to solve problems deductively. We learn about the problem, and we build solutions one step at a time. Loosely speaking, there is some deduction in GP; after all, highly fit programs are combined to obtain programs with hopefully higher fitness. But the computer programs generated by GP are not built up by logically by adding functionality one step at a time.

4. Engineers have been trained to solve problems deterministically. The more that we can remove randomness from our environment, the more control we can obtain. The more control we can obtain, the better we can proceed with our solution method. But GP, like other EAs, relies on randomness to find good solutions.

5. Engineers have been trained to solve problems economically. A short and simple solution is better than a complicated solution. But GP evolves computer programs with branches that never execute, with terminals that do not contribute to the final result, and with inefficient structures. This is like the problem solving processes that we see in nature. Many animals routinely have hundreds of babies for every one that survives. Evolution is a notoriously wasteful and inefficient process.

6. Engineers have been trained to solve problems with specific success criteria. We have tasks, subtasks, milestones, and schedules. Validation processes tell us if we have succeeded, and they tell us if we have failed. GP, however, does not have a well-defined termination point. This was discussed in more detail in Section 7.2.2.

Many of these factors apply to EAs other than GP, but they seem to be especially applicable to GP. The difference is that EAs find solutions, while GP finds solution methods. We seem to be better able to tolerate sloppiness in our solutions than in our solution methods. When we find a good solution to a problem, we are often not too worried about where the solution came from, as long as it works. However, when we find a solution *method*, we are likely to be distrustful of the method if we do not understand it, even if works.

We have seen in this chapter that GP can evolve computer programs. However, computer programs are still much more likely to be written by humans than by GP. This is because computer programs can usually be planned, structured, and organized in a way that is amenable to human experts. Computer programs can be modularized, and major tasks can be divided into subtasks and assigned to individual computer programmers. Large-scale software projects have too many degrees of freedom to expect GP, a process that is based on a random search, to succeed. Even if GP did succeed, the resulting program might be inefficient and hard to maintain. In summary, simple computer programming tasks are too easy for GP, because humans can complete such tasks without much effort; but difficult computer programming tasks are too hard for GP, and therefore require human ingenuity. This raises important questions about GP. What types of problems are suitable for GP that are truly difficult for humans? Does GP have any practical applications?

To discuss these questions, we look at the application areas of EAs. EAs are good at finding solutions to difficult, multidimensional, multimodal optimization problems. Computer programming can be difficult, multidimensional, and multimodal. However, computer programming is a task at which many humans excel. Parameter optimization is a task at which humans do *not* excel. Almost all nontrivial parameter optimization problems are solved by computer programs. We see that EAs have become widespread because they excel at problems which are difficult for humans.

Since many humans are skilled computer programmers, GP is not likely to be widely applied to real-world computer programming problems. However, GP could become widely applied to problems similar to computer programming at which humans do *not* excel. There are many engineering (and other) problems whose candidate solutions can be represented as tree-like structures, and at which humans do not excel. These problems include the design of lens systems [Koza et al., 2008], photonic crystal structures [Preble et al., 2005], algorithms for protein classification [Koza, 1997], cellular automata [Andre et al., 1996], algorithms to find numerical solutions to difficult equations [Balasubramaniam and Kumar, 2009], algorithms to solve puzzles and find game strategies [Hauptman and Sipper, 2007], [Hauptman et al., 2009], electric circuits [McConaghy et al., 2008], field programmable gate arrays [Koza et al., 1999], and antennas [Lohn et al., 2004].

The key feature of these problems is that they cannot be easily reduced to parameter optimization problems, and so they cannot be solved by GAs, ES, EP, or similar methods. They can be solved by a GP that evolves an algorithm, or that evolves a design program. The results from GP often improve on existing patents while using very little problem-specific information [Koza, 2010]. The term "computer intelligence" is used so much nowadays that it does not mean much; it has become a nearly content-less buzzword. But when a computer program creates a patentable invention, many people will probably agree that the program has a significant degree of intelligence.

7.6 MATHEMATICAL ANALYSIS OF GENETIC PROGRAMMING

We can mathematically analyze GPs, just as we can other EAs (see Chapter 4). This section extends GA schema theory (Section 4.1) to GP [Langdon and Poli, 2002, Chapter 4]. Our main approach in this section is to use simple examples to obtain a general idea for how GP schema theory works, and then use those examples to present a general GP schema formula.

7.6.1 Definitions and Notation

For our first example, suppose that our terminal set is $\{x, y\}$. We use # to indicate a "don't care" terminal. Consider the schema

$$H = (\ (+\ (-\ \#\ y\ )\ \#\ )\ . \tag{7.13}$$

If x and y are the only two available terminals, then this schema has four instances:

$$\begin{aligned} &(\ +\ (\ -\ x\ y\ )\ x\ ),\quad (\ +\ (\ -\ x\ y\ )\ y\ ), \\ &(\ +\ (\ -\ y\ y\ )\ x\ ),\quad (\ +\ (\ -\ y\ y\ )\ y\ ). \end{aligned} \tag{7.14}$$

We say that a schema matches an s-expression if the s-expression is an instance of the schema. For example, the schema $(+\ \#\ y)$ matches $(+\ x\ y)$, and it also matches $(+\ 2\ y)$, but it does not match $(-\ x\ y)$.

Now we define three important terms that are related to GP schemata.

1. The order of H, $o(H)$, is the number of defined symbols in H, including both functions and terminals. In Equation (7.13), $o(H) = 3$.

2. The length of H, $n(H)$, is the total number of symbols in H, including both defined functions and terminals, and "don't care" functions and terminals. In Equation (7.13), $n(H) = 5$.

3. The defining length of H, $L(H)$, is the minimum number of links in the syntax subtree that includes all of the defined symbols. Defining length is difficult to determine directly from an s-expression, but it can be easily seen by looking at the corresponding syntax tree.

Figure 7.14 shows some examples of syntax trees, their orders, their lengths, and their defining lengths.

Now consider how many schemata match an s-expression of length n. As an example, consider the s-expression

$$(\ (+\ (\ -\ 2\ x\ )\ (\ *\ 3\ y\ )\ )\ . \tag{7.15}$$

A schema can match this s-expression with either a + function or a # symbol in the top node. A similar statement can be made for all of the other nodes in the s-expression. Therefore, a schema matches the s-expression if the schema has either the given s-expression symbol, or a # symbol, at each node. We see that there are 2^n schemata that match an s-expression of length n. For example, there are $2^7 = 128$ schemata that match the s-expression of Equation (7.15).

Now we define the structure of a schema. We use G to denote the structure of a schema, and we obtain G by replacing all symbols in H with # symbols. For example, the schema of Equation (7.13) has the structure

$$G = (\ (\ \#\ (\ \#\ \#\ \#\ )\ \#\ )\ . \tag{7.16}$$

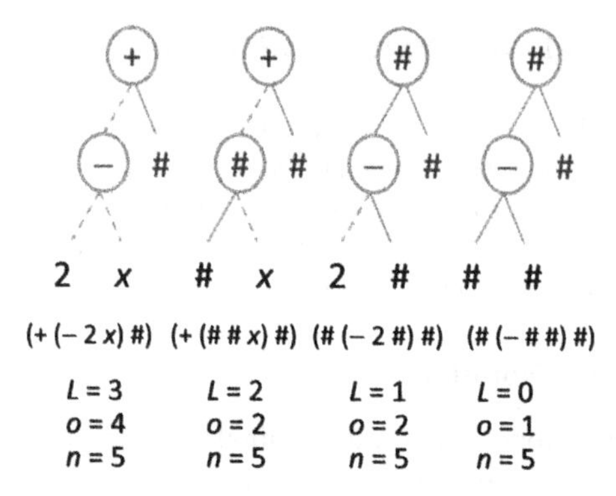

Figure 7.14 Four GP schemata in syntax tree form and in s-expression form. The defining length L, order o, and length n, is shown below each schema. The links that are used to determine the defining length are those that are in the smallest subtree that includes all of the non-# symbols; those links are shown as dashed lines.

7.6.2 Selection and Crossover

Consider a GP that uses roulette-wheel selection and crossover. We use $m(H,t)$ to denote the number of instances of schema H in a GP population at the t-th generation. We use $m(H,t+1/2)$ to denote the number of schema instances in the population after selection. Then $m(H,t+1)$ is the number of schema instances after selection, crossover, and mutation. If we use roulette-wheel selection, then on average,

$$m(H,t+1/2) = m(H,t)f(H,t)/f_{\text{ave}}(t) \tag{7.17}$$

where $f(H,t)$ is the average fitness of all of the instances of H at the t-th generation, and $f_{\text{ave}}(t)$ is the average fitness of all of the individuals at the t-th generation.

Now consider the effect of crossover on the population. Crossover might destroy an instance of H; that is, if a parent is an instance of H, it might cross over to produce children of which none are instances of H. This will result in one less instance of H at the next generation. Crossover can destroy an instance of H in two different ways. Consider a parent p_1 such that $p_1 \in H$ and $p_1 \in G$, where G is the structure of H. First, an instance of H might be destroyed by crossing p_1 with an individual $p_2 \notin G$; we call this event D_1. Second, an instance of H might be destroyed by crossing p_1 with an individual p_2 such that $p_2 \in G$ but $p_2 \notin H$; we call this event D_2. Since D_1 and D_2 are mutually exclusive events, the probability that crossover destroys an instance of H is given by

$$\Pr(D) = \Pr(D_1) + \Pr(D_2). \tag{7.18}$$

■ **EXAMPLE 7.1**

Figure 7.15 shows an example of event D_1. Parents $(+\ (-\ 2\ x)\ (-\ 3\ y))$ and $(+\ x\ y)$ are selected to cross over with each other. The crossover points are randomly chosen as the leftmost subtraction function in Parent 1, and the y terminal in Parent 2. The crossover results in children $(+\ y\ (-\ 3\ y))$ and

$(+\ x\ (-\ 2\ x))$. We see that neither child has the same structure as either of the parents. Crossover has destroyed all schemata H_1 of which Parent 1 is an instance, and all schemata H_2 of which Parent 2 is an instance.

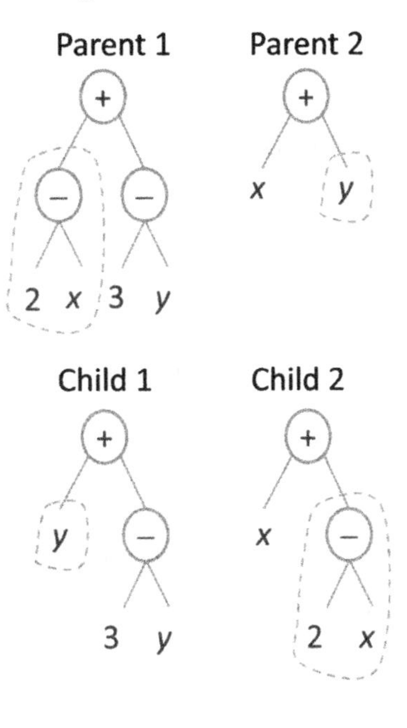

Figure 7.15 Crossover between these parents results in children that do not have the structure of either parent. This is an example of event D_1, in which crossover between individuals with different structures results in the destruction of schemata.

□

■ EXAMPLE 7.2

We consider the possibility of event D_2 by looking at an example. Consider the schema $H = (\ \#\ x\ y)$. Suppose that two parents are given as

$$\begin{aligned} p_1 &= (+\ x\ y) \in H \\ p_2 &= (-\ y\ x) \notin H. \end{aligned} \tag{7.19}$$

Both parents have the same structure G, but p_1 belongs to schema H, while p_2 does not. If the crossover points are chosen at the top link in both p_1 and p_2, then the children will be

$$\begin{aligned} c_1 &= (+\ y\ x) \notin H \\ c_2 &= (-\ x\ y) \in H. \end{aligned} \tag{7.20}$$

We see that the instance of H is preserved for the next generation.

Now consider the schema $H = (+ x \#)$ and the same two parents as shown in Equation (7.19). As before, $p_1 \in H$ and $p_2 \notin H$. However, in this case, neither c_1 nor c_2 from Equation (7.20) belong to H, so the instance of H is destroyed.

□

Now we consider the effect of D_1 on the number of schema instances in a population. Recall that D_1 is the destruction of an instance of schema H, which has structure G, due to crossover between a parent p_1 such that $p_1 \in H$ and $p_1 \in G$, with a parent $p_2 \notin G$. That is,

$$D_1 = D \cap (p_2 \notin G) \tag{7.21}$$

where D is the event that an instance of H is destroyed due to crossover. Bayes' theorem tells us that

$$\Pr(D_1) = \Pr(D \,|\, p_2 \notin G)\Pr(p_2 \notin G). \tag{7.22}$$

But the probability that $p_2 \notin G$ is proportional to the number of individuals in the population that do not belong to G after selection; that is,

$$\Pr(p_2 \notin G) = (N - m(G, t + 1/2))/N \tag{7.23}$$

where N is the population size. Combining this with Equation (7.22) gives

$$\Pr(D_1) = \Pr(D \,|\, p_2 \notin G)\frac{N - m(G, t + 1/2)}{N}. \tag{7.24}$$

Now consider the probability of event D_2. Recall that D_2 is the destruction of an instance of schema H, which has structure G, due to crossover between a parent p_1 such that $p_1 \in H$ and $p_1 \in G$, with a parent $p_2 \in G$. That is,

$$\begin{aligned} D_2 &= D \cap (p_2 \in G) \\ &= D \cap (p_2 \in G) \cap (p_2 \notin H) \end{aligned} \tag{7.25}$$

where the second equality comes from the fact that schema destruction cannot occur unless $p_2 \notin H$. Bayes' theorem tells us that

$$\Pr(D_2) = \Pr(D \,|\, p_2 \in G, p_2 \notin H)\Pr(p_2 \in G, p_2 \notin H). \tag{7.26}$$

The second term on the right side of Equation (7.26) is the probability that $p_2 \in G$ and $p_2 \notin H$. However, the event $p_2 \in H$ is a subset of $p_2 \in G$; therefore,

$$\Pr(p_2 \in G, p_2 \notin H) = \Pr(p_2 \in G) - \Pr(p_2 \in H) \tag{7.27}$$

as shown in Figure 7.16.

The probabilities on the right side of Equation (7.27) are respectively proportional to the number of G and H instances after selection; therefore,

$$\Pr(p_2 \in G, p_2 \notin H) = \frac{m(G, t + 1/2) - m(H, t + 1/2)}{N}. \tag{7.28}$$

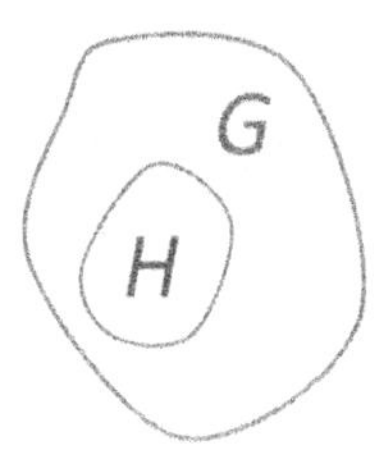

Figure 7.16 The set of programs belonging to schema H is a subset of the set of programs belonging to schema structure G. Therefore, $\Pr(p \in G, p \notin H) = \Pr(p \in G) - \Pr(p \in H)$.

Now consider the first term on the right side of Equation (7.26), which is the probability of D given that $p_2 \in G$ and $p_2 \notin H$. A schema instance can be destroyed only if crossover is at one of the defining links of the schema. For example, in Figure 7.14, the schema on the left will be destroyed only if crossover occurs at one of the three dashed links; this is true for all of the schemata depicted in the figure. Even if crossover does occur at one of those links, the schema might not be destroyed, depending on the contents of the subtree that is spliced into the schema instance. Since there are $L(H)$ links at which crossover must occur for a schema instance to be destroyed, and there are a total of $n(H) - 1$ links, there is a probability of $L(H)/(n(H) - 1)$ that crossover occurs at one of the defining links. Since crossover at one of these links is a necessary but not sufficient condition for the destruction of the schema instance, the probability of D given that $p_2 \in G$ is bounded from above by this probability; that is,

$$\Pr(D \mid p_2 \in G, p_2 \notin H) \leq \frac{L(H)}{n(H) - 1}. \tag{7.29}$$

The right side of the above equation is called the fragility of the node composition of schema H [Langdon and Poli, 2002, Section 4.4]. It gives an upper bound to the probability that an instance of schema H will be destroyed if the other parent has the same structure as H.

Combining Equations (7.18), (7.24), (7.28), and (7.29) gives an upper bound for the probability of the destruction of schema H:

$$\Pr(D) \leq \Pr(D|p_2 \notin G)\frac{N - m(G, t + 1/2)}{N} + \frac{L(H)}{n(H) - 1}\frac{m(G, t + 1/2) - m(H, t + 1/2)}{N}. \tag{7.30}$$

This gives us a general expression for the probability of the destruction of an instance of schema H due to crossover.

Considering the fact that crossover occurs with probability p_c, we assume that the GP population is large enough to use the law of large numbers [Grinstead and Snell, 1997] for $m(H, t + 1)$, which is the number of instances of H at generation $t + 1$. This quantity is given by the sum of two terms: (1) the number of instances of H after selection, multiplied by the probability that crossover does not occur; (2) the number of instances of H after selection, multiplied by the probability of crossover, multiplied by the probability that crossover does not destroy an instance of H. This gives

$$
\begin{aligned}
m(H,t+1) &= m(H,t+1/2)(1-p_c) + m(H,t+1/2)p_c(1-\Pr(D)) \\
&= m(H,t+1/2)(1-p_c\Pr(D)) \qquad (7.31)
\end{aligned}
$$

where $m(H, t + 1/2)$ is shown in Equation (7.17). Equation (7.31) gives us the approximate number of schema instances after crossover.

7.6.3 Mutation and Final Results

Now we consider the probability of schema destruction due to mutation. Suppose that the probability of mutation at each node is p_m. Then the probability that mutation does not occur at each node is $1-p_m$. The probability that mutation does not occur at any of the defined nodes is $(1 - p_m)^{o(H)}$, where $o(H)$ is the schema order (that is, the number of defined nodes). So the probability that mutation occurs at a defined node is

$$
\begin{aligned}
\Pr(D_m) &= 1-(1-p_m)^{o(H)} \\
&\approx p_m o(H) \qquad (7.32)
\end{aligned}
$$

where the approximation is based on a Taylor series expansion of $\Pr(D_m)$ around $p_m = 0$. Note that mutation at a defined node is a necessary but not sufficient condition for schema destruction. We therefore combine Equations (7.31) and (7.32) to obtain

$$
m(H,t+1) \geq m(H,t+1/2)(1-p_c\Pr(D))(1-p_m o(H)). \qquad (7.33)
$$

We combine this equation with Equations (7.17) and (7.30) to obtain

$$
\begin{aligned}
m(H,t+1) \geq\; & \frac{m(H,t)f(H,t)}{f_{\text{ave}}(t)}[1-p_m o(H)] \times \\
& \left\{1 - p_c\left[\Pr(D|p_2 \notin G)\left(1 - \frac{m(G,t)f(G,t)}{N f_{\text{ave}}(t)}\right) + \right.\right. \\
& \left.\left. \frac{L(H)}{n(H)-1}\frac{m(G,t)f(G,t)-m(H,t)f(H,t)}{N f_{\text{ave}}(t)}\right]\right\}. \qquad (7.34)
\end{aligned}
$$

This gives us a lower bound for the number of instances of schema H at generation $t+1$, where we have taken both crossover and mutation into account. As expected, this is slightly more complicated than the GA schema theory that we derived in Section 4.1. GP is more complicated than GAs because of the variable size and shape of GP individuals. However, we can make some simplifications in Equation (7.34). For example, early in the GP run we have a lot of diversity in the population, so it is unlikely that two individuals with different shapes will cross over at a point that results in the preservation of a schema. Therefore, $\Pr(D|g \notin G) \approx 1$ early in the GP run. Also, with high diversity, the total fitness of individuals belonging to a structure G will be small simply because there will be a small number of such programs; that is, $m(G,t)f(G,t)/(N f_{\text{ave}}) \ll 1$. Therefore, Equation (7.34) can be approximated early in the GP run as

$$
\begin{aligned}
m(H,t+1) \geq\; & \frac{m(H,t)f(H,t)}{f_{\text{ave}}(t)}[1-p_m o(H)] \times \\
& \left\{1 - p_c\left[1 + \frac{L(H)}{n(H)-1}\left(\frac{-m(H,t)f(H,t)}{N f_{\text{ave}}(t)}\right)\right]\right\}. \qquad (7.35)
\end{aligned}
$$

This gives an approximation to Equation (7.34) early in a GP run. We can make a further approximation for short schema, in which case $L(H)/(n(H)-1) \ll 1$:

$$m(H, t+1) \geq \frac{m(H,t)f(H,t)}{f_{\text{ave}}(t)} \left[1 - p_m o(H)\right](1 - p_c). \tag{7.36}$$

This GP schema approximation, which is valid for short schema early in a GP run, is similar to the GA schema expression in Equation (4.6). The schema theory that we derived in this section gives a lower bound for $m(H, t+1)$ rather than an equality. This is because we used approximations in our development. Therefore, this schema theory is called pessimistic.

Schema theory can be approached in many different ways. Other ways of modeling GP selection, crossover, and mutation give GP schema theories that are different than the one derived in this section. Some of those theories are exact instead of pessimistic, in which case we obtain an equation instead of inequality like Equation (7.34) [Altenberg, 1994], [Langdon and Poli, 2002, Chapters 3, 5]. Una-May O'Reilly developed a lower-bound schema theory based on John Koza's work, which defined schemata differently than in this chapter [Koza, 1992], [O'Reilly and Oppacher, 1995]. Justinian Rosca developed a GP schema theory using yet another schema definition [Rosca, 1997]. Schema theory has also been developed for GP systems that use program representations other than syntax trees [Whigham, 1995].

There are also other methods besides schema theory for mathematically analyzing GP. For example, Markov models for GP were introduced in 2001 [Poli et al., 2001], [Poli et al., 2004]. Also, Price's selection and covariance theorem can be used to mathematically model GP [Langdon and Poli, 2002, Chapter 3].

7.7 CONCLUSION

This chapter has been limited to GP using Lisp and syntax trees. GP has also been implemented with many other structures and in many other languages. For example, programs can be represented as a linear sequence of instructions, which is the programming format that most of us are used to in our everyday experience. This is called linear GP [Poli et al., 2008, Chapter 7], and is particularly suited for assembly code programs. It is difficult to evolve an assembly code program for an embedded system using tree syntax, because we would need to first build a tree-to-assembly compiler. However, if we evolve programs directly using assembly code, cost evaluation is much more straightforward.

Cartesian GP is a way of representing programs with a set of arrays. Each array includes an element that specifies the operation of that array, and elements that specify the arrays from which to obtain inputs [Miller and Smith, 2006]. Graph GP is a way of representing programs with nodes and edges [Poli et al., 2008]. Various other structures have also been used to represent computer programs in GP [Banzhaf et al., 1998, Chapter 9].

Genetic programming is becoming more widely used as its domain expands beyond computer programming to the more general evolution of engineering algorithms and designs. John Koza lists 76 results from GP that are competitive with human-generated results [Koza, 2010]. He also states that the production rate of GP-generated results that are competitive with human results is proportional to computing power. This portends a future dramatic increase in engineering designs that are generated by computer programs, and in meaningful collaborations between humans and computers.

There are some problems for which GP may not be suitable. The search space of a GP is the set of all computer programs, within user-defined syntax limitations. This broad search domain is both a strength and weakness of GP. The broad search domain allows GP to search for optima more thoroughly than other EAs, but also indicates that GP typically does not use much available domain information from the human programmer. In general, if we know the structure of the solution ahead of time, then a more standard EA should outperform GP, because we can more easily incorporate problem-specific information into a parameter optimization problem than a program optimization problem. However, if discovering the structure of the solution is a major challenge of the optimization problem, then a GP might be a suitable approach. Also, note that we can seed an initial GP population with known good candidate solutions. Then the GP will improve on these solutions as it progresses. Therefore, problems for which incremental improvements in existing solutions are highly desirable are also especially suitable for GP [Koza, 2010].

An interesting area for future work is the generation of computer programs that evolve computer programs that evolve computer programs. This might be called a meta-GP. A GP can evolve computer programs, but how do we find the best GP? Perhaps a meta-GP could evolve a GP. The meta-GP would presumably at least square the computing power required for a single GP. Jürgen Schmidhuber first proposed meta-GPs in his 1987 dissertation [Schmidhuber, 1987]. Meta-GPs are a type of meta-learning (that is, learning how to learn) [Anderson and Oates, 2007]. A meta-GP could be considered to be a search for a search and thus falls into the category of the vertical no-free-lunch theorem [Dembski and Marks, 2010].

Note that GP can be combined with other EAs. For example, we can combine GP and EDAs to find probabilistic descriptions of effective programs, which can in turn guide our search for better programs. This was first proposed with the name probabilistic incremental program evolution [Salustowicz and Schmidhuber, 1997]. In this algorithm, each node in a syntax tree has a probability of being equal to a specific function or terminal, and those probabilities depend on the fitness values of individual programs. New EAs are proposed in the literature quite often and it could be interesting to explore which of these new EAs are particularly suited for implementation as a GP (see Chapter 17).

Finally, we note that the serious student of GP should master the art of automatically defined functions (ADFs). ADFs are subroutines that evolve automatically and dynamically in a GP. Considering the fact that a human programmer naturally uses subroutines, it makes sense that a GP should also create and use subroutines. ADFs can significantly reduce the computational effort of a GP when applied to complex problems. ADFs are discussed in detail in [Koza, 1994, Chapter 4], [Koza et al., 1999], and in other GP books.

For further study, the reader can find several excellent books that are dedicated to GP. The book by Wolfgang Banzhaf et al. is a very readable introduction [Banzhaf et al., 1998]. John Koza's compendious volumes are standard references in the field that richly deserve their stellar reputation, especially his first book [Koza, 1992], [Koza, 1994], [Koza et al., 1999], [Koza et al., 2005]. *A Field Guide to Genetic Programming* is a freely available book that provides a nice overview of the topic of GP [Poli et al., 2008]. Detailed schema analyses can be studied in William Langdon and Riccardo Poli's book [Langdon and Poli, 2002].

PROBLEMS

Written Exercises

7.1 Write an s-expression and syntax tree for the positive solution to the quadratic equation: $(\sqrt{b^2 - 4ac} - b)/(2a)$. What is the depth of your syntax tree? Is your syntax tree full?

7.2 Write an s-expression that returns 8 if $x > 2$, and 9 otherwise.

7.3 Suppose you evaluate a GP candidate solution f on n different inputs $\{u_i\}$. Write a fitness function that gives twice as much weight to the average performance of f as it does to the worst-case performance of f.

7.4 Define a protected square root function in Lisp that returns 0 in case the input is negative.

7.5 Suppose we use use the grow method to generate a random s-expression with maximum depth D_c. Suppose there is a 50% chance at each node of selecting a terminal or a function. What is the probability that a given branch will reach its maximum possible depth?

7.6 Suppose we use the grow method to generate a random s-expression with maximum depth D_c. Suppose there is a 50% chance at each node of selecting a terminal or a function. Assume that each function takes two arguments. What is the probability that the s-expression will represent a full syntax tree with a depth of D_c?

7.7 List all of the programs that could be created by hoist mutation of the syntax tree of Figure 7.1.

7.8 List the unique programs that could be created by permutation mutation of the syntax tree of Figure 7.2.

7.9 Write the following equation in simplified form (for instance, (+ 1 1) can be replaced with 2). After obtaining the simplified version, write the equation in a more conventional form.

(defun Pgm (x) (+ (DIV (− (+ (+ 1 1) (+ (DIV x 1) (+ 1 1))) (− (* x 1) (+ 1 1))) (abs x)) (DIV (+ (− x 1) (abs x)) (− (− (* x 1) (+ 1 1)) (− (+ 1 1) (* 1 1)))))) .

7.10 What is the defining length, order, and length of the schema (if (# x #) 8 #)? What is the structure of the schema?

7.11 How does the lower bound for the number of schema instances vary with mutation probability? How does it vary with schema order? How does it vary with crossover probability? Explain your answers.

Computer Exercises

7.12 Lisp exercise:

a) Download and install the latest version of CLISP (Common Lisp).

b) Download and install an integrated development environment (IDE) for CLISP. Note: The minimum-time control problem in Section 7.3 was implemented with a program called LispIDE.

c) Run the Lisp IDE and type the following line at the command prompt:

(print (* 5 (+ 3 2)))

This will cause Lisp to print 25 to your terminal twice: once because of the print command, and once because of the value that is returned from the print function.

7.13 Minimum-time control exercise:

a) Download `GPCartControl.lisp` and associated files from the book web site and run it on your computer. This duplicates the minimum-time control GP results of Section 7.3. If you are using LispIDE, you can do this as follows.

- Run LispIDE.
- Open `GPCartControl.lisp` from LispIDE.
- Modify line 15 of `GPCartControl.lisp` so that the path points to the directory on your computer that contains the Lisp files.
- Select the entire `GPCartControl.lisp` (use your computer mouse, or type Ctrl-A).
- Select the "Edit → Send to Lisp" menu item. This defines the `GPCartControl` function in Lisp.
- Type (`GPCartControl`) at the LispIDE command prompt. This runs the program.

b) After `GPCartControl.lisp` is done, it will output two files. One file is [`DateTimeString`]`.txt`, which contains the generation number, best cost, and average cost. The other file is [`DateTimeString`]`.lisp`, which contains the best program found by the GP. (Note that [`DateTimeString`] is a text string representing the date and time that the file was created.)

c) Run the Lisp command (`setf LispPgm` [`BestProgram`]), where [`BestProgram`] is the text string that defines the best program found by the GP. You should get this text string from [`DateTimeString`]`.lisp`. For example,

```
(setf LispPgm
"(defun CartControl (x v) (if (> (* -1 x) (* v (abs v))) 1 -1))").
```

d) Run the Lisp function (`PhasePlane LispPgm`). This will create two files. One file is `PhasePlane.txt`, which is a list of (x, v, control) values generated by `LispPgm`. The other file is `PhasePlane1.txt`, which is a list of (x, v) values at which the control switches between -1 and $+1$. This assumes that the control generated by `LispPgm` is always saturated. If this assumption is not true, then `PhasePlane1.txt` will not be useful.

e) Run the Matlab program `PlotPhasePlane.m` with input string "PhasePlane." This will generate a phase plane plot of the control as a function of x and v, using the `PhasePlane.txt` and `PhasePlane1.txt` files generated above. However, the plot will not be useful unless the assumption stated above is satisfied – that is, unless the control generated by `LispPgm` is always saturated.

f) The fitness of a cart control Lisp program can be evaluated by running `EvalCartControl.lisp`. If you define `CartControl` as described in subproblem (c) above, then you can open `EvalCartControl.lisp`, evaluate it in the Lisp IDE so that `EvalCartControl` is a defined function, and then run it by typing the following command: (`EvalCartControl #'CartControl`).

7.14 Modify some of the parameters in `GPCartControl.lisp` to see what effect they have on performance. Some of the parameters you could modify include the following.

- `Dinitial`, the maximum initial tree depth
- `Dcreated`, the maximum tree depth
- `Pcross`, the probability of crossover
- `Preproduce`, the probability of reproduction
- `Pinternal`, the probability of crossover at internal (function) nodes
- `NumEvals`, the number of function evaluations per individual
- `NumElites`, the number of elites each generation
- `GenLimit`, the generation limit
- `PopSize`, the population size
- `SelectionMethod`, the selection method

7.15 Modify `GPCartControl.lisp` and its associated files so the GP can find a mapping $\hat{y}(x)$ that closely matches the target $y(x)$, where $y(x)$ is given as follows:

$$\begin{array}{lllll} y(0)=3, & y(1)=5, & y(2)=1, & y(3)=2, & y(4)=9 \\ y(5)=8, & y(6)=3, & y(7)=4, & y(8)=1, & y(9)=6. \end{array}$$

Hand in a GP convergence plot showing the progress of the minimum and average cost of the population as functions of generation number, the best program found by the GP, and a plot showing the target values $y(x)$ compared to the GP-approximated values $\hat{y}(x)$.

CHAPTER 8

Evolutionary Algorithm Variations

Many options exist.

—David Fogel [Fogel, 2000]

The previous chapters discussed four popular and foundational approaches to evolutionary computing. However, the previous chapters only presented the basic ideas and algorithms. There are many variations that we can implement in those algorithms. These variations have application not only to the EAs discussed in prior chapters, but also to those that will be discussed later in this book. Therefore, this chapter has broad applicability to a wide variety of EAs. Some of the variations that we discuss in this chapter can make a big difference in EA performance. When we compare two EAs, algorithms A and B, it is often *not* the essential difference between A and B that makes the difference between their performance levels, but it is often these supposedly minor variations and implementation details that make the difference.

Overview of the Chapter

Section 8.1 discusses different ways of initializing an EA population. Section 8.2 discusses various ways of deciding when to terminate an EA. Section 8.3 discusses how to represent candidate EA solutions, and how the chosen representation can make a significant difference in results. Section 8.4 discusses elitism, which was originally proposed for genetic algorithms but is now typically implemented in all EAs because of its inherent advantages. Section 8.5 discusses the difference between generational EAs and steady-state EAs.

EA populations often tend to converge to a single highly-fit individual; that is, the entire population becomes clones of a single candidate solution. This severely reduces the ability of the EA to search for an optimal solution, and so Section 8.6 discusses how to maintain diversity in an EA population.

Until now, we have focused on roulette-wheel selection for choosing parents, but there are also other approaches to selection, and we discuss some of them in Section 8.7. One important selection option is the stud option (Section 8.7.7), which was initially proposed for genetic algorithms but can also be used with other EAs. Section 8.8 discusses different ways of combining parents to obtain children, and Section 8.9 discusses different ways to implement mutation.

8.1 INITIALIZATION

We typically initialize an EA with a random population. This is the easiest and most popular initialization method. However, initialization can make a significant difference in the success of an EA. A little extra effort spent on initialization can pay big dividends.

Suppose we want to run an EA with N individuals. One initialization approach is to generate more than N individuals, and simply keep the best N as our initial population. For example, [Bhattacharya, 2008] generates $5N$ random individuals and keeps the best N as the initial population.

We could also randomly generate individuals and then locally optimize each individual to obtain our initial EA population. For example, we could generate N random individuals, perform gradient descent optimization on a subset of those individuals, and then use the resulting individuals as our initial EA population. We could implement this in several different ways; for example, we could use only the best individuals in our initial population, or we could perform gradient descent on only the best individuals but use all of the individuals in our initial population.

Another option is to use expert solutions to initialize the EA population. For example, suppose we want to use an EA to tune a control algorithm. We could use expert knowledge to estimate reasonable control solutions, and seed an initial EA population with those solutions. Or we could use candidate solutions that we find from any other source (other algorithms, other published results, and so on) to seed our initial EA population. The use of problem-dependent information to seed the initial EA population is often called directed initialization.

■ **EXAMPLE 8.1**

This example looks at the effect of additional initial individuals on an EP for the optimization of the 10-dimensional Rosenbrock function (see Appendix C.1.4). We use a population size of 10 and run the EP for 50 generations. In our standard EP implementation, we initialize the 10 initial individuals randomly. In our additional-initial-individuals implementation, we randomly generate 20 individuals and use the best 10 for our initial EP population. Figure 8.1

shows the results of the two algorithms, averaged over 20 Monte Carlo simulations. Initializing extra individuals requires twice as much computational effort during the initialization phase, but the extra effort results in much better performance. The figure shows that if we initialize additional individuals, we obtain significantly better results during the first 40 generations, although the two algorithms perform about the same by the time they reach the 50th generation.

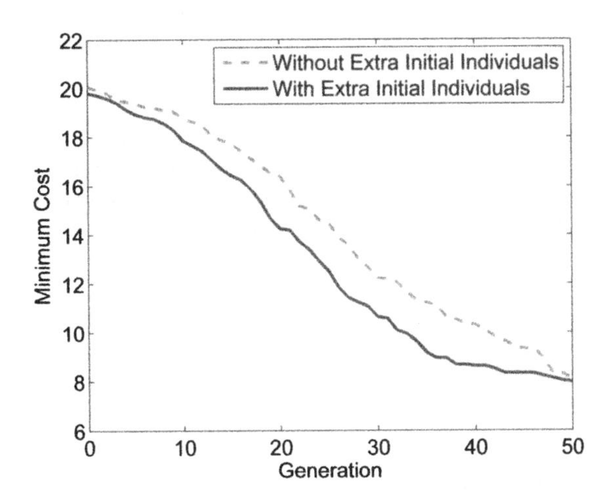

Figure 8.1 Example 8.1: Cost versus generation number for an EP that minimizes the 10-dimensional Rosenbrock function, averaged over 20 Monte Carlo simulations. The extra work of generating 10 more initial individuals appears to be worth the additional effort, at least during the first 40 generations.

□

Using directed initialization or additional initial individuals is an approach that makes sense for problems with few generations and few individuals; that is, problems that are restricted to few fitness function evaluations because of their high expense (see Section 21.1). With this type of problem, the extra effort expended during initialization can have benefits that last for many generations.

8.2 CONVERGENCE CRITERIA

One thing we have to decide when we implement an EA is when to stop. We have glossed over this issue in the algorithms of the previous chapters by using the generic phrase, "While not(termination criterion)" (for example, see Figures 3.6, 5.1, 6.1, and 7.5). What termination criterion should we use? How long should an EA run before we stop the program? There are several criteria that we can use to define convergence.

1. We can stop the EA after a preset number of generations. This has the advantage of simplicity, and run-time predictability, and is probably the most

commonly used termination criterion for EAs. However, if we are comparing different EAs, then we should stop the EAs after a preset number of objective function evaluations rather than a preset number of generations. This is because different EAs use a different number of function evaluations per generation, so we can obtain a fair comparison between EAs only by terminating after a common function evaluation limit.

2. We can stop the EA after the solution is "good enough." This termination criterion is problem-dependent because "good enough" varies from one problem to the next. This is an appealing termination criterion; if we find a solution that provides satisfactory performance, then why should we continue to search for a better solution? However, most solutions to real-world problems are never quite good enough. We always want to try to do better. On the other hand, doubling run time while improving performance by a miniscule amount is a waste of resources in many cases. The tradeoff between run time and performance is a problem-dependent issue that requires engineering judgment.

3. We can stop the EA after the best individual fitness does not change appreciably for a certain number of generations. This indicates that the EA may be stuck in a local minimum. The local minimum may also be the global minimum, but we will never know unless we can somehow find a better local minimum.

4. We can stop the EA after the average of the population's fitness values does not change appreciably for a certain number of generations. This is similar to the above criterion; it indicates that the population as a whole has stopped improving.

5. We can stop the EA after the standard deviation of the population's fitnesses stops decreasing, or drops below some threshold. This indicates that the population has reached a certain level of uniformity.

6. We can use some combination of the above criteria.

Items 3–5 imply (but do not guarantee) that the population is no longer improving, so we might as well stop the EA. However, there is a danger in this approach. Even when the EA has apparently converged, a statistically improbable mutation or recombination event could result in a significant improvement in the solution, as illustrated in Figure 8.2. We cannot keep running forever; we have to stop eventually. But regardless of when we stop, we take the chance that we are quitting just before a particularly fortunate mutation or recombination.

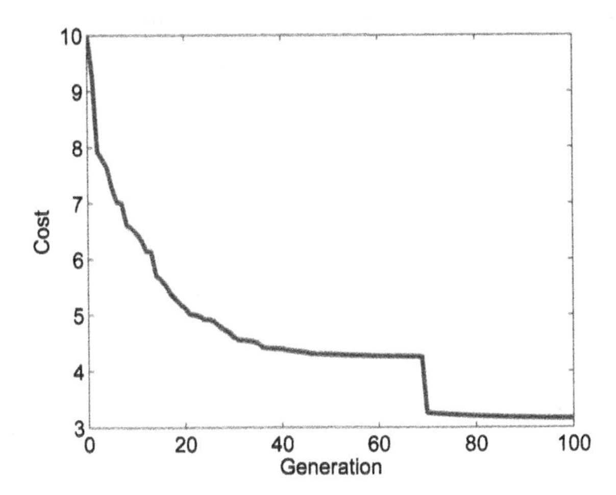

Figure 8.2 A plot of cost (best or mean) versus generation number for a hypothetical EA simulation. If we terminate the EA after the cost stops improving, we will miss a significant improvement at the 70th generation, which might be caused by a statistically improbable but serendipitous mutation. However, if a cost value of 5 satisfies the customer, then we might not care about the improvement at the 70th generation.

8.3 PROBLEM REPRESENTATION USING GRAY CODING

This section discusses how to implement binary EAs using gray coding. A gray code, also called a reflected binary code, is a way to represent numbers in such a way that the codes for neighboring numbers differ by only one bit [Doran, 2007]. Consider the representation of the numbers 0–7 in binary code:

$$\begin{aligned} 000 &= 0, & 001 &= 1, \\ 010 &= 2, & 011 &= 3, \\ 100 &= 4, & 101 &= 5, \\ 110 &= 6, & 111 &= 7. \end{aligned} \tag{8.1}$$

We see that the binary codes for neighboring numbers can differ by more than one bit. For example, the code for 3 is 011 while the code for 4 is 100. The binary codes for 3 and 4 differ in all three bit locations. The converse is also true; that is, binary codes that are very similar to each other sometimes represent numbers that are very far apart. For example, 000 represents the number 0; if we change one bit to obtain the code 100, we now have the representation for the number 4, which is far from 0 relative to the range of numbers that we are representing.

Now consider the representation of the numbers 0–7 in gray code:

$$\begin{aligned} 000 &= 0, & 001 &= 1, \\ 011 &= 2, & 010 &= 3, \\ 110 &= 4, & 111 &= 5, \\ 101 &= 6, & 100 &= 7. \end{aligned} \tag{8.2}$$

We see that the gray codes for neighboring numbers always differ by exactly one bit. For example, the code for 3 is 010 while the code for 4 is 110. The gray codes for 3 and 4 differ in only the left-most bit. Gray coding removes Hamming cliffs, which are large changes in integer values between representations that differ by only one bit [Deep and Thakur, 2007].

■ EXAMPLE 8.2

Consider the function $y = f(x)$ of Example 2.2. If we code the values $x \in [-4, +1]$ of 16 evenly-space values from the horizontal axis of Figure 2.1 with four-bit binary coding, we obtain

$$\begin{aligned} \text{Binary coding: } 0000 &= -4.00, & 0001 &= -3.67, \\ 0010 &= -3.33, & 0011 &= -3.00, \\ 0100 &= -2.67, & 0101 &= -2.33, \\ 0110 &= -2.00, & 0111 &= -1.67, \\ 1000 &= -1.33, & 1001 &= -1.00, \\ 1010 &= -0.67, & 1011 &= -0.33, \\ 1100 &= +0.00, & 1101 &= +0.33, \\ 1110 &= +0.67, & 1111 &= +1.00. \end{aligned} \tag{8.3}$$

On the other hand, we can also code the values with four-bit gray coding to obtain

$$\begin{aligned} \text{Gray coding: } 0000 &= -4.00, & 0001 &= -3.67, \\ 0011 &= -3.33, & 0010 &= -3.00, \\ 0110 &= -2.67, & 0111 &= -2.33, \\ 0101 &= -2.00, & 0100 &= -1.67, \\ 1100 &= -1.33, & 1101 &= -1.00, \\ 1111 &= -0.67, & 1110 &= -0.33, \\ 1010 &= +0.00, & 1011 &= +0.33, \\ 1001 &= +0.67, & 1000 &= +1.00. \end{aligned} \tag{8.4}$$

If we plot y versus x in such a way that neighboring values of x differ by one bit in their binary codes, we obtain the top plot of Figure 8.3. If we plot y versus x such that neighboring values of x differ by one bit in their *gray* codes, we obtain the bottom plot. We see that with gray coding, the plot retains its original shape (compare with Figure 2.1). This makes optimization easier for functions that are smooth, because small changes in codes result in small changes in function values.

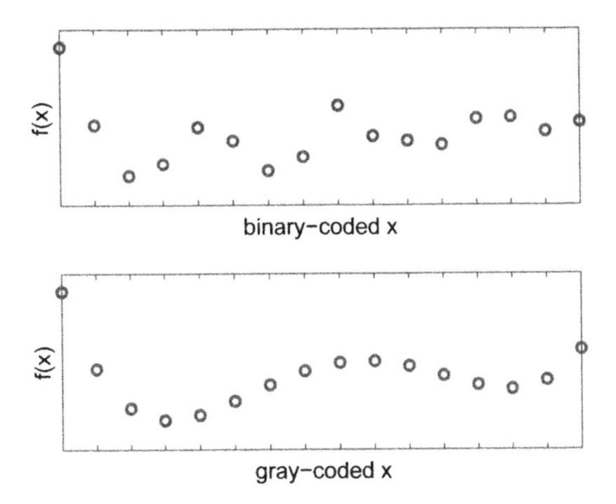

Figure 8.3 Example 8.2: In the top plot the x axis is arranged so that one-bit changes in x are adjacent when we use binary coding. The bottom plot shows the same thing when we use gray coding. A smooth function loses its smoothness if we use binary coding.

□

■ EXAMPLE 8.3

In this example, we test the effect of binary coding versus gray coding on GA performance. We use the GA to optimize the two-dimensional Ackley function described in Example 3.3, where each dimension is coded with six bits. We use a population size of 20, and a mutation rate of 2% per bit per generation. Figure 8.4 shows the average cost of the 20 GA individuals at each generation, averaged over 50 Monte Carlo simulations. Gray coding performs noticeably better than binary coding because of the smooth, regular surface of the Ackley function (see Figure 2.5).

□

Although gray coding seems to perform better in most practical applications, it can be proven that binary coding works better on worst-case problems [Whitley, 1999]. A worst-case problem is a discrete problem for which half of the points in the search space are local minima. Finally, we mention that we can use many other representations besides binary and gray coding in EAs. The representation that we use can have a significant impact on EA performance, and so although we do not discuss representations in much detail in this book, they should not be ignored in EA applications. The study of representations can be quite involved, and we point the reader to references such as [Choi and Moon, 2003] and [Rothlauf and Goldberg, 2003] for further study.

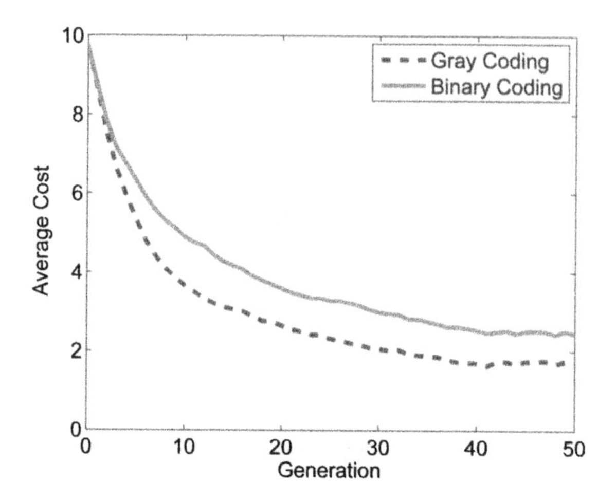

Figure 8.4 Example 8.3 results for the minimization of the two-dimensional Ackley function, where each dimension is coded with six bits. The plot shows the average cost of all GA individuals at each generation, averaged over 50 Monte Carlo simulations. A GA with gray coding performs noticeably better than a GA with binary coding.

■ EXAMPLE 8.4

Suppose that we have a worst-case problem for which even values of the binary coding have a cost of 1, and odd values have a cost of 2. If the individuals are represented with binary codes, the cost values for a three-bit worst-case problem are

$$\begin{aligned} f(000) &= 1, & f(001) &= 2, \\ f(010) &= 1, & f(011) &= 2, \\ f(100) &= 1, & f(101) &= 2, \\ f(110) &= 1, & f(111) &= 2. \end{aligned} \tag{8.5}$$

If the individuals are represented with gray codes, the cost values for a three-bit worst-case problem, written in the same order as the binary representations above, are

$$\begin{aligned} f(000) &= 1, & f(001) &= 2, \\ f(011) &= 1, & f(010) &= 2, \\ f(110) &= 1, & f(111) &= 2, \\ f(101) &= 1, & f(100) &= 2. \end{aligned} \tag{8.6}$$

If we look at the cost function values of Equation (8.5), we see that crossover between binary coded highly-fit individuals will result in children that are also highly fit. This is because all highly-fit individuals have a 0 at their right-most bit position, so any children from highly-fit individuals will also have a 0 at their right-most bit position. This means that they will be even, which means that they will be highly fit, just like their parents. However, Equation (8.6) shows that crossover between gray coded highly-fit individuals may result in low-fitness children. This simple example gives us an intuitive understanding that binary coding might be better than gray coding for problems with many local minima.

□

■ EXAMPLE 8.5

In this example, we test GA performance on a 20-bit worst-case problem for which even values of the binary coding have a cost of 1, and odd values have a cost of 2. We use a population size of 20, and a mutation rate of 2% per bit per generation. Figure 8.5 shows the average cost of the 20 GA individuals at each generation, averaged over 50 Monte Carlo simulations. Binary coding does much better than gray coding. This indicates that for problems with many local minima, binary coding might do better than gray coding at finding a wide variety of the local minima. Remember that for many practical optimization problems, we would like to find a variety of good solutions, rather than finding only one good solution.

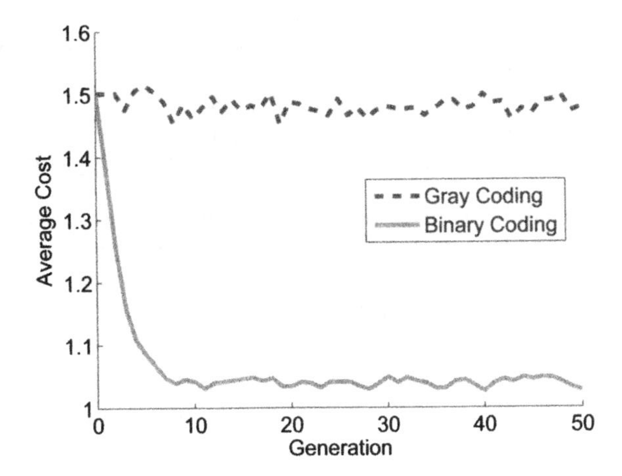

Figure 8.5 Example 8.5 results. The plot shows the average cost of all GA individuals at each generation of a 20-bit problem that has many local minima, averaged over 50 Monte Carlo simulations. Binary coding does better than gray coding at finding multiple local minima.

□

8.4 ELITISM

This section discusses elitism, which is a way of making sure that the best individuals in an EA are retained in the population from one generation to the next. Although the EA results that we have shown in this book look pretty good, there is the danger that we might lose some of our best individuals from one generation to the next. Elitism prevents this from happening.

Consider the GA outline of Figure 3.6. We see that the best parents recombine to produce children. However, if there is an excellent candidate solution x_e to our optimization problem at the i-th generation, there is no guarantee that the best individual at the $(i+1)$-st generation will be an improvement over x_e, or will even be as good as x_e. Individual x_e will recombine with other parents to produce children, but x_e will not be part of the next generation. How can we retain the nice results that arise from recombination, while avoiding the loss of the best individual in the population?

The answer to this question is to keep the best individuals in the EA from one generation to the next. This idea, first proposed in [De Jong, 1975], is called *elitism* and usually improves the performance of an EA. We can implement elitism in at least a couple of different ways.

1. We can implement elitism by producing only $(N-E)$ children each generation, where N is the population size and E is the user-defined number of elite individuals. Suppose we want to keep the best E individuals out of a total population of N from one generation to the next. In that case we would use recombination and mutation to produce $(N-E)$ children, and then we would merge the best E individuals with the children to obtain the next generation of N individuals. Figure 8.6, which is a modification of Figure 3.6, shows this option for an elitist GA. We can easily use this idea in other EAs also.

2. We can implement elitism by producing N children and replacing the worst children with the best E individuals of the previous generation. Figure 8.7 shows this option for an elitist GA, and we can easily use this idea in other EAs also. We can usually expect better performance with this option than with the above elitism option, but this option requires an additional sorting step.

3. There are other elitism options also. For example, we could produce N children and use some type of inverse-roulette-wheel selection algorithm to select E of them, where the worst children have the greatest probability of selection. We could then replace those children with the best E individuals of the previous generation. We could also implement other variations on this theme.

```
Parents ← {randomly generated population}
While not(termination criterion)
    Calculate the fitness of each parent in the population
    Elites ← Best E parents
    Children ← ∅
    While |Children| < |Parents| − E
        Use fitnesses to probabilistically select a pair of parents for mating
        Mate the parents to create children c1 and c2
        Children ← Children ∪ {c1, c2}
    Loop
    Randomly mutate some of the children
    Parents ← Children ∪ Elites
Next generation
```

Figure 8.6 Elitism option 1: A simple genetic algorithm modified for elitism. N is the population size, E is the number of elites that are retained from one generation to the next, and each generation produces $(N - E)$ children.

```
Parents ← {randomly generated population}
While not(termination criterion)
    Calculate the fitness of each parent in the population
    Elites ← Best E parents
    Children ← ∅
    While |Children| < |Parents|
        Use fitnesses to probabilistically select a pair of parents for mating
        Mate the parents to create children c1 and c2
        Children ← Children ∪ {c1, c2}
    Loop
    Randomly mutate some of the children
    Parents ← Children ∪ Elites
    Parents ← Best N Parents
Next generation
```

Figure 8.7 Elitism option 2: A simple genetic algorithm modified for elitism. N is the population size, E is the number of elites that are retained from one generation to the next, and each generation produces N children.

■ EXAMPLE 8.6

In this example, we test the effect of elitism on GA performance. We again optimize the two-dimensional Ackley function described in Example 3.3, where each dimension is coded with six bits. We use a population size of 20, and a mutation rate of 2% per bit per generation. Figure 8.8 shows the minimum cost of the 20 GA individuals at each generation, averaged over 20 Monte Carlo simulations. The elitist GA saves the best two individuals each generation, and uses the elitism option shown in Figure 8.6. We see that the incorporation of elitism makes a big improvement in GA performance.

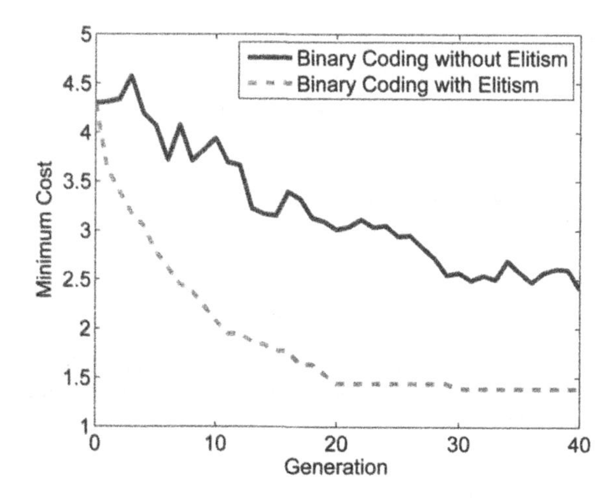

Figure 8.8 Example 8.6 results for the minimization of the two-dimensional Ackley function, where each dimension is coded with six bits. The plot shows the minimum cost of all GA individuals at each generation, averaged over 20 Monte Carlo simulations. An elitist GA does significantly better than a non-elitist GA.

□

We should almost always use elitism in our EAs because it costs very little but pays high dividends. However, there may be certain types of problems with expensive or dynamic cost functions for which a non-elitist EA performs better than an elitist EA (see Chapter 21).

8.5 STEADY-STATE AND GENERATIONAL ALGORITHMS

Most of the EAs that we have discussed so far are generational evolutionary algorithms. That means that the entire population is replaced each generation, with the possible exception of elite individuals as described in Section 8.4. However, this is not the way that evolution occurs in nature. Generations are staggered in nature, and death and birth occur continuously. This type of evolution is called steady state. Our observation of nature motivates us to implement steady-state versions of our EAs. Figure 8.9 gives an outline of a steady-state GA.

Figure 8.9 shows that we create only two children each generation, and the two children replace their two parents in the population. Compare this with the generational GA in Figure 3.6, in which we create all N children before they replace their parents. There are various options that we can implement in a steady-state EA.

```
Parents ← {randomly generated population}
Calculate the fitness of each parent in the population
While not(termination criterion)
    Use fitnesses to select a pair of parents, p1 and p2, for recombination
    Recombine the parents to create children c1 and c2
    Randomly mutate c1 and c2
    Calculate the fitness of c1 and c2
    p1 ← c1, and p2 ← c2
Next generation
```

Figure 8.9 A steady-state genetic algorithm.

- We could replace p_1 with c_1 only if c_1 has a better fitness than p_1, and we could do the same with p_2 and c_2. This would be similar to elitism as described in Section 8.4 because the best individual would never be lost from the population. It would also be similar to the $(\mu + \lambda)$-ES of Figure 6.10 in which a child survives to the next generation only if it is one of the best μ individuals out of $(\mu + \lambda)$ individuals total.

- We could create and replace more than just two individuals each generation. Note that the generational GA of Figure 3.6 replaces all N individuals each generation, while the steady-state GA of Figure 8.9 replaces only two individuals each generation. We could design a GA that lies somewhere between these extremes by replacing four individuals each generation, or six individuals, or a random number of individuals, or any number that we desire. Kenneth De Jong uses the term "generation gap" to refer to the number of individuals that we replace each generation [De Jong, 1975].

Note that we cannot compare the performance of the generational GA of Figure 3.6 with the performance of the steady-state GA of Figure 8.9 by running both algorithms for the same number of generations. A generation of Figure 3.6 produces N children, while a generation of Figure 8.9 produces only two children. Therefore, the generational GA will always outperform the steady-state GA for a given number of generations. This would not be a fair comparison. We can make a fair comparison only by running both algorithms for the same number of fitness function calculations. Therefore, $NG/2$ generations of the steady-state GA of Figure 8.9 would be computationally equivalent to G generations of the generational GA of Figure 3.6.

Figures 3.6 and 8.9 illustrate the generational and steady-state strategies for GAs, but we can easily extend the ideas to almost any other EA. As we have seen in the earlier pages of this book, and as we will also see later, some EAs seem to lend themselves more naturally to a generational approach, and others seem to fit more naturally with a steady-state approach. But the standard implementation of any EA can be modified as the user wishes to obtain either a generational or a steady-state algorithm.

8.6 POPULATION DIVERSITY

This section discusses how to handle duplicate individuals in the population, and how selection and recombination can be modified to encourage diversity in multimodal problems. First we consider the problem of duplicate individuals in Section 8.6.1. Then we discuss two methods for promoting diverse EA populations: restrictions on recombination in Section 8.6.2; and methods for the maintenance of population niches in Section 8.6.3, which include fitness sharing, clearing, and crowding.

8.6.1 Duplicate Individuals

In a population that is repeatedly recombined from one generation to the next, uniformity often results. This means that the entire population becomes a population of clones. Uniformity occurs more often in discrete-domain problems than in continuous-domain problems, but it can occur in both types of problems. Uniformity limits the EA from further exploration of the search space. Although the candidate solution to which an EA converges is usually a good solution, there may be much better solutions in other regions of the search space; therefore, even after the EA finds a good solution, we hope that it will keep exploring in an attempt to find even better solutions. When uniformity occurs before we have found a satisfactory solution to our optimization problem, it is called *premature convergence* [Ronald, 1998]. We can prevent this with higher mutation rates, but if we use too high of a mutation rate our EA will degenerate into a random search. One common way to prevent premature convergence is to continuously search for duplicate individuals and replace them in the population. We can do this in several different ways, as described below.

1. Whenever we create a child, we can scan the population to make sure that we are not creating a duplicate. If we have created a duplicate, then we can redo the recombination operation with different parents or different crossover parameters to obtain a different, non-duplicate child. Or we can mutate the child to obtain a non-duplicate.

2. Whenever we mutate an individual, we can scan the population to make sure that we are not creating a duplicate. If we have created a duplicate, then we can redo the mutation operation.

3. At the end of each generation, we can scan the population for duplicates. We can replace duplicates in a variety of ways. For example, we could replace duplicates with randomly-generated individuals, or we could mutate duplicates, or we could perform a recombination operation to replace each duplicate.

4. We could allow some duplicates in the population, but no more than a user-specified threshold D. Duplicates are likely to be highly-fit individuals, otherwise they would not be likely to occur in the population. Therefore, we might not mind duplicates because they provide a higher probability for high-fitness individuals to participate in recombination. So we could replace duplicates only if there are more than D of them. Or we could probabilistically replace them, depending on how fit they are, or how many duplicates there are.

Duplicate scanning can seem to be computationally expensive because it essentially requires a nested loop, and therefore requires computational effort on the order of N^2, where N is the population size. This can result in a significant portion of an EA's computational effort for benchmark problems. However, we should remember that real-world problems are typically orders of magnitude more complex than benchmark problems. For real-world problems, the fitness function evaluation comprises the vast majority of the computational effort (see Chapter 21), and the computational effort of a duplicate-search-and-replace operation will be insignificant. However, if computational effort on EA benchmark testing is a concern, we could reduce the effort by scanning the population for duplicates every G generations instead of every generation, where G is a user-defined parameter.

8.6.2 Niche-Based and Species-Based Recombination

The typical EA implementation selects parent individuals and combines them to obtain children, without any consideration of how similar or how different the parents are. In biology, however, we often see that parents are similar to each other, but not too similar. For example, we rarely see mating between individuals of different species, but also we rarely see mating between close relatives. Figure 8.10 illustrates the problem with recombination between two individuals that are very different from each other. First, the resulting child may be a poor solution to the optimization problem, because the midpoint between two highly-fit individuals may have poor fitness. Second, the crossover may result in the loss of genetic information that could be important for the problem solution.

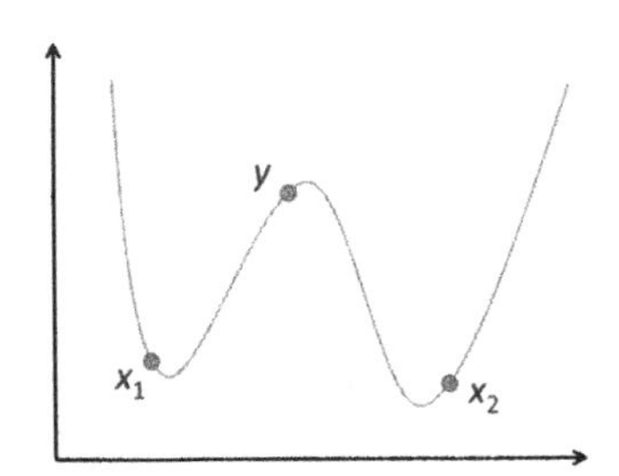

Figure 8.10 This figure illustrates a multimodal minimization problem. This example function illustrates the problem with recombining two individuals that are very different from each other. Parents x_1 and x_2 have low cost, but their offspring y has high cost. Also, if the child y replaces one of its parents (say, x_2), then it may be difficult for the EA to find the global optimum that is near x_2.

Figure 8.11 illustrates the problem with recombination between two individuals that are too similar to each other. If we do not allow recombination between individuals that are different from each other, the evolutionary search process may get stuck in a rut.

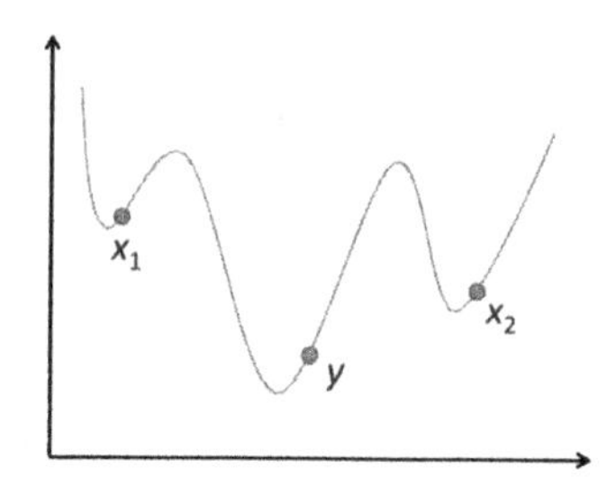

Figure 8.11 This figure illustrates a multimodal minimization problem. This example function illustrates the problem if individuals that are very different from each other are not allowed to recombine. If x_1 and x_2 are not allowed to recombine due to their dissimilarity, then it may be difficult to find the global minimum that is near y.

The problems discussed above motivate niche-based and species-based strategies for recombination.

- Niching strategies discourage recombination between individuals that are highly different from each other in domain space [Mahfoud, 1995b], [Mahfoud, 1995a]. Not only can this help find optimal solutions as shown in Figure 8.10, it can also help find multiple local optima, which is important for many problems. Niching can also be useful for multi-objective optimization (see Chapter 20) and for dynamic optimization (see Section 21.2).

- Species-based strategies discourage recombination between individuals that are highly similar to each other in domain space [Banzhaf et al., 1998, Section 6.4]. This encourages exploration by encouraging recombination between individuals that are much different than each other.

Note that niche-based and species-based strategies are opposite strategies for the same problem. The philosophy of niche-based recombination is that when we recombine fit individuals, we cannot expect the offspring to be fit unless the parents are similar to each other. The philosophy of species-based recombination is that when we recombine fit individuals, we must ensure that the parents are different from each other so that we can effectively explore the search space. Choosing which of these approaches to use is a problem-dependent decision.

8.6.3 Niching

In this section we used the term *niching* differently than in the previous section. Niching in this section is a method that allows EA individuals to survive in separate pockets of the search space. Niching in this section has the same motivation as in the previous section; however, the niching in the previous section specifically addressed the selection of parents, while the niching in this section involves the adjustment of fitness values.

Niching is motivated by multi-modal problems for which it may be important to maintain individuals near many local optima, or for which it may be important to find multiple good solutions. The earliest niching method is fitness sharing [Holland, 1975, page 164]. We discuss three niching strategies: fitness sharing in

Section 8.6.3.1, clearing in Section 8.6.3.2, and crowding in Section 8.6.3.3. Additional discussion of these ideas is provided in [Sareni and Krähenbühl, 1998].

8.6.3.1 Fitness Sharing Sometimes EA individuals in a good region of the search space can take over the population. This can lead to premature convergence. We would like to retain good individuals in the population, but we also want to maintain diversity so that we have a chance to explore new regions of the search space as the EA progresses from one generation to the next. Fitness sharing is especially useful for multimodal problems in which we want to find multiple solutions in different regions of the search space.

Consider Figure 8.12. We see that x^* is the global maximum, but individuals near x^* are unlikely to survive to the next generation because of their low fitness values relative to other individuals in the search space. To encourage diversity in the population, we can artificially increase the fitness values of individuals that are relatively unique, and decrease the fitness values of individuals that are relatively common.

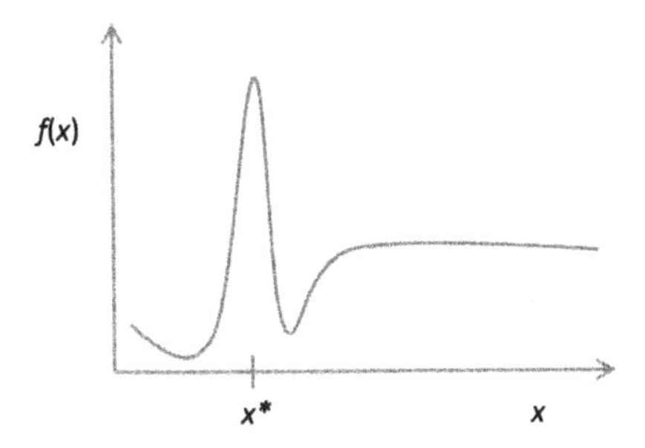

Figure 8.12 This function has a global maximum at x^*, but individuals near x^* are unlikely to be selected for recombination due to their low fitness values relative to other individuals in the search space.

Fitness sharing decreases the fitnesses of individuals that are close to each other in the search space [Sareni and Krähenbühl, 1998]. The biological motivation of this idea is the fact that similar individuals compete for similar resources. Therefore, even if an individual is highly fit, it may not be able to reproduce if there are many other similar individuals in the same geographic region.

Suppose that we have an EA population $\{x_i\}$ of N individuals, and that f_i is the fitness of x_i. Fitness sharing calculates modified fitness values as follows:

$$f_i' = f_i/m_i \tag{8.7}$$

where m_i, which is called the niche count of x_i, is related to the number of individuals that are similar to x_i. The niche count is computed as

$$m_i = \sum_{j=1}^{N} s(d_{ij}) \tag{8.8}$$

where $s(\cdot)$ is the sharing function, and d_{ij} measures the distance between individuals x_i and x_j. We often use the Euclidean distance to obtain d_{ij}. One commonly-used sharing function is

$$s(d) = \begin{cases} 1 - (d/\sigma)^\alpha & \text{if } d < \sigma \\ 0 & \text{otherwise} \end{cases} \tag{8.9}$$

where σ is a user-defined parameter called the dissimilarity threshold,, distance cutoff, or niche radius; and α is a user-defined parameter. We commonly use $\alpha = 1$, which gives a triangular sharing function. Researchers have suggested various methods to set the dissimilarity threshold [Deb and Goldberg, 1989]. For example,

$$\begin{aligned} \sigma &= rq^{-1/n} \\ r &= \frac{1}{2}\sum_{k=1}^{n}\left(\max_i x_i(k) - \min_i x_i(k)\right)^2 \end{aligned} \tag{8.10}$$

where n is the problem dimension, $x_i(k)$ is the k-th element of x_i, and q is the expected number of local optima in the fitness function. In fitness sharing, we use the modified fitness values of Equation (8.7) to select parents for recombination. Note that if we have a minimization problem, then we need to convert cost values to fitness values before applying Equation (8.7), and then we need to convert the modified fitness values to modified cost values (see Problem 8.7).

8.6.3.2 Clearing Clearing is similar to fitness sharing, but instead of sharing fitness values between individuals that share the same niche, we decrease the fitness of some of those individuals [Pétrowski, 1996], [Sareni and Krähenbühl, 1998]. There are several ways that we could implement this idea, including the following. First, we define the niche set D_i of each individual in the population:

$$D_i = \{x_j : d_{ij} < \sigma\} \tag{8.11}$$

where d_{ij} is the same distance as in Equation (8.8), and σ is a user-defined parameter. Next, we rank the individuals in each niche according to their fitness:

$$r_{ki} = \text{rank of } x_k \text{ in } D_i \tag{8.12}$$

where the best individual in each niche has a rank of 1, the second-best has a rank of 2, and so on. Finally, we define the parameter R as the number of individuals that we want to survive in each niche, and we obtain the modified fitness values as follows, where N is the population size.

```
For i = 1 to N
    For k = 1 to |D_i|
        If r_ki ≤ R then
            f'_k ← f_k
        else
            f'_k ← −∞
        End if
    Next niche
Next individual
```

The above algorithm ensures that the least fit individuals in each niche are not available for selection or recombination. However, it does not ensure that the most fit individuals are available for selection, because individuals may belong to more than one niche. For example, individual x_m might be the most fit in its niche, but it might also belong to another niche in which it is not the most fit, in which case its modified fitness could be set to $-\infty$. The set of individuals that survive the above algorithm depends on the order in which we process the niches $\{D_i\}$.

8.6.3.3 Crowding Crowding works by replacing individuals in the population with similar individuals that have recently been produced by recombination. Crowding was introduced by Kenneth De Jong [De Jong, 1975] to mimic resource competition in nature. Below we discuss three types of crowding: standard crowding, deterministic crowding, and restricted tournament selection.

Standard Crowding Standard crowding is used in conjunction with steady-state recombination (see Section 8.5). Standard crowding produces M children each generation and then compares those children with C_f randomly-selected parents, where M and C_f are user-specified parameters. C_f is called the crowding factor. Each child replaces the most similar individual from the group of C_f randomly-selected parent individuals. Commonly-used parameter values are $M = N/10$ and $C_f = 3$, where N is the population size [Mahfoud, 1992]. Figure 8.13 shows an implementation of standard crowding.

Parents $\{p_k\} \leftarrow$ {randomly generated population of N individuals}
Calculate the fitness of each parent p_k, $k \in [1, N]$
While not(termination criterion)
    Use fitnesses to probabilistically select M parents for recombination
    Recombine the parents to create M children c_i, $i \in [1, M]$
    Randomly mutate each child c_i, $i \in [1, M]$
    Calculate the fitness of each child c_i, $i \in [1, M]$
    For $i = 1$ to M
        Randomly select C_f individuals $\mathcal{I}$ from the parent population $\{p_k\}$
        $p_{\min} = \arg\min_p ||p - c_i|| : p \in \mathcal{I}$
        $p_{\min} \leftarrow c_i$
    Next child
Next generation

Figure 8.13 A steady-state evolutionary algorithm with standard crowding. M and C_f are user-selected parameters, and $||p - c_i||$ is a user-defined distance function.

Deterministic Crowding Deterministic crowding involves tournaments between children and parents [Mahfoud, 1995b]. Parents recombine to produce children, and each child replaces its most similar parent, but only if the child has a better fitness than that parent. Figure 8.14 shows an implementation of deterministic crowding.

Restricted Tournament Selection Restricted tournament selection has features in common with both standard crowding and deterministic crowding [Harik, 1995]. M parents recombine to produce two children. We then compare the children with

C_f randomly-selected individuals. Each child replaces the most similar individual from the group of randomly-selected individuals, but only if the child has a better fitness. See Figure 8.15 for an implementation of restricted tournament selection.

Parents $\leftarrow$ {randomly generated population}
Calculate the fitness of each parent in the population
While not(termination criterion)
    Use fitnesses to probabilistically select a pair of parents p_1 and p_2
    Recombine the parents to create children c_1 and c_2
    Randomly mutate c_1 and c_2
    Calculate the fitness of c_1 and c_2
    For $i = 1$ to 2
        If $||p_1 - c_i|| < ||p_2 - c_i||$ and fitness(c_i) > fitness(p_1) then
            $p_1 \leftarrow c_i$
        else if $||p_2 - c_i|| < ||p_1 - c_i||$ and fitness(c_i) > fitness(p_2) then
            $p_2 \leftarrow c_i$
        End if
    Next child
Next generation

Figure 8.14 A steady-state evolutionary algorithm with deterministic crowding. Each child replaces its nearest parent if the child is more fit than that parent.

Parents $\leftarrow$ {randomly generated population}
Calculate the fitness of each parent in the population
While not(termination criterion)
    Use fitnesses to probabilistically select M parents for recombination
    Recombine the parents to create M children c_i, $i \in [1, M]$
    Randomly mutate each child c_i, $i \in [1, M]$
    Calculate the fitness of each child c_i, $i \in [1, M]$
    Randomly select C_f individuals $\mathcal{I}$ from the parent population
    For $i = 1$ to M
        $p_{\min} = \arg\min_p ||p - c_i|| : p \in \mathcal{I}$
        If fitness(c_i) > fitness($p_{\min}$) then
            $p_{\min} \leftarrow c_i$
        End if
    Next child
Next generation

Figure 8.15 A steady-state evolutionary algorithm with restricted tournament selection. M and C_f are user-selected parameters, and $||p - c||$ is a user-defined distance function.

8.7 SELECTION OPTIONS

Before we can combine EA individuals to create children, we have to select which individuals to use as parents. Roulette-wheel selection, which we discussed in Section 3.4.2, is the standard selection method for GAs, and other EAs also use it. But there are also many other selection algorithms, and this section discusses seven of those algorithms. Almost all selection methods are biased toward fit individuals in the population. That is, no matter which selection method we use, a more fit individual will almost always have a greater chance of being selected than a less fit individual.

If a selection method is overly biased towards selecting fit individuals, then the population may converge too quickly to a uniform solution, while not exploring the search space widely enough. However, if a selection method is not biased strongly enough towards fit individuals, then the EA may not be able to properly exploit the information that is present in the most fit individuals.

A useful metric for quantifying the difference between various selection algorithms is selection pressure ϕ, which is defined as

$$\phi = \frac{\text{Pr(selection of most fit individual)}}{\text{Pr(selection of average individual)}} \tag{8.13}$$

where Pr(selection of x) is the probability that individual x is selected for recombination. Selection pressure quantities the relative probability that a highly-fit individual will take part in recombination. Below we discuss seven different types of selection.

8.7.1 Stochastic Universal Sampling

As noted above, the standard method for selection in GAs, and in many other EAs, is roulette-wheel selection, also called fitness-proportionate selection. Figure 3.2, which is reproduced below in Figure 8.16 for convenience, illustrates roulette-wheel selection for a four-member population.

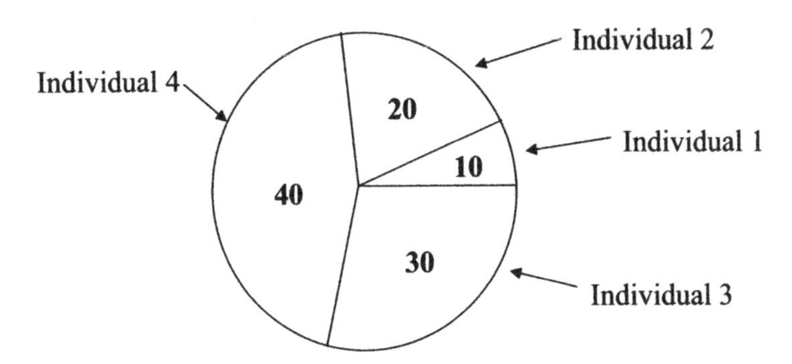

Figure 8.16 Illustration of roulette wheel selection for a four-member population. Each individual is assigned a slice whose area is proportional to its fitness. Each individual's selection as a parent is proportional to the area of its slice in the roulette wheel.

A potential problem with roulette-wheel selection is that there is a good chance that the best individual will not be selected for recombination. For example, suppose that we spin the roulette wheel of Figure 8.16 four times to select four parents for recombination. The probability that the best individual, individual #4, is not selected on any of the four spins is equal to $(0.6)^4 = 13\%$. We have a chance of about 1/7 that the best individual will not be selected for recombination. This may be an unacceptably high probability for losing the information of the best individual in the population.

Stochastic universal sampling [Baker, 1987] solves this problem while still using the roulette-wheel approach. Instead of spinning the roulette wheel of Figure 8.16 four times to select four parents, we instead use a spinner with four uniformly-spaced pointers, place it on the roulette wheel, and spin it once. This gives us four parents with a single spin, and guarantees that we get at least one selection of individual #3 and at least one selection of individual #4, since they both have fitness shares that are greater than 25% of the total summed fitness values. Figure 8.17 illustrates this idea.

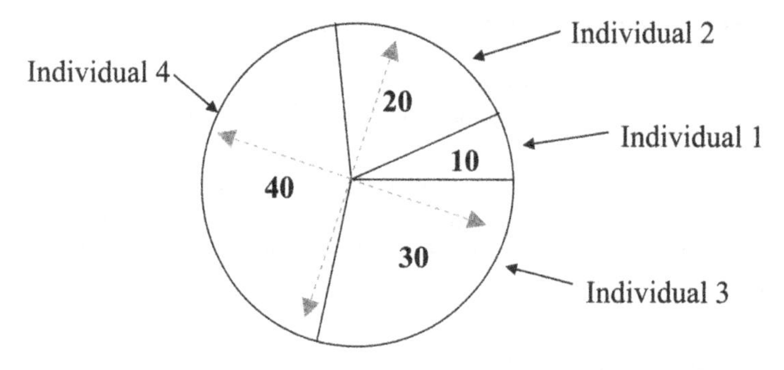

Figure 8.17 Stochastic universal sampling for a four-member population. Each individual is assigned a slice that is proportional to its fitness. A spinner with four evenly-spaced pointers is spun once to obtain the four parents.

Stochastic universal sampling applied to the fitness values of Figure 8.17 will give us one of the following selections of parents:

$$\begin{aligned} & \text{Individual } \#1, \#2, \#3, \text{ and } \#4 \\ \text{or } & \text{Individual } \#1, \#3, \#4, \text{ and } \#4 \\ \text{or } & \text{Individual } \#2, \#3, \#3, \text{ and } \#4 \\ \text{or } & \text{Individual } \#2, \#3, \#4, \text{ and } \#4. \end{aligned} \tag{8.14}$$

Figure 8.18 shows pseudo-code for stochastic universal sampling. Compare this with the roulette-wheel selection code of Figure 3.5. Figure 8.18 guarantees that individual x_i will be selected somewhere between $N_{i,\text{min}}$ and $N_{i,\text{max}}$ times, where

$$\begin{aligned} N_{i,\text{min}} &= \left\lfloor \frac{N f_i}{f_{\text{sum}}} \right\rfloor \\ N_{i,\text{max}} &= \left\lceil \frac{N f_i}{f_{\text{sum}}} \right\rceil \end{aligned} \tag{8.15}$$

where $\lfloor \alpha \rfloor$ is the largest integer that is less than or equal to α, and $\lceil \alpha \rceil$ is the smallest integer that is greater than or equal to α.

---

x_i = i-th individual in population, $i \in [1, N]$
$f_i \leftarrow$ fitness of x_i, for $i \in [1, N]$
$f_{\text{sum}} \leftarrow \sum_{i=1}^{N} f_i$
Generate a uniformly distributed random number $r \in [0, f_{\text{sum}}/N]$
$f_{\text{accum}} \leftarrow 0$
Parents $\leftarrow \emptyset$
$k \leftarrow 0$
While |Parents| $< N$
    $k \leftarrow k + 1$
    $f_{\text{accum}} \leftarrow f_{\text{accum}} + f_k$
    While $f_{\text{accum}} > r$
        Parents $\leftarrow$ Parents $\cup$ x_k
        $r \leftarrow r + f_{\text{sum}}/N$
    End while
Next parent

Figure 8.18 Pseudo-code for selecting N parents from N individuals using stochastic universal sampling. This code assumes that $f_i \geq 0$ for all $i \in [1, N]$.

8.7.2 Over-Selection

Over-selection is a method originally proposed by John Koza in the context of genetic programming [Koza, 1992, Chapter 6]. Over-selection modifies roulette-wheel selection by disproportionately weighting fitness values of highly fit individuals to increase their chances of selection. In Koza's version of over-selection, the best 32% of the population has an 80% chance of being selected, and the worst 68% of the population has a 20% chance of being selected. The exact percentages are not too important; the key feature of over-selection is that fit individuals have a disproportionately higher probability of being selected. This is a type of fitness scaling [Goldberg, 1989a].

Koza tested three different types of selection for genetic programming and found that roulette-wheel selection performed the worst, tournament selection performed better, and over-selection performed best [Koza, 1992, Chapter 25]. However, this may be due to the large population sizes that are typically used in GP. For smaller population sizes, over-selection might provide too much selection pressure during the early stages of evolution when the population's fitness variance is large (see Equation (8.13)), but its additional selection pressure might be beneficial during the later stages of evolution when the fitness variance is small.

8.7.3 Sigma Scaling

Sigma scaling normalizes fitness values relative to the standard deviation of the entire population's fitness. The scaled fitness values are

$$f'(x_i) = \begin{cases} \max\left[1 + (f(x_i) - \bar{f})/(2\sigma), \epsilon\right] & \text{if } \sigma \neq 0 \\ 1 & \text{if } \sigma = 0 \end{cases} \tag{8.16}$$

where $f(x_i)$ is the fitness of the i-th individual in the population, $\bar{f}$ is the mean of the fitness values, σ is the standard deviation of the fitness values, and ϵ is a non-negative user-defined minimum allowable scaled fitness value.

Note that the statement that σ is the standard deviation of the fitness values is ambiguous. If the fitness values are noise-free and we want to know the standard deviation of the specific fitness values that we have measured, then the standard deviation is defined as

$$\sigma = \left(\frac{1}{N} \sum_{i=1}^{N} (f(x_i) - \bar{f})^2 \right)^{1/2}. \tag{8.17}$$

However, if the fitness values include noise, or if we view the fitness values as samples from a probability distribution, then an unbiased estimate of the standard deviation of the fitness values is computed as follows [Simon, 2006, Problem 3.6]:

$$\sigma = \left(\frac{1}{N-1} \sum_{i=1}^{N} (f(x_i) - \bar{f})^2 \right)^{1/2}. \tag{8.18}$$

■ **EXAMPLE 8.7**

Suppose we have a four-member population with fitness values

$$\begin{aligned} f(x_1) &= 10, & f(x_2) &= 5, \\ f(x_3) &= 40, & f(x_4) &= 15. \end{aligned} \tag{8.19}$$

Roulette-wheel selection gives the individuals the following selection probabilities:

$$\begin{aligned} \Pr(x_1) &= 14\%, & \Pr(x_2) &= 7\%, \\ \Pr(x_3) &= 57\%, & \Pr(x_4) &= 22\%. \end{aligned} \tag{8.20}$$

The mean and standard deviation of the fitness values of Equation (8.19) are $\bar{f} = 17.5$ and $\sigma = 15.5$, where we use Equation (8.18) to estimate σ. Equation (8.16) gives scaled fitness values

$$\begin{aligned} f'(x_1) &= 0.76, & f'(x_2) &= 0.60, \\ f'(x_3) &= 1.72, & f'(x_4) &= 0.92. \end{aligned} \tag{8.21}$$

If we use these scaled fitness values in a roulette-wheel selection algorithm, we get the selection probabilities

$$\begin{aligned} \Pr(x_1) &= 19\%, & \Pr(x_2) &= 15\%, \\ \Pr(x_3) &= 43\%, & \Pr(x_4) &= 22\%. \end{aligned} \tag{8.22}$$

Comparing Equations (8.20) and (8.22), we see that sigma scaling tends to even out the selection probabilities of widely-differing fitness values.

□

■ **EXAMPLE 8.8**

As another example, suppose we have four individuals with the following fitness values:

$$\begin{aligned} f(x_1) &= 15, & f(x_2) &= 25, \\ f(x_3) &= 20, & f(x_4) &= 10. \end{aligned} \tag{8.23}$$

Roulette-wheel selection gives these individuals the following selection probabilities:

$$\begin{aligned} \Pr(x_1) &= 21\%, & \Pr(x_2) &= 36\%, \\ \Pr(x_3) &= 29\%, & \Pr(x_4) &= 14\%. \end{aligned} \tag{8.24}$$

The mean and standard deviation of the fitness values of Equation (8.23) are $\bar{f} = 17.5$ and $\sigma = 6.5$. Equation (8.16) gives scaled fitness values

$$\begin{aligned} f'(x_1) &= 0.81, & f'(x_2) &= 1.58, \\ f'(x_3) &= 1.19, & f'(x_4) &= 0.42. \end{aligned} \tag{8.25}$$

If we use these scaled fitness values in a roulette-wheel selection algorithm, we get the selection probabilities

$$\begin{aligned} \Pr(x_1) &= 20\%, & \Pr(x_2) &= 40\%, \\ \Pr(x_3) &= 30\%, & \Pr(x_4) &= 10\%. \end{aligned} \tag{8.26}$$

Comparing Equations (8.24) and (8.26), we see that sigma scaling tends to spread out out the selection probabilities of closely-spaced fitness values.

□

8.7.4 Rank-Based Selection

Rank-based selection, also called rank weighting, sorts individuals in the population from best to worst, and performs selection using the rankings rather than the absolute fitness values [Whitley, 1989]. For example, suppose that we have four individuals in a population with the fitness values of Equation (8.19) and the roulette-wheel selection probabilities of Equation (8.20). Rank-based selection ranks the individuals according to fitness values, giving the best individual a rank of N (where N is the population size), and giving the worst individual a rank of 1:

$$\begin{aligned} R(x_1) &= 2, & R(x_2) &= 1, \\ R(x_3) &= 4, & R(x_4) &= 3. \end{aligned} \tag{8.27}$$

Rank-based selection then performs selection on the basis of the rankings rather than on the basis of the fitness values.

■ EXAMPLE 8.9

Suppose that we use rank-based selection in conjunction with roulette-wheel selection for the ranks of Equation (8.27). This gives the following selection probabilities:

$$\begin{aligned} \Pr(x_1) &= 20\%, \qquad \Pr(x_2) = 10\%, \\ \Pr(x_3) &= 40\%, \qquad \Pr(x_4) = 30\%. \end{aligned} \tag{8.28}$$

We see that when fitness values are widely different from each other, as in Equation (8.19), rank-based selection evens out the selection probabilities. This prevents highly fit individuals from overtaking the population during the early stages of evolution; that is, it prevents premature convergence.

□

■ EXAMPLE 8.10

As another example, suppose we have four individuals with the fitness values of Equation (8.23) and the roulette-wheel selection probabilities of Equation (8.24). Rank-based selection ranks the individuals as follows:

$$\begin{aligned} R(x_1) &= 2, \qquad R(x_2) = 4, \\ R(x_3) &= 3, \qquad R(x_4) = 1. \end{aligned} \tag{8.29}$$

If we use rank-based selection in conjunction with roulette-wheel selection, then Equation (8.29) results in the following selection probabilities:

$$\begin{aligned} \Pr(x_1) &= 20\%, \qquad \Pr(x_2) = 40\%, \\ \Pr(x_3) &= 30\%, \qquad \Pr(x_4) = 10\%. \end{aligned} \tag{8.30}$$

We see that when fitness values are tightly clustered together, as in Equation (8.23), rank-based selection spreads out the selection probabilities. This gives a greater distinction between similar individuals late in an EA run after the population has begun to converge.

□

We can adjust the spread of selection probabilities by passing the rankings through a nonlinear function. For example, if we want more of a distinction between the selection probabilities of the individuals, we can square the ranking before using roulette-wheel selection. Equation (8.29) would then become

$$\begin{aligned} R^2(x_1) &= 4, \qquad R^2(x_2) = 16, \\ R^2(x_3) &= 9, \qquad R^2(x_4) = 1. \end{aligned} \tag{8.31}$$

Using the above values for roulette-wheel selection gives selection probabilities

$$\begin{aligned} \Pr(x_1) &= 13\%, \qquad \Pr(x_2) = 53\%, \\ \Pr(x_3) &= 30\%, \qquad \Pr(x_4) = 4\%. \end{aligned} \tag{8.32}$$

We see that squaring the rankings spreads out the selection probabilities compared to Equation (8.30). Other types of operations on the rankings (for example, the square root operation) might result in more uniform selection probabilities.

8.7.5 Linear Ranking

Linear ranking is a generalization of rank-based selection. In linear ranking, we set the probability of selection of individual x_i to

$$\Pr(x_i) = \alpha + \beta R(x_i) \tag{8.33}$$

where $R(x_i)$ is the rank of x_i as defined in Section 8.7.4, and α and β are user-defined parameters. Figure 8.19 shows the probability of selection with a population size of N. As the line becomes more steep, the selection pressure increases.

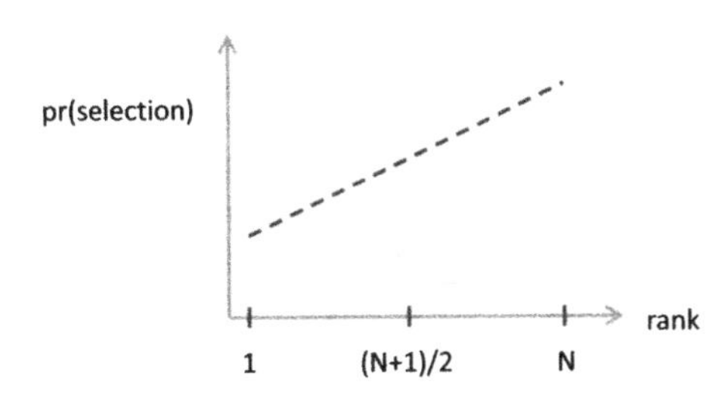

Figure 8.19 This figure illustrates the linear ranking method for selection in an EA with a population size of N. The worst individual has a rank of 1, and the best individual has a rank of N.

Since the best individual has a rank of N and the average individual has a rank of $(N+1)/2$, the selection pressure of Equation (8.13) is

$$\phi = \frac{\alpha + \beta N}{\alpha + \beta(N+1)/2}. \tag{8.34}$$

If we normalize the selection probabilities so that they sum to 1, we get

$$\sum_{i=1}^{N} (\alpha + \beta i) = \alpha N + \beta N(N+1)/2 = 1. \tag{8.35}$$

If we desire a certain selection pressure ϕ, then we can solve Equations (8.34) and (8.35) for α and β to obtain

$$\begin{aligned} \alpha &= \frac{2N - \phi(N+1)}{N(N-1)} \\ \beta &= \frac{2(\phi - 1)}{N(N-1)}. \end{aligned} \tag{8.36}$$

This tells us how to set α and β to obtain a desired selection pressure.

Since $\Pr(x_i)$ is a linear function of $R(x_i)$ as shown in Equation (8.33), we have

$$\begin{aligned} \Pr(\text{average } x) &= \frac{1}{2}[\Pr(\text{worst } x) + \Pr(\text{best } x)] \\ &\geq \frac{1}{2}\Pr(\text{best } x) \end{aligned} \tag{8.37}$$

assuming that all probabilities are non-negative. Combining this with the definition of selection pressure in Equation (8.13), we see that

$$\phi = \frac{\Pr(\text{best } x)}{\Pr(\text{average } x)} \leq 2. \tag{8.38}$$

If we try to set $\phi > 2$, then $\Pr(\text{worst } x)$ will be less than 0. Typically we want a low selection pressure during the early stages of an EA to avoid premature convergence, and a higher selection pressure during the later stages to exploit highly-fit individuals.

Linear Ranking and Roulette-Wheel Selection

Linear ranking has the advantage that even if we use it in conjunction with roulette-wheel selection, we do not need to use a loop in our computer program, as in Figure 3.5. To see how we can avoid looping, suppose that we generate a random number $r \sim U[0,1]$ for selection. This implies that we want to select the m-th individual, where

$$\begin{aligned} \sum_{i=1}^{m} \Pr(x_i) &\approx r \\ \sum_{i=1}^{m} (\alpha + \beta R(x_i)) &\approx r \\ \alpha m + \beta m(m+1)/2 &\approx r. \end{aligned} \tag{8.39}$$

But this is simply a quadratic equation for m, which we can solve as

$$m = \frac{-2\alpha - \beta + \sqrt{(2\alpha+\beta)^2 + 8\beta r}}{2\beta}. \tag{8.40}$$

Of course, since m is restricted to the set of integers, we need to round the right hand side of Equation (8.40) to the nearest integer to obtain m. We can thus implement linear ranking in conjunction with roulette-wheel selection without looping, in contrast to the standard roulette-wheel algorithm of Figure 3.5. Figure 8.20 shows an algorithm for roulette-wheel selection with linear ranking.

A disadvantage of linear ranking is that we need to sort the fitness values of the population, which we do not need to do for standard roulette-wheel selection. But if we use a steady-state EA in which we only generate a few children each generation, then it might be easy to maintain the EA population in order of fitness without a full-fledged sort each generation. In summary, there is no clear-cut advantage or disadvantage to using linear ranking and Equation (8.40) for roulette-wheel selection; it depends on the other aspects of the EA implementation, and the user's preference.

Choose user-specified selection pressure $\phi \in (1, 2)$
Solve Equation (8.36) for α and β
$\{x_i\}$ = EA population sorted in order of fitness, $i \in [1, N]$, where
    x_1 is the worst individual and x_N is the best individual
Generate a uniformly distributed random number $r \in (0, 1)$
Solve Equation (8.40) for m, the index of the selected parent

---

Figure 8.20 Pseudo-code for selecting one parent from N individuals using linear ranking and roulette-wheel selection.

8.7.6 Tournament Selection

Tournament selection reduces the computational cost associated with selection. Figure 3.5 shows that roulette-wheel selection of N parents from a population of N individuals requires nested loops, which can be computationally expensive for large populations. With tournament selection we randomly pick τ individuals from the population, where $\tau \geq 2$ is the user-defined tournament size. We then compare the fitness values of the selected individuals, and select the most fit for recombination.

To analyze tournament selection, consider selection pressure as defined in Equation (8.13). If the most fit individual is picked for a tournament, then it will be selected for recombination with a probability of 100%. If the average individual x is picked for a tournament, then it must be more fit than the $\tau - 1$ other individuals in the tournament to be selected for recombination. In this case there is a 50% probability that x is more fit than each individual with which it is compared,[1] so there is a $(1/2)^{\tau-1}$ probability that x will be selected for recombination. Equation (8.13) then becomes

$$\phi = 2^{\tau-1}. \tag{8.41}$$

We see that as τ increases, selection pressure also increases in tournament selection.

The above tournament selection method is called a *strict* tournament because the best individual in the tournament wins 100% of the time. A *soft* tournament is one in which the best individual in the tournament wins with probability $p < 1$ [Reeves and Rowe, 2003, Section 2.3]. The other, less-fit, individuals also have some probability of winning the tournament. Given the same tournament size, soft tournaments have less selection pressure than strict tournaments.

One advantage of tournament selection over other types of selection is that it can work with only subjective comparisons between individuals. That is, we do not need to calculate absolute fitness values to perform tournament selection; we only need to know the relative fitness values of the individuals in the tournament.

8.7.7 Stud Evolutionary Algorithms

Many EAs probabilistically choose individuals for recombination based on their relative fitness values. All of the methods we have discussed so far follow this prin-

[1] This is approximately correct, assuming that $\tau \ll N$.

ciple. In stud EAs, however, we always choose the best individual each generation for every recombination operation. The best individual each generation is called the *stud*. We then choose the other parents, with which the stud combines to create offspring, in the normal way (for example, fitness-based selection, rank-based selection, tournament selection, and so on). This idea was first applied to GAs, in which case it is called the stud GA [Khatib and Fleming, 1998]. The addition of stud logic to a GA results in the modification of the standard GA of Figure 3.6 and results in the stud GA of Figure 8.21.

Parents ← {randomly generated population}
While not(termination criterion)
    Calculate the fitness of each parent in the population
    Children ← $\emptyset$
    x_1 ← most fit parent
    While |Children| < |Parents|
        Use fitnesses to probabilistically select a second parent $x_2 : x_2 \neq x_1$
        Mate x_1 and x_2 to create children c_1 and c_2
        Children ← Children ∪ $\{c_1, c_2\}$
    Loop
    Randomly mutate some of the children
    Parents ← Children
Next generation

Figure 8.21 The above pseudo-code outlines the stud genetic algorithm.

■ EXAMPLE 8.11

In this example we simulate a continuous GA, with and without the stud option, on a set of 20-dimensional benchmark problems from Appendix C. We use a population size of 50, a generation limit of 50, and a mutation rate of 1% for each of the 20 features in each individual at each generation. We implement mutation by replacing an independent variable with one that we randomly choose from a uniform distribution between the minimum and maximum domain values. We also use an elitism parameter of two, which means that we keep the two best individuals from one generation to the next.

Table 8.1 shows the best performance of the standard GA and the stud GA, averaged over 50 Monte Carlo simulations. The table shows that the stud GA clearly outperforms the standard GA for the benchmarks shown. For some of the benchmarks, performance dramatically improves with the stud option.

□

To this point the stud EA has been mostly (perhaps exclusively) applied to GAs in the research literature, but we could easily apply it to many other EAs also. The addition of stud logic to an EA is a simple change in the basic EA, as we see from a comparison between Figures 3.6 and 8.21. Because of its ease of implementation, and the excellent performance shown in Table 8.1, we should give serious consideration to stud logic for our EAs. Another interesting area for future research would be to derive mathematical models for GAs (Chapter 4) and other EAs when they include stud logic.

Benchmark	Non-Stud GA	Stud GA
Ackley	1.44	1
Fletcher	3.26	1
Griewank	3.96	1
Penalty #1	1.05×10^5	1
Penalty #2	160.8	1
Quartic	9.14	1
Rastrigin	1.92	1
Rosenbrock	3.89	1
Schwefel 1.2	1.24	1
Schwefel 2.21	1.65	1
Schwefel 2.22	3.70	1
Schwefel 2.26	2.56	1
Sphere	4.47	1
Step	4.23	1

Table 8.1 Example 8.11 results, showing the relative performance of a GA with and without the stud option. The table shows the normalized minimum found by the two GA versions, averaged over 50 Monte Carlo simulations. See Appendix C for the definitions of the benchmark functions.

8.8 RECOMBINATION

The simple GA uses single-point crossover. This section discusses other types of recombination for both binary and continuous EAs. Note that we use the terms crossover and recombination interchangeably. Some of the recombination methods that we present in this section are further discussed in [Herrera et al., 1998].

Suppose that we have a population of individuals $\{x_1, x_2, \cdots, x_N\}$. Each individual has n features, and we denote the k-th feature of the i-th individual as $x_i(k)$ for $k \in [1, n]$. So we can represent x_i as the vector

$$x_i = \begin{bmatrix} x_i(1) & x_i(2) & \cdots & x_i(n) \end{bmatrix}. \tag{8.42}$$

We denote a child individual, which we also call an offspring and which is the result of recombination, as y. We denote the k-th feature of the offspring as $y(k)$, so

$$y = \begin{bmatrix} y(1) & y(2) & \cdots & y(n) \end{bmatrix}. \tag{8.43}$$

8.8.1 Single-Point Crossover (Binary or Continuous EAs)

Suppose that we have two parents, x_a and x_b, where $a \in [1, N]$ and $b \in [1, N]$. Single-point crossover, also called simple crossover or discrete crossover, is the type of crossover that was first used in binary GAs (see Chapter 3):

$$y(k) \leftarrow \begin{bmatrix} x_a(1) & \cdots & x_a(m) & x_b(m+1) & \cdots & x_b(n) \end{bmatrix} \tag{8.44}$$

where m is a randomly selected crossover point; that is, $m \sim U[0, n]$. If $m = 0$, then y is a clone of x_b. If $m = n$, then y is a clone of x_a. Single-point crossover is often implemented to obtain two children from a pair of parents. This is done by selecting each feature of the second child y_2 from the opposite parent than the one from which y_1 obtained its feature:

$$\begin{aligned} y_1(k) &\leftarrow \begin{bmatrix} x_a(1) & \cdots & x_a(m) & x_b(m+1) & \cdots & x_b(n) \end{bmatrix} \\ y_2(k) &\leftarrow \begin{bmatrix} x_b(1) & \cdots & x_b(m) & x_a(m+1) & \cdots & x_a(n) \end{bmatrix}. \end{aligned} \tag{8.45}$$

8.8.2 Multiple-Point Crossover (Binary or Continuous EAs)

Two-point crossover results in

$$\begin{aligned} y(k) \leftarrow \Big[& x_a(1) \quad \cdots \quad x_a(m_1) \\ & x_b(m_1+1) \quad \cdots \quad x_b(m_2) \\ & x_a(m_2+1) \quad \cdots \quad x_a(n) \Big] \end{aligned} \tag{8.46}$$

where the two crossover points are $m_1 \sim U[0, n]$ and $m_2 \sim U[m_1 + 1, n]$. If $m_1 = 0$ or $m_2 = n$, then two-point crossover reduces to single-point crossover. If $m_1 = n$, then y is a clone of x_a. Equation (8.46) can be extended to three-point crossover, or M-point crossover for any $M > 2$. As with single-point crossover, multiple-point crossover is often implemented to obtain two children from a pair of parents.

8.8.3 Segmented Crossover (Binary or Continuous EAs)

We can think of segmented crossover [Michalewicz, 1996, Section 4.6] as a generalization of multiple-point crossover. Child 1 gets its first feature from parent 1. Then, with a probability of ρ, we switch to parent 2 to obtain the second child 1 feature; and with a probability $(1 - \rho)$ we obtain the second child 1 feature from parent 1. Every time we obtain a feature for child 1, we switch to the other parent to obtain the next feature with a probability of ρ. Child 1 and child 2 obtain their features from different parents, so if child 1 gets feature k from parent 1, then child 2 gets feature k from parent 2, for $k \in [1, n]$. Similarly, if child 1 gets feature k from parent 2, then child 2 gets feature k from parent 1. Segmented crossover is equivalent to multiple-point crossover where the number of crossover points is a random number. Figure 8.22 shows an algorithm for segmented crossover. The switching probability ρ is often set to around 0.2.

8.8.4 Uniform Crossover (Binary or Continuous EAs)

Suppose that we have two parents, x_a and x_b. Uniform crossover [Ackley, 1987a], [Michalewicz and Schoenauer, 1996] results in the child y, where the k-th feature of y is

$$y(k) \leftarrow x_{i(k)}(k) \tag{8.47}$$

for each $k \in [1, n]$, where we randomly choose $i(k)$ from the set $\{a, b\}$. That is, we randomly choose each child feature from one of its two parents, each with a probability of 50%.

```
S ← true
For k = 1 to n
    If S then
        c1(k) ← p1(k)
        c2(k) ← p2(k)
    else
        c1(k) ← p2(k)
        c2(k) ← p1(k)
    End if
    r ← U[0, 1]
    If r < ρ then S ← not S
Next solution feature
```

Figure 8.22 Segmented crossover for n-dimensional individuals. p_1 and p_2 are the two parents, and c_1 and c_2 are their two children.

8.8.5 Multi-Parent Crossover (Binary or Continuous EAs)

The multi-parent crossover that we discuss here is a generalization of uniform crossover [Eiben, 2003], [Eiben and Bäck, 1998], [Eiben, 2000] and goes by several other names, including gene pool recombination [Bäck, 1996], [Bäck et al., 1997b], [Mühlenbein and Voigt, 1995], scanning crossover [Eiben and Schippers, 1996], and multi-sexual crossover [Schwefel, 1995] (in contrast to two-parent crossover, which is called bisexual crossover). In multi-parent crossover, we randomly choose each child feature from one of its parents, where the number of parents is greater than two. This was suggested as early as 1966 [Bremermann et al., 1966]. Multi-parent crossover gives

$$y_k \leftarrow x_{i(k)}(k) \tag{8.48}$$

for each $k \in [1, n]$, where we randomly choose $i(k)$ from a subset of $[1, N]$ (recall that we have N potential parents in the population). We need to make several choices when implementing multi-parent crossover. For example, how many individuals should be in the pool of potential parents? How should individuals be chosen for the pool? Once the pool has been determined, how should parents be selected from the pool? Finally, we note that other approaches to multi-parent crossover have also been proposed Eiben95.

8.8.6 Global Uniform Crossover (Binary or Continuous EAs)

One way of implementing multi-parent crossover is to randomly choose each child feature from one of its parents, where the parent pool is equivalent to the entire population. This gives global uniform recombination:

$$y_k \leftarrow x_{i(k)}(k) \tag{8.49}$$

for each $k \in [1, n]$, where $i(k)$ is randomly selected from the uniform distribution $[1, N]$ for each k. Alternatively, $i(k)$ could be chosen on the basis of fitness. That is, the probability that $i(k) = m$ could be proportional to the fitness of x_m for all $k \in [1, n]$ and $m \in [1, N]$.

8.8.7 Shuffle Crossover (Binary or Continuous EAs)

Shuffle crossover randomly rearranges the solution features in the parents [Eshelman et al., 1989]. We use the same rearrangement of solution features in all parents that contribute to a given child. We then perform one of the crossover methods above (usually single-point crossover) to obtain children. We then undo the rearrangment of solution features in the children. Figure 8.23 shows a shuffle crossover algorithm combined with single-point crossover.

$\{r_1, \cdots, r_n\} \leftarrow$ random permutation of $\{1, \cdots, n\}$
Crossover point $m \leftarrow U[1, n-1]$
For $k = 1$ to m
  $t_1(k) \leftarrow p_1(r_k)$
  $t_2(k) \leftarrow p_2(r_k)$
Next k
For $k = m + 1$ to n
  $t_1(k) \leftarrow p_2(r_k)$
  $t_2(k) \leftarrow p_1(r_k)$
Next k
For $k = 1$ to n
  $c_1(r_k) \leftarrow t_1(k)$
  $c_2(r_k) \leftarrow t_2(k)$
Next k

Figure 8.23 Shuffle crossover combined with single-point crossover for n-dimensional parents. p_1 and p_2 are the two parents, t_1 and t_2 are the two children before they are unshuffled, and c_1 and c_2 are the unshuffled children.

8.8.8 Flat Crossover and Arithmetic Crossover (Continuous EAs)

Flat crossover, also called arithmetic crossover, is described as follows:

$$\begin{aligned} y(k) &\leftarrow U[x_a(k), x_b(k)] \\ &= \alpha x_a(k) + (1-\alpha) x_b(k) \end{aligned} \tag{8.50}$$

where $\alpha \sim U[0, 1]$. That is, $y(k)$ is a random number taken from a uniform distribution between the k-th features of its two parents. This is equivalent to saying that the offspring is a linear combination of the features of its two parents. Sometimes flat crossover and arithmetic crossover are distinguished by saying that flat crossover gives one offspring while arithmetic crossover gives two offspring:

$$\begin{aligned} &\text{flat crossover:} && y(k) = \alpha x_a(k) + (1-\alpha) x_b(k) \\ &\text{arithmetic crossover:} && \begin{cases} y_1(k) = \alpha x_a(k) + (1-\alpha) x_b(k) \\ y_2(k) = (1-\alpha) x_a(k) + \alpha x_b(k). \end{cases} \end{aligned} \tag{8.51}$$

We could also use a triangular probability density function for α instead of a uniform density function:

$$\text{PDF}(\alpha) = \begin{cases} 1+\alpha & \text{if } -1 \leq \alpha < 0 \\ 1-\alpha & \text{if } 0 \leq \alpha \leq 1 \end{cases} \tag{8.52}$$

in which case Equation (8.51) is called fuzzy recombination [Eshelman and Schaffer, 1993].

8.8.9 Blended Crossover (Continuous EAs)

Blended crossover, which is also called BLX-α crossover and heuristic crossover [Houck et al., 1995], combines parents x_a and x_b as follows:

$$\begin{aligned} x_{\min}(k) &\leftarrow \min(x_a(k), x_b(k)) \\ x_{\max}(k) &\leftarrow \max(x_a(k), x_b(k)) \\ \Delta x(k) &\leftarrow x_{\max}(k) - x_{\min}(k) \\ y_k &\leftarrow U[x_{\min}(k) - \alpha\Delta x(k),\ x_{\max}(k) + \alpha\Delta x(k)] \end{aligned} \tag{8.53}$$

where α is a user-defined parameter. If $\alpha = 0$, then blended crossover is equivalent to flat crossover. If $\alpha < 0$ (with a lower limit of -0.5), then blended crossover shrinks the search domain, which is beneficial for exploitation of the current population. If $\alpha > 0$, then blended crossover expands the search domain, which is beneficial for exploration. [Herrera et al., 1998] recommends $\alpha = 0.5$.

8.8.10 Linear Crossover (Continuous EAs)

Linear crossover creates three offspring from parents x_a and x_b:

$$\begin{aligned} y_1(k) &\leftarrow (1/2)x_a(k) + (1/2)x_b(k) \\ y_2(k) &\leftarrow (3/2)x_a(k) - (1/2)x_b(k) \\ y_3(k) &\leftarrow (-1/2)x_a(k) + (3/2)x_b(k). \end{aligned} \tag{8.54}$$

We retain the most fit, or the two most fit, of the three offspring for the next generation, depending on the particular EA implementation.

8.8.11 Simulated Binary Crossover (Continuous EAs)

Simulated binary crossover (SBX) creates the following two offspring from parents x_a and x_b [Deb and Agrawal, 1995]:

$$\begin{aligned} y_1(k) &\leftarrow (1/2)[(1-\beta_k)x_a(k) + (1+\beta_k)x_b(k)] \\ y_2(k) &\leftarrow (1/2)[(1+\beta_k)x_a(k) + (1-\beta_k)x_b(k)] \end{aligned} \tag{8.55}$$

where β_k is a random number generated from the following density function:

$$\text{PDF}(\beta) = \begin{cases} \frac{1}{2}(\eta+1)\beta^{\eta} & \text{if } 0 \leq \beta \leq 1 \\ \frac{1}{2}(\eta+1)\beta^{-(\eta+2)} & \text{if } \beta > 1 \end{cases} \tag{8.56}$$

where η is any nonnegative real number. [Deb and Agrawal, 1995] includes a discussion of the effect of η on the SBX operator and generally recommends values between 0 and 5. We can generate β with the following algorithm:

$$\begin{aligned} r &\leftarrow U[0,1] \\ \beta &\leftarrow \begin{cases} (2r)^{1/(\eta+1)} & \text{if } r \le 1/2 \\ (2-2r)^{-1/(\eta+1)} & \text{if } r > 1/2. \end{cases} \end{aligned} \tag{8.57}$$

Note that SBX is equivalent to arithmetic crossover in Equation (8.51) if $\beta = 2\alpha - 1$. We could also implement SBX with β_k values that have distributions other than that of Equation (8.56).

8.8.12 Summary

The recombination methods discussed above were originally proposed for GAs, but they can also be used for other EAs. Researchers have also proposed other crossover methods [Herrera et al., 1998], but the above approaches give the main ideas. In addition, we could combine some of these approaches to create our own customized recombination algorithm. There is no clear winner among these crossover methods. One crossover method might work the best on one problem, while another works best on another problem. However, even though we cannot say which crossover method is best, we can usually say that single-point crossover is one of the worst.

8.9 MUTATION

In binary EAs, mutation is a straightforward operation. If we have a population of N individuals, where each individual has n bits, and our mutation rate is ρ, then at the end of each generation we flip each bit in each individual with a probability of ρ, as shown in Equation (3.6).

In continuous EAs, we have more options for mutation. We still call ρ the mutation rate, and we still modify $x_i(k)$ with a probability of ρ for each i and each k. But if we decide to modify $x_i(k)$, we then need to decide *how* to modify $x_i(k)$. One way is to generate $x_i(k)$ from a uniform or Gaussian distribution whose mean is at the center of the search domain. Another way is to generate $x_i(k)$ from a uniform or Gaussian distribution whose mean is at the non-mutated value of $x_i(k)$. We describe these options below, where we use $x_{\min}(k)$ and $x_{\max}(k)$ to denote the limits of the search domain of the k-th dimension in our optimization problem.

8.9.1 Uniform Mutation Centered at $x_i(k)$

Uniform mutation centered at $x_i(k)$ can be written as

$$\begin{aligned} r &\leftarrow U[0,1] \\ x_i(k) &\leftarrow \begin{cases} x_i(k) & \text{if } r \ge \rho \\ U[x_i(k) - \alpha_i(k),\, x_i(k) + \alpha_i(k)] & \text{if } r < \rho \end{cases} \end{aligned} \tag{8.58}$$

for $i \in [1, N]$ and $k \in [1, n]$, where $\alpha_i(k)$ is a user-defined parameter that determines the mutation magnitude. We often choose $\alpha_i(k)$ as large as possible while still ensuring that the mutation remains within the search domain:

$$\alpha_i(k) = \min(x_i(k) - x_{\min}(k),\, x_{\max}(k) - x_i(k)). \tag{8.59}$$

8.9.2 Uniform Mutation Centered at the Middle of the Search Domain

Uniform mutation centered at the middle of the search domain can be written as

$$\begin{aligned} r &\leftarrow U[0,1] \\ x_i(k) &\leftarrow \begin{cases} x_i(k) & \text{if } r \geq \rho \\ U[x_{\min}(k), x_{\max}(k)] & \text{if } r < \rho \end{cases} \end{aligned} \tag{8.60}$$

for $i \in [1, N]$ and $k \in [1, n]$.

8.9.3 Gaussian Mutation Centered at $x_i(k)$

Gaussian mutation centered at $x_i(k)$ can be written as

$$\begin{aligned} r &\leftarrow U[0,1] \\ x_i(k) &\leftarrow \begin{cases} x_i(k) & \text{if } r \geq \rho \\ \max\left[\min(x_{\max}(k), N(x_i(k), \sigma_i^2(k)), x_{\min}(k)\right] & \text{if } r < \rho \end{cases} \end{aligned} \tag{8.61}$$

for $i \in [1, N]$ and $k \in [1, n]$, where $\sigma_i(k)$ is a user-defined parameter that is proportional to the mutation magnitude. The min and max operations ensure that the mutated value of $x_i(k)$ remains within the search domain. This type of mutation is similar to the search operators that we use in EP and ES.

8.9.4 Gaussian Mutation Centered at the Middle of the Search Domain

Gaussian mutation centered at the middle of the search domain can be written as

$$\begin{aligned} r &\leftarrow U[0,1] \\ x_i(k) &\leftarrow \begin{cases} x_i(k) & \text{if } r \geq \rho \\ \max\left[\min(x_{\max}(k), N(c_i(k), \sigma_i^2(k)), x_{\min}(k)\right] & \text{if } r < \rho \end{cases} \end{aligned} \tag{8.62}$$

for $i \in [1, N]$ and $k \in [1, n]$, where $c_i(k) = (x_{\min}(k) + x_{\max}(k))/2$ is the center of the search domain, and where $\sigma_i(k)$ is a user-defined parameter that is proportional to the mutation magnitude. The min and max operations ensure that the mutated value of $x_i(k)$ remains within the search domain.

8.10 CONCLUSION

We have examined many EA variations in this chapter, but we actually restricted our discussion to only the most common variations. There are many other ways that EAs can be modified, such as the use of varying population sizes [Hu et al., 2010]; interacting sub-populations [Li et al., 2009]; diploidy or polyploidy, in which case each individual is associated with multiple candidate solutions [Wang et al., 2009]; and gender modeling, which restricts crossover to parents of the opposite gender [Mitchell, 1998]. Researchers have also proposed other modifications, but we do not have space here to discuss all of the variations that have been studied over the past few decades.

Considering all of the EA variations that are available, it may be useful to distinguish between an EA, and an EA instance [Eiben and Smit, 2011]. An EA is a general framework that defines the approach to optimization which includes a population of candidate solutions, selection, recombination, and mutation; and an EA instance is a realization of that framework which includes specific approaches to those tasks and specific tuning parameters (for example, a specifically-tuned GA, ES, or any of the other specific algorithms discussed in this book). This perspective views all EA instances as particular realizations of the general EA framework, which is useful for unifying the field, and for preventing the field from splintering into apparently disconnected fragments. This unified EA perspective is the basis of the books [De Jong, 2002], [Eiben and Smith, 2010].

Note that EA parameter adjustments can be considered from two different perspectives. First, we can tune parameters to optimize EA performance; and second, we can adjust parameters to study how performance varies with parameter settings [Eiben and Smit, 2011]. This second perspective is closely related to EA robustness, which we briefly discuss in Section 21.4.

The future will doubtless see the introduction of new and creative EA variations. The most challenging aspect of such studies is to first carefully explore past research to see if supposedly-new ideas have already been published. The current literature has many examples of the wheel being reinvented because of authors' and reviewers' ignorance of past research. Sometimes algorithms are reinvented and given different names by different authors, and sometimes new algorithms are invented but are given the same name as a completely different algorithm. In fairness, it is difficult to keep up with the explosion of EA literature over the past few decades, and we are all ignorant of past research to a certain extent. Nevertheless, when we document our research, we owe it to our readers and to past researchers to thoroughly study the published literature so that we can place our research in its proper context.

PROBLEMS

Written Exercises

8.1 Suppose you initialize a population of N individuals that are uniformly distributed in a one-dimensional search domain $[x_{\text{min}}, x_{\text{max}}]$, where $x_{\text{max}} = -x_{\text{min}}$.

a) What is the probability that at least one of those individuals will be within ϵ of the optimal point in the domain?

b) Suppose that $\epsilon/x_{\text{max}} \ll 1$ and that N is "not too large." Use Taylor series approximations to find the factor by which your answer to part (a) increases ifi the initial population is doubled.

8.2 Gray codes are not unique. Equation (8.2) shows a gray coding of the numbers 0–7. Give an alternate gray coding.

8.3 **Elitism and evolution strategies:**

a) Explain how an elitist GA is similar to a $(\mu + \lambda)$-ES.

b) What values would we use for N and E in Figure 8.7, and for μ and λ in Figure 6.10, to obtain an elitist GA and a $(\mu + \lambda)$-ES that are as similar as possible?

8.4 **Elitism and steady-state evolution:**

a) How could we combined elitism option 1 in Figure 8.6 with steady-state evolution?

b) How could we combined elitism option 2 in Figure 8.7 with steady-state evolution?

8.5 Suppose we have an EA with a generation gap of k. How many generations of this EA are computationally equivalent to G generations of a generational EA?

8.6 How many comparisons do we need to perform in a population of size N to completely check for duplicate individuals?

8.7 Write a sequence of equations to transform cost values (where lower is better) to modified cost values using fitness sharing.

8.8 Do we need to worry about the possibility of divide-by-zero in Equation (8.7)?

8.9 The clearing method of Section 8.6.3.2 may result in the most fit individual in a niche becoming unavailable for selection and recombination. Sketch a visual example of a situation where this could occur.

8.10 **Selection pressure:**

a) What is the selection pressure of roulette-wheel selection with the fitness values shown in Figure 8.16?

b) What is the selection pressure of stochastic universal sampling with the fitness values shown in Figure 8.17?

8.11 Stochastic universal sampling:

a) What is the probability that the most fit individual in Figure 8.17 will be selected twice using stochastic universal sampling?

b) What is the probability that the least fit individual in Figure 8.17 will be selected once using stochastic universal sampling?

8.12 Suppose we have four individuals in an EA population with fitness values 10, 20, 30, and 40.

a) What are the selection probabilities of each individual given one spin of a roulette-wheel?

b) Suppose we use over-selection so that the best 50% of the population has a 75% probability of selection, and the worst 50% of the population has a 25% probability of selection. What are the selection probabilities of each individual given one spin of a roulette-wheel?

c) What are the selection probabilities if we use sigma scaling?

d) What are the selection probabilities if we use rank-based selection?

8.13 Suppose we have four individuals in an EA population with fitness values 10, 20, 30, and 40. Use Equation (8.36) to calculate α and β to obtain the following selection pressures when linear ranking is used.

a) $\phi = 1.4$.

b) $\phi = 1.6$.

c) $\phi = 1.8$.

8.14 Suppose you use a soft tournament for selection with a tournament size of 3, where the probability of selecting the best individual in the tournament is 70%, the probability of selecting the second-best individual is 20%, and the probability of selecting the worst individual is 10%. What is the selection pressure of this tournament?

8.15 Suppose you are using an EA to solve a 20-dimensional problem.

a) What is the probability that children produced by single-point crossover will be clones of the parents?

b) What is the probability that children produced by two-point crossover will be clones of the parents?

c) What is the probability that children produced by segmented crossover with $\rho = 0.2$ will be clones of the parents?

d) What is the probability that children produced by uniform crossover will be clones of the parents?

Computer Exercises

8.16 Implement a continuous GA with elitism to minimize the 10-dimensional Ackley function. Run the GA for 50 generations with a population size of 50 and a mutation probability of 1%. Run the GA 20 times, and plot the average (over the 20 simulations) of the minimum cost as a function of generation number. Do this for 0 elites, 2 elites, 5 elites, and 10 elites. Put the four plots in a single figure to facilitate comparison.

8.17 Implement a continuous GA with the three types of crowding discussed in Section 8.6.3.3 to minimize the 10-dimensional Ackley function. Use the crowding parameters recommended in the text. Use a population size of 40, a mutation rate of 2%, an elitism parameter of 2, and replace duplicates at the end of each generation with randomly-generated individuals. Run each GA for 1,000 function evaluations. (Note that different crowding types give a different number of function evaluations per generation. Therefore, we need to run GAs for different generation limits to get a fair comparison between crowding types.) Report the minimum cost attained by the GA for no crowding, and for the three types of crowding discussed in the text, averaged over 20 Monte Carlo simulations.

8.18 Write a program to numerically confirm your answers to Problem 8.11.

8.19 Suppose you have the individuals of Problem 8.13 and linear ranking with selection pressure $\phi = 1.6$.

- **a)** Simulate Equation (8.40) a few thousand times and record the percentage of times that each individual is selected. Use rounding in Equation (8.40) to obtain the integer index of the selected individual. How do your simulation results compare with the theoretical selection probabilities?
- **b)** What is the largest possible value of Equation (8.40) (with no rounding) for this problem? How does this help explain the discrepancy between the simulated and theoretical results?

PART III

MORE RECENT EVOLUTIONARY ALGORITHMS

CHAPTER 9

Simulated Annealing

We conjecture that the analogy with thermodynamics can offer a new insight into optimization problems and can suggest efficient algorithms for solving them.

—V. Černý [Černý, 1985]

Simulated annealing (SA) is an optimization algorithm that is based on the cooling and crystallizing behavior of chemical substances. The literature often distinguishes SA from EAs because SA does not involve a population of candidate solutions. SA is a single-individual stochastic algorithm. However, the (1+1)-ES is actually a special case of an SA algorithm [Droste et al., 2002], so we can reasonably consider SA as an EA.

SA was first presented in its current form by Scott Kirkpatrick, Charles Gelatt, and Mario Vecchi in 1983 for the optimal solution of problems related to computer design, such as component placement and wire routing [Kirkpatrick et al., 1983]. SA was independently derived by Vlado Černý in 1985, who used it to solve the traveling salesman problem [Černý, 1985]. An optimization algorithm very similar to SA was developed in by Martin Pincus in the late 1960s [Pincus, 1968a], [Pincus, 1968b]. SA is sometimes called the Metropolis algorithm because it is closely related to the work of Nicholas Metropolis [Metropolis et al., 1953], whose development of an algorithm for investigating the properties of interacting particles formed the foundation for SA. Finally, SA is also sometimes called the Metropolis-Hastings algorithm due to the work of W. Keith Hastings [Hastings, 1970], who generalized the results of Metropolis et al.

Overview of the Chapter

Section 9.1 gives a brief discussion of statistical mechanics, which is the foundational principle of SA. Section 9.2 presents a simple SA algorithm. Section 9.3 discusses various cooling schedules, which is the primary tuning parameter for SA, and which has the strongest effect on its performance. Section 9.4 briefly discusses a few implementation issues, including ways that we can generate new candidate solutions in SA, when to reinitialize the cooling temperature, and why we need to keep track of the best candidate solution.

9.1 ANNEALING IN NATURE

Crystalline lattices are fascinating examples of the optimization ability of nature. A crystalline lattice is an arrangment of atoms or molecules in a liquid or solid. Some familiar examples that are common to most people's everyday experiences are the crystalline structures of quartz, ice, and salt. At high temperatures, crystalline materials don't exhibit much structure; high temperatures give the materials a lot of energy, which contributes to a lot of vibration and disorder. However, as the temperature decreases, the crystalline materials settle into a more ordered state. The particular state into which they settle is not always the same. A material that is heated and then cooled multiple times will settle into a different equilibrium state every time, but every equilibrium state tends to have low energy. Figure 9.1 compares a crystalline structure with a high entropy (a high level of disorder) at a high temperature, and one with a low entropy (a high level of order) at a low temperature. The process of heating and cooling a material to recrystallize it is called *annealing*.

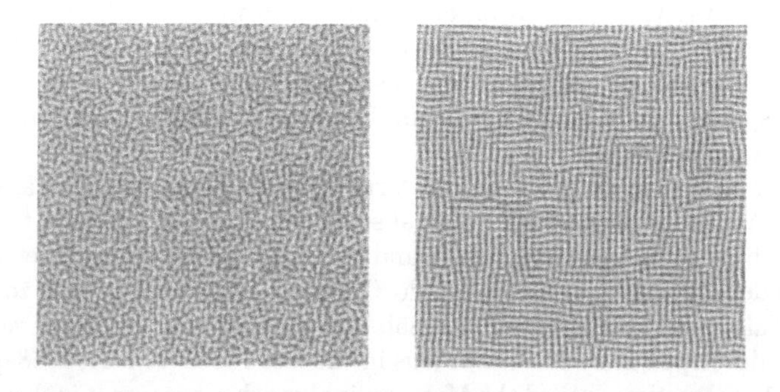

Figure 9.1 This figure gives a conceptual view of the annealing process. The figure on the left shows the disordered, high-energy state of a crystalline structure at a high temperature. The figure on the right shows the ordered, low-energy state of the same structure after it has cooled. (This figure was copied from http://en.wikipedia.org/wiki/Simulated_annealing and is distributed under the provisions of the GNU Free Documentation License.)

SA is based on statistical mechanics, which is the study of the behavior of large numbers of interacting particles, such as atoms in a gas. The number atoms in any material is on the order of 10^{23} per cubic centimeter, so when we examine the properties of a material, we only observe properties that are highly likely to occur. We also notice that equilibrium-energy configurations, and configurations that are similar to them, are observed very often, even though those configurations comprise only a tiny fraction of the possible configurations. This is because materials tend to converge to minimum-energy states; that is, nature is an optimizer.

Suppose that we use $E(s)$ to denote the energy of a specific configuration s of the atoms in some material. The probability that the system of atoms is in configuration s is given as

$$P(s) = \frac{\exp[-E(s)/(kT)]}{\sum_w \exp[-E(w)/(kT)]} \tag{9.1}$$

where k is Boltzmann's constant, T is the temperature of the system at equilibrium, and the sum in the denominator is taken over all posible configurations w [Davis and Steenstrup, 1987]. Now suppose that we have a system that is in configuration q, and we randomly select a configuration r that is a candidate for the system configuration at the next time step. If $E(r) < E(q)$, then we accept r as the configuration at the next time step with probability one:

$$P(r|q) = 1 \text{ if } E(r) < E(q). \tag{9.2}$$

That is, if our candidate configuration r has an energy that is less than that of s, we automatically move to r at the next time step. However, if $E(r) \geq E(q)$, then we move to r at the next time step with a probability that is proportional to the relative energy of q and r:

$$P(r|q) = \exp[(E(q) - E(r))/(kT)] \text{ if } E(r) \geq E(q). \tag{9.3}$$

That is, there is a nonzero probability $P(r|q)$ that the system moves to a configuration with higher energy. If $E(r) > E(q)$, then Equation (9.3) shows that the probability $P(r|q)$ that the system transitions from state q to state r is less than 1, but it increases as T increases. If we use the transition rules of Equations (9.2) and (9.3), then as time $\to \infty$, the probability that the system is in some configuration s converges to the Boltzmann distribution of Equation (9.1).

9.2 A SIMPLE SIMULATED ANNEALING ALGORITHM

Since annealing in nature results in low-energy configurations of crystals, we can simulate it in an algorithm to minimize cost functions. We start with a candidate solution s to some minimization problem. We also start with a high "temperature" so that the candidate solution is likely to change to some other configuration. We randomly generate an alternative candidate solution r and measure its cost, which is analogous to the energy of a crystalline structure. If the cost of r is less than that of s, then we update the candidate solution accordingly, as indicated by Equation (9.2). If the cost of r is greater than or equal to that of s, then we update the candidate solution with some probability less than or equal to one, as indicated by Equation (9.3). SA is sometimes called Boltzmann annealing because of its use of

Equations (9.2) and (9.3). As time progresses (that is, as the the iteration number increases), we decrease the temperature. This results in a tendency of the candidate solution to settle in a low-cost state. The analogies between annealing in nature and the SA algorithm are summarized as follows.

Annealing in Nature		Simulated Annealing
atomic configuration	$\longleftrightarrow$	candidate solution
temperature	$\longleftrightarrow$	tendency to explore search space
cooling	$\longleftrightarrow$	decreasing tendency to explore
changes to atomic configurations	$\longleftrightarrow$	changes to candidate solutions

We see that SA includes many standard EA behaviors. Although we have used annealing in nature to motivate SA in this chapter, SA can in fact be developed without any motivation from nature [Michiels et al., 2007]. There are advantages to both approaches. Appealing to nature may open up new avenues of SA research, but it may also limit the possible extensions to SA. A simple SA algorithm is shown in Figure 9.2.

T = initial temperature > 0
$\alpha(T)$ = cooling function: $\alpha(T) \in [0, T]$ for all T
Initialize a candidate solution x_0 to minimization problem $f(x)$
While not(termination criterion)
    Generate a candidate solution x
    If $f(x) < f(x_0)$
        $x_0 \leftarrow x$
    else
        $r \leftarrow U[0, 1]$
        If $r < \exp[(f(x_0) - f(x))/T]$ then
            $x_0 \leftarrow x$
        End if
    End if
    $T \leftarrow \alpha(T)$
Next iteration

Figure 9.2 A basic simulated annealing algorithm for the minimization of $f(x)$. The function $U[0, 1]$ returns a random number uniformly distributed on $[0, 1]$.

Figure 9.2 shows that the basic SA algorithm has features in common with most other EAs. First, it is simple and intuitively appealing. Second, it is based on an optimization process in nature. Third, it has several tuning parameters that can each have a significant impact on its performance.

- The initial temperature T provides an upper bound for the relative importance of exploration versus exploitation. If the initial temperature is too low, then the algorithm will not effectively explore the search space. If the initial temperature is too high, then the algorithm will take too long to converge.

- The cooling schedule $\alpha(T)$ controls the rate of convergence. At the beginning of the algorithm, exploration is high and exploitation is low. At the end of the algorithm, the converse is true: exploitation is high and exploration is low. The cooling schedule controls the transition from exploration to exploitation. If $\alpha(T)$ is too drastic, then, like a crystalline structure that cools too rapidly, the annealing process will converge to a disordered (high cost) state. If $\alpha(T)$ is too gradual, then the annealing process will take too long to converge. We discuss cooling schedules in Section 9.3.

- The strategy used to generate a candidate solution x at each iteration can have a significant impact on SA performance. Random generation of x may work, but a more intelligent method for trying to generate an x that is better than x_0 will probably give better performance. We discuss candidate generation strategies in Section 9.4.1.

A simplified SA algorithm can be implemented by replacing the acceptance test in Figure 9.2 as follows:

$$\text{Replace "If } r < \exp[(f(x_0) - f(x))/T]\text{" with "If } r < \exp[-c/T]\text{"} \tag{9.4}$$

where c is called the *acceptance probability constant*. This indicates that if the candidate solution x has a higher cost than x_0, then the probability of replacing x_0 with x is independent of its cost. The acceptance probability constant c controls exploration versus exploitation. If c is too large, then the algorithm will not explore the search space aggressively enough. If c is too small, then the algorithm will explore too aggressively without exploiting good solutions that it has previously discovered.

9.3 COOLING SCHEDULES

This section discusses different cooling schedules $\alpha(T)$ that can be used in the SA algorithm of Figure 9.2. The cooling schedule can have a significant impact on SA performance. If an SA implementation does not work on some problem, it may be because the cooling schedule is not appropriate for the problem. Some commonly use cooling schedules include linear cooling, exponential cooling, inverse cooling, logarithmic cooling, and inverse linear cooling, which we discuss in the following sections. We also note that optimization problems can have different scales along different dimensions, and so we discuss in dimension-dependent cooling in Section 9.3.6.

9.3.1 Linear Cooling

Linear cooling is the simplest type of cooling, and follows the schedule

$$\alpha(T) = T_0 - \eta k \tag{9.5}$$

where T_0 is the initial temperature, k is the SA iteration number, and η is a constant. We need to make sure that $T > 0$ for all k, so we should choose η such that the temperature at the maximum iteration number is positive. Alternatively, we could use the following modified form of linear cooling:

$$\alpha(T) = \max(T_0 - \eta k, T_{\min}) \tag{9.6}$$

where $T_{\min}$ is a user-specified minimum temperature.

9.3.2 Exponential Cooling

Exponential cooling follows the schedule

$$\alpha(T) = aT \tag{9.7}$$

where typically $a \in (0.8, 1)$. A larger value of a will give a slower cooling schedule. Figure 9.3 shows normalized temperature for the exponential cooling schedule for different values of a. We see that a should be quite close to 1, otherwise the cooling rate will be too drastic.

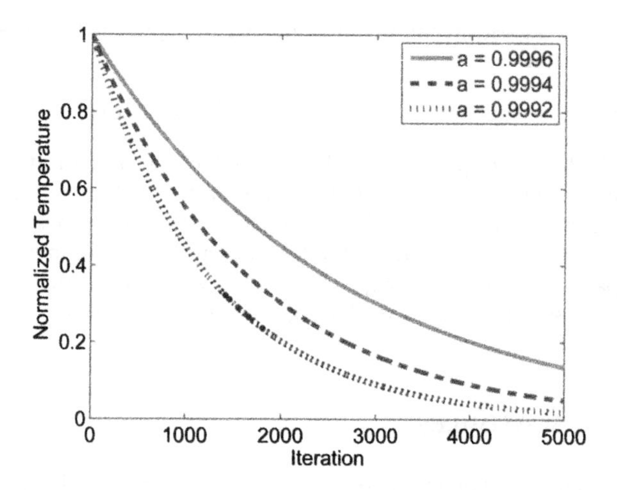

Figure 9.3 Normalized temperature as a function of a for the exponential cooling schedule. The parameter a is usually very close to 1 for this cooling schedule. The cooling rate is very sensitive to changes in a.

9.3.3 Inverse Cooling

Inverse cooling follows the schedule

$$\alpha(T) = T/(1 + \beta T) \tag{9.8}$$

where β is a small constant, typically on the order of 0.001. A smaller value of β will give a slower cooling schedule. This cooling schedule was first suggested in [Lundy and Mees, 1986]. Figure 9.4 shows normalized temperature for the inverse cooling schedule for different values of β. We see that β should be quite small, otherwise the cooling rate will be too drastic.

Comparing Figures 9.3 and 9.4, we see that the exponential cooling and inverse cooling schedules can be made to be very similar to each other by choosing appropriate values for a and β.

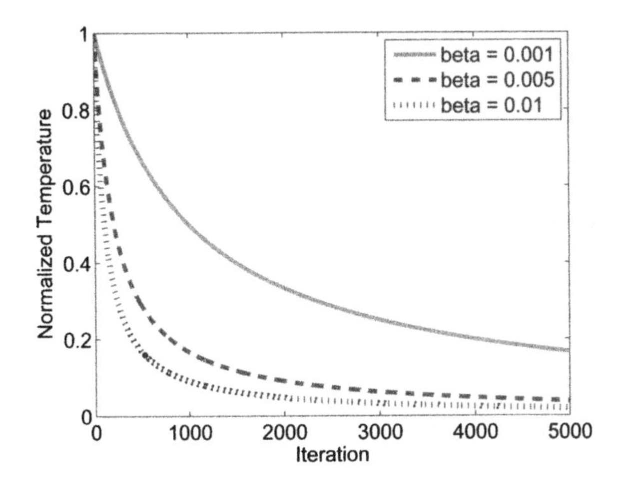

Figure 9.4 Normalized temperature as a function of β for the inverse cooling schedule. The parameter β is usually very small for this cooling schedule. The cooling rate is very sensitive to changes in β.

■ EXAMPLE 9.1

In this example, we optimize the 20-dimensional Ackley function, which is defined in Appendix C.1.2, with the SA algorithm of Figure 9.2. We use the inverse cooling function described in Equation (9.8): $T_{k+1} = T_k/(1+\beta T_k)$, where k is the iteration number, β is the cooling schedule parameter, T_k is the temperature at the k-th iteration, and $T_0 = 100$. We use a Gaussian random number centered at x_0 to generate a new candidate solution at each iteration:

$$x \leftarrow x_0 + N(0, T_k I) \tag{9.9}$$

where $N(0, T_k I)$ is a Gaussian random vector with a mean of 0 and a covariance of $T_k I$, and I is the 20×20 identity matrix. We use the simple acceptance test of Equation (9.4) with $c = 1$.

Figure 9.5 shows the best solution found as a function of the SA iteration number, averaged over 20 Monte Carlo simulations, and for three different values of β. We see that if β is too small (0.0002), then cooling occurs too slowly and the SA algorithm jumps around too aggressively in the search space without exploiting good solutions that it has already obtained. If β is too large (0.001), then cooling occurs too quickly and the SA algorithm tends to get stuck in local minima. If β is just right (0.0005), then cooling occurs at a rate that results in the best convergence. However, we note that toward the end of the plot, the $\beta = 0.0002$ trace appears to be rapidly overtaking the $\beta = 0.0005$ trace. This indicates that although $\beta = 0.0002$ is too small to give good convergence within the iteration number limit that we have used, it will eventually result in enough cooling if the iteration number continues to increase, and will eventually converge to a good result.

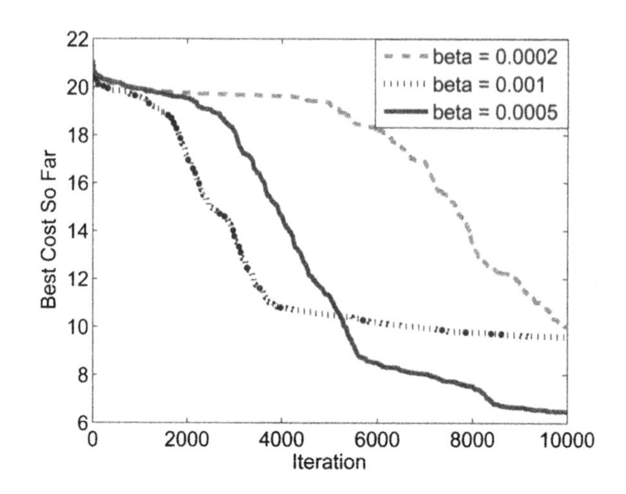

Figure 9.5 Example 9.1 simulation results of the SA algorithm for optimizing the 20-dimensional Ackley function. Results are averaged over 20 Monte Carlo simulations. The inverse cooling schedule parameter β has a significant impact on SA performance.

□

Example 9.1 shows that if cooling is too fast or too slow, then SA performance might not be good. The same conclusion can be drawn about the initial temperature T_0. In Example 9.1 we arbitrarily used $T_0 = 100$. Unfortunately, there are not any good guidelines for the selection of T_0; it depends entirely on the particular optimization problem.

9.3.4 Logarithmic Cooling

Logarithmic cooling follows the schedule

$$\alpha(T) = c/\ln k \tag{9.10}$$

where c is a constant, and k is the SA iteration number. This was first suggested in [Geman and Geman, 1984]. It is sometimes generalized to

$$\alpha(T) = c/\ln(k+d) \tag{9.11}$$

where d is a constant that is often set equal to 1 [Nourani and Andresen, 1998]. Logarithmic cooling is qualitatively different than exponential and inverse cooling, as seen from Figure 9.6. The temperature decreases very rapidly for the first few iterations, and then decreases extremely slowly. This slow decrease means that SA convergence is usually poor with the logarithmic cooling schedule. Therefore, the logarithmic cooling schedule is not recommended for practical applications.

However, the logarithmic cooling schedule is theoretically attractive and widely known in the SA community because it has been proven to give a global minimum under certain conditions [Geman and Geman, 1984]. As a simple demonstration

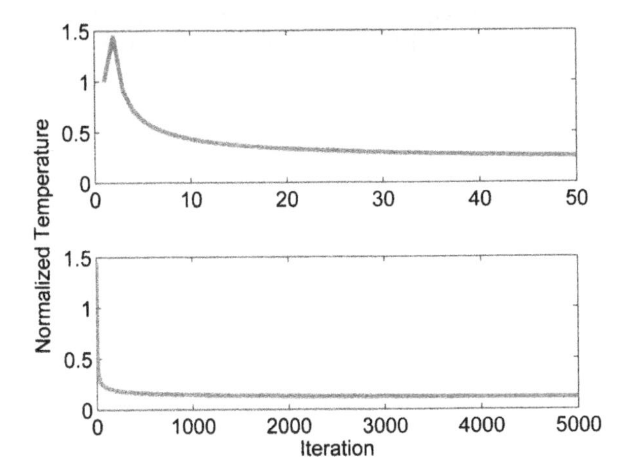

Figure 9.6 Normalized temperature for the logarithmic cooling schedule. The top figure shows the temperature for the first 50 iterations, and the bottom figure shows the temperature for the first 5,000 iterations. The temperature decreases very rapidly for the first few iterations, and then decreases so slowly that it is impractical for SA implementations.

of the proof [Ingber, 1996], suppose that we have a discrete problem so that the size of the search space is finite. We generate the candidate x from a Gaussian distribution so that the probability of generating x, given that x_k is the current candidate solution at the k-th iteration, is

$$g_k \equiv P(x|x_k) = (2\pi T_k)^{D/2} \exp\left[-||x - x_k||_2^2/(2T_k)\right] \tag{9.12}$$

where D is the problem dimension. In other words, the conditional probability of generating x given that x_k is the current candidate, is Gaussian with a mean of x_k and a covariance of $T_k I$, where T_k is the temperature at the k-th iteration, and I is the identity matrix. In order to visit every possible candidate solution in the search space, it suffices to show that as the iteration count approaches infinity, the probability of *not* visiting x approaches zero; that is,

$$\lim_{N\to\infty} \prod_{k=1}^{N} (1 - g_k) = 0. \tag{9.13}$$

Taking the natural log of the above equation gives

$$\ln\left[\lim_{N\to\infty} \prod_{k=1}^{N} (1 - g_k)\right] = \lim_{N\to\infty} \left[\ln \prod_{k=1}^{N} (1 - g_k)\right] = -\infty. \tag{9.14}$$

A Taylor series expansion of the logarithm about $g_1 = g_2 = \cdots = 0$ gives

$$\ln[(1 - g_1)(1 - g_2)\cdots] = \ln 1 - g_1 - g_2 - \cdots. \tag{9.15}$$

Combining the two previous equations gives the following sufficient condition for obtaining a 100% probability of visiting x as the iteration count approaches infinity:

$$\lim_{N\to\infty}\sum_{k=1}^{N} g_k = \infty. \tag{9.16}$$

If g_k is given by Equation (9.12), and if $T_k = T_0/\ln k$, then the left side of the above equation becomes

$$\begin{aligned}\lim_{N\to\infty}\sum_{k=1}^{N}(2\pi T_0/\ln k)^{D/2}\exp\left[-\|x-x_k\|_2^2/(2T_0/\ln k)\right] &\geq \\ \sum_{k=1}^{\infty}\exp(-\ln k) = \sum_{k=1}^{\infty} 1/k = \infty\end{aligned} \tag{9.17}$$

where the inequality is true if T_0 is large enough (see Problem 9.5).

9.3.5 Inverse Linear Cooling

Inverse linear cooling follows the schedule

$$\alpha(T) = T_0/k \tag{9.18}$$

where T_0 is the initial temperature, and k is the SA iteration number. The inverse linear cooling schedule exhibits the fast cooling of the logarithmic schedule during the first few iterations, but it avoids the nonzero temperatures and slow cooling of later iterations, as seen from Figure 9.7. The temperature decreases very rapidly and quickly reaches zero. This means that inverse linear cooling is not effective for problems that require a lot of exploration, but is more suitable for problems that can be initialized with a candidate solution that is known to be close to the optimal solution.

Inverse linear cooling, like logarithmic cooling, is theoretically attractive and widely known in the SA community because it has been proven to result in a global optimum under certain conditions [Szu and Hartley, 1987]. As a simple demonstration similar to that used in the previous section for logarithmic cooling [Ingber, 1996], suppose that we have a discrete problem so that the size of the search space is finite. We generate the candidate x from a Cauchy distribution so that the probability of obtaining x, given that x_k is the current candidate solution at the k-th iteration, is

$$g_k \equiv P(x|x_k) = \frac{T_k}{\left(\|x-x_k\|_2^2 + T_k^2\right)^{(D+1)/2}} \tag{9.19}$$

where D is the problem dimension. Note that in the previous section we used a Gaussian distribution to generate candidate solutions, while in this section we use a Cauchy distribution. Figure 9.8 compares a Cauchy PDF with a Gaussian PDF. We see that the Cauchy PDF has much fatter tails, which means that we are more likely to generate candidate solutions that are farther from the current candidate solution.

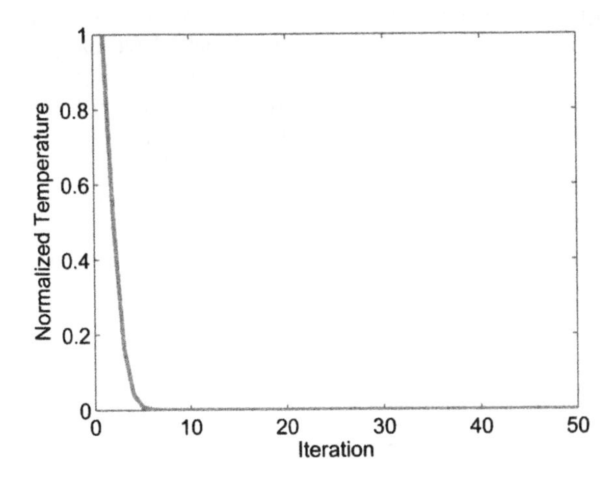

Figure 9.7 Normalized temperature for the inverse linear cooling schedule. The temperature reaches zero very quickly.

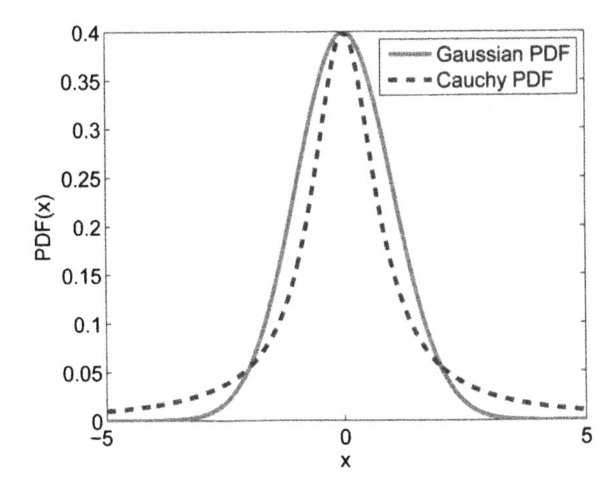

Figure 9.8 A comparison of the one-dimensional Cauchy and Gaussian probability density functions.

If g_k is given by Equation (9.19), and if $T_k = T_0/k$, then the left side of Equation (9.16) becomes

$$\lim_{N \to \infty} \sum_{k=1}^{N} \frac{T_0/k}{\left(||x - x_k||_2^2 + T_0^2/k^2\right)^{(D+1)/2}} \geq \sum_{k=1}^{\infty} 1/k = \infty \tag{9.20}$$

where the inequality is true for appropriate values of T_0 (see Problem 9.6).

Now we compare the convergence results obtained with the logarithmic and inverse linear cooling schedules. Recall that the Cauchy PDF has much fatter tails than the Gaussian PDF. Also recall that the candidate solution generation function of Equation (9.19), which uses the Cauchy PDF, is combined with the inverse linear cooling schedule of Figure 9.7. Finally, recall that the candidate solution generation function of Equation (9.12), which uses the Gaussian PDF, uses the logarithmic cooling schedule of Figure 9.6. Since the Cauchy generation function has fatter tails than the Gaussian generation function, it is guaranteed to converge with a much faster cooling schedule than the one used in conjunction with the Gaussian generation function.

9.3.6 Dimension-Dependent Cooling

In real-world applications, and even in some benchmark problems, the landscape of the cost function can look very different when viewed along different dimensions. For example, consider the function

$$\begin{aligned} f(x) &= 20 + e - 20\exp\left(-0.2\sum_{i=1}^{n} y_i^2/n\right) - \exp\left(\sum_{i=1}^{n}(\cos 2\pi y_i)/n\right) \\ y_i &= \begin{cases} x_i & \text{for odd } i \\ x_i/4 & \text{for even } i. \end{cases} \end{aligned} \tag{9.21}$$

This is simply a scaled version of the Ackley function, which is defined in Appendix C.1.2. The fact that x_i is scaled for even values of i means that the function is "stretched out" along those dimensions. Figure 9.9 shows a two-dimensional plot of this function. Because of the scaling of even dimensions, the function is much smoother along the x_2 dimension than along the x_1 dimension.

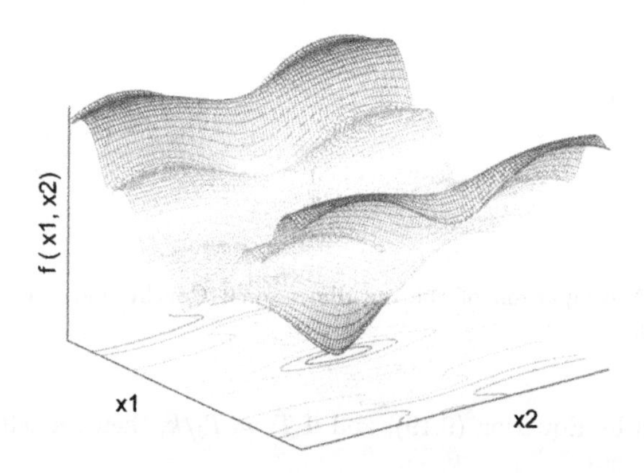

Figure 9.9 Scaled version of the two-dimensional Ackley function. The topology is much smoother along the x_2 direction, which indicates that the SA algorithm should use a slower cooling schedule along that dimension.

For functions that have different topologies along different dimensions, we might want to use a different cooling schedule for different dimensions. For the scaled Ackley function of Equation (9.21), we would want to use slower cooling along the even dimensions, and faster cooling along the odd dimensions. This will allow the SA algorithm to gradually converge to optimal values along the gradually-changing dimensions of the function. A fast cooling schedule along gradually-changing dimensions would prevent the SA algorithm from moving down gentle slopes. However, along the more dynamic dimensions of the function, a faster cooling schedule is needed. The SA algorithm will move downhill along dynamic dimensions even with a fast cooling rate, but a slow cooling rate will result in too much jumping around. As another way of looking at it, we can say that we need a more aggressive search (high temperatures) along low-sensitivity dimensions, and a less aggressive search (low temperatures) along high-sensitivity dimensions.

If we use the inverse cooling schedule of Equation (9.8), the above discussion implies a smaller value of β for the even dimensions and a larger value of β for the odd dimensions. This means that each dimension of the problem will have its own temperature. This results in a modification of the basic SA algorithm of Figure 9.2 to obtain the dimension-dependent SA algorithm of Figure 9.10.

T_i = initial temperature > 0, $i \in [1, n]$
$\alpha_i(T_i)$ = cooling function for i-th dimension, $i \in [1, n]$: $\alpha(T_i) \in [0, T_i]$ for all T_i
Initialize a candidate solution x_0 to minimization problem $f(x)$:
    $x_0 = [x_{01}, x_{02}, \cdots, x_{0n}]$
While not(termination criterion)
    Generate a candidate solution $x_1 = [x_{11}, x_{12}, \cdots, x_{1n}]$
    If $f(x_1) < f(x_0)$
        $x_0 \leftarrow x_1$
    else
        For $i = 1, \cdots, n$
            $r \leftarrow U[0, 1]$
            If $r < \exp[(f(x_0) - f(x_1))/T_i]$ then
                $x_{0i} \leftarrow x_{1i}$
            End if
        Next dimension i
    End if
    $T_i \leftarrow \alpha_i(T_i)$, $i \in [1, n]$
Next iteration

Figure 9.10 A dimension-dependent simulated annealing algorithm for the minimization of the n-dimensional function $f(x)$. The function $U[0, 1]$ returns a random number uniformly distributed on [0, 1]. This algorithm is a generalization of the basic SA algorithm of Figure 9.2; here we have allowed each dimension to have its own temperature and its own cooling schedule.

■ EXAMPLE 9.2

In this example we optimize the 20-dimensional scaled Ackley function, which is given by Equation (9.21), with the dimension-dependent SA algorithm of Figure 9.10. We use the inverse cooling function described in Equation (9.8) for each dimension: $T_{k+1,i} = T_{ki}/(1+\beta_i T_{ik})$, where i is the dimension number, k is the SA iteration number, β_i is the cooling schedule parameter for the i-th dimension, T_{ki} is the temperature at the k-th iteration of the i-th dimension, and $T_{0i} = 100$ for all i. We use a Gaussian random number centered at x_0 to generate a new candidate solution at each iteration:

$$x_{1i} \leftarrow x_{0i} + N(0, T_{ki}) \tag{9.22}$$

where $N(0, T_{ki})$ is a Gaussian random number with a mean of 0 and a variance of T_{ki}. We use the simple acceptance test of Equation (9.4) with $c = 1$.

Figure 9.11 shows the best solution found as a function of the SA iteration number, averaged over 20 Monte Carlo simulations, and for four different combinations of β. We see that if β is too small (0.001), then cooling occurs too slowly and the SA algorithm jumps around too aggressively in the search space without exploiting good solutions that it has already obtained. If β is too large (0.005), then cooling occurs too quickly and the SA algorithm tends to get stuck in local minima. However, if β is large for odd dimensions and small for even dimensions, then cooling occurs at a rate that results in the best convergence. This combination gives fast cooling for the highly dynamic odd dimensions, and slow cooling for the even dimensions.

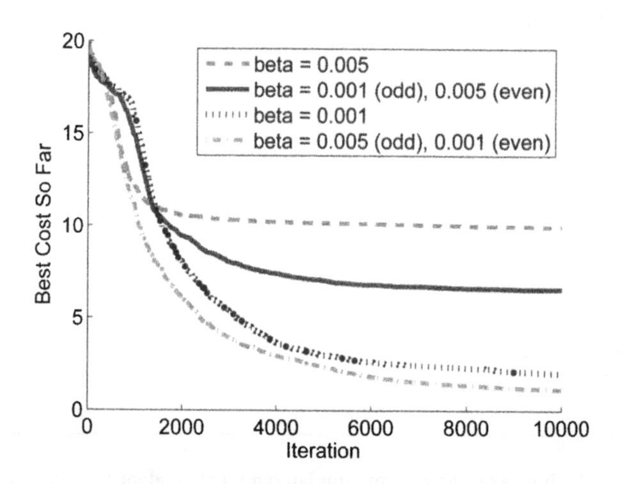

Figure 9.11 Example 9.2 simulation results of the dimension-dependent SA algorithm optimizing the 20-dimensional scaled Ackley function. Results are averaged over 20 Monte Carlo simulations. The cooling schedule parameters $\{\beta_i\}$ can be adjusted individually for each dimension to give the best results.

□

9.4 IMPLEMENTATION ISSUES

This section discusses a couple of implementation issues, including how to generate candidate solutions, when to reinitialize the cooling temperature, and why we need to keep track of the best candidate solution.

9.4.1 Candidate Solution Generation

The statement "Generate a candidate solution" in the SA algorithms of Figures 9.2 and 9.10 is deceptively simple. There are many different ways that this statement can be implemented, and the implementation choice can have a large impact on SA performance. One method of generating candidate solutions is to simply choose a random point in the search space. However, after the SA algorithm has begun converging to a good solution, we would expect that the current solution candidate x_0 is much better than most other points in the search space. Therefore, generating a random solution candidate will probably not be very effective. As a general rule, we should bias candidate solution generation toward the current candidate solution x_0. This is the reason that Equations (9.9) and (9.22) use a Gaussian random variable centered at x_0 for candidate solution generation. Furthermore, the variance of Gaussian random variable is equal to the temperature, which decreases with time, and so the search tends to narrow as the SA iteration count increases. Equation (9.19), the Cauchy distribution, can be used as a more aggressive method for generating candidate solutions while still being centered at x_0. Biasing candidate solution generation toward x_0 tends to exclude not only very poor candidates, but also very good candidates. However, very poor points in the search space are generally more common than very good candidates, so biasing the search toward x_0 is usually effective.

9.4.2 Reinitialization

As we discussed earlier in this chapter, the cooling schedule is an important contributor to SA performance. If we cool the temperature too quickly, then the SA will get stuck in a local optimum and performance will be poor. However, we usually do not know ahead of time what the appropriate cooling schedule is. Therefore, we often monitor the improvement of the SA algorithm, and if we do not find a better candidate solution within L iterations, we reinitialize the temperature to T_0 to increase exploration.

9.4.3 Keeping Track of the Best Candidate Solution

Recall from Figure (9.2) that a new candidate solution x might replace a current candidate solution x_0, even if x is worse than x_0. This is a necessary risk to sufficiently explore the search space, but it might result in the loss of a good candidate solution. Therefore, we usually want to implement an archive in SA so that we keep track of the best candidate solution obtained so far. This is similar to elitism in Section 8.4; however, in that section we actually retained the best candidate solutions in the population. We cannot do that directly in SA unless we increase the population size beyond 1, which is a possibility that we have not discussed in this chapter. However, regardless of the population size, we can always maintain an archive that contains the best candidate found so far. So even if we replace a good candidate solution with a poor candidate due to the exploratory nature of SA, we will still keep track of the best candidate found so far. The best candidate solution in the archive will never be replaced with a worse candidate. Then we can return the best candidate found when the SA algorithm is complete.

9.5 CONCLUSION

Simulated annealing is one of the older EAs, originating in 1983, but we have discussed in this part of the book because it is not always considered to be a classic EA. It is not population based, but it is clear that some of the classic EAs are not population based either, and so that is not a sufficient reason to remove it from the EA category. Since SA is based on a natural process, and since it is an iterative optimization algorithm, we generally consider it to be an EA. Its maturity and scientific roots have resulted in many papers, books, and applications. Readers who want a more comprehensive coverage of SA are recommended to the books [van Laarhoven and Aarts, 2010], [Otten and van Ginneken, 1989], and [Aarts and Korst, 1989]. Tutorial chapters are available at [Aarts et al., 2003] and [Henderson et al., 2003].

Like all of the EAs discussed in this book, SA can be useful for a wide variety of optimization problems, including both continuous-domain and discrete-domain problems. Current research directions in the area of SA mirrors current emphases in general EA research: SA for multi-objective problems [Bandyopadhyay et al., 2008], hybridizations of SA with other EAs [Cakir et al., 2011], parallelization [Zimmerman and Lynch, 2009], and constrained optimization [Singh et al., 2010].

This chapter has presented the background and implementation of SA, but there are other important aspects of SA that we have not had time to discuss. For instance, a Markov model and some theoretical convergence proofs are presented in [Michiels et al., 2007], although there is still much room for additional modeling and theoretical results. Practitioners are interested not only in convergence, but also in performance over finite time intervals, and this issue is discussed in [Henderson et al., 2003], [Vorwerk et al., 2009].

PROBLEMS

Written Exercises

9.1 What are the units of temperature in Equation (9.1)?

9.2 Draw a qualitatively correct plot of the probability $P(r|q)$ from Section 9.1 as a function of $\Delta E = E(r) - E(q)$.

9.3 What value of the acceptance probability constant c will give an acceptance probability of p when the new candidate solution x has a higher cost than the current candidate solution x_0?

9.4 Suppose you want to run SA for 10,000 iterations with linear cooling. What value of η should you use so that the temperature reaches 0 at the final iteration?

9.5 What is "large enough" in the convergence proof of the logarithmic cooling schedule?

9.6 What are "appropriate values of T_0" in the convergence proof of the inverse linear cooling schedule?

9.7 Write the linear and inverse linear cooling schedules in the form $T_{k+1} = \alpha(k, T_k)$, where k is the iteration number of the SA algorithm.

9.8 This problem compares exponential cooling and inverse cooling.

a) Write the exponential and inverse cooling schedules in the form $T_k = f(k, T_0)$, where k is the iteration number of the SA algorithm and T_0 is the initial temperature.

b) Find an expression for a in the exponential cooling schedule so that the temperature after N iterations is the same as it is in the inverse cooling schedule.

c) Given $T_0 = 100$, what value of a gives an equivalent temperature after 10,000 iterations when: (1) $\beta = 0.01$; (2) $\beta = 0.001$; (3) $\beta = 0.0001$?

Computer Exercises

9.9 This problem explores the reinitialization strategy discussed in Section 9.4.2. Implement an SA to minimize the 20-dimensional Ackley function. Use the following parameters:

- Inverse cooling with $\beta = 0.001$;
- Initial temperature $= 100$;
- Iteration limit $= 10{,}000$;
- Acceptance testing with Equation (9.4) with $c = 1$;

- Candidate solution generation using $x \leftarrow x_0 + r$, where r is a normally distributed zero-mean random number with variance T^2.

Keep track of the best-so-far solution x_k^* as a function of iteration number k, and plot the average of x_k^* over 20 Monte Carlo simulations. Compare the plots for the following reinitialization strategies:

- Reinitialize T whenever x is worse than x_0 for 10 consecutive iterations;
- Reinitialize T whenever x is worse than x_0 for 100 consecutive iterations;
- Reinitialize T whenever x is worse than x_0 for 1,000 consecutive iterations;
- Never reinitialize T.

What do you conclude from your results about the value of reinitializing T?

9.10 This problem explores methods used for generating candidate solutions. Implement an SA to minimize the 20-dimensional Ackley function. Use the same parameters as in Problem 9.9. Keep track of the best-so-far solution x_k^* as a function of iteration number k, and plot the average of x_k^* over 20 Monte Carlo simulations. Compare the plots for the following candidate solution generation strategies:

- $x \leftarrow x_0 + r_1$, where r_1 is a normally distributed zero-mean random number with variance T^2;
- $x \leftarrow r_2$, where r_2 is a random number uniformly distributed on the search domain.

What do you conclude from your results about the importance of candidate solution generation?

CHAPTER 10

Ant Colony Optimization

Go to the ant, you sluggard; consider its ways, and be wise!

—Proverbs 6:6

Ants are simple creatures but they can accomplish a lot by working together. The quote at the beginning of this chapter presents ants as a paradigm of hard work, but they can also be portrayed as the epitome of selfless cooperation. A single ant does not have much to offer. A solitary ant might wander aimlessly in circles until it dies of exhaustion [Delsuc, 2003]. The average ant has only 10,000 neurons in its brain, which doesn't seem like enough to accomplish much. But ants join together in colonies that can number in the millions. A one-million member ant colony has a collective neuron count of 10 billion, which begins to rival the neuron count of an average human. Ants seem to operate as a single entity and are therefore sometimes referred to as a superorganism [Hölldobler and Wilson, 2008]. An ant colony discovered on the Japanese island of Hokkaido was reported in 1979 to contain over 300 million ants living in 45,000 interconnected nests [Hölldobler and Wilson, 1990, page 1]. Ants thrive in almost every environment on earth, and are estimated to comprise over 15% of the mass of all land animals on earth [Schultz, 2000]. Myrmecologists (that is, those who study ants) tell us that the number of ant species is about 8,800 and have a global population of about one quadrillion, which means that there are about 150,000 ants for every person on earth. Why is it that ants, such tiny creatures, have been so successful for so long in so many different environments? Scientists attribute their adaptability and dominance to their social organization. Perhaps it is no coincidence that the most dominant mammal on Earth, humans, is also the clear leader in social organization.

Ants communicate mainly by using pheromones, which are chemical substances that they excrete. When ants travel along a path to a food source and bring food back to their colony, they leave a trail of pheromone. Other ants smell the pheromone with their antennas, follow the path, and bring more food back to the colony. In the process, ants continue to lay down pheromone, which other ants continue to smell, and the path to the food source is reinforced. The shortest path to the food thus becomes more attractive over time as it is strengthened by positive feedback.

Sometimes the food source is depleted or an obstacle prevents travel to the food source. When ants travel along a path and do not find food, they wander until they do find food. If they do not return to their colony using the original path, they do not deposit any additional pheromone on that path. As time passes, the pheromone on the original path evaporates, fewer ants take the original path, more ants take the new path to the new food source, and a new optimal path is discovered by the ants. This general process is depicted in Figure 10.1.

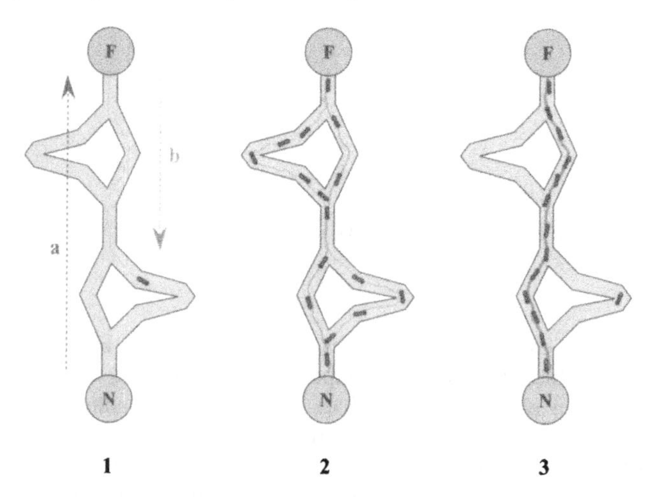

Figure 10.1 Ants depositing and following pheromone. (1) The first ant travels in the direction indicated by a, finds a food source F, and returns to the nest N in the direction indicated by b, laying a pheromone trail as it travels. (2) The ants follow one of four possible paths from N to F, but pheromone reinforcement makes the shortest path more appealing. (3) The ants tend to follow the path with the most pheromone, continuing to reinforce its desirability, while the pheromone on the longer paths evaporates. (This figure was created by Johann Dréo, was copied from `http://en.wikipedia.org/wiki/File:Aco_branches.svg`, and is distributed under the provisions of the GNU Free Documentation License.)

Ants can not only find the optimal path to a food source, but they can also perform many other impressive tasks by working together. They can build complex networks of tunnels, either underground, or, in the case of weaver ants, in trees. Their colonies have specialized rooms for storing, mating, and caring for larvae.

They can plant gardens to cultivate their own food source [Schultz, 1999]. They can form chains to cross gaps over the ground or over water (see Figure 10.2). They can form rafts to survive a flood, or to travel across water.

Figure 10.2 Ants form a bridge between leaves. Ants use bridges not only for transportation, but also to pull leaves together during nest construction. Many other photographs like this can be found in [Hölldobler and Wilson, 1994]. (This photograph was taken by Sean Hoyland, was copied from http://en.wikipedia.org/wiki/File:SSL11903p.jpg, and is distributed under the provisions of the GNU Free Documentation License.)

Overview of the chapter

In this chapter we discuss ant colony optimization (ACO), which is an algorithm that is motivated by the pheromone-depositing behavior of ants. Most ACO researchers emphasize that it is not an EA since candidate solutions do not directly exchange solution information with each other. We include ACO in this book not because we want to engage in the EA/non-EA debate, but simply because ACO is an interesting and effective biologically-motivated, population-based optimization algorithm.

ACO was developed by Martin Dorigo in his doctoral dissertation and was first published in 1991 [Colorni et al., 1991]. In Section 10.1 we discuss the pheromone deposition of biological ants, its evaporation, and mathematical models that describe these processes. In Section 10.2 we discuss the ant system (AS), which was proposed in the mid-1990s, and which was the first ACO algorithm. ACO was originally proposed to find optimal paths, but was quickly modified to deal with optimization problems with continuous domains, and that is what we discuss in Section 10.3. In Section 10.4 we discuss some popular modifications that have been made to the basic ant system algorithm, including the max-min ant system (MMAS) and the ant colony system (ACS). In Section 10.5 we give a brief overview of ACO research in the area of theory and modeling.

10.1 PHEROMONE MODELS

Suppose that we observe an ant nest and a food source, and that ants have two possible routes to obtain food. One route is long, and the other is short, as shown in Figure 10.3. Goss and his coworkers ran many experiments of this type with the Argentine ant, and they found that in 95% of their experiments, over 80% of the ant traffic was on the shorter path [Goss et al., 1989]. As ants reached the fork of the path, they made a random decision which path to take. The ants that chose the shorter path were able to return to their nest sooner than the ants that chose the longer path. This resulted in more ants taking the shorter path per unit time. This, in turn, resulted in more pheromone deposition on the shorter path. Finally, the larger amount of pheromone on the shorter path motivated later ants to take that path.

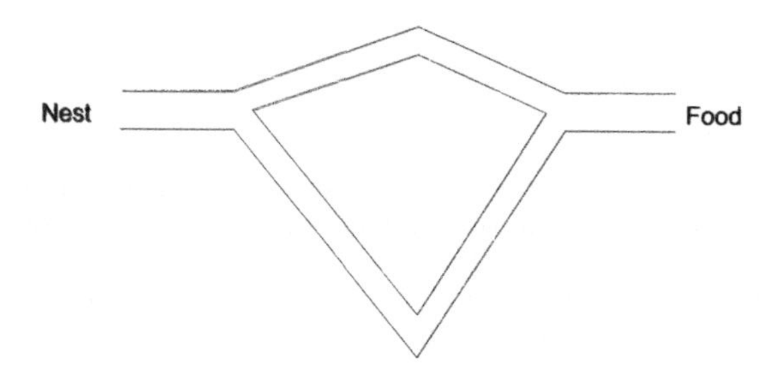

Figure 10.3 An experimental setup to explore how ants find the shortest distance to food. In 95% of the experiments, over 80% of the ant traffic was on the shorter path. Adapted from [Goss et al., 1989].

We see that ant travel is a positive feedback phenomenon, at least to a certain point. There were always some ants that chose the longer path because their choices are partially governed by random processes. However, in general, as more ants choose the shorter path, the shorter path receives more pheromone; and as the shorter path receives more pheromone, more ants choose it.

This positive feedback phenomenon is also a characteristic of EAs. For example, in GAs, individuals in the first generation with beneficial genetic features are more likely to be selected for recombination. This means that the second generation is more likely to possess those genetic features. The increased prevalence of those beneficial features in the second generation then makes it more likely that they will be passed on to the third generation. This positive feedback phenomenon is seen not only in ACO and GAs, but in all EAs.

Pheromones are not only deposited by ants, but they also evaporate. Goss performed another experiment in which only one path was available to the ant colony. The ants ran back and forth between their nest and the food source on this path because they did not have any other options. After a while, Goss added a short path to the food, as shown in Figure 10.4. The ants now had a choice, but all of the pheromone was on the long path. However, when ants reach a fork in the road, they do not automatically take the pheromone-saturated path. They are more likely to take a path with more pheromones, but there is also a random element to their behavior. Therefore, some of the ants took the newly-presented short path.

As they took this short path, they deposited pheromones on it, which made it more attractive to subsequent ant travelers. In about 20% of the experiments, the majority of ants ended up taking the short path, even though it did not have any initial pheromone deposits. This demonstrates the fact that pheromones evaporate. However, in this experiment, they did not evaporate quickly enough for the ants to take the shorter path more often than the longer path.

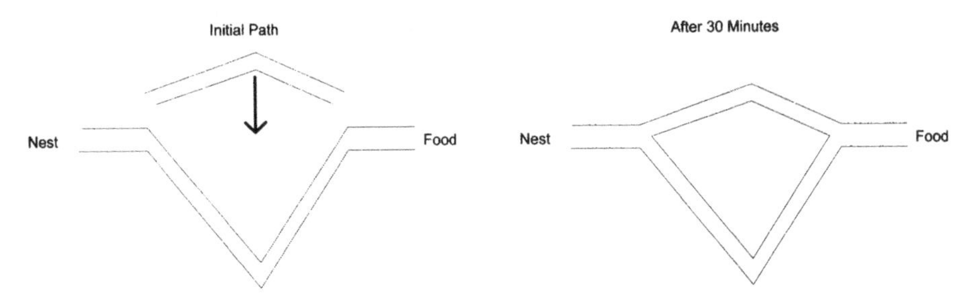

Figure 10.4 An experimental setup to explore how ants react when a shorter path to food is added. In 20% of the experiments, over 50% of the ant traffic converged to the shorter path. Adapted from [Goss et al., 1989].

In view of these experiments, and others like them, Deneubourg and his colleagues proposed a mathematical model for the deposition and evaporation of pheromone [Deneubourg et al., 1990]. Given two paths to choose from, the probability that an ant chooses path 1 is given by

$$p_1 = \frac{(m_1 + k)^h}{(m_1 + k)^h + (m_2 + k)^h} \tag{10.1}$$

where m_i is the number of ants that have previously chosen path i, and h and k are experimentally-determined parameters. This initial model does not take pheromone evaporation into account. Typical values for k and h are

$$k \approx 20, \quad h \approx 2. \tag{10.2}$$

Figure 10.5 shows the results of a simulation of Equation (10.1), which applies to two paths of equal length. The top plot shows that the behavior of the first 100 ants is not predictable. The ants' behavior is mostly random, and there is about a 50% chance that an ant will choose either path 1 or path 2. The bottom plot shows that as one path starts to receive the majority of pheromone deposits, it becomes more attractive, which results in the postive feedback phenomon that we discussed earlier in this section. Eventually 100% of the ant traffic will be on only one of the two paths.

We see that ants are able to find optimal solutions to the problems that they encounter in their everyday lives. This motivates us to simulate artificial ants to find optimal solutions to engineering problems.

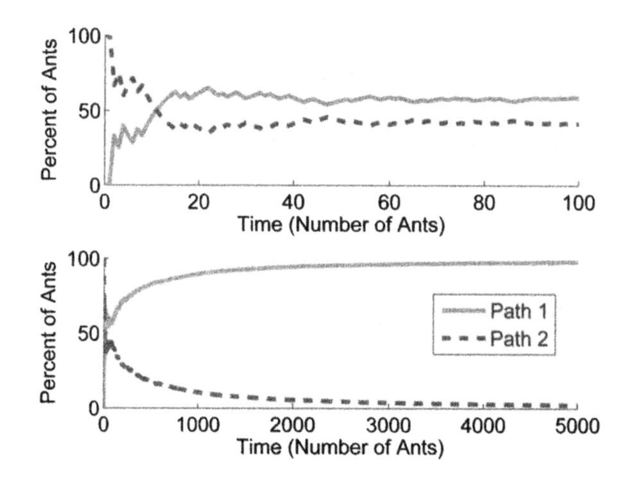

Figure 10.5 Simulation results of Equation (10.1). Initially the ants have about a 50% chance of choosing either path. After awhile, one path receives the predominance of pheromone, which results in a positive feedback phenomenon, and 100% of the ant traffic settles on one of the two paths.

10.2 ANT SYSTEM

The ant system was the first ACO algorithm that was published [Colorni et al., 1991], [Dorigo et al., 1996]. It can be illustrated on the traveling salesman problem (TSP; see Section 2.5 and Chapter 18). Each ant in the ACO simulation travels from one city to another, and the simulation deposits pheromone on the ants' paths after they complete their trip. Pheromones are not only deposited, but they also evaporate. The probability that an ant travels from its current city to some other city is proportional to the amount of pheromone between the cities. Ants are also assumed to have some knowledge about the problem that helps them make decisions during their travels. They know the distance from their current city to other cities, and they are more likely to travel to a close city than to a distant city, since the objective of the algorithm is to find the shortest path. The ant system algorithm is illustrated in Figure 10.6.

Figure 10.6 shows that the probability of each ant traveling from city i to city j is proportional to the amount of pheromone on the path between those cities, and is inversely proportional to the distance between those cities. The ratio α/β determines the relative importance of pheromone information to distance information when deciding which city to travel to. When an ant travels from city i to city j, the amount of pheromone on that path is increased in an amount proportional to the quality of that ant's solution (that is, inversely proportional to the ant's total travel distance).

Figure 10.6 is a fairly complete algorithm, but there are still some implementation details that are left to the programmer. For example, is $\tau_{ij} = \tau_{ji}$? In a biological ant system, the amount of pheromone between nodes i and j is the same as the

---

n = number of cities
α, β = relative importance of pheromones vs. heuristic information
Q = deposition constant
ρ = evaporation rate $\in (0, 1)$
$\tau_{ij} = \tau_0$ (initial pheromone between cities i and j) for $i \in [1, n]$ and $j \in [1, n]$
d_{ij} = distance between cities i and j for $i \in [1, n]$ and $j \in [1, n]$
While not(termination criterion)
    For $q = 1$ to $n - 1$
        For each ant $k \in [1, N]$
            Initialize the starting city c_{k1} of each ant $k \in [1, N]$
            Initialize the set of cities visited by ant k: $C_k \leftarrow \{c_{k1}\}$ for $k \in [1, N]$
            For each city $j \in [1, n]$, $j \notin C_k$
                probability $p_{ij}^{(k)} \leftarrow \left(\tau_{ij}^{\alpha}/d_{ij}^{\beta}\right) / \left(\sum_{m=1, m \notin C_k}^{n} \tau_{im}^{\alpha}/d_{im}^{\beta}\right)$
            Next j
            Let ant k go to city j with probability $p_{ij}^{(k)}$
            Use $c_{k,q+1}$ to denote the city selected in the previous line
            $C_k \leftarrow C_k \cup \{c_{k,q+1}\}$
        Next ant
    Next q
    $L_k \leftarrow$ total path length constructed by ant k, for $k \in [1, N]$
    For each city $i \in [1, n]$ and each city $j \in [1, n]$
        For each ant $k \in [1, N]$
            If ant k went from city i to city j
                $\Delta\tau_{ij}^{(k)} \leftarrow Q/L_k$
            else
                $\Delta\tau_{ij}^{(k)} \leftarrow 0$
            End if
        Next ant
        $\tau_{ij} \leftarrow (1 - \rho)\tau_{ij} + \sum_{k=1}^{N} \Delta\tau_{ij}^{(k)}$
    Next city pair
Next generation

Figure 10.6 A simple ant system (AS) for solving a TSP. Each generation, some of the pheromone between cities i and j evaporates, but the pheromone also increases due to ants that travel between the two cities.

amount between nodes j and i, but this is not necessary in ant system simulations. It can easily be imagined that travel from node i to j could lead to a good solution, while travel from node j to i could lead to a poor solution. This would result in $\tau_{ij} \neq \tau_{ji}$, which corresponds to the asymmetric TSP (see Figure 10.7).

Other implementation details that could be added to Figure 10.6 include intelligent initialization, elitism, and mutation. First, ACO performance, like EA performance, can strongly depend on proper initialization (see Section 8.1). For the TSP we may want to initialize certain individuals using a simple heuristic algorithm. For example, at the first generation we could force one of the ants to

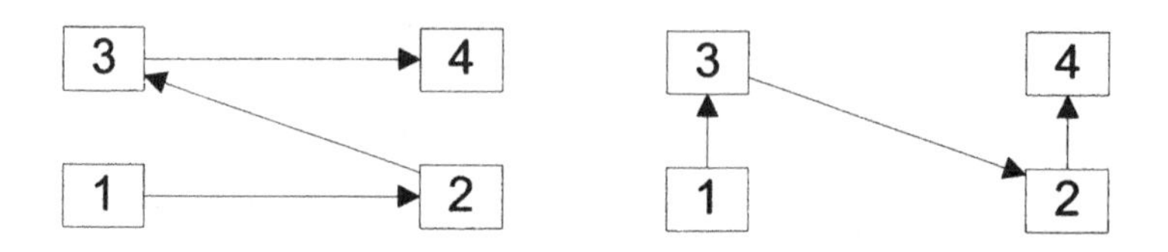

Figure 10.7 In this example, we assume that the tour begins at node 1. The tour on the left is much worse (that is, it has a longer distance) than the tour on the right, but both tours have paths between nodes 2 and 3. Since the tour on the left is long and the tour on the right is short, we would expect τ_{23} to be less than τ_{32} for an effective ACO algorithm; that is, it should be less attractive to go from node 2 to 3, than to go from node 3 to 2.

deterministically visit the closest city at each decision point. We discuss TSP initialization in more detail in Section 18.2. Second, elitism can be used in ACO, just as in any EA (see Section 8.4). Elitism could be incorporated by keeping track of the best few ants each generation, and forcing them to repeat the same route at the next generation. This ensures that the best route is not lost from one generation to the next. An ant system algorithm with elitism is sometimes called an elitist ant system [Dorigo et al., 1996], [Blum, 2005a]. Third, mutation could also be used in ACO, just like in any EA (see Section 8.9). Mutation could be incorporated by randomly altering routes with some mutation probability. Researchers have proposed several mechanisms for mutating TSP routes, which we discuss in Section 18.4.

Figure 10.6 shows that there are several parameters that need to be tuned for an ant system. These parameters include:

- The number of ants N, which is the population size;
- α and β, which are the relative importance of pheromone amounts and heuristic information;
- Q, which is the deposition constant;
- ρ, which is the evaporation rate;
- τ_0, which is the initial pheromone amount between each city.

The effects of these parameters have been studied by several researchers. As a typical example of recommended values [Dorigo et al., 1996]:

- $N = n$ (that is, number of ants = number of cities);
- $\alpha = 1$ and $\beta = 5$;
- $Q = 100$, although its effect is not significant;
- $\rho \in [0.5, 0.99]$;
- $\tau_0 \approx 10^{-6}$.

EXAMPLE 10.1

This example applies the ant system algorithm of Figure 10.6 to the Berlin52 TSP, which consists of 52 locations in Berlin, Germany [Reinelt, 2008]. Berlin52 is a symmetric TSP, which means that we are given a set of nodes and distances between each pair of nodes, and our goal is to find a round trip of minimal total length while visiting each node exactly once. In a symmetric TSP, such as the Berlin52 TSP, the distance from node i to j is the same as from node j to i. We use the following ant system parameters:

- $N = 53$;[1]
- $\alpha = 1$ and $\beta = 5$;
- $Q = 20$;
- $\rho = 0.9$;
- $\tau_0 = 10^{-6}$;
- In general, $\tau_{ij} \neq \tau_{ji}$;[2]
- Random initialization;
- Two elite ants each generation;
- No mutation.

Figure 10.8 shows the best tour of the initial population, with a total distance of 24,780. We see that the best initial tour appears to be quite poor. Figure 10.9 shows the convergence of the AS as it searches for the best tour. We see that the AS converges very quickly, and elitism ensures that the best tour never increases in total distance from one generation to the next. Figure 10.10 shows the best tour found by the AS after 10 generations, with a total distance of 7,796. We see that ACO has found a much better tour than the best one from the initial population, and the total distance has decreased by 69%.

Finally, Figure 10.11 shows the globally optimal tour, which has been proven to be optimal, and which has a total distance of 7,542. Comparing Figures 10.10 and 10.11, we see that ACO has found a tour that is similar to the optimal tour, and that is only 3% worse than the optimal tour. Every ACO simulation will find a different solution since ACO is a stochastic algorithm. But considering the fact that there are $51! = 1.6 \times 10^{66}$ potential solutions to this problem, ACO does quite well to find a solution that is only 3% worse than optimal.

□

[1]The standard population size for ACO varies from one paper to the next. Many ACO and TSP papers use $N = n$, while others use $N = n + 1$.

[2]Usually $\tau_{ij} = \tau_{ji}$ is recommended (but not required) for symmetric TSPs.

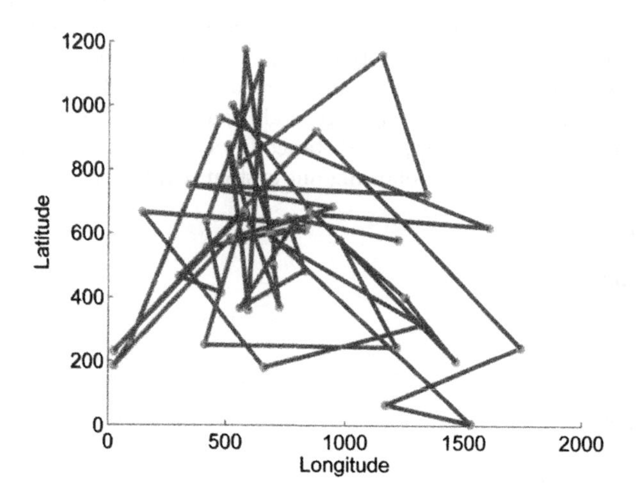

Figure 10.8 The best initial tour out of 53 random tours for Example 10.1, with a total distance of 24,780.

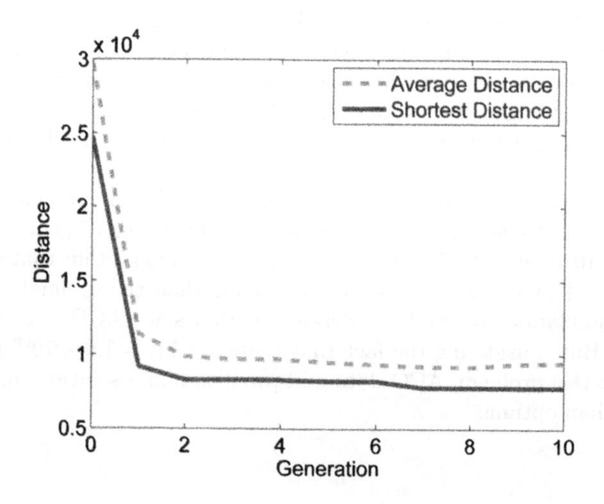

Figure 10.9 Ant system convergence of Example 10.1.

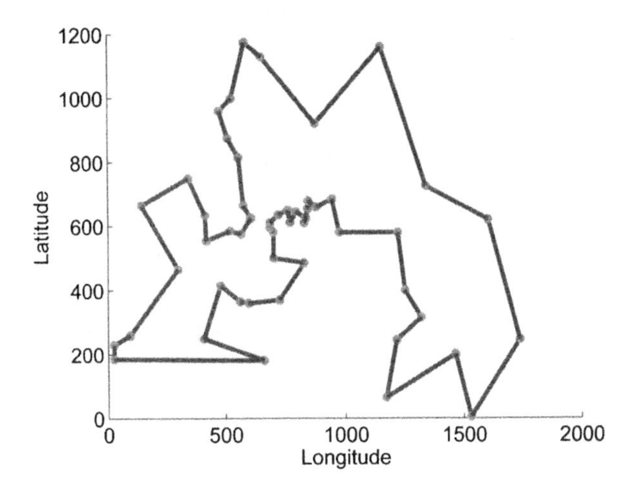

Figure 10.10 The best tour found after 10 ant system generations for Example 10.1, with a total distance of 7,796. This is 69% better than the best initial tour shown in Figure 10.8, and 3% worse than the globally optimal tour shown in Figure 10.11.

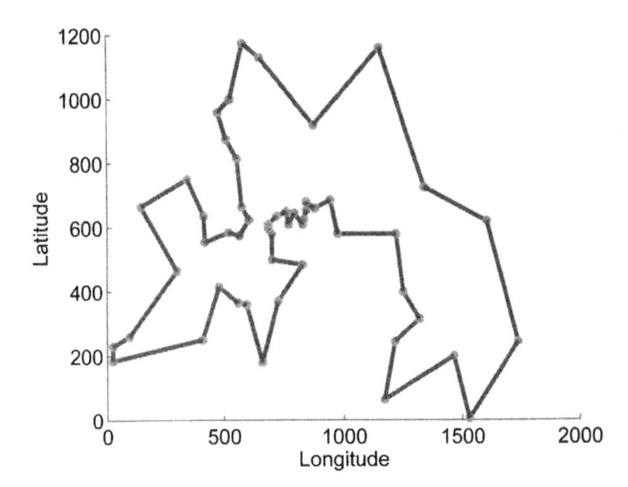

Figure 10.11 The globally optimal tour for Example 10.1, with a total distance of 7,542.

10.3 CONTINUOUS OPTIMIZATION

ACO was originally developed for TSP-like problems, but it has since been modified for optimization problems with continuous domains [Socha and Dorigo, 2008], [Tsutsui, 2004], [de Franca et al., 2008]. One simple approach for applying a discrete optimization algorithm like ACO to a continuous-domain problem is to divide each dimension i of the search space into discretized intervals. That is, we are trying to minimize the n-dimensional problem $f(x)$, where $x = [x_1, \cdots, x_n]$, and

$$\begin{aligned} x_i &\in [x_{i,\min}, x_{i,\max}] \\ x_{i,\min} = b_{i1} &< b_{i2} < \cdots < b_{i,B_i} = x_{i,\max} \end{aligned} \tag{10.3}$$

where $B_i - 1$ is the number of discrete intervals into which we divide the i-th domain. Each generation, if the i-th domain of a candidate solution is between b_{ij} and $b_{i,j+1}$, then we update the pheromone of that interval as in the standard ant system algorithm:

$$\text{if } x_i \in [b_{ij}, b_{i,j+1}] \text{ then } \tau_{ij} \leftarrow \tau_{ij} + Q/f(x) \tag{10.4}$$

where Q is the standard ant system deposition constant, and we assume that $f(x) > 0$ for all x. Equation (10.4) is analogous to the statement $\Delta\tau_{ij}^{(k)} \leftarrow Q/L_k$ in Figure 10.6. We use pheromone amounts to probabilistically construct new solutions at the beginning of each generation. If the interval $[b_{ij}, b_{i,j+1}]$ has a lot of pheromone, then there is a large probability that a candidate solution will be constructed such that its i-th dimension is in that interval. One way of doing this is to set the i-th dimension of the candidate solution to a random number $r \in [b_{ij}, b_{i,j+1}]$.

Figure 10.12 outlines a continuous ant system algorithm. Figure 10.12 assumes that the cost function, which is denoted as L_k, is positive for all k. If this property is not satisfied for a given problem, then the cost values of the population should be shifted so that it is satisfied. Figure 10.12 does not show the elitism option, but we can (and should) easily include elitism as described for EAs in Section 8.4. A continuous AS could also be combined with local search so that after an ant is placed in a discrete bin, local search is used to find the optimal solution within that bin.

The use of discretized intervals for each problem dimension is a simple way to extend the AS to continuous problems. A more rigorous implementation of a continuous ant system could use kernels to construct a continuous approximation to the discrete PDF that is represented by the pheromone amounts [Simon, 2006, Chapter 15], [Blum, 2005a].

---

n = number of dimensions
Divide the i-th dimension into $B_i - 1$ intervals as shown in Equation (10.3), $i \in [1, n]$
α = importance of pheromone amounts
Q = deposition constant
ρ = evaporation rate $\in (0, 1)$
$\tau_{i,j_i} = \tau_0$ (initial pheromone) for $i \in [1, n]$ and $j_i \in [1, B_i - 1]$
Randomly initialize a population of ants (candidate solutions) a_k, $k \in [1, N]$
While not(termination criterion)
    For each ant a_k, $k \in [1, N]$
        For each dimension $i \in [1, n]$
            For each discretized interval $[b_{ij}, b_{i,j+1}]$, $j \in [1, B_i - 1]$
                Probability $p_{ij}^{(k)} \leftarrow \tau_{ij}^{\alpha} / \sum_{m=1}^{B_i-1} \tau_{im}^{\alpha}$
            Next discretized interval
            $a_k(x_i) \leftarrow U[b_{ij}, b_{i,j+1}]$ with probability $p_{ij}^{(k)}$
        Next dimension
    Next ant
    $L_k \leftarrow$ cost of solution constructed by ant a_k, $k \in [1, N]$
    For each dimension $i \in [1, n]$
        For each discretized interval $[b_{ij}, b_{i,j+1}]$, $j \in [1, B_i - 1]$
            For each ant a_k, $k \in [1, N]$
                If $a_k(x_i) \in [b_{ij}, b_{i,j+1}]$
                    $\Delta\tau_{ij}^{(k)} \leftarrow Q/L_k$
                else
                    $\Delta\tau_{ij}^{(k)} \leftarrow 0$
                End if
            Next ant
            $\tau_{ij} \leftarrow (1 - \rho)\tau_{ij} + \sum_{k=1}^{N} \Delta\tau_{ij}^{(k)}$
        Next discretized interval
    Next dimension
Next generation

Figure 10.12 A simple ant system (AS) for solving a continuous-domain minimization problem. $a_k(x_i)$ is the i-th element of the k-th candidate solution. Each generation, pheromone in each bin evaporates, but the pheromone also increases in an amount proportional to the number of ants that construct a candidate solution in that bin. $U[b_{ij}, b_{i,j+1}]$ is a random number that is uniformly distributed between b_{ij} and $b_{i,j+1}$.

■ EXAMPLE 10.2

In this example we optimize the 20-dimensional Ackley function (see Appendix C.1.2). We use the algorithm of Figure 10.12 with the following parameters:

- $N = 50$;
- $\alpha = 1$;

- $Q = 20$;
- $\rho = 0.9$;
- $\tau_0 = 10^{-6}$;
- Two elite solutions each generation;
- Mutation rate = 1% per dimension per individual per generation;
- Number of intervals $B_i = 40$ or 80, for $i \in [1, n]$.

Figure 10.13 shows the best solution at each generation, averaged over 20 Monte Carlo simulations, for both 40 and 80 intervals per dimension. We see that convergence is better for more intervals per dimension, but computation time increases as the number of intervals increases. There are two reasons for this that can be seen from Figure 10.12. The first reason is the "for each discretized interval" loops. The second reason is that the decision of which interval in which to place $a_k(x_i)$ is more complicated. However, it should be noted that for most real-world optimization problems, the cost function calculation is the primary computational consideration, so the extra computation required for discretized domain intervals may not be a major problem (see Chapter 21).

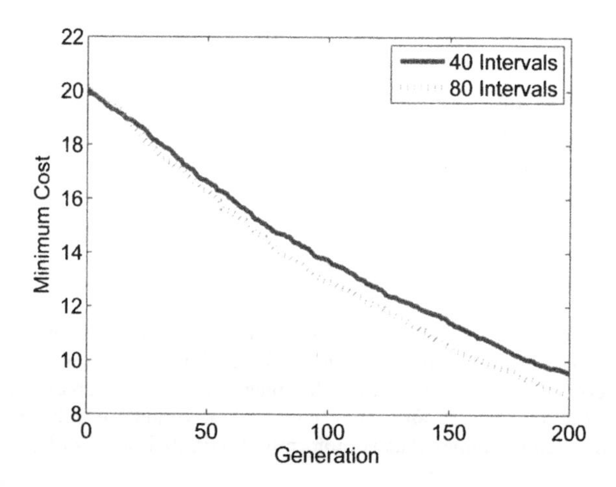

Figure 10.13 Example 10.2: Convergence of the continuous ant system applied to the 20-dimensional Ackley function. The plots show the best solution at each generation averaged over 20 Monte Carlo simulations. We get better performance if we distribute pheromone over more intervals per dimension.

□

10.4 OTHER ANT SYSTEMS

Many modifications have been made to the standard ant system algorithm that is described in the previous sections. This section describes two basic modifications: the max-min ant system in Section 10.4.1 and the ant colony system in Section 10.4.2.

10.4.1 Max-Min Ant System

The max-min ant system (MMAS) is a simple modification to the standard ant system algorithm [Dorigo et al., 2006], [Stützle and Hoos, 2000]. It is characterized by two main features. First, pheromone is increased only by the best ant each generation. This has the effect of reducing exploration and increasing exploitation of the best known solution. Second, the pheromone amount is bounded from above and below. This has the opposite effect; that is, it increases exploration because even the worst tours retain a nonzero amount of pheromone, and even the best tours cannot get so much pheromone that they completely dominate the ants' decisions.

The first difference between the standard ant system algorithm and MMAS can be seen in the following equations, which are replaced in Figures 10.6 and 10.12:

$$\begin{aligned} \text{Standard AS:} \quad & \tau_{ij} \leftarrow (1-\rho)\tau_{ij} + \sum_{k=1}^{N} \Delta\tau_{ij}^{(k)} \\ \text{MMAS:} \quad & \tau_{ij} \leftarrow (1-\rho)\tau_{ij} + \Delta\tau_{ij}^{(\text{best})} \end{aligned} \tag{10.5}$$

where *best* is the index of the best candidate solution. In the TSP of Figure 10.6, $\Delta\tau_{ij}^{(\text{best})}$ is given as

$$\Delta\tau_{ij}^{(\text{best})} \leftarrow \begin{cases} Q/L_{\text{best}} & \text{if city } i \to \text{city } j \text{ belongs to the best tour} \\ 0 & \text{otherwise} \end{cases} \tag{10.6}$$

and in the continuous AS of Figure 10.12, $\Delta\tau_{ij}^{(\text{best})}$ is given as

$$\Delta\tau_{ij}^{(\text{best})} \leftarrow \begin{cases} Q/L_{\text{best}} & \text{if the } i\text{-th dimension of the best individual} \in [b_{ij}, b_{i,j+1}] \\ 0 & \text{otherwise.} \end{cases} \tag{10.7}$$

The second difference between the standard ant system algorithm and MMAS is implemented with the following simple equations after τ_{ij} has been updated:

$$\begin{aligned} \tau_{ij} & \leftarrow \max(\tau_{ij}, \tau_{\min}) \\ \tau_{ij} & \leftarrow \min(\tau_{ij}, \tau_{\max}) \end{aligned} \tag{10.8}$$

where $\tau_{\min}$ and $\tau_{\max}$ are tuned for the specific problem that is being optimized.

With some imagination, we can see that MMAS could be generalized in several different ways. For example, instead of allowing only by the best ant to deposit pheromone, we could allow the best M ants to deposit pheromone, where M is a tuning parameter. Or we could allow the m-th best ant to deposit pheromone with probability p_m, where p_m decreases with increasing cost. Assuming that we want more exploration at the beginning of the optimization process and more exploitation at the end of the process, we could increase $(\tau_{\max} - \tau_{\min})$ as the generation count increases. With some imagination and experimentation, we could undoubtedly also find other extensions of MMAS that would improve performance on various kinds of problems.

EXAMPLE 10.3

In this example we repeat the minimization of the Ackley function with $n = 20$ dimensions as in Example 10.2. We use the following parameters:

- $N = 40$;
- $\alpha = 1$;
- $Q = 20$;
- $\rho = 0.9$;
- $\tau_0 = 10^{-6}$;
- Two elite candidate solutions each generation;
- Mutation rate 1% per dimension per individual per generation;
- Number of intervals $B_i = 20$ for $i \in [1, n]$;
- $\tau_{\min} = 0$ and $\tau_{\max} = \infty$.

We only allow M ants to deposit pheromone:

$$\tau_{ij} \leftarrow (1 - \rho)\tau_{ij} + \Delta\tau_{ij}^{(\text{best}_m)} \tag{10.9}$$

for $m \in [1, M]$, where best_m is the index of the m-th best individual each generation. That is, only the best M ants deposit pheromone on the domain that they have explored. Other than this change, the algorithm we use in this example is the same as the ant system algorithm in Example 10.2. Figure 10.14 shows the best solution at each generation, averaged over 20 Monte Carlo simulations, for $M = 4$ and $M = 40$. We see that convergence is much better when fewer ants are allowed to deposit pheromone. This makes sense intuitively. We do not want poor individuals to reinforce their solutions.

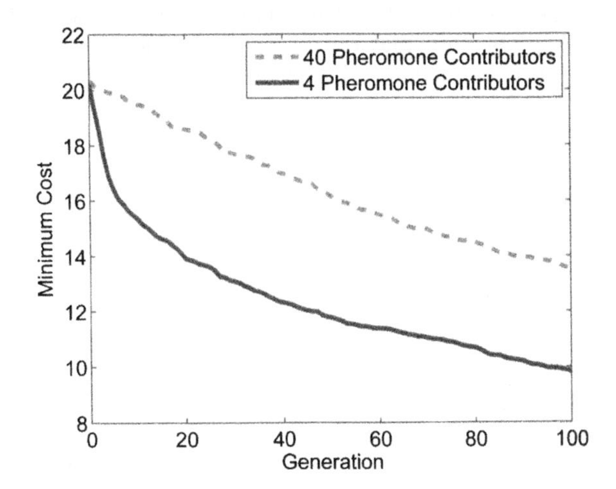

Figure 10.14 Example 10.3: Convergence of the continuous ant system applied to the 20-dimensional Ackley function. The plots show the best solution at each generation averaged over 20 Monte Carlo simulations. We get better performance if we allow only the best ants to deposit pheromone on their solutions.

□

10.4.2 Ant Colony System

The ant colony system (ACS) is an extension of AS [Dorigo and Gambardella, 1997a], [Dorigo and Gambardella, 1997b], [Dorigo et al., 2006]. In spite of their common roots, AS and ACS are quite different in their behavior and performance. ACS is characterized by two main extensions to AS. First, a local pheromone update is implemented by each ant as it constructs its solution. As soon as an ant travels from city i to city j, the pheromone along that path is updated as follows:

$$\tau_{ij} \leftarrow (1 - \phi)\tau_{ij} + \phi\tau_0 \tag{10.10}$$

where $\phi \in [0, 1]$ is the local pheromone decay constant, and τ_0 is the initial pheromone amount. If $\phi = 0$ then τ_{ij} does not change and we are back to the original ant system. Equation (10.10) indicates that pheromone between cities i and j decays as ants travel that path. This is not biologically accurate,[3] but it discourages other ants from following the same path and hence encourages exploration and diversity. After all ants have constructed a candidate solution we implement one of the standard global pheromone update rules of Equation (10.5).

The second extension that ACS makes to AS is the use of a pseudo-random proportional rule for candidate solution construction. Denote by $(a_k \to j)$ the event

[3]Extensions to ACO and EAs often stray from the algorithms' biological foundations, but our goal is primarily to develop effective optimization algorithms rather than to accurately model biology. The biological roots of ACO and EAs primarily serve as inspiration.

that the k-th ant goes to city j while constructing its candidate solution. Denote by $\Pr(a_k \to j)$ the probability that $(a_k \to j)$. The difference between standard AS candidate solution construction and ACS candidate solution construction is the following:

$$\text{AS:} \quad \Pr(a_k \to j) = p_{ij}^{(k)} \tag{10.11}$$

$$\text{ACS:} \quad \Pr(a_k \to j) = \begin{cases} \left.\begin{matrix} 1 & \text{if } j = \arg\max_J p_{iJ}^{(k)} \\ 0 & \text{otherwise} \end{matrix}\right\} & \text{if } r < q_0 \\ \quad p_{ij}^{(k)} & \text{if } r \geq q_0 \end{cases}$$

where r is a random number taken from a uniform distribution on $[0, 1]$, and $q_0 \in [0, 1]$ is a tuning parameter. In standard AS, probabilities are derived using pheromone amounts, and ant k decides which city to go to based on those probabilities (see Figures 10.6 and 10.12). However, in ACS, there is a q_0 probability that ant k goes to the city with the highest probability (that is, with the largest amount of pheromone leading from the current city to it, denoted by the arg max function in Equation (10.12)); and there is a $(1 - q_0)$ probability that ant k uses the standard AS rule to decide which city to go to. This biases the ants to explore highly promising options in their solution construction. This is conceptually equivalent to increasing the probability of high-pheromone paths, which is equivalent to increasing α in Figures 10.6 and 10.12.

The ACS probabilities of Equation (10.11) when $r \geq q_0$ are only approximately accurate. For better accuracy, they should be normalized so that they sum to 1 (see Problem 10.7).

■ **EXAMPLE 10.4**

In this example we investigate the use of the local pheromone decay constant ϕ in ACS. As in earlier examples in this chapter, we minimize the Ackley function with $n = 20$ dimensions. We use the following parameters:

- $N = 40$;
- $\alpha = 1$;
- $Q = 20$;
- $\rho = 0.9$;
- $\tau_0 = 10^{-6}$;
- Two elite solutions each generation;
- Mutation rate 1% per dimension per individual per generation;
- Number of intervals $B_i = 20$ for $i \in [1, n]$;
- $\tau_{\min} = 0$ and $\tau_{\max} = \infty$;
- The best four ants deposit pheromone each generation;
- Exploration constant $q_0 = 0$.

Figure 10.15 shows the best solution at each generation, averaged over 20 Monte Carlo simulations, for $\phi = 0$, 0.001, and 0.01. We see that performance is noticeably better with a nonzero value for the local pheromone decay constant. A positive value of ϕ encourages more exploration, which results in faster convergence. However, if ϕ is too large, other ants are discouraged too strongly from exploring previously used paths, and performance becomes worse. To make more firm conclusions, we should perform statistical significance tests on the results of Figure 10.15 (see Appendices B.2.4 and B.2.5).

□

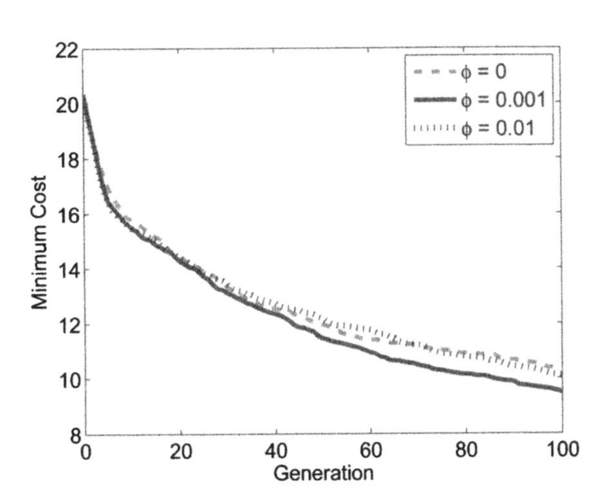

Figure 10.15 Example 10.4: Performance of the ant colony system (ACS) on the 20-dimensional Ackley function. The traces show the best solution at each generation averaged over 20 Monte Carlo simulations for various values of the local pheromone decay constant. We get better performance with $\phi > 0$, but if ϕ is too large then performance suffers.

■ EXAMPLE 10.5

In this example we investigate the use of the exploration constant q_0 in ACS. As in earlier examples in this chapter, we minimize the Ackley function with $n = 20$ dimensions. We use the same ACS parameters as in Example 10.4, except that we fix the local pheromone decay constant $\phi = 0$ and test various values of q_0. Figure 10.16 shows the best solution at each generation, averaged over 100 Monte Carlo simulations, for $q_0 = 0$, 0.001, and 0.01. We see that performance is slightly better with a nonzero value for the exploration constant. A positive value of q_0 provides a greater bias to the ants to use more favorable solution features. However, if q_0 is too large, then the ACS does not have enough exploration and performance becomes worse. To make more firm conclusions, we should perform statistical significance tests on the results of Figure 10.16 (see Appendices B.2.4 and B.2.5). Also note that these results are highly dependent on the particular problem that we solve, and on the other parameter settings listed in the previous example. Most ACS implementations use high values of q_0, such as $q_0 = 0.9$ [Dorigo and Gambardella, 1997b].

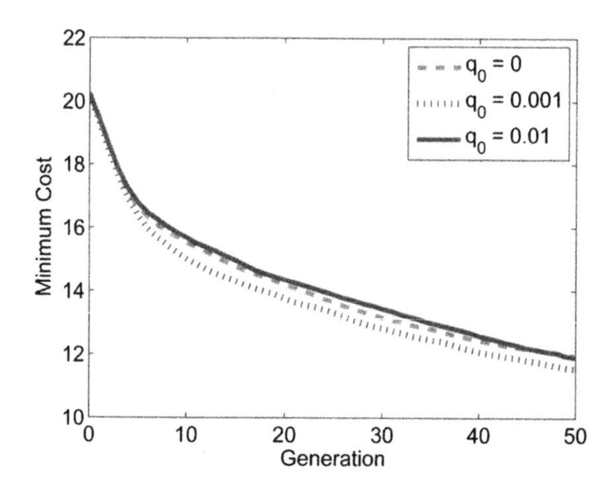

Figure 10.16 Example 10.5 performance of the ant colony system (ACS) on the 20-dimensional Ackley function. The traces show the best solution at each generation averaged over 100 Monte Carlo simulations for various values of the exploration constant. We get better performance with $q_0 > 0$, but if q_0 is too large then performance suffers.

□

10.4.3 Even More Ant Systems

Space prevents us from going into detail about other ant systems, but there are a few notable variations that we mention briefly here. In the elitist ant system, the best solution deposits pheromone every time the other ants deposit pheromone [Dorigo and Stützle, 2004, Chapter 3]. The $\Delta\tau$ calculation in Figure 10.12 is thus modified as follows:

$$\begin{aligned} \text{Standard AS: } \Delta\tau_{ij}^{(k)} &\leftarrow \delta_{ij}^{(k)} Q/L_k \\ \text{Elitist AS: } \Delta\tau_{ij}^{(k)} &\leftarrow \delta_{ij}^{(k)} Q/L_k + \delta_{ij}^{(\text{best})} Q/L_{\text{best}} \end{aligned} \tag{10.12}$$

where $\delta_{ij}^{(k)} = 1$ if the i-th dimension of the k-th candidate solution lies in the j-th discretized interval, and *best* is the index of the best individual in the population. We see that in the elitist ant system, every time an ant deposits pheromone, the best ant also does so.

Ant-Q is a hybrid of AS and Q-learning [Gambardella and Dorigo, 1995]. In rank-based AS, the amount of pheromone deposited depends not only on an ant's solution quality, but also on its rank relative to the other ants [Dorigo and Stützle, 2004,

Chapter 3]. Approximated non-deterministic tree search (ANTS) specifies certain mechanisms to define how attractive a move is, and how to update pheromone [Maniezzo et al., 2004]. The best-worst AS deposits extra pheromone on the best solution, applies extra evaporation on the worst solution, and also uses mutation to encourage exploration [Cordon et al., 2000]. The hypercube ACO algorithm limits pheromone amounts to the interval $[0, 1]$ to regularize the behavior of ACO on problems with different objectives, and to facilitate is theoretical investigation [Blum and Dorigo, 2004]. The population-based ACO maintains a population of pheromone histories rather than storing all information in a single pheromone map; it uses this population to modify the update algorithm [Guntsch and Middendorf, 2002]. Beam ACO is a hybrid of ACO and beam search, which is a popular tree search algorithm [Blum, 2005b].

10.5 THEORETICAL RESULTS

Ever since experimental results first began to show that ACO works, researchers have been working on developing ACO theory to explain when, why, and how it works. The first convergence proofs of ACO were given in [Gutjahr, 2000]. Since then various convergence proofs for various types of ACO algorithms have been published [Dorigo and Stützle, 2004]. Most of these proofs claim something like, "Given enough time, ACO will eventually find the best solution to a combinatorial optimization problem." Convergence results like this are mathematically interesting, but have limited practical interest. As long as the pheromone along each branch is maintained within lower and upper bounds as in the MMAS, there is always a nonzero probability of each ant exploring each possible branch of the solution space, so that given enough time, every branch will be explored. This means that eventually the optimal solution will be found. Of course, any stochastic search algorithm with a nonzero probability of searching each possible candidate solution will eventually converge. Even the simplest random search will eventually converge [Bäck, 1996].

More interesting theoretical results are along the lines of time to convergence [Gutjahr, 2008], [Neumann and Witt, 2009], probability of convergence within a given time, scalability with problem size, and descriptive mathematical models such as Markov models or dynamic system models (see Chapter 4 for mathematical models for GAs). Note that theoretical results for combinatorial problems are much different than theoretical results for continuous-domain problems. Also, if ACO could be shown to be equivalent to other optimization algorithms for which more interesting convergence proofs exist, then those convergence proofs might be able to be adapted to ACO to strengthen its theoretical foundations.

ACO has already been shown to be equivalent to the stochastic gradient ascent (SGA) and cross entropy (CE) optimization algorithms under certain conditions [Meuleau02 and Dorigo, 2002], [Zlochin et al., 2004], [Dorigo and Stützle, 2004]. SGA and CE are model-based optimization algorithms that construct solutions on the basis of a parameterized probability distribution over the search space. The evaluation of candidate solutions is used to modify the probability distribution so that it is biased toward better candidate solutions.

10.6 CONCLUSION

Some ACO researchers emphasize that ACO is not an algorithm but is instead a meta-heuristic because of its many variations. However, we could say that about any of the algorithms discussed in this book (GAs, EP, ES, GP, and so on); they all have many variations, and so they are all meta-heuristics. The difference between an algorithm and a meta-algorithm is one of degree, and so the difference is not black and white. Most ACO researchers emphasize that ACO is *not* an evolutionary algorithm because individuals do not exchange information with each other in the traditional sense of EAs. As we have seen in this chapter, although ACO solution construction parameters evolve over time, it is true that ACO individuals do not directly share information with each other.

Most of our discussion in this chapter has focused on trail pheromones. However, ants deposit other pheromones for purposes other than marking paths. The typical ant colony uses as many as 20 different pheromones [Hölldobler and Wilson, 1990, Chapter 7]. For example, ants deposit alarm pheromones when they are crushed. This can stimulate other ants to agressively fight the predator that crushed their colleague [Šobotník et al., 2008]. These types of pheromones could be simulated in an ACO algorithm by having a poor solution broadcast information that discourages other individuals from repeating its poor strategy. This is similar to the negative reinforcement PSO discussed in Section 11.6.

Female ants deposit epideictic pheromones when they lay their eggs to signal other females of the same species to lay their eggs elsewhere [Gómez et al., 2005]. Animals deposit territorial pheromones to mark their territory [Horne and Jaeger, 1988]. Territorial pheromones are present in the urine of cats and dogs, which they deposit on the boundaries of their claimed territory. Animals release sex pheromones to communicate their availability for breeding [Wyatt, 2003]. Ants release recruitment pheromones to attract other ants to some place where work is required [Hölldobler and Wilson, 1990]. These types of pheromones could be simulated in ACO by having individuals broadcast information about previously explored territory in the search space to prevent other individuals from searching in regions that have already been explored, or to encourage other individuals to explore promising regions of the search domain. Ants can also release task-specific pheromones [Greene and Gordon, 2007]. This could be simulated in ACO for multi-objective optimization with different individuals pursuing the optimization of different sub-problems. We see that there are many opportunities for biologically-motivated extensions of ACO.

Additional reading about ACO can be found in books [Bonabeau et al., 1999], [Dorigo and Stützle, 2004], [Solnon, 2010]; book chapters [Maniezzo et al., 2004], [Dorigo and Stützle, 2010]; and tutorial papers [Blum, 2005a], [Blum, 2007]. Other directions for future research in ACO are similar to research priorities for other optimization algorithms [Dorigo et al., 2006]. How can ACO be applied to dynamic optimization problems for which the search space changes with time, and how can ACO be applied to stochastic optimization problems with noisy fitness function evaluations (see Chapter 21)? How can ACO be applied to multi-objective optimization problems (see Chapter 20)? How can ACO be hybridized with other evolutionary algorithms?

PROBLEMS

Written Exercises

10.1 Give an example in a real-world problem when the cost of traveling from node A to node B would be different than the cost of traveling from node B to node A.

10.2 Let t be the total number of ants so that $m_1 \approx p_1 t$ and $m_2 \approx p_2 t$ in Equation (10.1).

a) What are the equilibrium ratios of p_1/p_2?

b) Which of the equilibrium ratios are stable, and which are unstable?

10.3 Suppose $\beta = 1$ in the ant system of Figure 10.6. If two path segments have equal amounts of pheromone and segment 1 is half as long as segment 2, how much more likely is an ant to travel on segment 1 than segment 2? What if $\beta = 2$? What if $\beta = 3$?

10.4 The ant system of Figure 10.6 sets the pheromone deposit of the k-th ant to $\Delta\tau_{ij}^{(k)} = \delta_{ij}^{(k)} Q/L_k$, where $\delta_{ij}^{(k)} = 1$ if the k-th ant went from city i to city j, and $\delta_{ij}^{(k)} = 0$ otherwise. Suppose we instead set it to $\delta_{ij}^{(k)} \epsilon \tau_{ij}$, where ϵ is a tuning parameter.

a) What range of ϵ makes the pheromone update equation stable?

b) What is the equilibrium value of τ_{ij} in this case? Is this a desirable equilibrium value?

10.5 In the standard continuous-domain AS of Figure 10.12, the m-th ant's pheromone deposit is $\Delta\tau_{ij}^{(m)} = Q/L_m$. Suppose we instead allow the m-th ant to deposit pheromone with probability p_m, where p_m decreases with increasing cost, as mentioned at the end of Section 10.4.1:

$p_m \leftarrow \frac{1}{L_m} \sum_{r=1}^{N} L_r$
$r \leftarrow U[0, 1]$ – that is, r is a random number uniformly distributed on $[0, 1]$
If $r < p_m$ then
    $\Delta\tau_{ij}^{(m)} \leftarrow Q_1/L_m$
else
    $\Delta\tau_{ij}^{(m)} \leftarrow 0$
End if

What value should we use for Q_1 in the above algorithm so that the average pheromone amount deposited by the m-th ant is equal to that deposited in the standard AS of Figure 10.12?

10.6 How does computational effort in the continuous-domain ant system of Figure 10.12 increase with the population size? How does it increase with the problem dimension? How does it increase with the number of discretized intervals in each dimension?

10.7 Ant colony system probabilities:

a) Suppose we have an ACS with four cities and $q_0 = 1/2$. Suppose the k-th ant is in city 1, and that

$$\begin{aligned} p_{11}^{(k)} &= 0 \\ p_{12}^{(k)} &= 1/4 \\ p_{13}^{(k)} &= 1/4 \\ p_{14}^{(k)} &= 1/2 \end{aligned}$$

According to Equation (10.11), what is the probability that the k-th ant will proceed to each of the four cities? Do these probabilities sum to 1?

b) Normalize the ACS probabilities $\Pr(a_k \to j)$ of Equation (10.11) so that the sum from $j = 1$ to n is 1.

c) Use your answer to part (b) to calculate new probabilities for the scenario described in part (a). Do the new probabilities sum to 1?

10.8 Propose a way to implement a rank-based AS such as the one mentioned in Section 10.4.3.

10.9 Propose a way to implement a best-worst AS such as the one mentioned in Section 10.4.3.

Computer Exercises

10.10 This problem explores the effect of β, which is the heuristic sensitivity of an ant system, on AS performance. Simulate the ant system of Example 10.1 20 times, recording the best cost among all ants at each generation. Plot the average of the 20 Monte Carlo simulations as a function of generation number. Do this for $\beta = 0.1$, 1, and 10. Discuss your results.

10.11 Repeat Example 10.3 with $M = 40$. Run 20 Monte Carlo simulations for each of the following values of τ_{min}: 0, 0.001, 0.01, and 0.1. Plot the results. Comment on the effect of τ_{min} on AS performance.

10.12 Repeat Example 10.3 with $M = 40$. Run 20 Monte Carlo simulations for each of the following values of τ_{max}: 1, 10, 100, and ∞. Plot the results. Comment on the effect of τ_{max} on AS performance.

CHAPTER 11

Particle Swarm Optimization

The particle swarm algorithm imitates human social behavior.
—James Kennedy and Russell Eberhart [Kennedy and Eberhart, 2001]

We observe collective intelligence in many natural systems. For example, ants exhibit an extraordinary level of collective intelligence, as we discussed at the beginning of Chapter 10. In such systems intelligence does not reside in individuals but is instead distributed among a group of many individuals. This can be seen in flocks of animals as they avoid predators, seek food, seek to travel more quickly, and other behaviors.

Animal groups can often avoid predators more effectively in a group than alone. For example, it might be easy for a lion to recognize a single zebra because of its contrast with the surrounding landscape, but a group of zebras blend together and are more difficult to recognize as individuals [Stone, 2009]. A group of animals might also appear to be larger, or sound louder, or be more threatening in other ways, than a solitary animal. Finally, it may be difficult for a predator to focus on a single animal when it is part of a large group. These phenomena are called the predator confusion effect [Milinski and Heller, 1978]. [Heinrich, 2002] gives an interesting description of how antelope use the predator confusion effect.

Another way that groups protect themselves from predators is described by the many-eyes hypothesis [Lima, 1995]. When a large group forages for food or drinks from a stream, random effects dictate that there will always be a few animals who are watching for predators. This collaboration not only provides more protection from predators, but also allows each individual more time for feeding and drinking.

Finally, groups protect themselves from predators because of the encounter dilution effect [Krause and Ruxton, 2002]. This can take several forms. First, individual animals might seek the cover and protection of a group as a type of selfish behavior to reduce their chance of being attached [Hamilton, 1971]. Second, as a predator wanders through its territory, it might be less likely to encounter a single group than one of many individuals scattered throughout the territory [Turner and Pitcher, 1986].

Animals also have more success in finding food when they are in groups than when they are alone. At first glance, this might not seem correct. After all, when an individual is in a group, it cannot approach its prey stealthily; and when it catches its prey, it has to share the food with others in the group. However, the success of groups in foraging is related to the many-eyes hypothesis in predator avoidance. With more eyes searching for food, the group has a disproportionately greater chance of success than a single animal that searches for food by itself [Pitcher and Parrish, 1993]. Also, a group increases its chances of success if it can surround its prey.

Animals can also move more quickly when in groups than when alone. This is seen in bicycle riders who ride in a line and draft off of each other. The trailing riders might expend as much as 40% less energy than the lead rider because of wind resistance [Burke, 2003]. The same type of effect, albeit to a lesser extent, can be seen in speed skating, running, swimming, and other sports. In the animal world, drafting can be seen in groups of geese as they fly [McNab, 2002], groups of ducks as they paddle [Fish, 1995], and groups of fish as they swim [Noren et al., 2008].

Particle swarm optimization (PSO) is based on the observation that groups of individuals work together to improve not only their collective performance on some task, but also each individual performance. The principles of PSO are clearly seen not only in animal behavior but also in human behavior. As we try to improve our performance at some task, we adjust our approach based on some basic ideas.

- Inertia. We tend to stick to the old ways that have proven to be successful in the past. "I've always done it this way, and so that is how I am going to continue doing things."

- Influence by society. We hear about others who have been successful and we try to emulate their approaches. We may read about the success of others in books, or on the internet, or in the newspaper. "If it worked for them, then maybe it will work for me too."

- Influence by neighbors. We learn the most from those who are personally close to us. We are influenced more by our friends than by society. In our conversations with others, we share stories of success and failure, and we modify our behavior because of those conversations. Investment advice from our millionaire neighbor or cousin will have a stronger influence on us than the more distant stories of billionaires that we read on the internet.

Overview of the Chapter

Section 11.1 gives a basic overview of PSO and some simple examples. Section 11.2 discusses ways of limiting the velocity of PSO particles, which is necessary for good optimization performance. Section 11.3 discusses inertia weighting and constriction coefficients, which are two features of PSO that indirectly limit particle velocities. Section 11.4 discusses the global PSO algorithm, which is a PSO generalization that uses the best individual at each generation to update each individual's velocity. Section 11.5 discusses the fully informed PSO algorithm, in which every individual's velocity contributes to every other individual's velocity each generation. Section 11.6 approaches PSO learning from the other direction – if we can learn from others' successes, then we can also learn from their mistakes.

11.1 A BASIC PARTICLE SWARM OPTIMIZATION ALGORITHM

Suppose that we have a minimization problem that is defined over a continuous domain of d dimensions. We also have a population of N candidate solutions, denoted as $\{x_i\}$, $i \in [1, N]$. Furthermore, suppose that each individual x_i is moving with some velocity v_i through the search space. This movement through search space is the essence of PSO, and it is the fundamental difference between PSO and other EAs. Most other EAs are more static than PSO because they model candidate solutions and their evolution from one generation to the next, but they do not model the dynamics of the movement of the candidate solutions through the search space.

As a PSO individual moves through the search space, it has some inertia and so it tends to maintain its velocity. However, its velocity can change due to a couple of different factors.

- First, it remembers its best position in the past, and it would like to change its velocity to return to that position. This is similar to the human tendency to remember the good old days, and to try to recapture the experiences of the past. In PSO, an individual travels through the search space, and its position in the search space changes from one generation to the next. However, the individual remembers its performance from past generations, and it remembers the search space location at which it obtained its best performance in the past.

- Second, an individual knows the best position of its neighbors at the current generation. This requires the definition of a neighborhood size, and it requires that all of the neighbors communicate with each other about their performance on the optimization problem.

These two effects randomly influence an individual's velocity and are similar to our own social interactions. Sometimes we feel more stubborn than at other times, and so we are not strongly influenced by our neighbors. Other times we feel more nostalgic than at other times, and so we are more strongly influenced by our past successes. We summarize the basic PSO algorithm in Figure 11.1.

Initialize a random population of individuals $\{x_i\}$, $i \in [1, N]$
Initialize each individual's n-element velocity vector v_i, $i \in [1, N]$
Initialize the best-so-far position of each individual: $b_i \leftarrow x_i$, $i \in [1, N]$
Define the neighborhood size $\sigma < N$
Define the maximum influence values $\phi_{1,\max}$ and $\phi_{2,\max}$
Define the maximum velocity $v_{\max}$
While not(termination criterion)
    For each individual x_i, $i \in [1, N]$
        $H_i \leftarrow \{\sigma \text{ nearest neighbors of } x_i\}$
        $h_i \leftarrow \arg\min_x \{f(x) : x \in H_i\}$
        Generate a random vector ϕ_1 with $\phi_1(k) \sim U[0, \phi_{1,\max}]$ for $k \in [1, n]$
        Generate a random vector ϕ_2 with $\phi_2(k) \sim U[0, \phi_{2,\max}]$ for $k \in [1, n]$
        $v_i \leftarrow v_i + \phi_1 \circ (b_i - x_i) + \phi_2 \circ (h_i - x_i)$
        If $|v_i| > v_{\max}$ then
            $v_i \leftarrow v_i v_{\max} / |v_i|$
        End if
        $x_i \leftarrow x_i + v_i$
        $b_i \leftarrow \arg\min\{f(x_i), f(b_i)\}$
    Next individual
Next generation

Figure 11.1 A basic particle swarm optimization algorithm for minimizing the n-dimensional function $f(x)$, where x_i is the i-th candidate solution and v_i is its velocity vector. The notation $a \circ b$ means element-by-element multiplication of the vectors a and b.

Figure 11.1 shows that there are several tuning parameters in the PSO algorithm.

- Not only do we have to initialize a population, as with every other EA, but we also have to initialize the population's velocity vectors. There are several ways to initialize velocities. For example, they could be initialized randomly, or they could be initialized to zero [Helwig and Wanka, 2008].

- We have to define the neighborhood size σ of the algorithm. Note that the term "neighborhood size" is ambiguous. Sometimes it means that each individual has σ close neighbors, and sometimes it means that since there are a total of σ individuals in the neighborhood, each individual has $(\sigma - 1)$ close neighbors. One of the initial PSO papers indicates that smaller neighborhoods (as small as two) provide better global behavior and avoid local minima, while larger neighborhoods provide faster convergence [Eberhart and Kennedy, 1995].

- We have to choose the maximum learning rates $\phi_{1,\max}$ and $\phi_{2,\max}$. The parameter ϕ_1, which is called the cognition learning rate, and ϕ_2, which is called the social learning rate, are random numbers distributed in $[0, \phi_{1,\max}]$ and $[0, \phi_{2,\max}]$ respectively. We discuss these further in Section 11.3.3, but for now we simply note the rule of thumb that $\phi_{1,\max}$ and $\phi_{2,\max}$ are often set to about 2.05.

- We have to choose the maximum velocity $v_{\max}$. Empirical evidence indicates that each element of $v_{\max}$ should be limited to the corresponding dynamic range of the search space [Eberhart and Shi, 2000]. This seems intuitive; if $v_{\max}$ were greater than the dynamic range of the search space, then a particle could easily leave the search space in a single generation. Other results suggest setting $v_{\max}$ to between 10% and 20% of the search space range [Eberhart and Shi, 2001]. There are some problems for which we do not have a search space range in mind; that is, we do not have any idea ahead of time about the location of the optimum for which we are searching. In this case we should still enforce a finite $v_{\max}$ for best performance [Carlisle and Dozier, 2001].

- We could simplify the velocity update of Figure 11.1 as follows:

$$v_i \leftarrow v_i + \phi_1(b_i - x_i) + \phi_2(h_i - x_i) \tag{11.1}$$

  where ϕ_1 and ϕ_2 are scalars instead of vectors with $\phi_1 \sim U[0, \phi_{1,\max}]$ and $\phi_2 \sim U[0, \phi_{2,\max}]$. In this option, called linear PSO [Paquet and Engelbrecht, 2003], each element of the velocity vector v_i is updated with the same ϕ_1 and ϕ_2 values. However, linear PSO is generally considered to provide worse performance than the standard algorithm of Figure 11.1.

- As with most other EAs, elitism often improves the performance of PSO. We have not shown elitism in Figure 11.1, but we can easily implement elitism as discussed in Section 8.4.

- The update equation $x_i \leftarrow x_i + v_i$ in Figure 11.1 may result in x_i moving outside of the search domain. We usually implement some type of limiting operation to keep x_i within the search domain. For instance, we could include the following two equations after the update equation:

$$\begin{aligned} x_i &\leftarrow \min(x_i, x_{\max}) \\ x_i &\leftarrow \max(x_i, x_{\min}) \end{aligned} \tag{11.2}$$

  where $[x_{\min}, x_{\max}]$ defines the limits of the search domain.

Particle Swarm Topologies

Figure 11.1 shows that each particle is influence by its σ nearest neighbors. The arrangement of the neighbors that influence a particle is call the *topology* of the swarm. Since the neighborhood of each particle in Figure 11.1 changes each generation, it is called a dynamic topology. Since the neighborhood is local (that is, it does not include the entire swarm), it is also called an *lbest* topology.

We can use many other methods to define the neighborhood of each particle [Akat and Gazi, 2008]. For instance, we could define neighborhoods at the beginning of the algorithm so that the neighborhoods are static and do not change from one generation to the next. Or, if the optimization process stagnates, we could at that time randomly redefine the neighborhoods [Clerc and Poli, 2006]. In the extreme case we can have a single neighborhood that encompasses the entire swarm, which means that H_i in Figure 11.1 is equal to the entire swarm for all i, and h_i is independent of i and is equal to the best particle among the entire population. This is called the *all* topology or the *gbest* topology. This is the topology with which PSO was originally developed, and it is still widely used. Another common topology is the*ring* topology, in which each particle is connected to two other particles. The *cluster* topology is one in which each particle is fully connected within its own cluster, while a few particles in each cluster are also connected to an additional particle in another cluster. The *wheel* topology is one in which a focal particle is connected to all other particles, while all of the other particles are connected only to the focal particle. The *square* topology, also called the *von Neumann* topology, is one in which each particle is connected to four neighbors. Figure 11.2 depicts some of these topologies. PSO performance can vary strongly with topology, and researchers have experimented with many other topologies besides the few that we mention here [Mendes et al., 2004], [del Valle et al., 2008].

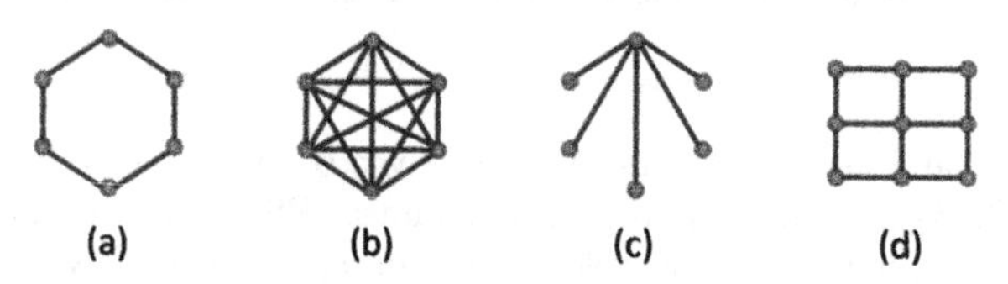

Figure 11.2 Some PSO topologies. (a) represents the *ring* topology, (b) represents the *all* topology, (c) represents the *wheel* topology, and (d) represents the *square* topology. The square topology wraps around from the top to the bottom, and from the left to the right, so that it forms a toroid with each particle connected to four neighbors. Each of these topologies can be either static or dynamic.

11.2 VELOCITY LIMITING

It has been found in many applications of PSO that if v_{max} is not used, PSO particles jump wildly around the search space [Eberhart and Kennedy, 1995]. To see why, consider the basic PSO algorithm of Figure 11.1, but with the simplification $\phi_2 = 0$. The position and velocity update in this case is

$$\begin{aligned} v_i(t+1) &= v_i(t) + \phi_1(b_i - x_i) \\ x_i(t+1) &= x_i(t) + v_i(t+1) \end{aligned} \tag{11.3}$$

where t is the generation number and ϕ_1 is the cognition learning rate. This can be written as

$$\begin{bmatrix} x_i(t+1) \\ v_i(t+1) \end{bmatrix} = \begin{bmatrix} 1-\phi_1 & 1 \\ -\phi_1 & 1 \end{bmatrix} \begin{bmatrix} x_i(t) \\ v_i(t) \end{bmatrix} + \begin{bmatrix} \phi_1 \\ \phi_1 \end{bmatrix} b_i. \tag{11.4}$$

The eigenvalues of the matrix on the right side of the above equation are

$$\lambda = \frac{2 - \phi_1 \pm \sqrt{\phi_1^2 - 4\phi_1}}{2} \tag{11.5}$$

and these eigenvalues govern the stability of the system.[1] If $\phi_1 \in [0, 4]$ then both eigenvalues have a magnitude of 1, which means that the system is marginally

[1]See any book on linear systems, or [Simon, 2006, Chapter 1], for a discussion of stability.

stable, and that $x_i(t)$ and $v_i(t)$ could become unbounded as $t \to \infty$, depending on initial conditions. If $\phi_1 > 4$, then one of the eigenvalues is greater than 1 in magnitude, which means that the system is unstable, and that $x_i(t)$ and $v_i(t)$ increase without bound for almost any initial conditions. This simple example shows why it could be important to use $v_{\max}$ to limit the magnitude of v_i, as shown in Figure 11.1.

However, this analysis assumes that b_i is not a function of x_i. Also, real implementations of PSO are more complicated than Equation (11.3), so our analysis may not be valid for general PSO algorithms. If $\phi_2 > 0$, or if an inertia weight less than 1 is used as we later discuss in Equation (11.9), then it may not be necessary to limit the velocity to get good performance [Carlisle and Dozier, 2001], [Clerc and Kennedy, 2002].

If we want to limit the velocity, then we could limit it in a couple of different ways. One way is to check the magnitude of v_i, and if it is greater than the scalar $v_{\max}$, then scale the components of v_i so that $|v_i| = v_{\max}$:

$$\text{If } |v_i| > v_{\max} \text{ then } v_i \leftarrow \frac{v_i v_{\max}}{|v_i|} \tag{11.6}$$

as shown in Figure 11.1. Another way is to limit the magnitude of each component of v_i. Recall that each individual in the population has n dimensions, so $v_i = \begin{bmatrix} v_i(1) & \cdots & v_i(n) \end{bmatrix}$.[2] With this approach, the maximum velocity of each dimension is specified, so we have $v_{\max}(j)$ defined for $j \in [1, n]$. Velocity limiting of the i-th particle is then performed as follows:

$$v_i(j) \leftarrow \begin{cases} v_i(j) & \text{if } |v_i(j)| \leq v_{\max}(j) \\ v_{\max}(j)\,\text{sign}(v_i(j)) & \text{if } |v_i(j)| > v_{\max}(j) \end{cases} \quad \text{for } j \in [1, n]. \tag{11.7}$$

Velocity limiting can be viewed as a control over the exploration-exploitation balance of PSO. A large $v_{\max}$ allows more change in each individual from one generation to the next, which emphasizes exploration. A small $v_{\max}$ restricts changes in individuals, which emphasizes exploitation.

11.3 INERTIA WEIGHTING AND CONSTRICTION COEFFICIENTS

To avoid velocity limiting, we can modify the velocity update equation in Figure 11.1 to prevent the velocity from increasing without bound. In this section, we first discuss the use of inertia weighting in Section 11.3.1. Then we discuss the equivalent but more commonly-used constriction coefficient in Section 11.3.2. Finally, we present some conditions for the stability of the PSO algorithm in Section 11.3.3.

11.3.1 Inertia Weighting

An inertia weight is often used in PSO applications. As we see from the velocity update equation of Figure 11.1, a particle tends to maintain its velocity from one

[2]Note that j in the term $v_i(j)$ here indicates a specific element of the vector v_i, while t in the term $v_i(t)$ in Equation (11.3) indicates the value of v_i at the t-th generation. This notation is not consistent, but its meaning should be clear from the context.

generation to the next, although some velocity changes are allowed due to the learning rates:

$$v_i(k) \leftarrow v_i(k) + \phi_1(b_i - x_i(k)) + \phi_2(k)(h_i(k) - x_i(k)) \text{ for } k \in [1, n] \tag{11.8}$$

where n is the problem dimension. However, it has been found empirically that decreasing inertia during the optimization process may provide better performance. Equation (11.8) is thus modified to the following equation:

$$v_i(k) \leftarrow wv_i(k) + \phi_1(k)(b_i(k) - x_i(k)) + \phi_2(h_i(k) - x_i(k)) \tag{11.9}$$

where w is the inertia weight, which often decreases from about 0.9 at the first generation to about 0.4 at the last generation [Eberhart and Shi, 2000]. This helps slow down the velocity of each particle as the generation count increases, which improves convergence.

[Clerc and Poli, 2006] recommend PSO parameters for velocity updates of the form of Equation (11.9). In that paper, the population size is set to 30, the neighborhood size is set to four and the neighborhoods are fixed until the PSO process stagnates, at which time the neighborhoods are randomly reinitialized. The velocity update is implemented as shown in Equation (11.9) with the following recommended parameters [Clerc and Poli, 2006, Equation 19]:

$$\begin{aligned} w &= 0.72 \\ \phi_1(k) &\sim U[0, 1.108] \text{ for } k \in [1, n] \\ \phi_2(k) &\sim U[0, 1.108] \text{ for } k \in [1, n]. \end{aligned} \tag{11.10}$$

Reference [Clerc and Poli, 2006] also proposes other PSO variations for improved performance. The stability of PSO with the velocity update of Equation (11.9) is discussed in [Poli, 2008].

A more extensive set of recommended PSO parameters are found for velocity updates of the form of Equation (11.9) in [Pedersen, 2010] and are shown in Table 11.1. These recommended parameters apply when the neighborhood for each particle is the entire swarm, so that h_i (for all i) in Equation (11.9) is equal to the best individual in the population.

Problem Dimension	Function Evaluations	N	w	ϕ_1	ϕ_2
2	400	25	0.3925	2.5586	1.3358
		29	−0.4349	−0.6504	2.2073
2	4,000	156	0.4091	2.1304	1.0575
		237	−0.2887	0.4862	2.5067
5	1,000	63	−0.3593	−0.7238	2.0289
		47	−0.1832	0.5287	3.1913
5	10,000	223	−0.3699	−0.1207	3.3657
		203	0.5069	2.5524	1.0056
10	2,000	63	0.6571	1.6319	0.6239
		204	−0.2134	−0.3344	2.3259
10	20,000	53	−0.3488	−0.2746	4.8976
20	40,000	69	−0.4438	−0.2699	3.3950
20	400,000	149	−0.3236	−0.1136	3.9789
		60	−0.4736	−0.9700	3.7904
		256	−0.3499	−0.0513	4.9087
30	600,000	95	−0.6031	−0.6485	2.6475
50	100,000	106	−0.2256	−0.1564	3.8876
100	200,000	161	−0.2089	−0.0787	3.7637

Table 11.1 Recommended PSO parameters for various problem dimensions and available fitness function evaluations [Pedersen, 2010]. N is the population size, and w, ϕ_1, and ϕ_2 are the recommended parameters for Equation (11.9) when each particle's neighborhood is comprised of the entire swarm. The table shows that some problem configurations have more than one recommended set of parameters because multiple sets of parameters give almost the same performance on the benchmarks.

11.3.2 The Constriction Coefficient

Instead of Equation (11.9), inertia weighting is often implemented with a constriction coefficient. This implementation, which accomplishes the same thing as the inertia weight, involves writing the velocity update equation as

$$v_i \leftarrow K\left[v_i + \phi_1(b_i - x_i) + \phi_2(h_i - x_i)\right] \tag{11.11}$$

where K is called the constriction coefficient [Clerc, 1999], [Eberhart and Shi, 2000], [Clerc and Kennedy, 2002]. We have used the linear velocity update in Equation (11.11) to simplify our analysis. Equation (11.11) is equivalent to the linear form of Equation (11.9) if $K = w$, and if ϕ_1 and ϕ_2 in Equation (11.11) are replaced with ϕ_1/K and ϕ_2/K respectively. To analyze this approach, we use t to denote the generation number and we write Equation (11.11) as follows:

$$\begin{aligned} v_i(t+1) &= K\left[v_i(t) + (\phi_1+\phi_2)\left(\frac{\phi_1 b_i(t) + \phi_2 h_i(t)}{\phi_1+\phi_2} - x_i(t)\right)\right] \\ &= K\left[v_i(t) + \phi_T(p_i(t) - x_i(t))\right] \end{aligned} \tag{11.12}$$

where ϕ_T and $p_i(t)$ are defined by the above equation. Now we define

$$y_i(t) = p_i(t) - x_i(t). \tag{11.13}$$

Assuming that $p_i(t)$ is constant with time, we can combine Equations (11.12) and (11.13) to get

$$\begin{aligned} v_i(t+1) &= Kv_i(t) + K\phi_T y_i(t) \\ y_i(t+1) &= p_i - x_i(t+1) \\ &= p_i - x_i(t) - v_i(t+1) \\ &= y_i(t) - Kv_i(t) - K\phi_T y_i(t) \\ &= -Kv_i(t) + (1-K\phi_T)y_i(t). \end{aligned} \tag{11.14}$$

These equations for $v_i(t+1)$ and $y_i(t+1)$ can be combined to give

$$\begin{bmatrix} v_i(t+1) \\ y_i(t+1) \end{bmatrix} = \begin{bmatrix} K & K\phi_T \\ -K & 1-K\phi_T \end{bmatrix} \begin{bmatrix} v_i(t) \\ y_i(t) \end{bmatrix}. \tag{11.15}$$

The matrix on the right side of the above equation, which governs the stability of the system, has eigenvalues

$$\begin{aligned} \lambda &= \frac{1}{2}\left[1 - K(\phi_T - 1) \pm \sqrt{1 + K^2(\phi_T-1)^2 - 2K(\phi_T+1)}\right] \\ &= \frac{1}{2}\left[1 - K(\phi_T-1) \pm \sqrt{\Delta}\right] \end{aligned} \tag{11.16}$$

where the discriminant Δ is defined by the above equation. We denote the eigenvalues as λ_1 and λ_2:

$$\begin{aligned} \lambda_1 &= \frac{1}{2}\left[1 - K(\phi_T-1) + \sqrt{\Delta}\right] \\ \lambda_2 &= \frac{1}{2}\left[1 - K(\phi_T-1) - \sqrt{\Delta}\right]. \end{aligned} \tag{11.17}$$

The dynamic system of Equation (11.15) is stable if $|\lambda_1| < 1$ and $|\lambda_2| < 1$. This analysis assumes that ϕ_T is constant. In the velocity update of Equation (11.11), the ϕ_i terms are random, but in this analysis we make the simplifying assumption that each ϕ_i is constant. See Problem 11.8 for a discussion of whether to use constant K or time-varying K in PSO. In the following section, we study the behavior of λ_1 and λ_2, which determines the stability of the PSO algorithm, as a function of the constriction coefficient K.

11.3.3 PSO Stability

We can use Equation (11.16) to make the following observation.

Observation 11.1 *When $K = 0$, we obtain $\Delta = 1$, $\lambda_1 = 1$, and $\lambda_2 = 0$.*

Next we consider the values of λ_1 and λ_2 as K increases from 0. Equation (11.16) shows that

$$\begin{aligned}
\lim_{|K|\to\infty} \Delta &= \infty \\
\Delta &= 0 \text{ for } K = \frac{\phi_T + 1 \pm 2\sqrt{\phi_T}}{(\phi_T - 1)^2} = \{K_1, K_2\} \\
\Delta &< 0 \text{ for } K \in (K_1, K_2)
\end{aligned} \tag{11.18}$$

where K_1 and K_2 are defined by the above equation. We assume that $\phi_T > 0$ so that $\sqrt{\phi_T}$ is real. Figure 11.3 shows a plot of Δ as a function of K. This leads us to the following observation.

Observation 11.2 *λ_1 and λ_2 are real for $K < K_1$.*

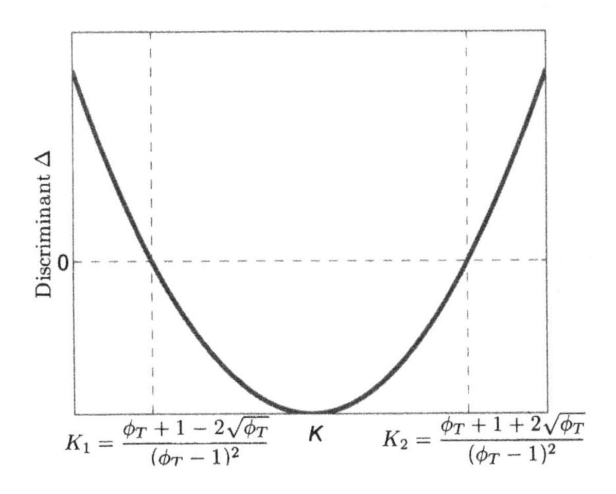

Figure 11.3 The discriminant Δ of Equation (11.16) as a function of the constriction coefficient K. $\Delta > 0$ for $K < K_1$ and $K > K_2$, which means that λ_1 and λ_2 are real. $\Delta < 0$ for $K \in (K_1, K_2)$, which means that λ_1 and λ_2 are complex.

When $K = K_1$, we see that $\Delta = 0$, which means that $\lambda_1 = \lambda_2$. In fact, it is easy to make the following observation from Equation (11.16).

Observation 11.3 *When $K = K_1$, we obtain $\lambda_1 = \lambda_2 = (1 - \sqrt{\phi_T})/(1 - \phi_T)$, which is between 0 and 1 for all $\phi_T \neq 1$.*

Now consider the behavior of λ_1 as K increases from 0 to K_1. Taking the derivative of λ_1 with respect to K, we see that

$$\frac{d\lambda_1}{dK} = \frac{1 - \phi_T}{2} - \frac{\phi_T - K(\phi_T - 1)^2 + 1}{\sqrt{K^2(\phi_T - 1)^2 - 2K(\phi_T + 1) + 1}}. \tag{11.19}$$

We can use basic algebraic manipulations to show that if $\phi_T > 1$, then this derivative is negative for $K < K_1$. We can similarly show that the derivative of λ_2 with respect to K is positive for $K < K_1$. This leads us to the following observation.

Observation 11.4 *If $\phi_T > 1$, then λ_1 and λ_2 are both between 0 and 1 for $K \in (0, K_1)$.*

Now consider the behavior of λ_1 and λ_2 as K increases from K_1. We see from Figure 11.3 that when $K \in (K_1, K_2)$, λ_1 and λ_2 are complex with the same magnitude, which can be derived as

$$|\lambda| = \frac{1}{2}\sqrt{[1 - K(\phi_T - 1)]^2 + 2K(\phi_T + 1) - 1 - K^2(\phi_T - 1)^2}. \tag{11.20}$$

After some algebraic manipulations, this reduces to $|\lambda| = \sqrt{K}$. The derivative of this expression is positive for all $K > 0$. This leads us to the following observation.

Observation 11.5 *When $K \in (K_1, K_2)$, λ_1 and λ_2 are complex and have the same magnitude, which monotonically increases with K.*

Now consider the value of λ_1 and λ_2 when $K = K_2$. From Figure 11.3, we know that λ_1 and λ_2 are real and equal when $K = K_2$. In fact, it is easy to see from Equation (11.16) that when $K = K_2$, $\lambda_1 = \lambda_2 = (1 + \sqrt{\phi_T})/(1 - \phi_T)$. This is between 0 and -1 for all $\phi_T > 4$, which we state as follows.

Observation 11.6 *When $K = K_2$, we obtain $\lambda_1 = \lambda_2 = (1 + \sqrt{\phi_T})/(1 - \phi_T)$, which is between 0 and -1 if $\phi_T > 4$.*

Now consider the values of λ_1 and λ_2 when $K > K_2$. Both λ_1 and λ_2 are real for this range of K. Equation (11.19) gives the derivative of λ_1 with respect to K when λ_1 is real. We can perform some basic algebraic manipulations of Equation (11.19) to show that if $\phi_T > 1$ and $K > K_2$, then the derivative of λ_1 is positive and the derivative of λ_2 is negative. Combining this reasoning with Observation 11.6, we see that λ_1 remains less than 1 in magnitude for all values of $K > K_2$. However, λ_2 approaches $-\infty$ as $K \to \infty$, as can be seen from Equation (11.17). This gives us the following observation.

Observation 11.7 *When $K > K_2$, λ_1 is real and negative and less than 1 in magnitude, and λ_2 is real and negative and approaches $-\infty$ as $K \to \infty$.*

The limit of λ_1 as $K \to \infty$ can be derived from Equation (11.16):

$$\begin{aligned}
\lambda_1 &= \frac{1}{2}\left[1 - K(\phi_T - 1) + \sqrt{1 + K^2(\phi_T - 1)^2 - 2K(\phi_T + 1)}\right] \\
&= \frac{1/K - (\phi_T - 1) + \sqrt{1/K^2 + (\phi_T - 1)^2 - 2(\phi_T + 1)/K}}{2/K} \\
&= \frac{N(K)}{D(K)}
\end{aligned} \tag{11.21}$$

where the numerator $N(K)$ and denominator $D(K)$ are defined by the above equation. The limit of both $N(K)$ and $D(K)$ is 0 as $K \to \infty$, so we can use l'Hopital's rule to evaluate the limit as follows.[3]

[3]Thanks to Steve Szatmary for deriving $\lim_{K\to\infty} \lambda_1$.

$$
\begin{aligned}
\frac{dN(K)}{dK} &= -K^{-2} + \frac{-2K^{-3} + 2(\phi_T + 1)K^{-2}}{2\sqrt{K^{-2} + (\phi_T - 1)^2 - 2(\phi_T + 1)K^{-1}}} \\
\frac{dD(K)}{dK} &= -2K^{-2} \\
\frac{dN(K)/dK}{dD(K)/dK} &= \frac{1}{2} + \frac{K^{-1} - \phi_T - 1}{2\sqrt{K^{-2} + (\phi_T - 1)^2 - 2(\phi_T + 1)K^{-1}}} \\
\lim_{K \to \infty} \lambda_1 &= \lim_{K \to \infty} \frac{dN(K)/dK}{dD(K)/dK} \\
&= \frac{1}{2} - \frac{\phi_T + 1}{2(\phi_T - 1)} \\
&= \frac{1}{1 - \phi_T}
\end{aligned} \tag{11.22}
$$

which is less than one in magnitude if $\phi_T > 2$. This leads us to the following observation, which is an expansion of Observations 11.6 and 11.7.

Observation 11.8 *As K increases from K_2 to ∞, λ_1 monotonically increases from $(1 + \sqrt{\phi_T})/(1 - \phi_T)$ to $1/(1 - \phi_T)$, and λ_2 monotonically decreases from $(1 + \sqrt{\phi_T})/(1 - \phi_T)$ to $-\infty$.*

Since λ_2 decreases from $(1 + \sqrt{\phi_T})/(1 - \phi_T)$, which is greater than -1, to $-\infty$, λ_2 must be equal to -1 at some value of K, which we denote as K_3. Therefore, from Equation (11.17) we have

$$
-1 = \frac{1}{2}\left[1 - K_3(\phi_T - 1) - \sqrt{1 + K^2(\phi_T - 1)^2 - 2K(\phi_T + 1)}\right]. \tag{11.23}
$$

Solving this equation for K_3 gives $K_3 = 2/(\phi_T - 2)$. Combining this with all of the above observations gives us the following theorem.

Theorem 11.1 *If b_i and h_i are constant in the velocity update of Equation (11.11), and if $\phi_T = \phi_1 + \phi_2 > 4$, then PSO is stable for*

$$
K < \frac{2}{\phi_T - 2}. \tag{11.24}
$$

Figure 11.4 illustrates how the eigenvalues of Equation (11.15) change in the complex plane as K increases from 0 to ∞. Figure 11.5 shows how their their magnitudes change with K.

We can write K for a stable PSO algorithm as

$$
K = \frac{2\alpha}{\phi_T - 2}, \text{ where } \phi_T = \phi_{1,\max} + \phi_{2,\max} \tag{11.25}
$$

with $\alpha \in (0, 1)$, so that α indicates how close the constriction coefficient K is to its theoretically maximum value before the PSO algorithm becomes unstable. A larger α allows more exploration, while a smaller α emphasizes more exploitation.

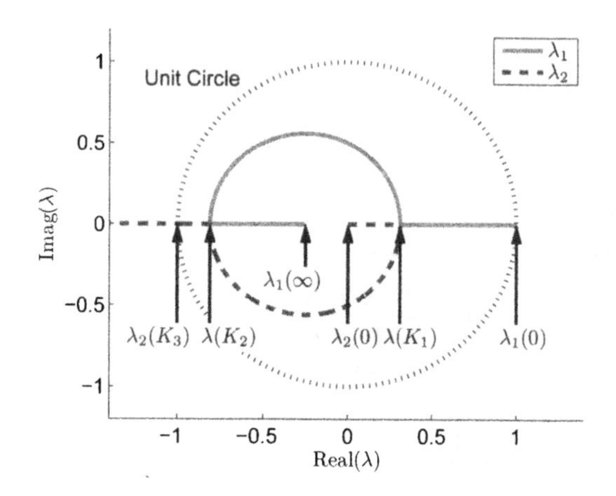

Figure 11.4 The eigenvalues of Equation (11.15) as the constriction coefficient K varies from 0 to ∞, illustrated for the case $\phi_T = 5$. At $K = 0$, $\lambda_1 = 1$ and $\lambda_2 = 0$. At $K = K_1$, $\lambda_1 = \lambda_2 > 0$. For $K \in (K_1, K_2)$, λ_1 and λ_2 are complex. At $K = K_2$, $\lambda_1 = \lambda_2 < 0$. At $K = K_3$, $\lambda_2 = -1$. As $K \to \infty$, $\lambda_1 \to 1/(1 - \phi_T)$ and $\lambda_2 \to -\infty$.

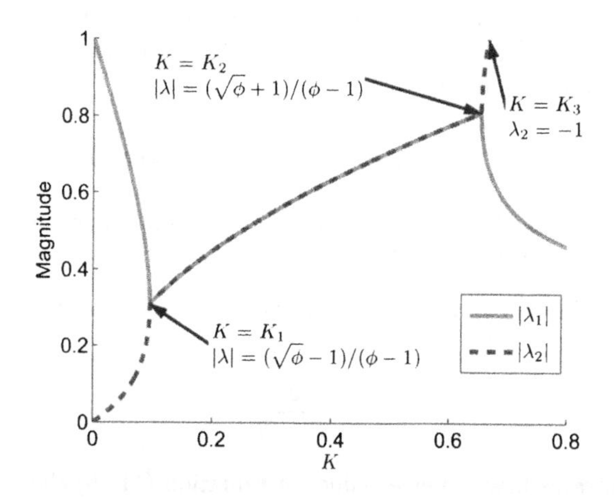

Figure 11.5 The magnitudes of the eigenvalues of Equation (11.15) as the constriction coefficient K increases from 0, illustrated for the case $\phi_T = 5$. This plot shows the magnitudes of the eigenvalues that are illustrated in Figure 11.4.

PSO algorithms are often presented in the books and research papers with the following recommendation [Carlisle and Dozier, 2001], [Clerc and Kennedy, 2002], [Eberhart and Shi, 2000], [Poli et al., 2007]:

$$\begin{aligned} \text{Common recommendation: } \phi_T &> 4 \\ K &< \frac{2}{\phi_T - 2 + \sqrt{\phi_T(\phi_T - 4)}}. \end{aligned} \tag{11.26}$$

This is equivalent to Theorem 11.1 as $\phi_T \to 4$, but Theorem 11.1 is more general for $\phi_T > 4$. Equation (11.26) does not provide any guidance for an upper bound for ϕ_T, or for how to allocate ϕ_T among $\phi_{1,\max}$ and $\phi_{2,\max}$. It is often recommended to set ϕ_T slightly larger than 4, and to allocate ϕ_T approximately equally among $\phi_{1,\max}$ and $\phi_{2,\max}$ – for example, $\phi_{1,\max} = \phi_{2,\max} = 2.05$. However, empirical results indicate there are some optimization problems for which better PSO performance can be obtained for values of ϕ_T that are much greater than 4.1, and for values of $\phi_{1,\max}$ and $\phi_{2,\max}$ that are far apart [Carlisle and Dozier, 2001]. Also note that our analysis takes only one specific approach, but other approaches with other assumptions lead to different stability conditions [Clerc and Poli, 2006].

11.4 GLOBAL VELOCITY UPDATES

One way that we can generalize the velocity update of Equation (11.11) is to write

$$v_i \leftarrow K\left[v_i + \phi_1(b_i - x_i) + \phi_2(h_i - x_i) + \phi_3(g - x_i)\right] \tag{11.27}$$

where g is the best individual found so far since the first generation. The analysis of the previous section is valid for Equation (11.27) if we define $\phi_T = \phi_{1,\max} + \phi_{2,\max} + \phi_{3,\max}$, and if we assume that $b_i + h_i + g$ is constant with time. The new term $\phi_3(g - x_i)$ adds a term to the velocity update equation that tends to drive each particle toward the best individual found so far since the first generation. This is conceptually similar to the stud EA, which uses the best individual at each generation for each recombination operation (Section 8.7.7). The difference is that g in Equation (11.27) is the best individual found since the first generation, while the stud in Section 8.7.7 is the best individual in the current generation. This similarity and difference could motivate the use of a more g-like operation in the stud EA, or the use of a more stud-like operation in the global PSO algorithm.

■ **EXAMPLE 11.1**

In this example we use PSO with the general velocity update of Equation (11.27) to optimize the 20-dimensional Ackley function. We use a population size of 50, an elitism parameter of 2, and a neighborhood size $\sigma = 4$. We use the nominal values

$$\begin{aligned} \phi_{1,\max} &= \phi_{2,\max} = \phi_{3,\max} = 2.1 \\ \phi_T &= \phi_{1,\max} + \phi_{2,\max} + \phi_{3,\max} \\ K &= \frac{2\alpha}{\phi_T - 2}, \quad \alpha = 0.9. \end{aligned} \tag{11.28}$$

Note that we can alternatively solve for ϕ_T in terms of K:

$$\phi_T = \frac{2(\alpha + K)}{K}. \tag{11.29}$$

Figures 11.6–11.9 show the average performance of PSO for various values of $\phi_{1,\max}$, $\phi_{2,\max}$, $\phi_{3,\max}$, and α, when the other parameters are equal to their nominal values. We see that the nominal values of Equation (11.28) are indeed approximately optimal for the 20-dimensional Ackley function.

Figures 11.6–11.8 show that when the $\phi_{\max}$ values are too small, the particles wander in an undirected manner. When they are too large, the particles are overly restricted and are unable to effectively explore the search space.

Figure 11.9 shows that when α (and hence K) is too small, the particles stagnate due to small velocities. When α (and hence K) is too large, the particles jump too aggressively through the search space.

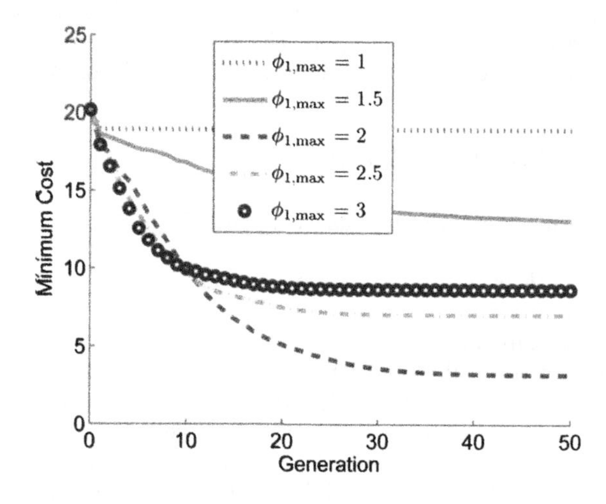

Figure 11.6 Example 11.1: Performance of PSO on the 20-dimensional Ackley function for various values of $\phi_{1,\max}$, averaged over 20 Monte Carlo simulations. $\phi_{1,\max} = 2$ is approximately optimal for this benchmark function.

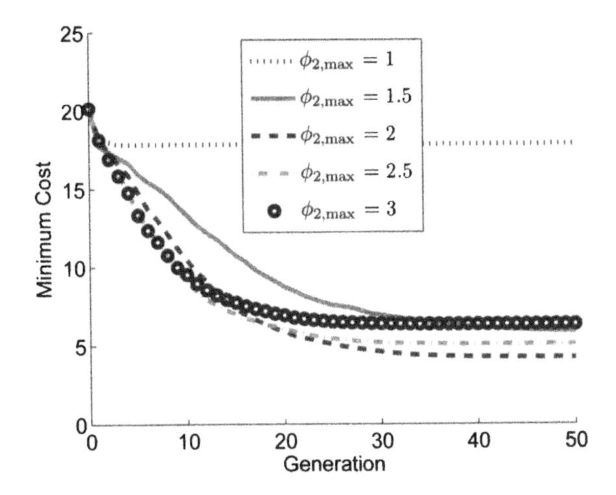

Figure 11.7 Example 11.1: Performance of PSO on the 20-dimensional Ackley function for various values of $\phi_{2,\max}$, averaged over 20 Monte Carlo simulations. $\phi_{2,\max} = 2$ is approximately optimal for this benchmark function.

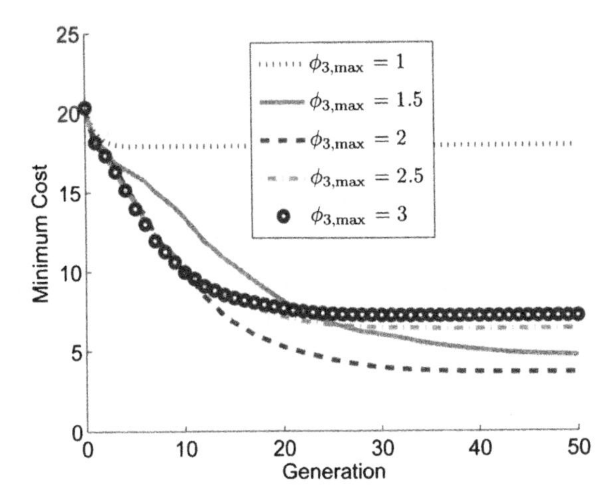

Figure 11.8 Example 11.1: Performance of PSO on the 20-dimensional Ackley function for various values of $\phi_{3,\max}$, averaged over 20 Monte Carlo simulations. $\phi_{3,\max} = 2$ is approximately optimal for this benchmark function.

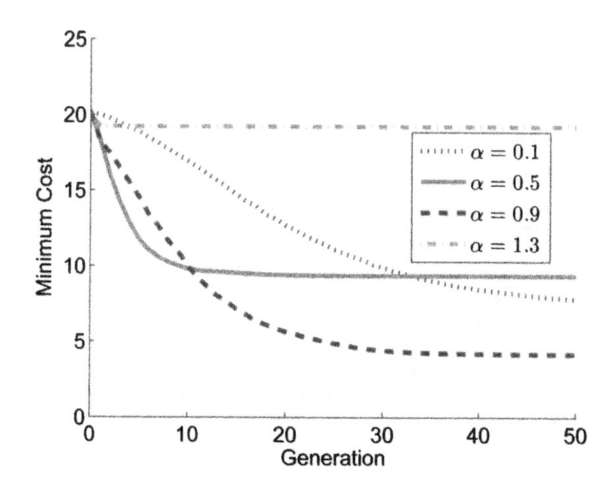

Figure 11.9 Example 11.1: Performance of PSO on the 20-dimensional Ackley function for various values of the constriction coefficient $K = \alpha K_{\max}$, averaged over 20 Monte Carlo simulations. $K = 0.9K_{\max}$ is approximately optimal for this benchmark function.

□

11.5 THE FULLY INFORMED PARTICLE SWARM

Equations (11.12) and (11.27) show that our most general (so far) form for the velocity update is

$$\begin{aligned} v_i(t+1) &= K\left[v_i(t) + \phi_T\left(p_i(t) - x_i(t)\right)\right] \\ \phi_T &= \phi_{1,\max} + \phi_{2,\max} + \phi_{3,\max} \\ p_i(t) &= \frac{\phi_1 b_i(t) + \phi_2 h_i(t) + \phi_3 g(t)}{\phi_1 + \phi_2 + \phi_3}. \end{aligned} \tag{11.30}$$

We see that three particle positions contribute to the velocity update: the current individual's best position so far $b_i(t)$, the neighborhood's best current position $h_i(t)$, and the population's best position so far $g(t)$. This leads to the idea of making the velocity update more general. Why not allow every individual in the population to contribute to the velocity update? A generalization of Equation (11.30) can be written as

$$\begin{aligned} v_i(t+1) &= K\left[v_i(t) + \phi_T\left(p_i(t) - x_i(t)\right)\right] \\ \phi_T &= \frac{1}{N}\sum_{j=1}^{N} \phi_{j,\max} \\ p_i(t) &= \frac{\sum_{j=1}^{N} w_{ij}\phi_j b_j(t)}{\sum_{j=1}^{N} w_{ij}\phi_j} \end{aligned} \tag{11.31}$$

where $b_j(t)$ is the best solution found so far by the j-th particle:

$$b_j(t) = \arg\min_x f(x) : x \in \{x_j(0), \cdots, x_j(t)\}. \tag{11.32}$$

Note the $1/N$ factor in the definition of ϕ_T in Equation (11.31), which is an ad-hoc approach to maintaining a reasonable balance between the contribution of $v_i(t)$ and $(p_i(t) - x_i(t))$ to the new velocity $v_i(t+1)$. The ϕ_j parameters in Equation (11.31) are random influence factors that are taken from the uniform distribution $U[0, \phi_{j,\max}]$. As indicated in Example 11.1, we often use

$$\begin{aligned} \phi_{j,\max} &\approx 2 \\ K &= 2\alpha/(3\phi_T - 2) \end{aligned} \tag{11.33}$$

where $\alpha \in (0,1)$. The factor of 3 in the value of K compensates for the fact that in Equation (11.27) ϕ_T is the sum of three $\phi_{j,\max}$ terms, while in Equation (11.31) it is the average of the $\phi_{j,\max}$ terms. The w_{ij} weights in Equation (11.31) are deterministic factors that describe the influence of the j-th particle on the velocity of the i-th particle. Sometimes we use w_{ij} = constant for all j. Other times, we want w_{ij} to be larger for values of j that correspond to better x_j particles, and also larger for values of j that correspond to x_j particles that are closer to x_i. For instance, if our problem is a minimization problem, we could use something like

$$w_{ij} = \left[\max_k f(x_k) - f(x_j)\right] + \left[\max_k |x_i - x_k| - |x_i - x_j|\right] \tag{11.34}$$

where $|\cdot|$ is a distance measurement. We might also need to weight the cost and fitness contributions appropriately so that they both contribute equal orders of magnitude to w_{ij}. For example,

$$\begin{aligned} S_i &= \frac{\max_k f(x_k) - \min_k f(x_k)}{\max_k |x_i - x_k|} \\ w_{ij} &= \left[\max_k f(x_k) - f(x_j)\right] + S_i \left[\max_k |x_i - x_k| - |x_i - x_j|\right]. \end{aligned} \tag{11.35}$$

S_i is a scale factor that makes the two terms that contribute to w_{ij} approximately equal. Since Equation (11.31) allows every particle to influence every other particle, it is called the fully informed particle swarm (FIPS) [Mendes et al., 2004]. This idea is reminiscent of global uniform recombination in EAs (Section 8.8.6).

■ EXAMPLE 11.2

In this example, we use the fully informed particle swarm of Equation (11.31) with the weights of Equation (11.35) to optimize the 20-dimensional Ackley function. We use a population size of 40 and an elitism parameter of 2. We use the nominal values

$$\begin{aligned} \phi_{j,\max} = \phi_{\max} &= 2, \quad \text{for } j \in [1, 20] \\ K &= 2\alpha/(3\phi_{\max} - 2), \quad \alpha = 0.9. \end{aligned} \tag{11.36}$$

Figures 11.10 and 11.11 show the average performance of PSO for various values of $\phi_{\max}$ and α, when the other parameter is equal to its nominal value.

Figure 11.10 shows that when $\phi_{\max}$ is small, the swarm converges very quickly, but it converges to a poor solution. As $\phi_{\max}$ increases, initial convergence slows, but the final converged solution becomes better. This may motivate us to use an adaptive $\phi_{\max}$ that is initially small and then gradually increases over time. Figure 11.11 shows that for small values of α, convergence is very slow. Convergence is fastest for $\alpha = 0.9$, but the final solution is better for $\alpha = 0.5$.

These results are very specific. They apply for a specific benchmark function with a specific dimension, a specific elitism parameter, and a specific form for the w_{ij} weighting parameters (Equation (11.35)). Additional experimentation is needed to see if the conclusions for this example can be generalized to a wider range of problems.

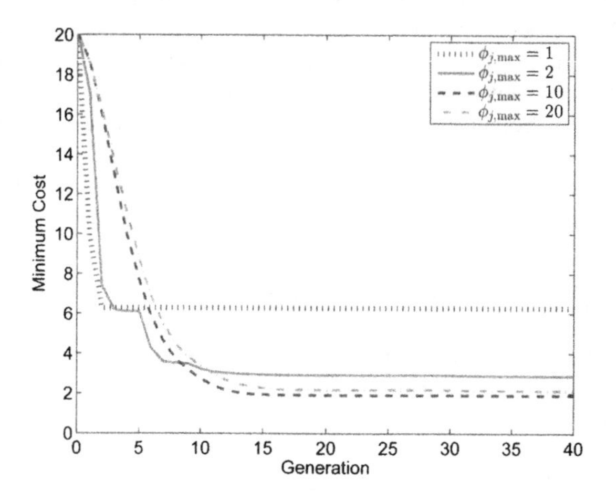

Figure 11.10 Example 11.2: Performance of the fully informed particle swarm on the 20-dimensional Ackley function for various values of $\phi_{\max}$, averaged over 20 Monte Carlo simulations. $\phi_{\max} = 1$ gives the best short-term performance, but larger values of $\phi_{\max}$ give better long-term performance.

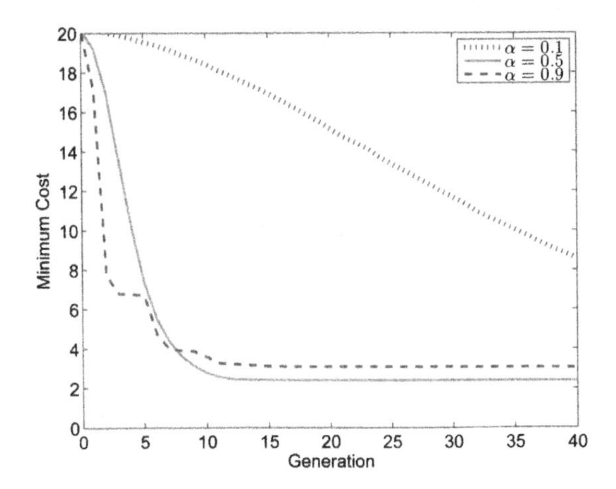

Figure 11.11 Example 11.2: Performance of the fully informed particle swarm on the 20-dimensional Ackley function for various values of α, averaged over 20 Monte Carlo simulations. $\alpha = 0.9$ gives the best short-term behavior, and $\alpha = 0.5$ gives the best long-term behavior.

□

Sometimes fully informed PSO is written differently than Equation (11.31). For example, Equation (11.31) can be replaced with the following [Poli et al., 2007]:

$$v_i(t+1) = K\left[v_i(t) + \frac{1}{n_i}\sum_{j=1}^{n_i}\phi_j\left(b_{i,j}(t) - x_i(t)\right)\right] \tag{11.37}$$

where n_i is the neighborhood size of the i-th particle, ϕ_j is taken from the uniform distribution $U[0, \phi_{\max}]$, and $b_{i,j}(t)$ is the best solution found so far by the j-th neighbor of the i-th particle. In this formulation, each particle has a certain fixed neighborhood, and each neighbor's best solution $b_{i,j}(t)$ has an equally weighted contribution (on average) to the velocity update of the i-th particle. Note that Equation (11.37) is equivalent to Equation (11.11) under certain conditions. Some papers have found that fully informed PSO performs poorly because the particles experience too many conflicting attractions, or because the search space of each particle decreases with increasing neighborhood size [de Oca and Stützle, 2008].

11.6 LEARNING FROM MISTAKES

PSO is based on the idea that biological organisms tend to repeat strategies that have proven successful in the past. This includes beneficial strategies that they have used themselves, and also beneficial strategies that they have observed in others. The basic equation for updating velocity, as we see in Equation (11.27), is

$$v_i \leftarrow K\left[v_i + \phi_1(b_i - x_i) + \phi_2(h_i - x_i) + \phi_3(g - x_i)\right] \tag{11.38}$$

where x_i and v_i are the position and velocity of the i-th particle; b_i is the previous best position of the i-th particle; h_i is the current best position of the i-th neighborhood; g is the previous best position of the entire swarm; and K, $\phi_{1,\max}$, $\phi_{2,\max}$, and $\phi_{3,\max}$ are positive tuning parameters.

However, biological organisms not only learn from successes, but also learn from mistakes. We tend to avoid strategies that have proven harmful in the past. This includes detrimental strategies that we have used ourselves, and also detrimental strategies that we have observed in others. A natural extension of PSO is to incorporate this avoidance of negative behavior in the basic PSO algorithm. This algorithm has been called "new PSO" in [Yang and Simon, 2005], [Selvakumar and Thanushkodi, 2007], but the term "new" is overused and nondescriptive, so we refer to it as "negative reinforcement PSO" (NPSO) in this section.

In NPSO, each particle adjusts its velocity not only in the direction of the best position of itself and its neighbors, but also away from the direction of the worst position of itself and its neighbors. Equation (11.38) is therefore modified to

$$\begin{aligned} v_i \leftarrow\; & K\left[v_i + \phi_1(b_i - x_i) + \phi_2(h_i - x_i) + \phi_3(g - x_i)\right. \\ & \left. -\phi_4(\bar{b}_i - x_i) - \phi_5(\bar{h}_i - x_i) - \phi_6(\bar{g} - x_i)\right] \end{aligned} \tag{11.39}$$

where $\bar{b}_i$ is the previous worst position of the i-th particle; $\bar{h}_i$ is the current worst position of the i-th neighborhood; $\bar{g}$ is the previous worst position of the entire swarm; each ϕ_j is taken from a uniform distribution on $(0, \phi_{j,\max})$; and each $\phi_{j,\max}$ is a positive tuning parameter.

We have to find a balance between the velocity adjustment towards beneficial solutions that comes from standard PSO, and the velocity adjustment away from detrimental solutions that we have added to NPSO. This balance is something that we all try to find in our everyday lives. How much do we focus on success and try to emulate it, compared to how much we focus on failure and try to avoid it? Most of us agree that positive reinforcement is more effective than negative reinforcement, but most of us also agree that both types of reinforcement are important for learning.

■ EXAMPLE 11.3

In this example, we use the NPSO of Equation (11.39) to optimize the 20-dimensional Schwefel 2.26 function. We use a population size of 20 and an elitism parameter of 2. We use the nominal values

$$\begin{aligned} \phi_{1,\max} = \phi_{2,\max} = \phi_{3,\max} &= 2 \\ \phi_{4,\max} = \phi_{5,\max} = \phi_{6,\max} &= 0 \\ K = \frac{2\alpha}{\phi_{1,\max} + \phi_{2,\max} + \phi_{3,\max} - 2}, \quad \alpha &= 0.9. \end{aligned} \tag{11.40}$$

Figures 11.12–11.14 show the average performance of NPSO for various values of $\phi_{4,\max}$, $\phi_{5,\max}$, and $\phi_{6,\max}$, when the other parameters are equal to their nominal values. Figure 11.12 shows that when $\phi_{4,\max}$, which determines how much each particle avoids its previous worst position, is increased above its nominal value of 0, it can result in a large improvement in performance. Figure 11.13 shows a similar but less dramatic improvement when $\phi_{5,\max}$, which determines how much each particle avoids the current worst position of its neighborhood, is increased beyond its nominal value of 0. Finally, Figure 11.14 shows that performance also improves when $\phi_{6,\max}$, which determines how much each particle avoids the previous worst position of the entire swarm, is increased beyond its nominal value of 0 . It appears from these figures that $\phi_{6,\max}$ has the greatest effect on NPSO performance.

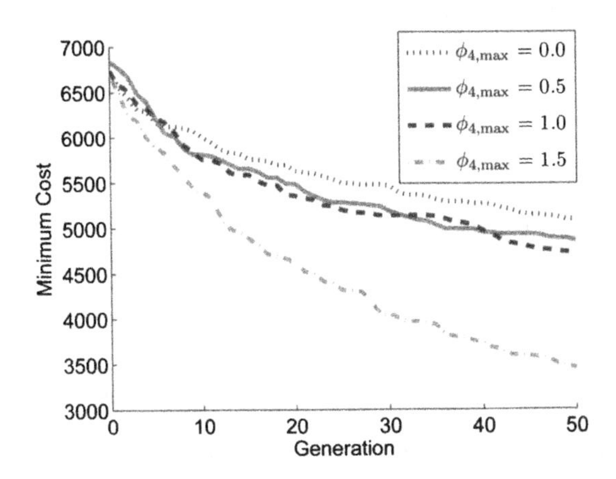

Figure 11.12 Example 11.3: Performance of NPSO on the 20-dimensional Schwefel 2.26 function for various values of $\phi_{4,\max}$, averaged over 20 Monte Carlo simulations. Particles that avoid their own previous worst position perform significantly better than particles that do not.

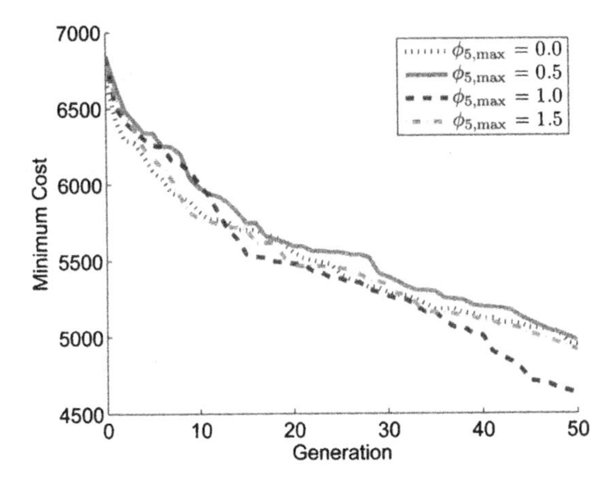

Figure 11.13 Example 11.3: Performance of NSPO on the 20-dimensional Schwefel 2.26 function for various values of $\phi_{5,\max}$, averaged over 20 Monte Carlo simulations. Particles that avoid the current worst position of their neighbors perform noticeably better than particles that do not.

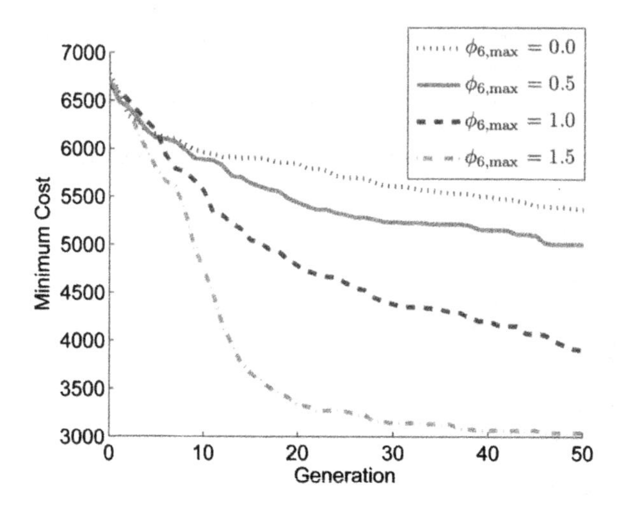

Figure 11.14 Example 11.3: Performance of NSPO on the 20-dimensional Schwefel 2.26 function for various values of $\phi_{6,\max}$, averaged over 20 Monte Carlo simulations. Particles that avoid the previous worst position of the swarm perform significantly better than particles that do not.

□

Example 11.3 indicates that the NPSO can perform much better than standard PSO. Notice in Example 11.3 that we changed only one of the negative reinforcement terms at a time while leaving the other two equal to zero. We have not tried combining nonzero values of $\phi_{4,\max}$, $\phi_{5,\max}$, and $\phi_{6,\max}$, but we leave this for future research by the reader. Also note that we could combine the idea of negative reinforcement with the fully informed PSO of Equation (11.31). We also leave this extension to the reader for further research. Finally, it would be interesting to rederive the stability results of Section 11.3 for NPSO; this is another area for future research.

11.7 CONCLUSION

PSO has proven itself to be an effective EA for a variety of problems. Any investigation of a newly proposed EA should include a comparison with PSO because of its good performance. Similar to ant colony optimization, some researchers do not consider PSO as an evolutionary algorithm, but instead consider it to be a type of swarm intelligence. It is true that PSO particles do not directly share candidate solution information with each other. However, PSO does include fitness-based selection, and PSO particles do share velocity information with each other, and velocity information directly affects the solutions. Therefore, we categorize PSO as an EA in this book.

Catfish PSO is a modification that was introduced to combat stagnation in PSO [Yang et al., 2011]. In a holding tank, sardines often settle into a locally optimal behavior and location, but then become lethargic and experience rapidly degrading health. If catfish are added to the tank, the sardines experience a renewed sense of stimulation and remain healthy for a longer period of time. Catfish PSO is based on this observation, and stimulates a PSO population when it stagnates. If the best individual in the PSO population has not improved for m consecutive generations (m is often between 3 and 7), then each independent variable of the worst 10% of the population is set equal to one of the boundaries of the search domain. The reason the particles are moved to the boundaries of the search domain is to maximize the search space. Also, optimization problems with constraints often have solutions that lie on a constraint boundary [Bernstein, 2006].

All of the discussion in this chapter has focused on PSO for continuous-domain problems. PSO has been extended in several different ways for combinatorial optimization [Kennedy and Eberhart, 1997], [Yoshida et al., 2001], [Clerc, 2004]. Other current research directions include simplifying the PSO algorithm [Pedersen and Chipperfield, 2010], hybridizing it with other EAs [Niknam and Amiri, 2010], adding mutation-like operators to avoid premature convergence [Xinchao, 2010], using multiple interacting swarms [Chen and Montgomery, 2011], removing randomness from the PSO algorithm [Clerc, 1999], using dynamic and adaptive topologies [Ritscher et al., 2010], exploring initialization strategies [Gutierrez et al., 2002], and adapting PSO parameters on-line [Zhan et al., 2009]. Also note that just as we model the velocity of each PSO particle, we could also model their accelerations [Tripathi et al., 2007]. Other future work could include particle swarm behavior and convergence analysis that takes the randomness of the algorithm into account, and that takes the relationships between the particles into account.

Additional recommended reading and study in the area of PSO includes books [Kennedy and Eberhart, 2001], [Clerc, 2006], [Sun et al., 2011]; and papers [Bratton and Kennedy, 2007], [Banks et al., 2007], [Banks et al., 2008]. Useful and extensive PSO web sites include [PSC, 2012] and [Clerc, 2012a].

PROBLEMS

Written Exercises

11.1 What are some arguments for having static neighborhoods in PSO? What are some arguments for having dynamic neighborhoods?

11.2 Acceleration in PSO:

a) How could you modify the PSO algorithm of Figure 11.1 to include acceleration?

b) Given this modification of the PSO algorithm, how would Equation (11.4) change, and what would be the eigenvalues?

11.3 Suppose that $\phi_1 = 4$ in Equation (11.4).

a) What are the eigenvalues of the matrix?

b) Is the system stable?

c) Give an initial condition and input b_i that will result in x_i and v_i being bounded as $t \to \infty$.

d) Give an initial condition and input b_i that will result in x_i and v_i being unbounded as $t \to \infty$.

11.4 Equation (11.35) uses the cost and distance of x_i to calculate the weight w_{ij}. What are some other features of x_i that we might consider using as part of the w_{ij} calculation?

11.5 Assuming that $p_i(t)$ is constant in Equation (11.30), write the dynamic state-space equations for $x_i(t+1)$ and $v_i(t+1)$. What are the eigenvalues of the system?

11.6 Under what conditions are Equations (11.11) and (11.37) equivalent?

11.7 Generalize the NPSO update of Equation (11.39) to obtain a fully-informed NPSO update equation.

11.8 Equation (11.25) recommends setting the constriction coefficient as follows:

$$K = \frac{2\alpha}{\phi_T - 2}$$

where $\alpha \in (0, 1)$. We can set ϕ_T to the sum of the maximum possible values of the ϕ_i terms, in which case ϕ_T is constant for the PSO algorithm; or we can set ϕ_T to the sum of the actual ϕ_i terms that are randomly computed for each velocity update, in which case ϕ_T is different for each velocity update. Assuming that we use Equation (11.27) for our velocity update, these two options can be written as follows:

$$\begin{aligned} K_1 &= \frac{2\alpha_1}{\phi_{1,\max} + \phi_{2,\max} + \phi_{3,\max} - 2} \\ K_2 &= \frac{2\alpha_2}{\phi_1 + \phi_2 + \phi_3 - 2} \end{aligned}$$

where each ϕ_i is uniformly distributed on $[0, \phi_{i,\max}]$. What value of α_2 in the above equations makes $K_1 = K_2$ on average? (See Problem 11.12 for the computer exercise counterpart to this problem.)

Computer Exercises

11.9 Neighborhood Sizes: Simulate the PSO algorithm of Figure 11.1 for 40 generations to minimize the 10-dimensional sphere function (see Appendix C.1.1 for the definition of the sphere function). Use a population size of 20, and use the global velocity update of Equation (11.27). Use $\phi_{1,\max} = \phi_{2,\max} = \phi_{3,\max} = 2$, use $v_{\max} = \infty$, and use $\alpha = 0.9$ to find the constriction coefficient K. Run 20 Monte Carlo simulations for neighborhood sizes $\sigma = 0$, 5, and 10. Plot the average performance of each Monte Carlo set as a function of generation number. What do you conclude about the importance of local neighborhoods in PSO?

11.10 Fully Informed Particle Swarm Distance Weighting: Equation (11.35) can be written as

$$\begin{aligned} w_{ij} &= w_{ij}(c) + S w_{ij}(d) \\ \text{where } w_{ij}(c) &= \max_k f(x_k) - f(x_j) \\ w_{ij}(d) &= \max_k |x_i - x_k| - |x_i - x_j|. \end{aligned}$$

$w_{ij}(c)$ is the cost contribution of x_j to w_{ij}, and $w_{ij}(d)$ is the distance contribution. The above equation can be generalized as follows:

$$w_{ij} = (w_{ij}(c) + D S w_{ij}(d))/(1 + D)$$

where D is the importance of the distance contribution relative to the cost contribution. Use this weight formula to simulate the fully informed PSO to optimize the 20-dimensional Rastrigin function (see Appendix C.1.11 for the definition of the Rastrigin function). Run 20 Monte Carlo simulations for $D = 0$, 0.5, 1, 2, and 1000. Plot the average performance of each Monte Carlo set as a function of generation number, and provide some general observations about your results.

11.11 Fully Informed Particle Swarm Neighborhood Sizes: Implement the velocity update of Equation (11.37) in a PSO simulation to minimize the 20-dimensional Rosenbrock function (see Appendix C.1.4 for the definition of the Rosenbrock function). Use a population size of 20 and a generation count limit of 40. Tune $\phi_{\max}$ and K for good performance. Run 20 Monte Carlo simulations with neighborhood sizes of 2, 5, 10, and 20. Plot the average performance of each Monte Carlo set as a function of generation number, and provide some observations about your results.

11.12 Constant vs. Time-Varying Constriction: Simulate the PSO algorithm of Figure 11.1 for 50 generations to minimize the 10-dimensional Ackley function (see Appendix C.1.2 for the definition of the Ackley function). Use a population size of 20, and use the global velocity update of Equation (11.27). Use $\phi_{1,\max} = \phi_{2,\max} = \phi_{3,\max} = 2.1$, use $v_{\max} = \infty$, and use $\alpha_1 = 0.9$ to find the constriction coefficient K_1 that is defined in Problem 11.8. Run 20 Monte Carlo simulations with the constant constriction coefficient K_1, and run 20 Monte Carlo simulations with the time-varying constriction coefficient K_2 that you found in Problem 11.8. Plot the average performance of each Monte Carlo set as a function of generation number, and comment on your results.

CHAPTER 12

Differential Evolution

> Compared to several existing EAs, DE is much simpler and straightforward to implement ... Simplicity of programming is important for practitioners from other fields, since they may not be experts in programming ....
>
> —S. Das, P. Suganthan, and C. Coello Coello [Das et al., 2011]

Differential evolution (DE) was developed by Rainer Storn and Kenneth V. Price around 1995. Like many new optimization algorithms, DE was motivated by real-world problems: the solution of Chebyshev polynomial coefficients, and the optimization of digital filter coefficients. DE made a quick and impressive entrance into the world of EAs by finishing as one of the top entries at the First International Contest on Evolutionary Computation [Storn and Price, 1996] and at the Second International Contest on Evolutionary Optimization [Price, 1997]. The first DE publications were in conference proceedings [Storn, 1996a], [Storn, 1996b], and the first journal publication was a year later [Storn and Price, 1997]. However, the first DE publication that was widely read was in a non-refereed magazine [Price and Storn, 1997]. DE is a unique evolutionary algorithm because it is not biologically motivated.

Overview of the Chapter

Section 12.1 outlines a basic DE algorithm for optimization over continuous domains. After the original introduction of DE, researchers introduced many variations, and we discuss some of these variations in Section 12.2. After DE proved to be successful for continuous-domain problems, researchers extended it to discrete domains, and so we discuss DE for discrete-domain problems in Section 12.3. DE was originally introduced *not* as a separate EA, but as a genetic algorithm variation, and so we look at DE from that perspective in Section 12.4.

12.1 A BASIC DIFFERENTIAL EVOLUTION ALGORITHM

DE is a population-based algorithm that is designed to optimize functions in an n-dimensional continuous domain. Each individual in the population is an n-dimensional vector that represents a candidate solution to the problem. DE is based on the idea of taking the difference vector between two individuals, and adding a scaled version of the difference vector to a third individual to create a new candidate solution. This process is depicted in Figure 12.1.

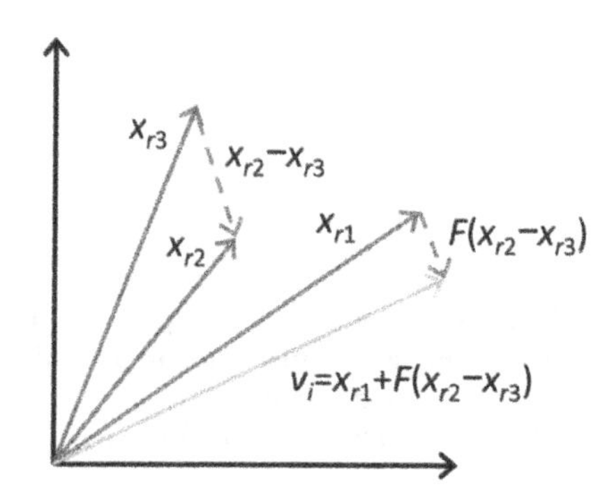

Figure 12.1 The basic idea of differential evolution, illustrated for a two-dimensional optimization problem ($n = 2$). x_{r1}, x_{r2}, and x_{r3} are candidate solutions. A scaled version of the difference between individuals x_{r2} and x_{r3} is added to x_{r1} to obtain a mutant vector v_i, which is a new candidate solution. Note that v_i is indexed with the subscript i because we generate n separate mutant vectors each generation, where n is the population size.

Figure 12.1 depicts DE in a two-dimensional search space. Two individuals, x_{r2} and x_{r3}, are randomly chosen with $r_2 \neq r_3$. A scaled version of the difference between those two individuals is added to a third randomly chosen individual, x_{r1}, where $r_1 \notin \{r_2, r_3\}$. This results in a mutant v_i that might be accepted into the population as a new candidate solution.

After the mutant vector v_i is created, it is combined (that is, crossed over) with a DE individual x_i, where $i \notin \{r_1, r_2, r_3\}$, to create a trial vector u_i. Crossover is implemented as follows:

$$u_{ij} = \begin{cases} v_{ij} & \text{if } (r_{cj} < c) \text{ or } (j = \mathcal{J}_r) \\ x_{ij} & \text{otherwise} \end{cases} \tag{12.1}$$

for $j \in [1, n]$, where n is the problem dimension and is also the dimension of u_i, v_i, and x_i; u_{ij} is the j-th component of u_i; v_{ij} is the j-th component of v_i; x_{ij} is the j-th component of individual x_i; r_{cj} is a random number taken from the uniform distribution $[0, 1]$; c is the constant crossover rate $\in [0, 1]$;[1] and $\mathcal{J}_r$ is a random integer taken from the uniform distribution $[1, n]$. We see that the trial vector u_i is a component-by-component combination of a current DE individual x_i and the mutant vector v_i. The purpose of $\mathcal{J}_r$ is to guarantee that u_i is not a clone of x_i, although this complication can be omitted for most problems (see Problem 12.3). The crossover rate c controls how likely it is that each component of u_i comes from the mutant vector v_i.

After N trial vectors u_i have been created as described above, where N is the population size, the u_i and x_i vectors are compared. The most fit vector in each (u_i, x_i) pair is kept for the next DE generation, and the least fit is discarded. The basic DE algorithm for an n-dimensional problem is summarized in Figure 12.2.

F = stepsize parameter $\in [0.4, 0.9]$
c = crossover rate $\in [0.1, 1]$
Initialize a population of candidate solutions $\{x_i\}$ for $i \in [1, N]$
While not(termination criterion)
  For each individual x_i, $i \in [1, N]$
    $r_1 \leftarrow$ random integer $\in [1, N] : r_1 \neq i$
    $r_2 \leftarrow$ random integer $\in [1, N] : r_2 \notin \{i, r_1\}$
    $r_3 \leftarrow$ random integer $\in [1, N] : r_3 \notin \{i, r_1, r_2\}$
    $v_i \leftarrow x_{r1} + F(x_{r2} - x_{r3})$ (mutant vector)
    $\mathcal{J}_r \leftarrow$ random integer $\in [1, n]$
    For each dimension $j \in [1, n]$
      $r_{cj} \leftarrow$ random number $\in [0, 1]$
      If $(r_{cj} < c)$ or $(j = \mathcal{J}_r)$ then
        $u_{ij} \leftarrow v_{ij}$
      else
        $u_{ij} \leftarrow x_{ij}$
      End if
    Next dimension
  Next individual
  For each population index $i \in [1, N]$
    If $f(u_i) < f(x_i)$ then $x_i \leftarrow u_i$
  Next population index
Next generation

Figure 12.2 A simple differential evolution (DE) algorithm for minimizing the n-dimensional function $f(x)$. This algorithm is called classic DE, or DE/rand/1/bin.

[1] Most DE literature uses the symbol Cr for the crossover rate. But two-letter symbols can be misinterpreted as separate symbols (for example, C multiplied by r), and so we use a more standard mathematical notation for the crossover rate in this chapter.

As seen from Figure 12.2, DE has several parameters that need to be tuned. As with any other EA, the population size needs to chosen. DE-specific parameters include the stepsize F, which is also called the scale factor, and the crossover rate c. These parameters are problem-dependent but are typically (but not always) chosen in the range $F \in [0.4, 0.9]$ and $c \in [0.1, 1]$. The optimal value of F generally decreases with the square root of the population size N. The optimal value of c generally decreases with the separability of the objective function [Price, 2013].

The algorithm in Figure 12.2 is often referred to as classic DE. It is also called DE/rand/1/bin because the base vector, x_{r1}, is randomly chosen; one vector difference (that is, $F(x_{r2} - x_{r3})$) is added to x_{r1}; and the number of mutant vector elements that are contributed to the trial vector closely follows a binomial distribution. It would exactly follow a binomial distribution if not for the "$j = \mathcal{J}_r$" test (see Problem 12.1).

Some thought about the classic DE algorithm of Figure 12.2 indicates why it works [Price, 2013]. First, perturbations of the form $(x_{r2} - x_{r3})$ decrease as the population narrows in on the problem solution. Second, perturbation magnitudes are different from one dimension to the next depending on the scale of the problem. That is, the magnitude of the p-th component of $(x_{r2} - x_{r3})$ is proportional to how close the population is to the problem solution along the p-th dimension. Third, perturbation steps are correlated between dimensions, which makes the search efficient even for highly nonseparable problems (see Appendix C.7.2). The result of these DE features is *contour matching*, which means that the DE population distributes itself along the contours of the objective function. A DE population tends to adapt to the objective function shape.

12.2 DIFFERENTIAL EVOLUTION VARIATIONS

In this section we look at some DE variations. Section 12.2.1 shows an alternative way of creating the trial vector u_i at each iteration, Section 12.2.2 shows some alternative ways of creating the mutant vector v, and Section 12.2.3 discusses some possibilities for using a random scale factor F.

12.2.1 Trial Vectors

Note that the method of Figure 12.2 does not include any mechanism for keeping solution features together from v_i or x_i. That is, the probability of copying v_{ij} to u_{ij} is the same whether or not $v_{i,j-1}$ was copied to $u_{i,j-1}$. However, there are many problems for which fitness depends on combinations of solution features rather than individual solution features, so it may be desirable to keep solution features together. DE/rand/1/L works by generating a random integer $L \in [1, n]$, copying L consecutive features from v_i to u_i, and then copying the remaining features from x_i to u_i [Storn and Price, 1996].

For example, suppose that we have a seven-dimensional problem ($n = 7$). The DE/rand/1/L algorithm works by first generating a random integer $L \in [1, n]$; suppose that $L = 3$. We then generate a random starting point $s \in [1, n]$; suppose that $s = 6$. Given these parameters, solution features from v_i and x_i are copied to trial vector u_i as follows:

$$
\begin{aligned}
u_{i1} &\leftarrow v_{i1} \\
u_{i2} &\leftarrow x_{i2} \\
u_{i3} &\leftarrow x_{i3} \\
u_{i4} &\leftarrow x_{i4} \\
u_{i5} &\leftarrow x_{i5} \text{ (ending point)} \\
u_{i6} &\leftarrow v_{i6} \text{ (starting point } s) \\
u_{i7} &\leftarrow v_{i7}.
\end{aligned} \tag{12.2}
$$

We see that since $s = 6$, we start copying elements of v_i to elements of u_i at the sixth dimension (that is, the sixth solution feature). Since $L = 3$, we copy three consecutive elements from v_i to u_i, where *consecutive* means that we wrap around to the beginning of the vectors after we reach the end. After copying three elements from v_i to u_i, we begin copying elements of x_i to u_i, stopping when u_i is completely defined. More formally, DE/rand/1/L works by replacing the "For each dimension" loop in Figure 12.2 with the loop in Figure 12.3.

$L \leftarrow$ random integer $\in [1, n]$
$s \leftarrow$ random integer $\in [1, n]$
$J \leftarrow \{s, \min(n, s + L - 1)\} \cup \{1, s + L - n - 1\}$
For each dimension $j \in [1, n]$
    If $j \in J$
        $u_{ij} \leftarrow v_{ij}$
    else
        $u_{ij} \leftarrow x_{ij}$
    End if
Next dimension

Figure 12.3 The DE/rand/1/L loop that copies elements from x_i and the mutant vector v_i to the trial vector u_i. The *a mod b* function returns the remainder of a/b. This loop replaces the "For each dimension" loop in Figure 12.2.

It is interesting to consider the average number of mutant vector elements v_{ij} that are copied to trial vector features u_{ij} for a given index i. For DE/rand/1/bin, there are n iterations of the "For each dimension" loop in Figure 12.2. One of those iterations has a 100% probability of copying v_{ij} to u_{ij}, and the other $(n - 1)$ iterations have a c probability of copying v_{ij} to u_{ij}. That means that the expected number of v_{ij} elements that are copied to the trial vector is

$$E(\text{number of } v_i \text{ elements copied}) = 1 + c(n - 1) \qquad \text{for DE/rand/1/bin.} \tag{12.3}$$

For DE/rand/1/L, L mutant vector features v_{ij} are copied to the trial vector. Since L is uniformly distributed in $[1, n]$,

$$E(\text{number of } v_i \text{ elements copied}) = n/2 \qquad \text{for DE/rand/1/L.} \tag{12.4}$$

Under what conditions is the expected number of mutant vector elements copied to the trial vector equal for the bin and L options? Equating Equations (12.3) and (12.4), we obtain

$$c = \frac{n-2}{2(n-1)}. \tag{12.5}$$

This value of the crossover parameter c in the DE/rand/1/bin algorithm of Figure 12.2, which is slightly less than 0.5, will result in the same average number of mutant vector elements copied to the trial vector as the DE/rand/1/L algorithm of Figure 12.3.

In general we could set L to a random integer between 1 and $L_{\max}$, with the user-specified constant $L_{\max} \in [1, n]$. We see that $L_{\max} = n$ in Figure 12.3, but values of $L_{\max}$ less than n might give better performance for some problems.

■ **EXAMPLE 12.1**

In this example we apply DE to the 20-dimensional Ackley function described in Appendix C.1.2. We use the following parameters:

- Population size = 50;
- Stepsize $F = 0.4$;
- Crossover rate $c = 0.49$ from Equation 12.5.

We will look at the difference between generating the trial vector using the bin option shown in Figure 12.2, and the L option shown in Figure 12.3. Figure 12.4 shows the best individual at each generation, averaged over 20 Monte Carlo simulations. We see that the L option converges more quickly at the beginning of the simulation, but the bin option gives significantly better performance in the long run. We would not expect to gain any improvement from using the L option because the solution features in the Ackley function are not coupled in any way; that is, the Ackley function is a separable problem. However, it is not clear why the bin option performs so much better than the L option.

□

12.2.2 Mutant Vectors

In this section we look at some alternatives for mutant vector creation. For example, instead of randomly choosing the base vector x_{r1}, it may be beneficial to always use the best individual in the population as the base vector. That way the entire set of trial vectors u_i for $i \in [1, n]$ is comprised of mutations of the best individual. This approach is called DE/best/1/bin [Storn and Price, 1996], [Storn, 1996b]. It is identical to Figure 12.2 except that the calculation of the mutant vector is replaced with

$$v_i \leftarrow x_b + F(x_{r2} - x_{r3}) \tag{12.6}$$

where x_b is the best individual in the population. This has the effect of increasing exploitation and reducing exploration. This idea is similar to the stud EA discussed in Section 8.7.7. If we use Equation (12.6) to create the mutant vector,

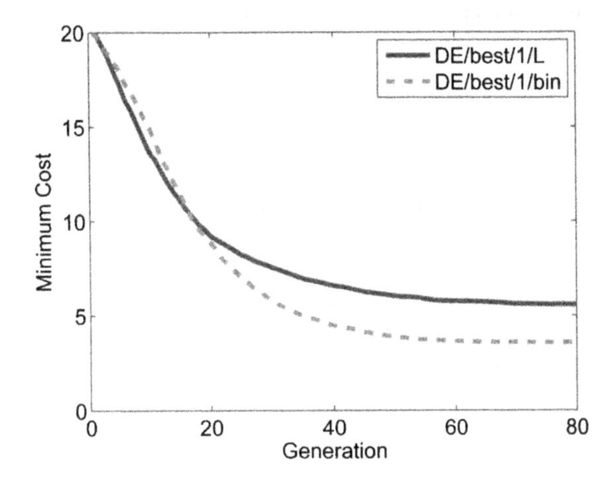

Figure 12.4 Example 12.1: DE performance on the 20-dimensional Ackley function for Example 12.1. The traces show the cost of the best individual at each generation, averaged over 20 Monte Carlo simulations. The bin option for trial vector generation performs noticeably better than the L option.

and Figure 12.3 to copy features to the trial vector, we obtain the DE/best/1/L algorithm.

Another option is to use two difference vectors to create the mutant vector [Storn and Price, 1996], [Storn, 1996b]. This can increase exploration because the total difference vector is not constrained to lie in the direction of the differences between pairs of vectors. The total difference vector has more degrees of freedom. This can be combined with a random selection of the base vector each x_i loop iteration as shown in Figure 12.2, or with the selection of the best individual as the base vector each x_i loop iteration as shown in Equation 12.6. This results in the following two options for generating the mutant vector:

$$\begin{aligned} r_4 &\leftarrow \text{random integer} \in [1, N] : r_4 \notin \{i, r_1, r_2, r_3\} \\ r_5 &\leftarrow \text{random integer} \in [1, N] : r_5 \notin \{i, r_1, r_2, r_3, r_4\} \\ v_i &\leftarrow \begin{cases} x_{r1} + F(x_{r2} - x_{r3} + x_{r4} - x_{r5}) & \text{DE/rand/2/?} \\ x_b + F(x_{r2} - x_{r3} + x_{r4} - x_{r5}) & \text{DE/best/2/?} \end{cases} \end{aligned} \tag{12.7}$$

Now we explain the question marks at the end of the above equation. If we use one of the options of Equation (12.7) to create the mutant vector and Figure 12.2 to copy features to the trial vector, we obtain the DE/rand/2/bin or DE/best/2/bin algorithm. If we use Equation (12.7) to create the mutant vector and Figure 12.3 to copy features to the trial vector, we obtain the DE/rand/2/L or DE/best/2/L algorithm.

Note that Equation (12.7) increases the effect of the difference vectors on the mutant vector. If $F = F_0$ is used in Figure 12.2 or Equation (12.6), then for a fair comparison, some $F < F_0$ should be used in Equation (12.7). The exact relationship between the two values of F depends on the shape of the objective function.

DE can also be implemented by using the current x_i as the base vector [Storn, 1996a]. For example,

$$v_i \leftarrow x_i + F\Delta x \tag{12.8}$$

where Δx is a difference vector. Depending on the method that we use to create the difference vector, and the method that we use to create the trial vector, this results in the DE/target/1/bin, DE/target/2/bin, DE/target/1/L, or DE/target/2/L algorithm.[2] In contrast to the DE/rand algorithms of Sections 12.1 and 12.2.1, DE/target algorithms seem to be much less sensitive to F [Price, 2013].

Yet another option is to create the difference vector by using the best individual in the population, x_b. This tends to create mutant vectors that all move toward x_b. The vector that is subtracted from x_b could be a random individual or the base individual. We can imagine many possibilities based on this idea. For example [Storn, 1996a],

$$\begin{aligned}
v_i &\leftarrow x_i + F(x_b - x_i) \\
v_i &\leftarrow x_{r1} + F(x_b - x_{r3}) \\
v_i &\leftarrow x_b + F(x_{r2} - x_{r3} + x_b - x_{r5}) \\
v_i &\leftarrow x_i + F(x_b - x_i + x_{r2} - x_{r3})
\end{aligned} \tag{12.9}$$

and so on. If the last equation above is used to generate v_i, the algorithm is called DE/target-to-best/1/bin [Price et al., 2005, Section 3.3.1].[3] Note that sometimes we use recombination to create v_i, sometimes we use mutation, and sometimes we use both. The first option in Equation (12.9) is a recombination operation because it involves a combination of x_i and another vector. The second and third options are mutation operations because x_i does not appear in the equations. The fourth option is a hybrid operation because it involves x_i, but it also involves a vector difference $(x_{r2} - x_{r3})$ in which x_i does not appear.

We could combine various methods by randomly deciding how to generate the mutant vector. For example, Figure 12.5 shows a method for generating the mutant vector that gives rise to the DE/rand/1/either-or algorithm [Price et al., 2005, Section 2.6.5]. If $a < p_f$, then the standard DE/rand/1/bin method is used to generate v. However, if $a \geq p_f$, then a special type of DE/rand/2 method is used to generate v.

At this point we are getting almost too many permutations of the DE algorithm to manage. Most of these options, though, are generally of secondary importance. The main idea of DE is depicted in Figures 12.1 and 12.2, and all of the possible variations are just details.

[2]The DE/target algorithms are also referred to in the literature as DE/current and DE/i.

[3]It seems that it should be called DE/target-to-best/2/bin to be consistent with other DE naming conventions. On the other hand, the equation can be interpreted as the addition of a single mutation to the base vector x_i, which justifies the terminology DE/target-to-best/1/bin.

p_f = mutation probability $\in [0, 1]$
$a \leftarrow$ random number $\in [0, 1]$
If $a < p_f$ then
    $v_i \leftarrow x_{r1} + F(x_{r2} - x_{r3})$
else
    $v_i \leftarrow x_{r1} + K(x_{r2} - x_{r1} + x_{r3} - x_{r1})$
End if

Figure 12.5 Mutant vector generation that results in the DE/rand/1/either-or algorithm. Generally, $K = (F + 1)/2$ gives good results in benchmark problems.

■ EXAMPLE 12.2

In this example we again apply DE to the 20-dimensional Ackley function described in Appendix C.1.2. Since Example 12.1 showed that the bin option performed better than the L option, we use the bin option in this example. In this example we will look at how performance changes depending on which vector we use as our base vector. We have three options. We can use a random vector x_{r1} as the base vector as shown in Figure 12.2, or we can use the best vector as the base vector as shown in Equation (12.6), or we can use the current population member as the base vector as shown in Equation (12.8). Figure 12.6 shows the best individual at each generation, averaged over 20 Monte Carlo simulations. We see that the random and current options perform about the same. This is expected since: (1) the current option results in each individual from the current population being used as the base vector exactly once per generation; and (2) the random option results in each individual being used as the base vector an average of once per generation. On the other hand, Figure 12.6 shows that the best option clearly outperforms the other two options. It is apparent that focusing the search around the best individual each generation is a beneficial strategy.

Since using the best individual as the base vector gives the best performance, we use that option in the rest of this example. For our last simulation in this example we look at how performance changes depending on how many difference vectors are used to generate the mutant vector. We can use a single difference vector as shown in Equation (12.6), or we can use two difference vectors as shown in Equation (12.7). Figure 12.7 shows the best individual at each generation, averaged over 20 Monte Carlo simulations. We see that using only one difference vector performs slightly better than using two difference vectors, even when F is corrected for the effect of using two difference vectors as discussed in the text following Equation (12.7). The reason for this is not clear, but it is consistent with the findings reported in [Price et al., 2005, Section 2.4.7], and it would be interesting to explore this issue in more detail.

□

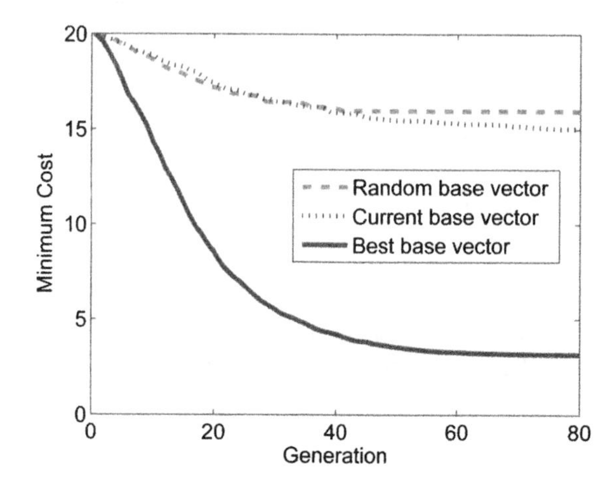

Figure 12.6 Example 12.2: DE performance on the 20-dimensional Ackley function. The traces show the cost of the best individual at each generation, averaged over 20 Monte Carlo simulations. Using the best individual as the base vector gives significantly better performance.

We could also implement other EA options in DE. For example, we could easily add a more standard EA mutation operation to the DE algorithm, such as Gaussian or uniform mutation centered at the candidate solution (see Section 8.9). However, this operation it might not have much impact since DE's mutant vector is already highly exploratory.

Recall that elitism is a common feature of EAs that ensures that we do not lose high-performing individuals, and that the performance of the best individual in the population never gets worse from one generation to the next (see Section 8.4). Elitism is an attractive option for all EAs and usually provides a significant improvement in performance. However, there is no need to implement elitism in DE, because DE automatically saves the best individuals each generation, as seen in the "for each population index" loop in Figure 12.2. But this raises the issue that there may be problems for which DE performs better with a less aggressive elitism strategy. Sometimes EAs need to tunnel through poor regions of the search space to reach good solutions. Non-elitist EAs may be better suited for certain problems with expensive or dynamic cost functions (see Chapter 21).

12.2.3 Scale Factor Adjustment

DE's scale factor F determines the effect that difference vectors have on the mutant vector. So far we have assumed that F is a constant. However, randomization is one of the hallmarks of EAs, so it makes sense to let F be a random variable. This allows a broader range of mutant vectors, which may lead to greater exploration by the DE algorithm. Also, making F random allows for the analysis of DE's convergence properties [Zaharie, 2002].

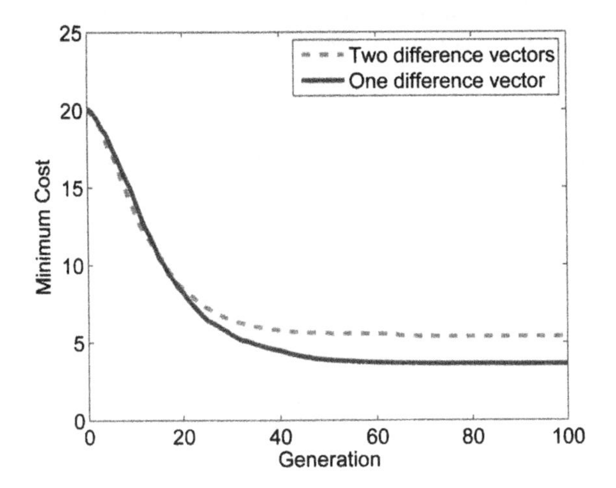

Figure 12.7 Example 12.2: DE performance on the 20-dimensional Ackley function for Example 12.2. The traces show the cost of the best individual at each generation, averaged over 20 Monte Carlo simulations. The use of only one difference vector for mutant vector generation is slightly better than the use of two difference vectors.

We can vary the DE scale factor two different ways. First, we can allow F to remain a scalar and randomly change it each time through the "for each individual" loop in Figure 12.2. This type of variation is called dither. Second, we can change F to an n-element vector and randomly change each element of F in the "for each individual" loop, so that each element of the mutant vector v is modified by a uniquely-scaled component of the difference vector. This type of variation is called jitter.

Dither replaces the mutant vector generation line in Figure 12.2 with the following:

$$\begin{aligned} F &\leftarrow U[F_{\min}, F_{\max}] \\ v_i &\leftarrow x_{r1} + F(x_{r2} - x_{r3}). \end{aligned} \tag{12.10}$$

That is, the scale factor is a random scalar uniformly distributed between $F_{\min}$ and $F_{\max}$. Other approaches to dithering allow F to be taken from a Gaussian distribution [Price et al., 2005, Section 2.5.2].

Jitter replaces the mutant vector generation line in Figure 12.2 with the following:

$$\begin{aligned} &\text{For each dimension } j \in [1, n] \\ &\quad F_j \leftarrow U[F_{\min}, F_{\max}] \\ &\quad v_{ij} \leftarrow x_{r1,j} + F_j(x_{r2,j} - x_{r3,j}) \\ &\text{Next dimension} \end{aligned} \tag{12.11}$$

That is, each element of the difference vector is scaled by a different amount to create the mutant vector.

In general, constant values of F seem to work better for simple functions (for example, the sphere function), and randomized values of F seem to work better for most multimodal functions. Jitter works best for functions that are mostly separable, and dither works best for highly nonseparable functions [Price, 2013].

■ EXAMPLE 12.3

In this example we explore the use of dithering and jittering. We use a population size of 50, as in previous examples. We use $F_{\min} = 0.2$ and $F_{\max} = 0.6$ in Equations (12.10) and (12.11). Figure 12.8 shows the average performance of DE on the 20-dimensional Ackley function with crossover rate $c = 0.9$ and with three different implementations of the scale factor: (1) Constant F; (2) Dithered F; and (3) Jittered F. We see that the dithering and jittering options perform about the same, while the constant F option performs the best. This indicates that randomizing F degrades performance. However, Figure 12.9 shows the same results on the Fletcher optimization benchmark. In this case jittering performs slightly but clearly better than constant F, which in turn performs better than dithered F.

These results indicate that the effects of dithering and jittering depend on the specific problem that is being solved, and also on the other parameters in the DE algorithm. The effect of varying F depends on what value of constant F we compare to, crossover rate, the range and type of distribution used for varying F, and so on.

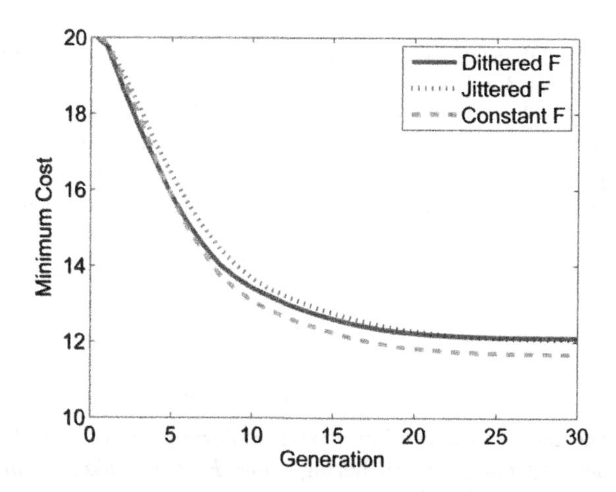

Figure 12.8 Example 12.3: DE performance on the 20-dimensional Ackley function with crossover rate $c = 0.9$. The traces show the cost of the best individual at each generation, averaged over 100 Monte Carlo simulations. The use of a constant scale factor F performs slightly better than dithering or jittering.

□

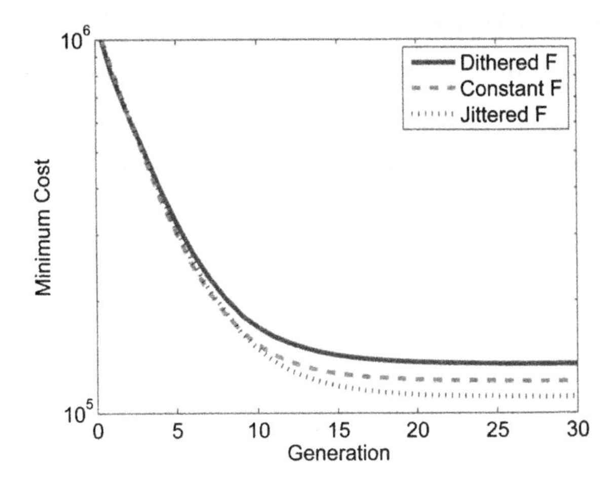

Figure 12.9 Example 12.3: DE performance on the 20-dimensional Fletcher function with crossover rate $c = 0.9$. The traces show the cost of the best individual at each generation, averaged over 100 Monte Carlo simulations. Jittering F performs slightly better than constant F, which in turn performs slightly better than dithered F.

Equations (12.10) and (12.11) both use a zero-mean uniform distribution to determine the scale factor variation ΔF from its nominal value $(F_{\text{min}} + F_{\text{max}})/2$. Other distributions, such as non-zero-mean uniform and log-normal, can also be used. These distributions have shown improvements in performance for some problems [Price et al., 2005, Section 2.5.2].

12.3 DISCRETE OPTIMIZATION

In this section we discuss how DE can be used to optimize functions over a discrete domain. The only place that discrete domains cause a problem in DE is in the generation of the mutant vector. Recall Figure 12.2, where we see that

$$v_i \leftarrow x_{r1} + F(x_{r2} - x_{r3}). \tag{12.12}$$

Since $F \in [0, 1]$, v_i might not belong the problem domain D. DE was originally designed for problems with a continuous domain, but it can be modified for discrete domains. There are two similar but fundamentally different ways to modify DE for discrete problems. First, we can generate the mutant vector v_i with standard DE methods, such as the method of Equation (12.12), and then modify it to lie in the problem domain D; we discuss one approach along these lines in Section 12.3.1. Second, we can modify the mutant vector generation method so that the mutant vector is directly generated so that it is in D; we discuss one approach along these lines in Section 12.3.2. See [Onwubolu and Davendra, 2009] for additional discussions of DE for discrete domains.

12.3.1 Mixed-Integer Differential Evolution

One obvious approach to ensure that $v_i \in D$ is to simply project it onto D. When DE is modified in this way to optimize over a discrete domain, it is often called mixed-integer DE [Huang and Wang, 2002], [Su and Lee, 2003]. For example, if D is the set of n-dimensional integer vectors, then we could replace Equation (12.12) with the following:

$$v_i \leftarrow \text{round}[x_{r1} + F(x_{r2} - x_{r3})] \tag{12.13}$$

where the *round* function operates element-by-element on a vector. A more general way to do this is

$$v_i \leftarrow P[x_{r1} + F(x_{r2} - x_{r3})] \tag{12.14}$$

where P is a projection operator such that $P(x) \in D$ for all x. Equation (12.13) gives a specific and simple possibility for P.

In general, P could be more complicated than Equation (12.13). For example, suppose again that the problem domain D is the set of N-dimensional integer vectors. Then we could define P as

$$P(x) = \arg\min_{\alpha} f(\alpha) : \alpha \in D, |x_j - \alpha_j| < 1 \text{ for all } j \in [1, n]. \tag{12.15}$$

This would project the real-valued vector x onto the integer-valued vector α which results in the lowest cost function, and where each element of α is within one unit of the corresponding element of x. This idea is illustrated in Figure 12.10 for two dimensions. Other possible forms for the projection operator could be used, and might depend on the specific problem.

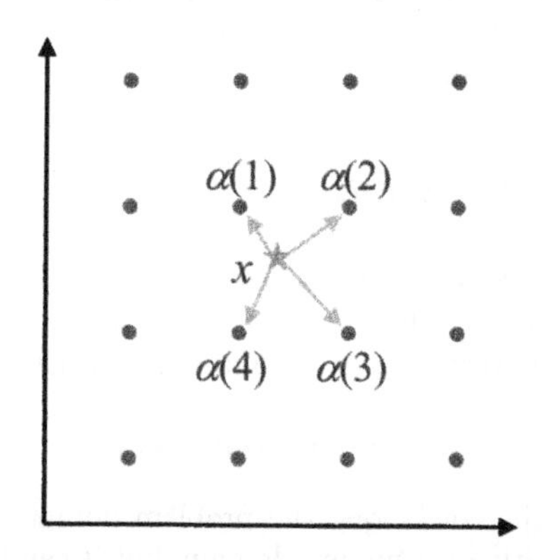

Figure 12.10 Projection of the continuous-valued vector x onto a discrete-valued vector α. This two-dimensional example shows that x is not in the problem domain of the discrete optimization problem. The cost function values of the four closest points to x in the problem domain are tested. The $\alpha(i)$ that results in the smallest value of the cost function is equal to $P(x)$, the projected value of x.

12.3.2 Discrete Differential Evolution

Another way to modify DE for discrete problems is to change the mutant vector generation method so that it directly creates mutant vectors that lie in the discrete domain D. When DE is modified in this way, it is often called discrete DE [Pan et al., 2008]. With this approach, we replace Equation (12.12) with

$$v_i \leftarrow G(x_{r1}, x_{r2}, x_{r3}) \tag{12.16}$$

where $G(\cdot) \in D$ if all of its arguments are in D. This is a generalization of Equation (12.14), and so we see that discrete DE is a generalization of mixed-integer DE. The function $G(\cdot)$ can be written to handle general discrete problems, or it can be formulated as a problem specific function. For example, suppose again that D is the set of n-dimensional integer vectors. Then we could use the following options for generating the mutant vector:

$$\begin{aligned} \text{Option 1:} \quad v_i &\leftarrow x_{r1} + \text{round}[F(x_{r2} - x_{r3})] \\ \text{Option 2:} \quad v_i &\leftarrow x_{r1} + \text{sign}(x_{r2} - x_{r3}) \end{aligned} \tag{12.17}$$

where the *round* and *sign* functions operate element-by-element on vectors.

Recall that the basic idea of mutant vector generation with discrete DE is to obtain v_i by modifying a candidate solution vector (x_{r1} in the above equation) using the difference between two other candidate solution vectors (x_{r2} and x_{r3} in the above equation). Any method of doing this that gives $v_i \in D$ is suitable for discrete DE. There are many possible methods that could be explored in future research.

12.4 DIFFERENTIAL EVOLUTION AND GENETIC ALGORITHMS

In this section we show that DE is a special type of continuous GA. Suppose that we do not know anything about DE, but that we want to develop an EA based on the material that we read in Part II of this book. In particular, suppose that we want to develop a modified version of a GA. For each individual x_i, we want to probabilistically copy independent variables from a randomly-selected individual v_i, which we call the mutant vector, to x_i, to obtain a child u_i. We use c, which we call the crossover rate, to denote the probability that an independent variable in x_i is replaced by the corresponding independent variable from v_i. This idea is very similar to uniform crossover in Section 8.8.4 if, in that section, we define $x_a = x_i$, $x_b = v_i$, and $y = u_i$. Furthermore, we want to replace x_i with the child u_i if the child is better; this idea is similar to (1+1)-ES of Section 6.1. With these ideas in mind, we propose the modified GA of Figure 12.11.

Now suppose that we want to tune our algorithm to get better performance. Instead of assigning a random individual to v_i, we decide to perturb a random individual to obtain v_i. In particular, we decide to perturb a random individual as shown in Figure 12.1. This is a conceptual change in the way that we obtain v_i, but it is, after all, still based on current population members. We also realize that because of the "If **rand**(0,1) $< c$" statement in Figure 12.11, it is possible that $u_{ij} = x_{ij}$ for all $j \in [1, n]$; that is, it is possible that the child u_i is a clone of its parent x_i. We want to prevent this, and so we think of a way to ensure that at least one independent variable in u_i is copied from v_i. We do this by adding another condition to the "If rand(0,1) $< c$" statement; we change the statement to "If (rand(0,1) $< c$) or (j = random index $\in [1, n]$)," where n is the dimension of the problem. With these ideas in mind, we obtain a generalization of Figure 12.11 as shown in the algorithm of Figure 12.12.

Initialize a population of candidate solutions $\{x_i\}$, $i \in [1, N]$
While not(termination criterion)
    For each individual x_i, $i \in [1, N]$
        $r_1 \leftarrow$ random integer $\in [1, N] : r_1 \neq i$
        $v_i \leftarrow x_{r1}$
        For each dimension $j \in [1, n]$
            If rand(0,1) $< c$ then
                $u_{ij} \leftarrow v_{ij}$
            else
                $u_{ij} \leftarrow x_{ij}$
            End if
        Next dimension
    Next individual
    For each $i \in [1, N]$, If $f(u_i) < f(x_i)$ then $x_i \leftarrow u_i$
Next generation

Figure 12.11 The above pseudo-code outlines Version 1 of a modified genetic algorithm for minimizing $f(x)$ where c is the crossover rate and **rand**(0,1) is a random number $\in [0, 1]$.

Now we note that Figure 12.12 is identical to the basic DE algorithm of Figure 12.2; that is, DE is a special type of genetic algorithm. This raises two questions.

1. Should a GA be called a GA, or should it be considered a special case of DE?

2. Should DE be called DE, or should it be considered a GA variant?

To answer the first question, we know that the GA label will never become obsolete because of its history and its foundational importance in the development of EAs. Furthermore, the GA label is useful because it encourages the incorporation of biological features into the algorithm (sexual reproduction, aging, island populations, and so on), which can lead to interesting and rewarding GA extensions.

To answer the second question above, the EA community has realized since the 1990s that DE is distinctive enough to be considered a separate EA, and should not be considered as a special case of some other EA. However, although these questions have been answered for DE, they have far-reaching implications for other EAs. Every year new EAs are proposed, some of which we discuss in Chapter 17. Which ones deserve their own class, and which ones should be considered as generalizations or special cases of already-established EAs? As more and more EAs arise in the literature, it will be more and more difficult for new EAs to find a niche. However, just as DE deserves its own class in spite of its similarity to GAs, some of these new EAs may also deserve their own class. We discuss this topic further in Chapter 17.[4]

Initialize a population of candidate solutions $\{x_i\}$, $i \in [1, N]$
While not(termination criterion)
    For each individual x_i, $i \in [1, N]$
        $r_1 \leftarrow$ random integer $\in [1, N] : r_1 \neq i$
        $r_2 \leftarrow$ random integer $\in [1, N] : r_2 \notin \{i, r_1\}$
        $r_3 \leftarrow$ random integer $\in [1, N] : r_3 \notin \{i, r_1, r_2\}$
        $v_i \leftarrow x_{r1} + F(x_{r2} - x_{r3})$
        $\mathcal{J}_r \leftarrow$ random integer $\in [1, n]$
        For each dimension $j \in [1, n]$
            If (**rand**(0,1) $< c$) or ($j = \mathcal{J}_r$) then
                $u_{ij} \leftarrow v_{ij}$
            else
                $u_{ij} \leftarrow x_{ij}$
            End if
        Next dimension
    Next individual
    For each $i \in [1, N]$, If $f(u_i) < f(x_i)$ then $x_i \leftarrow u_i$
Next generation

Figure 12.12 The above pseudo-code outlines Version 2 of a modified genetic algorithm for minimizing $f(x)$ where F is the stepsize, c is the crossover rate, and **rand**(0,1) is a random number $\in [0, 1]$.

12.5 CONCLUSION

Current research in DE mirrors current research in other EAs: simplification of the DE algorithm [Omran et al., 2009]; on-line adaptation of the DE control parameters [Qin et al., 2009]; hybridization with other search algorithms [Noman and Iba, 2008]; and the extension of DE to special types of optimization problems, like dynamic problems [Brest et al., 2009], multi-objective problems [Mezura-Montes et al., 2008], [Dominguez and Pulido, 2011], and constrained problems [Lampinen, 2002], [Mezura-Montes and Coello Coello, 2008]. As with many other EAs, there is a lot of room for theoretical and mathematical analyses of DE, so that would be a fruitful area for further research. It would be interesting to compare DE's approach to contour matching (recall the discussion at the end of Section 12.1) with that of CMA-ES (see the end of Section 6.5). Further reading on the topic of DE can be found in books [Price et al., 2005], [Feoktistov, 2006], [Qing, 2009], [Zhang and Sanderson, 2009]; and tutorial papers [Das and Suganthan, 2011], [Neri and Tirronen, 2010]; and book chapters [Syswerda, 2010].

[4]One of the first DE publications includes the subtitle, "A simple evolution strategy for fast optimization" [Price, 1997]. However, DE does not seem to have much in common with ES.

PROBLEMS

Written Exercises

12.1 Section 12.1 says that the number of mutant vector elements k that are contributed to the trial vector closely follows a binomial distribution. (See Problem 12.9 for the computer exercise counterpart to this problem.)

a) Given an experiment with a success probability c, what is the probability of obtaining k successes in n independent experiments?

b) Given the classic DE algorithm, what is the probability that the mutant vector will contribute k components to the trial vector?

12.2 The classic DE algorithm of Section 12.1 requires the generation of three random integers, but the random number generations might need to be repeated due to their restricted allowable values.

a) On average, how many random number generations are required to obtain acceptable values of r_1, r_2, and r_3? (Hint: Use the geometric distribution.)

b) Given $n = 20$, how many random number generations are required, on average, to obtain acceptable values of r_1, r_2, and r_3?

12.3 Suppose we omit the "$j = \mathcal{J}_r$" test in Figure 12.2. What is the probability that u_i is a clone of x_i? What is the probability if $c = 0.5$ and $n = 20$?

12.4 Suppose we want to implement DE/rand/1/L as described in Section 12.2.1, except that we do not want to allow elements of v to wrap around when copying them to u_i. In this case we could replace the J value in Figure 12.3 with the statement $J \leftarrow \{s, \min(n, s + L - 1)\}$. What would be the average number of v features copied to the trial vector?

12.5 How could you change the DE algorithm to be non-elitist?

12.6 Propose a stochastic projection operator for mixed-integer DE.

12.7 Propose a stochastic mutant vector generator for discrete DE.

12.8 Modify the statement "If $f(u_i) < f(x_i)$ then $x_i \leftarrow u_i$" in Figure 12.2 so that DE is more like a $(\mu + \lambda)$-ES.

Computer Exercises

12.9 Number of Mutant Contributions: In Problem 12.1 you obtained two probabilities: (1) the probability of k successes out of n independent trials, each of which have a success probability c; (2) the probability that the mutant vector contributes k components to the trial vector. Plot these two probabilities as a function of k for $n = 20$ and $c = 0.5$.

12.10 DE Step Size: Implement the classic DE algorithm of Figure 12.2 to minimize the 10-dimensional Rosenbrock function (see Appendix C.1.4 for the definition of the Rosenbrock function). Use a population size $N = 100$, a crossover rate $c = 0.9$, and a generation limit of 30. Run 40 Monte Carlo simulations for each of the following step size values F: 0.1, 0.3, 0.5, 0.7, and 0.9. For each set of Monte Carlo simulations, compute the average of the best cost of the 40 simulations at each generation. Plot the average performance of each Monte Carlo set as a function of generation number, and comment on your results.

12.11 DE Crossover Rate: Implement the classic DE algorithm of Figure 12.2 to minimize the 10-dimensional Rastrigin function (see Appendix C.1.11 for the definition of the Rastrigin function). Use a population size $N = 100$, a step size $F = 0.4$, and a generation limit of 50. Run 40 Monte Carlo simulations for each of the following crossover values CR: 0.1, 0.5, and 0.9. For each set of Monte Carlo simulations, compute the average of the best cost of the 40 simulations at each generation. Plot the average performance of each Monte Carlo set as a function of generation number, and comment on your results.

CHAPTER 13

Estimation of Distribution Algorithms

Estimation-of-distribution algorithms take a different approach to sample the search space. The population is used to estimate a probability distribution over the search space that reflects what are considered to be important characteristics of the population.

—Alden Wright [Wright et al., 2004]

An estimation of distribution algorithm (EDA) optimizes a function by keeping track of the statistics of the population of candidate solutions [Larrañaga and Lozano, 2002]. Since the statistics of the population are maintained, the actual population itself does not need to be maintained from one generation to the next. A population is created at each generation from the previous generation's population statistics, and then the statistics of the most fit individuals in the population are computed. Finally, a new population is created by using the statistics of the most fit individuals. This process repeats from one generation to the next. So EDAs are population-based algorithms that discard at least part of the population each generation and replace it using the statistical properties of highly-fit individuals. EDAs differ from most EAs in that they typically do not include recombination. EDAs are also called probabilistic model-building genetic algorithms (PMBGAs) [Pelikan et al., 2002], and iterated density estimation algorithms (IDEAs) [Bosman and Thierens, 2003].

Evolutionary Optimization Algorithms, First Edition. By Dan J. Simon

Overview of the Chapter

Section 13.1 begins this chapter by presenting the basic outline of a generic EDA, and by showing how meaningful statistics can be computed from a population of EA individuals. All EDAs use statistics such as those computed in Section 13.1.2 to create the next generation of individuals. Section 13.2 outlines some popular EDAs for discrete optimization problems that rely on only first-order statistics, including the univariate marginal distribution algorithm (UMDA), the compact genetic algorithm (cGA), and population based incremental learning (PBIL), which is a generalization of UMDA. Section 13.3 outlines some discrete EDAs that use second-order statistics, including mutual information maximization for input clustering (MIMIC), combining optimizers with mutual information trees (COMIT), and the bivariate marginal distribution algorithm (BMDA). Section 13.4 discusses multivariate EDAs, which are EDAs that use higher-order statistics, and gives an outline of the extended compact genetic algorithm (ECGA).

All of the EDAs mentioned above are designed for problems with binary domains. We conclude this chapter by showing how to extend those EDAs to problems with continuous domains. We illustrate this idea by presenting continuous UMDA and PBIL algorithms in Section 13.5.

13.1 ESTIMATION OF DISTRIBUTION ALGORITHMS: BASIC CONCEPTS

This section presents the basic outline of a generic EDA in Section 13.1.1, and shows how meaningful statistics can be computed from a population of EA individuals in Section 13.1.2.

13.1.1 A Simple Estimation of Distribution Algorithm

Figure 13.1 shows the basic outline of an EDA. Each EDA has its own unique approach to the three main steps of the algorithm of Figure 13.1. First, how are the M individuals selected from the total population of N candidate solutions? Second, what statistics are calculated from the M individuals, and how are those statistics calculated? Third, how are the statistics used to create a new population for the next generation? It is the answers to these questions that give rise to different types of EDAs, and this will occupy our attention for most of the remainder of this chapter.

The first step in the loop of Figure 13.1 is the selection of M individuals from a population of N candidate solutions, where $M < N$. This can be done in many different ways. There is nothing distinctive about the selection method in an EDA. Selection can be performed just as in any other EA, as discussed in Section 8.7, so we will not discuss selection any further in this chapter.

13.1.2 Computations of Statistics

This section describes how to compute the statistics of a population of individuals, which is the second step in the loop of Figure 13.1. We will explore this topic with a simple example. Suppose that we have a binary optimization problem, that we

Initialize a population of candidate solutions $\{x_i\}$, $i \in [1, N]$
While not(termination criterion)
    Select M individuals from $\{x_i\}$ according to fitness, where $M < N$
    Compute the statistics of the M individuals selected above
    Use the statistics to create a new population $\{x_i\}$, $i \in [1, N]$
Next generation

Figure 13.1 The basic outline of an estimation of distribution algorithm (EDA).

have N candidate solutions, that we evaluate their fitness values, and that we use some fitness-based method to select M of the individuals, where $M < N$. The selection should be biased toward the more fit individuals, as with any other EA (see Section 8.7). Suppose that $M = 10$ and we select the following 10 individuals:

$$\begin{array}{ll} x_1 = (0,1,1,1,1,0), & x_2 = (0,1,1,1,1,1), \\ x_3 = (1,0,0,1,1,0), & x_4 = (1,1,1,0,1,0), \\ x_5 = (0,1,0,0,0,1), & x_6 = (0,1,0,0,1,0), \\ x_7 = (0,0,1,1,1,0), & x_8 = (1,0,1,0,1,0), \\ x_9 = (0,1,0,0,0,0), & x_{10} = (0,1,1,1,1,1). \end{array} \tag{13.1}$$

The mean of these individuals can easily be computed as

$$\bar{x} = (0.3, 0.7, 0.6, 0.5, 0.8, 0.3). \tag{13.2}$$

The mean is a first-order statistic. We see that the first bit of this relatively fit subpopulation has only a 30% chance of being a 1. Therefore, when we generate the next population, the first bit of each individual should have a 30% chance of being a 1, and a 70% chance of being a 0. We also see that the second bit has a 70% chance of being a 1. So when we generate the next population, the second bit of each individual should have a 70% chance of being a 1.

However, we can also use second-order statistics. Note that if the first (leftmost) bit $x_i(1) = 1$ in Equation (13.1), then the second bit $x_i(2)$ has only a 1/3 chance of being a 1; also, if $x_i(1) = 0$ in Equation (13.1), then the second bit has a 6/7 chance of being a 1. It appears that there might be some correlation between the first and second bit values. So instead of giving the second bit a 70% chance of being a 1 in each member of the new population, maybe we should wait until after we generate the first bit. Then if the first bit is a 1 we should give the second bit a 1/3 chance of being a 1, and if the first bit is a 0 we should give the second bit a 6/7 chance of being a 1.

Finally, note that we could use also third-order or even higher-order statistics to create the next generation. For example, we see that if the fourth and fifth bits in Equation (13.1) are 0 and 1 respectively, the last bit is always 0.

13.2 FIRST-ORDER ESTIMATION OF DISTRIBUTION ALGORITHMS

This section presents three first-order EDAs, including the univariate marginal distribution algorithm (UMDA) in Section 13.2.1, the compact genetic algorithm (cGA) in Section 13.2.2, and population based incremental learning (PBIL) in Section 13.2.3.

13.2.1 The Univariate Marginal Distribution Algorithm (UMDA)

The univariate marginal distribution algorithm (UMDA) is the most basic EDA, and was introduced by Heinz Mühlenbein for binary problems in the late 1990s [Mühlenbein and Paaβ, 1996], [Mühlenbein and Schlierkamp-Voosen, 1997]. It uses only first-order statistics to generate the population at each generation. Figure 13.2 gives an outline of UMDA for binary optimization problems.

Although the standard UMDA does not include elitism, we can use elitism with UMDA just as with any other EA. An elitism parameter e means that we keep the best e individuals from one generation to the next. This ensures that the best individual at each generation is never worse than the best individual of the previous generation, and guarantees continuous improvement (see Section 8.4).

Initialize a population of candidate solutions $\{x_i\}$, $i \in [1, N]$
Note that each x_i includes n bits $x_i(1), \ldots, x_i(n)$
While not(termination criterion)
  Select M individuals from $\{x_i\}$ according to fitness, where $M < N$
  Index the M selected individuals as $\{x_i\}$, $i \in [1, M]$
  $\Pr(x(k) = 1) \leftarrow \sum_{i=1}^{M} \delta(x_i(k) - 1) \Big/ M$, for $k \in [1, n]$
  For $i = 1$ to N (population size)
    For $k = 1$ to n (number of bits in each candidate solution)
      $r \leftarrow U[0, 1]$
      If $r < \Pr(x(k) = 1)$
        $x_i(k) \leftarrow 1$
      else
        $x_i(k) \leftarrow 0$
      End if
    Next bit
  Next individual
Next generation

Figure 13.2 The basic outline of a univariate marginal distribution algorithm (UMDA) for optimization on an n-bit binary domain. $\delta(y)$ is the Kronecker delta function; that is, $\delta(y) = 1$ if $y = 0$, and $\delta(y) = 0$ if $y \neq 0$. $U[0, 1]$ is a random number generated from a uniform distribution between 0 and 1. $x_i(k)$ is the k-th bit in the i-th individual.

■ EXAMPLE 13.1

In this example, we use the UMDA of Figure 13.2 for the minimization of the 20-dimensional Ackley function, which is defined in Appendix C.1.2. We use six bits per dimension, so the minimization problem includes $n = 120$ bits. We use a domain of $[-5, +5]$ for each of the 20 dimensions, which gives us a resolution of $10/(2^6 - 1) = 0.16$ for each dimension. We use a population size $N = 100$ and an elitism parameter of two, which means that we always keep the best two individuals from one generation to the next. We tried four different values for M: 2, 10, 40, and 70. Figure 13.3 shows the performance of UMDA, averaged over 50 Monte Carlo simulations. Figure 13.3 shows that if we use too few or too many individuals to calculate probabilities, then performance is not good. We have to use just the right number of individuals in the probability calculation to get good performance.

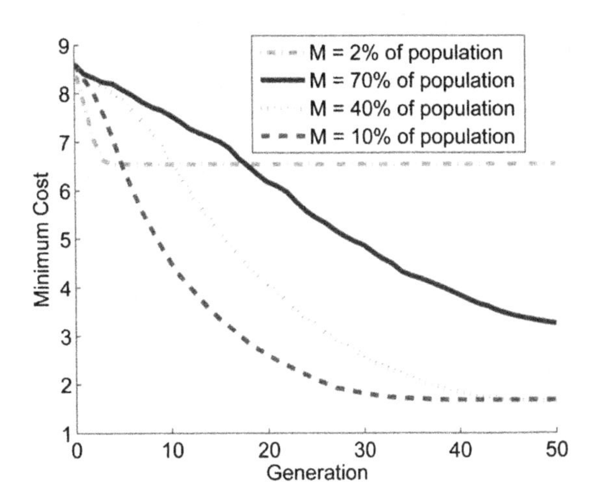

Figure 13.3 Example 13.1: UMDA results for the minimization of the 20-dimensional Ackley function with six bits per dimension. The results show the cost of the best individual at each generation, averaged over 50 Monte Carlo simulations.

□

Further UMDA research could include modifying the probability vector calculation to use weighted contributions from each individual. Figure 13.2 shows that the probability vector is calculated as

$$\Pr(x(k) = 1) \leftarrow \frac{1}{M} \sum_{i=1}^{M} \delta(x_i(k) - 1), \quad k \in [1, n]. \tag{13.3}$$

Each of the best M individuals in the population has the same contribution to the computation of the probability vector. But it makes sense to weight the best individuals more heavily than worse individuals. This extension would be similar to fully-informed particle swarm optimization. Standard PSO uses a neighborhood of a certain size to adjust each particle's velocity, but the fully-informed PSO of Section 11.5 uses the entire population to adjust each particle's velocity. This idea could result in the replacement of Equation (13.3) with something like the following:

$$\Pr(x(k) = 1) \leftarrow \sum_{i=1}^{N} w_i \delta(x_i(k) - 1) \bigg/ \sum_{i=1}^{N} w_i, \quad k \in [1, n] \tag{13.4}$$

where w_i is proportional to the fitness of x_i.

13.2.2 The Compact Genetic Algorithm (cGA)

The compact genetic algorithm (cGA) was developed by Georges Harik, Fernando Lobo, and David Goldberg in 1999 [Harik et al., 1999]. As its name indicates, it is a minimalist approach to evolutionary computation. Although the term *GA* is in its name, the cGA is more of an EDA than a GA. Like the UMDA, it uses only first-order statistics to generate individuals at each generation. Given an optimization problem on an n-element binary domain, we begin with an n-element probability vector p with each element initialized to 1/2. We then randomly generate two individuals x_1 and x_2, using p to determine the probability of the value for each bit in each of the two individuals. We then measure the fitness of the two individuals. If one individual is more fit than the other, and the i-th bit in the two individuals is different, then we adjust the i-th element of the probability vector p accordingly. We continue to the next generation by using the updated probability vector. Figure 13.4 shows a basic cGA algorithm.

```
Initialize the n-element probability vector p = [0.5, ···, 0.5]
Set p_min and p_max, the minimum and maximum values for each element of p
Set α, the probability update increment
While not(termination criterion)
    For i = 1 to 2 (population size)
        For k = 1 to n (number of bits in each candidate solution)
            r ← U[0, 1]
            If r < p(k)
                x_i(k) ← 1
            else
                x_i(k) ← 0
            End if
        Next bit
    Next individual
    Evaluate x_1 and x_2, and re-order them so that x_1 is more fit than x_2
    For k = 1 to n (number of bits in each candidate solution)
        If x_1(k) ≠ x_2(k) then
            If x_1(k) = 1 then
                p(k) ← p(k) + α
            else
                p(k) ← p(k) − α
            End if
        p(k) ← max(min(p(k), p_max), p_min)
        End if
    Next bit
Next generation
```

Figure 13.4 The basic outline of the compact genetic algorithm (cGA) for optimization on an n-bit binary domain. $U[0,1]$ is a random number generated from a uniform distribution between 0 and 1. $\alpha \in (0,1)$ governs the speed of convergence. $x_i(k)$ is the k-th bit in the i-th individual.

As with any other EA, elitism can be incorporated into the cGA. In this case, the best individual at the end of each generation is included with the two new individuals that are created at the next generation, and the best of the three individuals is used to adjust the probability vector.

EXAMPLE 13.2

In this example, we again attempt to minimize the 20-dimensional Ackley function, this time using the cGA of Figure 13.4. Our problem parameters are the same as in Example 13.1: six bits per dimension, resulting in a minimization problem with $n = 120$ bits; and a domain of $[-5, +5]$ for each of the 20 dimensions, giving a resolution of $10/(2^6 - 1) = 0.16$ for each dimension. We use a population size $N = 2$, as indicated by the algorithm of Figure 13.4. We used $p_{\min} = 0.05$ and $p_{\max} = 0.95$. We also use elitism, which means that we always keep the best individual from one generation to the next. We tried three different values for α: 0.001, 0.01, and 0.1. Figure 13.5 shows the performance of cGA, averaged over 50 Monte Carlo simulations. We see that if α is too large, then there is too much jumping around in the search space and convergence is poor. If α is too small, then the probability vector converges slightly more slowly during the early generations, but better performance is obtained in the long run. Even with the optimal value of α, though, convergence is not nearly as good as UMDA, as seen from Example 13.1. However, it should be be noted that cGA requires only two new individuals, and only one fitness function comparison, at each generation. Therefore, it is not really fair to compare UMDA and cGA with an equal number of generations; instead, they should be compared with an equal number of fitness function calculations (see Problems 13.3 and 13.12).

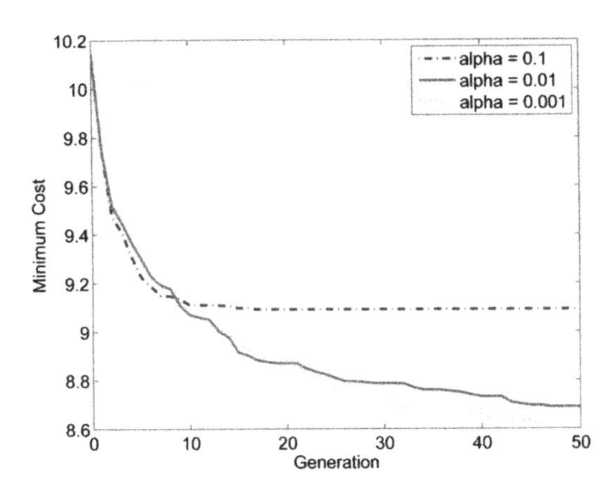

Figure 13.5 Example 13.2: cGA results for the minimization of the 20-dimensional Ackley function with six bits per dimension. The results show the best individual at each generation, averaged over 50 Monte Carlo simulations.

□

Example 13.2 showed only results with elitism, but the reader can confirm with his or her own experiments that removing elitism results in poor performance in the cGA. $p_{\min}$ and $p_{\max}$ can also have a strong influence on cGA performance. Finally, notice that there is nothing to prevent us from using more than two individuals per generation. If we create more than two individuals per generation, then we can modify the probability vector by comparing the best individual with the worst individual. See Figure 13.6 for a generalized version of the cGA.

Initialize the n-element probability vector $p = [0.5, \cdots, 0.5]$
Set $p_{\min}$ and $p_{\max}$, the minimum and maximum values for each element of p
Set α, the probability update amount
Set N, the population size
Initialize elite individual $x_e \leftarrow \emptyset$ (null vector)
While not(termination criterion)
    For $i = 1$ to N (population size)
        For $k = 1$ to n (number of bits in each candidate solution)
            $r \leftarrow U[0, 1]$
            If $r < p(k)$
                $x_i(k) \leftarrow 1$
            else
                $x_i(k) \leftarrow 0$
            End if
        Next bit
    Next individual
    $x_{\text{best}} \leftarrow$ best of $\{x_e, x_1, \cdots, x_N\}$
    $x_{\text{worst}} \leftarrow$ worst of $\{x_e, x_1, \cdots, x_N\}$
    For $k = 1$ to n (number of bits in each candidate solution)
        If $x_{\text{best}}(k) \neq x_{\text{worst}}(k)$ then
            If $x_{\text{best}}(k) = 1$ then
                $p(k) \leftarrow p(k) + \alpha$
            else
                $p(k) \leftarrow p(k) - \alpha$
            End if
        $p(k) \leftarrow \max(\min(p(k), p_{\max}), p_{\min})$
        End if
    Next bit
    $x_e \leftarrow x_{\text{best}}$
Next generation

Figure 13.6 A generalized version of the compact genetic algorithm (cGA) with elitism for optimization on an n-bit binary domain. $U[0, 1]$ is a random number generated from a uniform distribution between 0 and 1. $\alpha \in (0, 1)$ governs the speed of convergence. $x_i(k)$ is the k-th bit in the i-th individual.

EXAMPLE 13.3

In this example, we again attempt to minimize the 20-dimensional Ackley function, this time using the generalized cGA of Figure 13.6. Our problem parameters are the same as in Example 13.2, but we set $\alpha = 0.001$. We tried three different values for the population size: $N = 2$ (cGA default), $N = 5$, and $N = 20$. Figure 13.7 shows the performance of cGA, averaged over 50 Monte Carlo simulations. We see that as population size increases, performance improves. However, computational effort is directly proportional to population size. Instead of comparing with equal numbers of generations, a more fair comparison would use equal numbers of function evaluations (see Problems 13.4 and 13.13).

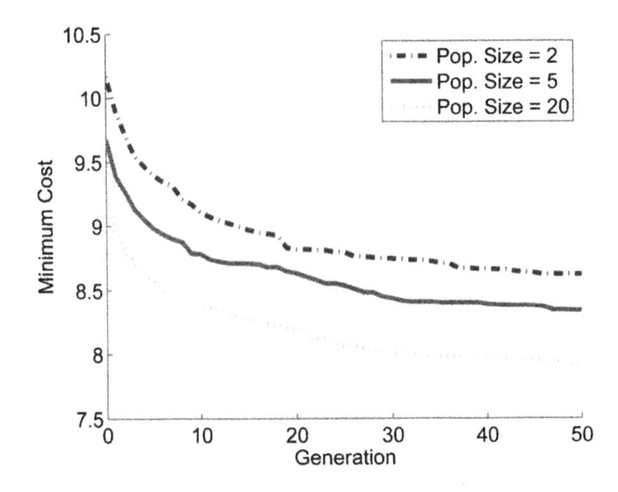

Figure 13.7 Example 13.3: cGA results for the minimization of the 20-dimensional Ackley function with six bits per dimension. The results show the best individual at each generation, averaged over 50 Monte Carlo simulations. cGA performance and computational effort are both directly proportional to population size.

□

13.2.3 Population Based Incremental Learning (PBIL)

This section presents population based incremental learning (PBIL), which is an EDA that uses first-order statistics. PBIL is a generalization of UMDA. PBIL was introduced in [Baluja, 1994], [Baluja and Caruana, 1995]. It is also called hill climbing with learning (HCwL) [Kvasnicka et al., 1996] and the incremental univariate marginal distribution algorithm (IUMDA) [Mühlenbein and Schlierkamp-Voosen, 1997]. Given an n-dimensional binary optimization problem, PBIL maintains an n-dimensional probability vector p. The k-th element of p specifies the probability that the k-th element of a candidate solution will be equal to 1. PBIL is motivated by competitive learning, which is a simple method of learning in artificial neural networks [Fausett, 1994].

At each generation we use the probability vector p to probabilistically generate a random set of candidate solutions. Then we test the fitness of each candidate solution. Next we adjust the probability vector so that the next generation is more likely to be similar to the most fit individuals, and less likely to be similar to the least fit individuals. Given this new probability vector, we proceed to the next generation by using p to create another random population of candidate solutions. This process continues until the user-defined convergence criterion is satisfied. Figure 13.8 outlines a basic PBIL algorithm for an n-dimensional binary optimization problem.

N = population size
N_{best} = number of good individuals that are used to adjust p
N_{worst} = number of bad individuals that are used to adjust p
$p_{\text{max}} \in [0,1]$ = maximum allowable value of p
$p_{\text{min}} \in [0,1]$ = minimum allowable value of p
η = learning rate $\in (0,1)$
Initialize the n-element probability vector $p = [0.5, \cdots, 0.5]$
While not(termination criterion)
    Use p to randomly generate N individuals $\{x_i\}$ as follows:
        For $i \in [1, N]$ (for each individual)
            For $k \in [1, n]$ (for each bit)
                $r \leftarrow$ random number in $U[0,1]$
                If $r < p_k$
                    $x_i(k) \leftarrow 0$
                else
                    $x_i(k) \leftarrow 1$
                End if
            Next dimension k
        Next individual i
    Sort the individuals so that $f(x_1) \le f(x_2) \le \cdots \le f(x_N)$
    For $i \in [1, N_{\text{best}}]$
        $p \leftarrow p + \eta(x_i - p)$
    Next i
    For $i \in [N - N_{\text{worst}} + 1, N]$
        $p \leftarrow p - \eta(x_i - p)$
    Next i
    Probabilistically mutate p
    $p \leftarrow \max(\min(p, p_{\text{max}}), p_{\text{min}})$
Next generation

Figure 13.8 A simple PBIL algorithm for minimizing $f(x)$, where the problem domain has n binary dimensions, and $x_i(k) \in \{0, 1\}$ is the kth element of the ith individual.

Figure 13.8 shows that there are several tuning parameters in the PBIL algorithm.

- We have to decide on a population size N, just as with all other EAs.
- We have to choose N_{best} and N_{worst}, which are the numbers of individuals used to modify the probability vector at each generation. Large values (close to $N/2$) for these parameters will result in a relatively stagnant, slow process of evolution. Small values (1 or slightly larger) will result in an aggressive learning process.
- We have to choose the learning rate η. This parameter has an effect that is the opposite to that of N_{best} and N_{worst}. A small η will result in slow optimization, and a large η will result in fast optimization. If optimization is too fast, then the PBIL algorithm will tend to overshoot the optimum.
- We have to decide on a mutation algorithm, just as with all other EAs (see Section 8.9).

We see from Figure 13.8 that we do not maintain the population from one generation to the next. We keep track of the probability vector, and we create a population at each generation, but the population is created anew at each generation.

Figure 13.8 shows that the probability vector is adjusted so that succeeding generations of individuals will be more likely to be similar to highly fit individuals. Conversely, the probability vector is adjusted so that later generations will be *less* likely to be similar to low-fitness individuals. This idea is illustrated for the two-dimensional case in Figure 13.9, where we suppose that x_1 is an individual with high fitness, and x_N is an individual with low fitness. Note from Figure 13.9 that if we add some multiple of $(x_1 - p)$ to p, then the result will be a new p that has moved closer to x_1:

$$\begin{aligned} p_{\text{new}} &\leftarrow p + \eta(x_1 - p) \\ \|p_{\text{new}} - x_1\|_2 &< \|p - x_1\|_2 \end{aligned} \tag{13.5}$$

where $\eta \in (0, 1)$ is the learning rate. Conversely, Figure 13.9 shows that if we subtract some multiple of $(x_N - p)$ from p, then the result will be a new p that has moved away from x_N:

$$\begin{aligned} p_{\text{new}} &\leftarrow p - \eta(x_N - p) \\ \|p_{\text{new}} - x_N\|_2 &> \|p - x_N\|_2. \end{aligned} \tag{13.6}$$

The PBIL algorithm combines Equations (13.5) and (13.6) for several candidate individuals (N_{best} good individuals and N_{worst} bad individuals) in a search space that is typically of high dimension. This results in the probability vector moving closer to the good individuals, and farther from the bad individuals. Subsequent generations are then more likely to be closer to the good individuals and farther from the bad individuals.

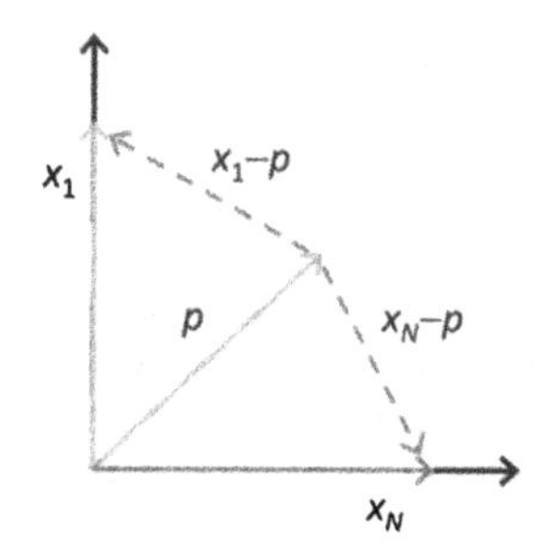

Figure 13.9 Probability vector adjustment for a two-dimensional optimization problem. We adjust p to move toward x_1, which is a good individual. We adjust p to move away from x_N, which is a poor individual.

13.3 SECOND-ORDER ESTIMATION OF DISTRIBUTION ALGORITHMS

Ideally we would like to use the entire probability distribution of the best M individuals when creating the next generation. However, that would be computationally infeasible, and so we relax rigor for the sake of simplicity and practicality. EDA algorithms like UMDA, cGA, and PBIL, which were discussed in the previous section, use only first-order statistics to generate the population; this emphasizes simplicity and practicality over rigor. The EDA algorithms in this section accept some additional complication for the sake of increased rigor, and thus use second-order statistics to generate the population. Section 13.3.1 discusses mutual information maximization for input clustering (MIMIC), Section 13.3.2 discusses combining optimizers with mutual information trees (COMIT), and Section 13.3.3 discusses the bivariate marginal distribution algorithm (BMDA).

13.3.1 Mutual Information Maximization for Input Clustering (MIMIC)

This section discusses mutual information maximization for input clustering (MIMIC), which is an EDA that uses second-order statistics, and which was developed by Jeremy De Bonet, Charles Isbell, and Paul Viola in 1997 [De Bonet et al., 1997]. The probability density function (PDF) of a random individual x can be written as

$$\begin{aligned} p(x) &= p(x(1), x(2), \cdots, x(n)) \\ &= p(x(1) \,|\, x(2), x(3), \cdots, x(n))\, p(x(2) \,|\, x(3), x(4), \cdots, x(n)) \cdots \\ &\quad\; p(x(n-1) \,|\, x(n))\, p(x(n)) \end{aligned} \tag{13.7}$$

where $x(k)$ is the k-th bit of a relatively fit candidate solution. For example, if we notice from the best M individuals that bit 4 has a 78% chance of being 1 when bits 5, 8, 9, 14, and 15, are equal to 0, 1, 1, 1, and 0 respectively, then we would like to use that information in the creation of the next generation. However for problems containing more than a few bits, it is not computationally realistic to obtain the complete distribution. That is why UMDA, cGA, and PBIL use only first-order statistics; they make the implicit assumption that the PDF of the population can be approximated with first-order statistics:

$$\hat{p}(x) = p(x(1))\,p(x(2))\cdots p(x(n)) \approx p(x). \tag{13.8}$$

MIMIC attempts to find a better approximation than the one in Equation (13.8):

$$\hat{p}(x) = p(x(k_1)\,|\,x(k_2))\,p(x(k_2)\,|\,x(k_3))\cdots p(x(k_{n-1})\,|\,x(k_n))\,p(x(k_n)) \tag{13.9}$$

where $(k_1, k_2, \cdots, k_n)$ is a permutation of $\{1, 2, \cdots, n\}$. Recall that a permutation is simply a re-ordering of integers. For example, if $n = 5$, then $(4, 5, 1, 3, 2)$ and $(5, 1, 2, 4, 3)$ are both permutations of $\{1, 2, 3, 4, 5\}$. The problem that MIMIC tries to solve is the determination of the permutation $(k_1, k_2, \cdots, k_n)$ that makes Equation (13.9) as close as possible to the true PDF $p(x)$; then MIMIC uses $\hat{p}(x)$ to create a new population of candidate solutions each generation.

Before we can find the permutation that minimizes the approximation error of $\hat{p}(x)$, we need to define approximation error. The similarity of two discrete PDFs $p(x)$ and $\hat{p}(x)$ can be quantified by the Kullback-Liebler divergence [Bishop, 2006]:

$$\begin{aligned} D(p, \hat{p}) &= \sum_x p(x) \log_2(p(x)/\hat{p}(x)) \\ &= \sum_x p(x)(\log_2 p(x) - \log_2 \hat{p}(x)) \\ &= \sum_x (p(x) \log_2 p(x) - p(x) \log_2 \hat{p}(x)) \end{aligned} \tag{13.10}$$

where the sum is taken over all points x where $p(x)$ or $\hat{p}(x)$ is nonzero. We want to minimize $D(p, \hat{p})$ with respect to $\hat{p}$. The first term on the right side of Equation (13.10) is not a function of $\hat{p}$, so our cost function can be written as

$$\begin{aligned} J(\hat{p}) &= -\sum_x p(x) \log_2 \hat{p}(x) \\ &= -\sum_x p(x) \log_2 \left[p(x(k_1)\,|\,x(k_2))\,p(x(k_2)\,|\,x(k_3))\cdots p(x(k_{n-1})\,|\,x(k_n))\,p(x(k_n))\right] \\ &= -\sum_x \left[p(x) \log_2 p(x(k_1)\,|\,x(k_2)) + p(x) \log_2 p(x(k_2)\,|\,x(k_3)) + \cdots + \right. \\ &\quad \left. p(x) \log_2 p(x(k_{n-1})\,|\,x(k_n)) + p(x) \log_2 p(x(k_n))\right] \\ &= -E\left[\log_2 p(x(k_1)\,|\,x(k_2))\right] - E\left[\log_2 p(x(k_2)\,|\,x(k_3))\right] - \cdots \\ &\quad -E\left[\log_2 p(x(k_{n-1})\,|\,x(k_n))\right] - E\left[\log_2 p(x(k_n))\right] \end{aligned} \tag{13.11}$$

where we substituted Equation (13.9) for $\hat{p}(x)$. Our cost can now be written as

$$J(\hat{p}) = h(k_1\,|\,k_2) + h(k_2\,|\,k_3) + \cdots + h(k_{n-1}\,|\,k_n) + h(k_n) \tag{13.12}$$

where the entropy terms $h(\cdot)$ are defined as follows [Gray, 2011]:

$$\begin{aligned} h(k_i) &= -E\left[\log_2 p(x(k_i))\right] \\ h(k_i\,|\,k_j) &= -E\left[\log_2 p(x(k_i)\,|\,x(k_j))\right]. \end{aligned} \tag{13.13}$$

Our problem is now more clear. We want to find the permutation $(k_1, k_2, \cdots, k_n)$ of $\{1, 2, \cdots, n\}$ so that the combined entropy on the right side of Equation (13.12) is minimized. The entropy of a single bit can be written as

$$h(k_i) = -\sum_{\alpha} \Pr(x_i = \alpha) \log_2 \Pr(x(i) = \alpha). \tag{13.14}$$

The conditional entropy of bit k_i given that bit k_j is equal to β can be written as

$$h(x(k_i) \,|\, x(k_j) = \beta) = -\sum_{\alpha} \Pr(x(k_i) = \alpha \,|\, x(k_j) = \beta) \log_2 \Pr(x(k_i) = \alpha \,|\, x(k_j) = \beta). \tag{13.15}$$

Finally, the conditional entropy of bit k_i given bit k_j can be written as

$$h(x(k_i) \,|\, x(k_j)) = \sum_{\beta} h(x(k_i) \,|\, x(k_j) = \beta) \Pr(x(k_j) = \beta). \tag{13.16}$$

■ EXAMPLE 13.4

Let us consider a few simple examples of entropy calculation. Suppose that we have four three-bit EA individuals:

$$x_1 = (0,0,0),\ x_2 = (0,0,0),\ x_3 = (1,0,0),\ x_4 = (1,1,0). \tag{13.17}$$

The entropy of the first (left) bit is

$$\begin{aligned} h(1) &= -E\left[\log_2 p(x(1))\right] \\ &= -\left[\Pr(x(1)=0)\log_2 \Pr(x(1)=0) + \Pr(x(1)=1)\log_2 \Pr(x(1)=1)\right] \\ &= -\left[0.5 \log_2 0.5 + 0.5 \log_2 0.5\right] = 1. \end{aligned} \tag{13.18}$$

The entropy of the second (middle) bit is

$$\begin{aligned} h(2) &= -E\left[\log_2 p(x(2))\right] \\ &= -\left[\Pr(x(2)=0)\log_2 \Pr(x(2)=0) + \Pr(x(2)=1)\log_2 \Pr(x(2)=1)\right] \\ &= -\left[0.75 \log_2 0.75 + 0.25 \log_2 0.25\right] = 0.81. \end{aligned} \tag{13.19}$$

The entropy of the third (right) bit is

$$\begin{aligned} h(3) &= -E\left[\log_2 p(x(3))\right] \\ &= -\left[\Pr(x(3)=0)\log_2 \Pr(x(3)=0) + \Pr(x(3)=1)\log_2 \Pr(x(3)=1)\right] \\ &= -\left[1 \log_2 1 + 0 \log_2 0\right] = 0 \end{aligned} \tag{13.20}$$

where we have used the convention $0 \log_2 0 = 0$, which is based on the fact that $\lim_{z \to 0}(z \log_2 z) = 0$. We see that the entropy of the first bit is the maximum possible value, which means that the first bit does not tell us anything about the most fit individual in the population. Based on the four individuals that we have, the most fit individual is equally likely to have either a 0 or a 1 as its left-most bit. On the other hand, the entropy of the third bit is the minimum possible value, which means that the third bit contains the maximum possible information about the most fit individual in the population. Based on the four individuals of Equation (13.17), the most fit individual has a 100% chance of having a 0 as its right-most bit.

The conditional entropies of bit 1 can be calculated as

$$\begin{aligned} h(x(1)\,|\,x(2)=0) &= -\Pr(x(1)=0\,|\,x(2)=0)\log_2 \Pr(x(1)=0\,|\,x(2)=0) \\ &\quad -\Pr(x(1)=1\,|\,x(2)=0)\log_2 \Pr(x(1)=1\,|\,x(2)=0) \\ &= -(2/3)\log_2(2/3) - (1/3)\log_2(1/3) = 0.92 \\ h(x(1)\,|\,x(2)=1) &= -\Pr(x(1)=0\,|\,x(2)=1)\log_2 \Pr(x(1)=0\,|\,x(2)=1) \\ &\quad -\Pr(x(1)=1\,|\,x(2)=1)\log_2 \Pr(x(1)=1\,|\,x(2)=1) \\ &= -0\log_2 0 - 1\log_2 1 = 0. \end{aligned} \tag{13.21}$$

We see that the conditional entropy of bit 1 given that bit 2 = 0 is relatively high, which means that knowing that bit 2 = 0 does not tell us much about the value of bit 1. On the other hand, the conditional entropy of bit 1 given that bit 2 = 1 is the lowest possible value, which means that knowing that bit 2 = 1 gives us 100% certainty about the value of bit 1. Combining these results gives the conditional entropy of bit 1 given bit 2 as

$$\begin{aligned} h(x(1)\,|\,x(2)) &= h(x(1)\,|\,x(2)=0)\Pr(x(2)=0) + h(x(1)\,|\,x(2)=1)\Pr(x(2)=1) \\ &= (0.92)(3/4) + (0)(1/4) = 0.69 \end{aligned} \tag{13.22}$$

which is a weighted sum of the two individual conditional entropy terms.

□

We want to find the permutation of $\{1, 2, \cdots, n\}$ that minimizes Equation (13.12). But there are many possible permutations of $\{1, 2, \cdots, n\}$. In general, there are $n!$ permutations of $\{1, 2, \cdots, n\}$. This number becomes extremely large for even small values of n, so a brute-force search for the optimal permutation is not possible. Instead we use a greedy algorithm [De Bonet et al., 1997] that approximately minimizes Equation (13.12) and quickly finds a good approximation for $\hat{p}(x)$. The greedy algorithm is shown in Figure 13.10. The first step is to find the bit k_n that has the lowest entropy (most information). The second step is to find the bit k_{n-1} that has the lowest conditional entropy given bit k_n. At each subsequent step, we find the bit that has the lowest conditional entropy given the previously discovered bit, making sure not to use the same bit more than once.

$k_n = \arg\min_j h(j)$
For $i = n-1, n-2, \cdots, 1$
  $k_i = \arg\min_j h(j\,|\,k_{i+1}) : j \notin \{k_{i+1}, k_{i+2}, \cdots, k_n\}$
Next i

Figure 13.10 A greedy algorithm for approximately minimizing Equation (13.12).

The MIMIC algorithm works by selecting a subset of highly-fit individuals from a population. It uses the greedy algorithm of Figure 13.10 to find a near-optimal solution to Equation (13.12). It then uses those probabilities to generate the next population of candidate solutions. Figure 13.11 gives a MIMIC algorithm for optimization on a binary domain.

Initialize a population of candidate solutions $\{x_i\}$, $i \in [1, N]$
Note that each x_i includes n bits $x_i(1), \cdots, x_i(n)$
While not(termination criterion)
    Select M individuals from $\{x_i\}$ according to fitness, where $M < N$
    Index the M selected individuals as $\{x_i\}$, $i \in [1, M]$
    $k_n \leftarrow \arg\min_j h(x(j))$
    For $m = n-1, n-2, \cdots, 1$
        $k_m \leftarrow \arg\min_j h(x(j) \mid x(k_{m+1})) : j \notin \{k_n, k_{n-1}, \cdots, k_{m+1}\}$
    Next m
    $\Pr(x(k_n) = 1) \leftarrow \sum_{i=1}^{M} \delta(x_i(k_n) - 1) \Big/ M$
    For $m = n-1, n-2, \cdots, 1$
        Define $\mathbf{1}_{m+1} = \{i \in [1, M] : x_i(k_{m+1}) = 1\}$
        Define $\mathbf{0}_{m+1} = \{i \in [1, M] : x_i(k_{m+1}) = 0\}$
        $\Pr(x(k_m) = 1 \mid x(k_{m+1}) = 1) \leftarrow \sum_{i \in \mathbf{1}_{m+1}} \delta(x_i(k_m) - 1) \Big/ |\mathbf{1}_{m+1}|$
        $\Pr(x(k_m) = 1 \mid x(k_{m+1}) = 0) \leftarrow \sum_{i \in \mathbf{0}_{m+1}} \delta(x_i(k_m) - 1) \Big/ |\mathbf{0}_{m+1}|$
    Next m
    For $i = 1$ to N (population size)
        $r \leftarrow U[0, 1]$
        If $r < \Pr(x(k_n) = 1)$ then $x_i(k_n) \leftarrow 1$; else $x_i(k_n) \leftarrow 0$
        For $m = n-1, n-2, \cdots, 1$
            $r \leftarrow U[0, 1]$
            If $x_i(k_{m+1}) = 0$
                If $r < \Pr(x(k_m) = 1 \mid x(k_{m+1}) = 0)$
                    $x_i(k_m) \leftarrow 1$
                else
                    $x_i(k_m) \leftarrow 0$
                End if
            else
                If $r < \Pr(x(k_m) = 1 \mid x(k_{m+1}) = 1)$
                    $x_i(k_m) \leftarrow 1$
                else
                    $x_i(k_m) \leftarrow 0$
                End if
            End if
        Next bit m
    Next individual i
Next generation

Figure 13.11 The basic outline of mutual information maximization for input clustering (MIMIC) for optimization on an n-bit binary domain. $h(y)$ is the entropy of the PDF of y, and is empirically estimated from candidate solutions. $\delta(y)$ is the Kronecker delta function; that is, $\delta(y) = 1$ if $y = 0$, and $\delta(y) = 0$ if $y \neq 0$. $U[0, 1]$ is a random number generated from a uniform distribution between 0 and 1.

The MIMIC algorithm includes a lot of probability calculations. During implementation we want to make sure that none of these probabilities are 0 or 1, because that would result in an inability to completely explore the search space of the optimization problem. Therefore, during the implementation of Figure 13.11, after the calculation of each probability p_i we might want to limit the probability value:

$$\begin{aligned} p_i &\leftarrow \max(p_i, \epsilon) \\ p_i &\leftarrow \min(p_i, 1-\epsilon) \end{aligned} \tag{13.23}$$

where ϵ is a small positive tuning parameter, often equal to about 0.01. An example of MIMIC is given later in Example 13.7.

13.3.2 Combining Optimizers with Mutual Information Trees (COMIT)

The COMIT algorithm was introduced in [Baluja and Davies, 1998]. COMIT is similar to MIMIC. However, in MIMIC we find a near-optimal permutation $(k_1, k_2, \cdots, k_n)$ by minimizing conditional entropy terms, as shown in Figure 13.10. In COMIT we instead find a near-optimal permutation by maximizing mutual information terms. Instead of finding the permutation that minimizes Equation (13.12), we find the permutation that maximizes

$$J_c(\hat{p}) = I(k_1 \,|\, k_2) + I(k_2 \,|\, k_3) + \cdots + I(k_{n-1} \,|\, k_n) - h(k_n) \tag{13.24}$$

The mutual information between bits k and m is defined as follows [Cover and Thomas, 1991]:

$$I(k,m) = \sum_{i,j} \Pr(x(k)=i, x(m)=j) \log_2 \left[\frac{\Pr(x(k)=i, x(m)=j)}{\Pr(x(k)=i)\Pr(x(m)=j)} \right] \tag{13.25}$$

where the summation is taken over $i \in [0,1]$ and $j \in [0,1]$.

■ **EXAMPLE 13.5**

In this example, which is based on [Chow and Liu, 1968], we illustrate the calculation of mutual information. Suppose that we are executing an EA algorithm on a four-bit optimization problem. We have many individuals from the algorithm, maybe 100 or so. We choose 20 relatively fit individuals. Suppose that these 20 individuals are given as follows:

$$\begin{array}{llllll} x_1 &=& (0,0,0,0), & x_2 &=& (0,0,0,0) \\ x_3 &=& (0,0,0,1), & x_4 &=& (0,0,0,1) \\ x_5 &=& (0,0,1,0), & x_6 &=& (0,0,1,1) \\ x_7 &=& (0,1,1,0), & x_8 &=& (0,1,1,0) \\ x_9 &=& (0,1,1,1), & x_{10} &=& (1,0,0,0) \\ x_{11} &=& (1,0,0,1), & x_{12} &=& (1,0,0,1) \\ x_{13} &=& (1,1,0,0), & x_{14} &=& (1,1,0,1) \\ x_{15} &=& (1,1,1,0), & x_{16} &=& (1,1,1,0) \\ x_{17} &=& (1,1,1,0), & x_{18} &=& (1,1,1,1) \\ x_{19} &=& (1,1,1,1), & x_{20} &=& (1,1,1,1). \end{array} \tag{13.26}$$

The bit numbers are indexed from 1 to 4 as we go from left to right. For example, $x_{13}(1) = x_{13}(2) = 1$, and $x_{13}(3) = x_{13}(4) = 0$. By counting the bits and following the procedure shown in Example 13.4, we find the following:

$$\begin{array}{lll} \Pr(x(1) = 1) = 0.55 & \rightarrow & h(1) = 0.993 \\ \Pr(x(2) = 1) = 0.55 & \rightarrow & h(2) = 0.993 \\ \Pr(x(3) = 1) = 0.55 & \rightarrow & h(3) = 0.993 \\ \Pr(x(4) = 1) = 0.50 & \rightarrow & h(4) = 1. \end{array} \tag{13.27}$$

We see that bits 1, 2, and 3 all have the same amount of information, while bit 4 has the least amount of information. Now we calculate the mutual information between bit 1 and the other bits. Equation (13.25) shows that before we calculate mutual information, we need to calculate the individual bit probabilities $\Pr(x_i)$. Consider the individuals in Equation (13.26). Equation (13.27) shows the following:

$$\begin{array}{ll} \Pr(x(1) = 0) = 0.45, & \Pr(x(1) = 1) = 0.55 \\ \Pr(x(2) = 0) = 0.45, & \Pr(x(2) = 1) = 0.55. \end{array} \tag{13.28}$$

Now note that there are six individuals in Equation (13.26) such that $x(1) = 0$ and $x(2) = 0$; there are three individuals such that $x(1) = 0$ and $x(2) = 1$; there are three individuals such that $x(1) = 1$ and $x(2) = 0$; and there are eight individuals such that $x(1) = 1$ and $x(2) = 1$. Therefore,

$$\begin{aligned} \Pr(x(1) = 0, x(2) = 0) &= 0.30 \\ \Pr(x(1) = 0, x(2) = 1) &= 0.15 \\ \Pr(x(1) = 1, x(2) = 0) &= 0.15 \\ \Pr(x(1) = 1, x(2) = 1) &= 0.40. \end{aligned} \tag{13.29}$$

Now we use Equation (13.25) to calculate the mutual information between bits 1 and 2 as follows:

$$\begin{aligned} I(1,2) &= \sum_{i,j} \Pr(x(1) = i, x(2) = j) \log_2 \left[\frac{\Pr(x(1) = i, x(2) = j)}{\Pr(x(1) = i)\Pr(x(2) = j)} \right] \\ &= \Pr(x(1) = 0, x(2) = 0) \log_2 \left[\frac{\Pr(x(1) = 0, x(2) = 0)}{\Pr(x(1) = 0)\Pr(x(2) = 0)} \right] + \\ &\quad \Pr(x(1) = 0, x(2) = 1) \log_2 \left[\frac{\Pr(x(1) = 0, x(2) = 1)}{\Pr(x(1) = 0)\Pr(x(2) = 1)} \right] + \\ &\quad \Pr(x(1) = 1, x(2) = 0) \log_2 \left[\frac{\Pr(x(1) = 1, x(2) = 0)}{\Pr(x(1) = 1)\Pr(x(2) = 0)} \right] + \\ &\quad \Pr(x(1) = 1, x(2) = 1) \log_2 \left[\frac{\Pr(x(1) = 1, x(2) = 1)}{\Pr(x(1) = 1)\Pr(x(2) = 1)} \right] \\ &= 0.30 \log_2[0.30/(0.45 \times 0.45)] + 0.15 \log_2[0.15/(0.45 \times 0.55)] + \\ &\quad 0.15 \log_2[0.15/(0.55 \times 0.45)] + 0.40 \log_2[0.40/(0.55 \times 0.55)] \\ &= 0.1146. \end{aligned} \tag{13.30}$$

We use the same method to calculate the mutual information between the other bits, which results in the following:

$$
\begin{aligned}
I(1,2) &= 0.1146 \\
I(1,3) &= 0.0001 \\
I(1,4) &= 0.0073 \\
I(2,3) &= 0.2727 \\
I(2,4) &= 0.0073 \\
I(3,4) &= 0.0073.
\end{aligned}
\tag{13.31}
$$

□

The mutual information $I(i, j)$ quantifies how much information is shared between bits i and j. It tells us how much we can know about the value of one bit if we know the value of the other bit. If bit j is given, maximizing the mutual information $I(i, j)$ over all i is similar to minimizing the conditional entropy $h(i \,|\, j)$ over all i. COMIT can therefore be performed with the same algorithm as MIMIC, but instead of using Figure 13.10 for selecting a permutation, we use Figure 13.12. The COMIT algorithm is therefore the same as the MIMIC algorithm of Figure 13.11, except that the first "For $m = n-1, n-2, \cdots, 1$" loop in Figure 13.11 is replaced with Figure 13.12.

$k_n = \arg\min_j h(j)$
For $i = n-1, n-2, \cdots, 1$
    $k_i = \arg\max_j I(j, k_{i+1}) : j \notin \{k_{i+1}, k_{i+2}, \cdots, k_n\}$
Next i

Figure 13.12 A greedy algorithm for approximately maximizing Equation (13.25). Compare with Figure 13.10.

■ EXAMPLE 13.6

In this example, which is a continuation of Example 13.5, we illustrate the greedy algorithm of Figure 13.12. The greedy algorithm first finds the bit with the most information, which is equivalent to finding the bit with the least entropy. In Equation (13.27) in Example 13.5 we saw that bits 1, 2, and 3 all have the same amount of information, while bit 4 has the least amount of information. Therefore, the solution of the problem $k_n = \arg\min_j h(j)$ in Figure 13.12 is either 1, 2, or 3. We arbitrarily choose bit 1 as the solution. Now we find the bit that shares the most information with bit 1; Equation (13.31) shows us that bit 2 shares the most information with bit 1. Now we find the bit (not including bit 1) that shares the most information with bit 2; Equation (13.31) shows us that bit 3 shares the most information with bit 2. Bit 4 is the only remaining bit, and so the greedy algorithm is complete and Equation (13.9) becomes

$$
\hat{p}(x) = p(x(1))\, p(x(2) \,|\, x(1))\, p(x(3) \,|\, x(2))\, p(x(4) \,|\, x(3)). \tag{13.32}
$$

We can use Equation (13.26) to calculate these probabilities. Table 13.1 shows the probability of each bit combination, the estimated probability using a first-order approximation of Equation (13.8), and the estimated probability from Equation (13.32).

$x(1)x(2)x(3)x(4)$	$p(x(1),x(2),$ $x(3),x(4))$	UMDA: $p(x(1))\,p(x(2))\cdot$ $p(x(3))\,p(x(4))$	COMIT: $p(x(1))\,p(x(2)\,\vert\,x(1))\cdot$ $p(x(3)\,\vert\,x(2))\,p(x(4)\,\vert\,x(3))$
0000	0.100	0.0456	0.1037
0001	0.100	0.0456	0.1296
0010	0.050	0.0557	0.0364
0011	0.050	0.0557	0.0303
0100	0.000	0.0557	0.0121
0101	0.000	0.0557	0.0152
0110	0.100	0.0681	0.0669
0111	0.050	0.0681	0.0558
1000	0.050	0.0557	0.0519
1001	0.100	0.0557	0.0648
1010	0.000	0.0681	0.0182
1011	0.000	0.0681	0.0152
1100	0.050	0.0681	0.0323
1101	0.050	0.0681	0.0404
1110	0.150	0.0832	0.1785
1111	0.150	0.0832	0.1488

Table 13.1 Example 13.6 results: True probabilities (second column) and estimated probabilities (third and fourth columns) of all possible bit combinations.

A cursory glance through the numbers in Table 13.1 shows that the true probability of the second column is estimated more accurately by the fourth column than by the third column. We can use Equation (13.10) to quantify the similarity between probability distributions. We find that the probabilities in the second and third columns have a closeness measure of 0.53, while those in the second and fourth columns have a closeness measure of 0.14.

Note that we do not use the probabilities of Table 13.1 in the COMIT algorithm. We only show them in this example to illustrate the effectiveness of the greedy algorithm of Figure 13.12. The COMIT algorithm uses the probabilities on the right side of Equation (13.32) to generate the population at the next generation.

□

Example 13.6 shows how to find a second-order approximation to a probability distribution using a mutual information criterion (COMIT) rather than a conditional entropy criterion (MIMIC). If we obtain a probability distribution estimate of highly-fit individuals in an EA, we can use the estimate to generate candidate solutions at each generation. This is the essence of COMIT, and is illustrated in the following example.

■ EXAMPLE 13.7

In this example, we use the MIMIC and COMIT algorithms of Figures 13.10, 13.11, and 13.12 to minimize the Ackley function, which is defined in Appendix C.1.2. We use six bits per dimension, so the minimization problem includes $n = 6D$ bits, where D is the dimension of the Ackley function. We use a domain of $[-5, +5]$ for each dimension, which gives us a resolution of $10/(2^6 - 1) = 0.16$ for each dimension. We use a population size $N = 100$, we use $M = 40$, and we use an elitism parameter of two, which means that we always keep the best two individuals from one generation to the next. We use $\epsilon = 0.01$ in Equation (13.23).

To study the performance of MIMIC, we compare it to UMDA (see Figure 13.2), which uses only first-order statistics. We also implement a MIMIC algorithm in which the permutation $(k_1, k_2, \cdots, k_n)$ is assigned randomly rather than using the greedy algorithm of Figure 13.10. Figure 13.13 shows the performance of MIMIC, UMDA, and random permutation MIMIC on the two-dimensional Ackley function, averaged over 20 Monte Carlo simulations. We see that the use of second-order statistics outperforms the first-order UMDA algorithm, even if the permutation is random. However, MIMIC performs the best because it uses the near-optimal greedy algorithm to determine the permutation of bit indices.

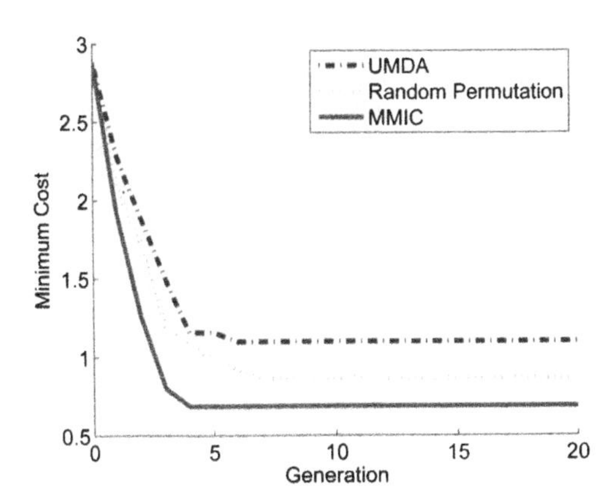

Figure 13.13 Example 13.7: UMDA and MIMIC results for the minimization of the two-dimensional Ackley function with six bits per dimension. The results show the cost of the best individual at each generation, averaged over 20 Monte Carlo simulations.

Figure 13.14 shows the performance of MIMIC, UMDA, and random permutation MIMIC on the 10-dimensional Ackley function, averaged over 20 Monte Carlo simulations. We see that MIMIC performs the best for the first few generations, but after a few generations UMDA and random-permutation MIMIC catch up and outperform MIMIC. This shows that there are no guarantees that MIMIC will perform better than first-order algorithms, but it could be a valuable optimization tool, depending on the problem.

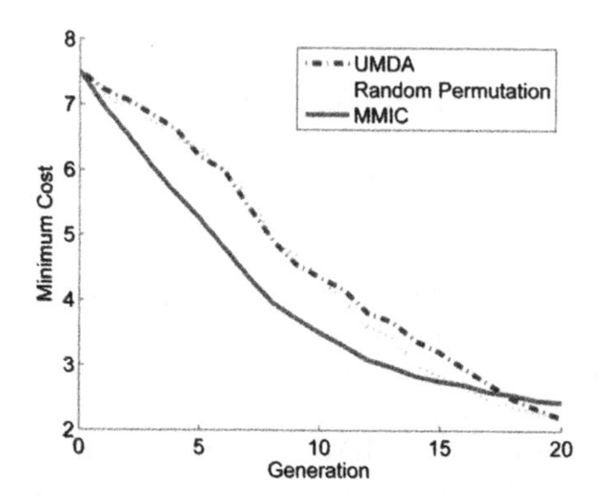

Figure 13.14 Example 13.7: UMDA and MIMIC results for the minimization of the 10-dimensional Ackley function with six bits per dimension. The results show the cost of the best individual at each generation, averaged over 20 Monte Carlo simulations.

Finally, we compare COMIT and MIMIC. Both algorithms are identically described by Figure 13.11, except MIMIC uses the greedy algorithm of Figure 13.10 and COMIT uses the greedy algorithm of Figure 13.12 to decide which bit indices to pair to generate candidate solutions at each generation. Figure 13.15 shows the performance of COMIT and MIMIC on the 10-dimensional Ackley function, averaged over 20 Monte Carlo simulations. We see that COMIT performs much better in the earlier generations. However, after about 15 generations, MIMIC catches up and surpasses the performance of COMIT.

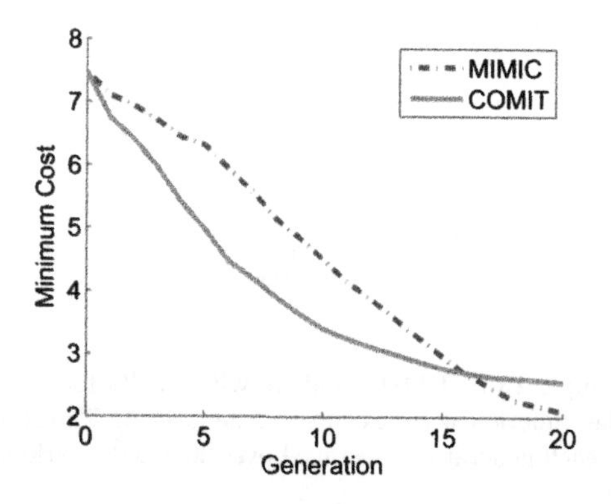

Figure 13.15 Example 13.7: MIMIC and COMIT results for the minimization of the 10-dimensional Ackley function with six bits per dimension. The results show the cost of the best individual at each generation, averaged over 20 Monte Carlo simulations.

□

13.3.3 The Bivariate Marginal Distribution Algorithm (BMDA)

The bivariate marginal distribution algorithm (BMDA) was developed in [Pelikan and Mühlenbein, 1998]. BMDA uses second-order statistics, like MIMIC and COMIT. However, it has a couple of notable differences. First, it uses Pearson's chi-square tests [Boslaugh and Watters, 2008] to establish links between interdependent bits. Second, it creates a PDF approximation that does not necessarily use all of the bits in a single, connected chain. Recall from Equation (13.9) that MIMIC and COMIT find a permutation $(k_1, k_2, \cdots, k_n)$ of $\{1, 2, \cdots, n\}$ (where n is the number of bits in x) such that

$$p(x) \approx p(x(k_1) \,|\, x(k_2))\, p(x(k_2) \,|\, x(k_3)) \cdots p(x(k_{n-1}) \,|\, x(k_n))\, p(x(k_n)). \tag{13.33}$$

The above approximation can be viewed as a single chain of k_i values:

$$k_1 \to k_2 \to \cdots \to k_{n-1} \to k_n. \tag{13.34}$$

BMDA, on the other hand, finds a more general approximation to $p(x)$:

$$p(x) \approx \prod_{x(r) \in R} p(x(r)) \prod_{x(i) \in V \setminus R} p(x(i) \,|\, x(m_i)). \tag{13.35}$$

In the approximation above, R is the set of root bit indices and is determined by BMDA, and V is the set of all bit indices; that is, $V = \{1, \cdots, n\}$. The $x(i)$ bits belong to $V \setminus R$, which is the set of all indices that do not belong to R; that is, $V \setminus R = \{i \in V : i \notin R\}$. Finally, $m(i)$ is a bit index determined by BMDA that has a high degree of dependence with bit i.

An example may clarify the interpretation of Equation (13.35). Suppose we have nine bits in our search domain. Equation (13.33) is used by MIMIC and COMIT, and may result in the chain

$$3 \to 9 \to 1 \to 5 \to 8 \to 2 \to 4 \to 7 \to 6 \tag{13.36}$$

which gives the approximation

$$\begin{aligned} p(x) \approx\; & p(x(3) \,|\, x(9))\, p(x(9) \,|\, x(1))\, p(x(1) \,|\, x(5))\, p(x(5) \,|\, x(8)) \times \\ & p(x(8) \,|\, x(2))\, p(x(2) \,|\, x(4))\, p(x(4) \,|\, x(7))\, p(x(7) \,|\, x(6))\, p(x(6)) \end{aligned} \tag{13.37}$$

Equation (13.35), which is used by BMDA, may result in the chains

$$\begin{aligned} & 3 \to 9 \to 1 \\ & 3 \to 5 \\ & 8 \to 2 \to 4 \to 7 \\ & 8 \to 6 \end{aligned} \tag{13.38}$$

which gives the approximation

$$\begin{aligned} p(x) \approx\; & p(x(3)) p(x(9) \,|\, x(3)) p(x(1) \,|\, x(9)) p(x(5) \,|\, x(3)) \times \\ & p(x(8)) p(x(2) \,|\, x(8)) p(x(4) \,|\, x(2)) p(x(7) \,|\, x(4)) p(x(6) \,|\, x(8)) \end{aligned} \tag{13.39}$$

The root bit indices 3 and 8 were determined by BMDA, and multiple bit index chains were determined by BMDA for each root.

BMDA works by first choosing a random bit index, r, as the first root bit index. Then the chi-square statistic is computed for the root bit index r and all of the remaining bits. The chi-square statistic between bits r and j is computed as

$$\chi_{rj}^2 = M \sum_{\alpha,\beta} \frac{[\Pr(x(r) = \alpha, x(j) = \beta) - \Pr(x(r) = \alpha)\Pr(x(j) = \beta)]^2}{\Pr(x(r) = \alpha)\Pr(x(j) = \beta)} \tag{13.40}$$

where M is the number of bit strings, and the summation is taken over all values of α such that $\Pr(x(r) = \alpha) \neq 0$, and over all values of β such that $\Pr(x(j) = \beta) \neq 0$. The χ_{rj}^2 statistic measures the amount of dependence between bits r and j. A high value of χ_{rj}^2 indicates that there is a high degree of correlation between bits r and j. In statistics, $\chi_{rj}^2 < 3.84$ is often used as a threshold for the independence of bits r and j. This value of χ^2 indicates that there is a 95% probability that the bit values are independent.

After BMDA computes χ_{rj}^2 for the root bit r and all remaining bits j, it selects the j with the highest value of χ_{rj}^2 as the next bit in the chain. Next, BMDA computes χ_{rk}^2 and χ_{jk}^2 for all $k \neq \{r, j\}$. Whichever k has the highest χ^2 statistic becomes the next bit in the chain, following either bit r or bit j. This process is continued until all of the χ^2 statistics are below some threshold. When that occurs, then another random root bit is chosen and the process repeats for the next chain. When all of the bits have been used, the probability approximation is complete. The BMDA algorithm is summarized in Figure 13.16 [Pelikan and Mühlenbein, 1998].

---

(1) $V \leftarrow \{1, 2, \cdots, n\}$
$A \leftarrow V$
(2) $v \leftarrow$ randomly chosen element from A
Add $\Pr(v)$ to the PDF approximation
(3) Remove v from A
If $A = \emptyset$, terminate
(4) Compute χ_{ij}^2 for all $i \in A$ and all $j \in V \backslash A$
If $\max_{i,j} \chi_{ij}^2 < 3.84$, go to (2)
(5) $\{v, v'\} = \arg\max_{i,j} \chi_{ij}^2 : i \in A, j \in V \backslash A$
Add $\Pr(v' \,|\, v)$ to the PDF approximation
Go to (3)

Figure 13.16 The basic outline of a bivariate marginal distribution algorithm (BMDA) for the generation of an n-dimensional PDF approximation. V is the constant set of all bit indices, and A is the set of available bit indices for inclusion in a chain of bits. The value 3.84 is used here as the 95% confidence level for independence.

■ EXAMPLE 13.8

Let us consider a few simple examples of the calculation of the χ^2 statistic. Suppose that we have four four-bit EA individuals:

$$x_1 = (0, 0, 0, 1),\ x_2 = (0, 0, 0, 1),\ x_3 = (1, 0, 0, 0),\ x_4 = (1, 1, 0, 0). \tag{13.41}$$

We find the marginal probabilities of bits 1 and 2 as

$$\begin{array}{ll} \Pr(x(1) = 0) = 1/2, & \Pr(x(1) = 1) = 1/2 \\ \Pr(x(2) = 0) = 3/4, & \Pr(x(2) = 1) = 1/4 \end{array} \tag{13.42}$$

where we define $x(1)$ (bit 1) as the left bit, $x(2)$ (bit 2) as the next bit, and so on. We find the joint probabilities of bits 1 and 2 as

$$\begin{array}{ll} \Pr(x(1) = 0, x(2) = 0) = 1/2, & \Pr(x(1) = 0, x(2) = 1) = 0 \\ \Pr(x(1) = 1, x(2) = 0) = 1/4, & \Pr(x(1) = 1, x(2) = 1) = 1/4. \end{array} \tag{13.43}$$

The χ^2_{12} statistic is computed from Equation (13.40) as

$$\chi^2_{12} = 4(1/24 + 1/8 + 1/24 + 1/8) = 4/3. \tag{13.44}$$

We see that there is some relationship between bits 1 and 2, but we have so few samples (four) that we cannot say with much confidence that the dependence is statistically signficant; that is, $\chi^2_{12} < 3.84$.

Now consider bits 1 and 3. In this case, the χ^2_{13} statistic is computed as

$$\chi^2_{13} = 4(0 + 0 + 0 + 0) = 0. \tag{13.45}$$

There is no relationship between bits 1 and 3. We can see this by looking at Equation (13.41); bit 1 is equal to 0 half the time and 1 half the time, while bit 3 is always equal to 0.

Finally, consider bits 1 and 4. In this case, the χ^2_{14} statistic is computed as

$$\chi^2_{14} = 4(1 + 1 + 1 + 1) = 4. \tag{13.46}$$

We see that there is a statistically significant relationship between the two bits. Even though we only have four individuals, since bits 1 and 4 are always complements of each other, we can be fairly certain that they are dependent on each other.

□

13.4 MULTIVARIATE ESTIMATION OF DISTRIBUTION ALGORITHMS

We have seen that first-order EDAs emphasize simplicity over mathematical rigor. Second-order EDAs place more emphasis on mathematical rigor, thus resulting in algorithms that are more complicated but potentially more effective. Multivariate EDAs take another step in this direction by using statistics that are even higher than second order, and we present one such algorithm in this section: the extended compact genetic algorithm (ECGA).

13.4.1 The Extended Compact Genetic Algorithm (ECGA)

As its name implies, the extended compact genetic algorithm (ECGA) is a generalization of the cGA of Section 13.2.2. The ECGA was proposed in [Harik, 1999] and is further explained in [Sastry and Goldberg, 2000], [Lima and Lobo, 2004], [Harik et al., 2010]. The ECGA attempts to find a probability distribution that satisfies two properties: first, it is simple; and second, it accurately approximates the probability distribution of a set of highly fit individuals. The approximate probability distribution is called a marginal product model (MPM). An example of an MPM $\hat{p}(x)$ for a 10-dimensional problem is the following:

$$\hat{p}(x) = p(x(1), x(3), x(6))\, p(x(2))\, p(x(4), x(5), x(7), x(10))\, p(x(8), x(9)). \quad (13.47)$$

The variables in each marginal distribution on the right side of the above equation is called a building block, so the MPM above has four building blocks: $(x(1), x(3), x(6))$, $(x(2))$, $(x(4), x(5), x(7), x(10))$, and $(x(8), x(9))$. The number of variables in the i-th building block B_i of an MPM is called the length L_i of the building block. The MPM of Equation (13.47) therefore has the building block lengths

$$L_1 = 3,\ L_2 = 1,\ L_3 = 4,\ L_4 = 2. \quad (13.48)$$

The variables in the building blocks are distinct, so if $x(k)$ belongs to B_i then it does not belong to B_j for any $j \neq i$. The complexity of an MPM is quantified as

$$C_m = (\log_2(M+1)) \sum_{i=1}^{N_b} (2^{L_i} - 1) \quad (13.49)$$

where M is the number of high-fitness individuals that are selected from the population, N_b is the number of building blocks, and L_i is the length of the i-th building block. The accuracy of an MPM is quantified as

$$C_p = \sum_{i=1}^{N_b} \sum_{j=1}^{2^{L_i}} N_{ij} \log_2(M/N_{ij}) \quad (13.50)$$

where N_{ij} is the number of individuals in the population that include the j-th bit sequence in the i-th building block. If the i-th building block has length L_i, then it includes 2^{L_i} possible bit sequences, and we order them as $(0, \cdots, 0, 0)$, $(0, \cdots, 0, 1)$, $\cdots$, $(1, \cdots, 1, 1)$. The ECGA consists of finding an MPM that minimizes the total cost

$$C_c = C_m + C_p. \quad (13.51)$$

The following example illustrates how to compute C_c for an MPM.

■ EXAMPLE 13.9

Suppose we have the following five highly-fit individuals ($M = 5$):

$$\begin{array}{lll} x_1 = (0,0,0,0), & x_2 = (0,0,1,0), & x_3 = (0,0,1,1), \\ x_4 = (1,0,1,0), & x_5 = (1,1,0,1). & \end{array} \quad (13.52)$$

One possible MPM is the univariate model

$$\hat{p}_1(x) = p(x(1))\, p(x(2))\, p(x(3))\, p(x(4)) \quad (13.53)$$

with $N_b = 4$ and $L_i = 1$ for $i \in [1, 4]$. The complexity of this model is

$$C_m = (\log_2 6) \sum_{i=1}^{4} \left(2^{L_i} - 1\right) = 10.4. \tag{13.54}$$

We order the building blocks as $B_i = p(x(i))$ for $i \in [1, 4]$. We order the bit sequences in binary order so that N_{i1} is the number of individuals for which $x(i) = 0$, and N_{i2} is the number of individuals for which $x(i) = 1$. The accuracy of the model is therefore given by

$$C_p = \sum_{i=1}^{4} \sum_{j=1}^{2} N_{ij} \log_2(M/N_{ij}) = 18.2. \tag{13.55}$$

Another possible MPM for the population of Equation (13.52) is the model

$$\hat{p}_2(x) = p(x(1), x(2))\, p(x(3))\, p(x(4)) \tag{13.56}$$

with $N_b = 3$, $L_1 = 2$, and $L_2 = L_3 = 1$. This appears to be more complex than Equation (13.53), but we can guess that it will be more accurate because it includes the joint distribution of $x(1)$ and $x(2)$, and the population of Equation (13.52) indicates that there is a significant correlation between those two bits; that is, in four of the five individuals, $x(1)$ and $x(2)$ have the same value. The complexity of $\hat{p}_2(x)$ is

$$C_m = (\log_2 6)\,(3 + 1 + 1) = 15.4 \tag{13.57}$$

which, as expected, is higher than the complexity of $\hat{p}_2(x)$ as shown in Equation (13.54). We order the building blocks of $\hat{p}_2(x)$ as shown in Equation (13.56). We order the bit sequences in binary order so that N_{11} is the number of individuals for which $(x(1), x(2)) = (0, 0)$, N_{12} is the number of individuals for which $(x(1), x(2)) = (0, 1)$, N_{13} is the number of individuals for which $(x(1), x(2)) = (1, 0)$, and N_{14} is the number of individuals for which $(x(1), x(2)) = (1, 1)$. Similarly, N_{21} is the number of individuals for which $x(3) = 0$, and N_{22} is the number of individuals for which $x(3) = 1$. Finally, N_{31} is the number of individuals for which $x(4) = 0$, and N_{32} is the number of individuals for which $x(4) = 1$. Given this convention, we can calculate the accuracy of the model as

$$C_p = \sum_{i=1}^{3} \sum_{j=1}^{2^{L_i}} N_{ij} \log_2(M/N_{ij}) = 16.6. \tag{13.58}$$

As expected, the accuracy of $\hat{p}_2(x)$ is better (that is, less) than that of $\hat{p}_1(x)$. We combine these results to obtain

$$\begin{aligned} C_m + C_p &= 10.4 + 18.2 = 28.6 \text{ for } \hat{p}_1(x) \\ C_m + C_p &= 15.4 + 16.6 = 32.0 \text{ for } \hat{p}_2(x). \end{aligned} \tag{13.59}$$

ECGA indicates that $\hat{p}_1(x)$ is a better model because of its lower complexity.

□

Now that we know how to quantify the cost of an MPM, we use a steepest descent algorithm to find the MPM that approximately minimizes C_c. Given an MPM with N_b building blocks, we can form $N_b(N_b-1)/2$ alternative sets of building blocks by merging each possible pair of building blocks. For example, given the four building blocks of Equation (13.47), we can form the following six alternative MPMs:

$$\begin{aligned}
&p(x(1), x(3), x(6), x(2))\, p(x(4), x(5), x(7), x(10))\, p(x(8), x(9)) \\
&p(x(1), x(3), x(6), x(4), x(5), x(7), x(10))\, p(x(2))\, p(x(8), x(9)) \\
&p(x(1), x(3), x(6), x(8), x(9))\, p(x(2))\, p(x(4), x(5), x(7), x(10)) \\
&p(x(1), x(3), x(6))\, p(x(2), x(4), x(5), x(7), x(10))\, p(x(8), x(9)) \\
&p(x(1), x(3), x(6))\, p(x(2), x(8), x(9))\, p(x(4), x(5), x(7), x(10)) \\
&p(x(1), x(3), x(6))\, p(x(2))\, p(x(4), x(5), x(7), x(10), x(8), x(9)).
\end{aligned} \tag{13.60}$$

ECGA works by selecting the MPM from this set that minimizes the cost of Equation (13.51). The ECGA algorithm is summarized in Figure 13.17. Note that Figure 13.17 executes each generation. To implement an ECGA, we need to select M, which is some number less than the population size N. We also need to select P_c, which is the proportion of children that we create using the best identified MPM. We create these children by randomly selecting MPM subsets from the M best individuals identified at the beginning of Figure 13.17. This is equivalent to $(N_b - 1)$-point crossover, so each child has N_b parents, some of which may be repeated.

Select the M best individuals from the current population
$\hat{p}_0(x) \leftarrow p(x(1))\, p(x(2)) \cdots p(x(n))$
While (true)
  $N_b \leftarrow$ number of building blocks in $\hat{p}_0(x)$
  Use $\hat{p}_0(x)$ to form alternative MPMs $\hat{p}_i(x)$ for $i \in [1, N_b(N_b-1)/2]$
  $\hat{p}(x) \leftarrow \arg\min\,(C_c(\hat{p}_i(x)) : i \in [1, N_b(N_b-1)/2])$
  If $\hat{p}(x) = \hat{p}_0(x)$ then exit this loop
Next iteration
Use the building blocks in $\hat{p}_0(x)$ to create NP_c individuals for the next generation
Randomly create $N(1 - P_c)$ individuals for the next generation

---

Figure 13.17 Marginal product model (MPM) construction using steepest descent in the extended compact genetic algorithm (ECGA). This algorithm executes each generation of the ECGA.

13.4.2 Other Multivariate Estimation of Distribution Algorithms

Researchers have proposed many other multivariate EDAs, including the factorized distribution algorithm (FDA) [Mühlenbein et al., 1999], the learning FDA [Mühlenbein et al., 1999], the estimation of Bayesian networks algorithm (EBNA) [Larrañaga et al., 1999a], [Larrañaga et al., 2000], the Bayesian optimization algorithm (BOA) [Pelikan et al., 1999], the Markov network factorized distribution algorithm (MN-FDA) [Santana, 2003], and the Markov network EDA (MN-EDA) [Santana, 1998]. Hierarchical BOA (hBOA) attempts to reduce the computational complexity of BOA by decomposing the optimization problem into subproblems [Pelikan, 2005].

13.5 CONTINUOUS ESTIMATION OF DISTRIBUTION ALGORITHMS

The preceding sections discussed various EDAs for discrete-domain problems. This section extends the EDA concept for continuous-domain problems. For discrete-domain problems, EA individuals are created from discrete probability distributions. The same idea is used in continuous-domain problems, except that the probability distributions are continuous instead of discrete.

To set the stage for continuous EDAs, first recall the procedure for discrete EDAs. Suppose that we have a discrete-domain problem where the probability of having a 0 bit in the i-th position of a candidate solution, $x(i)$, is 0.75, and the probability of having a 1 bit is 0.25. We can generate $x(i)$ with code like the following:

$$\begin{aligned} r &\leftarrow U[0,1] \\ x(i) &\leftarrow \begin{cases} 0 & \text{if } r < 0.75 \\ 1 & \text{otherwise} \end{cases} \end{aligned} \tag{13.61}$$

where r is a random number uniformly distributed in $[0,1]$. If, on the other hand, our problem has a continuous domain $[0,1]$, then the i-th position of a candidate solution is not a bit, but is a continuous variable. We can generate the variable by writing code like the following:

$$\begin{aligned} r &\leftarrow U[0,1] \\ x(i) &\leftarrow 3/2 - \sqrt{(9/4) - 2r}. \end{aligned} \tag{13.62}$$

This results in the probability density function (PDF) for $x(i)$ shown in Figure 13.18. This can be viewed as a continuous counterpart of Equation (13.61) because the probability of $x(i)$ linearly increases as $x(i)$ approaches 0. Note that functions that transform one PDF to another can be found with standard methods from probability texts (see Problems 13.11 and 13.15) [Grinstead and Snell, 1997].

EDAs for continuous-domain problems all operate on this same idea. Given a subpopulation of relatively fit individuals, we generate an approximate continuous PDF, and then use the PDF to create the next generation of candidate solutions.

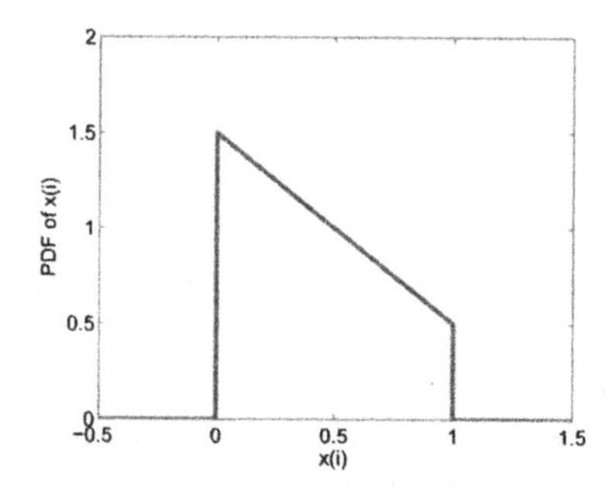

Figure 13.18 Sample probability density function (PDF) for the continuous variable described by Equation (13.62).

13.5.1 The Continuous Univariate Marginal Distribution Algorithm

We illustrate EDAs for continuous-domain problems in this section with a simple modification of the binary UMDA algorithm of Section 13.2.1. We use Gaussian distributions to create the next generation, and so this algorithm is denoted as UMDA_c^G [Gallagher et al., 2007]. UMDA_c^G may be the simplest continuous EDA, and is summarized in Figure 13.19. In UMDA_c^G we calculate the mean and variance of each element of the selected population subset, and then we use Gaussian random numbers to create the next generation. We could also modify Figure 13.19 to create the next generation from distributions other than Gaussian. See Equation (8.18) for the rationale for using $M-1$ instead of M in the estimate of σ_k.

Initialize a population of candidate solutions $\{x_i\}$, $i \in [1, N]$
Note that each x_i includes n continuous variables $x_i(1), \ldots, x_i(n)$
While not(termination criterion)
    Select M individuals from $\{x_i\}$ according to fitness, where $M < N$
    Index the M selected individuals as $\{x_i\}$, $i \in [1, M]$
    $\mu_k \leftarrow \frac{1}{M}\sum_{j=1}^{M} x_j(k)$
    $\sigma_k \leftarrow \left[\frac{1}{M-1}\sum_{j=1}^{M}(x_j(k)-\mu_k)^2\right]^{1/2}$
    For $i = 1$ to N (population size)
        For $k = 1$ to n (number of variables in each candidate solution)
            $x_i(k) \leftarrow N(\mu_k, \sigma_k^2)$
        Next variable
    Next individual
Next generation

Figure 13.19 Continuous Gaussian univariate marginal distribution algorithm (UMDA_c^G) for optimization on an n-dimensional continuous domain. $N(\mu_k, \sigma_k^2)$ is a Gaussian random variable with mean μ_k and variance σ_k^2. $x_i(k)$ is the k-th element of the i-th individual.

13.5.2 Continuous Population Based Incremental Learning

We illustrate continuous EDAs in this section by modifying the binary PBIL algorithm of Section 13.2.3 for continuous-domain problems [Sebag and Ducoulombier, 1998]. PBIL for continuous-domain problems is also called stochastic hill climbing with learning by vectors of normal distributions (SHCLVND) [Rudlof and Köppen, 1996], [Pelikan, 2005, Section 2.3]. Suppose that each independent variable $x_i(k)$ of a candidate solution x_i is constrained to lie within some domain:

$$x_i(k) \in [x_{\min}(k), x_{\max}(k)] \tag{13.63}$$

for $i \in [1, N]$ and $k \in [1, n]$, where N is the population size and n is the problem dimension. Suppose that we have an n-dimensional vector p such that $p_k \in [x_{\min}(k), x_{\max}(k)]$ for $k \in [1, n]$. We can then create the candidate solution element $x_i(k)$ for each individual x_i by generating a Gaussian random number that has a mean of p_k. As the generation count increases, we expect p to converge toward the optimal solution; therefore, we typically decrease the standard deviation of the Gaussian random number generator as the generation count increases. This idea is illustrated in Figure 13.20.

Figure 13.21 gives a simple PBIL algorithm for problems with continuous domains. It is very similar to the binary PBIL algorithm of Figure 13.8. The main difference is that the vector P is used to generate candidate solutions within the continuous domain of the problem, and the standard deviation that is used to generate those candidate solutions decreases at each generation as shown in Figure 13.20. This gives two additional tuning parameters, α and β, in Figure 13.21. Note that we should limit $x_i(k)$ to the domain $[x_{\min}(k), x_{\max}(k)]$ each time we update it in Figure 13.21.

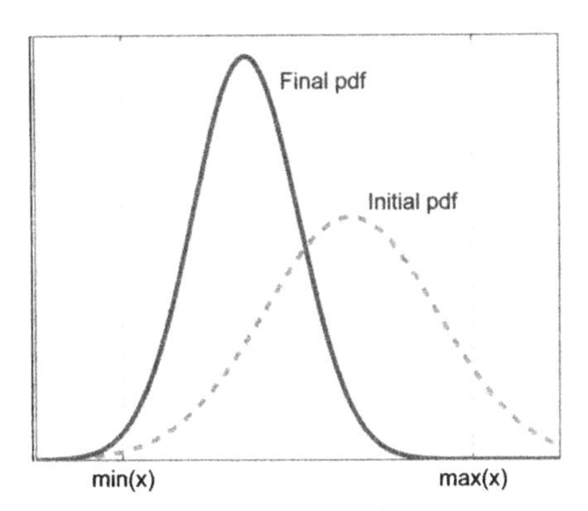

Figure 13.20 Illustration of PDF evolution in continuous PBIL. The probability density function (PDF) has a large variance at the beginning of the algorithm, which allows a lot of exploration in the search space. The PDF has a smaller variance in later generations, which allows the algorithm to narrow in on the optimal solution.

N = population size
N_{best}, N_{worst} = number of good and bad individuals used to adjust P
η = learning rate $\in (0,1)$
$[x_{\min}(k), x_{\max}(k)]$ = domain of the k-th element of the search space, $k \in [1,n]$
β = (initial standard deviation) $\div$ (parameter range) $\in (0,1)$
α = standard deviation contraction factor $\in (0,1)$
$\sigma_k \leftarrow \beta(x_{\max}(k) - x_{\min}(k))$ = initial standard deviations, $k \in [1,n]$
$p_k \leftarrow U[x_{\min}(k), x_{\max}(k)]$ for $k \in [1,n]$ (uniformly distributed random numbers)
While not(termination criterion)
    Use p to randomly generate N candidate solutions as follows:.
        For $i \in [1,N]$
            For $k \in [1,n]$
                $x_i(k) \leftarrow p_k + N(0,\sigma_k)$
            Next dimension k
        Next individual i
    Sort the individuals so that $f(x_1) \le f(x_2) \le \cdots \le f(x_N)$
    For $i \in [1, N_{\text{best}}]$
        $p \leftarrow p + \eta(x_i - p)$
    Next i
    For $i \in [N - N_{\text{worst}} + 1, N]$
        $p \leftarrow p - \eta(x_i - p)$
    Next i
    Probabilistically mutate p
    $\sigma_k \leftarrow \alpha\sigma_k$ for $k \in [1,n]$
Next generation

Figure 13.21 A PBIL algorithm for minimizing $f(x)$ on n continuous dimensions, and $x_i(k) \in [x_{\min}(k), x_{\max}(k)]$ is the k-th element of the i-th candidate solution. $N(0,\sigma_k)$ is a zero-mean Gaussian random variable with standard deviation σ_k.

■ EXAMPLE 13.10

In this example, we attempt to minimize the 20-dimensional Ackley function defined in Appendix C.1.2. We use the continuous PBIL algorithm of Figure 13.21 with the following settings.

- Population size $N = 50$.
- σ_k linearly decreases from 10% of the parameter range at the initial generation to 2% of the parameter range at the final generation. The parameter range is $[-30, 30]$ for each dimension, so σ_k linearly decreases from an initial value of 6 to a final value of 6/5. This does not exactly follow the σ_k profile shown in Figure 13.21, but it accomplishes the same essential purpose.
- We use the five best and the five worst individuals ($N_{\text{best}} = N_{\text{worst}} = 5$) to update the probability vector P at each generation.
- We do not use any mutation.

Figure 13.22 shows the effect of the learning rate η on PBIL performance. If the learning rate is too small, then the adjustment of P will not be aggressive enough and convergence will be slow. If the learning rate is too high, then P will jump agressively toward good solutions, which gives good initial performance. However, this could result in overshooting the optimal probability vector, and could lead P in misleading directions in the search space.

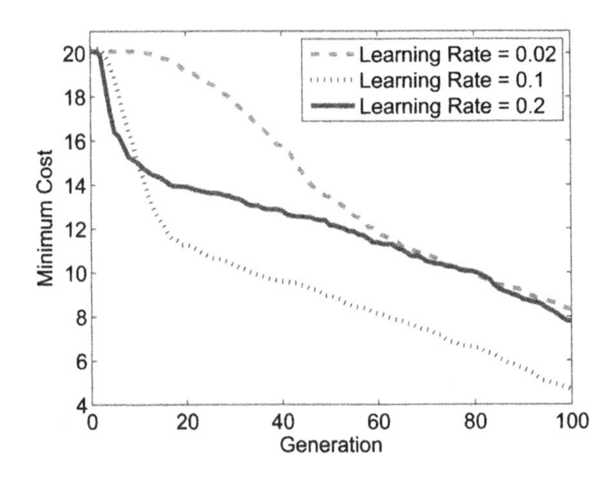

Figure 13.22 Example 13.10: Continuous PBIL results for the 20-dimensional Ackley function. The plot shows the cost of the best individual at each generation, averaged over 20 Monte Carlo simulations. We need to use an appropriate value for the learning rate η to get the best performance.

Next we explore the effect of N_{best} and N_{worst} on PBIL performance. We use $\eta = 0.1$ since that appears to be the best learning rate in Figure 13.22. Figure 13.23 shows PBIL performance for three different values of N_{best} and N_{worst}. We see that if these parameters are too small, then PBIL puts too much emphasis on a few individuals and does not obtain a broad enough picture of the performance of diverse individuals in the population. However, if these parameters are too large, then PBIL adjusts its probability vector using too many individuals, some of which may not be suitable for such use.

Finally, we explore the effect of σ_k on PBIL performance. We use $\eta = 0.1$ and $N_{\text{best}} = N_{\text{worst}} = 5$. We vary σ_k linearly from $k_0(x_{\max}(k) - x_{\min}(k))$ at the first generation to $k_f(x_{\max}(k) - x_{\min}(k))$ at the final generation. Figure 13.24 shows PBIL performance for three different combinations of k_0 and k_f. We see that if k_0 is too small, then initial convergence is slow due to the relative sluggishness of P. If k_f is too large, then PBIL does not converge well because the variation of candidate solutions is too large. We could run further experiments to explore the effect of k_0 being too large or k_f being too small.

□

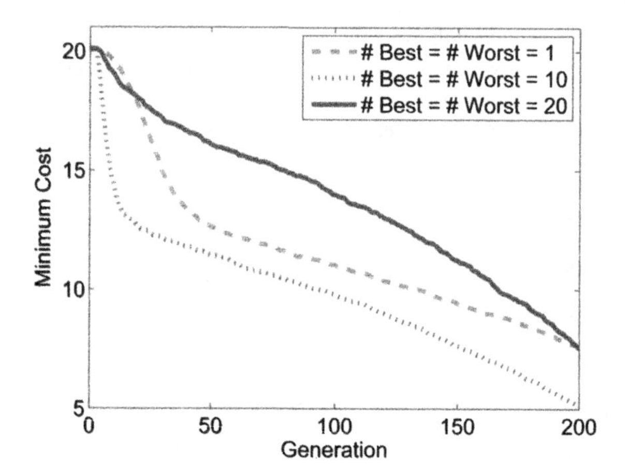

Figure 13.23 Example 13.10: Continuous PBIL results for the 20-dimensional Ackley function. The plot shows the cost of the best individual at each generation, averaged over 50 Monte Carlo simulations. We need to use appropriate values for N_{best} and N_{worst} to get the best performance.

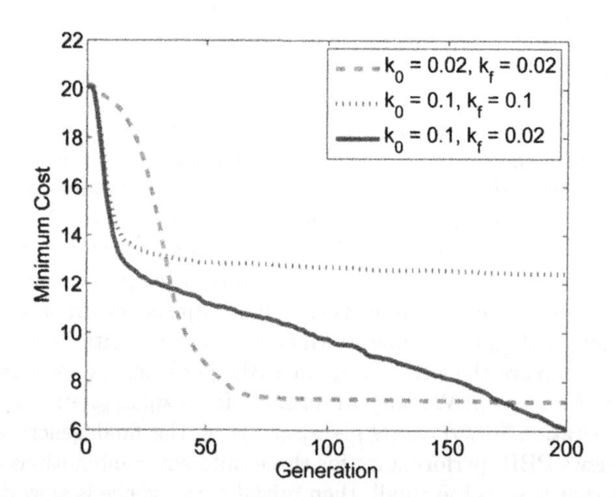

Figure 13.24 Example 13.10: Continuous PBIL results for the 20-dimensional Ackley function. The plot shows the cost of the best individual at each generation, averaged over 50 Monte Carlo simulations. k_0 and k_f control the standard deviation of candidate solution generation at the first and last generation. We need to use appropriate values for k_0 and k_f to get the best performance.

13.6 CONCLUSION

The EDAs in this chapter estimate probabilities to model the search space and find a global optimum. We could use maximum likelihood to estimate the sample mean and covariance of a population, and this is the approach taken in the estimation of multivariate normal algorithm (EMNA) [Larrañaga, 2002]. If we model the search space with Gaussian networks, we obtain estimation of Gaussian network algorithms (EGNAs) [Larrañaga, 2002], [Paul and Iba, 2003].

EDAs have been modeled mathematically use Markov chains [González et al., 2001], dynamic systems theory [González et al., 2000], [Mahnig and Mühlenbein, 2000], and other methods [González et al., 2002]. We discuss mathematical modeling of evolutionary algorithms in Chapter 4 for genetic algorithms and in Chapter 7 for genetic programming, but we do not discuss EDA math models in this book.

EDAs are relatively recent innovations, and so there is a lot of room of additional research and applications. Current directions in EDA research include multi-objective optimization [Bureerat and Sriworamas, 2007], dynamic optimization [Yang and Yao, 2008a], hybridizations of EDA with other algorithms [Peña et al., 2004], and on-line adaptation of EDA parameters [Santana et al., 2008].

This chapter has presented EDAs that use first-order and second-order statistics. The first paragraph in this conclusion mentions a few EDAs that use higher order statistics. This naturally raises the possibility of an EDA that gradually increases the order of statistics as the EDA gets closer to convergence. During the early stages of the EDA we could use first-order statistics to obtain a population that is reasonably close to local optima, and during the later stages of the EDA we could use higher order statistics to fine-tune our results.

Another promising direction for future work would be to combine EDAs with more traditional EAs. This would result in merging the ideas of recombination, mutation, and probability theory, to create the next generation of individuals. Other important research directions for continuous-domain EDAs include exploring methods for approximating continuous PDFs on the basis of a discrete set of EDA individuals. PDF approximation also needs to be performed in particle filtering, and so there is a lot of room for cross-fertilization between EDA research and particle filter research [Simon, 2006, Section 15.3]. Additional introductory, overview, and research material on EDAs can be found in [Larrañaga and Lozano, 2002], [Pelikan et al., 2002], [Kern et al., 2004], [Lozano et al., 2006], [Shakya and Santana, 2012], and [Larrañaga et al., 2012]. A MATLAB toolbox for EDA-based optimization is available on the internet at [Santana and Echegoyen, 2012].

PROBLEMS

Written Exercises

13.1 In Equation (13.1), what is the probability that bit 5 is a 1 if bits 3 and 4 are 1? What is the probability that bit 4 is a 1 if bits 3 and 5 are 1? What is the probability that bit 3 is a 1 if bits 4 and 5 are 1?

13.2 What are the values of w_i in Equation (13.4) that reduce it to Equation (13.3)?

13.3 How many generations in the cGA algorithm of Figure 13.4 give the same number of fitness function evaluations as one generation in the UMDA algorithm of Figure 13.2?

13.4 How many generations in the generalized cGA algorithm of Figure 13.6 with a population size N_1 are equivalent to one generation with a population size N_2 in terms of the number of fitness function evaluations?

13.5 In Example 13.4, calculate the conditional entropy of bit 2 given bit 1.

13.6 Given a set of binary EA individuals, what is the conditional entropy of bit k given bit k?

13.7 Suppose the entropy of five bits in an EA population are $h(1) = 0.3$, $h(2) = 0.4$, $h(3) = 0.5$, $h(2) = 0.5$, and $h(1) = 0.6$. Suppose also that the following table specifies the conditional entropy of bit j given bit k.

	$k = 1$	$k = 2$	$k = 3$	$k = 4$	$k = 5$
$j = 1$	0.0	0.1	0.4	0.3	0.4
$j = 2$	0.4	0.0	0.5	0.6	0.7
$j = 3$	0.9	0.8	0.0	0.7	0.6
$j = 4$	0.8	0.2	0.1	0.0	0.1
$j = 5$	0.2	0.5	0.2	0.5	0.0

a) Use the greedy algorithm of Figure 13.10 to minimize Equation (13.12). What value does the algorithm give for Equation (13.12)?

b) Start with $k_5 = 2$ and continue with the greedy algorithm to obtain the remaining values of k_i. What value does this approach give for Equation (13.12)?

13.8 Show that the mutual information between bits m and k is the same as that between bits k and m.

13.9 Verify the calculation of $I(1, 3)$ in Example 13.5.

13.10 Calculate χ^2_{23} for Example 13.8.

13.11 Given a random variable $x \sim U[0,1]$, find a function $y(x)$ with the PDF

$$g(y) = \begin{cases} 2a & \text{if } 0 < y < 3/4 \\ a & \text{if } 3/4 < y < 1. \end{cases}$$

What value of a is required to make $g(y)$ a valid PDF?

Computer Exercises

13.12 cGA versus UMDA: Repeat Example 13.2, but use a cGA generation limit that allows a fair comparison with the UMDA results of Example 13.1 (see Problem 13.3). What cGA generation limit should you use? How do your cGA results compare with the UMDA results?

13.13 cGA Population Size: Repeat Example 13.3 with $N = 2$ and $N = 20$, but use a larger generation limit with $N = 2$ to allow a fair comparison between the two different population sizes (see Problem 13.4). What generation limit should you use with $N = 2$? How do your cGA results compare for $N = 2$ and $N = 20$?

13.14 PBIL: Simulate the PBIL algorithm of Figure 13.8 to minimize the 20-dimensional Ackley function, using six bits per dimension. Run for 50 generations, use $N_{\text{best}} = N_{\text{worst}} = 5$, $P_{\min} = 0$, $P_{\max} = 1$, and do not mutate P. Run 20 Monte Carlo simulations. Plot the cost of the best individual each generation, averaged over the 20 Monte Carlo simulations. Do this for the following values of the learning rate η: 0.001, 0.01, and 0.1. Comment on your results.

13.15 PDF Transformation: Generate 100,000 random numbers $\{x_i\}$ that are uniformly distributed on $[0,1]$. Apply the function that you found in Problem 13.11 to $\{x_i\}$ to obtain $\{y_i\}$. Plot a histogram of $\{y_i\}$ to verify that you obtained the desired PDF.

CHAPTER 14

Biogeography-Based Optimization

"... the Zoology of Archipelagoes will be well worth examination ..."
—Charles Darwin [Keynes, 2001], [MacArthur and Wilson, 1967, page 3]

Biogeography is the study of the speciation, extinction, and geographical distribution of biological species. As Charles Darwin predicted in the above quote, biogeography has indeed been a fruitful area of examination. A recent search of Biological Abstracts, a biology research index, reveals that 37,847 papers were written in the year 2010 on the subject of biogeography, and there are several journals devoted to the subject. Popular science writer David Quammen has written a fascinating account of biogeography in his book *The Song of the Dodo* [Quammen, 1997].

Just as the behavior of biological ants gave rise to ant colony optimization, the science of genetics gave rise to genetic algorithms, and the study of animal swarms gave rise to particle swarm optimization, so the science of biogeography has given rise to biogeography-based optimization. BBO is a relatively recent addition to the stable of EAs, but we devote a full chapter to it in this book because of the following reasons.

- The popularity of BBO is growing rapidly. A search of Google Scholar shows the following:
  - 1 BBO paper in 2008;
  - 37 BBO papers in 2009;
  - 81 BBO papers in 2010;
  - 145 BBO papers in 2011.

  We are on pace to see over 200 BBO papers in 2012 (as of this writing). It remains to be seen whether this growth will continue, but these numbers indicate that BBO is rapidly gaining in popularity.
- In spite of its recent introduction, BBO has seen a lot of success in real-world applications, including biomedical problems, power optimization, antenna design, mechanical design, robotics, scheduling, navigation, military problems, and others. See the BBO web site [Simon, 2012] for more details.
- In contrast to many other recent EAs, there has been a relatively large amount of material written about BBO theory in the short time since its inception, including papers on Markov models [Simon et al., 2011a], dynamic system models [Simon, 2011a], and statistical mechanics models [Ma et al., 2013].
- The author of this book is also the inventor of BBO and thus has a natural interest in it.

Overview of the Chapter

This chapter gives an overview of natural biogeography in Section 14.1, and discusses its interpretation as an optimization process in Section 14.2. We then show how biogeography can be adapted to obtain the BBO algorithm in Section 14.3. We suggest some useful BBO modifications and extensions in Section 14.4.

14.1 BIOGEOGRAPHY

The science of biogeography can be traced to the work of 19th century naturalists, most notably Alfred Wallace [Wallace, 2006] and Charles Darwin [Keynes, 2001]. Wallace is usually considered the father of biogeography, although Darwin is much better known because of his theory of evolution.

Before the 1960s, biogeography was mostly descriptive and historical, with the notable exception of Eugene Munroe's quantitative doctoral thesis [Munroe, 1948]. In the early 1960s Robert MacArthur and Edward Wilson began working on mathematical models of biogeography, culminating with their classic 1967 book *The Theory of Island Biogeography* [MacArthur and Wilson, 1967]. They were mostly interested in the distribution of species between neighboring islands, and mathematical models for the extinction and migration of species. Since MacArthur and Wilson's work, biogeography has become a major subset of biology [Hanski and Gilpin, 1997].

Mathematical models of biogeography describe speciation (the evolution of new species), the migration of species between islands, and the extinction of species.

The term *island* here is descriptive rather than literal. An island is considered any habitat that is geographically isolated from other habitats. In the classic sense of the term, an island is isolated from other habitats by water. But islands can also be habitats that are isolated by stretches of desert, rivers, mountain ranges, predators, man-made artifacts, or other obstacles. An island could consist of a riverbank that supports herbs, a pond that supports amphibians, a rocky outcrop that supports snails, or a dead tree trunk that supports insects [Hanski and Gilpin, 1997].

Geographical areas that are friendly to life are said to have a high habitat suitability index (HSI) [Wesche et al., 1987]. Features that correlate with HSI include such factors as rainfall, vegetative diversity, topographic diversity, land area, and temperature. These variables which characterize habitability are called suitability index variables (SIVs). In terms of habitability, SIVs are the independent variables of the habitat and HSI is the dependent variable.

Islands with a high HSI tend to support many species, and islands with a low HSI can support only a few species. Islands with a high HSI have many species that emigrate to nearby habitats, simply by virtue of the large number of species that they host. Emigration from an island with a high HSI does not occur because species *want* to leave their home; after all, the home island is an attractive place to live. The reason that emigration occurs from these islands is due to the accumulation of random effects on a large number of species with large populations. Emigration occurs as animals ride flotsam, swim, fly, or ride the wind to neighboring islands. When a species emigrates from an island, it does not mean that the species completely disappears from the island; only a few representatives emigrate, so an emigrating species remains present on its home island while at the same time migrating to a neighboring island. However, in most of our discussion, we will assume that emigration from an island results in extinction from that island. This assumption will be necessary in our use of biogeography to develop BBO.

Islands with a high HSI not only have a high emigration rate, but they have a low immigration rate because they already support many species. The species that arrive at such islands will tend not to survive, even though the HSI is high, because there is too much competition for resources.

Islands with a low HSI have a high immigration rate because of their low populations. Again, this is not because species *want* to immigrate to such islands; after all, these islands are undesirable places to live. The reason that immigration occurs to these islands is because there is a lot of geographical room for additional species. Whether or not the immigrating species can survive in its new home, and for how long, is another question. However, species diversity is correlated with HSI, so the more species that arrive at a low HSI island, the greater chance that the island's HSI will increase [Wesche et al., 1987].

Figure 14.1 illustrates a model of species abundance on a single island [MacArthur and Wilson, 1967]. The immigration rate λ and the emigration rate μ are functions of the number of species on the island. We have depicted the migration curves as straight lines, but in general they might be more complicated curves, as we will discuss later.

Consider the immigration curve. The maximum possible immigration rate to the habitat is I, which occurs when there are zero species on the island. As the number of species increases, the island becomes more crowded, fewer species are able to successfully survive immigration, and the immigration rate decreases. The largest possible number of species that the habitat can support is $S_{\max}$, at which point the immigration rate is zero.

Now consider the emigration curve. If there are no species on the island, then the emigration rate is zero. As the number of species on the island increases, it becomes more crowded, more species are able to leave the island, and the emigration rate increases. The maximum emigration rate is E, when the island contains the largest number of species that it can support.

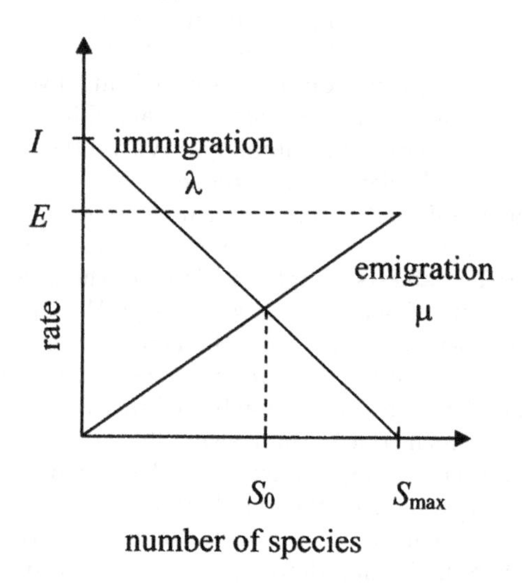

Figure 14.1 Species migration model of an island, based on [MacArthur and Wilson, 1967]. S_0 is the equilibrium species count.

A Mathematical Model of Biogeography

The remainder of this section presents a mathematical model of species counts in biogeography. This material is not necessary for an understanding of the BBO algorithm, and so the reader can safely skip to the next section if so inclined.

The equilibrium number of species in Figure 14.1 is S_0, at which point the immigration and emigration rates are equal. However, there will be occasional excursions from S_0 due to temporal, random causes. Positive excursions from S_0 could be due to an unusually large piece of flotsam arriving from a neighboring island, or a statistically unlikely high number of births. Negative excursions from S_0 could be due to disease, the temporary introduction of a new predator, or a natural catastrophe. It can take many years for the number of species to reach equilibrium after a large perturbation [Hanski and Gilpin, 1997], [Hastings and Higgins, 1994].

Now consider the probability P_s that the island contains S species. P_s changes from time t to $(t + \Delta t)$ as follows:

$$P_s(t + \Delta t) = P_s(t)(1 - \lambda_s \Delta t - \mu_s \Delta t) + P_{s-1}(t)\lambda_{s-1}\Delta t + P_{s+1}(t)\mu_{s+1}\Delta t \quad (14.1)$$

where λ_s and μ_s are the immigration and emigration rates when there are S species on the island. This equation holds if we assume that Δt is small enough so that the probability of more than one migration between time t and $(t+\Delta t)$ can be ignored. Therefore, to have S species at time $(t+\Delta t)$, one of the following conditions must hold:

1. There were S species at time t, and no immigration or emigration occurred between t and $(t+\Delta t)$; or,
2. There were $(S-1)$ species at time t, and one species immigrated; or,
3. There were $(S+1)$ species at time t, and one species emigrated.

Taking the limit of Equation (14.1) as $\Delta t \to 0$ gives

$$\dot{P}_s = \begin{cases} -(\lambda_s+\mu_s)P_s + \mu_{s+1}P_{s+1} & S=0 \\ -(\lambda_s+\mu_s)P_s + \lambda_{s-1}P_{s-1} + \mu_{s+1}P_{s+1} & 1 \le S \le S_{\max}-1 \\ -(\lambda_s+\mu_s)P_s + \lambda_{s-1}P_{s-1} & S = S_{\max}. \end{cases} \tag{14.2}$$

We define $n = S_{\max}$, and $P = \begin{bmatrix} P_0 & \cdots & P_n \end{bmatrix}^T$. Now we can arrange the $(n+1)$ equations of Equation (14.2) into the single matrix equation

$$\dot{P} = AP \tag{14.3}$$

where the matrix A is given as

$$A = \begin{bmatrix} -(\lambda_0+\mu_0) & \mu_1 & 0 & \cdots & 0 \\ \lambda_0 & -(\lambda_1+\mu_1) & \mu_2 & \ddots & \vdots \\ \vdots & \ddots & \ddots & \ddots & \vdots \\ \vdots & \ddots & \lambda_{n-2} & -(\lambda_{n-1}+\mu_{n-1}) & \mu_n \\ 0 & \cdots & 0 & \lambda_{n-1} & -(\lambda_n+\mu_n) \end{bmatrix}. \tag{14.4}$$

For the straight line migration rates of Figure 14.1, we have

$$\begin{aligned} \mu_k &= Ek/n \\ \lambda_k &= I(1-k/P). \end{aligned} \tag{14.5}$$

For the special case $E = I$, we have

$$\begin{aligned} \lambda_k + \mu_k &= E = I \text{ for all } k \in [0,n] \\ A &= E \begin{bmatrix} -1 & 1/n & 0 & \cdots & 0 \\ n/n & -1 & 2/n & \ddots & \vdots \\ \vdots & \ddots & \ddots & \ddots & \vdots \\ \vdots & \ddots & 2/n & -1 & n/n \\ 0 & \cdots & 0 & 1/n & -1 \end{bmatrix} \\ &= EA' \end{aligned} \tag{14.6}$$

where A' is defined by the above equation.

Theorem 14.1 *The (n+1) eigenvalues of A', for any natural n, are*

$$\begin{aligned} x(A') &= \{0, -2/n, -4/n, \cdots, -n\} \\ &= -2k/n, \quad k \in [0, n]. \end{aligned} \tag{14.7}$$

Furthermore, the eigenvector corresponding to the zero eigenvalue is

$$\begin{aligned} v(0) &= \left[\begin{array}{ccc} v_0(0) & \cdots & v_n(0) \end{array} \right]^T \\ \textit{where } v_k(0) &= \binom{n}{k} = \frac{n!}{k!\,(n-k)!}, \quad k \in [0, n]. \end{aligned} \tag{14.8}$$

The first part of the theorem was conjectured in [Simon, 2008] and proven in [Igelnik and Simon, 2011]. The second part of the theorem was proven in both references but in two different ways. After the publication of [Igelnik and Simon, 2011] we discovered that the basic idea of the first part of Theorem 14.1 had been previously stated without proof in [Clement, 1959] and [Gregory and Karney, 1969, Example 7.10].

In steady state we have $t \to \infty$, so $\dot{P}(\infty) = AP(\infty) = EA'P(\infty) = 0$; that is, $P(\infty)$ is the eigenvector that corresponds to the zero eigenvalue. Recall that eigenvectors are not defined uniquely, but are defined only up to a nonzero scaling factor. The elements of $P(\infty)$ must add up to 1 since they are probabilities. These facts give us the following [Simon, 2008], [Igelnik and Simon, 2011].

Theorem 14.2 *The steady state value for the probability of the number of each species is given by*

$$\begin{aligned} P(\infty) &= \frac{v(0)}{\sum_{k=0}^{n} v_k(0)} \\ &= 2^{-n} v(0). \end{aligned} \tag{14.9}$$

■ EXAMPLE 14.1

Consider an island that can support a maximum of four species. The maximum immigration and emigration rates are two species per unit of time. Therefore, $n = 4$ and $E = I = 2$. The A' matrix of Equation (14.6) is

$$A' = \left[\begin{array}{ccccc} -1 & 1/4 & 0 & 0 & 0 \\ 1 & -1 & 2/4 & 0 & 0 \\ 0 & 3/4 & -1 & 3/4 & 0 \\ 0 & 0 & 2/4 & -1 & 1 \\ 0 & 0 & 0 & 1/4 & -1 \end{array} \right]. \tag{14.10}$$

Theorem 14.1 tells us that the eigenvalues are $x = \{0, -1/2, -1, -3/2, -2\}$, and the eigenvector corresponding to $x = 0$ is $v(0) = \left[\begin{array}{ccccc} 1 & 4 & 6 & 4 & 1 \end{array} \right]^T$. Theorem 14.2 tells us that the steady-state probability for the number of each species count is

$$\begin{aligned} \Pr(S=0) = \Pr(S=4) &= 1/16 \\ \Pr(S=1) = \Pr(S=3) &= 4/16 \\ \Pr(S=2) &= 6/16. \end{aligned} \tag{14.11}$$

If we run a migration simulation for 5,000 time steps, we obtain the following probabilities for each species count:

$$\begin{aligned}
\Pr(S=0) &= 0.0714 \\
\Pr(S=1) &= 0.2605 \\
\Pr(S=2) &= 0.3734 \\
\Pr(S=3) &= 0.2358 \\
\Pr(S=4) &= 0.0544. \qquad (14.12)
\end{aligned}$$

These are fairly close to the analytical probabilities shown in Equation (14.11).

□

14.2 BIOGEOGRAPHY IS AN OPTIMIZATION PROCESS

We know that nature includes many processes that optimize [Alexander, 1996]. In fact, this premise is the foundational principle of most EAs. However, is biogeography an optimization process? At first glance it seems that biogeography simply maintains species count equilibria among islands, and that it is not necessarily optimal. This section discusses biogeography from the viewpoint of optimality.

As discussed earlier, biogeography is nature's way of distributing species, and it has often been studied as a process that maintains equilibrium in habitats. Equilibrium can be seen at the point S_0 in Figure 14.1 where the immigration and emigration curves intersect. One reason that biogeography has been viewed from the equilibrium perspective is that this viewpoint was the first to place biogeography on a firm mathematical footing [MacArthur and Wilson, 1963], [MacArthur and Wilson, 1967]. However, since then the equilibrium perspective has been increasingly questioned, or rather expanded, by biogeographers.

In engineering, we often view stability and optimality as competing objectives; for example, a simple system is typically easier to stabilize than a complex system, while an optimal system is typically complex and less stable than a simpler system [Keel and Bhattacharyya, 1997]. However, in biogeography, stability and optimality are two sides of the same coin. Optimality in biogeography involves diverse, complex communities that are highly adaptable to their environment. Stability in biogeography involves the persistence of existing populations. Field observations show that complex communities are more adaptable and stable than simple communities [Harding, 2006, page 82], and this observation has also been supported by simulation [Elton, 1958], [MacArthur, 1955].

Although the complementary nature of optimality and stability in biogeography has been challenged [May, 1973], those challenges have been adequately answered and the idea is generally accepted today [McCann, 2000], [Kondoh, 2006]. The equilibrium vs. optimality debate in biogeography thus becomes a matter of semantics, because equilibrium and optimality are simply two different perspectives on the same phenomenon in biogeography.

A dramatic example of the optimality of biogeography is Krakatoa, a volcanic island in the Indian Ocean which erupted in August 1883 [Winchester, 2008]. The eruption was heard from thousands of miles away and resulted in the death of over

36,000 people, mostly from tidal waves whose remnants were recorded as far away as England. The eruption threw dust particles 30 miles high which remained aloft for months and were visible all around the world. Rogier Verbeek, a geologist and mining engineer, was the first visitor to Krakatoa six weeks after the eruption, but the surface of the island was too hot to touch and showed no evidence of life. The island was completely sterilized [Whittaker and Bush, 1993]. The first animal life (a spider) was discovered on Krakatoa in May 1884, nine months after the eruption. By 1887, dense fields of grass were discovered on the island. By 1906, plant and animal life was abundant. Although volcanic activity continues today on Krakatoa, by 1983 (one century after its desolation) there were 88 species of trees and 53 species of shrubs [Whittaker and Bush, 1993], and the species count continues to increase linearly with time. Life immigrates to Krakatoa, and immigration makes the island more habitable, which in turn makes the island more friendly to additional immigration.

Biogeography is a positive feedback phenomenon, at least to a certain point. This is analogous to natural selection, also called survival of the fittest. As species become more fit, they are more likely to survive. As they survive longer, they disperse and become better able to adapt to their environment. Natural selection, like biogeography, entails positive feedback. However, the time scale of biogeography is much shorter than that of natural selection.

Another good example of biogeography as an optimization process is the Amazon rainforest, which is a typical case of a mutually optimizing life/environment system [Harding, 2006]. The rainforest has a large capacity to recycle moisture, which leads to less aridity and increased evaporation. This leads to cooler and wetter surfaces, which are more amenable to life. This suggests that a view of biogeography "based on *optimizing* environmental conditions for biotic activity seems more appropriate than a definition based on homeostasis" [Kleidon, 2004] (emphasis added). This view of the environment as a life-optimizing system was suggested as early as 1997 [Volk, 1997]. There are many other examples of the optimality of biogeography, such as Earth's temperature [Harding, 2006], Earth's atmospheric composition [Lenton, 1998], and the ocean's mineral content [Lovelock, 1990].

This is not to say that biogeography is optimal for any particular species. For example, investigations of the Bikini Atoll show that the high level of radioactivity resulting from nuclear tests had little effect on its natural ecology, but mammals were seriously affected [Lovelock, 1995, page 37]. This and similar studies indicate that the Earth "will take care of itself [and] environmental excesses will be ameliorated, but it's likely that such restoration of the environment will occur in a world devoid of people" [Margulis, 1996]. Interestingly, amid all of the current warnings about ozone depletion, we overlook the fact that for the first two billion years of life Earth had no ozone at all [Lovelock, 1995, page 109]. Life flourishes and evolves without ozone, but not in a human-centric way. Although global warming or an ice age might be disastrous for humans and many other mammals, it would be a minor event in the overall history of biogeography on our planet.

The premise that biogeography is optimization process motivates the development of BBO as an evolutionary optimization algorithm, which we discuss next.

14.3 BIOGEOGRAPHY-BASED OPTIMIZATION

Biogeography is nature's way of distributing species and optimizing environments for life, and is analogous to mathematical optimization. Suppose that we have an optimization problem and some candidate solutions, which we call individuals. Good individuals perform well on the problem, and poor individuals perform poorly. A good individual is analogous to an island with a high HSI, and a poor individual is analogous to an island with a low HSI. Good individuals resist change more than poor individuals, just like highly habitable islands have lower immigration rates than less habitable islands. By the same token, good individuals tend to share their features (that is, their independent variables) with poor individuals, just like highly habitable islands have high emigration rates. Poor individuals are likely to accept new features from good individuals, just like less habitable islands are likely to receive many immigrants from highly habitable islands. The addition of new features to poor individuals may raise the quality of those individuals. The EA that is based on this approach is called biogeography-based optimization (BBO).

We assume that each BBO individual is represented by an identical species count curve with $E = I$ for simplicity. Figure 14.2 illustrates the migration rates for a BBO algorithm with these assumptions. S_1 in Figure 14.2 represents a poor individual, while S_2 represents a good individual. The immigration rate for S_1 will be relatively high, which means that it will be likely to receive new features from other candidate solutions. The emigration rate for S_2 will be relatively high, which means that it will be likely to share its features with other individuals. Figure 14.2 is called a linear migration model since the μ and λ values are linear functions of fitness.

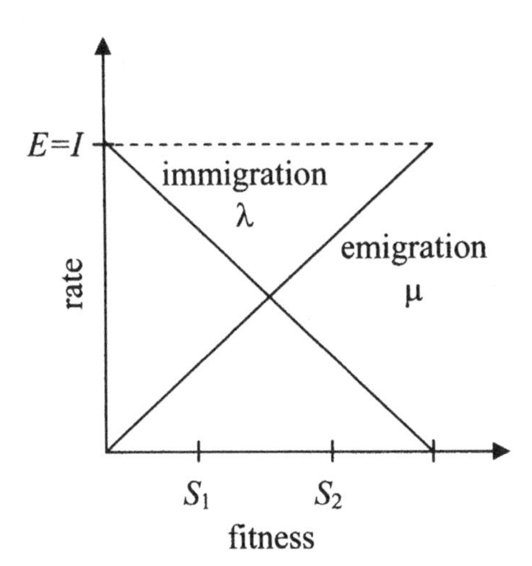

Figure 14.2 BBO feature-sharing relationship. S_1 represents a poor individual with a low probability of sharing features, but a high probability of receiving features from other individuals. S_2 represents a good individual with a high probability of sharing features, but a low probability of receiving features from other individuals.

We use the migration rates of each individual to probabilistically share information between individuals. There are several different ways to implement the details of BBO, but in this chapter we focus on the original BBO formulation [Simon, 2008], which is called partial immigration-based BBO [Simon, 2011b]. Using our standard notation, we suppose that we have a population size of N, that x_k is the k-th individual in the population, that the dimension of our optimization problem is n, and that $x_k(s)$ is the s-th independent variable in x_k, where $k \in [1, N]$ and $s \in [1, n]$. At each generation and for each solution feature in the k-th individual, there is a probability of λ_k (immigration probability) that it will be replaced:

$$\lambda_k = \text{Probability that } s\text{-th independent variable in } x_k \text{ will be replaced} \quad (14.13)$$

for $k \in [1, N]$ and $s \in [1, n]$. If a solution feature is selected to be replaced, then we select the emigrating solution with a probability that is proportional to the emigration probabilities $\{\mu_i\}$. We can use any fitness-based selection method for this step (see Section 8.7). If we use roulette-wheel selection, then

$$\Pr(x_j) \text{ is selected for emigration } = \frac{\mu_j}{\sum_{i=1}^{N} \mu_j}. \quad (14.14)$$

This gives the algorithm of Figure 14.3. Migration and mutation of each individual in the current generation occurs before any of the individuals are replaced in the population, which requires the use of the temporary population z in Figure 14.3. Borrowing from GA terminology [Vavak and Fogarty, 1996], we say that Figure 14.3 depicts a *generational* BBO algorithm as opposed to a *steady-state* algorithm. As with other EAs, we typically implement elitism in BBO (see Section 8.4), although this is not shown in Figure 14.3.

Figure 14.4 illustrates migration in BBO. The figure shows that individual z_k immigrates features. We use Equation (14.13) to decide whether or not to replace each feature in z_k. In the example of Figure 14.4, we see the following migration decisions:

1. Immigration is *not* selected for the first feature; that is why the first feature in z_k remains unchanged.

2. Immigration is selected for the second feature, and Equation (14.14) chooses x_1 as the emigrating individual; that is why the second feature in z_k is replaced by the second feature from x_1.

3. Immigration is selected for the third feature, and Equation (14.14) chooses x_3 as the emigrating individual; that is why the third feature in z_k is replaced by the third feature from x_3.

4. Immigration is selected for the fourth feature, and Equation (14.14) chooses x_2 as the emigrating individual; that is why the fourth feature in z_k is replaced by the fourth feature from x_2.

5. Finally, immigration is selected for the fifth feature, and Equation (14.14) chooses x_N as the emigrating individual; that is why the fifth feature in z_k is replaced by the fifth feature from x_N.

```
Initialize a population of candidate solutions {x_k} for k ∈ [1, N]
While not(termination criterion)
    For each x_k, set emigration probability μ_k ∝ fitness of x_k, with μ_k ∈ [0, 1]
    For each individual x_k, set immigration probability λ_k = 1 − μ_k
    {z_k} ← {x_k}
    For each individual z_k
        For each solution feature s
            Use λ_k to probabilistically decide whether to immigrate to z_k
                (see Equation (14.13))
            If immigrating then
                Use {μ_i}_{i=1}^N to probabilistically select emigrating individual x_j
                    (see Equation (14.14))
                z_k(s) ← x_j(s)
            End if
        Next solution feature
        Probabilistically mutate {z_k}
    Next individual
    {x_k} ← {z_k}
Next generation
```

Figure 14.3 Outline of the BBO algorithm with a population size of N. This algorithm is also known as partial immigration-based BBO. $\{x_k\}$ is the entire population of individuals, x_k is the k-th individual, and $x_k(s)$ is the s-th feature of x_k. Similarly, $\{z_k\}$ is the temporary population of individuals, z_k is the k-th temporary individual, and $z_k(s)$ is the s-th feature of z_k.

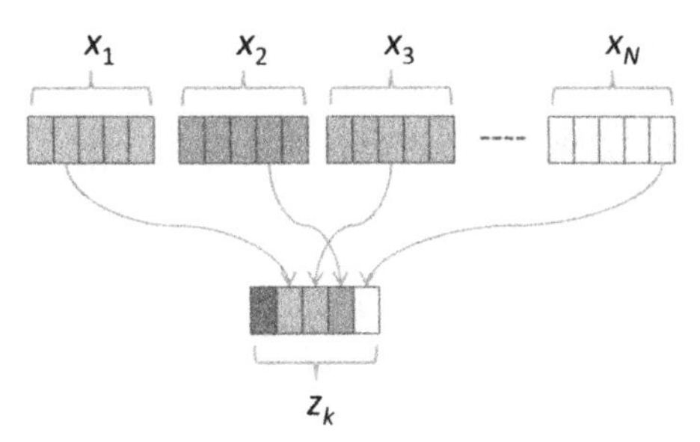

Figure 14.4 Illustration of BBO migration for a five-dimensional problem. Feature 1 is not selected for immigration, but features 2–5 are selected for immigration. Equation (14.14) is used to select the emigrating individuals.

■ EXAMPLE 14.2

This simple BBO experiment is motivated by David Goldberg's "GA simulation by hand" [Goldberg, 1989a]. Suppose that we want to maximize x^2, where x is encoded as a five-bit integer. We have to decide how many individuals we want in our population, and what mutation rate we want to use. We start with a randomly-generated population of four individuals, and a mutation rate of 1% per bit. For each individual, we compute the fitness value x^2, and then we assign migration rates in a linear manner as shown in Figure 14.2. Migration rates should be between 0 and 1, but we often set the smallest value to slightly greater than 0, and the largest value to slightly less than 1. This allows some randomness (non-determinism) even for the best and worst individuals in the population. For this example, we arbitrarily decide to use $1/N$ as the minimum values for λ and μ, and $(N-1)/N$ as the maximum values, where $N = 4$ is the population size. Suppose that our random initial population is created as shown in Table 14.1.

String number	x (binary)	x (decimal)	$f(x) = x^2$	μ	λ
1	01101	13	169	2/5	3/5
2	11000	24	576	4/5	1/5
3	01000	8	64	1/5	4/5
4	10011	19	361	3/5	2/5

Table 14.1 Example 14.2: Initial population for a simple BBO problem.

The first thing we do is copy the population x to temporary population z. Then we consider the possibility of immigration to each bit of the first individual in the temporary population, z_1, which is equal to x_1 (01101). We order bit numbers from left to right starting with index 1. We therefore see that

$$z_1(1) = 0, \quad z_1(2) = 1, \quad z_1(3) = 1, \quad z_1(4) = 0, \quad z_1(5) = 1. \tag{14.15}$$

Since z_1 is the third most fit individual, immigration rate $\lambda_1 = 3/5$, so there is a 60% chance of immigrating to each bit in z_1. We generate a random number $r \sim U[0,1]$ for each bit in z_1 to determine whether or not we should immigrate to that bit.

1. Suppose $r = 0.7$. Since $r > \lambda_1$, we will not immigrate to $z_1(1)$, so $z_1(1)$ remains equal to 0.

2. Suppose the next random number that we generate is $r = 0.3$. Since $r < \lambda_1$, we will immigrate to $z_1(2)$. We use roulette-wheel selection to choose the immigrating bit. $x_3(2)$ has the greatest probability of immigrating to $z_1(2)$, $x_1(2)$ has the second greatest probability, $x_4(2)$ has the third greatest probability, and $x_2(2)$ has the least probability. We could exclude $x_1(2)$ from consideration since z_1 is a copy of x_1, but this is an implementation detail that depends on the preference of the engineer. Suppose that this roulette-wheel selection process results in the choice of $x_3(2)$ for immigration. Then $z_1(2) \leftarrow x_3(2) = 1$. Even though we immigrated to $z_1(2)$, it did not change from its original value.

3. We continue this process for $z_1(3)$, $z_1(4)$, and $z_1(5)$. Suppose that the random numbers generated result in the following:

- $z_1(3) = 1$ (no immigration);
- $z_1(4) \leftarrow x_4(4) = 1$ (immigration); and
- $z_1(5) = 1$ (no immigration).

Now we have completed the migration process for z_1 and have obtained $z_1 = 01111$.

4. We repeat steps 1–3 for z_2, z_3, and z_4.

5. We next consider the possibility of mutation for each bit in each temporary individual z_1, z_2, z_3, and z_4. Mutation can be implemented as with any other EA (see Section 8.9).

6. Now that we have a modified population of $\{z_k\}$ individuals, we copy z_k to x_k for $k \in [1, 4]$, and the first BBO generation is complete.

The above process continues until some convergence criterion is met. For instance, we could continue for a specified number of generations, or continue until we achieve a satisfactory fitness value, or continue until the fitness value stops changing (see Section 8.2).

□

14.4 BBO EXTENSIONS

This section discusses some extensions that can be made to BBO to improve performance. We discuss migration curve shapes, blended migration, and alternative approaches to BBO implementation. We conclude this discussion with Section 14.4.4, which considers whether or not BBO should be viewed as a GA variation rather than a separate EA.

14.4.1 Migration Curves

Up to this point, we have assumed that the BBO migration curves are linear as shown in Figure 14.2. This is a convenient assumption, and it corresponds to linear rank-based selection (see Section 8.7.4); but in biogeography, migration curves are nonlinear. The exact shape of biogeography migration curves is difficult to quantify and changes from one island to the next. However, many curves in nature follow an S-shape. It was surmised in the original BBO paper [Simon, 2008] that nonlinear migration curves might give better performance than linear curves. This led to the investigation of several different migration curves in [Ma et al., 2009], [Ma, 2010]. Here we discuss the most promising curve: S-shaped migration curves.

Figure 14.2 models the unnormalized migration rates as

$$\begin{aligned} \mu_k &= r_k \\ \lambda_k &= 1 - r_k \end{aligned} \tag{14.16}$$

where r_k is the fitness rank of the k-th individual in the population, $r_k = 1$ for the least fit individual, and $r_k = N$ (where N is the population size) for the most fit individual. Sinusoidal migration rate modeling assigns the migration rates as

$$\begin{aligned} \mu_k &= \frac{1}{2}\left(1 - \cos(\pi r_k/N)\right) \\ \lambda_k &= 1 - \mu_k. \end{aligned} \tag{14.17}$$

These equations result in the S-shaped curves shown in Figure 14.5.

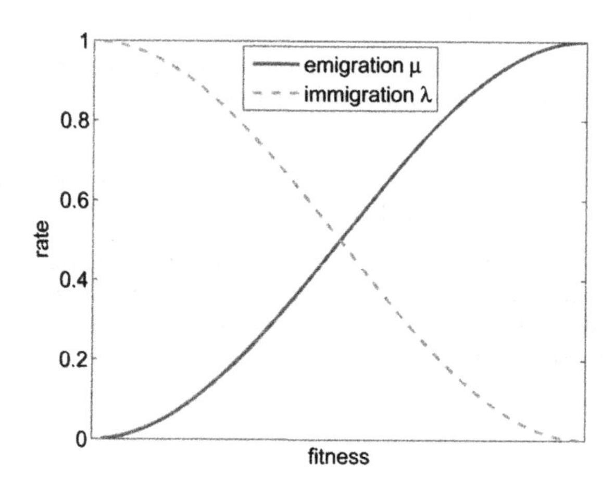

Figure 14.5 Sinusoidal BBO migration model. Compare with Figure 14.2.

■ EXAMPLE 14.3

If natural biogeography is really an optimization process, then it stands to reason that modeling BBO more closely after natural biogeography could result in better optimization performance. With this idea in mind, we simulate linear BBO and sinusoidal BBO on a set of 20-dimensional benchmark problems, obtaining the results shown in Table 14.2. We use a population size of 50, a generation limit of 50 for each BBO run, and a mutation rate of 1% per solution feature. We implement mutation by generating a solution feature that is uniformly distributed between the minimum and maximum domain values, with a 1% probability per individual per generation. We also use an elitism parameter of 2, which means that we keep the two best individuals from one generation to the next.

Table 14.2 shows that sinusoidal migration clearly outperforms linear migration for the standard benchmarks shown in the table. The average performance is 43% better for sinusoidal migration than for linear migration. This shows that migration models that are closer to nature outperform simpler migration models, and supports the hypothesis that natural biogeography is itself an optimization process.

□

Benchmark	Linear Migration	Sinusoidal Migration
Ackley	1.0373	1
Fletcher	1.2015	1
Griewank	1.2367	1
Penalty #1	1.4249	1
Penalty #2	4.3265	1
Quartic	1.6876	1
Rastrigin	1.0665	1
Rosenbrock	1.0759	1
Schwefel 1.2	1.0980	1
Schwefel 2.21	1.0468	1
Schwefel 2.22	1.0721	1
Schwefel 2.26	1.2471	1
Sphere	1.2582	1
Step	1.2683	1
Average	1.4319	1

Table 14.2 Example 14.3 results: Relative performance of BBO with linear and sinusoidal migration models. The table shows the normalized minimum found by the two BBO versions, averaged over 50 Monte Carlo simulations. See Appendix C for the definitions of the benchmark functions.

14.4.2 Blended Migration

Blended crossover has been shown to improve the performance of GAs and other EAs [McTavish and Restrepo, 2008], [Mezura-Montes and Palomeque-Oritiz, 2009], [Mühlenbein and Schlierkamp-Voosen, 1993] (see Section 8.8.9). In blended GA crossover, instead of copying a single parent's gene to a child gene, the child gene is obtained as a convex combination of two parent genes. This motivates the use of a blended migration operator for BBO [Ma and Simon, 2010], [Ma and Simon, 2011b]. In the standard BBO algorithm of Figure 14.3, a feature s of individual z_k is completely replaced by a feature from individual x_j:

$$z_k(s) \leftarrow x_j(s). \tag{14.18}$$

In blended migration in BBO, a feature of individual z_k is not simply replaced by a feature from individual x_j; instead, the feature of individual z_k is set equal to a convex combination of that of $z_k(s)$ and $x_j(s)$:

$$z_k(s) \leftarrow \alpha z_k(s) + (1-\alpha)x_j(s) \tag{14.19}$$

where $\alpha \in (0,1)$. If $\alpha = 0$, then blended BBO reduces to standard BBO; therefore, blended BBO is a generalization of standard BBO. The blend parameter α could be random, deterministic, or proportional to the relative fitness of z_k and x_j.

Blended migration is suitable for problems with continuous solution features. It could possibly be adapted for problems with discrete solution features, but we do not explore that idea here. There are a couple of justifications for blended migration compared to standard migration. First, good individuals will be less likely to be degraded due to migration, since they will retain a certain proportion of their original features in the migration process. Second, poor individuals will still accept at least part of the solution features from good individuals during migration.

■ **EXAMPLE 14.4**

To explore the effect of blended migration on BBO performance, we simulate standard BBO and blended BBO with $\alpha = 0.5$ on a set of 20-dimensional benchmark problems. We use the same BBO parameters as in Example 14.3, and we obtain the results shown in Table 14.3.

Benchmark	Standard BBO ($\alpha = 0$)	Blended BBO ($\alpha = 0.5$)
Ackley	1.6559	1.0
Fletcher	1.0	2.388
Griewank	3.4536	1.0
Penalty #1	701.47	1.0
Penalty #2	8817.7	1.0
Quartic	49.663	1.0
Rastrigin	1.0	1.6892
Rosenbrock	3.9009	1.0
Schwefel 1.2	12.63	1.0
Schwefel 2.21	4.0846	1.0
Schwefel 2.21	1.3280	1.0
Schwefel 2.26	1.0	4.8213
Sphere	5.4359	1.0
Step	4.5007	1.0
Average	686.34	1.4213

Table 14.3 Example 14.4 results: Relative performance of standard BBO and blended BBO with $\alpha = 0.5$. The table shows the normalized optimum found by the two BBO versions, averaged over 50 Monte Carlo simulations. See Appendix C for the definitions of the benchmark functions.

Table 14.3 shows that blended BBO performs better than standard BBO on 11 of the 14 benchmarks. The magnitude of the improvement is quite impressive. Standard BBO performs better on three benchmarks with an average factor of improvement of about 3. But blended BBO performs better on 11 benchmarks, with a factor of improvement as high as 8818 (Penalty #2 function).

□

14.4.3 Other Approaches to BBO

The algorithm presented in Figure 14.3 is called partial immigration-based BBO [Simon, 2011b]. The word *partial* in the name means that only one solution feature is considered for immigration at a time. That is, for individual z_k, λ_k is tested against a random number once for every solution feature to decide whether or not to replace that solution feature. The term *immigration-based* in the name means that λ_k is first used to decide whether or not to immigrate to z_k; it is only after immigration is decided upon that the $\{\mu_i\}$ variables are used to choose the emigrating solution, using some procedure like roulette-wheel selection.

However, there are also other ways that we could implement the idea of BBO. Instead of testing λ_k against a random number once for each solution feature, we could test λ_k against a random number only once for each individual, and then if immigration is decided upon, we replace all solution features in z_k. This could be called total immigration-based BBO.

Also, we could first use μ_k to decide whether or not to emigrate a feature from a given individual. Then, only if emigration is decided upon, would we use the $\{\lambda_i\}$ variables in a roulette-wheel process to select where to immigrate the chosen solution feature. This idea gives rise to emigration-based BBO.

Combining the above ideas results in four different BBO implementations. The first, partial immigration-based BBO, is the default implementation and is outlined in Figure 14.3. The other three are outlined in Figures 14.6–14.8. In addition, each of these approaches could be combined with sinusoidal migration curves as discussed in Section 14.4.1, and/or blended migration as discussed in Section 14.4.2. As with any other EA, we should also implement mutation and elitism, although these procedures are not shown in Figures 14.6–14.8. Theoretical and applied investigations of these BBO options are reported in [Ma and Simon, 2013].

Initialize a population of candidate solutions $\{x_k\}$ for $k \in [1, N]$
While not(termination criterion)
    For each x_k, set emigration probability $\mu_k \propto$ fitness of x_k, with $\mu_k \in [0, 1]$
    For each individual x_k, define immigration probability $\lambda_k = 1 - \mu_k$
    $\{z_k\} \leftarrow \{x_k\}$
    For each individual x_k
        For each solution feature s
            Use μ_k to probabilistically decide whether to emigrate $x_k(s)$
            If emigrating then
                Use $\{\lambda_i\}$ to probabilistically select the immigrating solution z_j
                $z_j(s) \leftarrow x_k(s)$
            End if
        Next solution feature
    Next individual
    Probabilistically mutate $\{z_k\}$
    $\{x_k\} \leftarrow \{z_k\}$
Next generation

Figure 14.6 The above algorithm outlines partial emigration-based BBO with a population size of N. $\{x_k\}$ is the entire population of individuals, x_k is the k-th individual, and $x_k(s)$ is the s-th feature of x_k. Similarly, $\{z_k\}$ is the temporary population of individuals, z_k is the k-th temporary individual, and $z_k(s)$ is the s-th feature of z_k.

```
Initialize a population of candidate solutions {x_k} for k ∈ [1, N]
While not(termination criterion)
    For each x_k, set emigration probability μ_k ∝ fitness of x_k, with μ_k ∈ [0, 1]
    For each individual x_k, define immigration probability λ_k = 1 − μ_k
    {z_k} ← {x_k}
    For each individual z_k
        Use λ_k to probabilistically decide whether to immigrate to z_k
        If immigrating then
            For each solution feature s
                Use {μ_i} to probabilistically select the emigrating solution x_j
                z_k(s) ← x_j(s)
            Next solution feature
        End if
    Next individual
    {x_k} ← {z_k}
Next generation
```

Figure 14.7 The above algorithm outlines total immigration-based BBO with a population size of N. $\{x_k\}$ is the entire population of individuals, x_k is the k-th individual, and $x_k(s)$ is the s-th feature of x_k. Similarly, $\{z_k\}$ is the temporary population of individuals, z_k is the k-th temporary individual, and $z_k(s)$ is the s-th feature of z_k.

```
Initialize a population of candidate solutions {x_k} for k ∈ [1, N]
While not(termination criterion)
    For each x_k, set emigration probability μ_k ∝ fitness of x_k, with μ_k ∈ [0, 1]
    For each individual x_k, define immigration probability λ_k = 1 − μ_k
    {z_k} ← {x_k}
    For each individual x_k
        Use μ_k to probabilistically decide whether to emigrate x_k
        If emigrating then
            For each solution feature s
                Use {λ_i} to probabilistically select the immigrating solution z_j
                z_j(s) ← x_k(s)
            Next solution feature
        End if
    Next individual
    {x_k} ← {z_k}
Next generation
```

Figure 14.8 The above algorithm outlines total emigration-based BBO with a population size of N. $\{x_k\}$ is the entire population of individuals, x_k is the k-th individual, and $x_k(s)$ is the s-th feature of x_k. Similarly, $\{z_k\}$ is the temporary population of individuals, z_k is the k-th temporary individual, and $z_k(s)$ is the s-th feature of z_k.

14.4.4 BBO and Genetic Algorithms

This section discusses the relationship between GAs and BBO. In GAs with uniform crossover we randomly choose each child gene from one of its two parents (see Section 8.8.4). In gene pool recombination, which is also known as multi-parent recombination and scanning crossover, we randomly choose each child gene from one of its parents, where the number of parents is greater than two (see Section 8.8.5). We need to make several choices when implementing gene pool recombination in GAs. For example, how many individuals should be in the pool of potential parents? How should individuals be chosen for the pool? Once the pool has been determined, how should parents be selected from the pool? One way of implementing gene pool recombination might be called global uniform recombination, in which we randomly choose each child gene from one of its parents, where the parent population is equivalent to the entire GA population, and the random selection is based on fitness values (for example, roulette-wheel selection).

If we use global uniform recombination, and if we also use fitness-based selection for each solution feature in each offspring, we obtain the algorithm shown in Figure 14.9, which we call the genetic algorithm with global uniform recombination (GA/GUR). Comparing Figures 14.3 and 14.9 we see that BBO is a generalization of a specific type of GA/GUR. This is because if, rather than setting $\lambda_k = 1 - \mu_k$ in the BBO algorithm of Figure 14.3, we instead set $\lambda_k = 1$ for all k, then the BBO algorithm of Figure 14.3 would be equivalent to the GA/GUR algorithm of Figure 14.9.

Initialize a population of candidate solutions $\{x_k\}$ for $k \in [1, N]$
While not(termination criterion)
    For $k = 1$ to N
        $\text{Child}_k \leftarrow \begin{bmatrix} 0 & 0 & \cdots & 0 \end{bmatrix} \in R^n$
        For each solution feature $s = 1$ to n
            Use fitness values to probabilistically select individual x_j
            $\text{Child}_k(s) \leftarrow x_j(s)$
        Next solution feature
        Probabilistically mutate Child_k
    Next child
    $\{x_k\} \leftarrow \{\text{Child}_k\}$
Next generation

Figure 14.9 Outline of genetic algorithm with global uniform recombination (GA/GUR) for an n-dimensional optimization problem. N is the population size, $\{x_k\}$ is the entire population of individuals, x_k is the k-th individual, and $x_k(s)$ is the s-th feature of x_k.

This discussion is similar to that in Section 12.4, where we showed that DE is a special type of continuous GA. In that section, we saw that even though DE and GAs have many similarities, DE is distinctive enough to be considered a separate EA rather than a special type of GA. We make a similar conclusion in this section. Even though BBO and GAs have many similarities, BBO is distinctive enough to be considered a separate EA rather than a special type of GA. There is also another even more important reason to consider BBO as a separate EA, and that is because the biogeography roots of BBO open up many avenues for extensions and modifications that would otherwise be unavailable to the researcher. We discussed one of these extensions in Section 14.4.1, and we discuss some others in the conclusion of this chapter.

14.5 CONCLUSION

We have seen how biogeography, the study of the geographical distribution of biological species, can be used to obtain the biogeography-based optimization (BBO) algorithm. BBO has been modeled using Markov theory [Simon et al., 2011a], dynamic systems [Simon, 2011a], and statistical mechanics [Ma et al., 2013]. Some of these models are analogous to those that we derived for GAs (see Chapter 4). GA and BBO Markov models are compared in [Simon et al., 2011b]. The BBO Markov model was extended to BBO with elitism in [Simon et al., 2009]. Like many other EAs, BBO has been applied to many real-world problems. A web site devoted to BBO is available at [Simon, 2012].

One shortcoming of the BBO algorithm presented here is that it migrates only one independent variable at a time between solutions. This works fine for separable problems, that is, problems whose fitness function $f(x)$ can be written as

$$f(x) = \sum_{i=1}^{n} f_i(x(s_i)) \tag{14.20}$$

where $x(s_i)$ is the i-th independent variable of x, and n is the problem dimension. However, most optimization problems are not separable. This means that if a candidate solution contains a group of independent variables that makes it highly fit, there is no easy way to migrate that group to another candidate solution. One remedy for this shortcoming might be to modify the BBO algorithm so that random groups of independent variables migrate, rather than simply one independent variable at a time. Something similar, although not identical, to this idea has been suggested in [Omran et al., 2013].

BBO is actually a family of algorithms, and so it could be called a metaheuristic. It includes the options shown in Table 14.4. A systematic study of the combinations of the options in Table 14.4, both theoretical and applied, remains as a task for future research.

Migration Approaches	Migration Curves	Migration Blending
Partial immigration-based	Linear	None ($\alpha = 0$)
Total immigration-based	Sinusoidal	$\alpha = 0.5$
Partial emigration-based	Other	α = some other constant
Total emigration-based		$\alpha \propto$ fitness

Table 14.4 BBO implementation options. BBO can be implemented with the combination of any choice from column 1, any choice from column 2, and any choice from column 3.

There are many other interesting possibilities for aligning BBO more closely with biogeography, including the following.

Habitat Similarity: In island biogeography, immigration rate is correlated with island isolation [Adler and Nuernberger, 1994]. Islands that are isolated are relatively well-buffered from immigration. This intuitive idea is called the distance effect [Wu and Vankat, 1995]. It also stands to reason that emigration rates are correlated with island isolation. In island biogeography, the environmental uniqueness of an island is related to island isolation because environmental conditions vary predictably with geographical distance [Lomolino, 2000a]. In BBO, candidate solution isolation would be related to candidate solution uniqueness. Similar islands could be viewed as clustered together in solution space, and dissimilar solutions as isolated in solution space. In biogeography language, similar solutions would belong to the same archipelago (island group). This would tend to increase immigration and emigration between similar solutions, and decrease those rates between dissimilar solutions. This could be implemented in BBO by probabilistically increasing solution feature sharing between similar solutions. This is analogous to species-based crossover, also called niching, in GAs [Stanley and Miikkulainen, 2002], and is also similar to the speciating island model [Gustafson and Burke, 2006] (see Section 8.6.2). It is also analogous to the idea of neighborhoods in particle swarm optimization [Kennedy and Eberhart, 2001] (see Chapter 11). However, the motivation and mechanism is entirely different. Niching in GAs is based on the likelihood of individuals to mate with similar individuals. Neighborhoods in PSO are based on the likelihood of individuals to congregate together in solution space. Archipelagos in BBO are formed on the likelihood of similar islands to cluster together. A quantitative way to determine the effect of island isolation on immigration rates is given in [Hanski, 1999].

Initial Immigration: Classic island biogeography theory indicates that immigration rate decreases as the number of species increases, as shown in Figures 14.1 and 14.5. In BBO this corresponds to a monotonic decrease in immigration rate as individual fitness increases. This means that as an individual becomes more fit, the probability of incorporating features from other individuals decreases. However, more recent advances in biogeography indicate that for some pioneer species (plants, for example), an initial increase in species count results in an initial increase in the immigration rate [Wu and Vankat, 1995]. This is because these early immigrants modify the island to make it more hospitable to other species. That is, the positive effect of increased diversity due to initial immigration overcomes the negative effect of increased population size. In BBO this would correspond to an initial increase in immigration rate as a very poor candidate solution initially improves its fitness. This can be viewed as a temporary positive feedback mechanism in BBO. A very poor individual accepts features from other individuals, increasing its fitness, which subsequently increases its likelihood of accepting even more features from other individuals. This is depicted in Figure 14.10. This idea can be incorporated into other EAs also, but its initial motivation comes from biogeography [Ma and Simon, 2011a].

Minimum Fitness Requirement: We could suppose that a habitat must have some minimum fitness rank to have a nonzero emigration rate. This is similar to supposing that an island must have a nonzero HSI to support any species [Hanski and Gilpin, 1997].

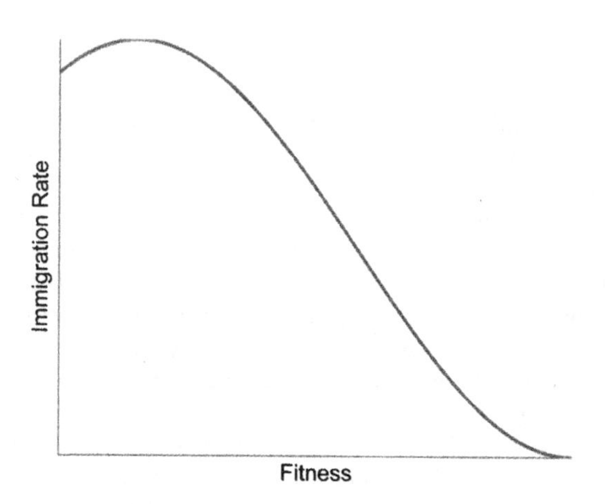

Figure 14.10 This proposed model shows that immigration initially increases with fitness. This gives poor but improving individuals momentum to continue improving. As an individual continues to become more fit after the initial increase in immigration, immigration begins to decrease to give less fit individuals relatively greater opportunities to immigrate good solution features.

Age Criterion: The reproductive value of an individual (that is, the expected number of offspring per unit time) is a triangular function of its age. Reproductive value is low at young ages due to immaturity, high at child-bearing ages, and low again at old ages due to loss of fertility. The same could be said of species. A young species might be poorly adapted to its environment and so has only a small chance of speciating, a middle-aged species is both mature enough and dynamic enough to speciate, and an old species is too stagnant to speciate. This could lead to the introduction of an age criterion in BBO, similar to that which has been used in GAs [Zhu et al., 2006].

Species Mobility: Classic island biogeography theory assumes that all species are equal in their migratory ability. In reality, some species are more mobile than others, and some species are better dispersers than others. Efforts are being made in biogeography to incorporate species-specific characteristics into island biogeography theory [Lomolino, 2000b]. BBO presently assumes that all species are equally mobile. BBO would be more consistent with its motivating framework if species mobility were proportional to the species' contribution to solution fitness. That is, given a population of individuals, statistical methods can be used to find the correlation of each solution feature with fitness. Mobility would then be defined as a solution feature, or a set of features, that is positively correlated with fitness. BBO species mobility would follow biogeography theory by assigning mobility values with a Gaussian distribution [Lomolino, 2000b]. Those solution features that tend to increase fitness would be more likely to emigrate. This should improve the mean fitness of the population.

Predator/Prey Relationships: In biology, certain species have adversarial relationships. These relationships do not necessarily harm the prey species. For instance, prey may respond to predators by reducing the exploitation of their resources, thus benefiting themselves in the long term [Hanski and Gilpin, 1997]. However, the more common scenario is the one in which predators reduce prey to such an extent that one or both populations face extinction. Predator/prey relationships can be inferred from a BBO population by examining individuals and noting which pairs of solution features have a low probability of coexisting. Those solution features would then be modeled as a predator/prey pair. Combining this information with the fitness contribution of each species would result in defining the predator solution feature as the adversary that is positively correlated with fitness, and the prey solution feature as the adversary that is negatively correlated with fitness. The predator/prey relationship might lead to a nonzero equilibrium population, or it might lead to the extinction of one or both populations [Gotelli, 2008], [Hanski and Gilpin, 1997]. This information could be used throughout the population of individuals to increase the likelihood of the presence of predator features, and reduce the likelihood of the presence of prey features. Most predator/prey models in biology are for two-species systems. These models could be used in BBO, but a more complete description would be obtained if existing predator/prey models could be extended to multi-species systems.

Resource Competition: In contrast to the predator/prey relationship described above, we note that similar species compete for similar resources. Therefore, it is unlikely that many similar species occupy the same island, especially if they have large populations [Tilman et al., 1994]. In BBO, this means that it is unlikely that solution features would emigrate to islands that already have large populations that are similar to each other. Alternatively, it could mean that emigration rate is not affected, but survival likelihood is lower. Resource competition in BBO also means that if two solution features have equal probability of extinction, then the feature most similar to other features in the solution is more likely to become extinct. This is a different type of interaction than the predator/prey relationship described above. However, both models are plausible, and competition is generally viewed in biology as a more significant driver of community composition than predator/prey interactions.

Time Correlation: In island biogeography, if a species migrates to an island in a given geographical direction, it is likely to continue moving in that direction to the next island. This is because migration is influenced by prevailing winds and currents, and those winds and currents have a positive time correlation. This is described by biodiffusion theory, the telegraph equation, and the equation of diffusion [Okubo and Levin, 2001]. If a species migrates from island A to island B, it is likely to continue in the same direction to the next island in the chain at the next time step. In BBO this means that if a solution feature migrates from one individual to the next, it is likely to continue migrating in that direction at the next evolutionary generation. The concept of "direction" in BBO could be defined in terms of solution location, where location is defined as a point in solution feature space.

Other aspects of biogeography could inspire other variations to BBO. The biogeography literature is so rich that there are many possibilities along these lines.

PROBLEMS

Written Exercises

14.1 We wrote Equation (14.1) under the assumption that Δt is small enough so that the probability of more than one migration between time t and $(t + \Delta t)$ can be ignored. Rewrite the equation under the assumption that no more than two migrations can occur between t and $(t + \Delta t)$.

14.2 We defined selection pressure for EAs in Equation (8.13). How could you change the migration curves of Figure 14.2 to increase the selection pressure?

14.3 How could you change the BBO algorithm of Figure 14.3 to be steady state rather than generational?

14.4 We typically set $\mu_k = r_k/(N + 1)$ and $\lambda_k = (N + 1 - r_k)/(N + 1)$, where r_k is the rank of the k-th individual, the best individual has rank N, and the worst individual has rank 1. This means that $\lambda_k \in [1/(N + 1), N/(N + 1)]$. What is the practical result of enforcing $\lambda_k > 0$ for the best individual?

14.5 This problem explores the initial immigration model of Figure 14.10. Let β denote the normalized fitness value at the peak value of λ.

- **a)** Find an equation for the immigration rate of Figure 14.10.
- **b)** Find the equilibrium species count for Figure 14.10.

14.6 Suppose we use linear migration rates with a population size N, so that $\lambda_1 = 1/(N+1)$ for the best individual, and $\lambda_N = N/(N+1)$ for the worst individual.

- **a)** What is the probability that the best individual x_b will receive an immigration of at least one solution feature in the partial immigration-based algorithm of Figure 14.3?
- **b)** What is the probability that the best individual x_b will receive an immigration of at least one solution feature from an individual other than itself in the partial emigration-based algorithm of Figure 14.6?

14.7 Look in Appendix C.1 for an example of a separable function and an example of non-separable function.

Computer Exercises

14.8 Repeat Example 14.1 for $n = 5$.

14.9 The standard BBO algorithm of Figure 14.3 indicates that for each individual x_k, we should set emigration probability $\mu_k \propto$ fitness of x_k, with $\mu_k \in [0, 1]$. We use the emigration probabilities to select the emigrating individual, and we can use the following options for this operation.

- Rank-based selection, as discussed in Section 8.7.4. This is the standard BBO option, as indicated in Example 14.2.
- Square ranking, which is also discussed in Section 8.7.4.
- Tournament selection, as discussed in Section 8.7.6.
- Stud selection, as discussed in Section 8.7.7.

Implement BBO to minimize the 10-dimensional Ackley function with $N = 50$, generation limit = 50, mutation rate = 1%, and elitism parameter = 2. Run BBO for 20 Monte Carlo simulations, keeping track of the lowest cost at each generation for each Monte Carlo simulation. Plot the lowest cost, averaged over the Monte Carlo simulations, as a function of generation number. Compare the BBO plots from each of the four emigration selection options mentioned above. Comment on your results.

14.10 Plot your answers to Problem 14.6 as a function of population size N for $N \in [10, 50]$ and problem dimension $n = 10$. Comment on your results.

14.11 One potential way to improve BBO performance is to select the next generation from the old individuals and the new individuals [Du et al., 2009]. That is, the $\{x_k\} \leftarrow \{z_k\}$ statement at the end of Figure 14.3 could be replaced with the following:

$$\{x_k\} \leftarrow \text{Best } N \text{ individuals from } \{x_k\} \cup \{z_k\}.$$

This is motivated by evolution strategy, and so we can call it BBO-ES. Implement BBO to minimize the 10-dimensional Ackley function with population size $N = 50$, generation limit = 50, mutation rate = 1%, and elitism parameter = 2. Run BBO for 20 Monte Carlo simulations, keeping track of the lowest cost at each generation for each Monte Carlo simulation. Plot the lowest cost, averaged over the Monte Carlo simulations, as a function of generation number. Compare the plot from the standard BBO algorithm with that from BBO-ES. Comment on your results.

CHAPTER 15

Cultural Algorithms

Culture optimizes cognition.

—James Kennedy [Kennedy, 1998]

The point of the above quote is that cognition (that is, the process of thinking) involves more than brain activity and neuronal behavior. Our thinking is influenced by our culture. Furthermore, this influence is beneficial (even optimal, according to the above quote). Without culture, our cognitive abilities would be impaired.

This idea is exemplified by the discovery of feral children [Newton, 2004]. Some of these children grow up in the wild, while others grow up in isolation as a result of abusive caretakers. Children who grow up without any social or cultural interaction typically never learn to assimilate society, never learn to speak, never learn to relate to others, and never learn to act in a socially acceptable way. Their inability to learn how to function in their new and civilized environment is not genetic; it is not due to a lack of innate intelligence. The lack of social and cultural influences during their upbringing severely compromises their intelligence. Feral children provide a strong argument for the *nurture* side in the *nature vs. nurture* debate.

Scientists used to believe that human culture originated at a high level and then later degenerated to lower levels. This degeneration resulted in low and uncivilized cultures scattered around the world. This belief was largely based on religious accounts of creation and the Biblical story of the Tower of Babel in the book of

Genesis. However, this view of cultural degeneration has testable implications. For instance, degenerationism should result in archeological records of increasing cultural sophistication as digs go deeper and farther back into the past. Although specific religious stories cannot be scientifically tested, the testability of degenerationism as a general principle was instrumental in its demise as a sociological theory.

Edward Tylor, a 19th-century anthropologist, showed that advanced cultures developed from primitive cultures, rather than the other way around [Tylor, 2011]. He showed that culture evolves from lower to higher forms just as biological organisms evolve. Tylor was the first to use the word *culture* in its modern sociological sense, and he defined it as "that complex whole which includes knowledge, belief, art, morals, law, custom, and any other capabilities and habits acquired by man as a member of society" [Tylor, 2009].

A society's culture is a complex entity that interacts with the environment, individuals, and other cultures. Individuals can act independently, but they also interact with each other, both directly and indirectly; individuals influence each other directly, and they influence each other indirectly through culture. Most individuals are constrained by the culture in which they live. Some individuals swim against the tide, but most individuals conform to society.

Overview of the Chapter

This chapter discusses some ways that culture can be modeled in evolutionary algorithms (EAs) to improve their performance. Section 15.1 is a preliminary section that discusses optimal strategies for human relationships at a high level; this provides some motivation and background for the remainder of the chapter. Section 15.2 discusses a particular model of culture called belief spaces, and discusses their co-evolution with candidate solutions in EAs. Section 15.3 uses belief spaces to develop a cultural evolutionary program (EP), and shows that it provides better performance than the standard EP. Section 15.4 takes a different perspective of culture and views it as more interpersonal and relationship oriented; this section discusses the adaptive culture model (ACM), and shows how ACM can solve the traveling salesman problem.

15.1 COOPERATION AND COMPETITION

This section discusses culture in the sense of interpersonal relationships. Modern society involves a lot of interpersonal communication, and the amount of communication is increasing at a rapid pace. Society also involves a lot of cooperation and a lot of competition. Sometimes we communicate for the purpose of cooperating, but sometimes we communicate for the purpose of competing. When we write a technical paper or a research proposal, we communicate the disadvantages of competing ideas and the advantages of our own ideas. Sometimes we exaggerate to make ourselves look better or to make someone else look worse. Sometimes our exaggerations are intentional, sometimes they are unintentional, and sometimes it is hard to discern our own intentions. In our communication with others, we sometimes tell the truth because we hope that others will in turn tell us the truth. But we often lie when the benefits or rewards outweigh the consequences. Consider typical answers to questions like the following.

1. How are you?
2. Did you like the meal I cooked for you?
3. How often do you lie?

We reason that the one who asks does not really want to know the answer to the question. The one who asks is only trying to make conversation, or is fishing for a specific answer, and so we willingly oblige even though our answer might technically be classified as a lie. Interestingly, everyone thinks that they lie less often than others [DePaulo et al., 1996]. Furthermore, everyone thinks that their own lies are more justifiable than those of others.

We can simulate interpersonal communication in evolutionary algorithms to study communication strategies, or to find solutions to optimization problems. The prisoner's dilemma (see Section 5.4) is one example of agents communicating with each other. It is an interesting example because it has many variations, the best strategy depends on the opposing player's strategy, and the best strategy is not always obvious.

El Farol

Another interesting example of interpersonal communication is the El Farol problem [Kennedy and Eberhart, 2001, Chapter 5]. This problem involves a man named Brian Arthur[1] who likes to go to a pub by the name of El Farol in downtown Sante Fe. He particularly likes to go on Thursdays when El Farol plays Irish music. However, he prefers to stay home if the place is crowded. In particular, he wants to go to El Farol if there will be fewer than 60 people there, but he wants to stay home if there will be 60 or more people. Brian's friends are in the same situation. They love to go to El Farol if there will be fewer than 60 people there, but they do not want to go if there will be 60 or more people.

Brian and his friends know that for the past 14 weeks, the number of people at El Farol have been

$$44, 78, 56, 15, 23, 67, 84, 34, 45, 76, 40, 56, 22, \text{and } 35. \tag{15.1}$$

Should he go this Thursday? In other words, based on the data from the past 14 weeks, will there be fewer than 60 people at El Farol this week? He could use various pattern recognition techniques and regression tests to try to predict the number of people who will be at El Farol this Thursday. If he found a good predictor, then his problem would be solved. However, if he told all of his friends about his predictor, then the predictor would not work any more. If all of his friends knew that his algorithm predicted fewer than 60 people, then they would all go to El Farol and there would be more than 60 people. If his algorithm predicted more than 60 people, then all of his friends would stay home and there would be fewer than 60 people. This is the paradox of a good prediction algorithm when the human element is taken into account. A good predictor becomes a poor predictor.

[1]Brian Arthur is an economist and co-founder of the Sante Fe Institute.

Now suppose that Brian and his friends talk with each other about whether or not they are going to El Farol on Thursday. If Brian decides to go to El Farol and tells all of his friends, then they will be more likely to stay home, and Brian will be rewarded with a small crowd at El Farol. If Brian decides to stay home and tells all of his friends, then they will be more likely to go, and his friends will be penalized with a large crowd.

However, Brian and his friends may not be completely honest with each other. They may all tell each other that they are going to El Farol in hopes that most of the others will stay home. After telling all of his friends that he is going, Brian may decide to stay home if he hears that everyone else is planning to go. In addition, Brian might tell his friends that he and 10 of his other friends are going. Brian might exaggerate the size of his party to encourage his friends to stay home. Of course, his friends may pursue the same strategy. That is, they might lie for their own benefit.

What is the optimal communication strategy for Brian? Should he always tell the truth? If he consistently lies, then his friends will eventually recognize his pattern of lies and will learn to ignore him. However, if his ultimate goal is to go to El Farol on uncrowded nights and avoid El Farol on crowded nights, then it seems that he may need to lie on occassion.

The El Farol problem is interesting because it involves truth, lies, trust, communication, and possibly conflicting objectives. If Brian's objective is to be liked by his friends, then he will probably tell the truth all the time. If his objective is to go to El Farol on uncrowded nights and avoid El Farol on crowded nights, then he might sometimes lie.

Other Examples

The other reason that El Farol is interesting is that, as mentioned earlier, a good predictor will become a poor predictor if everyone uses it. This characteristic arises in many real-world situations. For example, consider the romantic interest that a boy shows to a girl. If he shows too much interest then he will appear to be desperate, which will be unattractive to the girl. However, if his interest is not obvious enough, then he will appear to be uninterested, which will not be conducive to a relationship with the girl of his dreams. Furthermore, how should the girl interpret his apparent lack of interest? Should she interpret it as true disinterest, in which case she will turn her attention to other suitors? Or should she interpret his lack of interest as subdued passion, in which case she will respond? Courtship is a complex give-and-take activity between two individuals who behave according to their own goals but also according to their culture.

Baseball is another interesting example. When the count is three balls and two strikes, the pitcher needs to throw a strike to avoid walking the batter – especially if the bases are loaded late in a tie game. But the batter knows that the pitcher needs to throw a strike. This seems to give the advantage to the batter, because he knows approximately where the ball will be pitched. But since the pitcher is aware of the batter's thought process, the pitcher might deliver a pitch outside of the strike zone, trying to entice the batter to swing at a bad pitch. Of course, the batter is aware of the pitcher's thought process too. Baseball becomes not only a physical contest but also a mental battle. The batter needs to decide how aggressive he will be in his anticipation of a pitch in the strike zone. The pitcher needs to decide if he can risk a pitch outside of the strike zone. Players take into account not only the game situation when deciding their strategies, but also the history of previous encounters with their opponent.

Other examples arise in our business and in our research programs. What area should we focus on with our research proposals? Should we write proposals in highly funded areas? High funding amounts increase our chances, but everyone else is also writing proposals in those areas, which decreases our chances. Perhaps we should write proposals in areas with less funding so that we have less competition. If our proposal is the only one in a certain area, then it is more likely to be funded. But if our competitors follow that same strategy, then the strategy will fail. Deciding where to focus our proposal-writing efforts is complex and multidimensional, but the optimal strategy is probably to spread out our efforts among both high-risk and low-risk areas, and both high-funding and low-funding areas [Simon, 2005]. A similar mixed strategy can be used for investing (recall the investor's mantra of diversification), product marketing, and other applications.

Problems with interpersonal relationships, communication, cooperation, deception, and multiple objectives, have a lot in common with human culture. Humans have learned near-optimal ways of relating, structuring society, and developing culture. We are not aware of all of the optimization features that are inherent in human culture, but considering the possibility of such features would surely be a fascinating and rewarding study. Imitating and simulating the optimization behavior of human culture is another fascinating and rewarding study, and we turn our attention to that pursuit in the remaining sections of this chapter.

15.2 BELIEF SPACES IN CULTURAL ALGORITHMS

A cultural algorithm (CA) is similar to other EAs in viewing candidate solutions to an optimization problem as individuals. However, a CA also views the principles that guide the evolution of individuals as their culture. The CA models the influence between individuals and their culture to obtain an optimization algorithm. The cultural norms of the CA's virtual society are sometimes called a belief space. At each generation in a CA individuals recombine and mutate, just as in the other EAs that we have discussed earlier in this book. But in a CA this recombination and mutation is influenced by the belief space. The belief space can be designed by the programmer to impose constraints, or to favor preferred features in the population, or to avoid undesirable features.

■ EXAMPLE 15.1

In this example we discuss a general idea for how a belief space might be implemented in an EA. Recall the artificial ant problem in Section 5.5. An artificial ant is placed in a grid with some empty cells and some food-filled cells. The only sensory ability that the ant has is to sense the presence or absence of food in the cell directly in front of him. In each cell, the ant can take one of three actions: he can either move ahead, in which case he eats the food in the next cell, if food is present; or he can remain in his current cell and turn to the right; or he can remain in his current cell and turn to the left. We want to evolve a finite state machine (FSM) to help the ant navigate his way through the grid and consume as much food as possible. We know intuitively that if the ant senses food in the cell in front of him, he should probably move ahead and eat the food. However, we may not want to impose this action as a hard constraint. We know that evolution often requires the exploration of suboptimal solutions in its search for an optimum. So we would like to *encourage* each FSM in our EA population to move ahead whenever food is sensed, but we do not want to make this a strict requirement. This is the type of behavior that could be encoded in a belief space to encourage, but not require, a certain feature in an EA population.

The level of encouragement that we provide to EA individuals to move ahead when food is sensed represents the strength of the culture. In human society, some cultures are stronger than others and exert more pressure to conform. If our encouragement is mild, then we will allow quite a few individuals to explore other options rather than moving ahead when food is sensed. If our encouragement is strong, then we will allow only a very few individuals to explore alternate options. As the population evolves, we might want to modify our level of encouragement based on the fitness of the individuals who conform to the prevailing culture, relative to the fitness of those individuals who oppose the prevailing culture by taking an alternate action when sensing food.

□

Example 15.1 shows that CAs can implement dual inheritance: solution features are inherited by children from parents, and the belief space of one generation is inherited from the previous generation. Evolution still occurs on an individual level but the evolution is influenced by the belief space.

The belief space of a CA might be static or dynamic depending on how we implement it. If the belief space is static then it does not change with time. If the belief space is dynamic then it changes with time; that is, the culture can evolve. A CA with a dynamic belief space evolves not only a population of individuals from one generation to the next, but it also evolves a belief space from one generation to the next. In a CA with a dynamic belief space, the belief space not only influences the evolution of the population, but the population in turn influences the evolution of the belief space. Dynamic belief spaces are motivated by what we observe in human society. We see that culture evolves much faster than biology. We therefore hope that we can find optimal solutions more quickly in an EA with a dynamic belief space than in an EA without it.

Just as there are many theories about human culture [Welsch and Endicott, 2005], there are also several different types of CAs. Figure 15.1 shows a basic outline of a CA. As with any other EA, we initialize a population of candidate solutions, but we also initialize a belief space B. The belief space influences the evolution of the population; it can be said to guide the evolutionary process. The algorithm of Figure 15.1 proceeds like any other EA by evaluating the cost of each individual. But then it uses the individuals to modify B. There are many options for how we can implement the modification of B. For example, if the population indicates that most of the good individuals have a certain feature, then B might be updated to bias future candidate solutions toward that feature. Of course, future individuals are already biased toward that feature by virtue of the fact that fit individuals are more likely to recombine than unfit individuals. But B can also bias future individuals in more complex ways than standard recombination methods. For example, we can incorporate certain combinations of features, or types of behaviors, in B. (Recall the idea in Example 15.1 of biasing artificial ant FSMs toward solutions that move ahead if food is sensed.)

Initialize the population of candidate solutions $\{x_i\}$, $i \in [1, N]$
Initialize the belief space B
While not(termination criterion)
    Calculate the cost $f(x_i)$ of each individual in the population, $i \in [1, N]$
    Use the population $\{x_i\}$ to update B
    Incorporate B in the recombination and mutation of the population $\{x_i\}$
Next generation

Figure 15.1 Outline of a basic cultural algorithm, based on [Reynolds, 1994], [Engelbrecht, 2003, Chapter 14].

After B is updated in Figure 15.1, we perform recombination and mutation of the population $\{x_i\}$. This step can be performed with any of the EAs that we discuss in this book. Therefore, a CA should be viewed not as a separate EA but rather as a way of augmenting other EAs, or as a meta-EA. The distinctive feature of CAs is that recombination and mutation are influenced by a belief space B; individuals in the next generation tend to be consistent with B.

There are many details to be worked out in the CA of Figure 15.1. For example:

- What type of information will we encode in the belief space B?
- How will we update B?
- What type of recombination and mutation will we use? That is, what EA will we use as the baseline for the CA?
- How will we use B to influence recombination and mutation?

These questions all provide opportunities to the researcher to find answers that are effective for specific problems, or for general classes of problems. For instance, consider the first question in the above list. We can partially answer this question by noting that the belief space in a CA can represent the following aspects of an optimization problem.

- The belief space can represent constraints for the solution of optimization problems with either hard constraints or soft constraints [Coello Coello and Becerra, 2002], [Becerra and Coello Coello, 2004].
- The belief can represent domain-specific knowledge to bias the search in preferred directions that are based on human expertise [Sverdlik and Reynolds, 1993], [Alami and El Imrani, 2008].
- The belief space can include the importance of diversity to help preserve diversity in the search.

- The belief space can include the importance of cooperation to improve the performance of co-evolutionary systems. Co-evolution involves the development of distinct but interacting evolutionary systems in a common environment [Durham, 1992]. We do not discuss co-evolution in this book, but artificial co-evolution can find optimal solutions when fitness evaluations vary with time (see Section 21.2), or when the fitness evaluations of the population of candidate solutions depend on other populations which themselves change with time [Yang et al., 2008].

- The belief space can include the importance of creativity, which would bias an evolutionary algorithm's search toward novel candidate solutions or toward unexplored regions of the search space. These ideas are incorporated in opposition-based learning (see Chapter 16) and the search for novelty [Lehman and Stanley, 2011] but have not yet been incorporated in CA belief spaces.

15.3 CULTURAL EVOLUTIONARY PROGRAMMING

This section shows how a simple belief space can improve the performance of an evolutionary program (EP). The basic EP algorithm is outlined in Figure 5.1. In this section we include the option of implementing a belief space in the EP. The belief space indicates where the best-performing candidate solutions lie in the search space. The CA-influenced EP (CAEP) that we present in this section is similar to the one discussed in [Engelbrecht, 2003, Chapter 14]. The belief space B is encoded as $2n$ parameters, where n is the dimension of the optimization problem. The interval $[B_{\min}(k), B_{\max}(k)]$ indicates where the prevailing culture believes that the k-th dimension of good solutions lie in the search space. The belief space influences the mutation process of the EP. If $\beta = 0$ in the EP mutation method shown in Figure 5.1, we have

$$x_i'(k) \leftarrow x_i(k) + r_i(k)\sqrt{\gamma} \tag{15.2}$$

for $i \in [1, N]$ and $k \in [1, n]$, where N is the population size, n is the problem dimension, $r_i(k) \sim N(0, 1)$, and γ is the variance of the mutation. In the CAEP, Equation (15.2) is replaced with

$$\begin{aligned} \Delta_i(k) &\leftarrow \begin{cases} B_{\min}(k) - x_i(k) & \text{if } x_i(k) < B_{\min}(k) \\ 0 & \text{if } B_{\min}(k) \le x_i(k) \le B_{\max}(k) \\ B_{\max}(k) - x_i(k) & \text{if } B_{\max}(k) < x_i(k) \end{cases} \\ x_i'(k) &\leftarrow x_i(k) + r_i(k)\sqrt{\gamma} + \Delta_i(k). \end{aligned} \tag{15.3}$$

We see that if the k-th dimension of individual x_i is within the belief space, then its mutated version $x_i'(k)$ is a random variable with a mean of $x_i(k)$. However, if $x_i(k)$ is outside the belief space, then its mutated version is a random variable with a mean of either $B_{\min}(k)$ or $B_{\max}(k)$, whichever is closer to $x_i(k)$. Figure 15.2 illustrates this idea. The figure shows that when $x_i(k)$ is inside the belief space (the left part of the figure), then it is mutated in a standard way. However, when $x_i(k)$ outside the belief space (the right part of the figure), its mutated version is centered at the closest edge of the belief space. The mutated version could end up outside of the belief space – in fact, it has at least a 50% chance of being outside $[B_{\min}(k), B_{\max}(k)]$. However, it also has almost a 50% chance of being within the belief space, which is much higher than the chance would be if the mean of the mutation were not shifted.

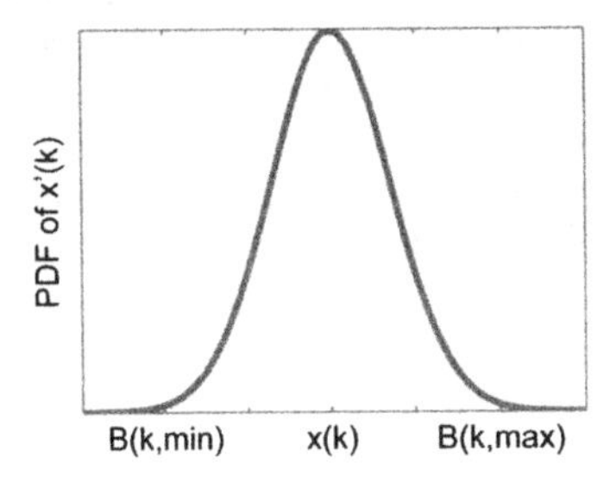

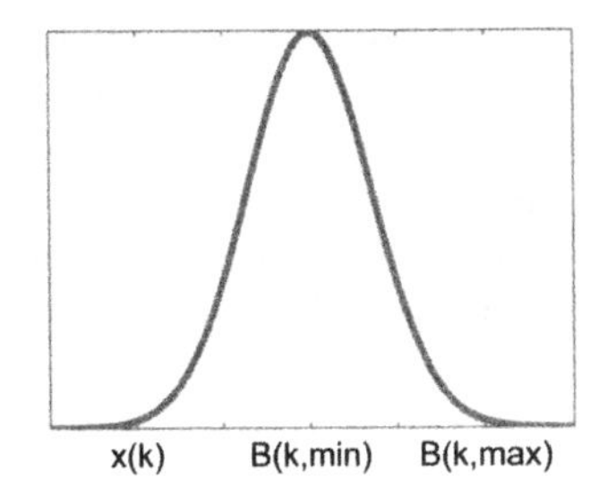

Figure 15.2 Mutation in the cultural evolutionary program. In the figure on the left, $x(k)$ is within the belief space, so the probability density function (PDF) of its mutated version has a mean of $x(k)$. In the figure on the right, $x(k)$ is outside the belief space, so the PDF of its mutated version has a mean that is equal to the closest edge of the belief space.

Now we discuss how to update the belief space in the CAEP. There are several ways that we could do this. For instance, we could use the best M individuals to update the belief space. First we find the minimum and maximum values of each dimension of the best M individuals:

$$\begin{aligned} x_{k,\min} &\leftarrow \min\{x_j(k) : j \in [1, M]\} \\ x_{k,\max} &\leftarrow \max\{x_j(k) : j \in [1, M]\} \end{aligned} \tag{15.4}$$

for $k \in [1, n]$, where the individuals are indexed from best to worst, so that $\{x_j(k) : j \in [1, M]\}$ comprises the best M individuals in the population. Now we use the minimum and maximum domain values to influence the belief space from one generation to the next:

$$\begin{aligned} B_{\min}(k) &\leftarrow \alpha B_{\min}(k) + (1 - \alpha)x_{k,\min} \\ B_{\max}(k) &\leftarrow \alpha B_{\max}(k) + (1 - \alpha)x_{k,\max} \end{aligned} \tag{15.5}$$

for $k \in [1, n]$. The parameter $\alpha \in [0, 1]$ is the belief space inertia, and it determines how stagnant the belief space is from generation to generation. Equation (15.5) shows that if $\alpha = 1$, then the belief space never changes. If $\alpha = 0$, then the belief space is entirely determined by the current population and is not influenced at all by the past generation's belief space.

■ EXAMPLE 15.2

This example shows how the incorporation of culture can improve the performance of EP. We use $N = 50$, $\beta = 0$, and $\gamma = 1$ in the basic EP of Figure 5.1. We use the EP to minimize the 20-dimensional Ackley function described in Appendix C.1.2, with each dimension of each individual randomly initialized in the domain $[-30, +30]$. For the CAEP, we use $M = 5$ in Equation (15.4), and $\alpha = 0.5$ in Equation (15.5). Figure 15.3 shows the optimization results of the standard EP and the cultural EP, averaged over 20 Monte Carlo simulations. We see that the CAEP far outperforms standard EP. Figure 15.4

shows how the belief space of the first dimension changes from one generation to the next. We see that the belief space converges pretty quickly to a small domain that includes the optimal solution, which is 0. There is no guarantee that the belief space will include the optimal solution. In fact, Figure 15.4 shows that the lower bound of the belief space sometimes slightly exceeds 0. But in general, the belief space gives a good indication of where good candidate solutions are likely to reside in the search space. A smaller value of α in Equation (15.5) would result in faster convergence, and a larger value of α would result in a slower convergence.

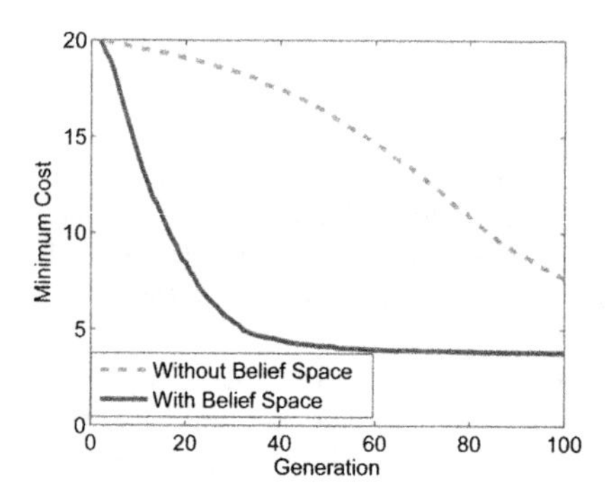

Figure 15.3 Example 15.2: Evolutionary programming without a belief space, and with a belief space. The figure shows the best cost of the population at each generation, averaged over 20 Monte Carlo simulations. CAEP far outperforms standard EP.

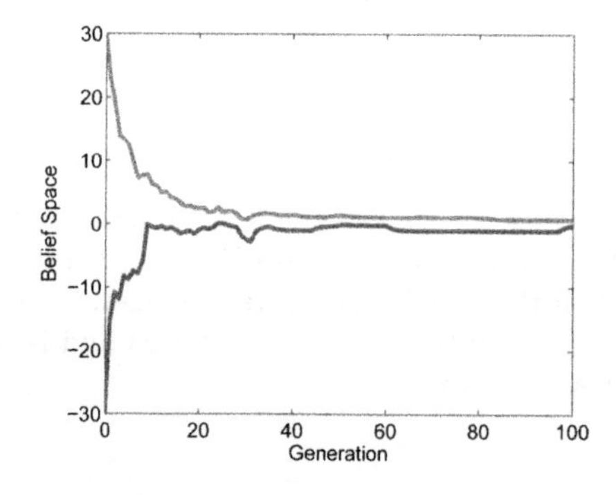

Figure 15.4 Example 15.2: The belief space of the first dimension of the CAEP for the 20-dimensional Ackley function. The belief space quickly converges to a small region around 0, which is the optimal solution.

□

15.4 THE ADAPTIVE CULTURE MODEL

This section discusses a cultural algorithm that is an alternative to the belief space approach of the previous sections. Th algorithm that we discuss in this section is called the adaptive culture model (ACM) [Axelrod, 1997], [Kennedy, 1998], [Kennedy and Eberhart, 2001, Chapter 6]. The ACM is based on the way that individuals in human societies interact with each other. For instance:

- Individuals are influenced more by those who are close to them, either geographically or relationally, than by those who are far from them [Latané et al., 1994]. This is reminiscent of the particle swarm neighborhoods in Chapter 11.
- Individuals are influenced more by those who are similar to them than by those who are different than them [Axelrod, 1997], [Kennedy, 1998].
- As a balance to the previous point, individuals are influenced more by those who are successful than by those who are not [Noel and Jannett, 2005]. A related point is that individuals are influenced more by those who are similar to their *ideal* selves than by those who are similar to their *actual* selves [Wetzel and Insko, 1982], [Kennedy, 1998].

The ACM can be simulated by laying out a grid of candidate solutions to some optimization problem. The candidate solutions are treated as individuals in a population. The proximity of two individuals can be measured in at least a couple of different ways. First, the individuals have a geographical proximity to each other since they are arranged in a grid. Second, the individuals have a behavioral proximity to each other depending on how similar they are with respect to their solution features.

Figure 15.5 shows an example of a grid of EA individuals in which each individual is encoded as an eight-character string. As the population evolves, the individuals maintain the same position in the grid, but their representations change from one generation to the next. Individuals that are closer to each other, either geographically or behaviorally, are more likely to exchange information with each other, and are therefore more likely to become even more similar to each other from a behavioral point of view. Also, when individuals exchange information with each other, the more fit individual is more likely to share information with the less fit individual, rather than vice versa.

We can use all of these ideas to obtain an algorithm for an ACM. Figure 15.6 shows a basic outline of an ACM. The population is initialized, and each candidate solution is assigned to a specific location in a grid. At each iteration of the ACM algorithm, we randomly select an individual and one of its neighbors. We randomly decide whether or not to share information between these two neighbors, using a higher probability if the neighbors are more similar. If we decide to share information, then we randomly replace one of the solution features in the less fit individual with a solution feature from the more fit individual.

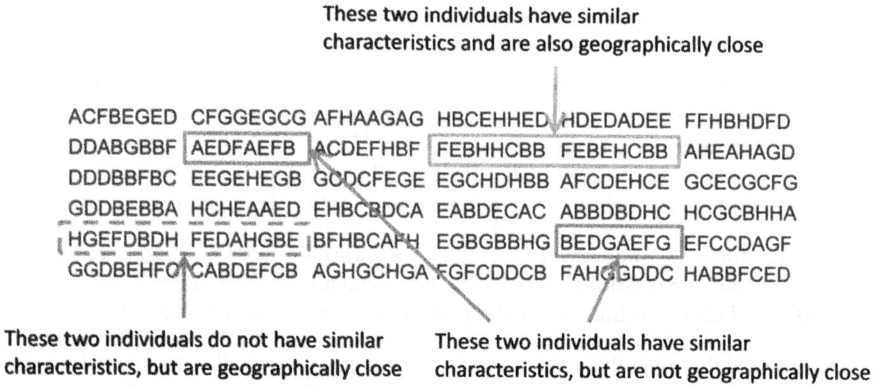

These two individuals have similar characteristics and are also geographically close

```
ACFBEGED  CFGGEGCG  AFHAAGAG  HBCEHHED  HDEDADEE  FFHBHDFD
DDABGBBF  AEDFAEFB  ACDEFHBF  FEBHHCBB  FEBEHCBB  AHEAHAGD
DDDBBFBC  EEGEHEGB  GCDCFEGE  EGCHDHBB  AFCDEHCE  GCECGCFG
GDDBEBBA  HCHEAAED  EHBCBDCA  EABDECAC  ABBDBDHC  HCGCBHHA
HGEFDBDH  FEDAHGBE  BFHBCAFH  EGBGBBHG  BEDGAEFG  EFCCDAGF
GGDBEHFC  CABDEFCB  AGHGCHGA  FGFCDDCB  FAHGGDDC  HABBFCED
```

These two individuals do not have similar characteristics, but are geographically close

These two individuals have similar characteristics, but are not geographically close

Figure 15.5 Example of a grid of ACM individuals. Some individuals are similar to each other but are not geographically close; they are not likely to share information with each other. Other individuals, like the two noted in the lower left portion of the grid, are geographically close but are not similar to each other; they are also unlikely to share information with each other. However, some individuals, like the two noted in the upper right portion of the grid, are geographically close and are also similar to each other. They are likely to share information with each other.

Initialize N individuals $\{x_i\}$, $i \in [1, N]$
Assign each individual to a random location in a grid
While not(termination criterion)
    Randomly select an individual x_i, where $i \in [1, N]$
    Randomly select a neighbor x_k of x_i
    Calculate the behavioral similarity $b_{i,k} \in [0, 1]$ between x_i and x_k
    $r \leftarrow U[0, 1]$
    If $r < b_{i,k}$
        Randomly select a solution feature index $s \in [1, n]$
        Comment: Begin Information Sharing
        If x_i is more fit than x_k then
            $x_k(s) \leftarrow x_i(s)$
        else
            $x_i(s) \leftarrow x_k(s)$
        End if
        Comment: End Information Sharing
    End if
Next interaction

Figure 15.6 The outline of a basic adaptive culture model (ACM). N is the population size, n is the problem dimension, and $U[0, 1]$ is a random number uniformly distributed between 0 and 1.

We see from Figure 15.6 that we always transmit information from the more fit individual to the less fit individual. In keeping with the spirit of stochasticity, we could instead make a probabilistic decision about who shares information with whom. We set the tuning parameter $p_1 \in [0.5, 1]$ equal to the probability of sharing from the better to the worse individual. We call p_1 the selection pressure. We always want $p_1 \geq 0.5$ because it does not make intuitive sense to bias the direction of information sharing from worse to better. Suppose that x_1 and x_2 are two neighboring candidate individuals. We then replace the code between "Comment: Begin Information Sharing" and "Comment: End Information Sharing" in Figure 15.6 with the information-sharing logic shown in Figure 15.7.

$\rho \leftarrow U[0,1]$
If $\rho < p_1$ then
    If x_i is more fit than x_k then
        $x_k(s) \leftarrow x_i(s)$
    else
        $x_i(s) \leftarrow x_k(s)$
    End if
else
    If x_i is more fit than x_k then
        $x_i(s) \leftarrow x_k(s)$
    else
        $x_k(s) \leftarrow x_i(s)$
    End if
End if

Figure 15.7 Adaptive cultural model with stochastic information sharing. $p_1 \in [0.5, 1]$ is the probability of sharing information from the better individual to the worse individual. This pseudo-code snippet replaces the code between "Comment: Begin Information Sharing" and "Comment: End Information Sharing" in Figure 15.6. This snippet results in a p_1 probability of sharing from the better individual to the worse individual. If $p_1 = 1$ then this code reduces to Figure 15.6.

■ EXAMPLE 15.3

In this example, which is motivated by [Kennedy and Eberhart, 2001, Chapter 6], we solve the traveling salesman problem (TSP) with the ACM. Suppose that we want to travel to eight locations in an order that minimizes the total travel distance. We suppose that the locations are arranged in a circle as shown in Figure 15.8, and that we begin at location A. We can easily see that there are two solutions to this TSP: A-B-C-D-E-F-G-H, and A-H-G-F-E-D-C-B. We randomly initialize an 18×8 grid of candidate solutions. We consider the grid to be toroidal so that individuals at the far right of the grid are neighbors to those on the far left, and individuals at the bottom of the grid are neighbors to those at the top. We use the ACM logic of Figures 15.6 and 15.7 to coordinate information sharing between candidate solutions. We implement the statement "Randomly select a neighbor x_k of x_i" in Figure 15.6 by randomly selecting one of the four closest individuals to x_i – that is, the individual immediately to the right, left, above, or below x_i. Use use $p_1 = 0.9$ in Figure 15.7.

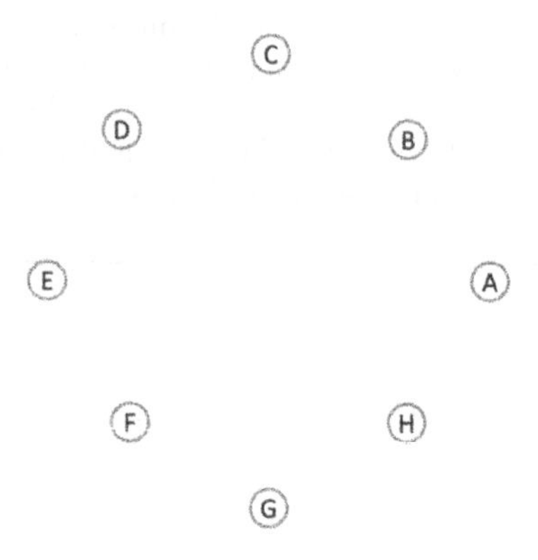

Figure 15.8 The TSP locations of Example 15.3. The goal is to visit all eight locations while minimizing the total travel distance. If we start at location A, there are two optimal solutions: A-B-C-D-E-F-G-H, and A-H-G-F-E-D-C-B.

Figure 15.9 shows the convergence of a typical ACM simulation. We find the first optimum solution after about 2500 iterations of the outer loop of Figure 15.6, and the average cost of the population steadily decreases with the number of interactions.

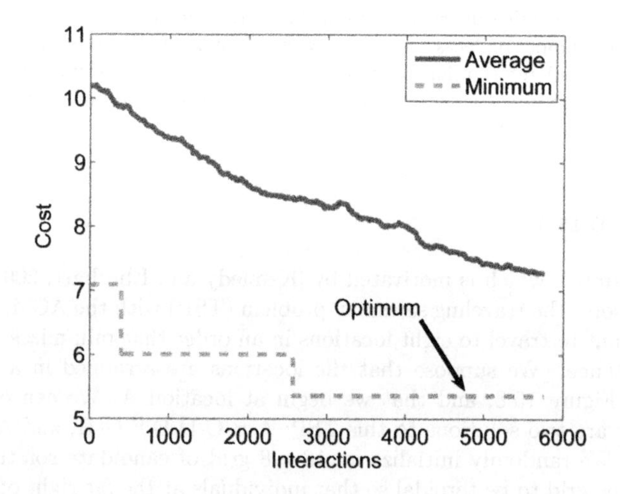

Figure 15.9 ACM convergence for the TSP of Example 15.3. If the eight cities of Figure 15.8 are arranged in a unit circle, the globally minimum cost is 5.3576.

As the individuals in the population continue to interact with each other, good individuals spread and poor individuals are gradually lost from the population. Figure 15.10 shows a typical progression of the spread of good individuals. The first optimal individual is found after about 2500 interactions, after which the prevalence of optimal solutions increases approximately linearly.

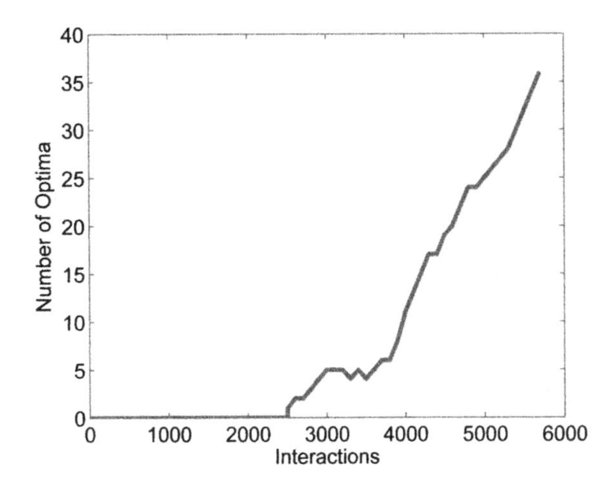

Figure 15.10 Example 15.3: The spread of optimal solutions in the population grid for the TSP. It takes about 2500 interactions for the first optimal solution to appear in the grid, after which the prevalance of optimal solutions increases approximately linearly.

Figure 15.11 shows the 18×8 population grid after 5760 interactions, which is an average of 80 interactions per individual. We see that 31 optimal solutions A-B-C-D-E-F-G-H are clustered together at the left and right edges of the grid (recall that the grid is a toroid, so the right and left edges are adjacent). We also see that there is a smaller cluster of five optimal solutions A-H-G-F-E-D-C-B near the bottom of the grid. A close look at Figure 15.11 reveals that there are also other clusters, which are suboptimal solutions to the TSP. This behavior is similar to how information and behavior spreads through a culture, how similar individuals tend to group together, and how we can simulate such cultural behavior to solve optimization problems.

```
ABHGFCDE ABHGFCDE ABHGFCDE ABHGFCDE ABHGFCDE ABHGFCDE ABHGFCDE ABHGFCDE
ABHGFCDE ABHGFCDE ABHGFCDE ABGHFCDE ABHGFCDE ABHGFCDE ABHGFCDE ABHGFCDE
ABHGFCDE ABHGFCDE ABHGFCDE ABHGFCDE ABHGFCDE ABHGFCDE ABHGFCDE ABHGFCDE
ABCGHFDE ABFGHCDE ABHGFCDE ABHGFCDE ABHGFCDE ABHGFCDE ABHGFCDE ABFDHCGE
ABCDEFGH ABCDHFGE ABHGFEDC ABHGFCDE ABHGFCDE ABHGFCDE ABCDEFGH ABCDEFGH
ABCDEFGH ABCDEFGH ABFGHEDC ABHGFCDE ABHGFCDE ABHDFCGE ABCDEFGH ABCDEFGH
ABCDEFGH ABCDEFGH ABCDEFGH ABFGHEDC ABHGFCDE ABCDEFGH ABCDEFGH ABCDEFGH
ABCDEFGH ABCDEFGH ABCDEFGH ABFDHEGC ABFEHGDC ABCDEFGH ABCDEFGH ABCDEFGH
ABCDEFGH ABCDEFGH ABCDEFGH AHDFCGBE AHDEFGBC ABCDEFGH ABCDEFGH ABCDEFGH
ABCDEFGH ABCDEFGH ABCDEFGH AHGFCBDE AHGFCBDE AHCDEFGB ABCDEFGH ABCDEFGH
AHFECBGD ABCDEFGH AHGFCBDE AHGFCBDE AHGFEBDC AHGFEBDC AHGFEBDC AHFECBGD
AHGFCBDE AHGFCBDE AHGFEBDC AHGFEBDC AHGFEBDC AHGFEBDC AHGFEBDC AHGFEBDC
ABCFEDGH ABEGFHDC ABDCHGEF AHGFEBDC AHGFEBDC AHGFEBDC ABHGFCDE ABCFEDGH
ABCFEDGH ABCFEDGH AHDFEGCB AHGFEBDC AHGFEBDC AHGFEBDC ABHGFCDE ABHGFCDE
ABCFEDGH ABCFEDGH AHGFEDCB AHGFEDCB AHGFECDB AHGFEBDC ABHGFCDE ABCFEDGH
ABHGFCDE ABCFEDGH AHGFEDCB AHGFEDCB AHGFEDCB ABHGFCDE AHGFEBDC AHGFEBDC
ABHGFCDE ABHGFCDE ABHGFCDE AHGFEDCB ABHGFCDE ABHGFCDE ABHGFCDE ABHGFCDE
ABHGFCDE ABHGFCDE ABHGFCDE ABHGFCDE ABHGFCDE ABHGFCDE ABHGFCDE ABHGFCDE
```

Figure 15.11 The population grid of Example 15.3 after 5760 interactions. Two optimal clusters (one with 31 individuals, and one with five individuals) are outlined in the grid. In the ACM, similar individuals group together, and highly fit solutions tend to spread throughout the population.

□

Example 15.3 shows how the ACM can find multiple solutions to a combinatorial optimization problem. The example can be extended to solve continuous optimization problems. We could make various generalizations to the ACM algorithms of Figures 15.6 and 15.7.

1. We allowed an individual to be influenced only by one of its four nearest neighbors. We could allow individuals to be influenced by more distant neighbors also. The probability or amount of interaction could be a decreasing function of distance.

2. We usually share information from more fit individuals to less fit individuals. This is in keeping with our desire to spread beneficial candidate solution features. However, we see in society that unsuccessful individuals can also exert an influence on others. We tend to avoid behaviors that we observe in unsuccessful individuals. We use this idea in the negative reinforcement particle swarm optimization in Section 11.6, but it may not yet have been explicitly used in CA, and so it is an open area for future research.

3. In Figure 15.6, we randomly select an individual x_i for an interaction. However, it might make more sense to select low-fitness individuals, since they are in more need of improvement. This idea is reminiscent of the immigration probabilities in BBO (see Chapter 14).

4. In Figure 15.6, we randomly select a neighbor x_k to interact with x_i. However, it might make more sense to randomly select a group of neighbors and decide

the information-sharing strategy based on relative fitness values. This idea is reminiscent of the emigration probabilities in BBO (see Chapter 14). This generalization, and the previous one, hint at interesting possibilities for a hybrid cultural BBO algorithm.

5. We could combine the idea of a belief space with the ACM. Individuals in human society are influenced by a combination of their neighbors and their culture. This idea is called the generalized other model (GOM), which can be loosely analogized with media influence [Shibani et al., 2001]. In Figure 15.6, we randomly select one of four neighbors to share information with x_i. In the GOM, we create a generalized neighbor that represents the consensus of the entire population. The generalized neighbor is a neighbor that does not actually exist in the population, but it is a pseudo-individual that is formed by taking the average of the entire population. The individual x_i could then receive information from either one of its four neighbors, or from the generalized neighbor. This idea is reminiscent of the fully informed particle swarm, which involves global information sharing (see Section 11.5). The generalized neighbor could also be obtained as a fitness-weighted average of the population, although this extension has apparently not yet been studied.

15.5 CONCLUSION

Cultural algorithms are a fascinating branch of evolutionary computation. They are different from typical EAs because they are not directly motivated by biology but are instead motivated by the social sciences. This motivation opens up a huge are of social science research that can be applied to self-organizing computational systems and optimization algorithms. Since cultural algorithms were first studied in the 1980s, it seems that most research in this area has been focused on simple applications or modifications of the basic CA ideas. However, there is a large body of research in the social sciences on many aspects of culture, including music, economics, language, nonveral communication, technology, family relationships, entertainment, education, sports, medicine, religion, art, literature, politics, war, and so on. Any engineering or computer science researcher who is interested in one of these aspects of culture has a virtually limitless reservoir of ideas to apply to CA research. Some other interesting and potentially important areas for future research include the following.

- Mathematical modeling of cultural algorithms seems to be a ripe area for future work. We see mathematical modeling work in the literature for other EAs (see Chapter 4 and Section 7.6), but there seems to be a dearth of mathematical modeling results for cultural algorithms.

- Cultures have multiple sets of beliefs, some of which are held by the majority of individuals, and others of which are held by the minority of individuals [Latané et al., 1994]. Sometimes these belief spaces conflict in what are called culture wars [Thomson, 2010]. We see this phenomenon in controversial areas such as religion, sports, and politics.

- Societies include multiple cultures. These cultures-within-cultures are called subcultures. Individuals within subcultures interact closely, and individuals from separate subcultures interact more loosely. Certain value systems are emphasized more in one subculture, and other value systems are emphasized more in other subcultures. This idea has applications to multi-objective optimization [Coello Coello and Becerra, 2003], [Alami et al., 2007].

How can these factors be modeled in a CA? How do these factors interact with each other? What other aspects of culture are important in human learning? How do cultural influences vary between individuals? These are all open research questions. Additional survey and tutorial reading in the area of CAs can be found in [Reynolds, 1994], [Reynolds and Chung, 1997], [Reynolds, 1999], and [Reynolds et al., 2011].

PROBLEMS

Written Exercises

15.1 Suggest a couple of different methods to predict the next number in the sequence of Equation (15.1). What values do your methods predict?

15.2 Consider Equation (15.3) and Figure 15.2.

a) If $x_i(k)$ is within the belief space, what is the probability that $x_i'(k)$ will be within the belief space?

b) If $x_i(k)$ is outside the belief space, what is the probability that $x_i'(k)$ will be within the belief space?

15.3 Consider Equation (15.3) and Figure 15.2.

a) What would be a more aggressive strategy of using the belief space when $x_i(k) \in B$? We use the term *more aggressive* here to indicate a higher probability that $x_i'(k) \in B$.

b) What would be a more aggressive strategy of using the belief space when $x_i(k) \notin B$?

15.4 How many fitness function evaluations are required each generation in the ACM algorithm of Figure 15.6? What does this imply for fair comparisons with other EAs?

15.5 The ACM algorithm of Figure 15.6 shares only one solution feature per interaction. What does this imply for its performance on non-separable problems? (Recall a similar discussion at the beginning of Section 14.5.) How could the ACM algorithm be modified to get better performance on non-separable problems?

15.6 Figure 15.9 shows that the ACM finds the optimal solution of the 8-city TSP in about 2500 iterations. Analyze the quality of that performance.

Computer Exercises

15.7 Repeat Example 15.2 with $\alpha = 0$, 0.25, 0.5, 0.75, and 1. Plot results similar to Figure 15.3 for each value of α. Comment on your results.

15.8 Repeat Example 15.2 with $M = 0$, 2, 5, 10, and 25. Plot results similar to Figure 15.3 for each value of M. Comment on your results.

15.9 Repeat Example 15.3 with selection pressure $p_1 = 0.5$, 0.7, 0.9, and 1. Limit the number of interactions to 2,000 for each simulation. Plot results similar to Figure 15.9 for each value of p_1; however, whereas Figure 15.9 shows the results of a typical ACM simulation, you should run 20 Monte Carlo simulations for each value of p_1, record the best cost at each generation for each value of p_1, and plot the average of the best cost as a function of generation number for each value of p_1. Comment on your results.

15.10 Repeat Example 15.3 with neighborhood sizes of 4 and 8. Limit the number of interactions to 2,000 for each simulation. Plot results similar to Figure 15.9 for each value of neighborhood size; however, whereas Figure 15.9 shows the results of a typical ACM simulation, you should run 20 Monte Carlo simulations for each neighborhood size, record the best cost at each generation for each neighborhood size, and plot the average of the best cost as a function of generation number for each neighborhood size. Comment on your results.

CHAPTER 16

Opposition-Based Learning

> Social revolutions are ... extremely fast changes in human society. They occur to establish, simply expressed, the *opposite* circumstances.
>
> —Hamid Tizhoosh [Tizhoosh, 2005]

Evolution is a slow process; change takes time. However, some types of change are rapid. One type of rapid change that almost all evolutionary algorithms (EAs) use is mutation. But there is also a type of rapid change that occurs in human society that we have not yet explored: social revolutions. A social revolution is a paradigm shift to the opposite of the currently accepted norm. Sometimes social revolutions have important and long-lasting effects, like when the United States fought against England in the revolutionary war and made the change from colonies to states. Other revolutions are less dramatic, like the introduction of synthetic materials in clothing, or the introduction of microwaves for cooking. However, all revolutions, by definition, result in significant lifestyle changes.

Opposition-based learning (OBL) was introduced as an attempt to increase the rate of learning in EAs. Since evolution is a slow process while revolution is a fast process, the simulation of revolution in EAs might speed up their convergence. OBL was originally introduced as an improvement to reinforcement learning, genetic algorithms, and neural network training [Tizhoosh, 2005]. It has also been implemented in many other optimization algorithms, including biogeography-based optimization (BBO) [Ergezer et al., 2009], particle swarm optimization [Omran, 2008], [Rashid and Baig, 2010], differential evolution [Rahnamayan et al., 2008], ant colony optimization [Malisia, 2008], and simulated annealing [Ventresca and Tizhoosh, 2007].

Overview of the Chapter

Section 16.1 presents some definitions of opposition as the term relates to numerical problems. Section 16.2 outlines how OBL can be incorporated into an EA, and particularly how it can be used to improve the performance of BBO. Section 16.3 mathematically studies the probability of EA improvement using various types of opposition. Section 16.4 introduces jumping ratio, which is a concept that is used in OBL. Although OBL was originally defined for continuous-domain problems, Section 16.5 discusses how it can be extended to combinatorial problems, and in particular to the traveling salesman problem. Section 16.6 reviews some concepts from dual learning, which preceded OBL, and shows the relationship between the two methods.

16.1 OPPOSITION DEFINITIONS AND CONCEPTS

This section discusses definitions and concepts related to the opposite of a scalar or vector. We begin by considering scalars. We begin by assuming that x is defined on the domain $[a, b]$, and the center of the domain is c:

$$\begin{aligned} x &\in [a, b] \text{ where } a < b \\ c &= (a + b)/2. \end{aligned} \tag{16.1}$$

16.1.1 Reflected Opposites and Modulo Opposites

We can think of several different ways to define the opposite of a scalar x [Tizhoosh et al., 2008]. For example, the reflected opposite of x is defined as

$$x_{o1} = a + b - x. \tag{16.2}$$

This means that x_{01} is the same distance as x from the center of the domain:

$$c - x = x_{o1} - c. \tag{16.3}$$

The modulo opposite of x is defined as

$$x_{o2} = (x - a + c) \bmod (b - a). \tag{16.4}$$

This views the domain $[a, b]$ as a circle, and defines the opposite of x as the number that lies on the opposite side of the circle. Figure 16.1 illustrates the reflected opposite and modulo opposite.

The reflected opposite and modulo opposite definitions can be extended to vectors in a simple but straightforward way. Suppose that x is an n-dimensional vector

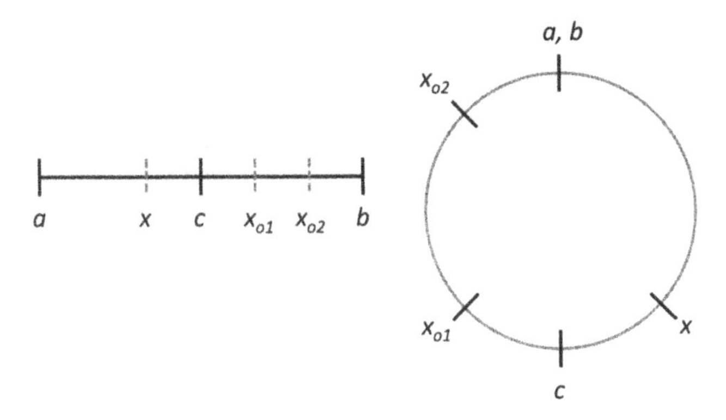

Figure 16.1 Illustration of the reflected opposite x_{o1} and the modulo opposite x_{o2} of a scalar x. The figure on the left illustrates the domain of x as linear segment, and the figure on the right illustrates it as circle. The scalar x is defined on the domain $[a, b]$, and c is the center of the domain. The reflected opposite x_{o1} is the same distance as x from c, and the modulo opposite x_{o2} is on the opposite side of the circle that defines the domain of x.

defined on a rectangular domain; that is, x_i is defined on the domain $[a_i, b_i]$, and the center of the domain of x_i is c_i:

$$\begin{aligned} x &= \begin{bmatrix} x_1 & \cdots & x_n \end{bmatrix} \\ \text{where } x_i &\in [a_i, b_i] \text{ and } a_i < b_i \text{ for } i \in [1, n] \\ c_i &= (a_i + b_i)/2 \text{ for } i \in [1, n]. \end{aligned} \tag{16.5}$$

The reflected opposite of x is defined as

$$\begin{aligned} x_{o1} &= \begin{bmatrix} x_{o1,1} & \cdots & x_{o1,n} \end{bmatrix} \\ \text{where } x_{o1,i} &= a_i + b_i - x_i \text{ for } i \in [1, n]. \end{aligned} \tag{16.6}$$

The modulo opposite of a vector x is defined as

$$\begin{aligned} x_{o2} &= \begin{bmatrix} x_{o2,1} & \cdots & x_{o2,n} \end{bmatrix} \\ \text{where } x_{o2,i} &= (x_i - a_i + c_i) \bmod (b_i - a_i). \end{aligned} \tag{16.7}$$

These definitions apply only to rectangular domains. The extension of these definitions to non-rectangular domains is left for future work but is probably not too difficult.

We will not use the modulo opposite any more in this chapter. In the remainder of this chapter we use the term "opposite of x" as shorthand for "reflected opposite of x," and we use the notation x_o as shorthand for x_{o1}.

16.1.2 Partial Opposites

Given a vector x, we can define x_p, a partial opposite of x, by taking the opposite of some of the dimensions of x while leaving other elements of x unchanged. For example:

$$\begin{aligned} x &= \begin{bmatrix} x_1 & \cdots & x_n \end{bmatrix} \\ \text{partial opposite } x_p &= \begin{bmatrix} x_{p1} & \cdots & x_{pn} \end{bmatrix} \\ \text{where } x_{pi} &= \begin{cases} x_{oi} & \text{for } i \in S \\ x_i & \text{for } i \in \bar{S} \end{cases} \end{aligned} \tag{16.8}$$

where S is some subset of $\{1, 2, \cdots, n\}$, and $\bar{S}$ is its complement; that is, $S \cup \bar{S} = \{1, 2, \cdots, n\}$, $S_j \notin \bar{S}$ for all $j \in \{1, \cdots, |S|\}$, and $\bar{S}_j \notin S$ for all $j \in \{1, \cdots, |\bar{S}|\}$. The *degree of opposition* of x_p is defined as

$$\tau(x_p) = |S|/n. \tag{16.9}$$

■ EXAMPLE 16.1

Suppose that $x = \begin{bmatrix} 0.5 & 0.5 \end{bmatrix}$, where both elements of x are defined on the domain $[0, 2]$. We can define four partial opposites of x:

$$\begin{aligned} x_p^{(1)} &= \begin{bmatrix} 0.5 & 0.5 \end{bmatrix} & \rightarrow & \quad \tau\left(x_p^{(1)}\right) = 0 \\ x_p^{(2)} &= \begin{bmatrix} 1.5 & 0.5 \end{bmatrix} & \rightarrow & \quad \tau\left(x_p^{(1)}\right) = 1/2 \\ x_p^{(3)} &= \begin{bmatrix} 0.5 & 1.5 \end{bmatrix} & \rightarrow & \quad \tau\left(x_p^{(1)}\right) = 1/2 \\ x_p^{(4)} &= \begin{bmatrix} 1.5 & 1.5 \end{bmatrix} & \rightarrow & \quad \tau\left(x_p^{(1)}\right) = 1. \end{aligned} \tag{16.10}$$

Figure 16.2 illustrates the partial opposites of x.

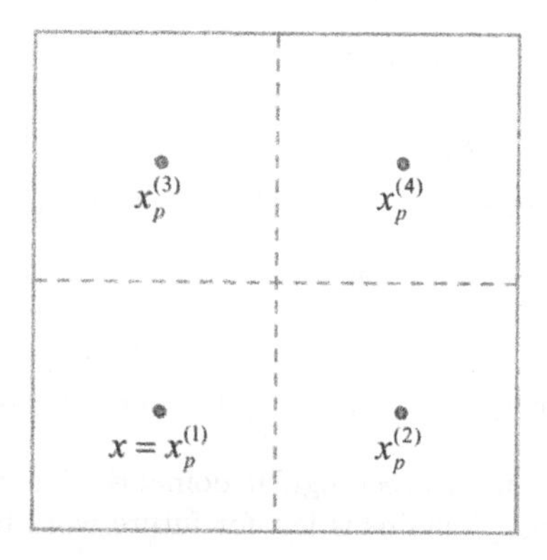

Figure 16.2 Example 16.1: Degree of opposition of partial opposites of the two-dimensional vector x. Vector $x_p^{(1)}$ is identical to x, so its degree of opposition is 0. Vectors $x_p^{(2)}$ and $x_p^{(3)}$ include one element that is opposite the corresponding element of x, and one element that is identical to the corresponding element of x, so their degree of opposition is 0.5. Each element of $x_p^{(4)}$ is opposite the corresponding element of x, so its degree of opposition is 1.

□

16.1.3 Type 1 Opposites and Type 2 Opposites

Up to this point, we have defined opposite in terms of the domain of a function; this is called type 1 opposition. We can also defined opposite in terms of the range of a function, and this is called type 2 opposition [Tizhoosh et al., 2008]. We begin with a scalar function $y(\cdot)$ of a scalar x, where x is defined on the domain $[a, b]$. The range $[y_{\min}, y_{\max}]$ is defined as

$$\begin{aligned} y_{\min} &= \min y(x) : x \in [a, b] \\ y_{\max} &= \max y(x) : x \in [a, b]. \end{aligned} \tag{16.11}$$

The center of the range is defined as

$$y_c = (y_{\max} - y_{\min})/2. \tag{16.12}$$

The type 2 reflected opposite of x is defined as

$$x_o^{(r)} = x' : y(x') = y_{\min} + y_{\max} - y(x). \tag{16.13}$$

This means that $y\left(x_o^{(r)}\right)$ is the same distance as $y(x)$ from y_c:

$$y_c - y\left(x_o^{(r)}\right) = y_c - y(x). \tag{16.14}$$

This definition can result in multiple values for $x_o^{(r)}$ unless $y(\cdot)$ is a one-to-one mapping. Figure 16.3 illustrates the difference between type 1 and type 2 opposites. Note that we can extend the definition of type 2 opposition to obtain the type 2 opposite of a vector, the type 2 modulo opposite of a vector, and the degree of type 2 opposition.

In the remainder of this chapter we restrict our discussion to type 1 opposition. Type 2 opposition deserves further study in the context of EAs, but we leave that to further research.

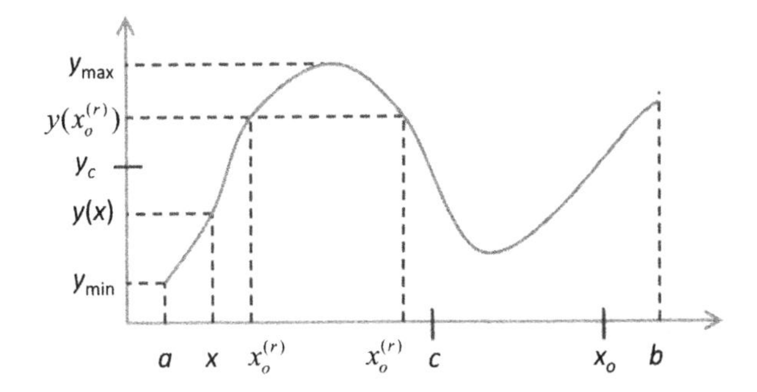

Figure 16.3 Consider the scalar x on the domain $[a, b]$, and the function $y(x)$. The type 1 opposite of x is x_o, and is obtained by reflecting x across the center of the domain c. The type 2 opposite of x is obtained by reflecting $y(x)$ across the center of the domain y_c to obtain $y(x_o^{(r)})$, and then computing the inverse of $y(x_o^{(r)})$ to obtain $x_o^{(r)}$. This results in two possible values of $x_o^{(r)}$ for this example.

16.1.4 Quasi Opposites and Super Opposites

Now we define three additional approaches to opposition. As before, we consider the scalar $x \in [a, b]$ with c as the center of its domain.

The quasi opposite of x is defined as follows [Tizhoosh et al., 2008]:

$$x_{qo} = \text{rand}(c, x_o) \tag{16.15}$$

where x_o is the standard reflected opposite defined in Equation (16.2). That is, x_{qo} is the realization of a random number that is uniformly distributed on $[c, x_o]$. Note that we define the *rand* function in such a way that its result is independent of the order of its arguments; that is, the notations $\text{rand}(c, x_o)$ and $\text{rand}(x_0, c)$ are equivalent.

The super opposite of x is defined as follows [Tizhoosh et al., 2008]:

$$x_{so} = \begin{cases} \text{rand}(x_o, b) & \text{if } x < c \\ \text{rand}(a, x_o) & \text{if } x > c. \end{cases} \tag{16.16}$$

That is, x_{so} is the realization of a random number that is uniformly distributed between x_o and the domain boundary that is farthest from x. This definition is not complete because it does not define x_{so} for the case $x = c$, but that special situation can be handled by arbitrarily changing one of the inequalities in Equation (16.16) so that it includes both equality and inequality.

The quasi reflected opposite of x is defined as follows [Ergezer et al., 2009]:

$$x_{qr} = \text{rand}(x, c). \tag{16.17}$$

That is, x_{qr} is the realization of a random number that is uniformly distributed between x and c. Note that the use of the word "reflected" in the term "quasi reflected" is not related to the word "reflected" in the term "reflected opposite" (see Equation (16.2)).

Figure 16.4 illustrates four different methods of opposition. We can extend these definitions to vectors, modulo opposites, and type 2 opposites, by following the procedures presented earlier in this section.

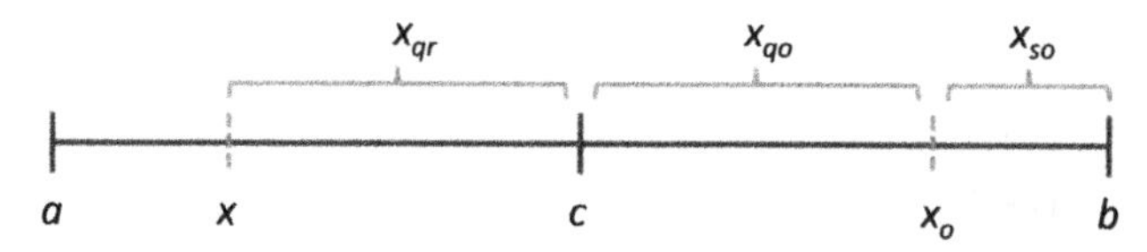

Figure 16.4 Suppose we have a scalar $x \in [a, b]$. The opposite of x is x_o, and is obtained by reflecting x across the center of the domain c. The quasi opposite of x is x_{qo}, and is obtained by generating a random number between c and x_o. The super opposite of x is x_{so}, and is obtained by generating a random number between x_o and the domain boundary that is farthest from x. The quasi reflected opposite of x is x_{qr}, and is obtained by generating a random number between x and c.

It is interesting to make connections between these opposition definitions and fuzzy logic. Figure 16.4 shows that the opposite x_o is crisp; given x, its opposite x_o is a crisp number or vector. However, x_{qr}, x_{qo}, and x_{so} could be defined as fuzzy quantities. Connections between these opposition definitions and fuzzy logic have not been presented in the literature, but the investigation of such connections seems to be a ripe area for further research.

16.2 OPPOSITION-BASED EVOLUTIONARY ALGORITHMS

This section presents a generic OBL algorithm and shows how it can augment an EA. One simple approach to using an OBL with any EA is to perform the following steps.

1. When the N individuals of the EA population are initialized, N opposite individuals are created, each opposite individual corresponding to one of the N original individuals. Given our $2N$ candidate solutions (N original individuals and N opposite individuals), we keep the best N as the starting population of the opposition-based EA. This general idea is discussed in Section 8.1.

2. We run a standard implementation of an EA. As we have seen earlier in this book, this involves a loop of cost function evaluations, recombinations, and mutations. By definition, the loop executes once per generation.

3. Once every few generations, we compute the opposite of each of the N individuals. Of these $2N$ candidate solutions (N standard EA individuals and N opposite individuals), we keep the best N for the next EA generation. At each generation, we perform this step with probability $J_r \in [0, 1]$, which is called the jumping rate.

We have to make some decisions in our opposition-based EA.

1. Which EA should we use? Answering this question also means that we must choose all of the tuning parameters of the EA.

2. What type of opposition should we use?

3. What value should we use for the jumping rate J_r?

The jumping rate is a tuning parameter. We don't have many guidelines for J_r, but we do not want to make it too high. The reason that we periodically create an opposite population is to explore uncharted areas of the search space. But we do not want to create an opposite population every generation because then we would just be repeatedly jumping back and forth in the search space, which would waste function evaluations. Results from opposition-based differential evolution indicate that $J_r \approx 0.3$ provides a good balance [Rahnamayan et al., 2008].

Note that a non-opposition-based EA that runs for G generations with N individuals requires a total of GN function evaluations. An opposition-based EA that runs for G' generations with N' individuals and a jumping rate J_r requires a total of $G'N'(1 + J_r)$ function evaluations, on average. To make a fair comparison between a non-opposition-based EA and its opposition-based version, we need to choose G', N', and J_r so that

$$GN = G'N'(1 + J_r). \tag{16.18}$$

We can do this by either setting $N' = N$ and reducing the opposition-based generation limit so that $G' = G/(1 + J_r)$, or by setting $G' = G$ and reducing the opposition-based population size so that $N' = N/(1 + J_r)$, or by reducing both G' and N' simultaneously to satisfy Equation (16.18).

Oppositional Biogeography-Based Optimization

Now we show how the OBL outline presented above can be used in biogeography-based optimization (BBO). We combine the standard BBO algorithm of Figure 14.3 with OBL to obtain oppositional BBO (OBBO) [Ergezer et al., 2009]. Figure 16.5 shows an outline of the OBBO algorithm. Note that the algorithm of Figure 16.5 is identical to that of Figure 14.3 except for the pseudo-code between the lines "Comment: Begin Opposition Logic" and "Comment: End Opposition Logic."

Initialize a population of candidate solutions $\{x_k\}$ for $k \in [1, N]$
While not(termination criterion)
    For each x_k, set emigration probability $\mu_k \propto$ fitness of x_k, with $\mu_k \in [0, 1]$
    For each individual x_k, set immigration probability $\lambda_k = 1 - \mu_k$
    $\{z_k\} \leftarrow \{x_k\}$
    For each individual z_k
        For each solution feature s
            Use λ_k to probabilistically decide whether to immigrate to z_k
            If immigrating then
                Use $\{\mu_i\}_{i=1}^N$ to probabilistically select emigrating individual x_j
                $z_k(s) \leftarrow x_j(s)$
            End if
        Next solution feature
        Probabilistically mutate $\{z_k\}$
    Next individual
    Comment: Begin Opposition Logic
    $r \leftarrow U[0, 1]$
    If $r < J_r$ then
        Use $\{z_k\}$ to create opposite population $\{\bar{z}_k\}$
        $\{z_k\} \leftarrow$ best N individuals from $\{z_k\} \cup \{\bar{z}_k\}$
    End if
    Comment: End Opposition Logic
    $\{x_k\} \leftarrow \{z_k\}$
Next generation

Figure 16.5 The oppositional biogeography-based optimization (OBBO) algorithm with a population size of N. $\{x_k\}$ is the entire population of individuals, x_k is the k-th individual, and $x_k(s)$ is the s-th feature of x_k. Similarly, $\{z_k\}$ is the temporary population of individuals, z_k is the k-th temporary individual, and $z_k(s)$ is the s-th feature of z_k.

■ EXAMPLE 16.2

In this example we optimize the 20-dimensional Griewank function. This function is defined in Appendix C.1.6, and is also listed here for convenience:

$$f(x) = 1 + \sum_{i=1}^{n} x_i^2/4000 - \prod_{i=1}^{n} \cos\left(x_i/\sqrt{i}\right) \tag{16.19}$$

where $x_i \in [-600, +600]$. The minimizing value of x is $x_i = 0$ for all $i \in [1, n]$. We use BBO with a population size $N = 50$ and a function evaluation limit of 2500. This results in 50 generations if we evaluate each BBO individual once per generation. We use a mutation probability of 1% per dimension per individual, and we use an elitism parameter of 2. We also add OBL to the BBO algorithm as shown in Figure 16.5. When implementing OBL we use a jumping rate $J_r = 0.2$, so the generation count decreases to about 41 or 42, depending on the random number sequence that controls the generation of the opposite population. After 20 Monte Carlo simulations, the average of the lowest costs found by BBO and OBBO are as follows:

$$\begin{aligned} \text{BBO} &: \quad 8.85 \\ \text{Reflected OBBO} &: \quad 9.69 \\ \text{Quasi OBBO} &: \quad 0.05 \\ \text{Super OBBO} &: \quad 11.82 \\ \text{Quasi Reflected OBBO} &: \quad 0.03 \end{aligned}$$

The meaning of the terms in the above list can be seen in Figure 16.4. Reflected OBBO refers to x_o, quasi OBBO refers to x_{qo}, super OBBO refers to x_{so}, and quasi reflected OBBO refers to x_{qr}. We see that reflected OBBO and super OBBO perform worse than BBO. However, quasi OBBO and quasi reflected OBBO perform amazingly better than BBO.

□

As we know from the no-free-lunch theorem (see Appendix B), the astounding performance of quasi OBBO and quasi reflected OBBO in Example 16.2 is not magic. The reason for their superior performance is that the solution of the Griewank problem lies at the exact center of its domain. Figure 16.4 shows us that quasi OBBO and quasi reflected OBBO both tend to move individuals closer to the center of the search domain. Reflected OBBO maintains individuals the same distance from the center (but on the opposite side of the search domain), which is why reflected OBBO performs worse than BBO. Reflected OBBO neither degrades nor improves an individual in the Griewank problem; it merely consumes function evaluations. Super OBBO does even worse. Figure 16.4 shows us that super OBBO always moves individuals farther from the center of the search domain, which degrades performance in the Griewank problem. So the results of Example 16.2 are exactly what we would have predicted from our understanding of OBL and the Griewank problem.

Of course, if we knew that the solution was near the center of the search domain, we would not need to use OBL; we could use any other method to bias BBO individuals toward the center of the domain. In this sense, the use of OBL for the Griewank problem is "cheating"; it relies on the fact that the Griewank problem solution is near the center of the domain. That is, it implicitly relies on problem-specific information. This is closely related to the no-free-lunch theorem that is discussed in Appendix B. A problem whose solution can lie *anywhere* in the search domain might provide a better test for OBL, and this leads us to the following example.

■ **EXAMPLE 16.3**

In this example we again optimize the Griewank function with $n = 20$ dimensions (see Appendix C.1.6). We use the same parameters as those that are used in Example 16.2. However, this time we randomly shift the solution of the Griewank problem:

$$f(x) = 1 + \sum_{i=1}^{n} (x_i - r_i)^2/4000 - \prod_{i=1}^{n} \cos\left((x_i - r_i)/\sqrt{i}\right) \tag{16.20}$$

where r_i is a random number uniformly distributed in the search domain $[-600, +600]$. The minimizing argument of $f(x)$ is $x_i^* = r_i$ for $i \in [1, n]$. After 20 Monte Carlo simulations, where we use a different set of $\{r_i\}$ values for each Monte Carlo sample, the average of the lowest cost found by BBO and OBBO are as follows:

BBO	:	10.4
Reflected OBBO	:	14.1
Quasi OBBO	:	13.8
Super OBBO	:	13.4
Quasi Reflected OBBO	:	13.9

All of the opposition-based BBO algorithms perform significantly worse than standard BBO. This is because the shifted Griewank solution is uniformly distributed in the search space, so an opposite point is no more likely than a BBO individual to be close to the optimal solution. In fact, as the BBO generation count increases, the opposite point is *less* likely to be close to the optimal solution. This is because as the generation count increases, BBO individuals move closer to the optimal solution by virtue of their information-sharing mechanism. Therefore, the opposition function is likely to move them farther away from the solution. The use of opposition in this case not only wastes function evaluations, but seems to do so in a counterproductive way.

□

Example 16.3 seems to show that after a more careful consideration, the initially exhilarating results of Example 16.2 turn out to be a mirage. However, all is not lost. When we try to solve a real-world optimization problem with an EA, we need to define the search domain. We typically define it so that we are reasonably sure that the solution lies within the search domain. That means that we often make the search domain larger than necessary. We want a large search domain because we are not sure where the solution lies. We tend to err more on the side of a larger-than-required search domain than on the side of a too-small search domain. But we probably suspect that the solution lies near the center of the domain. Therefore, a situation that is more realistic than either Example 16.2 or Example 16.3 might be to randomly shift the Griewank solution in a way that allows it to reach either extreme of the domain, but that tends to keep it near the center, and this leads us to the next example.

EXAMPLE 16.4

In this example we once again optimize the Griewank function with $n = 20$ dimensions. We use the same parameters as those that are used in Examples 16.2 and 16.3. However, this time we randomly shift the solution of the Griewank problem in such a way that the solution is the realization of a normally-distributed vector, each of whose elements have a standard deviation of 200:

$$\begin{aligned} r_i &\leftarrow 200N(0,1) \qquad \text{for } i \in [1,n] \\ r_i &\leftarrow \max(\min(r_i, 600), -600) \\ f(x) &= 1 + \sum_{i=1}^{n}(x_i - r_i)^2/4000 - \prod_{i=1}^{n}\cos\left((x_i - r_i)/\sqrt{i}\right). \end{aligned} \tag{16.21}$$

$N(0,1)$ is a normally distributed random number with zero mean and unity variance, which means that $200N(0,1)$ has a standard deviation of 200. The max/min operation in Equation (16.21) ensures that each element of the solution of the shifted Griewank function remains in the search domain $[-600, 600]$. After 20 Monte Carlo simulations, where we use a different set of $\{r_i\}$ values for each Monte Carlo sample, the average of the lowest cost found by BBO and OBBO are as follows:

BBO	:	9.5
Reflected OBBO	:	11.2
Quasi OBBO	:	9.9
Super OBBO	:	11.9
Quasi Reflected OBBO	:	6.0

The performance of BBO and quasi OBBO are statistically identical, while the performance of reflected OBBO and super OBBO are worse than standard BBO. However, quasi reflected OBBO performs noticeably better than standard BBO. This is because quasi reflected OBBO tends to move BBO individuals toward the center of the domain. It might be expected that quasi OBBO should also perform better than BBO, because quasi OBBO also moves individuals toward the center of the domain. However, quasi OBBO moves individuals toward the center of the domain while also moving them far away from the current individual x (see Figure 16.4). This tends to degrade performance in later generations when most of the individuals have a low cost. Quasi reflected OBBO performs better because it not only moves individuals toward the center but it also tends to keep individuals near their original location in the search space, which is beneficial after the first few generations.

□

16.3 OPPOSITION PROBABILITIES

Section 16.2 showed how OBL can be incorporated into BBO to improve its performance. This section studies the probability of getting closer to an optimization problem solution when we use various opposition types: reflected opposition, quasi opposition, and quasi reflected opposition. This section is highly mathematical, so the practice-oriented reader can safely skip this section or simply read the results at the end of this section in Table 16.1.

We make the following assumptions in this section.

1. We assume that the search space is one-dimensional. This is obviously very restrictive, but it is a starting point, and additional work should allow the extension of the one-dimensional case to higher dimensions.

2. We assume that the solution x^* of the optimization problem is not known, but that it is the realization of a random number that is uniformly distributed in the domain of x. This assumption is based on the principle of insufficient reason, which asserts that in the absence of prior knowledge we must assume that all events in the search space have equal probabilities [Dembski and Marks, 2009b], [Dembski and Marks, 2009a].

Suppose that we have an arbitrary EA individual x. We assume without loss of generality that x is in the lower half of the search domain. Let us consider the probability that its quasi opposite x_{qo} is closer than its opposite x_o to the optimal solution x^*. Figure 16.6 illustrates an arbitrary EA individual x, its opposite x_o, and its quasi opposite x_{qo}. The optimal solution x^* could be in one of the following three regions.

1. We define case 1 as the situation in which $x^* \in [a, c]$.

2. We define case 2 as the situation in which $x^* \in [c, x_o]$.

3. We define case 3 as the situation in which $x^* \in [x_o, b]$.

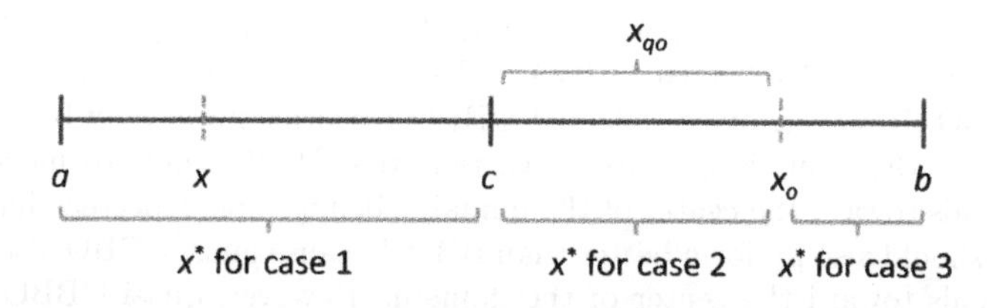

Figure 16.6 x is an EA individual, x_o is its opposite, and x_{qo} is its quasi opposite (taken from a uniform distribution between c and x_o). The solution x^* to an optimization problem is uniformly distributed on $[a, b]$, and so it could lie in one of three regions shown above.

Case 1

For case 1, it is clear that x_{qo} is closer than x_o to x^*. Therefore,

$$\Pr(|x_{qo} - x^*| < |x_o - x^*|) = 1 \text{ for case 1.} \tag{16.22}$$

Case 2

For case 2, x^* and x_{qo} are independent and uniformly distributed in $[c, x_o]$. We can use the total probability theorem [Mitzenmacher and Upfal, 2005], and the fact that $x_o - x^* > 0$ for case 2, to write the probability that x_{qo} is closer than x_o to x^* as follows:

$$\begin{aligned}
&\Pr(|x_{qo} - x^*| < |x_o - x^*|) \\
&= \Pr(|x_{qo} - x^*| < x_0 - x^* \,|\, x_{qo} - x^* < 0)\Pr(x_{qo} - x^* < 0) + \\
&\quad \Pr(|x_{qo} - x^*| < x_0 - x^* \,|\, x_{qo} - x^* > 0)\Pr(x_{qo} - x^* > 0) \\
&= \Pr(x_{qo} > 2x^* - x_0 \,|\, x_{qo} < x^*)\Pr(x_{qo} < x^*) + \\
&\quad \Pr(x_{qo} < x_0 \,|\, x_{qo} > x^*)\Pr(x_{qo} > x^*).
\end{aligned} \tag{16.23}$$

Consider the terms on the right side of the above equation. First, since x_{qo} and x^* are both uniformly distributed on $[c, x_o]$, we see that

$$\begin{aligned}
\Pr(x_{qo} < x^*) &= 1/2 \\
\Pr(x_{qo} > x^*) &= 1/2 \\
\Pr(x_{qo} < x_0 \,|\, x_{qo} > x^*) &= 1.
\end{aligned} \tag{16.24}$$

We can use Bayes' theorem to write the first expression on the right side of Equation (16.23) as

$$\begin{aligned}
&\Pr(x_{qo} > 2x^* - x_0 \,|\, x_{qo} < x^*)\Pr(x_{qo} < x^*) \\
&= \Pr(x_{qo} > 2x^* - x_0, x_{qo} < x^*) \\
&= \Pr(2x^* - x_o < x_{qo} < x^*) \\
&= \int_c^{x_o} \int_{x_{qo}}^{(x_{qo}+x_0)/2} f(x^*) f(x_{qo}) dx^* \, dx_{qo}
\end{aligned} \tag{16.25}$$

where we have assumed that x^* and x_{qo} are independent with PDFs $f(x^*)$ and $f(x_{qo})$. Assuming uniform PDFs, the above integration can be performed as

$$\begin{aligned}
&\Pr(x_{qo} > 2x^* - x_0 \,|\, x_{qo} < x^*)\Pr(x_{qo} < x^*) \\
&= \int_c^{x_o} \int_{x_{qo}}^{(x_{qo}+x_0)/2} \frac{1}{(x_o - c)^2} dx^* \, dx_{qo} \\
&= \int_c^{x_o} \frac{x_o - x_{qo}}{2(x_o - c)^2} dx_{qo} \\
&= 1/4.
\end{aligned} \tag{16.26}$$

Substituting Equations (16.24) and (16.26) into Equation (16.23) gives

$$\Pr(|x_{qo} - x^*| < |x_o - x^*|) = 3/4 \text{ for case 2.} \tag{16.27}$$

Case 3

For case 3, it is clear from Figure 16.6 that x_o is closer than x_{qo} to x^*. Therefore,

$$\Pr(|x_{qo} - x^*| < |x_o - x^*|) = 0 \text{ for case 3.} \tag{16.28}$$

Final Results

Let us use $\mathcal{E}$ as shorthand notation to indicate the event that x_{qo} is closer than x_o to the optimal solution x^*:

$$\mathcal{E} = \{|x_{qo} - x^*| < |x_o - x^*|\}. \tag{16.29}$$

Then we can combine the results from cases 1, 2, and 3 to obtain

$$\begin{aligned} \Pr(\mathcal{E}) &= \Pr(\mathcal{E} \,|\, x^* \in [a,c])\Pr(x^* \in [a,c]) + \\ &\quad \Pr(\mathcal{E} \,|\, x^* \in [c,x_o])\Pr(x^* \in [c,x_o]) + \\ &\quad \Pr(\mathcal{E} \,|\, x^* \in [x_o,b])\Pr(x^* \in [x_o,b]) \\ &= (1)\left(\frac{1}{2}\right) + \left(\frac{3}{4}\right)\left(\frac{x_o - c}{b-a}\right) + 0. \end{aligned} \tag{16.30}$$

If x is uniformly distributed in the lower half of the search domain, then x_o is uniformly distributed in the upper half of the search domain. Therefore, the expected value of x_o is

$$E(x_o) = (c+b)/2. \tag{16.31}$$

Taking the expected value of Equation (16.30) then gives

$$\begin{aligned} E\left[\Pr(|x_{qo} - x^*| < |x_o - x^*|)\right] &= \frac{1}{2} + \frac{3}{4}\frac{(b-c)/2}{b-a} \\ &= 1/2 + 3/16 = 11/16. \end{aligned} \tag{16.32}$$

The above derivation assumes that $x \in [a,c]$, but this does not affect the generality of the results; the same result holds if $x \in [c,b]$. We have thus obtained the following theorem.

Theorem 16.1 *Assume that an EA individual x and the solution x^* of a one-dimensional optimization problem are independent and uniformly distributed in the search space. Then the average probability that the quasi opposite of x is closer than the opposite of x to the solution x^* is 11/16.*

These results were first presented in [Ergezer et al., 2009], [Ergezer, 2011]. Some additional results are also available in those papers, and are summarized in Table 16.1. The first row of Table 16.1 shows that an EA individual and its opposite both are equally likely to be closer to the optimal solution. This is as expected from the symmetry between x and x_o.

Although Table 16.1 is restricted to one-dimensional problems, the extension of the approach in this section to higher dimensions should be conceptually straightforward and is left for further research. Some experimental results from higher dimensions are shown in [Ergezer et al., 2009], where it seems that the probabilities increase to an asymptote as the number of dimensions increases. Also see Problem 16.12.

Note that we arbitrarily defined x^* in this section as the solution to an optimization problem. We could just as well have defined it as the worst individual in the search space. The key to OBL is that after the opposite population is generated, the best N individuals from the original N individuals and the opposite N individuals are retained for the next generation. The reason that OBL works is that quasi

Event	Probability
$\lvert x_o - x^* \rvert < \lvert x - x^* \rvert$	1/2
$\lvert x_{qo} - x^* \rvert < \lvert x - x^* \rvert$	9/16
$\lvert x_{qr} - x^* \rvert < \lvert x - x^* \rvert$	11/16
$\lvert x_{qo} - x^* \rvert < \lvert x_o - x^* \rvert$	11/16
$\lvert x_{qr} - x^* \rvert < \lvert x_o - x^* \rvert$	9/16
$\lvert x_{qo} - x^* \rvert < \lvert x_{qr} - x^* \rvert$	1/2

Table 16.1 One-dimensional probabilities of certain oppositional points being closer to the optimal solution than other points.

opposite and quasi reflected points have a high probability of being closer than an arbitrary EA individual x to an arbitrary point in the search space.

Our derivation assumes that the EA individual x is uniformly distributed in the search space. We expect that as an EA progresses to later generations, most of the individuals will get closer to the optimal solution, which means that x will no longer be uniformly distributed. This seems to indicate that OBL should be more effective early in the search process. When implementing OBL, we may want to use a higher jumping rate early in the search process than later. This is the same type of logic that we often use in simulated annealing (see Chapter 9). We also often use similar reasoning in EA mutation; we use high rates of mutation early in the search process, and lower rates later [Haupt and Haupt, 2004, Section 5.9].

16.4 JUMPING RATIO

This section introduces the concept of a jumping ratio, which is a simple extension that can improve the performance of OBL. This idea is motivated by the realization that OBL requires computational resources. Every opposite individual that we generate requires an extra fitness evaluation, and fitness evaluations can be very computationally expensive in real-world problems (see Chapter 21). We do not want to arbitrarily generate opposite solutions during our EA implementation. We would prefer to generate opposite solutions only if we are reasonably confident that the extra computational effort will pay for itself in improved performance.

Note that the opposite of a highly-fit EA individual is less likely to be fit than the opposite of a low-fitness EA individual. That is, if an EA individual is close to the optimal solution, then it is not worth generating its opposite. Conversely, if an EA individual is far from the optimal solution, then it probably is worth generating its opposite. Of course, we do not know if a given individual is near to, or far from, the optimal solution. But we do know the relative fitness values of each individual in our EA population. Perhaps OBL should be implemented so that the probability of generating an opposite individual is a function of the fitness of that individual. The OBBO logic in Figure 16.5 could be replaced with something like that shown in Figure 16.7.

The parameter $\alpha \geq 0$ in Figure 16.7 controls the pressure to generate opposite individuals. Recalling that μ_k is proportional to the fitness of z_k, we see that fitness-based opposition logic makes it more likely that a low-fitness individual will have an opposite individual generated. A small value of α will result in the creation of a lot of opposite individuals. In the limit as $\alpha \to 0$, the fitness-based opposition logic is equivalent to the standard opposition logic in Figure 16.5 and we will generate an opposite for every EA individual. As α becomes larger, fewer opposite individuals will be created. As $\alpha \to \infty$, no opposite individuals will be created, and the OBBO algorithm will reduce to standard BBO. Creating opposite individuals is a risk; it requires extra computational effort because the fitness of each opposite individual needs to be evaluated. Is the potential payoff of new, highly fit opposite individuals worth the extra fitness evaluations? The parameter α provides the balance.

α = opposition pressure $\in [0, 1]$
$r_1 \leftarrow U[0, 1]$
If $r_1 < J_r$ then
    $m = 0$
    For each individual z_k
        $r_2 \leftarrow U[0, 1]$
        If $r_2 > \alpha\mu_k$ then
            $m \leftarrow m + 1$
            $\bar{z}_m \leftarrow$ opposite of z_k
        End if
    Next individual
    $\{z_k\} \leftarrow$ best N individuals from $\{z_k\} \cup \{\bar{z}_m\}$
End if

Figure 16.7 Fitness-based opposition logic. This logic can replace the standard OBBO logic in Figure 16.5.

Another way of implementing fitness-based opposition logic is to generate opposite individuals only for the least-fit proportion ρ of the individuals in the population. This is very similar to the idea outlined above, but is more deterministic, and can be implemented as shown in Figure 16.8, where $\rho \in [0, 1]$.

The ideas presented in this section are an attempt to make OBL more intelligent, more adaptive, and more effective. Creative researchers can develop other ideas along these lines to improve OBL. We might also be able to use ideas from guided mutation in general EA research to improve OBL [Zhang et al., 2005]. We demonstrate the jumping ratio logic outlined above in an example in the next section.

ρ = jumping ratio $\in [0, 1]$
$r \leftarrow U[0, 1]$
If $r < J_r$ then
    $m = 0$
    For each individual z_k
        If z_k is in the least-fit ρ proportion of the population then
            $m \leftarrow m + 1$
            $\bar{z}_m \leftarrow$ opposite of z_k
        End if
    Next individual
    $\{z_k\} \leftarrow$ best N individuals from $\{z_k\} \cup \{\bar{z}_m\}$
End if

Figure 16.8 Fitness-based proportional opposition logic. This logic can replace the standard OBBO logic in Figure 16.5. If $\rho = 1$ then this logic reduces to the standard opposition logic of Figure 16.5.

16.5 OPPOSITIONAL COMBINATORIAL OPTIMIZATION

This section extends OBL to combinatorial optimization problems. We clearly need to rethink the definitions of opposites in Section 16.1 if we want to extend OBL to combinatorial problems. Initial work in this area was presented in [Ergezer and Simon, 2011].

A combinatorial problem is one for which we want to find the best way to order a set of nodes. The traveling salesman problem (TSP) is a good example of a combinatorial problem (see Section 2.5 and Chapter 18). A TSP can either be a closed-path problem or an open-path problem. A closed-path problem is one for which the solution makes a close path; that is, the route begins and ends at the same city. An open-path problem is one for which the solution visits each city exactly one time, so the beginning and ending cities are different. We will consider open-path problems in this section.

Before we try to define the opposite of an individual in a combinatorial EA, we use a simple example to introduce some definitions. Suppose that we are trying to solve a four-city TSP with an EA. The cities are labeled A, B, C, and D. One of the candidate solutions is

$$A \rightarrow B \rightarrow C \rightarrow D. \tag{16.33}$$

1. We define one *leg* of a trip as the travel between two adjacent cities. We see that Equation (16.33) is comprised of three legs: $A \rightarrow B$, $B \rightarrow C$, and $C \rightarrow D$.

2. We define the *proximity* between two cities as the number of legs that it takes to get from one city to the other. In Equation (16.33), A and B have a proximity of one, A and C have a proximity of two, and A and D have a proximity of three.

3. We define the *total proximity* of a route as the sum of the proximities between each pair of adjacent cities. In Equation (16.33), the total proximity is three because $A \to B$, $B \to C$, and $C \to D$ each have a proximity of one. The total proximity of a route is always equal to $N - 1$, where N is the number of cities.

4. We define the *relative proximity* of route β as the sum of the proximities between each pair of adjacent cities in β, where the proximities are obtained from some other route α. For example, suppose that we have the following routes:

$$\begin{aligned} \alpha &: D \to C \to A \to B \\ \beta &: B \to D \to A \to C. \end{aligned} \tag{16.34}$$

   The proximity of β relative to α is six. This is because β consists of three legs: the first leg is $B \to D$, two cities that have a proximity of three in α; the second leg is $D \to A$, two cities that have a proximity of two in α; and the third leg is $A \to C$, two cities that have a proximity of one in α.

One way to define the opposite of a route α is to find a route β whose relative proximity is as large as possible. This is intuitive because the relative proximity of α relative to α is $N-1$, which is the minimum possible value. Using this definition, the opposite of the route of Equation (16.33) is

$$C \to A \to D \to B. \tag{16.35}$$

This route has a proximity of 7 relative to Equation (16.33), which is the maximum possible value.

However, the problem of finding a route to maximize relative proximity is itself a combinatorial optimization problem. That means that if we want to solve a problem like the TSP using OBL, we have to solve a combinatorial problem that consists of multiple combinatorial problems at each generation. This could quickly become computationally infeasible. Therefore, we define a greedy opposite of a combinatorial individual. The greedy opposite keeps the initial city unchanged, and then inserts the city with the greatest relative proximity as the second city. We set the new third city equal to the city with the greatest relative proximity from the new second city. We iterate this process to complete the greedy opposite route. This process is outlined in Figure 16.9.

Figure 16.9 gives the following greedy opposite of the route of Equation (16.33):

$$A \to D \to B \to C. \tag{16.36}$$

This route has a proximity of six relative to Equation (16.33), which is one less than the opposite route of Equation (16.35). Figure 16.10 shows another, simpler way of implementing the greedy opposite. These algorithms do not give the exact opposite of a TSP candidate solution, but hopefully they give a near-opposite with a reasonably low computational cost.

$\alpha = \{\alpha_1, \alpha_2, \cdots, \alpha_N\}$ = candidate solution
$p(\alpha_i, \alpha_j) = |i - j|$ = proximity between nodes α_i and α_j
$\beta_1 \leftarrow \alpha_1$
$\beta \leftarrow \{\beta_1\}$
For $k = 2$ to N
    $\beta_k \leftarrow \arg\max_a p(\alpha_{k-1}, a) : a \notin \beta$
    $\beta \leftarrow \beta \cup \beta_k$
Next k

Figure 16.9 The above pseudo-code outlines an algorithm to find the greedy opposite β of a candidate solution α to a combinatorial optimization problem, where N is the number of nodes in each candidate solution.

$\alpha = \{\alpha_1, \alpha_2, \cdots, \alpha_N\}$ = candidate solution
For $k = 1$ to N
    If k is odd then
        $m \leftarrow (k + 1)/2$
    else
        $m \leftarrow N + 1 - k/2$
    End if
    $\beta_k \leftarrow \alpha_m$
Next k

Figure 16.10 The above pseudo-code outlines a simple algorithm to find the greedy opposite β of a candidate solution α to a combinatorial optimization problem, where N is the number of nodes in each candidate solution. This algorithm is equivalent to Figure 16.9 but is simpler.

■ EXAMPLE 16.5

In this example we investigate the use of OBL for the TSP. We use inver-over crossover (see Section 18.3.1.5), and we use BBO to choose immigrating and emigrating population members. We use the Ulysses16 TSP benchmark, which consists of 16 cities (see Section C.6), and we use 10,000 function evaluations. Recall that a 16-city TSP has $16!/2 \approx 10^{13}$ possible solutions. Table 16.2 shows the average and standard deviation of the shortest route found by various BBO/OBL combinations after 40 Monte Carlo simulations. The results show that performance improves as jumping rate J_r and jumping ratio ρ increase. (See Figure 16.8 for the definition of jumping ratio ρ.) If J_r and ρ increase too much, then performance degrades, although we have not shown those results here.

□

	$\rho = 0.1$	$\rho = 0.2$	$\rho = 0.3$	$\rho = 0.4$
$J_r = 0.0$	7266 ± 353	7266 ± 353	7266 ± 353	7266 ± 353
$J_r = 0.1$	7153 ± 289	7284 ± 244	7122 ± 296	7127 ± 270
$J_r = 0.2$	7160 ± 297	7100 ± 324	7047 ± 251	6910 ± 315
$J_r = 0.3$	7180 ± 267	6976 ± 336	6945 ± 270	6869 ± 319
$J_r = 0.4$	7127 ± 201	7005 ± 326	6910 ± 265	6776 ± 207

Table 16.2 Example 16.5: Oppositional biogeography-based optimization results for the solution of the Ulysses16 TSP. The results show the average and standard deviation of the best solution found over 40 Monte Carlo simulations. $J_r = 0$ corresponds to standard BBO without any OBL. In general, performance improves as jumping rate J_r and jumping ratio ρ increase.

16.6 DUAL LEARNING

Opposition-based learning is similar to dual learning, which was first proposed in in the 1990s [Collard and Aurand, 1994], [Collard and Gaspar, 1996], and rediscovered in the early 2000s [Yang, 2003a], [Yang, 2003b]. Later, [Yang and Yao, 2005] suggested taking the dual of only the worst individuals in the population. Incorporating the ideas of dual learning in the OBBO algorithm of Figure 16.5 gives the duality logic of Figure 16.11. Note that Figure 16.5 can replace the "Opposition Logic" section of Figure 16.5.

$\{w_k\} \leftarrow \{N_d$ worst individuals in the population$\}$
Use $\{w_k\}$ to create a population of N_d opposites $\{\bar{w}_k\}$
For $i = 1$ to N_d
    If $\bar{w}_k$ is better than w_k, then replace w_k with $\bar{w}_k$ in the population
Next i

Figure 16.11 The above pseudo-code outlines duality logic. This logic can replace the "Opposition Logic" block of Figure 16.5. N_d is the number of duals that we create each generation, and can be adapted as described in the text.

The number of duals N_d that we create each generation can be adapted for optimal EA performance. [Yang and Yao, 2005] suggests the following adaptation scheme, which we execute each generation, and which we could add to the end of the duality logic of Figure 16.11:

$$\begin{aligned} N_v &\leftarrow |\bar{w}_k : f(\bar{w}_k) > f(w_k)| \\ s &\leftarrow (\delta N_d - N_v)/N \\ N_d &\leftarrow \beta^s N_d \\ N_d &\leftarrow \max(N_d, N_{d,\min}) \\ N_d &\leftarrow \min(N_d, N_{d,\max}). \end{aligned} \tag{16.37}$$

In Equation (16.37), $f(\cdot)$ is the fitness function, so a larger value of $f(\cdot)$ indicates a better-performing individual. N_v is the number of "valid" duals from the previous generation, which is the number of duals $\bar{w}_k$ that were better than the individuals from which they were obtained. The parameter $\delta \in (0, 1)$ is a decision threshold. If the proportion of valid duals is greater than δ, then we want to create more duals in the following generation; but if the proportion of valid duals is less than δ, then we want to create fewer duals in the next generation. $\beta \in (0, 1)$ is a constant that controls the adaptation speed. $N_{d,\min}$ and $N_{d,\max}$ are the minimum and maximum allowable values of N_d. The following values are suggested for the constants in Equation (16.37) [Yang and Yao, 2005]:

$$\begin{aligned}
\text{Initial } N_d &= 0.5N \\
\delta &= 0.9 \\
\beta &= 0.5 \\
N_{d,\min} &= 1 \\
N_{d,\max} &= 0.5N
\end{aligned} \tag{16.38}$$

where N is the population size. Dual learning can also be extended to PBIL for solving dynamic optimization problems [Yang and Yao, 2005], [Yang and Yao, 2008b]. In PBIL, a dual probability vector p_d is symmetric to probability vector p with respect to the 50% probability value: $p_d = 1 - p$.

16.7 CONCLUSION

Opposition-based learning (OBL) is a relatively new arrival on the optimization scene, and so there are a lot of possible extensions. Adaptive OBL might be an interesting avenue to pursue. Adaptation could be implemented in a few different ways. For example, since an EA population tends to converge to good solutions as the generation count increases, perhaps OBL should be implemented more often at the early stages of EA operation, and less often later in the EA. This could be done by making the jumping rate J_r and/or the jumping ratio ρ a decreasing function of the generation number. Also, we have restricted the opposition degree to 0 or 1 in this chapter (see Equation (16.9)). We could implement adaptive OBL by probabilistically decreasing the opposition degree with the generation count.

Other ways of implementing adaptation in an OBL algorithm might include changing the opposition type as the generation count increases, or changing the opposition type based on individual fitness. Low-fitness individuals should benefit more than high-fitness individuals from drastic opposition operations, so maybe super opposition should be reserved for low fitness individuals.

Although a lot of mathematical modeling has been done with OBL using probability theory, OBL as an optimization algorithm has not yet been mathematically modeled. Important areas for future OBL research include adapting mathematical EA models (see Chapter 4 and Section 7.6) for the incorporation of OBL.

Additional research could also focus on exploring the relationship between OBL and evolution through the search for novelty [Lehman and Stanley, 2011]. Also, since OBL is based on social revolutions, it would be interesting to incorporate more cultural models into OBL (see Chapter 15). Additional tutorial material on OBL can be found in [Tizhoosh, 2005] and [Tizhoosh et al., 2008].

PROBLEMS

Written Exercises

16.1 Equation (16.4) defines the modulo opposite. Give an equivalent definition that does not use the modulo function.

16.2 Give an example of a two-dimensional domain where the opposite of a point x in the domain could be outside of the domain.

16.3 Consider the point $(x, y) = (2, 2)$ where the domain of x is $[1, 5]$ and the domain of y is $[1, 7]$. What is the opposite, quasi opposite, super opposite, and quasi reflected opposite of this point?

16.4 Consider the point $(x, y) = (2, 2)$ where the domain of x is $[1, 5]$ and the domain of y is $[1, 7]$. What type of opposite is the point $(2, 5)$?

16.5 Explain how you could modify the OBBO algorithm of Figure 16.5 to use an adaptive jumping rate.

16.6 How does the assumption of Example 16.4 contradict the second assumption of Section 16.3? Which assumption do you think is more reasonable?

16.7 Suppose the BBO emigration rate μ_k in Figure 16.7 is a random variable uniformly distributed on $[0, 1]$.

- **a)** What is the probability that an opposite individual will be generated for a randomly selected individual z_k?
- **b)** Does the probability that you derive make intuitive sense in the limit as $\alpha \to 0$ and $\alpha \to \infty$?

16.8 In Figures 16.9 and 16.10 we arbitrarily defined the starting point of the greedy opposite β of route α to be the same as the starting point of α. However, the proximity of β relative to α depends on the starting point. Consider the route $\alpha = \{A \to B \to C \to D \to E\}$.

- **a)** What is the greedy opposite β if the starting city of β is A, and what is its proximity relative to α?
- **b)** What is the greedy opposite β if the starting city of β is B, and what is its proximity relative to α?

16.9 What are the minimum and maximum values of s in Equation (16.37)?

16.10 Suppose we use the dual adaptation logic of Equation (16.37) with the recommended constants, and that $N_v = 0.1N$ after the first generation. What will the value of N_d be during the second generation?

Computer Exercises

16.11 Write a program that sets $x^* \sim U[a,b]$ for some arbitrarily chosen values a and b such that $a < b$. Set $x \sim U[a,b]$, set x_o to the reflected opposite of x, and set x_{qo} to the quasi opposite of x. Check which opposite is closer to x^*. Run the program a few thousand times to confirm Theorem 16.1.

16.12 Solve Problem 16.11 for $n = 1$ to 20, where n is the number of dimensions. Plot the probability that $||x_{qo} - x^*||_2 < ||x_o - x^*||_2$ as a function of n. Comment on your results.

16.13 Repeat Example 16.4 with the fitness-based proportional opposition logic of Figure 16.8. Use $\rho = 0.1$, 0.5, and 1.0. What is the average (over 20 Monte Carlo simulations) of the lowest cost found by OBBO for each of these values of ρ? Comment on your results.

CHAPTER 17

Other Evolutionary Algorithms

> What has been will be again, what has been done will be done again; there is nothing new under the sun.
>
> —Ecclesiastes 1:9

This chapter gives an overview of some of the EAs that we have not had time to discuss in previous chapters. Some of the algorithms in this chapter are in the murky region between evolutionary and non-evolutionary algorithms, and so this chapter seems to be a good place to summarize them. Other algorithms in this chapter are clearly evolutionary but are also new, and so it is not clear how much of an impact they will have on future EA theory and practice. When deciding which EAs to include in this book it was clear that GAs, EP, ES, and GP should be covered in Part II because of their foundational importance and their history. Deciding which EAs to include in Part III was less clear. The EAs discussed in the previous chapters reflect the author's personal biases and his opinions of the importance of each algorithm.

There are several optimization algorithms that do not have their own chapter in this book but that we should discuss to at least some extent. That is the purpose of this chapter. The algorithms in this chapter are not necessarily less important, less effective, or less useful than those in previous chapters. Their placement in this chapter simply reflects the author's limited experience and subjective interests.

17.1 TABU SEARCH

Tabu search (TS) was introduced in [Glover and McMillan, 1986]. Tabu, or taboo, means forbidden, banned, or not allowed. Forbidden items, speech, or practices can be based on culture, religion, morality, or politics. TS is not strictly a population-based approach to optimization, but it can be considered an EA because it is based on the natural world, and it is an iterative search process. TS is based on the idea that if a certain region of search space has already been visited during the search process, then it is tabu and the search algorithm is discouraged from visiting it again. Similarly, if a certain search strategy has already been used during the search process, then that search strategy is tabu and the search algorithm is discouraged from using it again.

Figure 17.1 outlines a basic TS algorithm, where T is a list of tabu features, and x_0 is the current best candidate solution. When we create children from x_0, we do not allow the search process to include features from T. When an improved candidate solution x' is found, we add features from x' to the tabu list T. We periodically remove features from T, perhaps based on how long they have been in T. This simulates the gradual changing of tabu with time, as we see in human society. Note that the test for (features of x') $\notin T$ in Figure 17.1 is intentionally left ambiguous. The details of this test depends on the problem, on the method used to create neighbors of x_0, on user preference, and on other details.

Initialize a candidate solution x_0
$T \leftarrow \emptyset$
While not(termination criterion)
    Children $\leftarrow \emptyset$
    While |Children| $< M$
        Create a neighbor x' of x_0
        If (features of x') $\notin T$
            Children $\leftarrow$ Children $\cup\, x'$
        End if
    End while
    $x' \leftarrow \arg\min(f(x) : x \in \text{Children})$
    If $f(x') < f(x_0)$
        $T \leftarrow T \cup$ (features from x')
        $x_0 \leftarrow x'$
    End if
    Remove old features from T
Next generation

Figure 17.1 Outline of a tabu search (TS) algorithm to find the minimum of $f(x)$. Each iteration includes the creation of M children, where M is a user-specified parameter.

We can use many variations on the algorithm in Figure 17.1. For example, we could have varying degrees of tabu. We could also have a tabu list that contains not features to avoid, but that instead contains search strategies to avoid. TS is often used to augment other EAs. The brief outline in this section is intended to give the reader enough information to implement a simple TS algorithm, learn the basic idea of TS, and learn more details from other sources. Additional reading about TS can be found in [Reeves, 1993, Chapter 3], [Glover and Laguna, 1998], [Gendreau, 2003], and [Gendreau and Potvin, 2010].

17.2 ARTIFICIAL FISH SWARM ALGORITHM

The artificial fish swarm algorithm (AFSA), which was proposed in [Li et al., 2003] and is sometimes called the artificial fish school algorithm, is loosely based on the swarming behavior of fish. The position of an artificial fish in a search space is denoted as x_i, where $i \in [1, N]$ is the index of the fish, and N is the number of fish in the swarm. We denote the search domain for each dimension as $[l_k, u_k]$ for $k \in [1, n]$, where n is the dimension of the search space. Fish have a visual field within which they can see other fish, and beyond which they cannot see other fish. The visual range of the fish is defined as

$$v = \delta \max_k (u_k - l_k) \tag{17.1}$$

where δ is a tuning parameter that is often gradually decreased during the optimization process. [Fernandes et al., 2009] has found that values of δ between 1 and 10 give good performance for problems with between two and four dimensions, although this range may need to be adjusted for problems with more dimensions. The indices of the fish that are within visual range of fish x_i are denoted as follows:

$$V_i = \{j \neq i : ||x_i - x_j||_2 \leq v\}. \tag{17.2}$$

A fish is said to be in a crowded environment if there are relatively many fish within its visual range:

$$\begin{aligned} \frac{|V_i|}{N} > \theta \quad &\Longrightarrow \quad \text{The visual scope of } x_i \text{ is crowded} \\ \frac{|V_i|}{N} \leq \theta \quad &\Longrightarrow \quad \text{The visual scope of } x_i \text{ is not crowded} \end{aligned} \tag{17.3}$$

where θ is a tuning parameter. [Fernandes et al., 2009] has found that $\theta \approx 1$ gives good performance for low-dimensional problems. AFSA fish have five distinct behaviors which we discuss next: random, chasing, swarming, searching, and leaping.

17.2.1 Random Behavior

Sometimes fish behave randomly; that is, they move in a random direction in the search space. Figure 17.2 shows pseudo-code for a random move. Random behavior occurs if a fish does not have any other fish within its visual range, or if the optimization process has stagnated. Stagnation is defined as a failure of the best individual in the population to significantly improve during the previous m generations:

$$\arg\min_x f_{t-m}(x) - \arg\min_x f_t(x) < \eta \Longrightarrow \text{Stagnation} \tag{17.4}$$

where $f_t(x)$ is the optimization function value of individual x at the t-th generation, m is a positive integer-valued tuning parameter, and η is a non-negative tuning parameter. We assume in Equation (17.4) that our optimization problem is a minimization problem. [Fernandes et al., 2009] has found that $m \approx 10n$ and $\eta \approx 10^{-4}$ gives good performance for low-dimension benchmark problems, where n is the problem dimension.

For $k = 1$ to n
  $r \leftarrow U[0,1]$
  If $r < 1/2$ then
    $\rho \leftarrow U[0,1]$
    $y_i(k) \leftarrow x_i(k) + \rho \min(v, u_k - x_i(k))$
  else
    $\rho \leftarrow U[0,1]$
    $y_i(k) \leftarrow x_i(k) - \rho \min(v, x_i(k) - l_k)$
  End if
Next dimension

Figure 17.2 Random behavior in an artificial fish swarm algorithm. This code shows a random move of fish x_i to a new location y_i, where n is the number of dimensions in the optimization problem, $U[0,1]$ is a random number uniformly distributed in $[0,1]$, and v is the visual range defined in Equation (17.1).

17.2.2 Chasing Behavior

Sometimes a fish moves toward the fish that is at the location of highest food concentration within its visual range. Chasing behavior for fish x_i is described as follows:

$$\begin{aligned} j^* &\leftarrow \arg\min_j \{f(x_j) : j \in V_i\} \\ y_i &\leftarrow x_i + r(x_{j^*} - x_i) \end{aligned} \tag{17.5}$$

where $r \in [0,1]$ is a uniformly distributed random variable, and y_i is the new location of x_i. We again assume that our optimization problem is a minimization problem, so j^* is the index of the fish within the visual range of x_i that has the best performance on our optimization problem. If a fish is not within visual range of any other fish, then it cannot engage in chasing behavior. Also, x_i chases another fish only if the best fish x_{j^*} within its visual range has better optimization performance than x_i.

17.2.3 Swarming Behavior

Fish are social creatures, so sometimes they congregate. In this case, a fish x_i moves toward the centroid c_i of the fish that are within its visual range. Swarming behavior for fish x_i is described as follows:

$$\begin{aligned} c_i &\leftarrow \frac{1}{|V_i|} \sum_{j \in V_i} x_j \\ y_i &\leftarrow x_i + r(c_i - x_i) \end{aligned} \tag{17.6}$$

where $r \in [0,1]$ is a uniformly distributed random variable, and y_i is the new position of fish x_i. If a fish is not within visual range of any other fish, then it cannot engage in swarming behavior. Swarming occurs only if the fish's visual scope is not empty, and it is not crowded, and $f(c_i)$ is better than $f(x_i)$.

17.2.4 Searching Behavior

When a fish sees another fish that has more food, it moves toward that fish. Searching behavior for fish x_i is described as follows:

$$\begin{aligned} j &\leftarrow \text{random integer } \in V_i \\ y_i &\leftarrow x_i + r(x_j - x_i) \end{aligned} \tag{17.7}$$

where $r \in [0,1]$ is a uniformly distributed random variable, and y_i is the new location of fish x_i. Searching behavior is the movement of a fish toward a randomly selected fish that is within its visual range. If a fish is not within visual range of any other fish, then it cannot engage in searching behavior. Searching behavior occurs if the fish's visual scope is crowded, or if the fish's visual scope is not crowded and $f(c_i)$ in Equation (17.6) is worse than $f(x_i)$, or if the fish's visual scope is not crowded and $f(x_{j^*})$ in Equation (17.5) is worse than $f(x)$.

17.2.5 Leaping Behavior

Sometimes a fish randomly leaps through the search space. This is analogous to a fish leaping out of the water and randomly landing in a different location. Leaping occurs for a single randomly-selected fish if the optimization process has stagnated as indicated in Equation (17.4). Figure 17.3 shows pseudo-code for the leaping behavior of a fish.

For $k = 1$ to n
  $r \leftarrow U[0,1]$
  $\rho \leftarrow U[0,1]$
  If $r < 1/2$ then
    $x_i(k) \leftarrow x_i(k) + \rho(u_k - x_i(k))$
  else
    $x_i(k) \leftarrow x_i(k) - \rho(x_i(k) - l_k)$
  End if
Next dimension

Figure 17.3 Leaping behavior in an artificial fish swarm algorithm. This code shows a leap of individual x_i in an artificial fish swarm algorithm, where n is the number of dimensions in the optimization problem, and $U[0,1]$ is a random number uniformly distributed in $[0,1]$.

17.2.6 A Summary of the Artificial Fish Swarm Algorithm

AFSA uses a greedy selection method. That is, after random, chasing, swarming, and searching behavior, the fish x_i moves to its new position y_i only if that new position is better than its old position. Figure 17.4 shows pseudo-code for the AFSA, which appears to have behavior similar to particle swarm optimization. Researchers have proposed many variations and hybrids of the AFSA [Neshat et al., 2012]. Analyzing and modeling AFSA mathematically, incorporating additional features from biological fish behavior, and clarifying the relationship between AFSA and PSO, could all be important and fruitful areas for future AFSA research.

N = population size
Initialize a random population of candidate solutions $\{x_i\}$ for $i \in [1, N]$
While not(termination criterion)
    For each individual x_i
        Find the fish in the visual scope of x_i as shown in Equation (17.2)
        If $V_i = \emptyset$ then
            $y_i \leftarrow$ random move as shown in Figure 17.2
        else if the visual scope of x_i is crowded (see Equation (17.3)) then
            $y_i \leftarrow$ search move as shown in Equation (17.7)
        else
            If $f(c_i) < f(x_i)$ (see Equation (17.6)) then
                $y_i \leftarrow$ swarm move as shown in Equation (17.6)
            else
                $y_i \leftarrow$ search move as shown in Equation (17.7)
            End if
            If $f(x_{j^*}) < f(x_i)$ (see Equation (17.5) then
                $y_i \leftarrow$ chase move as shown in Equation (17.5)
            else
                $y_i \leftarrow$ search move as shown in Equation (17.7)
            End if
            $y_i \leftarrow \arg\min\{f(x_i), f(y_i)\}$
        End if
    Next individual
    $x_i \leftarrow \arg\min\{f(x_i), f(y_i)\}$ for $i \in [1, N]$
    If the algorithm has stagnated as indicated in Equation (17.4) then
        $j \leftarrow$ random integer $\in [1, N]$
        $x_j \leftarrow$ leap move as shown in Figure 17.3
    End if
Next generation

Figure 17.4 An artificial fish swarm algorithm (AFSA) for minimizing the n-dimensional function $f(x)$, where x_i is the i-th candidate solution.

17.3 GROUP SEARCH OPTIMIZER

The group search optimizer (GSO), also called group search optimization, is based on the food foraging behavior of animals [He et al., 2009]. This foundation is similar to that of the fish swarm algorithm (Section 17.2) and bacterial foraging optimization (Section 17.6), but GSO is based on the observed behaviors of land-based animals.

Some animals focus their efforts on searching for food; these animals are called producers. Other animals focus their efforts on following other animals and exploiting the food-finding success of others; these animals are called joiners, or scroungers. GSO includes a third type of animal called rangers, which perform a random walk to search for resources. Each individual has a location in the n-dimensional search space denoted as x_i, and a heading angle denoted as $\phi_i = \left[\begin{array}{ccc} \phi_{i,1} & \cdots & \phi_{i,n-1} \end{array} \right]$.

Producers

GSO assumes that there is only one producer in the population. The producer role is assumed at each generation by the individual with the lowest cost. Each generation, the producer scans three points in his immediate surroundings for a better cost function value than his current location in search space. This corresponds to local search. If we denote the producer as x_p, then the three points are

$$\begin{aligned} x_z &= x_p + r_1 l_{\max} D(\phi_p) \\ x_r &= x_p + r_1 l_{\max} D(\phi_p + r_2 \theta_{\max}/2) \\ x_l &= x_p + r_1 l_{\max} D(\phi_p - r_2 \theta_{\max}/2) \end{aligned} \tag{17.8}$$

where r_1 is a zero-mean, unity-variance, normally distributed random variable;[1] $r_2 \in [0,1]$ is a uniformly distributed random variable; ϕ_p is the heading angle of x_p; $l_{\max}$ is a tuning parameter that defines how far the producer can see; $\theta_{\max}$ is a tuning parameter that defines how far the producer can turn his head; and $D(\cdot)$ is a polar-to-Cartesian coordinate transformation defined as

$$\begin{aligned} D(\phi_p) &= \left[\begin{array}{ccc} d_1 & \cdots & d_n \end{array} \right] \\ d_1 &= \prod_{q=1}^{n-1} \cos \phi_{p,q} \\ d_j &= \sin \phi_{p,j-1} \prod_{q=j}^{n-1} \cos \phi_{p,q} \quad \text{for } j \in [2, n-1] \\ d_n &= \sin \phi_{p,n-1}. \end{aligned} \tag{17.9}$$

If the producer finds a better cost function value at one of the points defined by Equation (17.8), then it immediately moves to that point; otherwise, it remains where it is and randomly moves its heading angle ϕ_p to a new value. If the producer cannot find a better point after $a_{\max}$ searches, then it moves its heading angle back to the value that it was $a_{\max}$ generations ago. However, it is not clear why this last strategy should have any effect on optimization performance, and so we might be able to safely neglect it.

[1]Note that this definition of r_1 allows the producer to look backward as well as forward in the three specified directions.

Scroungers

Scroungers generally move toward the producer. But they do not move directly toward the producer; instead they move in a sort of zig-zag pattern toward the producer, which allows them to search for lower cost function values while they move. A scrounger's movement is modeled as

$$x_i \leftarrow x_i + r_3 \circ (x_p - x_i) \tag{17.10}$$

where r_3 is an n-dimensional vector of random variables, each of which is uniformly distributed on $[0, 1]$; and $\circ$ represents element-by-element multiplication.

Rangers

Rangers randomly travel through the search space looking for areas with low cost function values. Ranger movement is modeled as

$$\begin{aligned} \phi_i &\leftarrow \phi_i + \rho\alpha_{max} \\ x_i &\leftarrow x_i + a_{\max} l_{\max} r_1 D(\phi_i) \end{aligned} \tag{17.11}$$

where $\alpha_{\max}$ is a tuning parameter that defines how far a ranger can turn his head; $\rho \in [-1, 1]$ is a uniformly distributed random variable; $l_{\max}$ is a tuning parameter that is related to the maximum distance that a ranger can travel in one generation, and is the same as $l_{\max}$ in Equation (17.8); and r_1 is a zero-mean, unity-variance, normally distributed random variable.

Summary

Figure 17.5 outlines the GSO, and shows that GSO has several tuning parameters. Note that Figure 17.5 specifies that one individual is a producer, about 80% of the individuals are scroungers, and about 20% of the individuals are rangers. [He et al., 2009] studies the effect of these settings and the other tuning parameters, and recommends the following:

$$\begin{aligned} a_{\max} &= \text{round}\sqrt{n+1} \\ \theta_{\max} &= \pi / a_{\max}^2 \\ \alpha_{\max} &= \theta_{\max}/2 \\ l_{\max} &= \|U - L\|_2 \end{aligned} \tag{17.12}$$

where the n-dimensional vectors U and L are the upper and lower bounds of the search space, respectively.

GSO is similar to PSO. One difference is that in PSO, each individual retains a memory of its previous locations in search space. Another difference is that in PSO, each individual performs the same search strategy. One distinctive of GSO is its ranging behavior, although this behavior is also seen in the catfish PSO (see Section 11.7). Promising directions for future GSO research include mathematical modeling and analysis, on-line adaptation of tuning parameters, and the incorporation of additional nature-inspired features.

N = population size
Initialize a random population of candidate solutions $\{x_i\}$ for $i \in [1, N]$
Randomly initialize the heading angle ϕ_i of each candidate solution x_i
While not(termination criterion)
    Find the producer: $x_p \leftarrow \arg\min_{x_i}\{f(x_i) : i \in [1, N]\}$
    $\{x_z, x_r, x_l\} \leftarrow$ scanning result of Equation (17.8)
    If $\min\{f(x_z), f(x_r), f(x_l)\} < f(x_p)$ then
        $x_p \leftarrow \arg\min\{f(x_z), f(x_r), f(x_l)\}$
    else
        $\rho \leftarrow U[-1, 1]$
        $\phi(x_p) \leftarrow \phi(x_p) + \rho\alpha_{\max}$
    End if
    For each $x_i \neq x_p$
        $r_2 \leftarrow U[0, 1]$
        If $r_2 < 0.8$
            Let x_i scrounge using Equation (17.10)
        else
            Let x_i range using Equation (17.11)
        End if
    Next individual
Next generation

Figure 17.5 A group search optimizer (GSO) for minimizing the n-dimensional function $f(x)$, where x_i is the i-th candidate solution.

17.4 SHUFFLED FROG LEAPING ALGORITHM

The shuffled frog leaping algorithm (SFLA) was introduced in [Eusuff and Lansey, 2003], [Eusuff et al., 2006] as a hybrid of PSO and shuffled complex evolution (SCE). SCE is based on the idea of allowing sub-populations to evolve independently while periodically allowing interactions between the sub-populations [Duan et al., 1992], [Duan et al., 1993]. SCE uses probabilistic selection of parents in each sub-population and also randomly creates new individuals to prevent stagnation. SFLA is based on ideas from both SCE and PSO.

Figure 17.6 illustrates the global search strategy of the SFLA. We begin by randomly creating a set of N candidate solutions. We then divide these N individuals into m sub-populations, also called memeplexes. Usually N is a multiple of m so that each sub-population contains the same number of individuals. We then perform a local search algorithm in each sub-population. At the beginning of the next generation, we shuffle the population so that each individual is randomly assigned to a new sub-population. Common tuning parameters for the SFLA include a population size N of about 200, with about $m = 20$ sub-populations [Elbeltagi et al., 2005].

The statement "perform local search" in Figure 17.6 indicates execution of the algorithm of Figure 17.7. During local search, each sub-population independently performs an evolutionary search for $i_{\max}$ iterations. Each iteration, we update only

x_w, which is the sub-population individual with the worst cost:

$$x_w \leftarrow x_w + r(x_b - x_w) \tag{17.13}$$

where $r \in [0, 1]$ is a uniformly distributed random number, and x_b is the sub-population individual with the best cost. If Equation (17.13) does not improve x_w, then we update it again as follows:

$$x_w \leftarrow x_w + r(x_g - x_w) \tag{17.14}$$

where $r \in [0, 1]$ is a new random number, and x_g is the globally best individual from all m sub-populations. If Equation (17.14) does not improve x_w, then we replace x_w with a randomly-generated individual. The iteration limit $i_{\max} = 10$ is a common tuning parameter in Figure 17.7 [Elbeltagi et al., 2005]. Promising directions for future SFLA research include mathematical modeling and analysis, and the incorporation of additional nature-inspired features.

Initialize a random population $\{x_i\}$ for $i \in [1, N]$
While not(termination criterion)
    Randomly divide the population into m sub-populations
    For each sub-population $i = 1$ to m
        Perform local search in the i-th sub-population (Figure 17.7)
    Next sub-population
Next generation

Figure 17.6 The above pseudo-code outlines the global search strategy of the shuffled frog leaping algorithm (SFLA).

Find the best individual in the entire population, x_g
For $i = 1$ to $i_{\max}$
    Find the best and worst sub-population individuals, x_b and x_w
    Use Equation (17.13) to update x_w
    If the update did not improve x_w then
        Use Equation (17.14) to update x_w
        If the update did not improve x_w then
            $x_w \leftarrow$ randomly-generated individual
        End if
    End if
Next iteration

Figure 17.7 The above pseudo-code outlines the local search strategy of the shuffled frog leaping algorithm (SFLA).

17.5 THE FIREFLY ALGORITHM

The firefly algorithm was introduced in [Yang, 2008b, Chapter 8], [Yang, 2010b]. The firefly algorithm is based on the attraction of fireflies to one another. Attraction is based on the perceived brightness of a firefly, which exponentially decreases with distance. A firefly is attracted only to those fireflies that are brighter than itself.

Figure 17.8 shows pseudo-code for the firefly algorithm. As $\gamma \to 0$, all fireflies are attracted to each other equally, which corresponds to zero dispersion of light in the atmosphere. We would observe this type of behavior in a vacuum. As $\gamma \to \infty$, fireflies are not attracted to each other at all, which corresponds to random flight and a random search. We would observe this type of behavior in a dense fog. The parameters β_0 and α determine the tradeoff between exploitation (attraction to other fireflies) and exploration (random search). Typical tuning parameters are as follows:

$$\begin{aligned} \gamma_i &= \frac{\gamma_0}{\max_j ||x_i - x_j||_2}, \text{ where } \gamma_0 = 0.8 \\ \alpha &= 0.01 \\ \beta_0 &= 1. \end{aligned} \tag{17.15}$$

Each firefly x_i compares its brightness with every other firefly x_j, one at a time. If x_j is brighter than x_i, then x_i will make a move that includes both a component that is random, and a component that is directed toward x_j. The quantity αr in Figure 17.8 is the random component. This is usually relatively small due to the small value of α (see Equation 17.15). The quantity $\beta_0 e^{-\gamma_i r_{ij}^2}(x_j - x_i)$ is the directed component; as stated earlier, its magnitude is an exponential function of the distance r_{ij} between x_j and x_i. Although the exponential function is biologically motivated, we might want to try some other functions that decrease with increasing distance.

One thing that we notice from Figure 17.8 is that the best individual in the population is never updated. We might be able to improve the algorithm's performance if we periodically update the best individual to search for a better one. However, this approach might be a high-risk, low-payoff operation in that it could require many function evaluations before finding a location in search space that is better than the currently best position.

Additional variations on the firefly algorithm are discussed in [Łukasik and Żak, 2009], [Yang, 2009b], [Yang, 2010a]. For example, the parameter α is often a decreasing function of time, which serves to reduce exploration as the population becomes more optimized. A version of the algorithm for combinatorial problems has been proposed in [Sayadi et al., 2010]. The firefly algorithm, like the AFSA discussed of Section 17.2, is very similar to PSO (see Chapter 11). We can make some simple modifications to the firefly algorithm of Figure 17.8 to make it equivalent to a special case of PSO (see Problem 17.5).

Initialize a random population $\{x_i\}$ for $i \in [1, N]$
While not(termination criterion)
    For each individual x_i
        For each individual $x_j \neq x_i$
            If $f(x_j) < f(x_i)$
                For each dimension $k \in [1, n]$
                    $\rho \leftarrow U[0, 1]$
                    If $\rho < 1/2$
                        $r_k \leftarrow (u_k - x_i(k))U[0, 1]$
                    else
                        $r_k \leftarrow (x_i(k) - l_k)U[0, 1]$
                    End if
                Next dimension k
                $r_{ij} \leftarrow$ distance between x_i and x_j
                $x_i \leftarrow x_i + \beta_0 e^{-\gamma_i r_{ij}^2}(x_j - x_i) + \alpha r$ (this is a vector operation)
            End if
        Next x_j
    Next x_i
Next generation

Figure 17.8 A firefly algorithm for minimizing the n-dimensional function $f(x)$. In this algorithm, x_i is the i-th candidate solution, and $x_i(k)$ is the k-th element of x_i. $U[0, 1]$ is a random number uniformly distributed on $[0, 1]$, and l_k and u_k are the lower and upper bounds of the k-th dimension of the search space, respectively.

17.6 BACTERIAL FORAGING OPTIMIZATION

The bacterial foraging optimization algorithm (BFOA) was introduced in [Passino, 2002] and is based on the behavior of escherichia coli bacteria, commonly known as E. coli. BFOA is based on the premise that natural selection favors the propagation of genetics that lend themselves to successful food foraging behaviors. Food foraging is, of course, common among all species, not only bacteria. Sometimes animals forage cooperatively, and sometimes they forage alone. If they forage alone, they have the advantage of keeping the food that they find entirely to themselves. But if they forage in teams, they have the advantage of being able to more easily fight off predators. Animals need to balance the probability of foraging success with risks from predators.

If an animal finds a geographical area with a lot of food, the animal needs to balance its exploitation of those known resources with the possibility of finding better resources in another location; this is another type of risk-balancing behavior. As an animal depletes its resources in a given location, there is an optimal time to leave the known resources to search for regions with more resources.

Foraging also includes behaviors besides searching for food. Foraging includes pursuing and attacking the prey, and also consuming the prey. If the prey is larger than the predator, then the predators need to coordinate with each other to attack and consume the prey. If the prey is smaller than the predator, then it may be more optimal for the predator to forage on its own. Some foragers continually move through their environment while searching for prey. Other foragers remain hidden in a stationary location and wait for prey to come within striking distance. Other foragers practice a combination of these approaches. BFOA is specifically modeled on the foraging behavior of bacteria, but foraging theory is a widely studied discipline [Stephens and Krebs, 1986], [Giraldeau and Caraco, 2000] that has many potential applications to optimization theory [Quijano et al., 2006].

BFOA is based on three behaviors of bacteria. First, bacteria propel themselves through their environment; this behavior is called chemotaxis. Second, bacteria reproduce. Third, bacteria are eliminated from, and dispersed throughout, their habitat due to environmental events.

Chemotaxis

The first behavior of bacteria, self-propulsion or chemotaxis, can be further divided into two behaviors. First, bacteria can tumble in random directions. Second, they can propel themselves in the direction of an increasing food supply. This second type of self-propulsion is influenced not only by the food supply, but also by the presence of other bacteria. Other bacteria serve to both attract and repel each other. They have a certain level of attraction because the presence of a bacterium at a certain location implies that there is food at that location. They have a certain level of repulsion because the presence of a bacterium at a certain location indicates that there is competition for food at that location.

Suppose that we want to find the minimum of a function $f(x)$. In BFOA, the self-propulsion of a bacterium is modeled as

$$x \leftarrow x + c\Delta \tag{17.16}$$

where x is the location in the search space of an individual in the population, c is the step size, and Δ is a unit vector in some direction in the search space. When tumbling, Δ is a random unit vector. The combination of attraction and repulsion by other bacteria results in an effective cost function $f'(x)$ perceived by an individual x as follows:

$$f'(x) = f(x) + \sum_{i=1}^{N} \left[h \exp\left(-w_r \|x - x_i\|_2^2\right) - d \exp\left(-w_a \|x - x_i\|_2^2\right) \right] \tag{17.17}$$

where N is the population size, and h, w_r, d, and w_a are tuning parameters related to the repulsive and attractive forces that bacteria exert on each other. If an individual x tumbles in a direction that decreases $f'(x)$, then it continues moving in that direction, although an upper limit N_s (another tuning parameter) is placed on the number of moves in a given direction.

Reproduction

Each individual repeats the self-propulsion described above for N_c iterations. That is, it first tumbles in a random direction. If the random tumble decreases $f'(x)$, it continues moving in that direction. N_c movements defines the lifetime of each bacterium. After this, the health of each bacterium is measured as its average $f'(x)$ value over the previous N_c iterations. The most healthy half of the bacteria reproduce by spawning two clones per bacterium, thus providing N new bacteria for the next generation.

Elimination and Dispersal

After reproduction, the elimination-dispersal step takes place. Each bacterium is dispersed to a random location in the search space with probability p_e (another tuning parameter).

Summary

Figure 17.9 outlines a basic BFOA. Typical tuning parameters are given as follows [Passino, 2002]:

$$\begin{aligned}
\text{step size } c &= 0.1 \\
\text{population size } N &= 50 \\
\text{number of chemotaxis steps } N_c &= 100 \\
\text{number of cost reduction steps } N_s &= 4 \\
\text{number of reproduction steps } N_r &= 4 \\
\text{number of elimination-dispersal steps } N_e &= 2 \\
\text{attraction force depth } d &= 1 \\
\text{repulsion force depth } h &= 1 \\
\text{attraction force width } w_a &= 0.2 \\
\text{repulsion force width } w_r &= 10 \\
\text{probability of elimination-dispersal } p_e &= 0.25. \qquad (17.18)
\end{aligned}$$

The number of generations in BFOA is not as well-defined as in other EAs. The outermost loop of Figure 17.9 executes N_e times, which is typically only twice. The best way to measure EA computational effort is not with generations but with function evaluations.

We note from Figure 17.9 that individuals reproduce by cloning. We might be able to improve performance by using a more sophisticated recombination operation (see Section 8.8), even though this would stray from the bacterial foundations of BFOA. Also, we arbitrarily clone the best half of the population in Figure 17.9 to create the next generation. We could instead clone the best B individuals, where B is a tuning parameter.

BFOA is a research area with many possibilities. There are many aspects of bacterial foraging, and animal foraging in general, that could be modeled to obtain improved optimization performance. Automatic adaptation of the tuning parameters might be especially important for BFOA since there are so many parameters, and this could give improved performance [Dasgupta et al., 2009]. BFOA could be hybridized with other EAs, as it already has been with PSO [Biswas et al., 2007b] and DE [Biswas et al., 2007a]. Some mathematical analysis of BFOA is provided in [Das et al., 2009], although much more remains to be done in that area. Note that the bacterial chemotaxis model is a separate but similar EA that is based on the behavior of bacteria [Muller et al., 2002].

```
Initialize the parameters shown in Equation (17.18)
Initialize a random population {x_i} for i ∈ [1, N]
For l = 1 to N_e (elimination-disperal steps)
    For k = 1 to N_r (reproduction steps)
        For j = 1 to N_c (chemotaxis steps)
            For each individual x_i, i ∈ [1, N]
                Compute effective cost as shown in Equation (17.17)
                Generate a random n-dimensional unit vector Δ
                For m = 1 to N_s (cost reduction steps)
                    x̂_i ← x_i + cΔ
                    If f'(x̂_i) < f'(x_i) then
                        x_i ← x̂_i
                    else
                        m ← N_s (exit the cost reduction loop)
                    End if
                Next m
            Next individual
        Next j
        For each individual x_i, i ∈ [1, N]
            F_i ← average value of f'(x_i) during N_c steps of chemotaxis loop
        Next individual
        Eliminate the worst N/2 individuals based on {F_i}
        Clone the best N/2 individuals based on {F_i}
    Next k
    For each individual x_i, i ∈ [1, N]
        Random number r ← U[0, 1]
        If r < p_e then
            x_i ← random point in the search space
        End if
    Next individual
Next l
```

Figure 17.9 A bacterial foraging optimization algorithm (BFOA) for minimizing the n-dimensional function $f(x)$, where x_i is the i-th candidate solution.

17.7 ARTIFICIAL BEE COLONY ALGORITHM

The artificial bee colony (ABC) algorithm is based on the behavior of bees and was first published in [Basturk and Karaboga, 2006], [Karaboga and Basturk, 2007]. ABC is based on the search by bees for an optimal food source. The location of a food source is analogous to a location in the search space of an optimization problem. The amount of nectar at a location is analogous to the fitness of a candidate solution. ABC simulates three different types of bees.

First, forager bees, also called employed bees, travel back and forth between a food source and their hive. Each forager is associated with a specific location, and remembers that location as it travels back and forth between the hive. When a forager takes its nectar to the hive, it returns to its food source, but it also engages in local exploration as it searches in the nearby vicinity for a better source.

Second, onlooker bees are not associated with any particular food source, but they observe the behavior of the foragers when they return to the hive. Onlookers observe the amount of nectar that is returned by the foragers (that is, the fitness of each forager's location in search space), and use that information to decide where to search for nectar. The onlookers' search location is decided probabilistically based on their observations of the foragers.

Third, scout bees are explorers and, like onlookers, are not associated with any particular food source. If a scout sees that a forager has stagnated and is not progressively increasing the amount of nectar that it returns to the hive, then the scout randomly searches for a new nectar source in the search space. Stagnation is indicated when the explorer fails to increase the amount of nectar it brings to the hive after a certain number of trips.

These ideas lead to the ABC algorithm, which is summarized in Figure 17.10. The figures shows that the division between forager, onlooker, and scout bees is simply an analogy, and is not pressed too far in the ABC algorithm. The key idea of the ABC algorithm is that foraging, onlooking, and scouting behaviors are simulated in the search for a global optimum.

Figure 17.10 shows that each forager randomly modifies its position in the search space. If the random modification results in an improvement, then the forager moves to the new position. The onlooker bees also randomly modify the position of a forager, where the forager that is modified is randomly chosen using roulette-wheel selection. Again, if the random modification improves the forager, then the forager moves to the new position. Finally, a scout replaces a forager if the forager has not improved after a preset number of random modifications. The $T(x_i)$ counters in Figure 17.10 are forager trial counters that keep track of how many consecutive unsuccessful modifications have been performed for each forager. Figure 17.10 shows that ABC includes several tuning parameters. Typical ABC parameters are

$$P_f = P_o = N/2, \text{ stagnation limit } L = Nn/2. \tag{17.19}$$

The literature discusses several variations of the ABC algorithm [Karaboga and Basturk, 2007], [Karaboga and Basturk, 2008], [Karaboga and Akay, 2009], [Karaboga et al., 2013], and we could think of many modifications by examining Figure 17.10. For instance, Figure 17.10 shows that each forager updates its position deterministically if it finds a better position; we could instead make this update stochastic. Figure 17.10 also shows that onlookers choose which forager to follow based on roulette-wheel selection; we could instead use another fitness-based selection method (see Section 8.7).

Several other algorithms similar to ABC have been proposed; see [Tereshko, 2000], [Teodorović, 2003], [Benatchba et al., 2005], [Wedde et al., 2004], and the references in [Karaboga and Basturk, 2008]. One algorithm that is very similar to ABC is the bees algorithm [Pham et al., 2006]. The evolutionary algorithm in this book with which ABC seems to have the most in common is differential evolution (DE; see Chapter 12). Future research in the area of ABC could include the investigation of its commonality with DE, its commonality with other bee-oriented algorithms such as those referenced above, and the incorporation of more biologically-inspired features.

N = population size
Initialize the positive integer L, which is the stagnation limit
Initialize the forager population size $P_f < N$
Initialize the onlooker population size $P_o = N - P_f$
Initialize a random population of foragers $\{x_i\}$ for $i \in [1, P_f]$
Initialize the forager trial counters $T(x_i) = 0$ for $i \in [1, P_f]$
While not(termination criterion)
    Forager Bees:
    For each forager x_i, $i \in [1, P_f]$
        $k \leftarrow$ random integer $\in [1, N]$ such that $k \neq i$
        $s \leftarrow$ random integer $\in [1, n]$
        $r \leftarrow U[-1, 1]$
        $v_i(s) \leftarrow x_i(s) + r(x_i(s) - x_k(s))$
        If $f(v_i)$ is better than $f(x_i)$ then
            $x_i \leftarrow v_i$
            $T(x_i) \leftarrow 0$
        else
            $T(x_i) \leftarrow T(x_i) + 1$
        End if
    Next forager
    Onlooker Bees:
    For each onlooker v_i, $i \in [1, P_o]$
        Select a forager x_j, where $\Pr(x_j) \propto \text{fitness}(x_j)$ for $j \in [1, P_f]$
        $k \leftarrow$ random integer $\in [1, P_f]$ such that $k \neq j$
        $s \leftarrow$ random integer $\in [1, n]$
        $r \leftarrow U[-1, 1]$
        $v_i(s) \leftarrow x_j(s) + r(x_j(s) - x_k(s))$
        If $f(v_i)$ is better than $f(x_j)$ then
            $x_j \leftarrow v_i$
            $T(x_j) \leftarrow 0$
        else
            $T(x_j) \leftarrow T(x_j) + 1$
        End if
    Next onlooker
    Scout Bees:
    For each forager x_i, $i \in [1, P_f]$
        If $T(x_i) > L$ then
            $x_i \leftarrow$ randomly-generated individual
            $T(x_i) \leftarrow 0$
        End if
    Next forager
Next generation

Figure 17.10 An artificial bee colony (ABC) algorithm for optimizing the n-dimensional function $f(x)$, where x_i is the i-th candidate solution.

17.8 GRAVITATIONAL SEARCH ALGORITHM

The gravitational search algorithm (GSA) was introduced in [Rashedi et al., 2009] and is based on the law of gravity. GSA is similar to central force optimization, which is a deterministic evolutionary optimization algorithm that is also based on gravity [Formato, 2007], [Formato, 2008]. Other similar algorithms include space gravitational optimization [Hsiao et al., 2005], and integrated radiation optimization [Chuang and Jiang, 2007]. GSA is similar to particle swarm optimization and operates on the principle that each individual in the population has a position and velocity in the search space, but it also includes an acceleration. The particles attract each other based on their mass values, which are proportional to their fitness values (that is, inversely proportional to their cost values). Figure 17.11 depicts the GSA algorithm.

Initialize a random population of individuals $\{x_i\}$, $i \in [1, N]$
Initialize each individual's velocity v_i, $i \in [1, N]$
Initialize the gravitational constant G_0 and the decay rate α
Initialize the generation number $t = 0$ and the generation limit $t_{\max}$
While not(termination criterion)
    Gravitational constant $G \leftarrow G_0 \exp(-\alpha t / t_{\max})$
    For each individual x_i, $i \in [1, N]$
        $m_i \leftarrow \frac{f(x_i) - \max_k f(x_k)}{\min_k f(x_k) - \max_k f(x_k)} \in [0, 1]$
        Normalized fitness $M_i \leftarrow \frac{m_i}{\sum_{k=1}^{N} m_k}$
    Next individual
    For each individual x_i, $i \in [1, N]$
        Distance $R_{ik} \leftarrow ||x_k - x_i||_2$ for $k \in [1, N]$
        Force vector $F_{ik} \leftarrow \frac{G M_i M_k}{R_{ik} + \epsilon}(x_k - x_i)$ for $k \in [1, N]$
        Random number $r_k \leftarrow U[0, 1]$ for $k \in [1, N]$
        Acceleration vector $a_i \leftarrow \frac{1}{M_i} \sum_{k=1, k \neq i}^{N} r_k F_{ik}$
        Random number $r \leftarrow U[0, 1]$
        Velocity vector $v_i \leftarrow r v_i + a_i$
        Position vector $x_i \leftarrow x_i + v_i$
    Next individual
    Increment generation number: $t \leftarrow t + 1$
Next generation

Figure 17.11 A gravitational search algorithm for minimizing $f(x)$, where x_i is the i-th candidate solution. ϵ is small positive constant to prevent division by zero.

In the generational loop of Figure 17.11, we first update the value of the gravitational constant G. The time invariance of G in nature is a matter of debate, with some physicists arguing that it is time varying [Jofré et al., 2006]. The gradual reduction of G in GSA reduces the exploration component of the algorithm as time progresses. Next, we set the fitness values so that the worst individual has a fitness $m_i = 0$ and the best individual has a fitness $m_i = 1$; the fitness values correspond to gravitational masses. Next, we obtain normalized fitness values $\{M_i\}$ that sum to 1. Next, for each pair of individuals, we calculate the attractive force, which is a vector whose magnitude is proportional to their fitness values and the distances between them. Next, we take a random combination of the force vectors to obtain the acceleration vector of each individual. We finally use the acceleration vector to update the velocity and position of each individual. Typical tuning values in Figure 17.11 are $G_0 = 100$ and $\alpha = 20$.

Researchers have proposed various modifications and extensions to GSA, including alternative ways to adjust G each generation. The acceleration equation can be updated so that only the best individuals attract each particle:

$$a_i \leftarrow \frac{1}{M_i} \sum_{k \in B, k \neq i} r_k F_{ik} \tag{17.20}$$

where B is the set containing the best individuals, and the size of B is a tuning parameter. In addition, we can use different types of effective mass values for active gravitational force, passive gravitational force, and inertia [Rashedi et al., 2009]. An extension of GSA to discrete search domains is given in [Rashedi et al., 2010]. Given the similarity between GSA and PSO, it seems that many of the extensions proposed for PSO could also be implemented in GSA (see Chapter 11).

17.9 HARMONY SEARCH

Harmony search (HS) was introduced in [Geem et al., 2001] and is further explained in [Lee and Geem, 2006]. HS is based on musical processes. Each musician in a choir or band sounds a note within some allowable domain. If all of the notes result in good harmony, the positive experience is saved in the choir's collective memory and the possibility of achieving continued good harmony is increased. In HS, a choir or band is analogous to a candidate problem solution, and a musician is analogous to an independent variable or candidate solution feature.

Figure 17.12 outlines the HS algorithm [Omran and Mahdavi, 2008]. Harmony search often uses alternative notation rather than standard EA notation. For example, *harmony vector* is used to refer to the EA individual or candidate solution x, *harmony memory size* is used to refer to the population size N, *harmony memory considering rate* is similar to the crossover rate c in GAs, *pitch adjusting rate* is used to refer to the mutation rate p_m, and *distance bandwidth* is used to refer to the standard deviation σ of the Gaussian mutation. Typical values for these parameters are $c = 0.9$; p_m increases linearly from 0.01 at the first generation to 0.99 at the last generation; and σ decreases exponentially from 5% of the search domain to 0.01% of the search domain.

We see from Figure 17.12 that HS creates one child each generation. For each solution feature, we generate a random number r_c. If r_c is less than the crossover rate c, then that solution feature in the child is set equal to a randomly selected solution feature from the population; this step is similar to global uniform recombination (see Section 8.8.6). However, if r_m is greater than the crossover rate, then that solution feature in the child is set equal to a random number in the search domain; this step is similar to uniform mutation centered at the middle of the search domain (see Section 8.9.2). If the child's solution feature is obtained from the population rather than a random number, then we perform Gaussian mutation centered at the solution feature (see Section 8.9.3). Finally, if the child is better than the

p_m = mutation rate $\in [0, 1]$
σ^2 = Gaussian mutation variance
c = crossover rate $\in [0, 1]$
Initialize a population of candidate solutions $\{x_k\}$ for $k \in [1, N]$
While not(termination criterion)
  Child = $[\begin{array}{cccc} 0 & 0 & \cdots & 0 \end{array}] \in R^n$
  For each solution feature $s = 1, \cdots, n$
    $r_c \leftarrow U[0, 1]$
    If $r_c < c$ then
      $j \leftarrow$ random integer $\in [1, n]$
      Child$(s) \leftarrow x_j(s)$
      $r_m \leftarrow U[0, 1]$
      If $r_m < p_m$ then
        Child$(s) \leftarrow$ Child$(s) + N(0, \sigma^2)$
      End if
    else
      Child$(s) \leftarrow U[x_{\min}(s), x_{\max}(s)]$
    End if
  Next solution feature
  $m \leftarrow \arg\max_k (f(x_k) : k \in [1, N])$
  If f(Child) $< f(x_m)$ then
    $x_m \leftarrow$ Child
  End if
Next generation

Figure 17.12 Outline of the harmony search (HS) algorithm with a population size of N for minimizing the n-dimensional function $f(x)$. $\{x_k\}$ is the entire population of individuals, x_k is the k-th individual, and $x_k(s)$ is the s-th feature of x_k.

worst individual in the population, then the child replaces that individual in the population; this last step is the same strategy that we use in EP (see Section 5.1).

In summary, it appears that there are no fundamentally new ideas in the HS algorithm. HS is an amalgamation of previously established EA ideas, including global uniform recombination, uniform mutation, Gaussian mutation, and replacement of the worst individual each generation. The contribution of HS lies in two areas. First, the way that HS combines these ideas is novel. Second, the musical motivation of HS is novel. However, very few publications in the area of HS discuss musical motivations or extensions of HS. Most publications deal with hybridizing HS with other EAs, tuning HS parameters, or applying HS to specific problems. If more musically motivated extensions could be applied to HS, this would help set it apart as its own distinctive EA. Such research would require studying music theory, studying the process of musical composition and arrangement, studying educational theories of music, and creatively applying those theories to the HS algorithm. Further reading in the area of HS can be found in two edited volumes [Geem, 2010a], [Geem, 2010b], and one book [Geem, 2010c].

17.10 TEACHING-LEARNING-BASED OPTIMIZATION

Teaching-learning-based optimization (TLBO) was introduced in [Rao et al., 2011] and is further explained in [Rao et al., 2012], [Rao and Patel, 2012], [Rao and Savsani, 2012, Chapter 6]. TLBO is based on the teaching and learning process in a classroom. Each generation, the best candidate solution in the population is considered the teacher, and the other candidate solutions are considered learners. The learners mostly accept instruction from the teacher, but also learn from each other. In TLBO, an academic subject is analogous to an independent variable or candidate solution feature. The teacher phase consists of modifying each independent variable $x_i(s)$ in each candidate solution x_i as follows:

$$\begin{aligned} c_i(s) &\leftarrow x_i(s) + r(x_t(s) - T_f \bar{x}(s)) \\ \text{where } \bar{x}(s) &= \frac{1}{N}\sum_{k=1}^{N} x_k(s) \end{aligned} \tag{17.21}$$

for $i \in [1, N]$ and $s \in [1, n]$, where N is the population size, n is the problem dimension, x_t is the best individual in the population (that is, the teacher), r is a random number taken from a uniform distribution on $[0, 1]$, and T_f is called the teaching factor and is set equal to either 1 or 2 with equal probability. The child c_i replaces the parent x_i if the child is better than the parent. In general, Equation (17.21) adjusts $x_i(s)$ in a direction toward the best individual $x_t(s)$. We can see this by taking the expected value of Equation (17.21), which gives

$$\begin{aligned} \bar{c}_i(s) &= \bar{x}(s) + \frac{1}{2}\left(x_t(s) - \frac{3}{2}\bar{x}(s)\right) \\ &= \frac{x_t(s)}{2} + \frac{\bar{x}(s)}{4}. \end{aligned} \tag{17.22}$$

That is, on average, $c_i(s)$ is closer to $x_t(s)$ than it is to $x_i(s)$. Equation (17.22) also shows that $c_i(s)$ is, on average, closer to zero than the parent population, which may give TLBO an unfair advantage on problems whose solution is zero. Many optimization benchmarks have solutions at $x^* = 0$, so further research on TLBO should carefully investigate its performance on problems with nonzero solutions, and adjust the algorithm as needed to remove this inherent bias.

After the teacher phase completes, the learner phase begins. The learner phase entails adjusting each individual based on another randomly selected individual:

$$c_i(s) \leftarrow \begin{cases} x_i(s) + r(x_i(s) - x_k(s)) & \text{if } x_i \text{ is better than } x_k \\ x_i(s) + r(x_k(s) - x_i(s)) & \text{otherwise} \end{cases} \tag{17.23}$$

for $i \in [1, N]$ and $s \in [1, n]$, where k is a random integer in $[1, N]$ such that $k \neq i$, and r is a random number taken from a uniform distribution on $[0, 1]$. The learner phase adjusts $x_i(s)$ away from x_k if x_k is worse (the first case above), and toward x_k if x_k is better (the second case above).

Figure 17.13 outlines the TLBO algorithm. An inspection of Figure 12.2 reveals that of all the EAs discussed in this book, TLBO has the most in common with differential evolution (DE). Suppose in Figure 12.2 that we set the crossover rate $c = 1$. Further suppose that instead of a constant stepsize parameter F we use a

Initialize a population of candidate solutions $\{x_k\}$ for $k \in [1, N]$
While not(termination criterion)
    For each individual x_i where $i \in [1, N]$
        Comment: Teacher Phase
        $x_t \leftarrow \arg\min_x \left(f(x) : x \in \{x_k\}_{k=1}^N\right)$
        $T_f \leftarrow$ random integer $\in \{1, 2\}$
        For each solution feature $s \in [1, n]$
            $\bar{x}(s) \leftarrow \frac{1}{N} \sum_{k=1}^N x_k(s)$
            $r \leftarrow U[0, 1]$
            $c_i(s) \leftarrow x_i(s) + r(x_t(s) - T_f \bar{x}(s))$
        Next solution feature
        $x_i \leftarrow \arg\min_{x_i, c_i}(f(x_i), f(c_i))$
        Comment: Learner Phase
        $k \leftarrow$ random integer $\in [1, N] : k \neq i$
        If $f(x_i) < f(x_k)$ then
            For each solution feature $s \in [1, n]$
                $r \leftarrow U[0, 1]$
                $c_i(s) \leftarrow x_i(s) + r(x_i(s) - x_k(s))$
            Next solution feature
        else
            For each solution feature $s \in [1, n]$
                $r \leftarrow U[0, 1]$
                $c_i(s) \leftarrow x_i(s) + r(x_k(s) - x_i(s))$
            Next solution feature
        End if
        $x_i \leftarrow \arg\min_{x_i, c_i}(f(x_i), f(c_i))$
    Next individual
Next generation

Figure 17.13 Outline of the teaching-learning-based optimization (TLBO) algorithm with a population size of N for minimizing the n-dimensional function $f(x)$. x_t is the best individual in the population and is called the *teacher*.

random stepsize parameter that is different for each independent variable, so that one independent variable at a time is set in the mutant vector v. Further suppose that we replace each individual x_i with its child immediately after the child is created, rather than waiting until after we have created all of the children. These are not necessarily insignificant changes to DE, but they are straightforward. With these changes, the DE algorithm of Figure 12.2 becomes the modified DE algorithm of Figure 17.14.

Now suppose that instead of randomly generating r_1, r_2, and r_3 in Figure 17.14, we set them as follows:

$$\begin{aligned} r_1 &= i \\ r_2 &= \arg\min_i\{f(x_i) : i \in [1, N]\} \\ x_{r3}(s) &= \text{the average of the } s\text{-th solution feature of the population.} \end{aligned} \tag{17.24}$$

Initialize a population of candidate solutions $\{x_i\}$, $i \in [1, N]$
While not(termination criterion)
  For each individual x_i, $i \in [1, N]$
    $r_1 \leftarrow$ random integer $\in [1, N] : r_1 \neq i$
    $r_2 \leftarrow$ random integer $\in [1, N] : r_2 \notin \{i, r_1\}$
    $r_3 \leftarrow$ random integer $\in [1, N] : r_3 \notin \{i, r_1, r_2\}$
    For each solution feature $s \in [1, n]$
      $r \leftarrow U[0, 1]$
      $v(s) \leftarrow x_{r1}(s) + r(x_{r2}(s) - x_{r3}(s))$
    Next solution feature
    $x_i \leftarrow \arg\min_{x_i, v}[f(x_i), f(v)]$
  Next individual
Next generation

Figure 17.14 A modified differential evolution algorithm for minimizing $f(x)$.

With these additional changes, the DE algorithm of Figure 17.14 becomes the algorithm of Figure 17.15, which is equivalent to the teacher phase of TLBO in Figure 17.13 with $T_f = 1$.

Similarly, suppose that instead of randomly generating r_1, r_2, and r_3 in Figure 17.14, we set them as follows:

$$\begin{aligned} r_1 &= i \\ r_2 &= \arg\min_{i,k}\{f(x_i), f(x_k)\} \\ r_3 &= \arg\max_{i,k}\{f(x_i), f(x_k)\} \end{aligned} \tag{17.25}$$

where k is a random integer $\in [1, N]$ such that $k \neq i$. Then we obtain the learner phase of TLBO.

In summary, it appears that there are no fundamentally new ideas in the TLBO algorithm. TLBO is modification of DE, which is itself a genetic algorithm variation (see Section 12.4). The contribution of TLBO is to execute DE in two distinct phases, one called the teacher phase, and the other called the learner phase. Also, the teaching-learning motivation of TLBO is novel. However, TLBO publications to this date have not exploited teaching-learning theories to improve the TLBO algorithm. If more teaching-learning based extensions could be applied to TLBO, this would help set it apart as its own distinctive EA and would also provide the potential for improved performance. Such research would require studying learning theory, learning styles, and teaching styles, and creatively applying those theories to the TLBO algorithm. Another important research topic in TLBO, as mentioned earlier, is to explore its performance on problems whose solutions are not at the center of the search domain, and to adjust the algorithm to remove its center-of-domain bias. See [Črepinšek et al., 2012] for additional critiques of TLBO, and see [Waghmare, 2013] for responses to critiques.

Initialize a population of candidate solutions $\{x_i\}$, $i \in [1, N]$
While not(termination criterion)
    For each individual x_i, $i \in [1, N]$
        $x_t \leftarrow \arg\min_x f(x) : x \in \{x_k\}_{k=1}^N$
        For each solution feature $s \in [1, n]$
            $\bar{x}(s) \leftarrow \frac{1}{N} \sum_{k=1}^N x_k(s)$
            $r \leftarrow U[0, 1]$
            $v(s) \leftarrow x_i(s) + r(x_t(s) - \bar{x}(s))$
        Next solution feature
        $x_i \leftarrow \arg\min[f(x_i), f(v)]$
    Next individual
Next generation

Figure 17.15 Another modified differential evolution algorithm for minimizing $f(x)$. This algorithm is equivalent to the teacher phase of teaching-learning-based optimization.

17.11 CONCLUSION

Researchers have proposed many EAs since Nils Barricelli's first genetic algorithm in 1953 [Dyson, 1998, page 111]. It seems that virtually every natural process can be interpreted as an optimization algorithm [Alexander, 1996]. We have seen in this chapter, and elsewhere, that many of these optimization processes have similar algorithmic features. It is therefore difficult to know where one EA ends, and another begins. When does a new EA belong in its own class, and when should it instead be classified as a variation of an existing EA? One of the challenges for the research community is to find this balance, and to encourage new research while still maintaining high standards for the introduction and development of purportedly new algorithms.

We have covered several additional EAs in this chapter. Some of these are popular and useful, but have not fit well elsewhere in this book. Others are relatively new, and the level of their adoption by engineers and computer scientists is still to be determined. There are many other EAs that we have not had time to discuss, some of which are the following:

- Society and civilization algorithm [Ray and Liew, 2003];
- Charged system search [Kaveh and Talatahari, 2010];
- Invasive weed optimization [Mehrabian and Lucas, 2006];
- Cuckoo search [Yang, 2009a];
- Intelligent water drops [Shah-Hosseini, 2007];
- River formation dynamics [Rabanal et al., 2007];
- Stochastic diffusion search [Bishop, 1989];

- Gaussian adaptation [Kjellström, 1969];
- Big bang big crunch algorithm [Erol and Eksin, 2006];
- Imperialist competitive algorithm [Atashpaz-Gargari and Lucas, 2007];
- Squeaky wheel optimization [Joslin and Clements, 1999];
- Grammatical evolution [O'Neill and Ryan, 2003];
- Glowworm swarm optimization [Krishnanand and Ghose, 2009];
- Chemical reaction optimization [Lam and Li, 2010];
- Krill herd [Gandomi and Alavi, 2012];
- Bat-inspired algorithm [Yang, 2010c];
- Threshold accepting [Dueck and Scheuer, 1990];
- Great deluge algorithm and record-to-record travel [Dueck, 1993];
- Bacterial chemotaxis model [Muller et al., 2002], which we also briefly mentioned at the end of Section 17.6;
- Several artificial bee algorithms, which we briefly mentioned in Section 17.7;
- Several gravity-based and force-based algorithms, which we briefly mentioned in Section 17.8.

There are doubtless other EAs that belong in the above list, or that deserve further discussion, and that are omitted only because of the author's lack of awareness. The algorithms in the above list, along with those discussed in the earlier sections of this chapter, could provide a lifetime of productivity to the interested student and researcher. There are also other computational methods that are not usually classified as EAs, but sometimes the dividing line between machine learning and optimization is fuzzy. This is why we do not discuss algorithms in this book such as neural networks [Fausett, 1994], fuzzy logic [Ross, 2010], artificial immune systems [Hofmeyr and Forrest, 2000], artificial life [Adami, 1997], membrane computing [Păun, 2003], and many other computing paradigms.

PROBLEMS

Written Exercises

17.1 Rewrite the definition of stagnation in Equation (17.4) for the case that the optimization problem is a maximization problem.

17.2 Write an algorithm that is simpler but functionally equivalent to the artificial fish leaping behavior of Figure 17.3.

17.3 Is the group search optimizer elitist?

17.4 How many function evaluations does the shuffled frog leaping algorithm perform each generation?

17.5 Give some specific conditions under which the firefly algorithm of Figure 17.8 could be considered a special case, or a generalization, of the particle swarm optimization algorithm of Figure 11.1.

17.6 Section 8.7 discusses several selection options for EAs. What type of selection does bacterial foraging optimization use?

17.7 Write some pseudo-code showing how you could generate a random unit vector for bacterial foraging optimization.

17.8 Give some specific conditions under which the artificial bee colony algorithm of Figure 17.10 could be considered a special case, or a generalization, of the differential evolution algorithm of Figure 12.2.

17.9 Give some specific conditions under which the gravitational search algorithm of Figure 17.11 could be considered a special case, or a generalization, of the particle swarm optimization algorithm of Figure 11.1.

17.10 What is the probability that a specific child in harmony search is comprised entirely of pre-existing, non-mutated solution features from the parent population?

Computer Exercises

17.11 Simulate one of the EAs in this chapter. Vary a couple of the parameters to see what effect they have on optimization performance.

17.12 The teaching-learning-based optimization (TLBO) algorithm of Figure 17.13 has a teacher phase and a learner phase. Write simulations for three variations of TLBO: the first variation is the original algorithm as shown in Figure 17.13, the second variation uses only the teacher phase, and the third variation uses only the learner phase. Optimize the 10-dimensional Ackley function with a population size of 100 and a function evaluation limit of 10,000. Report the best performance obtained by the three TLBO variations, averaged over 20 Monte Carlo simulations. Repeat for the 10-dimensional Rosenbrock function. Comment on your results.

PART IV

SPECIAL TYPES OF OPTIMIZATION PROBLEMS

CHAPTER 18

Combinatorial Optimization

> We designated as the Messenger Problem (since this problem is encountered by every postal messenger, as well as by many travellers) the task of finding, for a finite number of points whose pairwise distances are known, the shortest path connecting the points.
> —Karl Menger [Gutin and Punnen, 2007, page 1]

Until now, this book has emphasized continuous optimization problems. This chapter discusses discrete optimization problems: that is, $\min_x f(x)$ where the domain of x is discrete. A discrete optimization problem, also called a combinatorial optimization problem, can be thought of as finding the optimal object from among a finite set of candidate objects:

$$\min_x f(x) \text{ where } x \in \{x_1, x_2, \cdots, x_{N_x}\}. \tag{18.1}$$

N_x is called the cardinality of the search space. We could theoretically solve Equation (18.1) by evaluating $f(x)$ for all N_x possible solutions. This approach to combinatorial optimization is called exhaustive search or brute force. However, combinatorial problems often have such a large search space that it is infeasible to check every possible solution.

The knight's tour problem is a classic combinatorial optimization problem. The knight's tour problem was first discussed from a mathematical perspective by Leonhard Euler in 1759 [Ball and Coxeter, 2010]. How can a knight move on an otherwise empty chess board so that it visits each square exactly one time? A closed tour (also called a re-entrant tour) is one in which the knight finishes its tour in the same square in which it started, and thus requires 64 moves; otherwise, the tour is open and thus requires 63 moves.

The open knight's tour problem can be posed as $\min_x f(x)$, where x consists of an initial position and a sequence of 63 moves, and $f(x)$ measures how many squares the knight failed to visit. The cardinality N_x of x (that is, the size of the search space) is over 3.3×10^{13} [Löbbing and Wegener, 1995]. We would not want to try to solve this problem using brute force, but by using human insight and ingenuity we can solve the knight's tour without much difficulty. We see that the cardinality of a combinatorial optimization problem is not necessarily indicative of its difficulty. Figure 18.1 shows a solution to the open knight's tour problem. The knight's tour problem has also been studied using chessboards with a size other than 8×8.

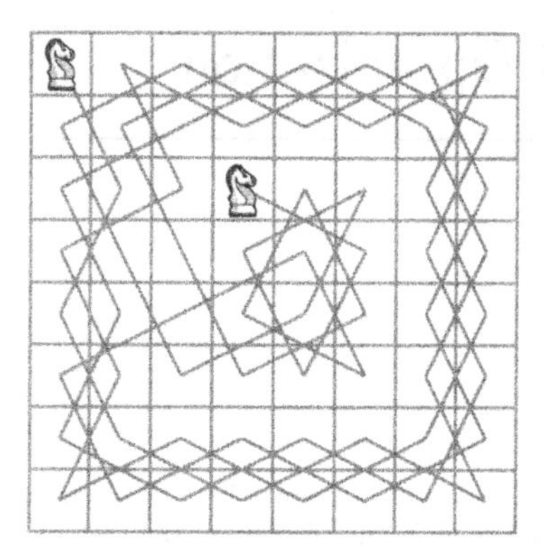

Figure 18.1 A solution to the open knight's tour problem [Fealy, 2006, page 237].

Overview of the Chapter

Most of this chapter is devoted to the traveling salesman problem (TSP), which is perhaps the most famous, applicable, and widely studied combinatorial optimization problem. Section 18.1 gives an overview of the TSP. Section 18.2 discusses a few simple and popular non-evolutionary heuristics for solving TSPs; it has relevance to EAs because we can use the heuristics discussed there to initialize or improve our EA population. Section 18.3 discusses various ways to represent TSP candidate solutions, and how to combine candidate solutions in an EA to obtain child solutions. Section 18.4 discusses some ways to mutate TSP candidate solutions. Section 18.5 ties together the material of the preceding sections and presents a basic EA that we can use to solve the TSP. Section 18.6 discusses the graph coloring problem, which is another popular combinatorial optimization problem. Note that Appendix C.6 discusses TSP benchmark problems and other combinatorial benchmark problems.

18.1 THE TRAVELING SALESMAN PROBLEM

The knight's tour that we discussed above is not a difficult problem, but it leads to the TSP, which is a very hard problem: What is the minimal-length tour that visits each of N cities exactly once? As with the knight's tour, a closed TSP is one in which the tour ends in the same city in which it began; otherwise, the TSP is open. The TSP first appears in written form in a German pamphlet that was published in 1832 with the title, "The traveling salesman – how he should be and what he has to do, to obtain orders and to be sure of a happy success in his business – by an old traveling salesman." Austrian mathematician Karl Menger called the TSP "the messenger problem," and he was the first to discuss it in the technical literature during the late 1920s and early 1930s [Schrijver, 2005].

The TSP has applications in robotics, circuit board drilling, welding, manufacturing, transportation, and many other areas. As we discussed in Section 2.5, an open n-city TSP has $(n-1)!$ possible solutions. This number becomes impossibly large for even moderate values of n. For example, the number of possible solutions to a 50-city TSP is $49! = 6.1 \times 10^{62}$. And 50 cities is not very many for a TSP. A circuit board could have tens of thousands of holes, and a drill needs to be programmed to visit each of those holes while minimizing some cost function (time or energy, for example).

In general, we will assume that an n-city TSP has cities denoted as city 1, city 2, $\cdots$, city n. We assume that there is a given distance $D(i,j)$ between cities i and j for all $i \in [1,n]$ and $j \in [1,n]$, and that $D(i,j) = D(j,i)$. This is called the symmetric TSP because the distance (or cost) from city i to city j is the same as the distance from city j to city i. We could imagine scenarios where $D(i,j) \neq D(j,i)$ (for example, it might cost more to go uphill than downhill) – such problems are called asymmetric TSPs, but we do not discuss them further in this chapter.

A valid tour in an open TSP is one in which all n cities are visited exactly once. A valid tour in a closed TSP is one that begins and ends at the same city, and that visits the other $(n-1)$ cities exactly one. An example of a valid tour in an open four-city TSP is the following:

$$\text{Valid four-city open tour: } 3 \rightarrow 2 \rightarrow 4 \rightarrow 1. \tag{18.2}$$

An example of a valid tour in a closed four-city TSP is the following:

$$\text{Valid four-city closed tour: } 3 \rightarrow 2 \rightarrow 4 \rightarrow 1 \rightarrow 3. \tag{18.3}$$

An edge, or leg, is one segment of a tour. Equation (18.2) consists of three edges: $3 \rightarrow 2$ is the first edge, $2 \rightarrow 4$ is the second edge, and $4 \rightarrow 1$ is the third edge. Equation (18.3) consists of four edges. In general, an open n-city tour includes $(n-1)$ edges, and a closed n-city tour includes n edges.

In the TSP, we try to minimize total distance. Suppose that n cities in an open TSP are listed in the order $x_1 \rightarrow x_2 \rightarrow \cdots \rightarrow x_n$. Then the total distance is

$$D_T = \sum_{i=1}^{n-1} D(x_i, x_{i+1}). \tag{18.4}$$

Note that we use the term "distance" in a general sense. It might refer to physical distance, financial cost, or any other quantity that we want to minimize in a combinatorial problem.

18.2 TSP INITIALIZATION

This section presents a few popular and simple non-evolutionary heuristics that we can use to try to solve TSPs [Nemhauser and Wolsey, 1999]. We can use these heuristics not only to search for TSP solutions, but also to initialize the population of an EA that is designed to solve a TSP. If we initialize an EA intelligently rather than randomly, we can greatly increase our chances of finding a good solution. This applies not only to the TSP but also to any other problem that we want to solve with an EA (see Section 8.1).

The initialization algorithms in this section are called *greedy* because they all make the incremental change to their candidate solutions which promises the best immediate change in performance. That is, they all build candidate solutions based on the highest immediate payoff. Section 18.2.1 builds a candidate solution by iteratively adding the city that is closest to the previously-added city. Section 18.2.2 builds a candidate solution by iteratively adding the next shortest edge. Section 18.2.3 builds a candidate solution by iteratively adding the city that is closest to any of the previously-added cities. Finally, Section 18.2.4 discusses the use of randomness in greedy initialization methods.

18.2.1 Nearest-Neighbor Initialization

One simple and intuitive way to initialize a candidate solution is with a nearest-neighbor strategy. This strategy is described as follows for an n-city TSP.

1. Initialize $i = 1$.
2. Randomly select a city $s(1) \in [1, n]$ as the starting city.
3. $s(i+1) \leftarrow \arg\min_\sigma \{D(s(i), \sigma) : \sigma \notin s(k) \text{ for } k \in [1, i]\}$. That is, find the city that is closest to $s(i)$ that has not yet been assigned to an element of s, and assign it to $s(i+1)$.
4. Increment i by one.
5. If $i = n$, terminate; otherwise, go to step 3.

At the end of this process we have an open tour $s(1) \to s(2) \to \cdots \to s(n)$ that gives us a reasonable guess for a TSP solution. If we want a closed tour then we simply add $s(1)$ to the end of the open tour.

Because of the random selection of the starting city we will generally obtain different candidate solutions if we perform this algorithm more than once. For instance, consider the distance matrix

$$D = \begin{bmatrix} \times & 3 & 2 & 9 & 3 \\ 3 & \times & 5 & 8 & 11 \\ 2 & 5 & \times & 4 & 6 \\ 9 & 8 & 4 & \times & 10 \\ 3 & 11 & 6 & 10 & \times \end{bmatrix} \tag{18.5}$$

where D_{ij}, which we can also write as $D(i, j)$, represents the distance between city i and city j. If we start at city 1, the nearest-neighbor algorithm gives the tour $1 \to 3 \to 4 \to 2 \to 5$, which has a total cost of 25. If we start at city 2, the algorithm gives the tour $2 \to 1 \to 3 \to 4 \to 5$, which has a total cost of 19.

Note that a general $n \times n$ distance matrix D is symmetric since the distance between city i and city j is the same as that between city j and city i. Also, an $n \times n$ matrix has $n(n-1)/2$ terms above the diagonal. Therefore, a symmetric n-city TSP has $n(n-1)/2$ unique edges.

If we want to intelligently initialize an EA to solve the TSP, we could use nearest-neighbor initialization for just one individual in the population, or for a few individuals in the population, or for the entire population. However, if we initialize too many individuals this way then we will probably obtain many duplicate individuals. We could also use a stochastic nearest-neighbor initialization algorithm – in this case, the probability of assigning a given city to $s(i+1)$ at each iteration would be inversely proportional to its distance from $s(i)$. Finally, we could take the nearest-neighbor algorithm to another level by performing a "nearest two-neighbor" algorithm. In this approach, given $s(i)$, we could assign a city to $s(i+1)$ that results in the smallest combined distance $D(s(i), s(i+1)) + D(s(i+1), \sigma)$, where σ is allowed to be equal to any city $\neq s(k)$ for $k \leq i+1$.

It is not difficult to find an example where nearest-neighbor initialization performs poorly. Figure 18.2 shows a simple five-city TSP. In the figure on the left, we start at city 1 and get a poor result with nearest-neighbor initialization. In the figure on the right, we start at city 3 and get the globally optimal solution with nearest-neighbor initialization. In Figure 18.2, the performance of nearest-neighbor initialization strongly depends on the starting city. But in general, nearest-neighbor initialization often fails due to the fact that it does not look more than one city ahead when planning its route.

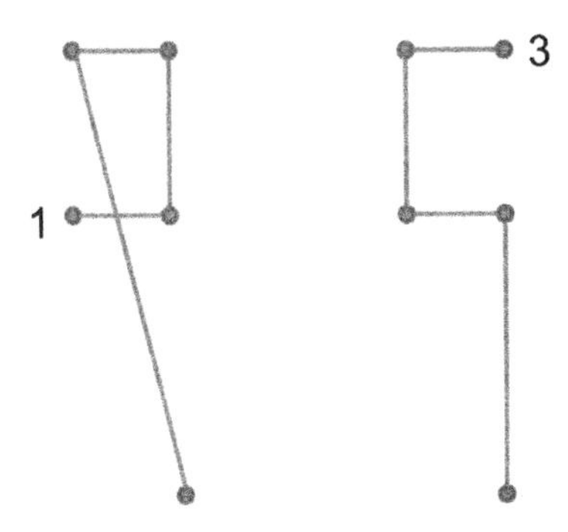

Figure 18.2 Nearest-neighbor initialization results for a five-city open TSP. Depending on the starting city, we either obtain a poor result (left) or a good result (right).

18.2.2 Shortest-Edge Initialization

Another simple way to initialize a candidate TSP solution is with a greedy shortest-edge algorithm. Suppose that we have an n-city TSP with a distance matrix D as shown in Equation (18.5). We define L_k as the edge that is associated with the k-th smallest number in D. That is, $\{L_k\}$ consists of $n(n-1)/2$ edges that are sorted in ascending order of distance. Shortest-edge initialization for an n-city closed TSP proceeds as follows.

1. Define T as the set of edges in the tour. Initialize T to the empty set.

2. Find the shortest edge in $\{L_k\}$ that satisfies the following constraints: (a) It is not in T; (b) If added to T, it will not result in a closed tour with less than n edges; (c) If it joins cities i and j and it is added to T, then T will not have more than two edges associated with city i or city j.

3. If T has n edges, then we are done; otherwise, go to step (2).

Shortest-edge initialization includes in the tour the edges from the cities that are nearest to each other, and it continues that process until it obtains a valid tour. Unlike nearest-neighbor initialization, shortest-edge initialization is not stochastic so it results in the same tour every time it executes. Therefore, in general, shortest-edge initialization should be used to initialize only one individual in an EA. The only exception to this statement is if more than one pair of cities have the same distance, in which case a random process can be used to break the tie in step (2) above, and in which case a different tour will result (in general), depending on the result of the random process.

As an example of shortest-edge initialization, consider the distance matrix of Equation (18.5). Shortest-edge initialization proceeds as follows.

1. The shortest edge is between cities 1 and 3, so we include that edge in T.

2. The shortest remaining edge is between cities 1 and 2, and cities 1 and 5. We randomly choose the edge between cities 1 and 5 to include in T.

3. The shortest remaining edge is between cities 1 and 2, but but city 1 already has two edges in T. So we look for the next shortest edge, which is between cities 3 and 4, and include that edge in T.

4. The shortest remaining edge is between cities 2 and 3, but city 3 already has two edges in T. So we look for the next shortest edge, which is between cities 3 and 5, but again, city 3 already has two edges in T. So we look for the next shortest edge, which is between cities 2 and 4, so we include that edge in T.

5. The only remaining edge that satisfies the constraints of the shortest-edge algorithm is the one between cities 2 and 5, so we include that edge in T to complete the closed tour.

The above algorithm gives the closed tour $1 \to 3 \to 4 \to 2 \to 5 \to 1$. If we want to use shortest-edge initialization to find an open tour, we would simply stop the above algorithm after obtaining $(n-1)$ edges in T, which would give the tour $5 \to 1 \to 3 \to 4 \to 2$.

18.2.3 Insertion Initialization

Insertion initialization begins with a subtour and then adds one city at a time to the tour such that the addition of the selected city gives the smallest increase in distance [Rosenkrantz et al., 1977]. The initial subtour is often a single edge, which is usually the shortest. In this case we have the nearest insertion algorithm, which is given as follows for the open TSP.

1. Define T as the set of edges in the tour. Initialize T to the shortest edge in the distance matrix.

2. $c \leftarrow \arg\min_c\{D(c,k) : (c \notin T) \text{ and } (k \in T)\}$. That is, among all cities that are not in T, select the one that is closest to T.

3. $\{k,j\} \leftarrow \arg\min_{k,j}\{(D(k,c) + D(c,j)) - D(k,j) : D(k,j) \in T\}$. That is, select the edge $D(k,j)$ from T such that the difference between the subtour distance $k \to c \to j$ and the distance $k \to j$ is minimized.

4. Remove $D(k,j)$ from T, and add $D(k,c)$ and $D(c,j)$ to T.

5. If T includes $(n-1)$ edges, then we are done; otherwise, go to step (2).

If we want a closed tour, then we simply add one more edge to T to complete the tour. We can modify the nearest insertion algorithm by changing the initialization in step (1) so that T is initialized to the convex hull of the cities, or initialized randomly, or initialized with a variety of other options. This would allow us to initialize more than one EA individual with an insertion algorithm.

As an example of nearest insertion initialization, consider the distance matrix of Equation (18.5). Nearest insertion proceeds as follows.

1. The shortest edge is between cities 1 and 3, so we include that edge in T.

2. The cities that are not yet in T are cities 2, 4, and 5. Among those cities, the one that is closest to T (that is, closest to either city 1 or city 3) is city 2 or city 5, which are both 3 units from city 1. we randomly select city 5 to include in T. We then remove the 1/3 edge from T, and add the 1/5 and 5/3 edges to T.

3. The cities that are not yet in T are cities 2 and 4. Among those cities, the one that is closest to T (that is, closest to either city 1, city 3, or city 5) is city 2, which is 3 units from city 1. This gives us two options.

   (a) We could remove the 1/5 edge from T, and replace it with the 1/2 and 2/5 edges. This would increase the subtour distance from 3 units (the 1/5 edge distance) to 14 units (the sum of the 1/2 and 2/5 edge distances). This is an increase of 11 units.

   (b) We could remove the 5/3 edge from T, and replace it with the 5/2 and 2/3 edges. This would increase the subtour distance from 6 units (the 5/3 edge distance) to 16 units (the sum of the 5/2 and 2/3 edge distances). This is an increase of 10 units.

   We choose option (b) since it results in the smallest increase. T now includes the 1/5, 5/2, and 2/3 edges.

4. The only city that is not yet in T is city 4. We thus need to add an edge to T that includes city 4. This gives us three options.

   (a) We could remove the 1/5 edge from T, and replace it with the 1/4 and 4/5 edges. This would increase the subtour distance from 3 units (the 1/5 edge distance) to 19 units (the sum of the 1/4 and 4/5 edge distances). This is an increase of 16 units.

   (b) We could remove the 5/2 edge from T, and replace it with the 5/4 and 4/2 edges. This would increase the subtour distance from 11 units (the 5/2 edge distance) to 18 units (the sum of the 5/4 and 4/2 edge distances). This is an increase of 7 units.

   (c) We could remove the 2/3 edge from T, and replace it with the 2/4 and 4/3 edges. This would increase the subtour distance from 5 units (the 2/3 edge distance) to 12 units (the sum of the 2/4 and 4/5 edge distances). This is an increase of 7 units.

   We could choose either option (b) or option (c) since they result in the smallest increase. We randomly choose option (c). T now includes the 1/5, 5/2, 2/4, and 4/3 edges.

The above algorithm gives the open tour $1 \to 5 \to 2 \to 4 \to 3$.

18.2.4 Stochastic Initialization

Nearest-neighbor initialization is the only initialization method that we have discussed so far that is stochastic. However, we could modify any of the other initialization options to include a random component. In addition, we could modify nearest-neighbor initialization to be more random than the algorithm that we presented in Section 18.2.1. The addition of randomness to an initialization method maintains the attractive features of the method, while also incorporating one of the fundamental components of EAs. It also allows us to use the initialization methods for more than one EA individual.

In nearest-neighbor initialization (Section 18.2.1), we could replace step (3), "Find the city that is closest to $s(i)$," with the selection of a city with a probability that is inversely proportional to the distance from $s(i)$. This would give the greatest probability of selection to the city that is closest to $s(i)$, but it would also give nonzero selection probabilities to all of the other cities in the TSP.

In shortest-edge initialization (Section 18.2.2), we could replace the selection of the shortest edge that satisfies the given constraints, with the selection of an edge (among those that satisfy the given constraints) with a probability that is inversely proportional to the edge length. This would give the greatest probability to the shortest edge, but it would also give nonzero selection probabilities to all of the other edges in the TSP.

In insertion initialization (Section 18.2.3), we could add randomness to two steps. First, in step (2), instead of selecting the city that is closest to T, we could select a city with a probability that is inversely proportional to the distance from T. This would give the greatest probability to the nearest city, but it would also give nonzero selection probabilities to all of the other cities that are not yet in T. Second, in step (3), instead of selecting the edge such that the difference in the subtour distances is minimized, we could select an edge with a probability that is inversely proportional to the differences in the subtour distances. This would give the greatest probability to the minimum distance-difference edge, but it would also give nonzero selection probabilities to all of the other edges in T.

18.3 TSP REPRESENTATIONS AND CROSSOVER

This section discusses different ways to represent TSP candidate solutions. We discuss path (Section 18.3.1), adjacency (Section 18.3.2), ordinal (Section 18.3.3), and matrix (Section 18.3.4) representations. We also discuss how candidate solutions can be combined via crossover using these various representations.

18.3.1 Path Representation

Path representation is the most natural way of representing a TSP tour. In path representation, the vector

$$x = \begin{bmatrix} x_1 & x_2 & \cdots & x_n \end{bmatrix} \tag{18.6}$$

represents the n-city tour $x_1 \to x_2 \to \cdots \to x_n$. The following sections discuss some ways that we can combine parent individuals that are represented in this way to obtain child individuals.

18.3.1.1 Partially Matched Crossover Partially matched crossover (PMX) [Goldberg and Lingle, 1985] is based on classic single-point crossover as is often used in GAs (see Section 8.8). As an example, consider the two parent vectors

$$\begin{aligned} P_1 &= \begin{bmatrix} 2 & 3 & 4 & 5 & 6 & 1 \end{bmatrix} \\ P_2 &= \begin{bmatrix} 3 & 2 & 6 & 1 & 4 & 5 \end{bmatrix}. \end{aligned} \tag{18.7}$$

If we perform single-point crossover at the midpoint of the two vectors, we obtain the children

$$\begin{aligned} c_1 &= \begin{bmatrix} 2 & 3 & 4 & 1 & 4 & 5 \end{bmatrix} \\ c_2 &= \begin{bmatrix} 3 & 2 & 6 & 5 & 6 & 1 \end{bmatrix}. \end{aligned} \tag{18.8}$$

These children are invalid because c_1 visits city 4 twice and does not visit city 6 at all. c_2 has the opposite problem; it visits city 6 twice and does not visit city 4 at all. We can easily repair the children by replacing one of the 4 elements in c_1 with a 6, and by replacing one of the 6 elements in c_2 with a 4. For example, this might result in the children of Equation (18.8) being modified to

$$\begin{aligned} c_1 &= \begin{bmatrix} 2 & 3 & \mathbf{6} & 1 & 4 & 5 \end{bmatrix} \\ c_2 &= \begin{bmatrix} 3 & 2 & 6 & 5 & \mathbf{4} & 1 \end{bmatrix} \end{aligned} \tag{18.9}$$

where the bold-faced cities are the ones that were randomly changed to give valid tours.

18.3.1.2 Order Crossover Order crossover (OX) copies a section of a tour from one parent to the child [Davis, 1985]. This results in a child that has a partial tour. Order crossover then completes the child by copying the remaining required cities from the second parent to the child, while maintaining the relative order of those cities from Parent 2. For example, suppose that we have the parents

$$\begin{aligned} P_1 &= \begin{bmatrix} 9 & 2 & 3 & 8 & 4 & 5 & 6 & 1 & 7 \end{bmatrix} \\ P_2 &= \begin{bmatrix} 4 & 5 & 2 & 1 & 8 & 7 & 6 & 9 & 3 \end{bmatrix}. \end{aligned} \tag{18.10}$$

We randomly select a subtour from P_1; suppose that we select subtour $\begin{bmatrix} 8 & 4 & 5 & 6 \end{bmatrix}$ from P_1. This gives the partial child

$$c_1 = \begin{bmatrix} - & - & - & 8 & 4 & 5 & 6 & - & - \end{bmatrix}. \tag{18.11}$$

We see that c_1 still needs cities 1, 2, 3, 7, and 9. Those cities occur in the order $\{2, 1, 7, 9, 3\}$ in P_2. We therefore copy those cities in that order into c_1 to obtain

$$c_1 = \begin{bmatrix} 2 & 1 & 7 & 8 & 4 & 5 & 6 & 9 & 3 \end{bmatrix}. \tag{18.12}$$

In order crossover, we often create a second child by using the above process with the roles of P_1 and P_2 reversed. In our example, this would give a preliminary second child with a subtour copied from P_2 as

$$c_2 = \begin{bmatrix} - & - & - & 1 & 8 & 7 & 6 & - & - \end{bmatrix}. \tag{18.13}$$

We would then copy the remaining cities 9, 2, 3, 4, and 5 in order from P_1 to complete the second child:

$$c_2 = \begin{bmatrix} 9 & 2 & 3 & 1 & 8 & 7 & 6 & 4 & 5 \end{bmatrix}. \tag{18.14}$$

18.3.1.3 Cycle Crossover Cycle crossover (CX), introduced in [Oliver et al., 1987], creates a child from two parents in a way that preserves as much sequence information as possible from the first parent, while completing the child with information from the second parent. Cycle crossover is best explained with an example. Suppose that we have parents

$$\begin{aligned} P_1 &= \begin{bmatrix} 2 & 3 & 4 & 5 & 6 & 1 \end{bmatrix} \\ P_2 &= \begin{bmatrix} 4 & 5 & 2 & 1 & 6 & 3 \end{bmatrix}. \end{aligned} \tag{18.15}$$

We create a child as follows.

1. Select a random index between 1 and n. Suppose that we select 4. $P_1(4) = 5$, so the child is initialized as $c = \begin{bmatrix} - & - & - & 5 & - & - \end{bmatrix}$.
2. $P_2(4) = 1$, and city 1 occurs in the sixth position of P_1, so the child is augmented to become $c = \begin{bmatrix} - & - & - & 5 & - & 1 \end{bmatrix}$.
3. $P_2(6) = 3$, and city 3 occurs in the second position of P_2, so the child is augmented to become $c = \begin{bmatrix} - & 3 & - & 5 & - & 1 \end{bmatrix}$.
4. $P_2(2) = 5$, but the child already includes city 5. Therefore we copy the remaining required cities to the child from P_2, which gives $c = \begin{bmatrix} 4 & 3 & 2 & 5 & 6 & 1 \end{bmatrix}$.

We often create a second child by reversing the roles of P_1 and P_2. Figure 18.3 shows how cycle crossover operates. Cycle crossover always results in valid children.

P_1 = Parent 1, P_2 = Parent 2
$s \leftarrow$ random integer from $[1, n]$
$r \leftarrow P_1(s)$
Initialize the child to an empty tour: $C(i) = 0$ for $i \in [1, n]$
$C(s) \leftarrow r$
While $C(i) = 0$ for some $i \in [1, n]$
    $r \leftarrow P_2(s)$
    If $C(i) \neq r$ for all $i \in [1, n]$ then
        $s \leftarrow \{i : P_1(i) = r\}$
        $C(s) \leftarrow r$
    else
        For $i = 1$ to n
            If $C(i) = 0$ then $C(i) \leftarrow P_2(i)$
        Next i
    End if
Next city

Figure 18.3 Cycle crossover for the n-city TSP.

18.3.1.4 Order-Based Crossover Order-based crossover (OBX) is a modification of cycle crossover [Syswerda, 1991]. Order-based crossover randomly selects several positions in the first parent P_1, finds the cities in the corresponding positions in P_2, and then re-orders those cities in P_1 with their order from P_2. The result is the child. For example, suppose that we have the parents

$$\begin{aligned} P_1 &= \begin{bmatrix} 2 & 3 & 4 & 5 & 6 & 1 \end{bmatrix} \\ P_2 &= \begin{bmatrix} 4 & 5 & 2 & 1 & 6 & 3 \end{bmatrix}. \end{aligned} \tag{18.16}$$

Order-based crossover would proceed by randomly selecting a certain number of positions in P_1. Suppose that we select positions 1, 3, and 4. The cities in those positions in P_2 are cities 4, 2, and 1. The child is initialized with all cities from P_1 except for the cities 4, 2, and 1:

$$c_1 = \begin{bmatrix} - & 3 & - & 5 & 6 & - \end{bmatrix}. \tag{18.17}$$

Next, we copy cities 4, 2, and 1 to the child in the same order that they occur in P_2. This gives the child

$$c_1 = \begin{bmatrix} 4 & 3 & 2 & 5 & 6 & 1 \end{bmatrix}. \tag{18.18}$$

We often create a second child by reversing the roles of P_1 and P_2, which in the above example results in c_2 being initialized with all cities from P_2 except for the cities 2, 4, and 5 (since those cities are in positions 1, 3, and 4 in P_1):

$$c_2 = \begin{bmatrix} - & - & - & 1 & 6 & 3 \end{bmatrix}. \tag{18.19}$$

We then copy cities 2, 4, and 5 to the child in the same order that they occur in P_1, which gives

$$c_2 = \begin{bmatrix} 2 & 4 & 5 & 1 & 6 & 3 \end{bmatrix}. \tag{18.20}$$

18.3.1.5 Inver-Over Crossover Given two parents P_1 and P_2, inver-over crossover works as follows [Tao and Michalewicz, 1998].

1. Randomly select a position s from P_1. Suppose that $P_1(s) = r$.
2. Suppose that r is in the k-th position in P_2; that is, $P_2(k) = r$. Set the end-point city as $e = P_2(k+1)$.
3. Reverse the order of the cities between $P_1(s+1)$ and e in P_1 to obtain the child.

For example, suppose that we have the parents

$$\begin{aligned} P_1 &= \begin{bmatrix} 2 & 3 & 4 & 5 & 6 & 1 \end{bmatrix} \\ P_2 &= \begin{bmatrix} 4 & 5 & 2 & 1 & 6 & 3 \end{bmatrix}. \end{aligned} \tag{18.21}$$

We randomly select a position s from P_1; suppose that we select $s = 4$. We see that $P_1(4) = 5$. We see that city 5 is in the second position in P_2; that is, $P_2(2) = 5$. So we set $e = P_2(3) = 2$ as the end-point city. We then reverse the order of the cities between $P_1(5)$ and city 2 in P_1 to obtain

$$c = \begin{bmatrix} 6 & 5 & 4 & 3 & 2 & 1 \end{bmatrix}. \tag{18.22}$$

If $k = n$ in Step 2, then $P_2(k+1)$ is not defined. In this case we can use some ad-hoc method to continue the crossover process; for example, we could set $e = P_2(k-1)$, or we could go back to Step 1 and select a new random s.

18.3.2 Adjacency Representation

In adjacency representation [Grefenstette et al., 1985], if a tour represented by vector x includes a direct path from city i to city j, then $x(i) = j$ – that is, the i-th element of x is equal to j. For example, consider the vector

$$x = \begin{bmatrix} 2 & 4 & 8 & 3 & 9 & 7 & 1 & 5 & 6 \end{bmatrix}. \tag{18.23}$$

Vector x is interpreted as follows.

- $x(1) = 2$, so the tour includes an edge from city 1 to city 2.
- $x(2) = 4$, so the tour includes an edge from city 2 to city 4.
- $x(3) = 8$, so the tour includes an edge from city 3 to city 8.
- $x(4) = 3$, so the tour includes an edge from city 4 to city 3.
- $x(5) = 9$, so the tour includes an edge from city 5 to city 9.
- $x(6) = 7$, so the tour includes an edge from city 6 to city 7.
- $x(7) = 1$, so the tour includes an edge from city 7 to city 1.
- $x(8) = 5$, so the tour includes an edge from city 8 to city 5.
- $x(9) = 6$, so the tour includes an edge from city 9 to city 6.

Putting it all together, we see that the tour represented by x is

$$1 \to 2 \to 4 \to 3 \to 8 \to 5 \to 9 \to 6 \to 7 \to 1. \tag{18.24}$$

A vector x that uses the adjacency representation for an n-city TSP has the following properties.

- $x(i) \neq i$ for all $i \in [1, n]$.
- For all $j \in [1, n]$, there exists exactly one $i \in [1, n]$ such that $x(i) = j$.

The above properties are necessary but not sufficient properties for x to represent a valid tour. For example, the vector

$$x = \begin{bmatrix} 2 & 1 & 8 & 3 & 9 & 7 & 4 & 5 & 6 \end{bmatrix} \tag{18.25}$$

is invalid. If we start in city 1, we repeat the sequence $1 \to 2 \to 1 \to 2 \to \cdots$ indefinitely, and we never visit the rest of the cities. The following sections discuss some ways that we can combine parent individuals that are represented in this way to obtain child individuals.

18.3.2.1 Classic Crossover First we discuss a crossover method that does *not* work with the adjacency representation, and that is single-point crossover as used in GAs (see Section 8.8). For example, consider the two parent vectors

$$\begin{aligned} P_1 &= \begin{bmatrix} 2 & 4 & 1 & 3 \end{bmatrix} \\ P_2 &= \begin{bmatrix} 4 & 3 & 1 & 2 \end{bmatrix}. \end{aligned} \tag{18.26}$$

If we perform single-point crossover at the midpoint of the two vectors, we obtain the children

$$\begin{aligned} c_1 &= \begin{bmatrix} 2 & 4 & 1 & 2 \end{bmatrix} \\ c_2 &= \begin{bmatrix} 4 & 3 & 1 & 3 \end{bmatrix}. \end{aligned} \tag{18.27}$$

c_1 represents the tour $1 \to 2 \to 4 \to 2$, which is invalid because it never visits city 3. Note that c_1 includes two "2" entries and no "3" entries. c_2 represents the tour $1 \to 4 \to 3 \to 1$, which is invalid because it never visits city 2. Note that c_2 includes two "3" entries and no "2" entries.

18.3.2.2 Alternating Edges Crossover Alternating edges crossover starts with classic crossover and repairs invalid tours [Grefenstette et al., 1985]. For instance, we know that c_1 in Equation (18.27) is invalid because it has two "2" entries and no "3" entries. We can try to repair it by replacing one of the "2" entries with a "3." If we replace the first "2" with a "3" we obtain

$$c_1' = \begin{bmatrix} 3 & 4 & 1 & 2 \end{bmatrix}. \tag{18.28}$$

This does not work because it leads to the cycle $1 \to 3 \to 1 \to \ldots$ We can therefore try to repair c_1 by instead replacing the second "2" with a "3" to obtain

$$c_1'' = \begin{bmatrix} 2 & 4 & 1 & 3 \end{bmatrix} \tag{18.29}$$

which represents the valid tour $1 \to 2 \to 4 \to 3$. The above example shows single-point crossover, but alternating edges crossover can also be used with two-point crossover, or a larger number of crossover points. Although alternating edges crossover works in the sense that it creates valid tours, it often disrupts good tours, and so it usually does not work well in practice.

18.3.2.3 Heuristic Crossover Heuristic crossover is so called because it uses common sense to combine two candidate solutions [Grefenstette et al., 1985]. It works by combining the best edges from two parents to create the child. Heuristic crossover is described as follows.

1. Choose a random city r as the starting point.
2. Compare the edges from the parents that leave city r. Select the shorter edge for the child.
3. The city on the other side of the edge selected above is the starting point for the selection of the next city.
4. If the selected city is already in the child, then replace the selected city with a random one that is not in the child.
5. Continue with Step 2 until the child tour is complete.

Figure 18.4 presents pseudocode for heuristic crossover. Note that we can use heuristic crossover with any vector-based TSP representation. We can easily extend Figure 18.4 for more than two parents. Also, we could experiment with other modifications of Figure 18.4. For example, instead of deterministically choosing $r_{\min}$ from $\{r_1, r_2\}$ to minimize $[d(r, r_1), d(r, r_2)]$, we could select $r_{\min}$ stochastically. This could entail, for instance, setting $r_{\min} = r_i$ with a probability that is inversely proportional to $d(r, r_i)$ for $i \in [1, 2]$.

P_1 = Parent 1, P_2 = Parent 2
$r \leftarrow$ random city in $[1, n]$
Child $C \leftarrow \{r\}$
While $|C| < n$
    Use r_i to indicate the city that follows r in Parent i, for $i \in [1, 2]$
    $d(r, r_i)$ = distance from r to r_i, for $i \in [1, 2]$
    $r_{\min} \leftarrow \min_{\{r_1, r_2\}} [d(r, r_1), d(r, r_2)]$
    $r \leftarrow r_{\min}$
    If $r \in C$ then
        $r \leftarrow$ random city in $[1, n]$ such that $r \notin C$
    End if
    $C \leftarrow \{C, r\}$
Next city

Figure 18.4 Heuristic crossover for the n-city TSP.

As an illustration of heuristic crossover, suppose that we have a four-city TSP with the distance matrix

$$D = \begin{bmatrix} - & 13 & 9 & 15 \\ 13 & - & 4 & 7 \\ 9 & 4 & - & 12 \\ 15 & 7 & 12 & - \end{bmatrix} \tag{18.30}$$

where D_{ij} represents the distance between city i and city j. Suppose we have parents

$$\begin{aligned} P_1 &= \begin{bmatrix} 2 & 4 & 1 & 3 \end{bmatrix} \\ P_2 &= \begin{bmatrix} 4 & 3 & 1 & 2 \end{bmatrix}. \end{aligned} \tag{18.31}$$

Heuristic crossover proceeds as follows.

1. We randomly select a starting city $r \in [1, 4]$; suppose that we select $r = 2$.
2. We see that P_1 has the edge $2 \to 4$, and $d(2, 4) = 7$. We see that P_2 has the edge $2 \to 3$, and $d(2, 3) = 4$. Therefore, we select the edge $2 \to 3$ for the child, which gives $C = \{2, 3\}$.
3. We see that both parents have the edge $3 \to 1$, so the child is augmented to become $C = \{2, 3, 1\}$.
4. We see that P_1 has the edge $1 \to 2$, and $d(1, 2) = 13$. We see that P_2 has the edge $1 \to 4$, and $d(1, 4) = 15$. Therefore, we select the edge $1 \to 2$ for the child. But city 2 is already in C, so we choose a random city that is not in C. Suppose we choose city 4 (which is, in fact, the only city that is not yet in C). This gives $C = \{2, 3, 1, 4\}$.
5. C is now complete, and its adjacency representation is $C = \begin{bmatrix} 4 & 3 & 1 & 2 \end{bmatrix}$.

18.3.3 Ordinal Representation

In ordinal representation [Grefenstette et al., 1985], an n-city tour is represented as a vector

$$x = \begin{bmatrix} x_1 & x_2 & \cdots & x_n \end{bmatrix} \tag{18.32}$$

where $x_i \in [1, n - i]$. That is

$$\begin{aligned} x_1 &\in [1, n] \\ x_2 &\in [1, n-1] \\ x_3 &\in [1, n-2] \\ &\vdots \\ x_n &= 1. \end{aligned} \tag{18.33}$$

Suppose that we have an ordered list of cities:

$$L = \{\ 1 \quad 2 \quad \cdots \quad n\ \}. \tag{18.34}$$

That is, $L(i) = i$ for $i \in [1, n]$. In ordinal representation, x_1 represents the first city of the tour. x_2 gives the index in the set $L_2 = L \backslash \{x_1\}$ of the second city of the tour.[1] x_3 gives the index in the set $L_3 = L \backslash \{x_1, x_2\}$ of the third city of the tour. In general, x_k gives the index in the set $L_k = L \backslash \cup_{i=1}^{k-1} x_i$ of the k-th city of the tour. Note that any vector of the form of Equation (18.33) represents a valid tour.

[1]We use the notation $A \backslash B$ to indicate the set $\{x : x \in A \text{ and } x \notin B\}$. That is, $A \backslash B$ means the set of all elements in A that are not in the set B.

For example, suppose that a six-city tour is represented as

$$x = \begin{bmatrix} 5 & 2 & 4 & 1 & 2 & 1 \end{bmatrix}. \tag{18.35}$$

Given the ordered list $L = \{1, 2, 3, 4, 5, 6\}$, we construct the tour represented by x as follows.

1. $x_1 = 5$, and $L(5) = 5$, so city 5 is the first city in the tour. Removing 5 from L gives $L_2 = \{1, 2, 3, 4, 6\}$.
2. $x_2 = 2$, and $L_2(2) = 2$, so city 2 is the second city in the tour. Removing 2 from L_2 gives $L_3 = \{1, 3, 4, 6\}$.
3. $x_3 = 4$, and $L_3(4) = 6$, so city 6 is the third city in the tour. Removing 6 from L_3 gives $L_4 = \{1, 3, 4\}$.
4. $x_4 = 1$, and $L_4(1) = 1$, so city 1 is the fourth city in the tour. Removing 1 from L_4 gives $L_5 = \{3, 4\}$.
5. $x_5 = 2$, and $L_5(2) = 4$, so city 4 is the fifth city in the tour. Removing 4 from L_5 gives $L_6 = \{3\}$.
6. $x_6 = 1$, and $L_6(1) = 3$, so city 3 is the sixth city in the tour.

This gives the tour $5 \to 2 \to 6 \to 1 \to 4 \to 3$.

Suppose that we want to try single-point crossover with ordinal representation to combine two parents and obtain a child. Consider the parents

$$\begin{aligned} P_1 &= \begin{bmatrix} 5 & 2 & 4 & 1 & 2 & 1 \end{bmatrix} \\ P_2 &= \begin{bmatrix} 1 & 5 & 3 & 3 & 1 & 1 \end{bmatrix}. \end{aligned} \tag{18.36}$$

If we select the crossover point as the midpoint of the parents, we obtain the children

$$\begin{aligned} c_1 &= \begin{bmatrix} 5 & 2 & 4 & 3 & 1 & 1 \end{bmatrix} \\ &= 5 \to 2 \to 6 \to 4 \to 1 \to 3 \\ c_2 &= \begin{bmatrix} 1 & 5 & 3 & 1 & 2 & 1 \end{bmatrix} \\ &= 1 \to 6 \to 4 \to 2 \to 5 \to 3. \end{aligned} \tag{18.37}$$

Both children represent valid tours. Although ordinal representation seems a little awkward at first, it has the advantage that single-point crossover always results in a valid tour.

18.3.4 Matrix Representation

In matrix representation, an open n-city tour is represented by an $n \times n$ matrix M containing only zeros and ones [Fox and McMahon, 1991]. $M_{ik} = 1$ if and only if city i occurs before city k in the tour. For instance, consider the matrix

$$M = \begin{bmatrix} 0 & 1 & 0 & 1 & 1 \\ 0 & 0 & 0 & 1 & 1 \\ 1 & 1 & 0 & 1 & 1 \\ 0 & 0 & 0 & 0 & 1 \\ 0 & 0 & 0 & 0 & 0 \end{bmatrix}. \tag{18.38}$$

The ones in the first row indicate that city 1 is before cities 2, 4, and 5. The ones in the second row indicate that city 2 is before cities 4 and 5. The ones in the third row indicate that city 3 is before cities 1, 2, 4, and 5. The one in the fourth row indicates that city 4 is before city 5. Finally, the fact that the fifth row is comprised of all zeros indicates that city 5 is the last city in the tour. Therefore, Equation (18.38) represents the tour $3 \to 1 \to 2 \to 4 \to 5$.

Another way to interpret Equation (18.38) is to note that the row with the most ones is the first city, the row with the second most ones is the second city, and so on. The row with the k-th most ones is the k-th city in the tour.

We note several properties for any $n \times n$ matrix M that represents a valid tour.

- Exactly one row of M has $(n-1)$ ones, exactly one row of M has $(n-2)$ ones, and so on.
- The above property allows us to find the number of ones in M:

$$\text{Number of ones} = \sum_{i=1}^{n}(n-i) = n(n-1)/2. \tag{18.39}$$

- No city occurs before itself in the tour, so $M_{ii} = 0$ for all $i \in [1, n]$.
- If city i occurs before city j, and city j occurs before city k, then city i occurs before city k. That is,

$$(M_{ij} = 1 \text{ and } M_{jk} = 1) \Longrightarrow M_{ik} = 1. \tag{18.40}$$

The following sections discuss a couple of ways that we can combine parent matrices to obtain children: we can take the intersection of the two parent matrices, or the union of the two matrices.

18.3.4.1 Intersection Crossover We illustrate intersection crossover with an example. Suppose that Equation (18.38) represents parent M_1, and that the second parent is given as

$$M_2 = \begin{bmatrix} 0 & 1 & 1 & 0 & 1 \\ 0 & 0 & 1 & 0 & 1 \\ 0 & 0 & 0 & 0 & 0 \\ 1 & 1 & 1 & 0 & 1 \\ 0 & 0 & 1 & 0 & 0 \end{bmatrix}. \tag{18.41}$$

M_2 represents the tour $4 \to 1 \to 2 \to 5 \to 3$. We obtain the intersection of M_1 and M_2 by performing an element-by-element logical AND operation on the two matrices. This gives the partially-defined child

$$M_c = M_1 \wedge M_2 = \begin{bmatrix} 0 & 1 & 0 & 0 & 1 \\ 0 & 0 & 0 & 0 & 1 \\ 0 & 0 & 0 & 0 & 0 \\ 0 & 0 & 0 & 0 & 1 \\ 0 & 0 & 0 & 0 & 0 \end{bmatrix}. \tag{18.42}$$

This does not represent a valid tour, but it does indicate that city 1 is before cities 2 and 5, that city 2 is before city 5, and that city 4 is before city 5. This ordering occurs because the same ordering occurs in both parents – in fact, this is the only ordering that is common in both parents. At this point we can pseudo-randomly add ones to M_c until it is valid tour (that is, until it satisfies all of the properties enumerated above). For example, we might choose to add ones to M_c to obtain

$$M_c = \begin{bmatrix} 0 & 1 & \mathbf{1} & \mathbf{1} & 1 \\ 0 & 0 & 0 & 0 & 1 \\ 0 & 0 & 0 & 0 & 0 \\ 0 & \mathbf{1} & \mathbf{1} & 0 & 1 \\ 0 & \mathbf{1} & \mathbf{1} & 0 & 0 \end{bmatrix} \tag{18.43}$$

where the added ones are denoted in bold font. M_c now satisfies all of the properties of a valid tour, and represents the tour $1 \to 4 \to 5 \to 2 \to 3$.

18.3.4.2 Union Crossover We now illustrate union crossover with an example. Suppose that Equations (18.38) and (18.41) represent parents M_1 and M_2. We obtain the union of M_1 and M_2 by performing an element-by-element logical OR operation on the two matrices. This gives the partially-defined child

$$M_c = M_1 \vee M_2 = \begin{bmatrix} 0 & 1 & 1 & 1 & 1 \\ 0 & 0 & 1 & 1 & 1 \\ 1 & 1 & 0 & 1 & 1 \\ 1 & 1 & 1 & 0 & 1 \\ 0 & 0 & 1 & 0 & 0 \end{bmatrix}. \tag{18.44}$$

We next select a random "cut point" that divides M_c into four quadrants (not necessarily of equal size). Suppose that we generate a random cut point at the second row and the second column. We write M_c with the upper-left and lower-right quadrants unchanged, but with the lower-left and upper-right quadrants replaced with undefined terms:

$$M_c = \begin{bmatrix} 0 & 1 & \times & \times & \times \\ 0 & 0 & \times & \times & \times \\ \times & \times & 0 & 1 & 1 \\ \times & \times & 1 & 0 & 1 \\ \times & \times & 1 & 0 & 0 \end{bmatrix}. \tag{18.45}$$

We next make necessary changes to M_c to remove contradictions. For example, $M_{c34} = 1$ and $M_{c43} = 1$, so one of those elements needs to be changed to a 0. Similarly, $M_{c35} = 1$ and $M_{c53} = 1$, so one of those elements needs to be changed to a 0. This gives the corrected but still partially-defined child as

$$M_c = \begin{bmatrix} 0 & 1 & \times & \times & \times \\ 0 & 0 & \times & \times & \times \\ \times & \times & 0 & 0 & 0 \\ \times & \times & 1 & 0 & 1 \\ \times & \times & 1 & 0 & 0 \end{bmatrix}. \tag{18.46}$$

Finally we pseudo-randomly add ones to the off-diagonal blocks in M_c until it is valid tour (that is, until it satisfies all of the properties enumerated earlier). For example, we might choose to add ones to M_c to obtain

$$M_c = \begin{bmatrix} 0 & 1 & 1 & 1 & 1 \\ 0 & 0 & 1 & 1 & 1 \\ 0 & 0 & 0 & 0 & 0 \\ 0 & 0 & 1 & 0 & 1 \\ 0 & 0 & 1 & 0 & 0 \end{bmatrix}. \quad (18.47)$$

M_c now satisfies all of the properties of a valid tour, and represents the tour $1 \to 2 \to 4 \to 5 \to 3$.

18.4 TSP MUTATION

This section discusses a few ways to mutate TSP solutions. We restrict our discussion to path representations (see Section 18.3.1). Mutations for other representations can be found in the literature. Also, any representation could be converted to path representation, and then we could use one of the mutation methods discussed in this section

18.4.1 Inversion

Inversion reverses the order of the tour between two randomly-selected indices [Fogel, 1990]. For example, x could be mutated to become x_m as follows:

$$\begin{aligned} x &= 1 \to \underbrace{5 \to 4 \to 7 \to 6} \to 2 \to 3 \\ x_m &= 1 \to \overbrace{6 \to 7 \to 4 \to 5} \to 2 \to 3 \end{aligned} \quad (18.48)$$

where we randomly selected the start and end point of the mutated segment. Inversion is also called 2-opt mutation [Beyer and Schwefel, 2002]. There are $n(n-1)/2$ unique ways to implement inversion to an n-city TSP tour. The lowest cost solution that results from all possible inversions of an n-city TSP tour always results in a tour without any crossed edges [Bäck et al., 1997a].

18.4.2 Insertion

Insertion moves the city in position i to position k, where i and k are randomly selected [Fogel, 1988]. For example, suppose that we have the tour x shown in Equation (18.48). Suppose that we randomly select $i = 4$ and $k = 2$. We then move city 7, which is in position 4, to position 2 to obtain the mutated tour

$$x_m = 1 \to 7 \to 5 \to 4 \to 6 \to 2 \to 3. \quad (18.49)$$

Insertion is also called or-opt mutation [Beyer and Schwefel, 2002].

18.4.3 Displacement

Displacement is a generalization of insertion [Michalewicz, 1996, Chapter 10]. Displacement takes the sequence of q cities beginning at the i-th position and moves them to the k-th position in the tour, where q, i, and k are randomly selected. For example, suppose that we have the tour x shown in Equation (18.48). Suppose that we randomly select $q = 2$, $i = 4$, and $k = 2$. We then take the two-city sequence beginning at position 4 (cities 7 and 6), and move it to position 2 to obtain the mutated tour

$$x_m = 1 \to 7 \to 6 \to 5 \to 4 \to 2 \to 3. \tag{18.50}$$

Displacement is also called shifting [Beyer and Schwefel, 2002]. We could combine displacement with inversion by reversing the order of the selected cities before we move them to their new position.

18.4.4 Reciprocal Exchange

Reciprocal exchange swaps the cities in the i-th and k-th positions, where i and k are randomly selected [Banzhaf, 1990]. For example, suppose that we have the tour x shown in Equation (18.48). Suppose that we randomly select $i = 5$ and $k = 1$. We then take the swap the cities in the first and fifth positions to obtain the mutated tour

$$x_m = 6 \to 5 \to 4 \to 7 \to 1 \to 2 \to 3. \tag{18.51}$$

Reciprocal exchange is also called 2-exchange mutation [Beyer and Schwefel, 2002]. We could generalize this method by swapping sequences of cities rather than single cities. We could then combine this generalization with inversion by reversing the order of one or more of the swapped sequences.

18.5 AN EVOLUTIONARY ALGORITHM FOR THE TRAVELING SALESMAN PROBLEM

Given the background of the preceding sections, we can now present a basic EA to solve the TSP, which is shown in Figure 18.5. We have many options in the implementation of Figure 18.5.

- We have several options for population initialization, as we discussed in Section 18.2.
- We have several options for crossover, as we discussed in Section 18.3. We could also use more than one crossover method, probabilistically switching back and forth between various methods from one generation to the next. Furthermore, we could keep track of which crossover method gives the best results, and adapt the frequency of the crossover methods depending on the fitness of their offspring.
- We have several options for mutation, as we discussed in Section 18.4. As with crossover, we could use more than one mutation method, probabilistically switching back and forth between various methods from one generation to the next. Also as with crossover, we could keep track of which mutation method gives the best results, and adapt the frequency of the mutation methods depending on the fitness of their results.
- We need to specify the "Select parents" statement in Figure 18.5 when implementing the algorithm. We could use any of the selection methods in

Section 8.7: fitness-proportionate selection, rank-based selection, tournament selection, and so on.

```
p_m = mutation rate
Initialize N candidate solutions {x_i} (see Section 18.2)
Represent the candidate solutions using the desired representation:
    path, adjacency, ordinal, or matrix representation
Calculate the tour distance for each candidate solution
While not(termination criterion)
    For k = 1 to N
        Select parents from {x_i} to create a child
        Create a child C_k using one of the crossover methods discussed earlier:
        If using path representation then
            Use a crossover method from Section 18.3.1 to create C_k
        else if using adjacency representation then
            Use a crossover method from Section 18.3.2 to create C_k
        else if using ordinal representation then
            Use a crossover method from Section 18.3.3 to create C_k
        else if using matrix representation then
            Use a crossover method from Section 18.3.4 to create C_k
        End if
        r ← U[0, 1] (random number uniformly distributed between 0 and 1)
        If r < p_m then
            Mutate C_k using one of the methods from Section 18.4
        End if
        Calculate the tour distance for C_k
    Next child
    Replace duplicate individuals in {x_i} ∪ {C_i}
    {x_i} ← best N individuals from {x_i} ∪ {C_i}
Next generation
```

Figure 18.5 An evolutionary algorithm to solve the traveling salesman problem.

- With problems like the TSP where problem-specific information (distances between cities) is readily available, we can often get much better results by combining the EA with an algorithm that uses distance information. For instance, after obtaining each child C_k in Figure 18.5, we could select a random sub-tour from C_k and re-arrange it using either one of the heuristics from Section 18.2, or, if the subtour is small enough, using brute force [Jayalakshmi et al., 2001]. These types of approaches are called hybrid EAs because they combine standard EA operators with non-evolutionary algorithms that are specifically designed for TSPs. The literature proposes many hybrid EA variations for the TSP.

- The line "Replace duplicate individuals" in Figure 18.5 is often necessary because in combinatorial optimization, the best few candidate solutions tend to dominate the entire population since the search space is discrete rather than continuous. That is, the population converges so that it consists of only a few (sometimes only one) unique individual. We discussed diversity for continuous optimization in Section 8.6, and diversity requires even more consideration in combinatorial optimization. We can use several methods to replace duplicate individuals. We could replace them with randomly-generated individuals, we could replace them with individuals generated by one of the heuristics of Section 18.2, we could mutate them using one of the methods of Section 18.4, or we could use some combination of these methods.

■ EXAMPLE 18.1

In this example, we investigate the Berlin52 TSP, which is a set of 52 locations in Berlin, Germany. This problem is available on the TSPLIB web site (see Appendix C.6). Figure 18.6 shows a plot of the 52 cities and the minimum-distance closed tour. The latitude and longitude units are normalized. The minimum-distance tour is 7542 units.

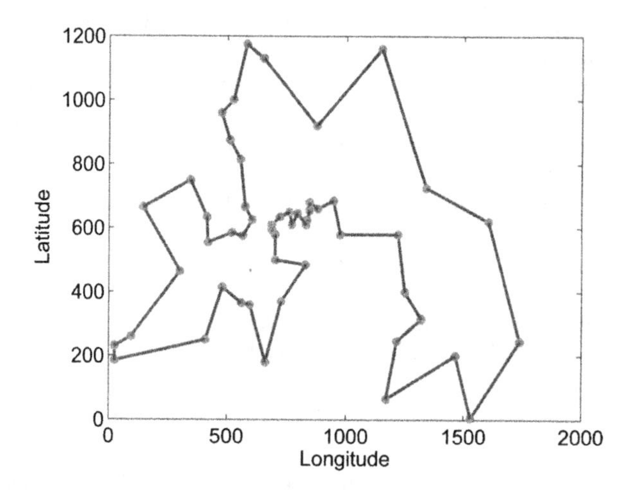

Figure 18.6 Example 18.1: The Berlin52 TSP cities and the minimum-distance tour, which is 7542 units.

In this example, we implement the evolutionary TSP algorithm of Figure 18.5 on the Berlin 52 TSP with the following parameters.

- We use a population size $N = 53$ (one more than the number of cities).
- We initialize the population by generating N random tours.
- We use path representation.
- At each generation we define the fitness $f(x)$ of tour x as

$$f(x) = \max_z D(z) + \min_z D(z) - D(x) \tag{18.52}$$

where the maximum and minimum are taken over the entire population of candidate solutions, and $D(z)$ is the distance of tour z. This equation transforms distance to fitness so that a low-distance solution has a high fitness value, and vice versa. The equation maps distance to fitness in such a way that all fitness values are positive.

- We select parents in the algorithm of Figure 18.5 using roulette-wheel selection.
- We use partially matched crossover (PMX) (see Section 18.3.1).
- We use a mutation rate $p_m = 5\%$ and inversion mutation (see Section 18.4).
- We replace duplicate individuals with a two-step process. First, we scan the parent/child population and mutate duplicate individuals. Second, we again scan the parent/child population and replace duplicates with random tours.
- We run 20 Monte Carlo simulations, each for 300 generations, and find the average distance D^* of the best tour of the 20 simulations.

We run several experiments in this example. First, we try five different crossover methods and obtain the following results.

$$\begin{aligned}
\text{partially matched crossover: } D^* &= 8724 \\
\text{order crossover: } D^* &= 8393 \\
\text{cycle crossover: } D^* &= 9493 \\
\text{order-based crossover: } D^* &= 17109 \\
\text{inver-over crossover: } D^* &= 10595.
\end{aligned} \tag{18.53}$$

We see that order crossover works the best.

Second, we use order crossover and try three different mutation methods, obtaining the following results.

$$\begin{aligned}
\text{inversion mutation: } D^* &= 8393 \\
\text{insertion mutation: } D^* &= 9776 \\
\text{reciprocal exchange mutation: } D^* &= 10036.
\end{aligned} \tag{18.54}$$

We see that inversion mutation works the best.

Third, we use order crossover and inversion mutation and try three different initialization methods. In the first method, we initialize the entire population to random tours. In the second method, we initialize two individuals using nearest-neighbor initialization (see Section 18.2.1), and initialize the remaining $(N - 2)$ cities randomly. In the third method, we initialize the entire population using nearest-neighbor initialization. We obtain the following results.

$$\begin{aligned}
N \text{ random and 0 nearest-neighbor tours: } D^* &= 8393 \\
(N-2) \text{ random and 2 nearest-neighbor tours: } D^* &= 8140 \\
0 \text{ random and } N \text{ nearest-neighbor tours: } D^* &= 8115.
\end{aligned} \tag{18.55}$$

We see that initializing the entire population with the nearest-neighbor algorithm works the best. However, we can obtain nearly the entire benefit of nearest-neighbor initialization by initializing only a couple of individuals with the nearest-neighbor algorithm. Figure 18.7 shows the results of a typical EA simulation with order crossover, inversion mutation, and all individuals initialized with the nearest-neighbor algorithm.

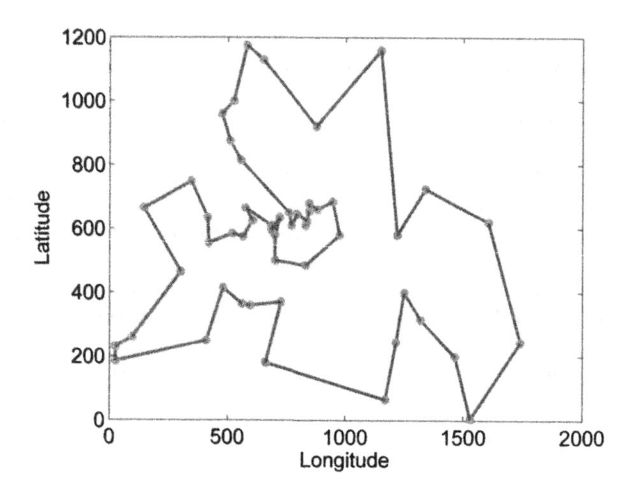

Figure 18.7 Example 18.1: Typical EA result for the Berlin52 problem. The distance of the tour shown here is 8036 units, which is 6.5% worse than the optimal solution.

□

Example 18.1 investigated several different EA options, but we could still not obtain the globally-optimal solution. In one sense this is not surprising since the search space cardinality is on the order of $51!/2 = 10^{66}$. We ran 300 generations each EA simulation with a population size of 53 individuals. We therefore evaluated $300 \times 53 = 15,900$ potential solutions each simulation, which is a miniscule, almost negligible portion of the search space, and we still got to within less than 10% of the optimal solution. However, the Berlin52 benchmark is considered to be a pretty easy TSP. Figure 18.6 shows the layout of the cities, and it does not look like it should be very difficult for a human, or for a good computer program, to find the optimal tour. These results underscore one of the points that we made at the end of the previous section, and which we restate here for emphasis.

> With problems like the TSP where problem-specific information (distances between cities) is readily available, we can often get much better results by combining the EA with an algorithm that uses distance information.

Any serious EA implementation for the TSP needs to take this advice seriously. The EA researcher needs to study non-evolutionary state-of-the-art TSP heuristics and carefully incorporate them into the EA to get competitive results.

Finally, we need to run many more than the 300 generations of each simulation in Example 18.1 to get good results. Figure 18.8 illustrates a typical graph of the minimum-distance solution in the EA population as a function of generation number. That figure indicates that the best candidate solution is continuing to improve even after 300 generations, and that we could expect much better performance if we allowed the EA to run for a few hundred more generations. EA implementations that hope to get competitive results usually need to run at least tens-of-thousands of generations. Of course, processing-power-constrained EAs are interesting and important for obtaining real-time TSP solutions, but the price to be paid for faster performance is worse results.

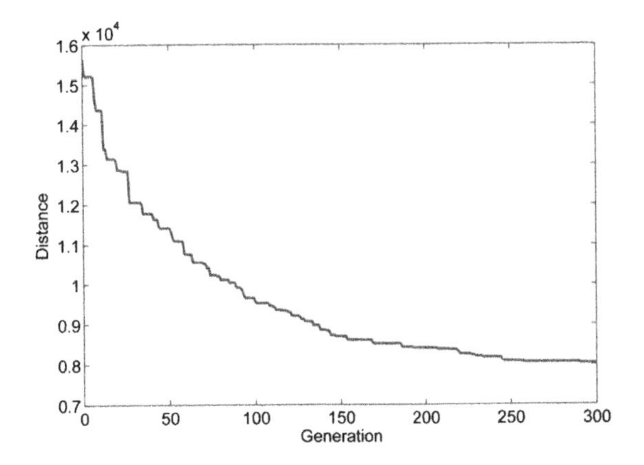

Figure 18.8 Example 18.1: Typical EA convergence behavior for the Berlin52 problem. The EA has mostly converged after 300 generations, but it appears that the best candidate solution would continue to improve for a few hundred more generations.

18.6 THE GRAPH COLORING PROBLEM

A graph is a set of partially-connected nodes. Each node has a unique index, and a weight that is generally not unique [Pardalos and Mavridou, 1998]. Figure 18.9 shows an example of a graph.

There are many related but distinct graph coloring problems. The classical graph coloring problem is defined as either:

1. Determine the smallest number of colors n such that each node of a connected graph can be colored with one of these n colors, under the constraint that linked nodes are not assigned the same color; or

2. Color each node in a connected graph with one of n colors, where n is given, under the constraint that linked nodes are not assigned the same color.

Note that the first problem listed above, which is also called the *n-coloring problem*, can be solved by repeatedly solving the second problem for successively smaller values of n.

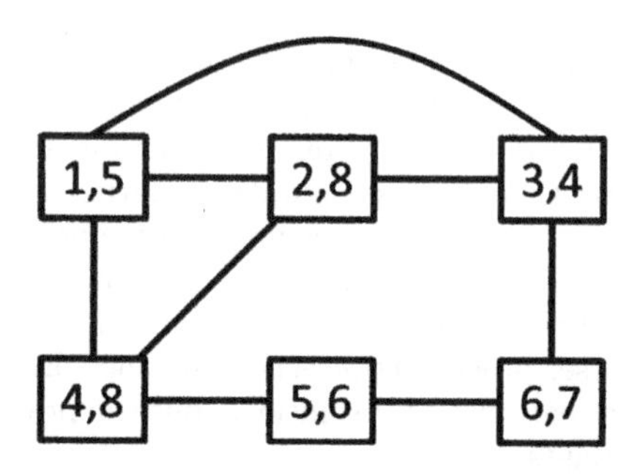

Figure 18.9 An example of a connected graph. Each node is labeled (i, w), where i is the unique node index and w is its weight.

The *weighted graph coloring problem* is a generalization of the second definition above. The weighted graph coloring problem consists of assigning one of n colors to each node, under the constraint that linked nodes are not assigned the same color, in such a way as to maximize the sum of the weights of the colored nodes. This is the problem that we emphasize in this section. Note that the second classical graph coloring problem above can be solved by assigning each node a weight of 1 and solving the weighted graph coloring problem.

In the weighted graph coloring problem, the fitness of a candidate solution is the sum of the weights of the colored nodes, and the problem is to color the nodes in such a way as to maximize fitness. The weighted graph coloring problem has applications in scheduling, computer networks, fault detection and diagnosis, pattern matching, communication theory, games, and many other areas [Ufuktepe and Bacak, 2005]. When we are faced with a practical optimization problem, if we can convert it to an equivalent graph coloring problem, then we can use a wealth of tools that are available for graph coloring problems to solve our practical optimization problem.

The reason that these problems are called *graph coloring problems*, or sometimes *map coloring problems*, is because a map can be represented as a connected graph.[2] As an example of a conversion from a map to a graph, Figure 18.10 shows a map where each region is labeled with an index. To convert the map to a graph, we first note that the map shows that region 1 shares a boundary with regions 2, 4, and 5; therefore, the graph on the right shows that node 1 is connected to nodes 2, 4, and 5. Next, we note that the map shows that region 2 shares a boundary with regions 1, 3, and 4; therefore, the graph on the right shows that node 2 is connected to nodes 1, 3, and 4. We continue this process to convert the map to an equivalent connected graph. We see that the problem of coloring the map with n colors so that neighboring regions do not have the same color, is equivalent to the graph coloring problem. Note that the converse is not true; that is, a connected graph cannot necessarily be converted to a planar map. For example, a completely connected graph with five nodes cannot be converted to a planar map.

Consider the one-color graph coloring problem for the graph of Figure 18.9. Nodes 2 and 4 have the highest weights, but we cannot color both of them the

[2]The famous four-color theorem states that no more than four colors are required to color any map in such a way that no two connected regions are the same color.

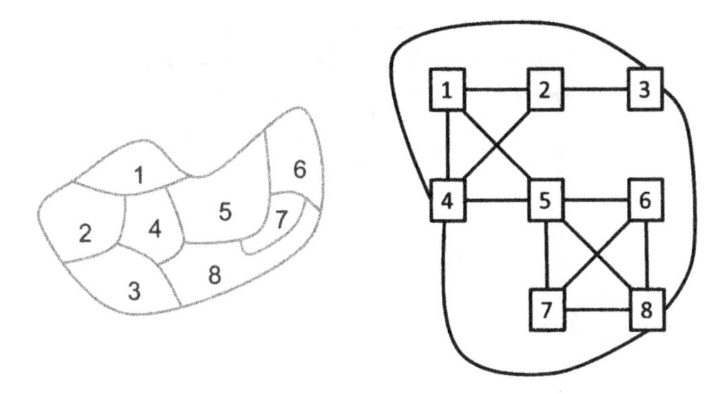

Figure 18.10 The figure on the left shows a map. The figure on the right shows the equivalent connected graph. A map can always be converted to an equivalent connected graph, but the converse is not true.

same color because they are connected. Which nodes should we color to obtain the highest fitness? One popular algorithm is the greedy algorithm, which is shown in Figure 18.11. The greedy algorithm is simple; it sorts the nodes in order of decreasing weight, and then assigns the first legal color to the nodes in their sorted order.

Given the graph of Figure 18.9, the greedy algorithm sorts the nodes in the order $\{2, 4, 6, 5, 1, 3\}$, although nodes 2 and 4 could be interchanged since they have the same weight. For the one-color problem, we color nodes 2 and 6, which gives a fitness of 15. For the two-color problem (red and green, for example), we assign red to node 2, green to node 4, red to node 6, and green to node 3, which gives a fitness of 27.

The greedy algorithm is simple and quick, and it often performs pretty well. However, it is not too difficult to find a case where the greedy algorithm fails. For instance, consider the graph of Figure 18.12. When we use the greedy algorithm for the one-color problem for this graph, we color only node 7, which gives a fitness of 5. It is clear from looking at the graph that we can obtain a better fitness by coloring nodes 1, 3, and 5, which gives a fitness of 9.

$\{x_i\}$ = N nodes sorted in order of decreasing weight
$\{C_k\}$ = n colors
For $i = 1$ to N
    For $k = 1$ to n
        If legal, then assign C_k to x_i and exit the "For k" loop
    Next color
Next node

Figure 18.11 A greedy algorithm for the N-node, n-color graph coloring problem.

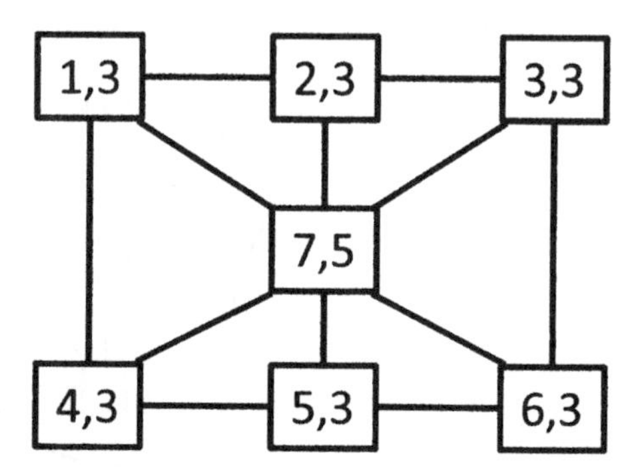

Figure 18.12 When we use the greedy algorithm for the one-color graph coloring problem for this graph, we color only node 7, which gives a suboptimal solution.

Evolutionary Algorithms and Graph Coloring Problems

How could we use an EA to solve the graph coloring problem? One way is to define an individual in an EA population as an ordered list of nodes. Then we can use the greedy algorithm to assign colors, based on the order of the nodes. Each individual then has a fitness value. We can use any type of fitness-dependent selection, and then we can use any of the recombination methods of Section 18.3 and any of the mutation methods of Section 18.4 to create children. This approach transforms the graph coloring problem into TSP format, which allows us to use all of the TSP results of the previous sections.

As with the TSP, any serious graph coloring EA needs to incorporate non-evolutionary heuristics to get good results. There is a lot of literature on the graph coloring problem. [Jensen and Toft, 1994] provides a good background on the problem, analysis, theoretical results, and some heuristic algorithms. In addition, there are many other EA-based approaches for solving graph coloring problems. Hybrid EAs that combine evolutionary search with local optimization are among the best-performing graph coloring algorithms. See [Galinier et al., 2013] for a survey.

■ **EXAMPLE 18.2**

This example shows how a scheduling problem can be represented as a graph coloring problem. Suppose that we want to schedule events 1, 2, 3, 4, 5, and 6 so that the following pairs of events do not occur at the same time:

(1 and 2), (1 and 3), (3 and 5), (3 and 6), and (4 and 6).

We can represent this problem with the graph of Figure 18.13. Each node has the same weight, and so the weights are not shown in the figure. Nodes that correspond to events that cannot be scheduled at the same time are connected in the graph. This graph does not have a one-color solution. However, we can obtain a two-color solution by coloring nodes 1, 5, and 6 one color, and node 2 another color, and nodes 3, and 4 a third color. In other words, we can schedule events 1, 5, and 6 during the first time slot, and event 2 during the second time slot, and events 3, and 4 during the third time slot. We see that we can use graph coloring algorithms to schedule operations in a manufacturing plant where certain tasks use the same resources, to schedule classes in a school where certain students and teachers are involved in multiple classes, and to solve a variety of other scheduling problems.

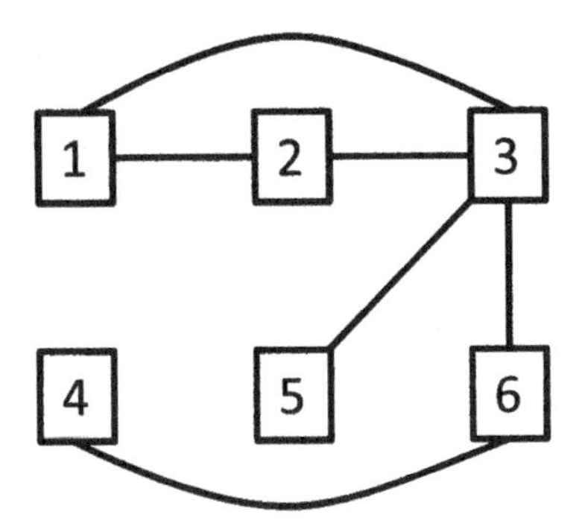

Figure 18.13 We can represent the scheduling problem of Example 18.2 with this graph.

□

18.7 CONCLUSION

We have summarized the traveling salesman problem (TSP), and discussed some of the most commonly-used TSP representations and operators. We have also discussed the graph coloring problem and have shown how it can be converted to a TSP. Researchers have proposed many TSP operators that we have not had time to cover in this chapter. [Larrañaga et al., 1999b] gives a good overview of TSP representations and operators. The TSP has a long history, and researchers have solved it using many methods other than EAs. Many good books are devoted to the TSP, including [Applegate et al., 2007] and [Lawler et al., 1985]. See [Hao and Middendorf, 2012] for the proceedings of a conference devoted to evolutionary algorithms for combinatorial problems.

This chapter has discussed only two combinatorial optimization problems (the TSP and the graph coloring problem), but there are also many other popular and widely applicable combinatorial optimization problems. These include the minimum spanning tree problem, the job shop scheduling problem, the knapsack problem, and the bin packing problem. EAs have been applied to all of these problems but there is plenty of room for additional research. Some of the newer EAs and EA variations have yet to be applied to some of these combinatorial optimization problems. Efficient ways to hybridize EAs with non-evolutionary combinatorial heuristics could be a fruitful area for future research. Finally, we need more theoretical results that quantify the performance of EAs on combinatorial problems and that can provide guidance for practical applications.

Our discussion of TSPs in this chapter has covered only symmetric TSPs, that is, problems where the distance from city i to city k is the same as that from city k to city i. Appendix C.6 discusses a few other types of TSP-related problems, including the asymmetric TSP, the sequential ordering problem, the capacitated vehicle routing problem, and the Hamiltonian path problem.

Another interesting variation is the close-enough TSP. In this problem, we are given a connected graph in which each node i has a "close-enough" radius r_i associated with it. The objective is to find the minimum distance Hamiltonian cycle that passes within r_i units of each node i [Yuan et al., 2007]. This problem is closely related to the TSP but it is actually a continuous optimization problem, although it does include combinatorial elements. Finally we mention the Dubins TSP, which is a TSP for a vehicle with kinematic constraints. For instance, a vehicle might need to visit a set of locations while traveling the minimum possible distance with the constraint that it cannot instantaneously change its direction of travel [Savla et al., 2008].

PROBLEMS

Written Exercises

18.1 What TSP tour and distance results from nearest-neighbor initialization given the distance matrix of Equation (18.5) if we start at city 3?

18.2 Suppose you have an n-city TSP and you want to determine M tours in the initial population using the nearest-neighbor strategy described in Section 18.2.1. What is the probability that you will select M different starting cities with this strategy? What is the probability if $n = 100$ and $M = 10$?

18.3 Formulate a five-city open TSP such that the nearest two-neighbor algorithm described in Section 18.2.1 performs better than the nearest-neighbor algorithm.

18.4 In the second step of the shortest-edge initialization example in Section 18.2.2, we had to make a random selection since two edges had the same length. Suppose we chose the other option in that example. What would be the final closed tour, and how would the total distance compare to the example in Section 18.2.2?

18.5 In the second step of the insertion initialization example in Section 18.2.3, we had to make a random selection since two cities were both the same distance to T. Suppose we chose the other option in that example. What would be the final closed tour?

18.6 Consider the open tour $1 \rightarrow 2 \rightarrow 3 \rightarrow 4 \rightarrow 5$. What is the path representation, adjacency representation, ordinal representation, and matrix representation of this path?

18.7 What is the rank of the matrix representation of a TSP tour?

18.8 We used the greedy graph coloring algorithm to solve the two-color problem of Figure 18.9 by sorting the nodes in the order $\{2, 4, 6, 5, 1, 3\}$, assigning the first color to nodes 2 and 6, and assigning the second color to nodes 4 and 3, which gave a fitness of 27. However, we could also sort the nodes in the order $\{4, 2, 6, 5, 1, 3\}$ since nodes 2 and 4 have the same weight. What color assignment would result with this order, and what would the fitness be?

18.9 Explain how a 9×9 Sudoku puzzle can be formulated as a graph coloring problem. Hint: the graph will have 81 nodes.

18.10 Consider the graph coloring problem for the graph on the left side of Figure 18.14, where each node has the same weight.

- **a)** Use the greedy graph coloring algorithm to find the minimum number of colors required to color all of the nodes, where the nodes are ordered as shown.
- **b)** Repeat for the graph on the right side of the figure, which is the same graph except that the nodes are ordered differently.

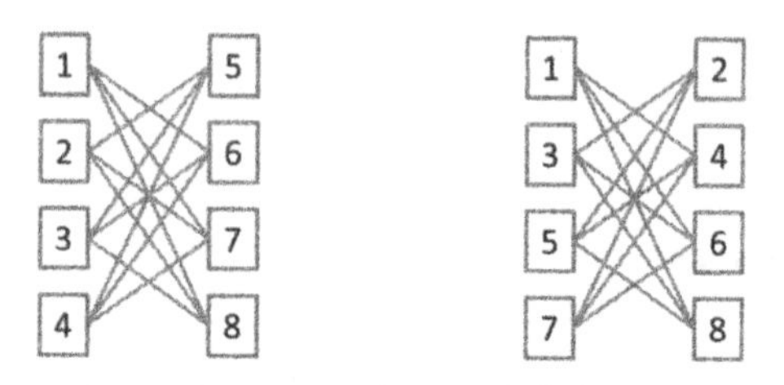

Figure 18.14 Problem 18.10: The connected graphs on the left and right are equivalent. The only difference is the ordering of the nodes.

Computer Exercises

18.11 Repeat Example 18.1 using one of the other TSPs from the TSPLIB web site.

18.12 Repeat Example 18.1 with order crossover, random initialization of the entire population, and the opposition-based logic outlined in Figure 16.10 for mutation. What is the distance of the best solution found, averaged over 20 Monte Carlo simulations? How does this compare with the results obtained with the three mutation methods used in Example 18.1.

CHAPTER 19

Constrained Optimization

[It] is necessary to find ways of incorporating the constraints (normally existing in any real-world application) into the fitness function.
—Carlos A. Coello Coello [Coello Coello, 2002]

All real-world optimization problems are constrained, at least implicitly if not explicitly. This chapter discusses various approaches to handling constraints in optimization problems. A constrained optimization problem can be written as

$$\begin{aligned} \min_x f(x) \quad \text{such that} \quad & g_i(x) \leq 0 \text{ for } i \in [1, m] \\ \text{and} \quad & h_j(x) = 0 \text{ for } j \in [1, p]. \end{aligned} \tag{19.1}$$

This problem includes $(m + p)$ constraints, m of which are inequality constraints, and p of which are equality constraints. The set of x that satisfies all $(m + p)$ constraints is called the feasible set, and the set of x that violates one or more constraints is called the infeasible set:

$$\begin{aligned} \text{feasible set } \mathcal{F} &= \{x : g_i(x) \leq 0 \text{ for } i \in [1, m] \text{ and } h_j(x) = 0 \text{ for } j \in [1, p]\} \\ \text{infeasible set } \bar{\mathcal{F}} &= \{x : x \notin \mathcal{F}\}. \end{aligned} \tag{19.2}$$

We use the term *constrained evolutionary algorithms* to refer to evolutionary algorithms that are designed to solve problems of the form of Equation (19.1).[1]

Overview of the Chapter

Constrained EAs can be broadly classified into various categories.

1. *Penalty function approaches* modify the cost function of an EA individual x based on some measure of its constraint violation. Penalty function approaches that allow, and sometimes even encourage, infeasible solutions in the population are called exterior approaches. In this case they penalize the cost or selection of infeasible candidate solutions. Penalty function approaches that do not allow infeasible solutions in the population are called interior point methods. We discuss general ways to implement penalty function approaches in Section 19.1. We show how to implement various penalty function approaches in constrained EAs in Section 19.2.

2. *Special representations* are problem-dependent approaches for representing constrained problems in such a way that the representation is unconstrained while the candidate solutions remain constrained. *Special operators* are problem-dependent approaches for performing selection, recombination, and mutation, in such a way that the constraints are always satisfied by child individuals. These two approaches do not allow infeasible candidate solutions in the population. We discuss these approaches in Section 19.3.

3. *Repair algorithms* modify infeasible EA individuals so that they become feasible. These algorithms are largely problem-dependent. They may allow some infeasible individuals to remain in the population, while repairing other infeasible individuals. The only repair method that we discuss in this chapter is the Genocop algorithm of Section 19.3.2.

4. *Hybrid methods* combine features from the above methods, or from non-evolutionary constrained optimization algorithms. For example, many constrained EAs use one of the above methods along with local search. In this chapter we present some basic approaches to constrained EAs, but we do not discuss how they can be hybridized. However, the literature includes many examples of hybridization. After the reader becomes familiar with the basic constrained EA approaches in this chapter, he should be well-prepared to explore hybrid algorithms in the literature, or to take the best features of various constrained EAs to develop his own hybrid algorithm.

We do not pretend to cover all of the constraint-handling methods that have been proposed over the years, but Section 19.4 outlines a few other approaches to constrained optimization, including the use of cultural algorithms, and the use of multi-objective optimization.

[1]The term *constrained evolutionary algorithms* is not quite correct from a grammatical viewpoint. A grammatically strict interpretation of the phrase would indicate that it refers to evolutionary algorithms that are constrained, rather than evolutionary algorithms that are designed to solve constrained problems. But the term is convenient, concise, and popular, so with this caveat we are confident that the reader will not confuse its meaning.

One of the main problems that we need to solve during a constrained optimization algorithm is how to rank candidate solutions. Some of the solutions have a high cost but satisfy the constraints, while other solutions have a low cost but violate the constraints. Section 19.5 summarizes the ranking approaches presented earlier in the chapter, and discusses a few alternative ranking methods. Section 19.6 ties together all of the material in this chapter and presents a comparison of various constrained BBO algorithms on some benchmarks.

19.1 PENALTY FUNCTION APPROACHES

Penalty function approaches penalize candidate solutions that violate constraints or that come close to violating constraints. Penalty function approaches for general constrained optimization problems were first proposed by Richard Courant in 1943 [Courant, 1943]. They are often cited as being the most popular algorithms for constrained optimization, but other approaches for constrained EAs are rapidly gaining in popularity.

We can design a penalty function method in two different ways. First, we could penalize feasible individuals as they move closer to the constraint boundary; these are called interior point methods, or barrier methods. This approach, which we briefly discuss in Section 19.1.1, does not allow any infeasible individuals in the population.

Second, we could allow infeasible individuals in the population, but penalize them in terms of cost, or in terms of selection for contributing to the next generation. This approach, several examples of which we present in Section 19.1.2, generally does not penalize feasible individuals, no matter how close they are to the constraint boundary. Such approaches are called exterior methods.

19.1.1 Interior Point Methods

Interior point methods allow only feasible individuals in the population. These methods penalize the cost of individuals as they move close to the constraint boundary, thus encouraging those individuals to remain within the constraints. We illustrate the idea of interior point methods with a simple example.

■ EXAMPLE 19.1

Consider the scalar problem

$$\min f(x) \text{ such that } x \geq c, \text{ where } f(x) = x^2. \tag{19.3}$$

We can modify this problem so that feasible values of x are penalized as they approach the constraint. The modified function is called a barrier function. For example, we could convert the constrained problem of Equation (19.3) to the unconstrained problem

$$\min f'(x), \text{ where } f'(x) = x^2 + (x - c + \delta)^{-\alpha} \tag{19.4}$$

where $\delta > 0$ is a small constant, and $\alpha > 0$ is another constant. As $\alpha \to 0$, $\arg\min_x f'(x) \to \arg\min_x f(x)$, but we also obtain a more poorly-behaved $f'(x)$; that is, $f'(x)$ becomes less smooth.

Figure 19.1 shows $f(x)$ and $f'(x)$ for $c = 1$, $\delta = 0.01$, and $\alpha = 1$. Of course, we still have to make sure that $x \geq c$ for this simple example, so our interior point approach did not really help much. But this example illustrates how interior point methods can prevent feasible individuals in an EA from violating constraints after recombination or mutation.

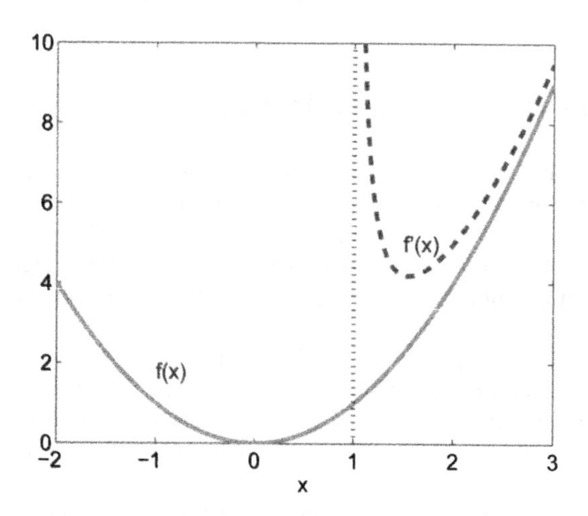

Figure 19.1 Example 19.1. We want to minimize $f(x)$ such that $x \geq 1$. We use an interior point method to convert the constrained minimization of $f(x)$ to the unconstrained minimization of $f'(x)$.

□

Interior point methods are not used very often in constrained EAs. This is because for many constrained optimization problems, finding candidate solutions that satisfy all of the constraints is itself a challenging problem. Also, infeasible solutions may include information that is valuable in the search for a constrained optimum. For example, a problem with a small feasible region might be solvable by combining two infeasible individuals (see Figure 19.2).

However, there are also many optimization problems for which it is relatively easy to find feasible candidate solutions. The paucity of interior point methods for constrained EAs is therefore surprising in view of the vast literature on interior point methods for general-purpose optimization algorithms [Wright, 1987]. This may indicate a neglected area of constrained EA research.

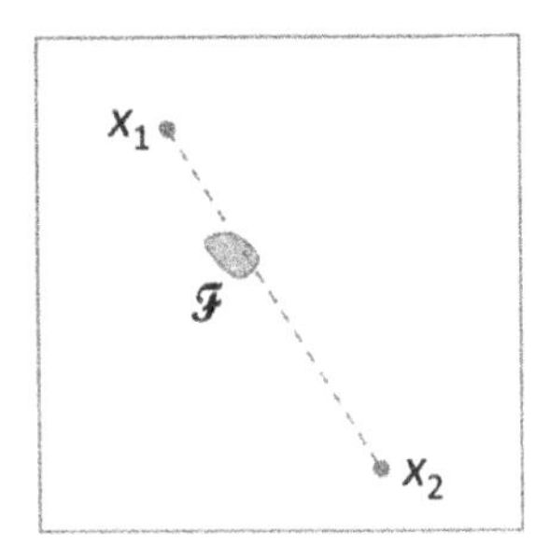

Figure 19.2 Example of a small feasible set $\mathcal{F}$ in a large search space. It may be difficult to directly find a candidate solution $x \in \mathcal{F}$, but it may be much easier to find two infeasible individuals x_1 and x_2 that can combine to produce a feasible individual.

19.1.2 Exterior Methods

Exterior methods allow infeasible individuals in the population, but penalize their cost or selection probabilities. This section gives a broad overview of exterior methods for constrained optimization.

19.1.2.1 Death Penalty Approaches Death penalty approaches are exterior methods that allow infeasible individuals in the population, but only for brief periods of time. A death penalty approach takes the penalty function approach to an extreme. With this approach we immediately remove any infeasible individual $\bar{x}$ from the population. If we obtain $\bar{x}$ by recombination, then we reject it, and we repeat the recombination operation until we obtain a feasible individual. If we obtain $\bar{x}$ by mutation, then we reject it, and we repeat the mutation operation until we obtain a feasible individual.

The death penalty is a convenient approach to constrained optimization. It has the advantage of not requiring cost evaluations of infeasible individuals, which can save computational effort. However, for many problems, it is difficult to obtain feasible individuals in the first place, so rejecting infeasible individuals might be overly strict. Instead of completely rejecting infeasible individuals, we might need to retain them in the population while giving a relatively lower cost to those that violate the constraints less severely, thus encouraging the population to move toward the feasible region. In summary, the effectiveness of the death penalty approach is problem-dependent.

19.1.2.2 Non-Death-Penalty Approaches The remainder of this chapter discusses non-death-penalty constraint-handling approaches. These approaches are more forgiving exterior methods than death penalty approaches because they allow infeasible individuals to remain in the population for the entire duration of the EA. We transform the standard constrained optimization problem of Equation (19.1) into the following unconstrained problem:

$$\begin{aligned}
&\min_x \phi(x), \text{ where } \phi(x) = f(x) + \sum_{i=1}^{m} r_i G_i(x) + \sum_{j=1}^{p} c_j L_j(x) \\
&G_i(x) = [\max(0, g_i(x))]^\beta \\
&L_j(x) = |h_j(x)|^\gamma
\end{aligned} \tag{19.5}$$

where r_i and c_j are positive constants that are called penalty factors, and β and γ are positive constants that are often set equal to 1 or 2. $\phi(x)$ is called the penalized cost function, and we obtain $\phi(x)$ as a weighted sum of the original cost function $f(x)$ and the constraint violation magnitudes $\{G_i(x)\}$ and $\{L_j(x)\}$. We see that if $x \in \mathcal{F}$, then $\phi(x) = f(x)$. However, if $x \notin \mathcal{F}$, then $\phi(x) > f(x)$ by an amount that increases with the amount of constraint violation.

Now that we have a penalized cost function $\phi(x)$, we can run an EA that uses $\phi(x)$ as the cost function to select individuals for the next generation. We can therefore extend any of the unconstrained EAs discussed in this book to constrained optimization. We simply use $\phi(x)$ instead of $f(x)$ as the cost function.

The constraints $h_j(x) = 0$ are very unforgiving. If we randomly generate an initial population in a continuous search domain, we have an essentially zero chance of obtaining individuals that satisfy equality constraints. Therefore, we often change the hard equality constraints to soft constraints that require $h_j(x)$ to be approximately zero, rather than exactly zero. This results in

$$|h_j(x)| \leq \epsilon \tag{19.6}$$

where ϵ is a small positive constant. This is equivalent to the two constraints

$$\begin{aligned}
h_j(x) - \epsilon &\leq 0 \\
-h_j(x) - \epsilon &\leq 0.
\end{aligned} \tag{19.7}$$

Depending on the value of ϵ, we have a reasonable chance of obtaining individuals that satisfy the soft constraint of Equation (19.6). One of way of assigning ϵ is to use relatively large values of ϵ early in the EA so that we can obtain some feasible individuals, and then gradually decrease ϵ as the generation count increases [Brest, 2009], [Zavala et al., 2009]. Many research papers that compare constrained optimization algorithms on benchmark functions use $\epsilon = 0.0001$ [Liang et al., 2006].

The conversion of equality constraints to inequality constraints transforms Equation (19.5) to

$$\begin{aligned}
&\min_x \phi(x), \text{ where } \phi(x) = f(x) + \sum_{i=1}^{m+p} r_i G_i(x) \\
&G_i(x) = \begin{cases} [\max(0, g_i(x))]^\beta & \text{for } i \in [1, m] \\ [\max(0, |h_i(x)| - \epsilon)]^\beta & \text{for } i \in [m+1, m+p] \end{cases}
\end{aligned} \tag{19.8}$$

where we have simplified the problem by setting $\gamma = \beta$. Problems like the one in Equation (19.8) can be solved with static methods or dynamic methods. Static methods use values of r_i, β, and ϵ that are independent of the EA generation number.

In contrast, dynamic methods use values of r_i, β, and ϵ that depend on the EA generation number. Static methods are simpler to implement, but dynamic methods may perform better because of their flexibility. Dynamic methods may be able to intelligently adapt their weights, based on the population distribution or the problem characteristics, to improve performance. Dynamic methods often increase r_i and β, and decrease ϵ, as the generation count increases. This increases the weight given to constraint violation, which results in a gradual attraction of more and more infeasible individuals toward the feasible region.

19.2 POPULAR CONSTRAINT-HANDLING METHODS

This section discusses several popular constraint-handling approaches that are used in EAs. These are all non-death-penalty approaches.

19.2.1 Static Penalty Methods

Equation (19.8) is proposed in [Homaifar et al., 1994] with $\beta = 2$, and r_i a function of the constraint violation magnitude. That is, r_i is a nondecreasing function of $G_i(x)$. Sometimes the penalty factor r_i is set equal to one of a set of discrete values depending on the amount of the constraint violation:

$$r_i = \begin{cases} R_{i1} & \text{if } G_i(x) \in (0, T_{i1}] \\ R_{i2} & \text{if } G_i(x) \in (T_{i1}, T_{i2}] \\ \vdots & \\ R_{iq} & \text{if } G_i(x) \in (T_{i,q-1}, \infty) \end{cases} \tag{19.9}$$

where q is the user-specified number of constraint levels, the R_{ij} values are user-defined weights, and the T_{ij} values are user-defined constraint thresholds. This is a static approach because the penalty on the constraints is not a function of the generation count. The research literature often criticizes this well-known approach because it requires many tuning parameters. It requires $(2q-1)(m+p)$ tuning parameters, although we can reduce this number by combining some of the weight levels and and thresholds to simplify the algorithm.

19.2.2 Superiority of Feasible Points

The method of the superiority of feasible points [Powell and Skolnick, 1993] modifies the penalized cost function of Equation (19.8) as follows:

$$\begin{aligned} \min_x \phi'(x), \text{ where } \phi'(x) &= \phi(x) + \theta(x) \\ &= f(x) + \sum_{i=1}^{m+p} r_i G_i(x) + \theta(x) \end{aligned} \tag{19.10}$$

where $\theta(x)$ is an additional term that is designed to guarantee that $\phi'(x) \le \phi'(\bar{x})$ for all $x \in \mathcal{F}$ and for all $\bar{x} \notin \mathcal{F}$. That is, $\phi'(x) \le \phi'(\bar{x})$ for all feasible x and for all infeasible $\bar{x}$. This can be accomplished by setting $\theta(x)$ as follows:

$$\theta(x) = \begin{cases} 0 & \text{if } x \in \mathcal{F} \\ \max f(y) : y \in \mathcal{F} & \text{if } x \notin \mathcal{F} \end{cases} \tag{19.11}$$

assuming that $f(x) \geq 0$ for all x. A less conservative way [Michalewicz and Schoenauer, 1996] to implement this method, which does not assume that $f(x) \geq 0$, is to set $\theta(x)$ as follows:

$$\theta(x) = \begin{cases} 0 & \text{if } x \in \mathcal{F} \\ \max\left[0, \max_{y \in \mathcal{F}} f(y) - \min_{y \notin \mathcal{F}} \phi(y)\right] & \text{if } x \notin \mathcal{F}. \end{cases} \tag{19.12}$$

This definition of $\theta(x)$ gives $\phi'(x) = \phi(x)$ for all x, under the condition that $\phi(x) \leq \phi(\bar{x})$ for all $x \in \mathcal{F}$ and for all $\bar{x} \notin \mathcal{F}$. That is, if the penalized cost function of Equation (19.8) results in all feasible individuals being ranked better than all infeasible individuals, then we do not make any changes to Equation (19.8). However, if Equation (19.8) results in $\phi(x) > \phi(\bar{x})$ for some $x \in \mathcal{F}$ and for some $\bar{x} \notin \mathcal{F}$, then Equation (19.12) shifts the penalized cost function values of all the infeasible individuals so that $\min_{\bar{x}} \phi'(\bar{x}) = \max_x \phi'(x)$; that is, the best infeasible penalized cost is equal to the worst feasible penalized cost.

The method of the superiority of feasible points may be an attractive approach if the optimization problem includes difficult constraints. If the constraints are hard to satisfy, then this method provides a lot of selection pressure for feasible points to remain in the population, which allows their information to carry on to the next generation.

19.2.3 The Eclectic Evolutionary Algorithm

The eclectic EA proposes another approach to enforcing the superiority of feasible points [Morales and Quezada, 1998]. The eclectic EA defines the penalized cost function as

$$\phi(x) = \begin{cases} f(x) & \text{if } x \in \mathcal{F} \\ K\left(1 - \frac{s(x)}{m+p}\right) & \text{if } x \notin \mathcal{F} \end{cases} \tag{19.13}$$

where K is a large constant, $m + p$ is the total number of constraints, and $s(x)$ is the number of constraints that are satisfied by x. The user-defined constant K needs to be large enough to guarantee that $\phi(\bar{x}) > \phi(x)$ for all $\bar{x} \notin \mathcal{F}$ and for all $x \in \mathcal{F}$. If we use a ranking method to select individuals for recombination, then there is no upper bound for K. However, if we use a roulette-wheel method, or some other method that uses absolute values of $\phi(\cdot)$ for selection, then we need to be careful not to set K too large; we want to make sure that although infeasible are ranked worse than feasible individuals, infeasible individuals still have a reasonable chance of being selected for recombination.

The eclectic EA differs from Equation (19.10) because the eclectic EA does not evaluate the cost $f(x)$ for infeasible individuals. This could result in significant computational savings. Also, the eclectic EA considers only the number of constraint violations in determining the penalized cost function, and it does not consider the magnitude of the constraint violations. Equation (19.10), on the other hand, considers only the magnitude of constraint violation, and it does not consider the number of constraint violations. This could provide another computational advantage to the eclectic EA because in real-world problems it is often much easier to count the number of constraint violations than to quantify the exact level of those violations.

19.2.4 Co-evolutionary Penalties

It would be interesting to combine the approaches of Equations (19.10) and (19.13), because sometimes the magnitude of constraint violation may be important to us, but other times the number of constraint violations may be more important. A co-evolutionary approach that incorporates this idea is proposed in [Coello Coello, 2000b], [Coello Coello, 2002], and uses the penalized cost

$$\phi(x) = f(x) + w_1 \sum_{i=1}^{m+p} G_i(x) + w_2 \left(1 - \frac{s(x)}{m+p}\right) \tag{19.14}$$

where w_1 and w_2 are weights. This is a co-evolutionary approach because it involves two populations. One population, P_1, consists of candidate solutions x and evolves according to the penalized cost $\phi(x)$. A second population, P_2, consists solely of (w_1, w_2) pairs. An individual in P_1 evolves using a specific individual from P_2 (that is, a specific (w_1, w_2) pair). The cost of a (w_1, w_2) pair is evaluated as

$$\begin{aligned} \psi(w) &= \frac{1}{M_1(w)} \sum_{x \in \mathcal{F}} \phi(x) - M_1(w) \\ M_1(w) &= |x : x \in \mathcal{F}| \end{aligned} \tag{19.15}$$

where w refers to a specific (w_1, w_2) pair from P_2. Note that $M_1(w)$ is the number of feasible individuals in P_1 after it has finished evolving using w. The cost $\psi(w)$ of an individual w depends on the average penalized cost of all feasible individuals that it evolves in P_1, and also depends on the number of feasible individuals that it evolves in P_1. Equation (19.15) is undefined if $M_1(w) = 0$. If $M_1(w) = 0$, then $\psi(w)$ can presumably be set equal to an arbitrarily high cost.

For each individual and generation in P_2, an EA evolves a population P_1. This co-evolutionary approach can be computationally demanding because of the nested evolutionary algorithms, but it is amenable to parallel implementation, which would decrease computational effort.

Figure 19.3 gives an outline of the co-evolutionary penalty algorithm. The outer loop evolves the P_2 population. For each P_2 generation (that is, each outer loop iteration), $|P_2|$ EAs run in the inner loop to evolve candidate solutions x using the penalized cost of Equation (19.14).

We could modify Figure 19.3 in several ways to try to improve performance. For example, we could use various types of elitism to preserve the best P_1 individuals from one P_1 evolution to the next. We could also perform more than one P_1 evolution for each P_2 individual to obtain the average or best performance for a given w.

Figure 19.3 is an example of co-evolution. Here we use it for constrained optimization, but co-evolution also has many other interesting applications. We see co-evolution many places in nature. For example, flowers and bees have evolved in such a way that they depend on one another for their mutual survival [Pyke, 1978]. Co-evolution has also been studied in many different ways in EAs [Paredis, 2000], and it will surely provide an active and fruitful area for future research.

$P_2 = \{w\} \leftarrow$ randomly initialized population of candidate weights
While not(termination criterion)
    For each $w \in P_2$
        $P_1 = \{x\} \leftarrow$ randomly initialized population of candidate solutions
        Run an EA to minimize Equation (19.14) with respect to x
        Use Equation (19.15) to compute $\psi(w)$
    Next w
    Use the $\psi(w)$ costs for selection, recombination, and mutation of P_2
Next P_2 generation

Figure 19.3 Outline of the co-evolutionary penalty algorithm for the minimization of $f(x)$ subject to $G_i(x) = 0$ for $i \in [1, m+p]$.

19.2.5 Dynamic Penalty Methods

The penalized cost function of Equation (19.8) is proposed in [Joines and Houck, 1994] with $\beta = 1$ or 2, and $r_i = (ct)^\alpha$, where c and α are constants, and where t is the generation count:

$$\begin{aligned} \phi(x) &= f(x) + (ct)^\alpha M(x) \\ M(x) &= \sum_{i=1}^{m+p} G_i(x). \end{aligned} \tag{19.16}$$

This is a dynamic approach because the penalty on the constraints increases with the generation count. However, in order to be successful with this approach, the cost $f(\cdot)$ and the constraint violation magnitude $M(\cdot)$ should be normalized so that the penalized cost function $\phi(\cdot)$ is written as follows:

$$\begin{aligned} \phi(x) &= f'(x) + (ct)^\alpha M'(x) \\ M'(x) &= \begin{cases} M(x)/\max_x M(x) & \text{if } \max_x M(x) > 0 \\ 0 & \text{if } \max_x M(x) = 0 \end{cases} \\ f'(x) &= f(x)/\max_x |f(x)| \end{aligned} \tag{19.17}$$

assuming that $f(x) > 0$ for all x. This ensures that the components of the penalized cost $\phi(x)$ have approximately the same magnitude. Another option is to combine a dynamic penalty method with the superiority of feasible points method that is described in Section 19.2.2. With this approach, the penalized cost is written as

$$\phi(x) = \begin{cases} f'(x) & \text{if } x \in \mathcal{F} \\ f'(x) + (ct)^\alpha M'(x) + \theta(x) & \text{if } x \notin \mathcal{F} \end{cases} \tag{19.18}$$

where $\theta(x)$ is defined such that all feasible points have a lower penalized cost than all infeasible points. The literature [Joines and Houck, 1994] often reports typical constant values of $c = 1/2$ and $\alpha = 1$ or 2, but appropriate values of c depend on the maximum generation count. For shorter EA simulations (a couple hundred generations or fewer), c should be larger than $1/2$ by one or two orders of magnitude. If c is too small, then the constraint violation penalty will be too small and the EA will place too high of a value on individuals with low costs but large constraint violations.

19.2.5.1 Exponential Dynamic Penalties An exponential dynamic penalty function is proposed in [Carlson and Shonkwiler, 1998] as[2]

$$\phi(x) = f(x)\exp(M(x)/T) \tag{19.19}$$

where $M(x)$ is the constraint violation magnitude defined in Equation (19.16), and T is a monotonically nonincreasing function of the generation count t. $T = 1/\sqrt{t}$ is proposed in [Carlson and Shonkwiler, 1998]. This gives $\lim_{t\to\infty} T = 0$, so the penalized cost of infeasible individuals tends to infinity as the generation count tends to infinity.

Equation (19.19) assumes that $f(x) \geq 0$ for all x; otherwise the constraint penalty would serve to decrease the cost (that is, make it more negative). If this assumption is not satisfied, then we should shift the cost function before we penalize it. We can also add a tuning parameter to the penalty part of $\phi(x)$.

$$\begin{aligned} \phi(x) &= f'(x)\exp(\alpha M'(x)/T) \\ f'(x) &= f(x) - \min_x f(x) \end{aligned} \tag{19.20}$$

where the normalized constraint violation magnitude $M'(x)$ is defined in Equation (19.17), and α is a tuning parameter to adjust the relative weight of the constraint violation. We find that values of α around 10 usually work pretty well.

As with the additive penalty method described in Equation (19.17), we could combine the exponential dynamic penalty method with the superiority of feasible points method that is described in Section 19.2.2. With this approach, the penalized cost is written as

$$\phi(x) = \begin{cases} f'(x) & \text{if } x \in \mathcal{F} \\ f'(x)\exp(M(x)/T) + \theta(x) & \text{if } x \notin \mathcal{F} \end{cases} \tag{19.21}$$

or

$$\phi(x) = \begin{cases} f'(x) & \text{if } x \in \mathcal{F} \\ f'(x)\exp(\alpha M'(x)/T) + \theta(x) & \text{if } x \notin \mathcal{F} \end{cases} \tag{19.22}$$

where $\theta(x)$ is large enough to ensure that all feasible points have a lower cost than all infeasible points.

19.2.5.2 Other Dynamic Penalty Approaches More complicated forms for dynamic penalty functions are proposed in [Coit and Smith, 1996], [Coit et al., 1996], [Joines and Houck, 1994], [Kazarlis and Petridis, 1998], and [Smith and Tate, 1993]. Dynamic penalty functions are surveyed in [Coello Coello, 2002]. Dynamic penalty methods often work better than static methods, but they require additional tuning. Tuning is problem-dependent. Penalties that are too high discourage exploration of the infeasible set, but sometimes we need to use infeasible individuals to find good

[2]Note that [Carlson and Shonkwiler, 1998] uses $\phi(x) = f(x)\exp(-M(x)/T)$ because the optimization problem there is a maximization problem.

solutions that satisfy the constraints (recall Figure 19.2). However, penalties that are too low result in too much exploration of the infeasible set, and poor convergence to feasible solutions. These considerations lead us to a discussion of adaptive penalty methods in the following section.

19.2.6 Adaptive Penalty Methods

The problems with static and dynamic penalty methods motivate the development of special types of dynamic methods that are called adaptive methods. Adaptive methods use feedback from the population to adjust the penalty weights. One adaptive approach is proposed in [Hadj-Alouane and Bean, 1997], and sets the penalty weights of Equation (19.8) as follows:

$$r_i(t+1) = \begin{cases} r_i(t)/\beta_1 & \text{if case 1} \\ \beta_2 r_i(t) & \text{if case 2} \\ r_i(t) & \text{otherwise} \end{cases} \tag{19.23}$$

where t is the generation number, β_1 and β_2 are constants satisfying $\beta_1 > \beta_2 > 1$, *case 1* means that the best individual was feasible for each of the past k generations, and *case 2* means that there were no feasible individuals in any of the past k generations. The generation window k is a tuning parameter that affects the speed of adaptation. We see that if the best individual in the population is feasible, we decrease the constraint weight to allow more infeasible individuals in the population. If there are no feasible individuals in the population, we increase the constraint weight to try to obtain some feasible individuals. The goal is to obtain a balanced mix of feasible and infeasible individuals to thoroughly explore the search space, and to exploit information from infeasible individuals even though they do not satisfy the constraints. Typical constant values for this method are $r_i(1) = 1$, $\beta_1 = 4$, $\beta_2 = 3$, and $k = n$, where n is the problem dimension (that is, the number of independent variables in $f(x)$) [Hadj-Alouane and Bean, 1993].

19.2.7 Segregated Genetic Algorithm

The segregated GA [Le Riche et al., 1995] is a clever approach for handling the difficulty of tuning the penalty function parameters. The r_i parameters in Equation (19.8) are hard to tune. If they are too large, then the constrained EA focuses too much on satisfying the constraints, and not enough on minimizing the cost function. If they are too small, then the constrained EA focuses too much on minimizing the cost function, and not enough on satisfying the constraints. The segregated GA solves this problem by creating two ranked lists of individuals: the first list uses small penalty weights r_{1i}, and the second list uses large penalty weights r_{2i}. We select individuals for the next generation by choosing alternately from the two lists. This is roughly equivalent to the use of two subpopulations, one with small penalty weights, and one with large penalty weights.

This approach appears to provide a lot of room for additional research. For example, we could use more than two penalty weights. We could also use the segregated GA concept to combine multiple constraint-handling methods.

19.2.8 Self-Adaptive Fitness Formulation

An approach called the self-adaptive fitness formulation [Farmani and Wright, 2003] penalizes the cost values of infeasible individuals in two stages. First, if any infeasible individual $\bar{x}$ has an unpenalized cost that is better than the best feasible individual x (that is, if $f(\bar{x}) < f(x)$ for some $\bar{x} \notin \mathcal{F}$ and for the best $x \in \mathcal{F}$), then the cost value of each infeasible individual is penalized. However, if $f(\bar{x}) > f(x)$ for all $\bar{x} \notin \mathcal{F}$ and for the best $x \in \mathcal{F}$, then none of the infeasible individuals are penalized. This prevents the unnecessary penalization of infeasible individuals. It allows infeasible individuals to have reasonably low penalized cost values so that they can remain in the population, and so that their information can be exploited.

Next we implement a second penalization phase. All of the infeasible individuals are penalized in such a way that the individual with the greatest constraint violation has the worst penalized cost. We do this by first defining the total infeasibility for each individual x as follows:

$$\iota(x) = \frac{1}{m+p} \sum_{i=1}^{m+p} G_i(x) / \max_{\bar{x} \notin \mathcal{F}} G_i(\bar{x}) \tag{19.24}$$

where $G_i(\cdot)$ is given in Equation (19.8) with $\beta = 1$. Next we define the individuals that are best (x_b), worst in terms of feasibility (x_{wf}), and worst in terms of cost (x_{wc}), as follows:

$$\begin{aligned} x_b &= \begin{cases} \arg\min_x f(x) : x \in \mathcal{F} & \text{if } \mathcal{F} \neq \emptyset \\ \arg\min_x \iota(x) & \text{otherwise} \end{cases} \\ x_{wf} &= \begin{cases} \arg\max_x \iota(x) : f(x) < f(x_b) & \text{if } \exists\, \bar{x} \notin \mathcal{F} \text{ such that } f(\bar{x}) < f(x_b) \\ \arg\max_x \iota(x) & \text{otherwise} \end{cases} \\ x_{wc} &= \arg\max_x f(x). \end{aligned} \tag{19.25}$$

Note that if $\mathcal{F} \neq \emptyset$, then $x_b \in \mathcal{F}$ even if there is some $\bar{x} \notin \mathcal{F}$ with a lower cost. With these definitions, the infeasibility metric is normalized to $[0, 1]$ as

$$\tilde{\iota}(x) = \frac{\iota(x) - \iota(x_b)}{\iota(x_{wf}) - \iota(x_b)}. \tag{19.26}$$

Now we can mathematically define the first penalization phase as

$$\phi(x) = \begin{cases} f(x) + \tilde{\iota}(x)(f(x_b) - f(x_{wf})) & \text{if } \exists\, \bar{x} \notin \mathcal{F} \text{ such that } f(\bar{x}) < f(x_b) \\ f(x) & \text{otherwise.} \end{cases} \tag{19.27}$$

The second penalty maps $\phi(x)$ to an additionally penalized cost $\phi'(x)$:

$$\begin{aligned} \phi'(x) &= \phi(x) + \gamma|\phi(x)| \left(\frac{\exp(2\tilde{\iota}(x)) - 1}{\exp(2) - 1} \right) \\ \gamma &= \begin{cases} (f(x_{wc}) - f(x_b))/f(x_b) & \text{if } f(x_{wf}) < f(x_b) \\ 0 & \text{if } f(x_{wf}) = f(x_b) \\ (f(x_{wc}) - f(x_{wf}))/f(x_{wf}) & \text{if } f(x_{wf}) > f(x_b). \end{cases} \end{aligned} \tag{19.28}$$

The exponential function in Equation (19.28) results in only a small penalty for individuals with small constraint violations. The scaling factor γ ensures that the individual with the greatest constraint violation has the the greatest penalized cost:

$$\phi'(x_{wf}) \geq \phi'(x) \text{ for all } x. \tag{19.29}$$

This two-stage penalization method is somewhat involved, but the basic goal is that after penalization, the best individuals in the population include some that are feasible and some that are infeasible. Individuals with low cost and small constraint violations are competitive, in terms of penalized cost, with feasible individuals. This idea seems to be pretty effective, and it has been applied to a lot of constrained optimization problems. Additional work could focus on better tuning, and simplification of the penalization approach while still accomplishing its fundamental goals.

19.2.9 Self-Adaptive Penalty Function

The self-adaptive penalty function (SAPF) algorithm [Tessema and Yen, 2006] adapts penalty functions based on the distribution of the population. If there are only a few feasible individuals, then we want to assign low penalized costs $\phi(\cdot)$ to individuals with small constraint violations, even though they may have high costs $f(\cdot)$. On the other hand, if there are many feasible individuals, then we want to assign low penalized costs $\phi(\cdot)$ only to individuals with low costs $f(\cdot)$. The SAPF algorithm consists of the following steps.

1. Normalize the cost function value for each individual x:

$$f_1(x) = \frac{f(x) - \min_x f(x)}{\max_x f(x) - \min_x f(x)}. \tag{19.30}$$

   This gives the normalized cost $f_1(x) \in [0, 1]$ for all x, where the best individual in terms of cost has a normalized cost equal to 0, and the worst individual in terms of cost has a normalized cost equal to 1.

2. Compute the normalized constraint violation magnitude $\iota(x)$ of each individual as shown in Equation (19.24). This gives $\iota(x) \in [0, 1]$ for all x. Note that $\iota(x) = 0$ for all $x \in \mathcal{F}$, and $\iota(\bar{x}) > 0$ for all $\bar{x} \notin \mathcal{F}$. Also, there may or may not exist an $\bar{x}$ such that $\iota(\bar{x}) = 1$.

3. Compute the distance value for each individual x:

$$d(x) = \begin{cases} \iota(x) & \text{if } \mathcal{F} = \emptyset \\ \sqrt{f_1^2(x) + \iota^2(x)} & \text{if } \mathcal{F} \neq \emptyset. \end{cases} \tag{19.31}$$

   If there are no feasible individuals in the population, then the distance value of x is equal to the total constraint violation of x, without any consideration for the cost of x. Among two feasible individuals, the one with the lower cost will have a smaller distance. If there are feasible individuals in the population, then the distance of an infeasible individual is a combination of its cost and its constraint violation. Therefore, when we compare a feasible individual x with an infeasible individual $\bar{x}$, either one may have a smaller distance, depending on their relative cost values.

4. Compute two additional penalized cost functions:

$$\begin{aligned} X(x) &= \begin{cases} 0 & \text{if } \mathcal{F} = \emptyset \\ \iota(x) & \text{if } \mathcal{F} \neq \emptyset \end{cases} \\ Y(x) &= \begin{cases} 0 & \text{if } x \in \mathcal{F} \\ f_1(x) & \text{if } x \notin \mathcal{F}. \end{cases} \end{aligned} \tag{19.32}$$

   The penalized cost $X(x)$ is equal to 0 if there are no feasible individuals in the population, and it is equal to $\iota(x)$ if there are feasible individuals in the population. This penalized cost serves to penalize infeasible individuals based on the magnitude of their constraint violations, but only if the population contains feasible individuals. The penalized cost $Y(x)$ is equal to 0 if x is feasible, and it is equal to the normalized cost $f_1(x)$ if x is infeasible. This penalized cost serves to further penalize infeasible individuals by an amount that is proportional to their cost values.

5. Compute the penalized cost function

$$\phi(x) = d(x) + (1 - r)X(x) + rY(x) \tag{19.33}$$

   where $r \in [0, 1]$ is the proportion of feasible individuals in the population. If there are a lot of feasible individuals in the population, $\phi(x)$ emphasizes $Y(x)$, which includes cost-based penalties on infeasible individuals. On the other hand, if there are few feasible individuals in the population, $\phi(x)$ emphasizes $X(x)$, which includes constraint violation penalties on infeasible individuals.

SAPF has also been adapted to constrained multi-objective optimization problems [Yen, 2009].

19.2.10 Adaptive Segregational Constraint Handling

The adaptive segregational constraint handling evolutionary algorithm (ASCHEA) is proposed in [Hamida and Schoenauer, 2000], [Hamida and Schoenauer, 2002]. ASCHEA is based on two ideas. First, we try to maintain a specified ratio of feasible individuals to infeasible individuals. This is similar to the adaptive approach that we discussed Section 19.2.6. This allows us to explore the entire search space, including both the feasible portion and the infeasible portion.

Second, if there are few feasible individuals in the population, then we allow feasible individuals to recombine only with infeasible individuals. This is based on the idea that constrained optimization problem solutions often lie on, or near, the constraint boundary [Leguizamón and Coello Coello, 2009], [Ray et al., 2009b]. Therefore, to solve constrained optimization problems, it makes sense to push both feasible and infeasible individuals toward the constraint boundary. Recombining feasible individuals with infeasible individuals tends to bring their offspring closer to the constraint boundary.

ASCHEA uses the penalty approach of Equation (19.8) with $\beta = 1$ and the following update method for the penalty weight:

$$\begin{aligned} \phi(x) &= f(x) + \sum_{i=1}^{m+p} r_i G_i(x) \\ r_i(t+1) &= \begin{cases} r_i(t)/\gamma & \text{if } \tau(t) > \tau_d \\ \gamma r_i(t) & \text{otherwise} \end{cases} \end{aligned} \tag{19.34}$$

where t is the generation number, $\gamma > 1$ is a tuning parameter, τ_d is the target ratio of the number of feasible individuals to infeasible individuals, and $\tau(t)$ is the ratio at generation t:

$$\tau(t) = \frac{|\{x : x \in \mathcal{F}\}|}{|\{\bar{x} : \bar{x} \notin \mathcal{F}\}|} \text{ at generation } t. \tag{19.35}$$

ASCHEA is often implemented with a target value $\tau_d = 1/2$.

The second idea implemented in ASCHEA is to allow feasible individuals to recombine only with infeasible individuals if $\tau(t) < \tau_d$. This encourages the offspring of infeasible individuals to move toward the feasible region, which hopefully increases the number of feasible individuals.

However, if $\tau(t) \geq \tau_d$, then we already have enough feasible individuals in the population, so we perform selection with the following two steps: first, we select a specified number of individuals from the feasible set $\mathcal{F}$; second, we select the remaining individuals for recombination on the basis of penalized cost values $\phi(\cdot)$ without any explicit consideration of feasibility. The minimum number of feasible individuals that we select for recombination is typically about 30% of the total number of individuals that we need to select. For example, if we want to select 100 individuals for recombination, we first select 30 feasible individuals on the basis of cost, and then we select the remaining 70 individuals from the entire population on the basis of penalized cost.

A similar algorithm, called the infeasibility driven evolutionary algorithm (IDEA), is proposed in [Ray et al., 2009b] for both single-objective and multi-objective optimization.

19.2.11 Behavioral Memory

Behavioral memory uses a divide-and-conquer method to solved constrained optimization problems [Michalewicz et al., 1996], [Schoenauer and Xanthakis, 1993]. Given the problem of Equation (19.8) with $m + p$ constraints, we first evolve a population that minimizes $G_1(x)$; that is, we minimize the violation of the first constraint without considering any of the other constraints or the cost function. We terminate this EA after a user-specified fraction of the population satisfies the first constraint. After this evolution is complete, we use its final population to initialize an EA that minimizes $G_2(x)$. During this second EA, we remove any individuals from the population that violate the $G_1(x)$ constraint, but we do not consider any of the other constraints or the cost function.

We repeat these steps for each constraint. We initialize the i-th EA with the results of the $(i-1)$-st EA. The i-th EA evolves a population that minimizes $G_i(x)$, and during this evolution we remove any individuals that stray outside any of the $G_j(x)$ constraints for $j \in [1, i-1]$. Finally after all $m + p$ constraints have been satisfied by the successive implementation of $m + p$ EAs, we use the resulting population to initialize an EA that minimizes the cost function subject to all $m + p$ constraints. Figure 19.4 gives an outline of the behavioral memory algorithm.

Behavioral memory can be classified as a penalty function approach because it uses the death penalty. However, the line in Figure 19.4 that begins with the statement "Run an EA to minimize ..." may or may not involve penalty function approaches. That EA could include any EA and any constraint-handling approach,

---

$\{x\}_0 \leftarrow$ randomly initialized population
For $i = 1, \cdots, m+p$
    Initialize the i-th EA: $\{x\} \leftarrow \{x\}_{i-1}$
    Run an EA to minimize $G_i(x)$ subject to $G_j(x) = 0$ for $j < i$, using the death penalty to ensure that all individuals satisfy the G_j constraints, and denote the final population of this EA as $\{x\}_i$
Next i
Initialize the final EA: $\{x\} \leftarrow \{x\}_{m+p}$
Run an EA to minimize $f(x)$ subject to $G_j(x) = 0$ for $j \in [1, m+p]$, using the death penalty to ensure that all individuals satisfy the G_j constraints

---

Figure 19.4 Outline of a behavioral memory algorithm for the minimization of $f(x)$ subject to $G_i(x) = 0$ for $i \in [1, m+p]$.

as long as it eventually removes individuals that violate previously-considered constraints.

Behavioral memory is actually a generalization of unconstrained optimization algorithms that gradually increase the number of cost function evaluations [de Garis, 1990], [Gathercole and Ross, 1994]. For example, suppose that we want to minimize $f_i(x)$ with respect to x, where i could be any of a large set of values $\mathcal{I} = \{1, 2, \cdots, i_{\max}\}$. One way to approach this is to minimize $f_1(x)$ with an EA. Then, using that final population, we minimize $f_1(x) + f_2(x)$. Again, using that second final population, we minimize $f_1(x) + f_2(x) + f_3(x)$. We continue this process until we are satisfied that we have minimized $f_i(x)$, averaged over all $i \in \mathcal{I}$. Another approach is to minimize the combination of a random selection of $f_i(x)$ instances, gradually increasing the number of instances as the generation count increases. This approach is called stochastic sampling [Banzhaf et al., 1998, Section 10.1.5]. We often see this approach in genetic programming since evaluating the cost of a single computer program requires computer runs for many input cases (see Chapter 7).

This minimization of $f_i(x)$ for $i \in \mathcal{I}$ appears to have the same form as a multi-objective optimization problem (see Chapter 20), but the problem discussed here is actually a single-objective optimization problem; each cost function $f_i(x)$ is the same function, but it is evaluated with different parameters for different values of i. However, the dividing line between single-objective optimization problems with multiple parameters, and multi-objective optimization problems, is fuzzy. The same problem might be considered or treated as a single-objective problem by one person, but as a multi-objective problem by another person.

19.2.12 Stochastic Ranking

Stochastic ranking [Runarsson and Yao, 2000] adds a stochastic component to constrained EAs. Since randomness is such an important component of EAs, it makes sense to include randomness in the constraint-handling approach of EAs. Stochastic ranking sometimes ranks candidate solutions according to their cost $f(\cdot)$, and sometimes ranks them according to their constraint violation magnitude. The decision of how to rank individuals is stochastic. When we compare two individuals x_1 and x_2, we consider individual x_1 to be better than x_2 if:

- Both solutions are feasible and $f(x_1) < f(x_2)$; or,
- A randomly-generated number $r \sim U[0, 1]$ is less than a user-defined probability P_f, and $f(x_1) < f(x_2)$; or,
- Neither of the above conditions are satisfied, and x_1 has a smaller constraint violation than x_2.

Otherwise, we consider x_2 to be better than x_1. We see that we might compare x_1 and x_2 on the basis of their costs, or we might compare them on the basis of their constraint violations, depending the outcome of a random number generator. After we have compared and sorted all of the individuals in the population, we then perform selection and recombination for the next generation. Probability values $P_f \in (0.4, 0.5)$ give good results for many benchmark problems [Runarsson and Yao, 2000].

19.2.13 The Niched-Penalty Approach

The niched-penalty approach [Deb and Agrawal, 1999], [Deb, 2000] is motivated by the difficulty of tuning the parameters of penalty methods. It uses tournament selection to select individuals for recombination according to the following rules.

- Given two feasible individuals, the one with the lower cost wins the tournament.
- Given one feasible individual and one infeasible individual, the feasible individual wins the tournament.
- Given two infeasible individuals, the one with the smaller constraint violation wins the tournament.

This method is attractive because of its simplicity; it does not require any tuning of penalty parameters. A comparison of two infeasible individuals does not require any cost function evaluations, which can reduce computational effort. The niched-penalty approach often obtains good results on constrained optimization problems. However, its simplicity may also be a disadvantage because it considers a feasible individual with a very high cost to be better than a slightly infeasible individual with a very low cost. Therefore, it may not work well for problems whose solutions are on the constraint boundary, which is the case for many real-world optimization problems [Leguizamón and Coello Coello, 2009], [Ray et al., 2009b].

The "niched" part of this approach is not integral to its constraint-handling capability, but is intended to preserve diversity in the population, and is described as follows. We do not allow individuals to participate in a tournament with each other (for selection) if they are far from each other in domain space. After we randomly choose individuals for tournament selection, we then compute their distance from each other. If the individuals are too far apart from each other, then we randomly choose different individuals for the tournament. This prevents distant clusters of individuals from disappearing from the population and thus maintains diversity.

The multimembered evolution strategy (MMES) [Mezura-Montes and Coello Coello, 2005] is similar to the niched-penalty approach. The distinctive feature of MMES is that every so often (around 3% of the generations), the infeasible individuals with the lowest cost function values, or the infeasible individuals with the smallest constraint violations, are guaranteed to be selected for the next generation. This elitist-like approach ensures that good infeasible individuals contribute their features to the EA search process.

19.3 SPECIAL REPRESENTATIONS AND SPECIAL OPERATORS

A special representation is a way of formulating a problem so that the constraints are automatically satisfied by candidate solutions. A special operator is a way of defining an EA's recombination and mutation operators in such a way that child individuals automatically satisfy the constraints. Both of these approaches are largely problem-dependent. We cannot write special representation code, or special operator code, that applies to a broad class of problems. However, although the problem-dependent nature of special representations and special operators creates more work for the EA designer, that work often pays high dividends. This is because the resulting EA uses problem-specific information, which often results in better performance than we can expect from a more general-purpose EA. This concept is directly related to the no free lunch theorem (see Appendix B.1).

Section 19.3.1 discusses special representations, including a method called the decoder approach. Section 19.3.2 discusses special operators, including a popular constrained EA called Genocop.

19.3.1 Special Representations

As a simple example of the special representation approach to constrained optimization, consider the two-dimensional problem

$$\min_x f(x) \text{ such that } x_1^2 + x_2^2 \leq K \tag{19.36}$$

where K is some constant. This can be transformed into the equivalent unconstrained problem

$$\begin{aligned}&\min_{\rho,\theta} f(x) \text{ such that } \rho \in [0, K] \text{ and } \theta \in [0, 2\pi]\\ &\text{where } x_1 = \rho\cos\theta \text{ and } x_2 = \rho\sin\theta.\end{aligned} \tag{19.37}$$

This simple rectangular-to-polar transformation converts the constrained problem of Equation (19.36) into the unconstrained problem of Equation (19.37). The original problem of Equation (19.36) has a nonlinear constraint, but we have transformed it into the problem of Equation (19.37), whose only constraints are simple limits on the search domain.

Decoders

One approach to solving constrained optimization problems is to encode instructions that determine candidate solutions, ensuring that the instruction set always results in a feasible individual. That is, instead of directly encoding a candidate solution to the problem, we encode an instruction set that we use to obtain a candidate solution. Each individual in the EA population then consists of an instruction set for building a candidate solution. We can also use this approach for unconstrained optimization problems, but it seems to be particularly applicable to constrained optimization problems, because depending on the specific problem, we may be able to determine a set of instructions that always satisfies the problem constraints.

An instruction set for building a candidate solution is called a decoder [Palmer and Kershenbaum, 1994], [Koziel and Michalewicz, 1998], [Koziel and Michalewicz, 1999], and should satisfy several properties.

- For each feasible solution to the optimization problem, there should exist at least one decoder.
- Each feasible solution should correspond to the same number of decoders so as not to introduce bias in the search.
- Each decoder should correspond to a feasible solution.
- The transformation between decoder and candidate solution should be computationally fast relative to cost evaluation.
- Small changes in a decoder should correspond to small changes in the candidate solution.

These rules can be relaxed for specific applications in case they are too difficult to satisfy [Koziel and Michalewicz, 1999], but they at least provide useful guidelines. For example, we may have a problem whose constraints are very difficult to satisfy, but we can find a set of decoders that gives candidate solutions that are feasible a high percentage of the time. Even though such a decoder set does not strictly satisfy the above conditions, it may be preferable to encoding candidate solutions directly in our EA.

Once we obtain a set of instructions for building candidate solutions, we redefine the EA population to consist of a set of instructions (decoders). We then perform selection, recombination, and mutation on the decoders, thus guaranteeing constraint satisfaction.

EXAMPLE 19.2

Consider the convex feasible region $\mathcal{F}$ of the two-dimensional constrained optimization problem represented by Figure 19.5. The reference point r is an arbitrary point in $\mathcal{F}$. Any point $x \in \mathcal{F}$ is uniquely represented by $\alpha b + (1-\alpha)r$ for some $\alpha \in [0, 1]$ and some b on the boundary of the search domain.[3]

A decoder to solve this optimization problem might proceed as follows.

1. We somehow find a feasible point r.
2. We find the boundary of $\mathcal{F}$ by moving b around the entire search domain boundary. For each b, we find the maximum α such that $\alpha b + (1-\alpha)r \in \mathcal{F}$. This maximum is denoted as $\alpha_{\max}(b) = \max\{\alpha : \alpha b + (1-\alpha)r \in \mathcal{F}\}$.

[3]There is one exception: r has an infinite number of such representations since $r = (0)(b) + (1)(r)$ for all b.

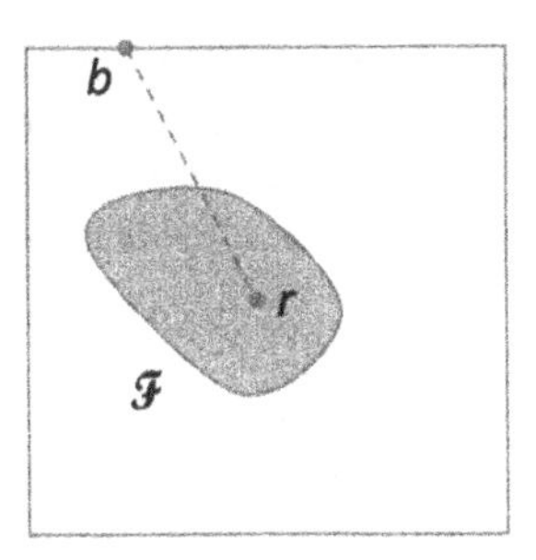

Figure 19.5 Example of a decoder algorithm for optimization over $\mathcal{F}$. Given any point $r \in \mathcal{F}$, any point $x \in \mathcal{F}$ is uniquely represented by $\alpha b + (1 - \alpha)r$ for some $\alpha \in [0, 1]$ and some b on the boundary of the search domain.

3. We define a population of EA individuals so that each individual is comprised of a (b, α) pair. The parameter b can be any point on the search domain boundary, and α can be any real number in $[0, \alpha_{\max}(b)]$.

4. We run an EA by performing recombination on (b, α) pairs. The only feasibility check that we need to worry about is $\alpha \in [0, \alpha_{\max}(b)]$.

□

Example 19.2 would need to be modified for nonconvex feasible regions, but it illustrates how a decoder can be used to simplify constraints.

19.3.2 Special Operators

In many real-world optimization problems, the solution lies on the constraint boundary [Leguizamón and Coello Coello, 2009], [Ray et al., 2009b]. Consider the problem of optimizing the purchase of some equipment. If we were given the assignment of purchasing the best possible equipment, that assignment might be constrained by an upper bound on price. We would probably spend the maximum allowable amount of money, because better equipment typically costs more than worse equipment. That is, the constrained optimum would lie on the price constraint boundary. Constraints related to the color of the equipment would typically *not* be part of the equipment purchase assignment, because equipment optimality and equipment color are typically unrelated.

Similarly, if we wanted to travel from point A to point B in minimum time with an upper limit on fuel use, we would probably use the maximum allowable amount of fuel, because we can typically travel faster using more fuel. That is, the constrained optimum would lie on the fuel constraint boundary. As we think about real-world problems, we realize that most optimization problem solutions lie on the constraint boundary.[4]

[4]This statement applies to real-world problems but not necessarily to the benchmark problems that we see in the literature. This difference between benchmark problems and real-world problems is related to the discussion of the no free lunch theorem in Appendix B.

This leads to the idea of solving an optimization problem by exploring the constraint boundary. We might be able to safely ignore the interior of the constraint boundary because we expect optimization solutions to lie on the constraint boundary. For example, consider an optimization problem with the product constraint

$$x_1 x_2 x_3 x_4 \geq 0.75. \tag{19.38}$$

If we have domain knowledge that leads us to expect that the constrained optimum lies on the constraint boundary, then we can try to solve the optimization problem by searching (x_1, x_2, x_3, x_4) combinations that satisfy the following [Michalewicz and Schoenauer, 1996]:

$$x_1 x_2 x_3 x_4 = 0.75. \tag{19.39}$$

We can initialize an individual in an EA population as follows:

$$\begin{aligned} x_1 &= U[-x_{\min}, x_{\max}] \\ x_2 &= U[-x_{\min}, x_{\max}] \\ x_3 &= U[-x_{\min}, x_{\max}] \\ x_4 &= 0.75/(x_1 x_2 x_3). \end{aligned} \tag{19.40}$$

This ensures that (x_1, x_2, x_3, x_4) satisifies Equation (19.38). Now suppose that we have another individual (y_1, y_2, y_3, y_4) such that $y_1 y_2 y_3 y_4 = 0.75$. Then if we create a child individual z such that $z_i = x_i^\alpha y_i^{1-\alpha}$ for $i \in [1, 4]$, where the real number $\alpha \in [0, 1]$, the child will always satisfy the constraint $z_1 z_2 z_3 z_4 = 0.75$. This is true because

$$\begin{aligned} z_1 z_2 z_3 z_4 &= x_1^\alpha y_1^{1-\alpha} x_2^\alpha y_2^{1-\alpha} x_3^\alpha y_3^{1-\alpha} x_4^\alpha y_4^{1-\alpha} \\ &= (x_1 x_2 x_3 x_4)^\alpha (y_1 y_2 y_3 y_4)^{1-\alpha} \\ &= (0.75)^\alpha (0.75)^{1-\alpha} = 0.75. \end{aligned} \tag{19.41}$$

We see that our specialized crossover operator ensures that the child of two feasible parents will always be feasible.

We can also design a specialized mutation operator for this problem:

$$\begin{aligned} x'_i &\leftarrow q x_i \\ x'_j &\leftarrow x_j / q \end{aligned} \tag{19.42}$$

where $i \in [1, 4]$ and $j \in [1, 4]$ are distinct random integers, and q is a random number. Any feasible individual x that is mutated this way will result in a feasible individual x'.

These simple recombination and mutation operators illustrate the use of specialized operators for constrained EAs. Note that specialized operators are problem-specific; for any constrained optimization problems, the EA designer has to formulate his own problem-specific operators to preserve feasibility. An additional example of a specialized operator is given in [Michalewicz and Schoenauer, 1996].

19.3.3 Genocop

Next we discuss an algorithm known as the genetic algorithm for numerical optimization of constrained problems (Genocop) [Michalewicz and Janikow, 1991].

This algorithm includes several special features and techniques for constrained optimization. We include Genocop in this section because it was first proposed with the idea of implementing special operators on EA individuals to ensure that they satisfy linear constraints.

The idea behind Genocop is that sometimes, depending on the form of the constraint, we can use problem-specific operators to transform infeasible individuals into feasible individuals. We could do this with a constraint in the form of Equation (19.38). If we have an individual that does not satisfy the constraint, then we can replace its fourth component as shown in the last line of Equation (19.40). This would be a repair approach. Alternatively, if we create the first three elements of a child individual, we could then create the fourth element as shown in the last line of Equation (19.40). This would be a special operator approach.

Suppose that we have the linear constraint

$$-8x_1 + x_3 \leq 0. \tag{19.43}$$

If we have an individual that does not satisfy this constraint, we can easily repair it by setting x_3 equal to any number less than or equal to $8x_1$. The modified individual will then satisfy the constraint. This example shows that any linear constraint can be easily satisfied using repair algorithms or special operators.

Genocop is efficient but its design is problem-specific. As shown above, Genocop is limited to linear constraints, and special forms of nonlinear constraints in which one variable can be solved in terms of the others. This is a disadvantage from the viewpoint of user effort, but an advantage from the viewpoint of EA efficiency.

19.3.4 Genocop II

Genocop II [Michalewicz and Attia, 1994] combines Genocop as described above with a dynamic penalty similar to the one in Section 19.2.5. Genocop II uses specialized operators to maximize the feasibility of an EA population. First, we satisfy all linear constraints by repairing infeasible individuals as proposed in Genocop. Second, we handle nonlinear constraints by minimizing $\phi(x)$ in Equation (19.8), where all of the constraints in that equation are nonlinear, since we already satisifed the linear constraints with special operators. The r_i weight in Equation (19.8) is $1/\tau$. Genocop II maintains a constant value of τ for several generations. After a while (for example, after a specific generation count, or after a specified fraction of the population is feasible), Genocop II decreases τ. This increases the constraint pressure, which results in a gradual attraction of more and more individuals to the feasible set. Early Genocop II papers suggest decreasing τ by a factor of 10 each time it is decreased [Michalewicz and Attia, 1994], [Michalewicz and Schoenauer, 1996].

19.3.5 Genocop III

Genocop III [Michalewicz and Nazhiyath, 1995] is a further modification of Genocop. In this method, a co-evolutionary algorithm maintains a population $P_r = \{x_r\}$ of reference points that satisfy all the constraints, and a population $P_s = \{x_s\}$ of search points that satisfy the linear constraints due to the approach used in Genocop. P_r and P_s may be different sizes. We assign the cost function value of each individual x_s from P_s using its repaired version. That is, we use information from P_r to repair x_s and to obtain an individual x'_s that satisfies all constraints. We then assign $f(x_s) \leftarrow f(x'_s)$. We create x'_s by generating a sequence of points $x = ax_s + (1-a)x_r$ for a set of random numbers $a \in [0,1]$, and for a randomly selected $x_r \in P_r$. That is, we search through a random set of points that are on the straight line connecting x_s and x_r. After this search gives us a feasible x, we assign $x'_s \leftarrow x$, and we assign $f(x_s) \leftarrow f(x'_s)$. Also, if $f(x'_s) < f(x_r)$, we replace x_r with x'_s in P_r. Finally, we also replace x_s with x'_s in P_s with some user-defined probability of replacement ρ.

The question of whether or not to replace an individual x_s with its repaired version x'_s is related to Lamarckian inheritance: that is, can an organism pass on traits that it acquires during its lifetime to its offspring? Some researchers never replace individuals with their repaired versions ($\rho = 0$), others always replace individuals with their repaired versions ($\rho = 1$), and others recommend that values of ρ between 5% and 20% give good results [Michalewicz and Schoenauer, 1996], [Orvosh and Davis, 1993].

Figure 19.6 outlines the Genocop III algorithm. The first step in Genocop III is the random initialization of P_s. We generate this population without any regard for feasibility, except for the satisfaction of linear constraints as proposed by Genocop. The second step is the evaluation of $f(x_s)$ for each $x_s \in P_s$. We perform this with the "Evaluate $f(x_s)$" function at the bottom of Figure 19.6. The third step is the random initialization of P_r. We generate this population in such a way that each individual satisfies all constraints.[5] The fourth step is the evaluation of $f(x_r)$ for each $x_r \in P_r$. Finally we perform the evolutionary algorithm loop to evolve the P_s and P_r populations. We can use any EA to implement selection, recombination, and mutation as we modify the P_s and P_r populations.

Genocop III seems to give good results and is thus an attractive algorithm for applications and future research. For example, we could try various EAs for the evolution of the P_s and P_r populations, and we do not have to use the same EA for the two evolutions. We could also experiment with self-adaptation of ρ. We note that [Michalewicz and Schoenauer, 1996] indicates that a new P_r population does not have to be created each generation; for example, we might need to perform the "generate a new population P_r" step in Figure 19.6 once every few times through the loop. This would decrease computational effort. Finally, hybrids of Gencop III with other constrained EAs might improve performance.

[5]Finding individuals that satisfy all constraints can be a challenging task in itself, but Genocop III assumes that we have some method to do this. The discussion of constraint programming in Section 19.7 is relevant to this task.

$P_s \leftarrow$ randomly initialized population of search points
Evaluate $f(x_s)$ for each $x_s \in P_s$ as shown in the algorithm below
$P_r \leftarrow$ randomly initialized population of feasible points
Evaluate $f(x_r)$ for each $x_r \in P_r$
While not(termination criterion)
    Use an EA with the $f(x_s)$ values to generate a new population P_s
    Evaluate $f(x_s)$ for each $x_s \in P_s$ as shown in the algorithm below
    Use an EA with the $f(x_r)$ values to generate a new population P_r
    Evaluate $f(x_r)$ for each $x_r \in P_r$
Next generation
~~~~~~~~~~
Evaluate $f(x_s)$:
If $x_s \in \mathcal{F}$ then
    Compute $f(x_s)$ using the cost function
else
    $x'_s \leftarrow x_s$
    While $x'_s \notin \mathcal{F}$
        Randomly select an x_r from P_r
        Randomly generate $a \sim U[0,1]$
        $x'_s \leftarrow a x_s + (1-a) x_r$
    End while
    $f(x_s) \leftarrow f(x'_s)$
    If $f(x'_s) < f(x_r)$ then $x_r \leftarrow x'_s$
    Randomly generate $\alpha \sim U[0,1]$
    If $\alpha < \rho$ then $x_s \leftarrow x'_s$
End if

Figure 19.6 Outline of the Genocop III algorithm for the minimization of $f(x)$ subject to constraints.

19.4 OTHER APPROACHES TO CONSTRAINED OPTIMIZATION

This section briefly discusses a couple of other approaches to constrained optimization. These approaches are not penalty function approaches, and they do not involve special representations or special operators, so we discuss them in this separate section. Section 19.4.1 discusses cultural algorithms, and Section 19.4.2 discusses multi-objective optimization for constrained problems.

19.4.1 Cultural Algorithms

Cultural algorithms (CAs) are EAs that use belief spaces to guide their evolution. That is, as a CA tries to solve an optimization problem, its search is biased in certain directions. Constrained CAs are not penalty methods (in general), because penalty methods increase the evaluated cost function of infeasible solutions, whereas constrained CAs bias the search so that infeasible solutions are less likely to exist
~~~~~~~~~~

in the population in the first place. However, using belief spaces in a CA opens up a wide area of possible approaches and implementations because of the cultural foundations of CAs. We discussed CAs in Chapter 15, and so we do not discuss them any further here, but we leave it to the reader to explore the use of CAs for constrained optimization [Becerra and Coello Coello, 2004], [Coello Coello and Becerra, 2002].

19.4.2 Multi-Objective Optimization

Chapter 20 discusses multi-objective optimization problems (MOPs). A MOP is a problem for which we want to simultaneously minimize M cost functions:

$$\min_x [f_1(x), \cdots, f_M(x)]. \tag{19.44}$$

A constrained optimization problem can be viewed as a MOP by defining the first objective as the cost, and definining the remaining objectives as the constraints. Consider a constrained optimization problem that is written in the standard form of Equation (19.1):

$$\begin{aligned} \min_x f(x) \quad \text{such that} \quad & g_i(x) \leq 0 \text{ for } i \in [1, m] \\ \text{and} \quad & h_j(x) = 0 \text{ for } j \in [1, p]. \end{aligned} \tag{19.45}$$

This problem is equivalent to the MOP of Equation (19.44) if

$$\begin{aligned} f_1(x) &= f(x) \\ f_2(x) &= G_1(x) \\ &\vdots \\ f_M(x) &= G_{m+p}(x) \end{aligned} \tag{19.46}$$

where $G_i(x)$ is given in Equation (19.8). Therefore, we can use any MOP algorithm to solve a constrained optimization problem. Chapter 20 discusses EAs for MOPs. Research on the use of MOP algorithms for constrained optimization can be found in [Aguirre et al., 2004], [Cai and Wang, 2006], [Coello Coello, 2000a], [Coello Coello, 2002], and [Mezura-Montes and Coello Coello, 2008], among many other references.

19.5 RANKING CANDIDATE SOLUTIONS

The preceding sections discussed several ways of ranking candidate solutions for constrained optimization problems. This section summarizes the previously-discussed ranking approaches, and presents a couple of alternative approaches. First we summarize the previously-discussed approaches.

- Equation (19.8) penalizes the cost function with a function of the constraint violation magnitudes.
- Equation (19.10) modifies Equation (19.8) so that all feasible individuals have a better rank than all infeasible individuals, while infeasible individuals are ranked according to the magnitude of their constraint violations. Section 19.2.13 also takes this approach.

- Equation (19.13) ranks all feasible individuals better than all infeasible individuals, while ranking infeasible individuals on the basis of the number of constraint violations rather than on the magnitude of constraint violations.
- Equation (19.14) penalizes the cost function with both the magnitude of constraint violations and the number of constraint violations.
- Equation (19.19) penalizes the cost function with a constraint violation penalty that increases as the generation count increases.
- Equations (19.23) and (19.34) impose a constraint violation penalty that is a function of the number of feasible individuals in the population.
- Sections 19.2.7 and 19.2.8 adjust the cost penalty based on a combination of the number of feasible individuals and the relative costs of various individuals.
- Section 19.2.12 uses a random process to determine how to rank candidate solutions.

We have already discussed quite a few ranking approaches for constrained optimization, and the literature includes several others. Next we present three additional ranking approaches.

19.5.1 Maximum Constraint Violation Ranking

Instead of using the sum of constraint violation magnitudes, or the number of constraint violations, we could rank individuals using their maximum constraint violation magnitude [Takahama and Sakai, 2009]. In this case, we replace the penalized cost function of Equation (19.8) with

$$\phi(x) = f(x) + \max_i G_i(x). \tag{19.47}$$

We could also rank candidate solutions using a combination of the sum of the constraint violation magnitudes, the number of constraint violations, and the maximum constraint violation magnitude.

19.5.2 Constraint Order Ranking

[Ray et al., 2009b] proposes a way of combining the magnitude of constraint violations with the number of constraint violations. Suppose that x_k is the k-th individual in a population of N individuals. Suppose that we have a constrained optimization problem with $m+p$ constraints. We use $G_i(x_k)$ to denote the magnitude of the i-th constraint violation of x_k, with $G_i(x_k) \geq 0$. We then use $c_i(x_k)$ to denote the rank of the i-th constraint violation of x_k, where a lower rank means less constraint violation, and we set $c_i(x_k) = 0$ if $G_i(x_k) = 0$. Note that $c_i(x_k) \in [0, N]$. Here is a simple example with a population size of five:

$$\left.\begin{array}{l} G_1(x_1) = 3.5 \\ G_1(x_2) = 5.7 \\ G_1(x_3) = 0.0 \\ G_1(x_4) = 1.3 \\ G_1(x_5) = 0.0 \end{array}\right\} \rightarrow \left\{\begin{array}{l} c_1(x_1) = 2 \\ c_1(x_2) = 3 \\ c_1(x_3) = 0 \\ c_1(x_4) = 1 \\ c_1(x_5) = 0. \end{array}\right. \tag{19.48}$$

We then define the constraint violation measure as

$$v(x_k) = \sum_{i=1}^{m+p} c_i(x_k). \tag{19.49}$$

19.5.3 ϵ-Level Comparisons

ϵ-level comparisons are similar to the static penalty approach of Section 19.2.1 in which different penalty function weights are used depending on the level of constraint violation. However, ϵ-level comparisons use only two levels of constraint violations for ranking [Takahama and Sakai, 2009].

First, we quantify the constraint violation $M(x)$ of each individual x by either combining all constraint violations, or by finding the maximum constraint violation:

$$M(x) = \begin{cases} \sum_{i=1}^{m+p} G_i(x) & \text{constraint sum method} \\ \max_i G_i(x) & \text{maximum constraint method.} \end{cases} \tag{19.50}$$

As mentioned in Section 19.5.1, we could also combine the constraint sum method and the maximum constraint method to obtain $M(x)$.

Second, we rank two individuals x and y as follows:

$$x \text{ is better than } y \text{ if: } \begin{cases} f(x) < f(y) \text{ and } M(x) \le \epsilon \text{ and } M(y) \le \epsilon, \text{ or} \\ f(x) < f(y) \text{ and } M(x) = M(y), \text{ or} \\ M(x) < M(y) \text{ and } M(y) > \epsilon \end{cases} \tag{19.51}$$

where $\epsilon \ge 0$ is a user-defined constraint violation threshold.[6] We see that a constraint violation that is less than ϵ is considered to be a feasible solution for the purpose of ranking. Note that if $\epsilon = \infty$, then individuals are ranked solely on the basis of cost. If $\epsilon = 0$, then feasible individuals are ranked on the basis of cost, infeasible individuals are ranked solely on the basis of their constraint violation, and feasible individuals are always ranked better than infeasible individuals. We typically decrease ϵ as the generation count increases, which gradually increases the importance of constraint satisfaction:

$$\begin{aligned} \epsilon(0) &= M(x_p) \\ \epsilon(t) &= \begin{cases} \epsilon(0)\,(1 - t/T_c)^c & \text{if } 0 < t < T_c \\ 0 & \text{if } t \ge T_c \end{cases} \end{aligned} \tag{19.52}$$

where $\epsilon(t)$ is the value of ϵ during the t-th generation, x_p is the individual with the p-th smallest constraint violation, $p = N/5$, N is the population size, and c and T_c are tuning parameters that are often set to values of about $c = 100$ and $T_c = t_{\max}/5$ [Takahama and Sakai, 2009]. We could also try other tuning parameters and other profiles for decreasing ϵ as a function of t.

19.6 A COMPARISON BETWEEN CONSTRAINT-HANDLING METHODS

This section presents a comparison between nine inequailty-constraint-handling methods. We use Equation (19.16) to measure the constraint violation magni-

[6] This ϵ is not the same as the one in Equation (19.6). The same variable is used in the literature for both Equation (19.6) and (19.51), and so we follow that convention in this chapter also.

tude of a candidate solution, where $G_i(x)$ is given by Equation (19.8) with $\beta = 1$. The constraint-handling methods that we test include the following.

1. EE: The eclectic EA of Section 19.2.3.
2. DP: The dynamic penalty method of Equation (19.16) in Section 19.2.5 with $c = 10$ and $\alpha = 2$.
3. DS: The dynamic penalty method combined with the superiority of feasible points, as defined by Equation (19.18) in Section 19.2.5, with $c = 10$ and $\alpha = 2$.
4. EP: The exponential dynamic penalty method of Equation (19.20) in Section 19.2.5.1 with $\alpha = 10$.
5. ES: The exponential dynamic penalty method combined with the superiority of feasible points, defined by Equation (19.22) in Section 19.2.5.1, with $\alpha = 10$.
6. AP: The adaptive penalty method of Equation (19.23) in Section 19.2.6 with $\beta_1 = 4$, $\beta_2 = 3$, and $k = n$, where n is the problem dimension.
7. SR: The stochastic ranking method of Section 19.2.12 with $P_f = 0.45$.
8. NP: The niched-penalty approach of Section 19.2.13.
9. ϵC: The ϵ-level comparison method of Section 19.5.3 with $c = 100$, $T_c = 200$, and $p = N/5$, where N is the population size.

Since the constraint-handling methods listed above are solely concerned with how to rank candidate solutions, we can use any EA in conjunction with any of these constraint-handling methods. In this section we use the BBO algorithm of Figure 14.3, while calculating the fitness of each individual with one of the nine constraint-handling methods listed above.

We test the constraint-handling methods on the 2010 Congress on Evolutionary Computation (CEC) benchmarks listed in Appendix C.2 with $n = 10$ dimensions. However, we only test on the benchmarks that do not include equality constraints: C01, C07, C08, C13, C14, and C15.

Equality-constrained problems require special handling. As mentioned in Section 19.1.2.2, equality constraints are very unforgiving. If we randomly generate an initial population, we have an essentially zero probability of obtaining any individuals that satisfy equality constraints. There are two basic approaches to generate individuals that satisfy equality constraints: (1) Use problem-specific information, as discussed in Section 19.3; (2) Use Equation (19.8) to convert the equality constraints to inequality constraints, use large values of ϵ early in the EA, and gradually decrease ϵ as the generation count increases. Although it is possible to formulate general-purpose equality-constrained optimization algorithms, this section is focused on inequality-constraint-handling; we leave it to the reader to use dynamic adaptations of ϵ to perform similar comparisons for equality-constrained problems.

We use a population size of 100 and a generation count limit of 100. This gives a total of 10,000 function evaluations during the EA simulation. Note that many studies in the literature use hundreds of thousands of function evaluations (that is, hundreds of individuals and thousands of generations) when benchmarking EA performance. We believe that such a high number of function evaluations is not realistic for most real-world problems. When solving real-world problems for which function evaluations are relatively expensive computationally, it is not reasonable to perform hundreds of thousands of function evaluations. We encourage students and researchers to focus more on achieving good convergence with a relatively low number of function evaluations, rather than trying to achieve excellent convergence with an unreasonably high number of function evaluations. This will shorten the path from academic theory to practical application in EA research. See Section 21.1 for more details.

Because of the relatively low number of function evaluations that we use, the results in this section are not comparable with many of the published results for the CEC 2010 benchmarks. But the point here is not to try to achieve the best possible performance with an unreasonably high number of function evaluations. The point is instead to compare constraint-handling methods on an even playing field. Our choice of 100 individuals and 100 generations is a good tradeoff because it does not take too long to run such simulations, but it still gives different EAs and constraint-handling methods enough time to differentiate themselves.

We implement mutation by replacing a feature in an individual with a value randomly chosen from a uniform distribution in the search domain. Each feature in each individual has a 1% probability of mutation each generation. We also use an elitism parameter of two, which means that we keep the two best individuals from one generation to the next.

For the purpose of elitism, we define the best individual as the feasible individual with the lowest cost. If there are not any feasible individuals, then we define the best individual as the one with the lowest penalized cost, where we obtain penalized cost using one of the nine methods listed above. We take this approach because some of the constraint-handling methods listed above rank infeasible individuals better than feasible individuals. We can afford this approach when there are a relatively large number of individuals and the ranking leads to selection for recombination. But if we are saving only two elite individuals from one generation to the next, we need to make sure that feasible individuals are always preferred above infeasible ones. This ensures that once the EA finds a feasible individual, it will always have at least one feasible individual for the rest of the simulation.

Table 19.1 shows a comparison of the nine constraint-handling methods on the six CEC 2010 constrained optimization benchmarks that do not have equality constraints. The results in the table are averaged over 20 Monte Carlo simulations. We randomly generated the offset values $\{o_i\}$ for the benchmarks in Appendix C.2, but we used the same $\{o_i\}$ for all constraint-handling methods for a given Monte Carlo trial. Some of the benchmarks use a rotation matrix M, which we randomly generated in the same way as the offset values.

Table 19.1 shows some interesting features. First, we notice that all algorithms perform similarly for C01. This indicates that C01 is either very easy, meaning that any method works well, or it is very hard, meaning that no method works well. Second, we note that all algorithms except EE perform about the same for C13; EE cannot find a feasible solution for C13, even after 20 Monte Carlo runs, and so its cost function value is written as ∞. EE performs the worst on C13, C14, and C15, which is interesting because these are the three benchmarks whose constraints are the most difficult to satisfy (see Table C.1 in Appendix C.2).

Table 19.1 indicates that, on average, the exponential penalty approach of Equation (19.20) in Section 19.2.5.1 performs best. However, there are many constrained optimization problems that have been discussed in the literature, and the constraint-handling mathods above have several tuning parameters. Different conclusions might be obtained if we ran tests with other tuning values and other benchmark functions.

	C01	C07	C08	C13	C14	C15
EE	−0.46	**19800**	2.39×10^5	∞	5.78×10^{13}	6.911×10^{13}
DP	−0.46	31900	1.51×10^5	−600	1.64×10^{13}	0.066×10^{13}
DS	−0.45	24700	1.46×10^5	−601	2.34×10^{13}	0.228×10^{13}
EP	−0.44	22000	$\mathbf{1.03 \times 10^5}$	−593	$\mathbf{0.01 \times 10^{13}}$	$\mathbf{0.001 \times 10^{13}}$
ES	**−0.47**	56800	1.32×10^5	**−606**	$\mathbf{0.01 \times 10^{13}}$	0.005×10^{13}
AP	−0.46	21000	1.64×10^5	−599	0.13×10^{13}	0.072×10^{13}
SR	−0.46	50500	1.25×10^5	−592	4.93×10^{13}	0.231×10^{13}
NP	−0.46	30900	1.97×10^5	−596	0.99×10^{13}	0.353×10^{13}
ϵC	−0.46	30900	7.68×10^5	−604	2.74×10^{13}	0.160×10^{13}

Table 19.1 Comparison of the best feasible cost function values found by nine constraint-handling BBO algorithms on six 10-dimensional benchmark problems, averaged over 20 Monte Carlo simulations. See the list at the beginning of Section 19.6 for the definitions of the acronyms. The best cost for each benchmark is in **bold font**.

19.7 CONCLUSION

We see from this chapter that there are many constraint-handling methods that can be used with EAs. Many of these algorithms have similar performance levels. Rather than trying all of these algorithms in a search for the best method, we would do better to remember some basic principles that may be important when solving constrained optimization problems.

Important Principles for Constrained Optimization

- As with unconstrained optimization problems, the more problem-specific information that we can incorporate into the EA, the better our chances for success. Handling constraints with special representations or special operators is usually more effective than using a general-purpose approach. Black-box optimization tools are easy to use, and sometimes they are necessary, but we can almost always get better performance by using difficult-to-obtain domain expertise.

- Given a constrained optimization problem, we should quantify its difficulty of constraint satisfaction. This can be measured with the parameter

$$\rho = |\mathcal{F}|/|\mathcal{S}| \tag{19.53}$$

where $|\mathcal{F}|$ is the size of the feasible set, and $|\mathcal{S}|$ is the size of the search space. We can approximate ρ by randomly generating many individuals in the search domain, and testing how many of them satisfy the constraints. We do not need to evaluate the cost function at this point; the purpose of this exercise is to see how difficult the constraints are. If only a tiny percentage of random individuals satisfy the constraints, then the constraints are difficult and our constrained EA should focus on constraint satisfaction. If a reasonably high percentage of random individuals satisfy the constraints, then the constraints are fairly easy and our constrained EA can focus more on cost function minimization.

- We should quantify the difficulty of *individual* constraint satisfaction. As above, this can be done by randomly generating many individuals in the search domain. Constraints that are satisfied by very few random individuals are difficult constraints and so the constrained EA should focus on satisfying those constraints. Constraints that are satisfied by a relatively large number of individuals are easy constraints and so the constrained EA does not need to focus on them as much. Alternatively, we may be able to normalize the constraints so that the satisfaction of each constraint is equally difficult.

- One of the biggest challenges in many constrained optimization problems is finding feasible solutions. This is especially true for problems with equality constraints. In this case, we might want to run constraint satisfaction algorithms before, or instead of, running a constrained EA. Constraint satisfaction algorithms fall in the field of study known as constraint programming. Constraint programming is outside of the scope of this book but it is an important field of study related to constrained optimization. Anyone who is seriously interested in constrained optimization should study constraint programming. Some good introductions to this topic are available in [Dechter, 2003], [Marriott and Stuckey, 1998], and [Rossi et al., 2006].

- In spite of the above points, the difficulty of constraint satisfaction is not necessarily an indication of the difficulty of the constrained optimization problem. Some problems with a relatively small area of feasibility are not difficult for constrained EAs. For example, [Michalewicz and Schoenauer, 1996] report two problems with $\rho = 0.0111\%$ and $\rho = 0.0003\%$ as being relatively easy for constrained EAs.

- When running a constrained EA, we should keep track of how many individuals satisfy the constraints from one generation to the next. Populations in which all the individuals are feasible are often inefficient, so we should take care to include both feasible and infeasible individuals in our search for a constrained optimum.

Current and Future Research in Constrained Evolutionary Algorithms

Constrained evolutionary optimization is an active research area because: (1) it is a relatively new area; (2) it is lacking in theoretical results (as typified by this chapter, which does not include any theoretical results); and (3) real-world optimization problems are almost always constrained. We conclude this chapter by mentioning some popular and important directions of current research.

- Much current research includes the incorporation of standard constraint-handling methods, such as those discussed in this chapter, into newer EAs. The literature continually introduces new EAs. These new EAs are often nothing more than modifications of older EAs, but sometimes they have distinctive new features and capabilities (see Chapter 17). It is important to explore how well current constraint-handling methods perform when incorporated into different types of EAs. The relative performances of different EAs on unconstrained problems does not necessarily correlate with their relative performances on constrained problems.

- This chapter discussed constrained optimization problems, and Chapter 20 discusses multi-objective optimization problems. Current research is beginning to combine these two fields to find algorithms for the solution of constrained multi-objective optimization problems [Yen, 2009].

- Theoretical results would be a highly fruitful area for future research in constrained optimization. This book discusses Markov models, dynamic system models, and schema theory for GAs and GP. Perhaps those tools, or others, could also be used to analyze constrained EAs.

- Related to the above discussion of constraint programming is the idea of searching the constraint boundary to solve constrained optimization problems [Leguizamón and Coello Coello, 2009]. Boundary search is related to constraint programming, but constraint programming focuses on finding feasible solutions while boundary search focuses on evolving a population that lies on the constraint boundary.

- As mentioned earlier in this chapter, some problems have constraints that are difficult to satisfy, so it is challenging just to find feasible regions in the search space. However, beyond the problem of finding feasible regions is the challenge of designing an EA that can effectively oscillate between feasible and infeasible regions. This type of behavior is often desirable for problems whose solution lies on the constraint boundary [Schoenauer and Michalewicz, 1996].

- Just as EAs can be combined in various ways, constraint handling methods can also be combined. For example, an ensemble of constraint handling methods could all use the same cost function results, and the best method at each generation would dominate the next generation [Mallipeddi and Suganthan, 2010]. This is similar to some of the multi-objective algorithms of Chapter 20, in which different cost functions are used at different stages of the optimization process.

- As a further level of abstraction beyond ensembles, hyper-heuristics combine multiple EAs and multiple constraint handling methods into a single algorithm. Recall that a heuristic is a family of algorithms (for example, a family of ACO variations, or a family of DE variations). A hyper-heuristic is a family of families of algorithms (for example, a family containing an ACO heuristic, a DE heuristic, and other heuristics). Hyper-heuristics can be used for any type of optimization problem, but we mention them here because of their promise for constrained problems [Tinoco and Coello Coello, 2013].

Constrained optimization surveys can be found in [Eiben, 2001], [Coello Coello, 2002], and [Coello Coello and Mezura-Montes, 2011]. The reader who is interested in further research should note that Carlos Coello Coello maintains a bibliography of papers related to constrained evolutionary optimization, which includes 1036 references as of August 2012 [Coello Coello, 2012a].

PROBLEMS

Written Exercises

19.1 Many equality-constrained benchmarks use $\epsilon \approx 0.0001$ in Equation (19.7). What is the probability of satisfying the scalar constraint $|x| \leq \epsilon$ with this value of ϵ and with a randomly generated $x \in [-1000, +1000]$?

19.2 This problem shows how the superiority of feasible points method that uses Equation (19.11) works if $f(x) \geq 0$ for all x, but how it may fail if that assumption is not satisfied. Use Equations (19.10) and (19.11) to find $\phi'(x)$ for a two-element population with the following characteristics.

a)

$$f(x_1) = 0, \quad \sum_i r_i G_i(x_1) = 1$$
$$f(x_2) = 10, \quad \sum_i r_i G_i(x_2) = 0$$

b)

$$f(x_1) = -10, \quad \sum_i r_i G_i(x_1) = 1$$
$$f(x_2) = 0, \quad \sum_i r_i G_i(x_2) = 0$$

19.3 This problem shows how the superiority of feasible points method that uses Equation (19.12) works. Use Equations (19.10) and (19.12) to find $\phi'(x)$ for the two-element populations shown in Problem 19.2.

19.4 Give an analytical expression for the smallest value of K in the eclectic EA of Equation (19.13) that guarantees that $\phi(\bar{x}) > \phi(x)$ for all $\bar{x} \notin \mathcal{F}$ and for all $x \in \mathcal{F}$.

19.5 Suppose you have four individuals in an EA population with the following cost values and constraint violation levels:

$$\begin{aligned}
f(x_1) = 3, \quad G_1(x_1) = 0, \quad G_2(x_1) = 0 \\
f(x_2) = 2, \quad G_1(x_2) = 1, \quad G_2(x_2) = 0 \\
f(x_3) = 1, \quad G_1(x_3) = 1, \quad G_2(x_3) = 1 \\
f(x_4) = 4, \quad G_1(x_4) = 0, \quad G_2(x_4) = 0.
\end{aligned}$$

Use the self-adaptive fitness formulation of Section 19.2.8 to find penalized cost values for these individuals. Give an intuitive explanation of your answer.

19.6 Suppose you have four individuals in an EA population with the cost values and constraint violation levels shown in Problem 19.5. Use the self-adaptive penalty function method of Section 19.2.9 to find penalized cost values for these individuals. Give an intuitive explanation of your answer.

19.7 This problem deals with the adaptive segregational constraint handling algorithm of Section 19.2.10.

a) Explain how the r_i update algorithm of Equation (19.34) attains the target ratio of feasible individuals to infeasible individuals.

b) Explain the effect of increasing γ in Equation (19.34).

19.8 Several constrained EAs, including the stochastic ranking algorithm of Section 19.2.12 and the niched-penalty approach of Section 19.2.13, include comparing individuals to see which one has a "smaller constraint violation." Suggest three ways that the size of a constraint violation might be measured.

19.9 The traveling salesman problem (TSP) is a constrained problem: a candidate solution must visit each city exactly once to be considered a valid tour. Crossover operators for the path representation of TSP individuals are discussed in Section 18.3.1. Which of these operators preserve the TSP constraint, and which ones do not?

19.10 Suppose you have four individuals in an EA population with the cost values and constraint violation levels shown in Problem 19.5. Suppose we use the ϵ-level comparison of Equation (19.51) in conjunction with the constraint sum method.

a) For what values of ϵ will x_2 be ranked better than x_1?

b) For what values of ϵ will x_3 be ranked better than x_1?

c) For what values of ϵ will x_2 be ranked better than x_3?

19.11 How many references are listed on Carlos Coello Coello's web site "List of References on Constraint-Handling Techniques used with Evolutionary Algorithms"?

Computer Exercises

19.12 Recreate Figure 19.1 with $\alpha = 0.5$. What difference do you see between your figure and Figure 19.1?

19.13 Suppose you have a circular search domain with a radius of 1 unit that is centered at the origin. Suppose individuals are constrained to lie in a circle with a radius of $\rho_c = 0.1$ units, also centered at the origin.

a) Use the Genocop III algorithm to generate a random feasible x_r, a random infeasible x_s, and random parameter $a \in [0, 1]$, to generate a potentially repaired individual x'_s. Perform this experiment many times to estimate the probability that x'_s is feasible. Repeat for $\rho_c = 0.5$ units.

b) Repeat part (a) for a spherical domain.

19.14 Section 19.6 compared nine constraint-handling methods in conjunction with BBO. Compare some of the constraint-handling methods in this chapter on one or more constrained optimization problems using an EA other than BBO.

CHAPTER 20

Multi-Objective Optimization

Multiple, often conflicting objectives arise naturally in most real-world optimization scenarios.

—Eckart Zitzler [Zitzler et al., 2004]

All real-world optimization problems are multi-objective, at least implicitly if not explicitly. This chapter discusses how to modify EAs for multi-objective optimization problems (MOPs). As the quote at the beginning of this chapter asserts, real-world optimization problems typically (perhaps always) include multiple goals, and those goals usually are in conflict. For example:

- When designing a bridge, we might want to minimize its cost and maximize its strength. The minimum-cost bridge might be made from styrofoam and would be very weak. The maximum-strength bridge might be made from titanium and would be very expensive. What is the best tradeoff between cost and strength?
- When purchasing a car, we might want to maximize comfort and minimize cost. The maximum-comfort car would be too expensive, but the minimum-cost car would be too uncomfortable.
- When designing a consumer product, we might want to maximize profit and maximize market share. The maximum-profit product would not provide

enough market penetration to position our company for future products, but the maximum market-share product would not result in enough profit.

- When designing a control system, we might want to minimize rise time and minimize overshoot. The minimum rise-time controller would have too much overshoot, but a critically damped (zero overshoot) controller would not have a fast enough rise time.

- When designing a control system, we might want to maximize input sensitivity and minimize disturbance sensitivity. The maximum input sensitivity controller would be too sensitive to noise, but the minimum disturbance sensitivity controller would not be responsive enough to control inputs.

Multi-objective optimization is also called multi-criteria optimization, multi-performance optimization, and vector optimization. In this chapter we assume that the independent variable x is n-dimensional, and we assume that our MOP is a minimization problem. An MOP can be written as follows:

$$\min_x f(x) = \min_x [f_1(x), f_2(x), \cdots, f_k(x)]. \tag{20.1}$$

That is, we want to minimize a vector $f(x)$ of functions. Of course, we cannot minimize a vector in the typical sense of the word *minimize*. Nevertheless, our goal in an MOP is to simultaneously minimize all k functions $f_i(x)$. We see that we must redefine our definition of optimality for MOPs.

Multi-objective optimization has been studied by the operations research community by many years [Ehrgott, 2005]. It appears that [Rosenberg, 1967] was the first to suggest using EAs for MOPs, [Ito et al., 1983] was the first implementation, and [Schaffer, 1985] was the first widely-known publication on the topic.

MOPs often include constraints, but as we see from the problem statement of Equation (20.1), we do not deal with constrained MOPs in this chapter. We can incorporate constraints into multi-objective evolutionary algorithms (MOEAs) in the same way that we incorporate them into single-objective EAs (see Chapter 19). Some researchers have proposed constraint-handling techniques that are unique to MOPs, but we do not discuss them in this chapter.

Overview of the Chapter

Section 20.1 discusses the concept of Pareto optimality, which is an extension of optimality to MOPs that are in the form of Equation (20.1). Since an MOP has multiple objectives, there are many ways to measure the performance of an MOEA and we discuss some of these ways in Section 20.2. We follow that discussion with a presentation of several popular MOEAs. Section 20.3 discusses MOEAs that do *not* explicitly use the concept of Pareto optimality, and Section 20.4 discusses MOEAs that *do* explicitly use the concept of Pareto optimality. Section 20.5 shows how we can combine biogeography-based optimization (BBO; see Chapter 14) with some of the MOEA approaches in this chapter, and presents a comparative study on some multi-objective benchmarks. The concluding section of this chapter provides references to additional resources, and suggests several important topics for current and future MOEA research.

20.1 PARETO OPTIMALITY

This section outlines some basic concepts and examples that are related to MOPs. We first list some definitions that are often used in multi-objective optimization.

1. *Domination:* A point x^* is said to dominate x if the following two conditions hold: (1) $f_i(x^*) \leq f_i(x)$ for all $i \in [1, k]$, and (2) $f_j(x^*) < f_j(x)$ for at least one $j \in [1, k]$. That is, x^* is at least as good as x for all objective function values, and it is better than x for at least one objective function value. We use the notation
$$x^* \succ x \tag{20.2}$$
to indicate that x^* dominates x. This notation can be confusing because the symbol $\succ$ looks like a "greater than" symbol, but since we deal mainly with minimization problems in this chapter, the symbol $\succ$ means the function values of x^* are less than or equal to those of x. However, this notation is standard in the literature, so this is the notation that we use. The statement "x^* is superior to x" is identical to the statement "x^* dominates x."

2. *Weak Domination:* A point x^* is said to weakly dominate x if $f_i(x^*) \leq f_i(x)$ for all $i \in [1, k]$. That is, x^* is at least as good as x for all objective function values. Note that if x^* dominates x, then it also weakly dominates x. Also note that if $f_i(x^*) = f_i(x)$ for all $i \in [1, k]$, then x^* and x weakly dominate each other. We use the notation
$$x^* \succeq x \tag{20.3}$$
to indicate that x^* weakly dominates x. Some authors use the equivalent terminology that x^* covers x.

3. *Nondominated:* A point x^* is said to be nondominated if there is no x that dominates it. Noninferior, admissible, and efficient, are synonyms for nondominated.

4. *Pareto optimal points:* A Pareto optimal point x^*, also called a Pareto point, is one that is not dominated by any other x in the search space. That is,
$$\begin{aligned} & x^* \text{ is Pareto optimal } \iff \\ & \nexists x : (f_i(x) \leq f_i(x^*) \text{ for all } i \in [1, k], \text{and} f_j(x) < f_j(x^*) \text{ for some } j \in [1, k]) . \end{aligned} \tag{20.4}$$

5. *Pareto optimal set:* The Pareto optimal set, also called the Pareto set and denoted as P_s, is the set of all x^* that are nondominated.
$$\begin{aligned} P_s &= \{x^* : [\nexists x : (f_i(x) \leq f_i(x^*) \text{ for all } i \in [1, k], \text{ and} \\ & \qquad f_j(x) < f_j(x^*) \text{ for some } j \in [1, k])]\} . \end{aligned} \tag{20.5}$$
The Pareto set is also called the efficient set, and it is sometimes called the admissible set, although this term usually implies constraint satisfaction rather than Pareto optimality.

6. *Pareto front:* The Pareto front, also called the nondominated set and denoted as P_f, is the set of all function vectors $f(x)$ corresponding to the Pareto set.
$$P_f = \{f(x^*) : x^* \in P_s\}. \tag{20.6}$$

The literature sometimes uses the terms Pareto set and Pareto front incorrectly or interchangeably, but the above list gives the technically correct definitions. Note that the statement that x^* is nondominated does not necessarily mean that x^* dominates all x that are not equal to x^*. It may be true that $f_i(x^*) = f_i(x)$ for all $i \in [1, k]$. In this case both x and x^* are nondominated with respect to each other, yet neither one dominates the other. It may also be the case, for example, in a two-objective problem, that $f_1(x) < f_1(x^*)$ and $f_2(x^*) < f_2(x)$. Again, in this case both x and x^* are nondominated with respect to each other, yet neither one dominates the other.

This idea of Pareto optimality for MOPs is often attributed to Francis Edgeworth, who introduced it in 1881 [Edgeworth, 1881], and to Vilfredo Pareto, who generalized Edgeworth's work in 1896 [Pareto, 1896]. However, the idea of trade-offs is a common one for anyone who has ever tried to balance conflicting objectives.

■ EXAMPLE 20.1

Suppose that we have a MOP for which the independent variable x is two dimensional ($n = 2$), and that x can take one of six discrete values $x^{(i)}$ where $i \in [1, 6]$. Further suppose that we have two objectives ($k = 2$) with function values

$$\begin{aligned} f_1(x^{(1)}) &= 1, \quad f_2(x^{(1)}) = 3, \\ f_1(x^{(2)}) &= 1, \quad f_2(x^{(2)}) = 4, \\ f_1(x^{(3)}) &= 2, \quad f_2(x^{(3)}) = 2, \\ f_1(x^{(4)}) &= 2, \quad f_2(x^{(4)}) = 3, \\ f_1(x^{(5)}) &= 3, \quad f_2(x^{(5)}) = 1, \\ f_1(x^{(6)}) &= 3, \quad f_2(x^{(6)}) = 3. \end{aligned} \tag{20.7}$$

If $x = x^{(1)}$ or $x = x^{(2)}$, then $f_1(x)$ is minimized. If $x = x^{(5)}$, then $f_2(x)$ is minimized. There is not a single value of x that minimizes both $f_1(x)$ and $f_2(x)$. The optimal value of x is the one that provides the best tradeoff between the multiple objectives, where *best* is based on our problem-dependent judgment. Another interesting point in Equation (20.7) is $x^{(3)}$, because any point $x \neq x^{(3)}$ gives either $f_1(x) > f_1(x^{(3)})$ or $f_2(x) > f_2(x^{(3)})$.

A good way of visualizing this problem is to view a plot of f_2 vs. f_1, as shown in Figure 20.1. This clearly shows that for $x \in \{x^{(1)}, x^{(3)}, x^{(5)}\}$, no other x exists that simultaneously decreases all objective function values. $x^{(1)}$, $x^{(3)}$, and $x^{(5)}$ are therefore good tradeoff values for this MOP, and they comprise the Pareto set. If we connect all of the optimal points in the f_1/f_2 plane in Figure 20.1, we obtain a curve that forms a lower bound for all of the other points in the f_1/f_2 plane.

Another way of visualizing this problem is to view a plot of the search space, with the Pareto set indicated by some special notation. Figure 20.2 shows one possibility for the search space for this example, with the Pareto points indicated with stars. This shows the search space region that corresponds to the Pareto front of Figure 20.1.

□

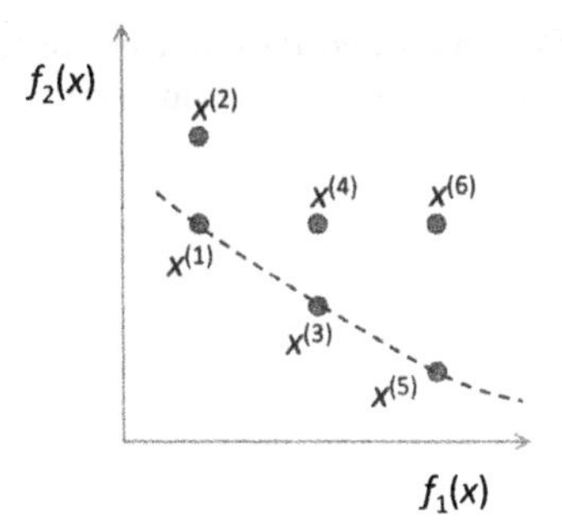

Figure 20.1 Example 20.1: $\{x^{(1)}, x^{(3)}, x^{(5)}\}$ form the Pareto set for this multi-objective minimization problem. The function vectors that correspond to the Pareto set form the Pareto front.

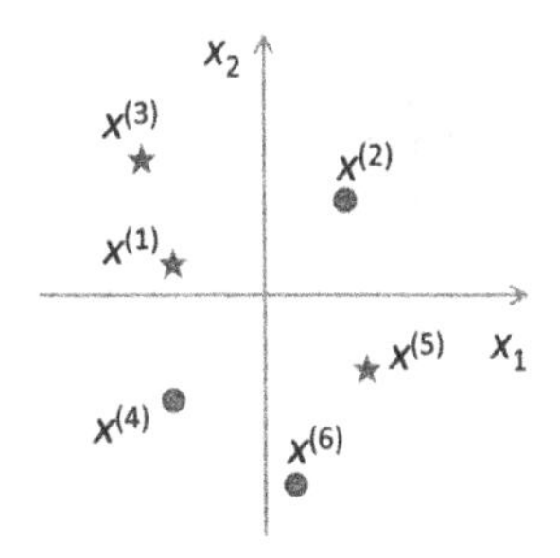

Figure 20.2 Example 20.1: This figure shows a possible two-dimensional search space for Example 20.1. The search space consists of six two-dimensional vectors. The stars indicate the Pareto set.

■ EXAMPLE 20.2

Consider the MOP

$$\min f(x) = \min[f_1(x),\ f_2(x)] = \min[x_1^2 + x_2^2,\ (x_1 - 2)^2 + (x_2 - 2)^2] \tag{20.8}$$

where $x_1 \in [0, 2]$ and $x_2 \in [0, 2]$. This is a two-dimensional MOP ($n = 2$) because each x in the search space has two elements. This MOP also has two objectives ($k = 2$). The point $x^{(1)} = (0, 0)$ minimizes $f_1(x)$, and so $(0, 0)$ is one of the Pareto points. The point $x^{(2)} = (2, 2)$ minimizes $f_2(x)$, and so $(2, 2)$ is also one of the Pareto points. If we use a brute-force search to find all of the Pareto points, we find that the Pareto set is

$$P_s = \{x : x_1 = x_2,\ \text{where } x_1 \in [0, 2]\}. \tag{20.9}$$

That is, the Pareto set forms a straight line in the search space. We can find the Pareto front by substituting the Pareto points into Equation (20.8):

$$P_f = \{(f_1, f_2) : f_1 = 2x_1^2, f_2 = 2(x_1 - 2)^2,\ \text{where } x_1 \in [0, 2]\}. \tag{20.10}$$

Figure 20.3 shows the Pareto set and Pareto front for this example. Any point other than a Pareto point maps to a function vector that is above and to the right of the Pareto front.

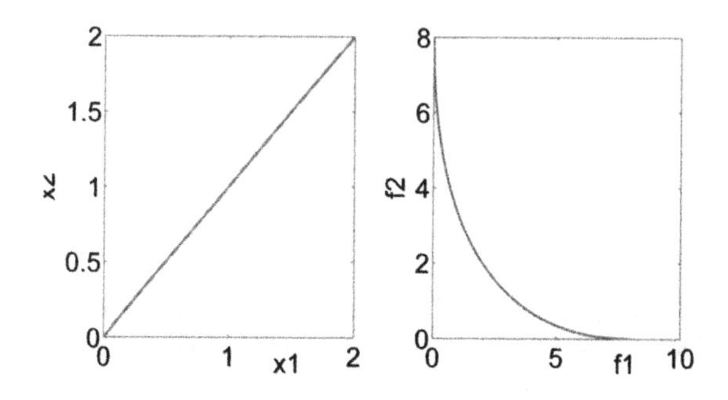

Figure 20.3 Example 20.2: The figure on the left shows the Pareto set. The figure on the right shows the corresponding Pareto front. Any point in the Pareto set provides a reasonable tradeoff for the multiple-objective optimization problem.

□

ϵ Dominance

One limitation of the concept of Pareto dominance is its either-or, black-and-white nature. For example, consider the following three sets of cost function values:

$$\begin{aligned} f_1(x) &= 200, \quad f_2(x) = 300 \\ f_1(y) &= 201, \quad f_2(y) = 301 \\ f_1(z) &= 500, \quad f_2(z) = 600. \end{aligned} \tag{20.11}$$

x dominates both y and z, but the concept of Pareto dominance does not allow for any distinction between the level of domination, and it does not recognize two candidate solutions that are very close to each other in the objective function space. In Equation (20.11), x dominates y, but since x and y are so similar, they are almost nondominated with respect to each other. In fact, we can almost say that y dominates x. This gives rise to the concept of ϵ dominance.

1. *Additive ϵ Dominance:* A point x^* is said to additively ϵ-dominate x if $f_i(x^*) \leq f_i(x) + \epsilon$ for some $\epsilon \geq 0$ and for all $i \in [1, k]$. That is, x^* is "close" to dominating x, where "closeness" is quantified additively with the parameter ϵ.

2. *Multiplicative ϵ Dominance:* A point x^* is said to multiplicatively ϵ-dominate x if $f_i(x^*) \leq f_i(x)(1 + \epsilon)$ for some $\epsilon \geq 0$ and for all $i \in [1, k]$. That is, x^* is "close" to dominating x, where "closeness" is quantified multiplicatively with the parameter ϵ.

We use the notation

$$x^* \succ_\epsilon x \tag{20.12}$$

to indicate that x^* ϵ-dominates x, where the type of epsilon domination (additive or multiplicative) should be clear from the context. Note that if $\epsilon = 0$ then ϵ dominance is equivalent to weak dominance:

$$(x^* \succ_\epsilon x) \text{ for } \epsilon = 0 \Longleftrightarrow (x^* \succeq x). \tag{20.13}$$

Also note that ϵ dominance for $\epsilon > 0$ is an even weaker type of dominance than weak dominance. That is, if x^* weakly dominates x, then x^* also ϵ-dominates x for all $\epsilon \geq 0$. Conversely, if x^* ϵ-dominates x for some $\epsilon > 0$, then x^* may or may not weakly dominate x:

$$(x^* \succeq x) \Longrightarrow (x^* \succ_\epsilon x) \text{ for all } \epsilon > 0. \tag{20.14}$$

The ϵ-dominance relationship between two individuals depends on the value of ϵ that we use in our definition. In Equation (20.11), $x \succ_\epsilon y$ for all $\epsilon \geq 0$. Also, $y \succ_\epsilon x$ holds in the additive sense if $\epsilon \geq 1$, and it holds in the multiplicative sense if $\epsilon \geq 0.005$.

20.2 THE GOALS OF MULTI-OBJECTIVE OPTIMIZATION

The goal of a single-objective optimization algorithm is usually straightforward: find the minimum value of the cost function and its corresponding decision vector. However, even in single-objective optimization, we might be interested in several different performance metrics for an EA. We might be interested not only in finding the minimum cost function value, but also in quickly finding a "good" solution that is not necessarily the best. We might also be interested in finding many good solutions in diverse regions of the search space. So even in the apparently straightforward problem of single-objective optimization, we may have several performance metrics. This complication increases with multiple-objective optimization. Some potential goals of an MOEA might be the following.

1. Maximize the number of individuals that we find within a certain distance of the true Pareto set.

2. Minimize the average distance between the MOEA-approximated Pareto set and the true Pareto set.

3. Maximize the diversity of the individuals that we find in the approximated Pareto set.

4. Minimize the distance of a candidate solution in objective function space to an ideal point, also called a utopia point.[1]

Goals 1 and 2 are concerned with finding the "best" approximation of the true Pareto set. Goal 3 is concerned with finding a diverse set of solutions so that the human decision maker has enough resources to make an informed decision among

[1]Some papers define the terms "ideal point" and "utopia point" (or "utopian point") slightly differently from each other, but for the purposes of this chapter we consider them to be synonymous.

the possible trade-offs. In contrast to the other goals, Goal 4 is concerned with finding a solution that is as close as possible to the decision maker's ideal solution, which may not exist. However, most current MOEAs are primarily concerned with finding the best approximation to the true Pareto set.

Goals 1 and 2 above assume that we know the true Pareto set in the first place, so those criteria might be useful when testing MOEAs on well-understood benchmarks, but the criteria are useless when running an MOEA on a real-world optimization problem. But if we know the true Pareto set P_s, and an MOEA gives us an approximate Pareto set $\hat{P}_s$, the average distance $M(P_s, \hat{P}_s)$ between them can be computed as

$$M_1(P_s, \hat{P}_s) = \frac{1}{|\hat{P}_s|} \sum_{x \in \hat{P}_s} \min_{x^* \in P_s} ||x^* - x|| \tag{20.15}$$

where $||\cdot||$ is any user-specified distance metric.

Goal 3 above can be measured in a few different ways. First, we could measure the average distance of each individual to its nearest neighbor in the approximated Pareto set. Second, we could measure the distance between the two extreme individuals in the approximated Pareto set. Third, we could compute the average number of individuals that are farther than some threshold from each element in the approximate Pareto set [Zitzler et al., 2000]:

$$M_2(\hat{P}_s) = \frac{1}{|\hat{P}_s|} \sum_{x \in \hat{P}_s} \left| x' \in \hat{P}_s : ||x' - x|| > \sigma \right| \tag{20.16}$$

where σ is a user-specified distance threshold. In general, M_2 increases as the number of elements in $\hat{P}_s$ increases, and also as the diversity of the elements in $\hat{P}_s$ increases. [Khare et al., 2003] discusses some additional diversity metrics for MOPs.

Goal 4 above is called target vector optimization [Wienke et al., 1992], goal attainment [Wilson and Macleod, 1993], or goal programming. It assumes that the user is thinking of some ideal point in objective function space, and it requires a definition of "distance." Usually we use the Euclidean distance D_2, also called the two-norm distance, between an objective function vector f and an ideal point f^*. The distance between f and f^* is defined as follows:

$$D_2^2(f^*(x), f(x)) = ||f^*(x) - f(x)||_2^2 = \sum_{i=1}^{k} (f_i^*(x) - f_i(x))^2. \tag{20.17}$$

However, we can also use other distance measures, such as the weighted two-norm, the one-norm, or the infinity-norm.

Recall Example 20.2. The user might think that it would be ideal to achieve both $f_1(x) = 0$ and $f_2(x) = 0$. After obtaining the Pareto front of Figure 20.3, we see that the closest we can get to the ideal point in terms of Euclidean distance is $x_1 = x_2 = 1$, which gives $f_1(x) = 2$ and $f_2(x) = 2$. On the other hand, if the user's ideal solution is $f_1(x) = 2$ and $f_2(x) = 0$, the closest we can get is $x_1 = x_2 = 1.20$, which gives $f_1(x) = 2.87$ and $f_2(x) = 1.29$. Using Goal 4 to quantify MOEA performance incorporates the user's preferences into the final solution of the MOP.

Note that we can pursue Goals 1–3 in terms of either the Pareto front or the Pareto set. For example, in Goal 1, instead of maximizing the number of individuals that are within a certain distance of the true Pareto set, we could maximize the number of individuals whose function vectors are within a certain distance of the true Pareto front. In summary, we see that there are many possible performance criteria for an MOEA. In other words, optimizing the performance of an MOEA is itself a MOP. This seems appropriate, but it complicates the evaluation of MOEAs.

EXAMPLE 20.3

Figure 20.4 shows the performance of three different EAs on an MOP. Figure 20.4(a) shows a solution that is fairly diverse and reasonably close to the true Pareto front. The Figure 20.4(b) shows a solution that is more diverse than Figure 20.4(a) in the sense that the distance between the extreme solutions is farther, but Figure 20.4(b) includes only three solutions while Figure 20.4(a) includes four solutions. Figure 20.4(c) shows solutions that are closer to the true Pareto front than Figures 20.4(a) or (b), but the diversity is not as good. Which of the three solutions is "best"? It depends on the priorities of the decision maker.

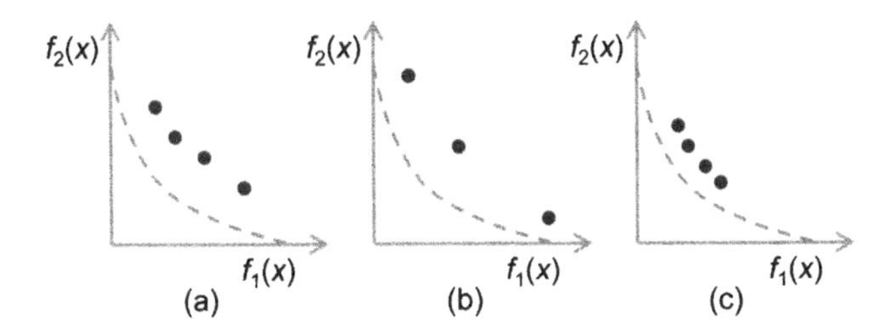

Figure 20.4 Example 20.3: This figure shows three potential EA solutions to a two-objective MOP. The true Pareto front is the dotted line, and the circles are the approximations that were found by each EA. Which solution is "best"? It depends on the priorities of the decision maker with respect to solution diversity and closeness to the true Pareto front.

□

20.2.1 Hypervolume

Another metric that researchers often use to measure the quality of a Pareto front is its hypervolume. Suppose that an MOEA has found M points in an approximate Pareto front $\hat{P}_f = \{f(x_j)\}$ for $j \in [1, M]$, where $f(x_j)$ is a k-dimensional function. The hypervolume can be computed as

$$S(\hat{P}_f) = \sum_{j=1}^{M} \prod_{i=1}^{k} f_i(x_j). \tag{20.18}$$

Given two MOEAs that compute two Pareto front approximations to a given MOP, we can use the hypervolume measure to quantify how good the two approximations are relative to each other. For a minimization problem, a smaller hypervolume indicates a better Pareto front approximation.

■ EXAMPLE 20.4

Suppose that we have two MOEAs, each of which are designed to approximate the Pareto front of a two-objective minimization problem. Figure 20.5 shows their Pareto front approximations. Figure 20.5(a) has the Pareto front approximation points

$$\hat{P}_f(1) = \{[f_1(x_j), f_2(x_j)]\} = \{[1,5], [2,3], [5,1]\} \tag{20.19}$$

which gives the hypervolume 5+6+5=16. Figure 20.5(b) has the Pareto front approximation points

$$\hat{P}_f(2) = \{[f_1(x_j), f_2(x_j)]\} = \{[1,4], [3,3], [4,1]\} \tag{20.20}$$

which gives the hypervolume 4+9+4=17. According to the hypervolume measure of Equation (20.18), Figure 20.5(a) gives a slightly better $\hat{P}_f$ than Figure 20.5(b).

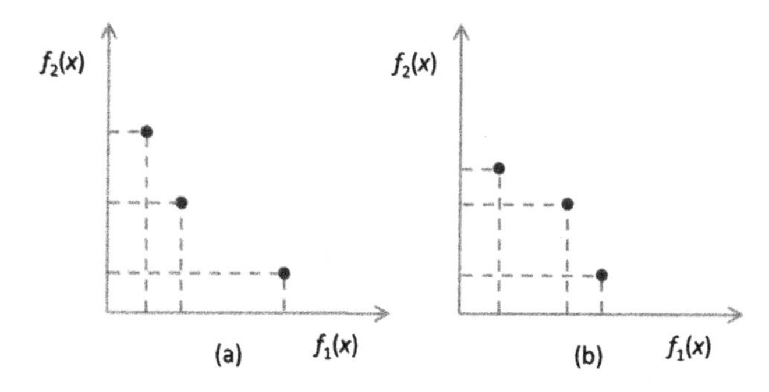

Figure 20.5 Example 20.4: This figure shows two Pareto front approximations to a two-objective MOP. A hypervolume measurement is used to quantify the goodness of the approximations. The approximation on the left has a hypervolume of 16, and the approximation on the right has a hypervolume of 17.

□

Hypervolume cannot be blindly used as an indicator of Pareto front quality. Equation (20.18) shows that an empty Pareto front approximation ($M = 0$) gives the smallest possible value of S. Therefore, a more accurate measure might be the normalized hypervolume $S_n(\hat{P}_f) = S(\hat{P}_f)/M$. However, even this quantity may not be a good metric for a Pareto front approximation. We can see this by considering the possibility that a certain Pareto front approximation $\hat{P}_f(1)$ has normalized hypervolume $S_n(\hat{P}_f(1))$. Now suppose that we add a single new point to $\hat{P}_f(1)$ to obtain $\hat{P}_f(2)$. This might result in $S_n(\hat{P}_f(2)) > S_n(\hat{P}_f(1))$ even though the only difference between $\hat{P}_f(1)$ and $\hat{P}_f(2)$ is that $\hat{P}_f(2)$ has an additional point. $\hat{P}_f(2)$ is clearly better than $\hat{P}_f(1)$, but $S_n(\hat{P}_f(2))$ is greater than $S_n(\hat{P}_f(1))$, which is counterintuitive.

This leads us to modify the hypervolume measure by computing it *not* with respect to the origin of the objective function space, but instead with respect to a reference point that lies outside the Pareto front. Suppose that we want to compare Q Pareto front approximations $\hat{P}_f(q)$ for $q \in [1, Q]$. We compute the reference point vector $r = [r_1, \cdots, r_k]$, where

$$r_i > \max_q \left[\max_{x \in \hat{P}_s(q)} f_i(x) \right] \tag{20.21}$$

and then we compute the hypervolumes S' with respect to the reference point:

$$S'(\hat{P}_f(q)) = \sum_{j=1}^{M(q)} \prod_{i=1}^{k} (r_i - f_i(x_j(q))) \tag{20.22}$$

where $M(q)$ is the number of points in the q-th Pareto front approximation, and $x_j(q)$ is the j-th point in the q-th Pareto set approximation. A larger reference-point hypervolume S' indicates that we have a better Pareto front for a minimization problem. We can either use the normalized reference-point hypervolume

$$S'_n(\hat{P}_f(q)) = S'(\hat{P}_f(q))/M(q) \tag{20.23}$$

or we can use the total reference-point hypervolume measurement $S'(\hat{P}_f(q))$ if we want the metric to take the number of Pareto points into account. Figure 20.6 illustrates the reference-point hypervolume in two dimensions.

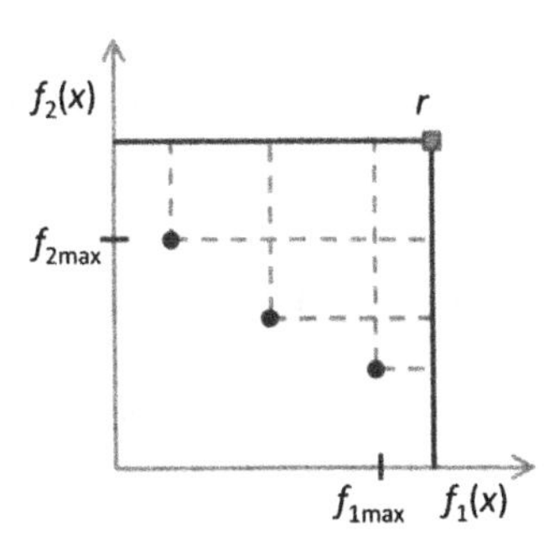

Figure 20.6 Reference-point hypervolume computation $S'(\hat{P}_f(q))$ of Equation (20.22). The reference point r is an arbitrary reference point whose i-th component is larger than that of each of the points in the Pareto front approximation. A larger $S'(\hat{P}_f(q))$ indicates a better Pareto front approximation to a minimization problem.

Many MOEA discussions in the literature convert MOPs into maximization problems, which means that larger hypervolumes in Equation (20.18) are more desirable. This is consistent with more points in $\hat{P}_f$ (larger M) being more desirable. Equations (20.18) and (20.22) give the basic idea of hypervolume calculation, but the literature includes several other methods, definitions, and algorithms for hypervolume calculation [Auger et al., 2012], [Bringmann and Friedrich, 2010], [Zitzler et al., 2003]. Most papers use the *union* of the M hyperboxes of Equations (20.18) and (20.22) to calculate hypervolume. For example, if we compute the hypervolume of the union of the hyperboxes of Figure 20.5, both $\hat{P}_f$ approximations have a hypervolume of 11. Other ways of computing hypervolume may lead to different conclusions about the relative merits of the two $\hat{P}_f$ approximations. However, Equations (20.18) and (20.22) are simpler to implement than computing the volume of the union of hyperboxes. There is clearly a high correlation between the sum of M hypervolumes, and the hypervolume of the union of M hyperboxes, although the correlation is not perfectly linear (see Problem 20.7).

20.2.2 Relative Coverage

Another way to compare Pareto front approximations is by computing the average number of individuals in one approximation that are weakly dominated by at least one individual in the other approximation [Zitzler and Thiele, 1999]. Suppose that we have two $\hat{P}_f$ approximations denoted as $\hat{P}_f(1)$ and $\hat{P}_f(2)$. We define the coverage of $\hat{P}_f(1)$ relative to $\hat{P}_f(2)$ as the average number of individuals in $\hat{P}_f(2)$ that are weakly dominated by at least one individual in $\hat{P}_f(1)$:

$$C(\hat{P}_f(1), \hat{P}_f(2)) = \frac{\left| a_2 \in \hat{P}_f(2) \text{ such that } \exists \left[a_1 \in \hat{P}_f(1) \text{ such that } a_1 \succeq a_2 \right] \right|}{|\hat{P}_f(2)|}. \tag{20.24}$$

Note that $C(\hat{P}_f(1), \hat{P}_f(2)) \in [0, 1]$. If $C(\hat{P}_f(1), \hat{P}_f(2)) = 0$, then for each individual $a_2 \in \hat{P}_f(2)$, there is no individual in $\hat{P}_f(1)$ that weakly dominates a_2. If $C(\hat{P}_f(1), \hat{P}_f(2)) = 1$, then for each individual $a_2 \in \hat{P}_f(2)$, there is at least one individual in $\hat{P}_f(1)$ that weakly dominates a_2. Although we cannot use the coverage equation to obtain an absolute measure of the goodness of a Pareto front approximation, it can be valuable in comparing several approximations.

20.3 NON-PARETO-BASED EVOLUTIONARY ALGORITHMS

This section discusses several MOEAs that do not explicitly use the concept of Pareto dominance. Section 20.3.1 discusses aggregation methods, Section 20.3.2 discusses the vector evaluated genetic algorithm (VEGA), Section 20.3.3 discusses lexicographic ordering approaches, Section 20.3.4 discusses the ϵ-constraint method, and Section 20.3.5 discusses gender-based approaches.

20.3.1 Aggregation Methods

Aggregation methods combine the objective function vector of the MOP into a scalar objective function. For example, we can convert the k-objective MOP of Equation (20.1) into the problem

$$\min_x f(x) \Longrightarrow \min_x \sum_{i=1}^{k} w_i f_i(x), \text{ where } \sum_{i=1}^{k} w_i = 1. \tag{20.25}$$

$\{w_i\}$ is a set of positive weight parameters whose elements sum to 1. Equation (20.25) is called the weighted sum approach, but other aggregation methods can also be used. For example, we can combine the objectives in a product:

$$\min_x f(x) \Longrightarrow \min_x \prod_{i=1}^{k} f_i(x). \tag{20.26}$$

If we use the product aggregation method, then we should make sure that all of the objectives $f_i(x) > 0$ for all x. But whatever aggregation method we use, the main point is to convert the MOP into a single-objective optimization problem.

■ EXAMPLE 20.5

Consider the MOP of Example 20.2. If we use Equation (20.25) to convert it to a single-objective problem, we obtain

$$\min_x \{w_1 f_1(x) + w_2 f_2(x)\} = \min_x \left\{w_1(x_1^2 + x_2^2) + (1 - w_1)[(x_1 - 2)^2 + (x_2 - 2^2)]\right\}. \tag{20.27}$$

We can minimize this equation by taking its partial derivative with respect to x_1 and x_2 to obtain

$$\begin{aligned} \frac{\partial[w_1 f_1(x) + w_2 f_2(x)]}{\partial x_1} &= 2x_1 + 4(w_1 - 1) \\ \frac{\partial[w_1 f_1(x) + w_2 f_2(x)]}{\partial x_2} &= 2x_2 + 4(w_1 - 1). \end{aligned} \tag{20.28}$$

Setting these two equations to zero and solving for x gives the Pareto set

$$x_1^* = x_2^* = 2(1 - w_1). \tag{20.29}$$

Substituting the Pareto set into the equations for $f_1(x)$ and $f_2(x)$ gives the Pareto front

$$\begin{aligned} f_1(x^*) &= 8(1 - w_1)^2 \\ f_2(x^*) &= 2w_1^2. \end{aligned} \tag{20.30}$$

Plotting Equations (20.29) and (20.30) as w_1 varies from 0 to 1 gives plots that are identical to Figure 20.3.

□

Example 20.5 shows that the aggregation method can find the Pareto set and Pareto front, at least for some MOPs. In fact, the solution of Equation (20.25) for any set of weights results in a Pareto-optimal point. However, if the Pareto front is concave, then the aggregation method cannot find the complete Pareto set and Pareto front, as the following example illustrates.

■ EXAMPLE 20.6

Consider the problem

$$\min_x f(x) = \min_x [x^2, \cos^3 x] \tag{20.31}$$

where $x \in [0, 4]$. This is a single-dimensional MOP with two objectives. We can aggregate the two objectives into the scalar objective

$$\min f(x) = \min \left[wx^2 + (1 - w) \cos^3 x \right] \tag{20.32}$$

where $w \in [0, 1]$, and we can solve Equations (20.31) and (20.32) with exhaustive search. However, the solutions of these two equations are not the same. Figure 20.7 shows the solutions of the two problems. We see that the true Pareto front is concave. The aggregation method correctly gives the convex part of the Pareto front, but it does not correctly give the concave part.

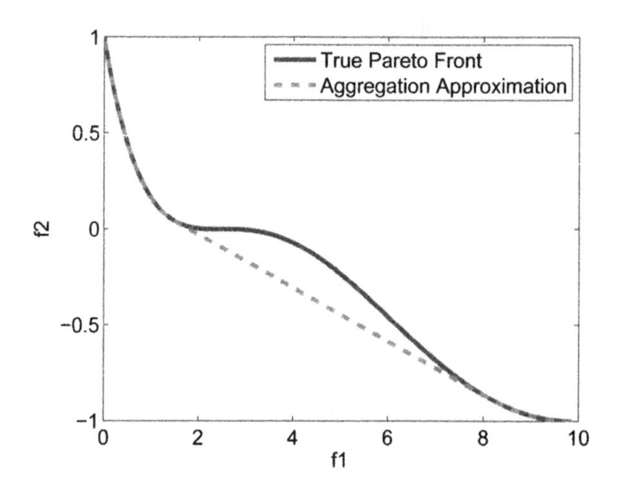

Figure 20.7 Example 20.6: The aggregation method of Equation (20.32) correctly gives the convex part of the Pareto front, but it does not correctly give the concave part.

□

Goal Attainment for Concave Pareto Fronts

It is not just the particular problem of Example 20.6 for which aggregation fails to find the Pareto front. In fact, it is impossible to find any concave Pareto front using an aggregation method [Fleming et al., 2005]. However, we can find concave Pareto fronts using an extension of the goal attainment approach of Section 20.2. Goal attainment is sometimes approached by solving the following problem:

$$\min \alpha \text{ such that } f_i(x) \leq f_i^* + w_i \alpha \text{ for all } i \in [1, k] \text{ and for some } x \tag{20.33}$$

where f_i^* is the ideal value of the i-th objective, and $\{w_i\}$ is a set of positive weights that indicate the relative importance of each objective. Equation (20.33) allows for the possibility that solutions may exist that are better than the user's ideal solution. If the optimal $\alpha > 0$, then the ideal point is not attainable, but the solution to Equation (20.33) finds the solution vector that is as close as possible to the ideal point. If the optimal $\alpha < 0$, then we can find a solution that is better than the ideal point for each objective. It can be shown [Chen and Liu, 1994], [Coello Coello, 1999] that solving Equation (20.33) for all weight combinations $\{w_i\}$ such that $\sum_i w_i = 1$ gives the Pareto set to the original MOP, even if the Pareto front is concave.

Note that Equation (20.33) is an optimization problem with a single objective, k constraints, and $n+1$ independent variables (the scalar α and the original decision vector x). We can thus use constrained optimization algorithms (see Chapter 19) to solve Equation (20.33).

20.3.2 The Vector Evaluated Genetic Algorithm (VEGA)

VEGA was the original MOEA [Schaffer, 1985]. VEGA operates by performing selection on the population using one objective function at a time. This gives a set of subpopulations, one set for each objective function. We then select individuals from the subpopulations to obtain the parents for the next generation, and combine the parents using standard EA recombination methods to obtain children. Figure 20.8 gives an outline of VEGA.

Initialize a population of candidate solutions $P = \{x_j\}$ for $j \in [1, N]$
$M \leftarrow \lceil N/k \rceil$
While not(termination criterion)
  Compute the cost $f_i(x_j)$ for each objective i and for each individual $x_j \in P$
  For $i = 1$ to k
    $P_i \leftarrow M$ individuals probabilistically selected from P using $f_i(\cdot)$
  Next i
  $P \leftarrow N$ individuals selected from $\{P_1, \cdots, P_k\}$
  $C \leftarrow N$ children created from recombining the individuals in P
  Probabilistically mutate the children in C
  $P \leftarrow C$
Next generation
$\hat{P}_s \leftarrow$ nondominated elements of P

Figure 20.8 Outline of the vector evaluated genetic algorithm (VEGA) for solving an optimization problem with k objectives.

Figure 20.8 shows that VEGA begins with a population of N individuals that we usually generate randomly. At each generation, we compute the value of all k cost function values for all N individuals. We then use any desired selection scheme (see Section 8.7) to select M individuals, where $M = \lceil N/k \rceil$ is the smallest integer that is greater than or equal to N/k. We perform this selection probabilistically, first using $f_1(\cdot)$ to create population P_1, then using $f_2(\cdot)$ to create population P_2, and so on. After we have created the P_i subpopulations, we combine them to obtain a parent population P. We then recombine the individuals in P to create a set of children C. We can perform recombination using any of the EAs discussed in this book (genetic algorithms, differential evolution, biogeography-based optimization, and so on). We see that the name VEGA is somewhat of an anachronism; depending on the recombination method that we use, we could call it VEDE, or VEBBO, or whatever other acronym seems appropriate for the type of recombination that we use. This point also applies to many other popular MOEAs that we discuss in this chapter (NSGA, MOGA, and so on).

As we have seen in single-objective EAs, elitism can greatly improve optimization performance, and this point also applies to MOEAs. There are several ways that we could implement elitism in Figure 20.8. For example, at each generation we could find the best individuals with respect to each objective function, and make sure that they are preserved from one generation to the next. Or we could find the nondominated individuals at each generation, and make sure that at least a few of them are preserved from one generation to the next. Although VEGA is not generally defined as an elitist algorithm, the addition of elitism to Figure 20.8 would not change its essence, and would probably improve its performance.

An algorithm similar to VEGA, sometimes referred to as the Hajela-Lin genetic algorithm (HLGA), uses the weighted sum approach. HLGA includes the weight vector of Equation (20.25) as part of the decision variable of each individual [Hajela and Lin, 1997]. This approach uses single-objective optimization based on the weighted sum of Equation (20.25). HLGA uses fitness sharing to achieve a diversity of weights (see Section 8.6.3.1).

20.3.3 Lexicographic Ordering

Lexicographic ordering is similar to VEGA, but allows the user to prioritize objectives [Fourman, 1985]. We perform tournament selection by comparing individuals on the basis of prioritized objectives. Instead of using prioritized objectives, we can also use randomly selected objectives for each tournament [Kursawe, 1991]. Lexicographic ordering is similar to the behavioral memory approach for constrained optimization (see Section 19.2.11) in its sequential handling of objectives.

Figure 20.9 outlines the lexicographic ordering method. In the original lexicographic ordering method, the outer loop in Figure 20.9 executes for $i \in [1, k]$ in the prioritized order of the objective functions. In the randomized variation, the outer loop executes until a user-defined termination criterion is satisfied, and the index i varies randomly at each generation. As with VEGA, we could implement many variations in Figure 20.9, including elitism, and including the incorporation of any of the EAs described in this book or elsewhere.

$P \leftarrow$ randomly-generated population
While not(termination criterion)
    Set the objective function index i
    Initialize an EA with the population P
    Use the EA to minimize $f_i(x)$, denoting the final population as P
Next EA

Figure 20.9 Lexicographic ordering for an optimization problem with k objectives. The objective $f_i(x)$ each iteration can depend on user prioritization, or can be selected randomly. Also, the objective $f_i(x)$ can change from one generation to the next during each EA.

20.3.4 The ϵ-Constraint Method

The ϵ-constraint method [Ritzel et al., 1995] minimizes one objective function at a time while constraining the other objective function values to be below a given threshold. First we find the minimum of each objective function value $f_i(x)$ for $i \in [1, k]$ by minimizing it with a single-objective EA. This gives a lower bound for the objective function constraints ϵ_i:

$$\epsilon_i > \min_i f_i(x) \text{ for } i \in [1, k]. \tag{20.34}$$

Then we minimize the first objective while constraining the other objectives to be smaller than some threshold:

$$\min_x f_1(x) \text{ such that } f_i(x) < \epsilon_i \text{ for } i \in [1, k],\ i \neq 1. \tag{20.35}$$

Using the final population of the EA that resulted from Equation (20.35) as the initial population of the next EA, we minimize the next objective:

$$\min_x f_2(x) \text{ such that } f_i(x) < \epsilon_i \text{ for } i \in [1, k],\ i \neq 2. \tag{20.36}$$

We repeat this process for all k objectives. We then decrease the ϵ_i values and repeat the sequential minimization process. This sequential approach is similar to lexicographic ordering (see Section 20.3.3), and to the behavioral memory approach for constrained optimization (see Section 19.2.11). Figure 20.10 outlines the ϵ-constraint MOEA. The most challenging part of implementing the ϵ-constraint MOEA is deciding exactly how to "set ϵ_i to some number greater than $f_i^*(x)$" and how to "decrease ϵ_i for $i \in [1, k]$" in Figure 20.10. As with other MOEAs, we could implement many variations in Figure 20.10, including elitism, and including the incorporation of any of the EAs described in this book or elsewhere.

For each objective function $f_i(x)$, where $i \in [1, k]$
    Use an EA to find $f_i^*(x) = \min_x f_i(x)$
    Set ϵ_i to some number greater than $f_i^*(x)$
Next objective function
Initialize the EA population P for the MOP
While not(termination criterion)
    For each objective function $f_i(x)$, where $i \in [1, k]$
        Initialize an EA population with the result of the previous EA
        Use an EA to minimize $f_i(x)$ such that $f_r(x) < \epsilon_r$ for $r \in [1, k]$, $r \neq i$
        $P \leftarrow$ final EA population
    Next objective function
    Decrease ϵ_i for $i \in [1, k]$
Next iteration

Figure 20.10 Outline of the ϵ-constraint method for solving an optimization problem with k objectives.

20.3.5 Gender-Based Approaches

Gender-based approaches assign a gender to each individual based on the objective function with which they are to be evaluated [Allenson, 1992], [Lis and Eiben, 1997]. Gender-based approaches also use a secondary population called an archive. The archive is a collection of nondominated individuals, and similar to elitism, it prevents the loss of good individuals. Many other MOEAs also use archives, including some that we discuss below.

In a gender-based approach for a k-objective MOP, we have k different genders, each corresponding to a different objective. We create an initial population with an equal number of individuals for each gender. We then select an individual from each gender i based on $f_i(x)$. We then use the selected individuals for recombination to obtain a child individual. We can assign the child individual's gender on the basis of which objective for which it performs the best. We can perform recombination using any of the methods discussed in this book. For standard single-point GA crossover, we would use only two individuals for recombination. For multi-parent crossover (see Section 8.8.5), we could use one or more individuals from each gender. At the end of each generation we usually compare the population with the archive and store the nondominated individuals in the archive while removing dominated individuals from the archive. Figure 20.11 outlines a gender-based approach for multi-objective optimization.

N_g = desired population size for each gender
Initialize k EA populations P_i, where $|P_i| = N_g$ for $i \in [1, k]$
EA population size $N \leftarrow kN_g$
While not(termination criterion)
    For $m = 1$ to N
        For $i = 1$ to k
            Use $f_i(x)$ to probabilistically select one parent from P_i
        Next i
        Use the k selected individuals to create a child individual c_m
    Next m
    Assign the genders of the children $\{c_m\}$
    Randomly mutate the children $\{c_m\}$
    Store the nondominated children to the archive
    Remove dominated individuals from the archive
    Replace the EA populations $\{P_i\}$ with children of the appropriate gender
Next generation

Figure 20.11 Outline of a gender-based algorithm for solving an optimization problem with k objectives.

We can see many opportunities for modifying the algorithm of Figure 20.11. For example, depending on the dimension of x, we may want to select more than one parent from each gender for recombination. Also, the line "Assign the genders of the children" leaves a lot of details to be determined. Should we restrict each child to a single gender? Should we make sure that we maintain an equal number of each gender in the population? The approach outlined in Figure 20.11 creates N children, but we may want to create more than that so that we can obtain a child population with highly-fit children for each gender.

The statement "store the nondominated children to the archive" in Figure 20.11 also leaves out a lot of details. Should we allow the archive to grow without bound? Should we place an upper limit on the size of the archive? Should we add individuals to the archive if they are nondominated with respect to the current population, or only if they are nondominated with respect to the archive? We discuss some of these archive-related issues in general terms in Section 20.4.

20.4 PARETO-BASED EVOLUTIONARY ALGORITHMS

The MOEA approaches of the previous section attempt to find a diverse Pareto-optimal set of solutions to a MOP. However, none of them directly use the concept of Pareto optimality to compute the relative dominance between individuals or groups of individuals (except when adding individuals to the archive in Figure 20.11). The following sections discuss approaches that directly use Pareto dominance.

- Section 20.4.1 discusses the simple evolutionary multi-objective optimizer (SEMO) and the diversity evolutionary multi-objective optimizer (DEMO).
- Section 20.4.2 discusses the ϵ-based MOEA (ϵ-MOEA).
- Section 20.4.3 discusses the nondominated sorting genetic algorithm (NSGA) and an updated version of it (NSGA-II).
- Section 20.4.4 discusses the multi-objective genetic algorithm (MOGA).
- Section 20.4.5 discusses the niched Pareto genetic algorithm (NPGA).
- Section 20.4.6 discusses the strength Pareto evolutionary algorithm (SPEA) and an updated version of it (SPEA2).
- Section 20.4.7 discusses the Pareto archived evolution strategy (PAES).

20.4.1 Evolutionary Multi-Objective Optimizers

This section discusses two evolutionary multi-objective optimizers: the simple evolutionary multi-objective optimizer (SEMO), and the diversity evolutionary multi-objective optimizer (DEMO). As will be seen in this section, these algorithms are motivated by the basic ideas of EP and ES.

The Simple Evolutionary Multi-Objective Optimizer (SEMO)

SEMO was originally proposed for binary optimization [Laumanns et al., 2003], but is easily extended to continuous optimization. In SEMO we begin with a randomly-generated population of individuals. The original SEMO algorithm begins with a population size of one. The population grows as the algorithm finds more and more nondominated solutions. At each generation we mutate one randomly-selected individual from the population to create a child. We add the child to the population if the child is nondominated by the population, and we also remove any dominated individuals from the population. Figure 20.12 illustrates the SEMO algorithm. The "random mutation" in Figure 20.12 could be any of the mutation methods discussed in Chapter 5, Chapter 6, or Section 8.9.

SEMO provides a useful starting point for multi-objective optimization. We can modify SEMO based on ideas from other MOEAs. For example, instead of using a "randomly selected individual from P" as the parent of y, we could instead use fitness-proportional selection. Also, we could generate y from multiple parents, or we could prune the population periodically as in SPEA2 (see Section 20.4.6).

Initialize a population of candidate solutions $P = \{x_j\}$
Compute the cost $f_i(x_j)$ for each objective $i \in [1, k]$ and for each individual $x_j \in P$
While not(termination criterion)
    $y \leftarrow$ mutation of randomly selected individual from P
    If y is not dominated by any individuals in P then
        $P \leftarrow \{P, y\}$
        Remove all individuals from P that are dominated by y
    End if
Next generation

Figure 20.12 Outline of the simple evolutionary multi-objective optimizer (SEMO).

The Diversity Evolutionary Multi-Objective Optimizer (DEMO)

One problem with SEMO is the unbounded growth of its population. We can handle this problem by using ϵ-dominance instead of dominance in the test for including y in the population P. This results in the diversity evolutionary multi-objective optimizer (DEMO). DEMO uses the same algorithm as the one in Figure 20.12, except that it uses ϵ dominance as the criterion for including y in P [Horoba and Neumann, 2010]. This raises the standard for including a child individual y in the current population; we include y in the population only if it is not ϵ-dominated by any other individuals in P. Since ϵ-dominance is a weaker type of dominance than Pareto dominance (see Section 20.1), the criterion for including y in the population is more strict in DEMO than in SEMO. DEMO essentially divides the objective space into hyperboxes, and does not allow the population to contain any more than one individual per hyperbox. We usually use additive ϵ dominance in DEMO.

Figure 20.13 illustrates the DEMO concept for a two-objective optimization problem. We would not include child y_1 in the population because it is ϵ-dominated by the individual that is in the same hyperbox. However, we would include y_2 in the population because it is not ϵ-dominated by any of the current population members. Note that Figure 20.13 is not precisely correct because DEMO hyperboxes are defined relative to the current population. Nevertheless, the figure gives a conceptual illustration of the use of ϵ-dominance in DEMO. Although Figure 20.13 shows that $\epsilon_1 = \epsilon_2$ (that is, the boxes are square), the user can choose a different value ϵ_i for each objective index $i \in [1, k]$ based on the desired solution accuracy for each objective.

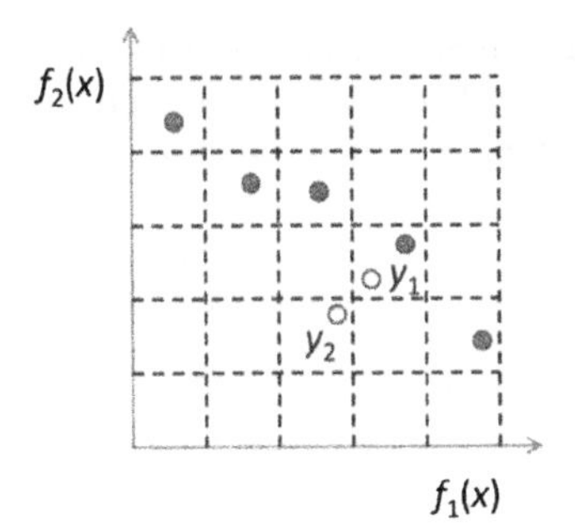

Figure 20.13 ϵ-dominance in the diversity evolutionary multi-objective optimizer (DEMO). Current population members are indicated with solid circles, and children are indicated with unfilled circles. We add a child to the population only if it is not ϵ-dominated by any of the current population. In this figure, we would not add y_1 to the population, but we would add y_2.

20.4.2 The ϵ-Based Multi-Objective Evolutionary Algorithm (ϵ-MOEA)

The ϵ-MOEA uses the concept of ϵ dominance in a similar manner to the ϵ dominance that DEMO uses as described above [Deb et al., 2005]. The ϵ-MOEA includes a population and an archive. At each generation we select one individual from the population and one from the archive, and we use some recombination method to obtain a child.

If the child dominates an individual in the population, then we replace the dominated individual with the child. If the child dominates more than one individual, then we replace a randomly-selected individual.

Next we compare the child with the archive. There are four situations that could result from this comparison. (1) If the child is dominated by any of the archived individuals, then the child is not placed into the archive. (2) If the child dominates any of the archived individuals, then the child is added to the archive, and the dominated individuals are removed from the archive.

If neither of these two conditions hold, then we calculate the ϵ-box $B(x)$ of the child x:

$$B_j(x) = \lfloor f_j(x)/\epsilon_j \rfloor \tag{20.37}$$

for $j \in [1, k]$, where k is the number of objectives, ϵ_j is the desired resolution of the j-th objective, and $\lfloor \cdot \rfloor$ returns the largest integer that is less than or equal to its argument. This brings us to the third situation that could result from the comparison of the child with the archive. (3) If the child x is in the same ϵ-box as an archived individual a, then the child replaces a if the child is closer to the origin of the objective function space:

$$x \text{ replaces } a \text{ if } \left[B(x) = B(a), \text{ and } \sum_{j=1}^{k} f_j^2(x) < \sum_{j=1}^{k} f_j^2(a) \right]. \tag{20.38}$$

The above condition assumes that the objective functions are normalized so that the magnitudes of each objective function value are commensurate with each other.

Equation (20.38) also uses the Euclidean norm to measure distance, although we could use any other norm. (4) If neither of three previous conditions occurs, then we add the child to the archive, which increases the size of the archive by one individual.

Figure 20.14 illustrates these four possibilities. Note that the logic described here ensures that no more than one individual is in each ϵ-box of the archive. Although the archive can grow from one generation to the next, its size is limited by this logic. Figure 20.14(a) illustrates the case that the child is dominated by one or more of the archived individuals; in this case, the child is not added to the archive. Figure 20.14(b) illustrates the case that the child dominates one or more archived individuals; in this case, the child replaces the dominated individuals. Figure 20.14(c) illustrates the case that the child and the archive are nondominated with respect to each other, and the child is in the same ϵ-box as one of the archived individuals; in this case, the child replaces the archived individual that is in the same ϵ-box if the child is closer to the origin of the objective function space. Finally, Figure 20.14(d) illustrates the case that the child and the archive are nondominated with respect to each other and the child does not share the same ϵ-box as any of the archived individuals; in this case, the child is added to the archive, which increases the size of the archive by one.

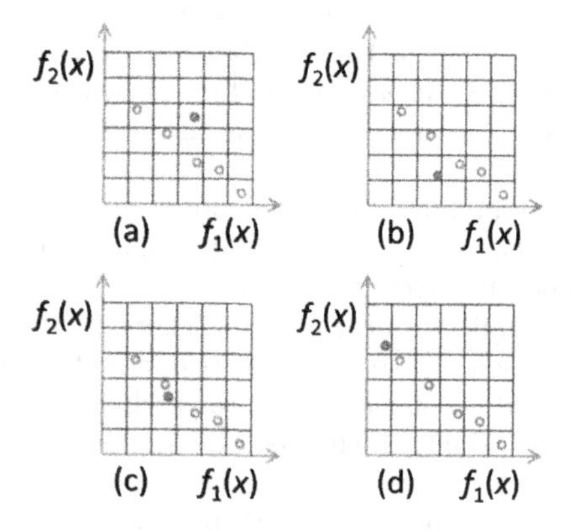

Figure 20.14 The ϵ-MOEA logic for adding a child to the archive. The empty circles are archived individuals, the solid circle is a child individual, and the grid in the objective function space defines ϵ boxes. In case (a), the child is not added to the archive. In case (b), the child replaces the individuals that it dominates, which in this figure decreases the archive size by one. In case (c), the child replaces the individual that is in the same ϵ-box, provided that the child is closer to the origin of the f_1/f_2 plane. In case (d), the child is added to the archive, and the size of the archive increases by one.

Figure 20.15 outlines the ϵ-MOEA. We could experiment with several variations on this algorithm. For example, usually we use tournament selection with a tournament size of two to select the parent x from P; however, we could use any other selection method. Usually we randomly select the parent a from the archive A; again, we could use any other selection method. The recombination method that we use to create the child could be any method from Section 8.8. If we use multiple-parent recombination, then we need to decide how many parents to choose from P and how many to choose from A.

Initialize a population of candidate solutions $P = \{x_j\}$ for $j \in [1, N]$
Copy the nondominated individuals from P to the archive A
While not(termination criterion)
    Select one parent x from P and one parent a from A
    Create a child c by recombining x and a
    $D_P \leftarrow \{x \in P : c \text{ dominates } x\}$
    If $D_P \neq \emptyset$ then
        Replace a random $x \in D_P$ with c
    End if
    $D_A \leftarrow \{a \in A : c \text{ dominates } a\}$
    If $D_A \neq \emptyset$ then
        Add c to A
        Remove D_A from A
    else if Equation (20.38) is satisfied then
        Add c to A
        Remove a from A
    else if (c is nondominated with respect to A) and ($B(c) \neq B(a)$ for all a) then
        Add c to A
    End if
Next generation

Figure 20.15 The above pseudocode outlines the ϵ-MOEA for solving an optimization problem with k objectives.

20.4.3 The Nondominated Sorting Genetic Algorithm (NSGA)

NSGA was proposed in [Srinivas and Deb, 1994] and is based on ideas from [Goldberg, 1989a]. NSGA assigns the cost of each individual based on how dominant it is. First we copy all individuals to a temporary population T. Then we find all nondominated individuals in T; these individuals, which we denote as the set B, are assigned the lowest cost value. Next we remove B from T and find all nondominated individuals in the reduced set T. These individuals are assigned the second-lowest cost value. We repeat this process, obtaining a cost for each individual that is based on its level of nondomination. Figure 20.16 outlines NSGA.

Figure 20.16 shows that we begin with a population of N individuals, usually generated randomly. At each generation, we compute the value of all k cost function values for all N individuals. We copy the individuals to a temporary population T. We assign all individuals that are nondominated a cost value of 1. We remove all of those individuals from T, find all the individuals in the reduced set T that are nondominated, and assign them a cost value of 2. We repeat this process until all individuals have been assigned a cost value based on their level of domination. We then use the cost values $\phi(\cdot)$ in Figure 20.16 to perform selection, and we recombine the individuals in P using any desired EA and any desired recombination method.

Finally we mutate the child population, replace the parents with the children, and continue to the next generation.

NSGA is sometimes criticized for its inefficiency due to the inner loop for the calculation of the cost $\phi(\cdot)$ in Figure 20.16. But for real-world problems, the function evaluations $f_i(\cdot)$ comprise the overriding computational burden, and the nondomination loop of NSGA adds a trivial amount of computational overhead.

Initialize a population of candidate solutions $P = \{x_j\}$ for $j \in [1, N]$
While not(termination criterion)
    Temporary population $T \leftarrow P$
    Nondomination level $c \leftarrow 1$
    While $|T| > 0$
        $B \leftarrow$ nondominated individuals in T
        Cost $\phi(x) \leftarrow c$ for all $x \in B$
        Remove B from T
        $c \leftarrow c + 1$
    Next nondomination level
    $C \leftarrow N$ children created from recombining the individuals in P
    Probabilistically mutate the children in C
    $P \leftarrow C$
Next generation

Figure 20.16 Outline of the nondominated sorting genetic algorithm (NSGA) for solving an optimization problem with k objectives. We use the cost function values $\phi(x_j)$ to select parents for recombination.

NSGA-II

NSGA-II is a modification of NSGA [Deb et al., 2002a]. NSGA-II computes the cost of an individual x by taking into account not only the individuals that dominate it, but also the individuals that it dominates. For each individual, we also compute a crowding distance by finding the distance to the nearest individuals along each objective function dimension. We use the crowding distance to modify the fitness of each individual. NSGA-II does not use an archive, but instead uses a $(\mu + \lambda)$ evolution strategy approach to implement elitism (see Chapter 6).

NSGA sets the crowding distance of each individual x equal to its average distance to its nearest neighbors along each objective function dimension. For example, suppose that we have N individuals in the NSGA. Further suppose that individual x has the objective function vector

$$f(x) = [f_1(x), \cdots, f_k(x)]. \tag{20.39}$$

For each objective function dimension, we find the closest larger value and the closest smaller value in the population, as follows:

$$f_i^-(x) = \max_y [f_i(y) \text{ such that } f_i(y) < f_i(x)]$$
$$f_i^+(x) = \min_y [f_i(y) \text{ such that } f_i(y) > f_i(x)]. \quad (20.40)$$

We then compute the crowding distance of x as

$$d(x) = \sum_{i=1}^{k} (f_i^+(x) - f_i^-(x)). \quad (20.41)$$

Individuals that are in more crowded regions of the objective function space tend to have a smaller crowding distance. Individuals at the extreme values of the objective function space have an infinite crowding distance:

$$d(x) = \infty \text{ for } x \in \left\{ \arg\min_y f_i(y) \cup \arg\max_y f_i(y) \text{ for all } i \in [1, k] \right\}. \quad (20.42)$$

The crowding distance corresponds to half of the perimeter of the largest hypercube, called a cuboid in [Deb et al., 2000], whose boundaries do not extend beyond the objective function space coordinates of the nearest neighbors of x in each dimension. Figure 20.17 illustrates the hypercube in a two-dimensional objective function space, which is a rectangle.[2] In Figure 20.17, the nearest neighbors of x in the f_1 direction are A and C, and the nearest neighbors of x in the f_2 direction are A and B.

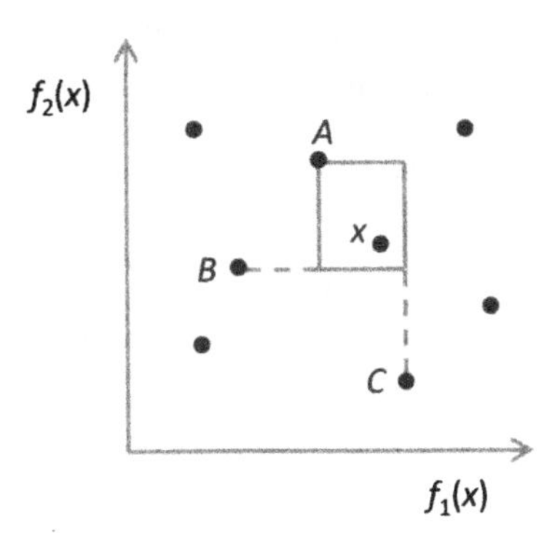

Figure 20.17 Illustration of the crowding distance calculation in NSGA-II. The crowding distance of x is obtained as half the perimeter of the largest hypercube (which is a rectangle in two-dimensional space) whose boundaries do not extend beyond the objective function space coordinates of the nearest neighbors of x.

Now that each individual in the population has a crowding distance, we use it as a secondary sorting parameter for obtaining the rank of each individual. As in the NSGA algorithm of Figure 20.16, we rank each individual on the basis of its

[2]Contrary to [Deb et al., 2000], the hypercube is not the largest one that encloses x without including any other points. As seen from Figure 20.17, that definition would give a different hypercube and would generally result in different NSGA-II performance.

nondomination level, but we also include a more fine-grained ranking level on the basis of crowding distance. That is, x is ranked better than y if $\phi(x) < \phi(y)$, or if $\phi(x) = \phi(y)$ and $d(x) > d(y)$. Whereas NSGA uses $\phi(x)$ to select parents for recombination in Figure 20.16, NSGA-II instead uses the ranks described above to select parents for recombination.

20.4.4 The Multi-Objective Genetic Algorithm (MOGA)

MOGA was introduced in [Fonseca and Fleming, 1993]. Like NSGA, it assigns cost values on the basis of domination, but MOGA approaches cost assignment from the opposite direction. Whereas NSGA assigns the cost of x based on how many levels of individuals need to be removed from the population before x is nondominated, MOGA assigns the cost of x based on how many individuals dominate it. We assign the same cost to all nondominated individuals. For each dominated individual x, we assign its cost based on how many individuals dominate it, and also based on how many individuals are near it. Similar to the use of crowding distance in NSGA-II, this encourages diversity in the population.

In MOGA, x is ranked better than y (that is, $\phi(x) < \phi(y)$) if it is dominated by fewer individuals in the population P (that is, $d(x) < d(y)$),[3] or if it is dominated by the same number of individuals and there are fewer individuals near x than there are near y in the objective function space (that is, $s(x) < s(y)$). This ranking approach can be expressed as follows.

$$\begin{aligned} d(x) &= |x' \in P : x' \text{ dominates } x| \\ s(x) &= |x' \in P : ||f(x) - f(x')|| < \sigma| \\ \phi(x) < \phi(y) &\text{ if } \{d(x) < d(y), \text{ or } d(x) = d(y) \text{ and } s(x) < s(y)\} \end{aligned} \tag{20.43}$$

where σ is a user-defined sharing parameter and $|| \cdot ||$ is some distance metric. Sharing can also be automatically implemented so that the user does not need to define the sharing parameter [Ahn and Ramakrishna, 2007]. Figure 20.18 gives an outline of MOGA. As with the other MOEAs described in this chapter, we could implement many variations in Figure 20.18, such as various recombination methods and various elitism approaches.

20.4.5 The Niched Pareto Genetic Algorithm (NPGA)

NPGA was proposed in [Horn et al., 1994]. It is similar to NSGA and MOGA in its assignment of cost on the basis of domination. NPGA is an attempt to reduce the computational effort of NSGA and MOGA. We randomly select two individuals from the population, x_1 and x_2. We then randomly select a subset S of the population, which is typically around 10% of the population. If one of the individuals x_1 or x_2 is dominated by any of the individuals in S, and the other is not, then the nondominated individual, denoted as r, wins the tournament and is selected for recombination. If both individuals x_1 and x_2 are dominated by at least one individual in S, or both individuals are not dominated by any individuals in S, then we use fitness sharing to decide the tournament winner; that is, the

[3]Note the change in terminology from NSGA-II; $d(x)$ is crowding distance in NSGA-II, but domination level in MOGA.

Initialize a population of candidate solutions $P = \{x_j\}$ for $j \in [1, N]$
While not(termination criterion)
    Use Equation (20.43) to find the rank $\phi(x_j)$ for each $x_j \in P$
    $C \leftarrow N$ children created from recombining the individuals in P
    Probabilistically mutate the children in C
    $P \leftarrow C$
Next generation

Figure 20.18 Outline of the multi-objective genetic algorithm (MOGA) for solving an optimization problem with k objectives. We use the ranks $\phi(x_j)$ to select parents for recombination.

individual that is in the least crowded region of the objective function space wins the tournament. This selection process can be described as follows:

$$\begin{aligned} d_i &= |y \in S : y \succ x_i| \text{ for } i \in [1, 2] \\ s_i &= \text{Crowding distance of } x_i \text{ for } i \in [1, 2] \\ r &= \begin{cases} x_1 & \text{if } \begin{cases} (d_1 = 0) \text{ and } (d_2 > 0), \text{ or} \\ (d_1 > 0) \text{ and } (d_2 > 0) \text{ and } (s_1 < s_2), \text{ or} \\ (d_1 = 0) \text{ and } (d_2 = 0) \text{ and } (s_1 < s_2) \end{cases} \\ x_2 & \text{otherwise.} \end{cases} \end{aligned} \tag{20.44}$$

d_i is the number of individuals that dominate x_i, s_i is the crowding distance of x_i, and r is the individual (either x_1 or x_2) that we finally select for recombination. The crowding distance s_i could be computed with a method from Section 8.6.3.1, or it could use Equation (20.43), or it could be any other calculation that quantifies the crowdedness of x_1 and x_2. The crowding distance is smaller for individuals that are in more crowded regions of the search space or the objective function space. The use of crowding distance in NPGA encourages diversity; like other such algorithms, this makes it especially suitable for multi-modal problems, or problems in which the user is interested in finding good potential solutions in widely separated regions of the function space or search space. Note that the crowding distance of Equation (20.44)) could be computed in either function space or search space, depending on the priorities of the user.

Figure 20.19 gives an outline for NPGA. NPGA, with its randomly-selected population subset S in Equation (20.44), was the first MOEA to save computational effort by reducing the number of individuals involved in the ranking process [Coello Coello, 2009]. As with the other MOEAs in this chapter, we could customize Figure 20.19 to include elitism, an archive, various recombination strategies, or various selection strategies.

Initialize a population of candidate solutions $P = \{x_j\}$ for $j \in [1, N]$
While not(termination criterion)
    $R \leftarrow \emptyset$
    While $|R| < N$
        Randomly select two individuals x_1 and x_2 from P
        Randomly select a population subset $S \subset P$
        Use Equation (20.44) to select r from $\{x_1, x_2\}$
        $R \leftarrow \{R, r\}$
    End while
    Recombine the individuals in R to obtain N children
    Randomly mutate the children
    $P \leftarrow$ children
Next generation

Figure 20.19 Outline of the niched Pareto genetic algorithm (NPGA) for solving an optimization problem with k objectives.

20.4.6 The Strength Pareto Evolutionary Algorithm (SPEA)

SPEA was the first MOEA to explicitly use elitism [Zitzler and Thiele, 1999], [Zitzler et al., 2004]. Of course, any of the previously-discussed MOEAs can be implemented with elitism, but for some reason most of them did not incorporate elitism when originally introduced. Elitism is usually a common sense option in both single-objective and multi-objective EAs. Also, elitism is theoretically necessary to guarantee convergence in MOEAs [Rudolph and Agapie, 2000]. However, if a user-preference-based approach is used in an MOEA and the preferences change over time, then elitism may result in a degradation of performance [Zitzler et al., 2000]. This is similar to the drawbacks of elitism for dynamic optimization problems, where fitness functions are time varying (see Chapter 21).

SPEA works by maintaining all nondominated individuals that are found during the learning process in an archive. Whenever we find a nondominated individual we copy it to the archive. We assign a strength value $S(\alpha)$ to each archived individual α based on how many individuals in the population that α dominates:

$$S(\alpha) = \frac{|x \in \{P\} \text{ such that } \alpha \succ x|}{N+1} \quad \text{for all } \alpha \in A \qquad (20.45)$$

where P is the set of candidate solutions, N is the size of P, and A is the archive set. Note that $S(\alpha) \in [0, 1)$. For each individual x in P, we find the set $\alpha(x)$ of all archived individuals that dominate it. We then compute the raw cost of x, denoted as $R(x)$, as the sum of the strengths of the individuals in $\alpha(x)$:

$$\begin{aligned} R(x) &= 1 + \sum_{y \in \alpha(x)} S(y), \text{ for all } x \in P \\ \text{where } \alpha(x) &= \{y \in A : y \succ x\}. \end{aligned} \qquad (20.46)$$

Adding one in the above equation ensures that $R(x) \geq 1$, which in turn ensures that $R(x) > S(\alpha)$ for all $x \in P$ and all $\alpha \in A$. Note that if x has a low raw cost, then x is a high-performing individual.[4]

Figure 20.20 illustrates strength and raw cost calculations for an archive size $|A| = 3$ and a population size $|P| = 6$. Figure 20.20 shows the strength values of the Pareto front points as the normalized number of individuals that they dominate. The figure also shows the raw cost value of each dominated point as 1 plus the sum of the strengths of the Pareto front points that dominate it. Note that the raw cost values become larger as the individuals move farther away from the Pareto front (that is, as the individuals are dominated by more Pareto front points). Also, note the dominated individual toward the upper left of the figure with the raw cost 9/7, and compare it with the two dominated individuals at the lower right with the raw cost values 10/7. The individuals in the lower right have a higher raw cost because they are in a more crowded region of objective function space. Since they are in a crowded area, the strength of the Pareto front point that dominates them is larger, which results in their raw cost being higher.

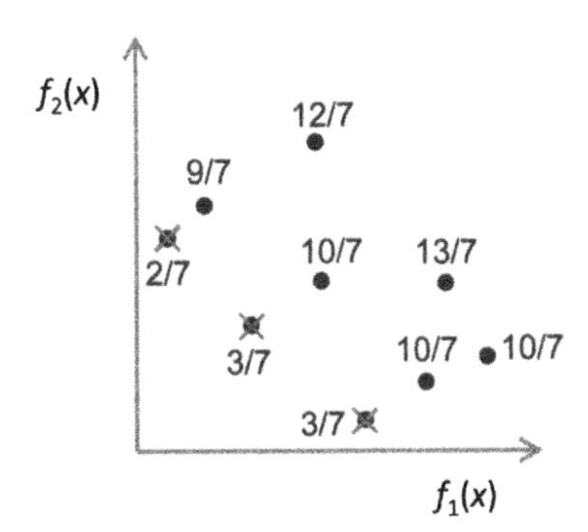

Figure 20.20 Illustration of SPEA strength and raw cost calculations for a two-objective minimization problem. The Pareto front individuals are marked with the symbol ×, and their strength values are shown beside them. The non-Pareto-front individuals are shown as filled circles, and their raw cost values are shown beside them.

As mentioned above, at each generation, all individuals in $\{P, A\}$ that are non-dominated are added to the archive A. However, this can result in unbounded growth of the archive. SPEA handles this potential problem with a clustering method [Zitzler and Thiele, 1999]. If the archive has $|A|$ individuals, we begin by defining each individual as a cluster. We then merge the two closest clusters into a single cluster so that the cluster count of A is reduced by one. We repeat this process until the archive contains N_A clusters, which is the desired archive size. Finally, we retain only one point from each cluster, usually the one that is closest to the cluster center.

[4]The SPEA literature often refers to $R(x)$ as raw fitness, but we use the terminology raw cost to be consistent with the intuition that low cost is good and low fitness is bad.

SPEA2

An improved version of SPEA, denoted as SPEA2, proposes some improvements to the original algorithm [Zitzler et al., 2001]. First, we assign a strength value $S(\alpha)$ not only to the individuals in the archive, but also to the individuals in the population:

$$S(\alpha) = |x \in \{P, A\} \text{ such that } \alpha \succ x| \text{ for all } \alpha \in \{P, A\}. \tag{20.47}$$

We also see from comparing the above equation with Equation (20.45) that we do not normalize the strength values.

Second, we calculate the raw cost of each individual in P slightly differently by summing the strengths of the dominating individuals in both the population and the archive:

$$\begin{aligned} R(x) &= \sum_{y \in \alpha(x)} S(y), \text{ for all } x \in P \\ \text{where } \alpha(x) &= \{y \in \{P, A\} : y \succ x\}. \end{aligned} \tag{20.48}$$

We also see from comparing the above equation with Equation (20.46) that we do not add one in the raw cost calculations.

Third, we modify the raw cost of each $x \in P$ based on how many individuals it is near; that is, we penalize the cost of individuals that are near many other individuals in objective function space. We do this by finding the distance between $f(x)$ and $f(y)$, for all $x \in P$, and for all $y \in \{P, A\}$ such that $y \neq x$. This distance metric can be any vector norm that the user thinks is appropriate, although we usually use the Euclidean norm. For each $x \in P$ we sort the distances between it and each $y \in \{P, A\}$ in increasing order, so we have an ordered distance list for each x with $(|P| + |A|)$ elements. We then select the j-th element in the distance list, which gives the distance between x and its j-th nearest neighbor, denoted as $\sigma_j(x)$. We can use various strategies to select j; for example, some researchers have had good success with $j = \sqrt{|P| + |A|}$, but others simply set $j = 1$ [Zitzler et al., 2004]. We define the density of x as

$$D(x) = \frac{1}{\sigma_j(x) + \gamma} \tag{20.49}$$

where we choose the constant γ in the denominator to ensure that $D(x) < 1$. The original SPEA2 paper suggests $\gamma = 2$ [Zitzler et al., 2001] Finally, we obtain the modified cost of x by adding the raw cost to the density:

$$C(x) = R(x) + D(x). \tag{20.50}$$

Since all nondominated individuals have a raw cost value of 0, as seen from Equation (20.48), and since $D(x) < 1$ for all x, we see that all nondominated individuals have a cost $C(x) < 1$.

The fourth modification of SPEA2 involves controlling the archive size. In SPEA there is not a lower bound on the archive size, but in SPEA2 the archive size is maintained at a constant value. If at any point during the SPEA2 process the archive size becomes too small, we add the lowest-cost individuals from the population, even though they are dominated, to the archive until the archive size reaches the desired value. SPEA uses a clustering method to reduce the archive size in case it becomes too large, but SPEA2 uses a different approach. In SPEA2, if the archive size becomes too large we remove individuals by finding the distance from each $x \in A$ to its nearest neighbor in objective function space:

$$D_{\min}(x) = \min_y \left[\sum_{i=1}^{k} (f_i(x) - f_i(y))^2 \text{ where } y \in A \text{ and } y \neq x \right], \text{ for } x \in A. \quad (20.51)$$

This gives us $|A|$ values of $D_{\min}(x)$, where $|A|$ is the archive size. Next we use D to denote the set of individuals that have the smallest $D_{\min}(x)$ value:

$$D = \{x : D_{\min}(x) \leq D_{\min}(y) \text{ for all } y \in A\}. \quad (20.52)$$

D will have at least two individuals in it, since the distance between any two individuals x and y is the same as the distance between y and x. Among all individuals in D we find the individual, denoted as $x_{\min}$, that is the nearest to any archived individual not in D:

$$x_{\min} = \arg\min_x \left[\min_y \sum_{i=1}^{k} (f_i(x) - f_i(y))^2 \text{ where } y \in A \text{ and } y \notin D \right]. \quad (20.53)$$

We remove $x_{\min}$ from the archive, which reduces the archive size by one. If $|A|$ is too large, we repeat Equations (20.51)–(20.53), removing one individual during each iteration, until the archive reaches the desired size.

The fifth and final modification of SPEA2 is that only members of A participate in selection and recombination to create the population at the next generation. The literature includes several variations and modifications of SPEA and SPEA2, but Figure 20.21 outlines the basic algorithm.

Figure 20.21 includes the SPEA principles, but leaves many details to the creativity of the researcher. Here we clarify a few points about Figure 20.21 and mention several possibilities for modifications.

- The reader needs to choose the population size N and the archive size N_A. Usually $N_A < N$.

- The statement "Copy nondominated individuals from P to A" indicates that all individuals $x \in P$ need to be compared with all individuals $y \in \{P, A\}$. Any individual x that is nondominated by all individuals y is copied to A. However, this statement leaves unanswered the question of whether or not to remove the nondominated individuals from P. Since the nondominated individuals are in A, it might not make sense to keep them as duplicates in P. But this raises a follow-on question about the population size of P. If we remove nondominated individuals from P, then should we replace them with some other individuals? We could leave P in a reduced state and always create N children regardless of the size of P, or we could replace the nondominated individuals that that we remove from P with some randomly-created individuals.

N = population size
N_A = maximum archive size
Initialize a population of candidate solutions $P = \{x_j\}$ for $j \in [1, N]$
Initialize the archive A as the empty set
While not(termination criterion)
    Copy nondominated individuals from P to A:
        $A \leftarrow \{A \cup \{x \in P : \nexists (y \in \{P, A\} : y \succ x)\}$
    Remove dominated individuals from A
    While $|A| > N_A$
        Use a clustering method (SPEA), or Equations (20.51)–(20.53) (SPEA2),
            to remove an individual from A
    End while
    If SPEA2 then
        While $|A| < N_A$
            Add the lowest-cost non-duplicate individual from P to A:
                $A \leftarrow \{A, (\arg\min_x C(x)$ such that $x \in P, x \notin A\}$
        End while
    End if
    Use Equation (20.46) (SPEA), or Equation (20.50) (SPEA2),
        to calculate the cost of each individual in P
    Select parents from $\{P, A\}$ (SPEA), or from A (SPEA2)
    Use a recombination method to create children C from the parents
    Probabilistically mutate the child population C
    Use a replacement method to replace individuals in P with individuals in C
Next generation

Figure 20.21 Outline of the strength Pareto evolutionary algorithm (SPEA and SPEA2).

- The loop "While $|A| > N_A$" removes individuals from crowded regions of objective function space in case the archive is too large. SPEA and SPEA2 each have their own methods for accomplishing this, and the reader can undoubtedly find other methods with which to experiment.

- The loop "While $|A| < N_A$" adds low-cost individuals to the archive in case it is too small, but only for SPEA2. In SPEA, this step is omitted; that is, $|A|$ does not have a lower bound.

- The statement "Select parents from $\{P, A\}$ (SPEA), or from A (SPEA2)" leaves a lot of room for flexibility. We can use any type of selection for this step (see Section 8.7). The success of SPEA and SPEA2 strongly depends on the implementation of this statement.

- The statement "Use a recombination method to create children C from the parents" also leaves a lot of room for flexibility. We can use any of the EAs discussed in this book, and any type of recombination (see Section 8.8), to create children. As with selection, the success of SPEA and SPEA2 strongly depends on the implementation of recombination.

- The statement "Use a replacement method to replace individuals in P with individuals in C" can be implemented in several different ways. If $|C| = |P|$ we can simply replace P with C. If $|C| < |P|$ we can select the best N individuals from $C \cup P$ to replace P, or we can select N individuals from $C \cup P$ using a fitness-proportionate algorithm. If $|C| > |P|$ we can select the best N individuals from C, or the best N individuals from $C \cup P$, or we can select N individuals from either C or $C \cup P$ using a fitness-proportionate algorithm.

EXAMPLE 20.7

This example illustrates SPEA2 archive pruning as described in Equations (20.51)–(20.53). Figure 20.22 shows an archive of nondominated individuals in a two-dimensional objective function space. Suppose that we need to reduce the archive size from the eight individuals shown to five individuals. First we find the individuals that are closest to each other, which are individuals f and g in the figure. We remove g because it is closer to its next-nearest neighbor (h) than f is to its next-nearest neighbor (e). Now that we have removed g, we find the next two individuals that are closest to each other, which are d and e in the figure. We remove d because it is closer to its next-nearest neighbor (c) than e is to its next-nearest neighbor (f). After removing d, we find the next two individuals that are closest to each other, which are a and b. We remove b because it is closer to its next-nearest neighbor (c) than a is to its next-nearest neighbor (c). We now have reduced the archive size to five individuals as desired. Note that this method always retains the extreme individuals in the archive (a and h in this example).

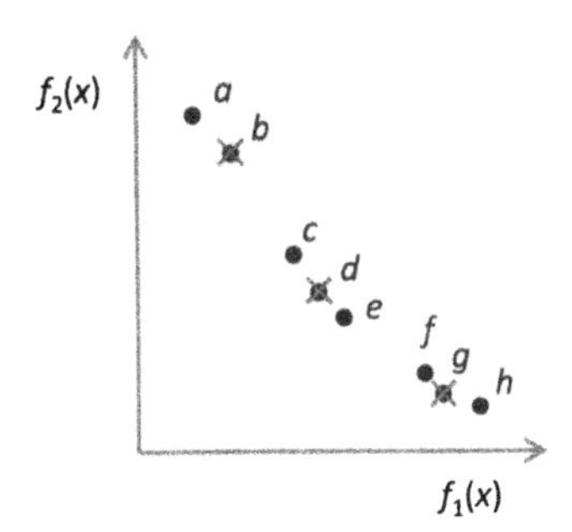

Figure 20.22 Example 20.7: This figure illustrates how an archive of nondominated solutions is reduced to fewer individuals in SPEA2. In this figure, adapted from [Zitzler et al., 2004], the archive size is reduced from eight to five individuals by removing g, d, and b in that order.

□

■ EXAMPLE 20.8

The SPEA2 approach for limiting the archive size might result in the deterioration of the Pareto front approximation. Figure 20.23 shows an example of how this could happen. Figure 20.23(a) shows a Pareto front approximation containing four points. If we want to keep the archive size equal to three, then we discard point c since it is the most crowded individual. At some later generation, however, the EA might find the nondominated solution e, which it adds to the archive as shown in Figure 20.23(b). Now, since individual b is the most crowded individual in Figure 20.23(b), SPEA2 removes b from the archive while retaining e. We see that, in hindsight, we should have retained c in the population since it dominates the new archive point e. This indicates that although the SPEA2 distance approach may be a good method for pruning a set of individuals, it might be better to never discard nondominated individuals [Zitzler et al., 2004].

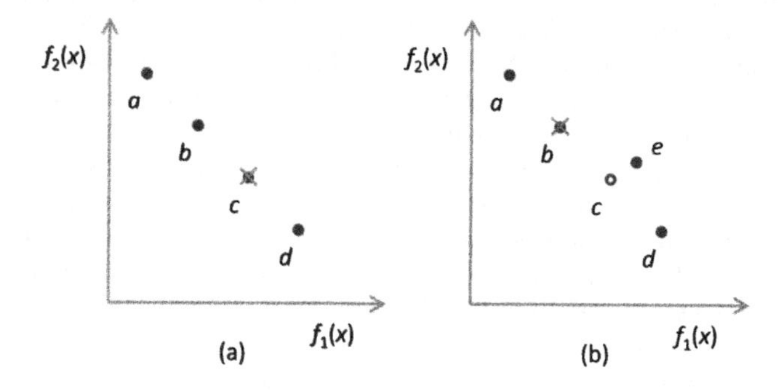

Figure 20.23 Example 20.8: This figure illustrates how the SPEA2 archive pruning process might inadvertently damage the Pareto front approximation. Individual c is discarded from the archive due to its crowdedness, and individual e is added to the archive later even though it would have been dominated by c, if c had been retained.

□

In SPEA, SPEA2, and any other MOEA that includes an archive, we can use the archive in a few different ways. First, we could simply use it to store nondominated solutions; this approach provides the highest degree of segregation between the MOEA population and its archive. Second, we could copy individuals from the archive to the child population at the end of each generation; this approach allows some interaction between the population and the archive. Third, we could allow the archive to participate in the selection process so that not only the population, but also the archive, is involved in recombination; this approach, which SPEA and SPEA2 use, provides the highest degree of interaction between the population and the archive.

20.4.7 The Pareto Archived Evolution Strategy (PAES)

PAES was introduced in [Knowles and Corne, 2001] and is motivated by the (1+1) evolution strategy (see Chapter 6). At each generation a single individual produces a single child using mutation. Each time a parent produces a child, we add the child to the archive if it is not dominated by any of the individuals in the archive. If the archive size exceeds a threshold, then we prune the archive by removing the individual that has the smallest crowding distance (that is, the individual that is in the most crowded region of the search space or objective function space). The original PAES uses grids in objective function space to compute the crowding distance. This is similar to the ϵ boxes of ϵ-MOEA (see Section 20.4.2). It is also similar to the clustering approach of [Parks and Miller, 1998], which does not add individuals to the archive unless they are sufficiently different from the individuals currently in the archive. Figure 20.24 shows an outline of a general PAES.

N_A = upper bound on archive size
Randomly generate a population of candidate solutions $P = \{x_j\}$ for $j \in [1, N]$
Initialize the archive A as the empty set
While not(termination criterion)
  Select a parent x from P
  Mutate x
  If x is not dominated by any individuals in A then
    Add x to A: $A \leftarrow \{A \cup x\}$
  End if
  If $|A| > N_A$ then
    Compute a crowding distance $s(\alpha)$ for all $\alpha \in A$
    $\alpha_{\min} \leftarrow \arg\min_\alpha s(\alpha)$
    Remove $\alpha_{\min}$ from A
  End if
Next generation

Figure 20.24 Outline of the Pareto archived evolution strategy (PAES). $s(\alpha)$ is the crowding distance of α, and is small for individuals in crowded regions of the search space or objective function space.

20.5 MULTI-OBJECTIVE BIOGEOGRAPHY-BASED OPTIMIZATION

This section shows how biogeography-based optimization (BBO), which is discussed in Chapter 14, can be combined with some of the previously-discussed MOEA approaches of this chapter. The combination of the BBO approach with various MOEA approaches results in several multi-objective biogeography-based optimization (MOBBO) algorithms. We then present a comparative study of these MOBBO algorithms on some standard multi-objective benchmark problems. This section could serve as a template for the extension of any other EA to multi-objective optimization.

20.5.1 Vector Evaluated BBO

In this section we discuss how to combine BBO with VEGA (Section 20.3.2). Recall that Figure 20.8 presents VEGA for a k-objective optimization problem. Since BBO is based on migration, we propose basing multi-objective BBO immigration on the k_i-th objective function value of each individual, where k_i is a random objective function index at the i-th migration trial. We then propose basing emigration on the k_e-th objective function value of each individual, where k_e is also a random objective function index. This results in the vector evaluated biogeography-based optimization (VEBBO) algorithm of Figure 20.25.

Initialize a population of candidate solutions $P = \{x_j\}$ for $j \in [1, N]$
While not(termination criterion)
  Compute the cost $f_i(x_j)$ for each objective i and for each individual $x_j \in P$
  $r_{ji} \leftarrow$ rank of x_j with respect to i-th objective function, $j \in [1, N]$, $i \in [1, k]$
  Immigration rates $\lambda_{ji} \leftarrow r_{ji} \Big/ \sum_{q=1}^{N} r_{qi}$ for $j \in [1, N]$, $i \in [1, k]$
  Emigration rates $\mu_{ji} \leftarrow 1 - \lambda_{ji}$ for $j \in [1, N]$, $i \in [1, k]$
  For each individual x_j, where $j \in [1, N]$
    For each independent variable $s \in [1, n]$
      $k_i \leftarrow \text{rand}(1, k) =$ uniformly distributed integer
      $r \leftarrow \text{rand}(0, 1)$
      If $r < \lambda_{j,k_i}$ then perform immigration
        $k_e \leftarrow \text{rand}(1, k) =$ uniformly distributed integer
        Probabilistically select emigrant x_e, where
          $\Pr(x_e = x_m) = \mu_{m,k_e} \Big/ \sum_{q=1}^{N} \mu_{q,k_e}$ for $m \in [1, N]$
        $x_j(s) \leftarrow x_e(s)$
      End immigration
    Next independent variable
  Next individual x_j
  Probabilistically mutate the population P
Next generation
$\hat{P}_s \leftarrow$ nondominated elements of P

Figure 20.25 Outline of vector evaluated biogeography-based optimization (VEBBO) for solving an n-dimensional optimization problem with k objectives. At each generation, the best individual x_b with respect to the i-th objective value has rank $r_{bi} = 1$, and the worst individual x_w has rank $r_{wi} = N$.

20.5.2 Nondominated Sorting BBO

Now we discuss how to combine BBO with NSGA (see Section 20.4.3). Recall that Figure 20.16 presents NSGA. To modify Figure 20.16 for BBO, we only need to change the recombination statement, "$C \leftarrow N$ children created from recombining ..." to a BBO migration operation. This results in the nondominated sorting biogeography-based optimization (NSBBO) algorithm of Figure 20.26.

Immigration rates $\lambda_j \leftarrow \phi(x_j) \Big/ \sum_{q=1}^{N} \phi(x_q)$ for $j \in [1, N]$
Emigration rates $\mu_j \leftarrow 1 - \lambda_j$ for $j \in [1, N]$
For each individual x_j, where $j \in [1, N]$
    For each independent variable $s \in [1, n]$
        $r \leftarrow \text{rand}(0, 1)$
        If $r < \lambda_j$ then
            Probabilistically select emigrant x_e, where
                $\Pr(x_e = x_m) = \mu_m \Big/ \sum_{q=1}^{N} \mu_q$ for $m \in [1, N]$
            $x_j(s) \leftarrow x_e(s)$
        End immigration
    Next independent variable
Next individual x_j
Child population $C \leftarrow \{x_j\}$

Figure 20.26 Outline of the migration portion of nondominated sorting biogeography-based optimization (NSBBO) for solving an optimization problem with n independent variables, k objectives, and a population size of N. This pseudo-code replaces the line "$C \leftarrow N$ children created from recombining" in Figure 20.16.

20.5.3 Niched Pareto BBO

This section proposes a simple way to combine BBO with NPGA (see Section 20.4.5). Recall that Figure 20.19 presents NPGA. Similar to the NSBBO algorithm of the previous section, to modify Figure 20.19 for BBO we only need to change the recombination statement "Recombine the individuals in R" to a BBO migration operation. Since NPGA already selects the individuals in R on the basis of non-domination, we can simply use equal migration rates for each individual in R to select migration operations. This results in the niched Pareto biogeography-based optimization (NPBBO) algorithm of Figure 20.27.

At this point we mention that we could combine the MOBBO algorithms discussed in this section with all of the possible BBO variations discussed in Chapter 14. For example, we could use emigration-based BBO rather than immigration-based BBO. We could use nonlinear migration curves. We could use blended migration. We could migrate groups of independent variables rather than one at a time (see Section 14.5). We could use a temporary population for migration so that we do not change any of the emigrating individuals until all of the migrations are completed. In general, all of the BBO extensions, variations, and hybridizations discussed in the literature could be combined with all of the MOBBO approaches discussed in this section to obtain many MOBBO algorithms. We could also make such efforts using any of the other EAs discussed in this book. An especially fruitful research direction would be to extend the many new EAs that have been recently introduced, including those discussed in Chapter 17, to multi-objective optimization.

For each individual $x_j \in R$, where $j \in [1, N]$
    For each independent variable $s \in [1, n]$
        $r \leftarrow \text{rand}(0, 1)$
        If $r < 1/N$ then
            Probabilistically select emigrant x_e, where
                $\Pr(x_e = x_m) = 1/N$ for $m \in [1, N]$
            $x_j(s) \leftarrow x_e(s)$
        End immigration
    Next independent variable
Next individual x_j
Child population $\leftarrow \{x_j\}$

Figure 20.27 Outline of the migration portion of niched Pareto biogeography-based optimization (NPBBO) for solving an optimization problem with n independent variables, k objectives, and a population size of N. This pseudo-code replaces the line "Recombine the individuals in R to obtain N children" in Figure 20.19.

20.5.4 Strength Pareto BBO

This section proposes a method for combining BBO with SPEA or SPEA2 (see Section 20.4.6). Recall that Figure 20.21 presents SPEA and SPEA2. To modify Figure 20.21 for BBO, we need to change the "Select parents" statement and the "Use a recombination method" statement. We can do this by calculating migration rates with the raw cost of Equation (20.46) for SPEA, or with the modified cost of Equation (20.50) for SPEA2. We can then implement BBO migration using these rates. In this section we take the SPEA approach in which parents can be selected from both the population P and the archive A. This results in the strength Pareto biogeography-based optimization (SPBBO) algorithm of Figure 20.28.

20.5.5 Multi-Objective BBO Simulations

Here we present simulation results for the MOBBO algorithms presented in the previous subsections. For each algorithm we use a population size of 100 and a generation limit of 1000. We use a mutation rate of 1% per independent variable per generation, and we use uniform mutation centered at the middle of the search domain (see Section 8.9). We check the population for duplicates every 100 generations and replace any duplicates that we find with randomly-generated individuals (see Section 8.6.1).

We incorporate elitism in VEBBO, NSBBO, and NPBBO by examining the population at each generation for nondominated individuals. If we find nondominated individuals, we replace the worst individuals in the population (in terms of nondomination level as shown in Figure 20.16) with two randomly chosen nondominated individuals from the previous generation.

We do not use elitism with SPBBO because SPBBO stores nondominated individuals in the archive (see Figure 20.28). For SPBBO, when we move nondominated individuals from the population P to the archive A, we replace those individuals

---

Use Equation (20.45) to calculate the strength $S(\alpha)$ of each individual $\alpha \in A$
Calculate the cost $R(\alpha) \leftarrow 1 - S(\alpha)$ for each $\alpha \in A$
Use Equation (20.46) to calculate the cost $R(x)$ of each individual $x \in P$
Immigration rates: $\lambda_j \leftarrow R(x_j) \Big/ \sum_{q=1}^{|P|} R(x_q)$ for all $x_j \in P$
Emigration rates: $\mu_j \leftarrow 1 - R(x_j) \Big/ \sum_{q=1}^{|P|+|A|} R(x_q)$ for all $x_j \in \{P, A\}$
For each individual $x_j \in P$
    For each independent variable $s \in [1, n]$
        $r \leftarrow \text{rand}(0, 1)$
        If $r < \lambda_j$ then
            Probabilistically select emigrant x_e, where
                $\Pr(x_e = x_m) = \mu_m \Big/ \sum_{q=1}^{|P|+|A|} \mu_q$ for $x_m \in \{P, A\}$
            $x_j(s) \leftarrow x_e(s)$
        End immigration
    Next independent variable
Next individual x_j

Figure 20.28 Outline of the migration portion of strength Pareto biogeography-based optimization (SPBBO) for solving an optimization problem with n independent variables. This pseudo-code replaces the six lines starting with "Use Equation (20.46)" and ending with "Use a replacement method" in Figure 20.21. Note that immigration occurs in the individuals in P, while emigration occurs from the individuals in $P \cup A$.

in P with randomly-generated individuals so that $|P|$ is maintained at 100. We do not use a lower bound on $|A|$, but we limit the maximum value of $|A|$ to 100 by using a simple clustering algorithm as described in Section 20.4.6.

We test the four MOBBO algorithms on some of the unconstrained multi-objective benchmarks of Appendix C.3, each with 10 dimensions, and selected because of their variety. We use problem U01 because of its convex Pareto front; problem U04 because of its concave Pareto front; problem U06 because of its discontinuous Pareto front; and problem U10 because it has three objectives (the other benchmarks that we test have only two objectives).

We evaluate the performance of the algorithms with two metrics: the reference-point hypervolume S' of Equation (20.22), and the normalized reference-point hypervolume S'_n of Equation (20.23). These metrics do not take into account diversity, but they do take into account the closeness of the approximated Pareto front to the true Pareto front. The reference-point hypervolume S' also takes into account the number of points in the approximated Pareto front.

Table 20.1 shows the results, averaged over 10 Monte Carlo simulations. We see that, in general, SPBBO performs the best. However, for the discontinuous U06 function, VEBBO performs the best in terms of total hypervolume, while NSBBO performs the best in terms of normalized hypervolume. That is, VEBBO finds the best combination of Pareto front quality and quantity, while NSBBO finds the best quality. For the more complex U10 function, although SPBBO finds the best total hypervolume, NSBBO finds better quality Pareto front points. In summary, we can say that SPBBO generally performs the best because of its archive, but there are no guarantees that it will perform the best for any specific problem.

		U01	U04	U06	U10
VEBBO	Hyper	(8.02, 8.93)	(2.73, 2.51)	(**63.53**, 59.99)	(299.68, 444.24)
	Norm	(1.28, 1.98)	(0.19, 0.18)	(21.18, 19.95)	(76.72, 103.07)
NSBBO	Hyper	(5.76, 8.01)	(3.27, 3.51)	(56.51, 48.44)	(373.51, 599.38)
	Norm	(1.26, 1.69)	(0.21, **0.22**)	(**21.94**, 20.09)	(101.40, **118.89**)
NPBBO	Hyper	(9.23, 18.78)	(2.95, 3.05)	(57.92, 48.31)	(419.96, 577.28)
	Norm	(1.18, 2.02)	(0.15, 0.15)	(20.01, 20.09)	(58.65, 57.33)
SPBBO	Hyper	(13.87, **33.10**)	(4.48, **4.60**)	(14.82, 18.33)	(934.29, **3884.32**)
	Norm	(0.90, **2.24**)	(**0.22**, **0.22**)	(3.47, 4.08)	(90.65, 108.08)

Table 20.1 Multi-objective BBO results on four 10-dimensional benchmark functions. The table shows the relative hypervolume, and normalized relative hypervolume, using linear BBO migration (the first number in each pair) and using sinusoidal migration (the second number in each pair). The best performance for each benchmark with respect to relative hypervolume and normalized relative hypervolume is shown in **boldface** font. See Section 14.4.1 for a discussion of linear migration vs. sinusoidal migration in BBO.

20.6 CONCLUSION

This chapter is not intended to provide a complete exposition of the subject of MOEAs, but has only covered some of the most popular MOEAs and associated ideas. Many other MOEAs have been proposed, and new ones are continually appearing in the literature. MOEA books include [Sakawa, 2002], [Collette and Siarry, 2004], [Coello Coello et al., 2007], [Deb, 2009], and [Tan et al., 2010]. Also, swarm-based approaches such as particle swarm optimization (see Chapter 11) are becoming popular for MOPs [Banks et al., 2008].

[Coello Coello, 2006] gives an interesting, high-level, historical view of MOEAs. The earliest technical survey on MOEAs is [Fonseca and Fleming, 1995], and additional surveys are provided in [Coello Coello, 1999], [Van Veldhuizen and Lamont, 2000], [Zitzler et al., 2004], and [Konak et al., 2006]. Although some of these papers are quickly aging due to the rapid expansion of MOEA research, they are all still very helpful and full of useful insights into the fundamental issues related to MOEAs.

We have seen in this chapter that diversity is an important consideration in MOEAs. Some approaches for increasing diversity include fitness sharing, grids, clustering, crowding, entropy, and mating restriction [Fonseca and Fleming, 1995]. Diversity can be an important consideration in single-objective EAs also, as discussed in Section 8.6. All of the diversity-seeking mechanisms of that section can be applied to the MOEAs discussed in this chapter.

Some important topics for future MOEA research include the following.

- Automatic on-line adaptation of MOEA tuning parameters;
- Hybridization of MOEAs with local search strategies;
- MOEAs that can provide good performance with few function evaluations;
- MOEAs for many objectives (more than just two or three);
- The incorporation of user preferences in MOEAs;
- Conceptually new approaches to MOEA design that do not rely on standard Pareto ranking methods; and
- MOEA theory and mathematical models.

We discuss a few of these topics in the following paragraphs.

The second topic listed above, the incorporation of local search strategies in MOEAs, is an important topic. In particular, MOEAs can be hybridized with derivative-based algorithms (or other local search methods) to fine-tune the optimization results. Such algorithms are called memetic algorithms because they involve (at least implicitly) the use of problem-specific information in the hybridized algorithm. Memetic strategies seem to be used a lot in single-objective optimization [Ong et al., 2007], but they have not yet been used much in MOPs, although there are a few exceptions [Jaszkiewicz and Zielniewicz, 2006].

The problem of expensive fitness function evaluations for MOPs is important because those types of fitness functions often arise in real-world problems, and because MOPs often require many more fitness function evaluations than single-objective problems. Section 21.1 discusses expensive fitness functions in general, but there is also research on EAs that are specifically designed for MOPs with expensive fitness functions [Chafekar et al., 2005], [Eskandari and Geiger, 2008], [Knowles, 2005], [Santana-Quintero et al., 2010]. [Goh and Tan, 2007] discusses MOEAs for problems with noisy fitness function evaluations.

The design of MOEAs for many objectives (10 or more) is also an important area for future research. Some results have been published in this area, but the more challenging problem is not necessarily the approximation of the Pareto set but rather how to help human decision makers choose a solution from an MOEA's Pareto set approximation. Some research on many-objective problems emphasizes their special challenges [Fleming et al., 2005], but other research shows that it is actually easier to find a good Pareto set approximation for problems with many objectives [Schütze et al., 2011]. After some thought, this makes intuitive sense because the more conflicting objectives that we have, the more likely it is that some random candidate solution will give good performance on at least one of those objectives. [Van Veldhuizen and Lamont, 2000] shows that the more objectives we add to an MOP, the larger the Pareto set becomes.

However, even though a Pareto set approximation may be easier to find with more objectives, it will also require more candidate solutions. For example, if we suppose that 10 candidate solutions can give a good Pareto set approximation for a two-objective problem, then we probably need at least 100 individuals in the two-objective MOEA. This means that we might need 10^k individuals for a k-objective MOP, which means that we need, for example, 100,000 individuals for a relatively small five-objective MOP. So the problem with many-objective problems is not the theoretical difficulty of approximating the Pareto set, but the practical difficulties of computational effort, and of approximating high-dimensional surfaces with only a few points. [Schütze et al., 2011] provides a good review of current research on many-objective problems.

This leads us to the issue of user preferences. We often have predefined preferences when we solve an MOP. For example, we may assign greater importance to certain objectives than others, or we may assign greater importance to certain combinations of objectives. The user is not always interested in obtaining the entire Pareto set. If we could somehow incorporate user preferences into an MOEA, then we could guide the evolution to a user-preferred region of the search space or objective space. We would then be able to reduce the MOEA population size for many-objective problems since we would not need to approximate the entire Pareto set. Another approach for dealing with many-objective problems is to simply reduce the number of objectives, since such problems often have objectives that are correlated with each other [López Jaimes et al., 2009].

Pareto set approximation is difficult enough, but even if an EA can obtain a good approximation, how can a human decision maker choose from among the large set of potential solutions? Some of the MOEAs discussed in this chapter incorporate user preferences (see Section 20.3.1), but we have not dealt explicity with this topic. The first attempt to incorporate user preferences in an MOEA was proposed in [Tanaka and Tanino, 1992]. Since then, many other approaches have been proposed; see [Thiele et al., 2009] for a good review.

Theoretical results for MOEAs are sparse, and so there is a lot of room for contributions in this area. [Rudolph and Agapie, 2000] provides a preliminary Markov model for MOEAs, and a few other researchers have studied MOEA theory [Zitzler et al., 2010], but compared to single-objective EAs theoretical studies for MOEAs are sparse.

Finally, the reader who is interested in additional MOEA results and research should note that Carlos Coello Coello maintains an exhaustive and useful web-based bibliography of papers related to multi-objective evolutionary optimization. His bibliography included 4861 references as of August 2012 [Coello Coello, 2012b].

PROBLEMS

Written Exercises

20.1 Given a set of points and a MOP, is it always true that one point dominates the others?

20.2 Does every MOP have a Pareto set?

20.3 Figure 20.1 shows a sketch of a Pareto front for a MOP in which we desire to minimize both objectives. Sketch and explain a sample convex Pareto front for a MOP in which we desire to: (a) minimize f_1 and maximize f_2; (b) maximize f_1 and minimize f_2; (c) maximize both f_1 and f_2.

20.4 Consider the following points and objective function values for a multi-objective minimization problem:

$$\begin{array}{ll} f_1(x^{(1)}) = 1, & f_2(x^{(1)}) = 1, \\ f_1(x^{(2)}) = 1, & f_2(x^{(2)}) = 2, \\ f_1(x^{(3)}) = 2, & f_2(x^{(3)}) = 1, \\ f_1(x^{(4)}) = 2, & f_2(x^{(4)}) = 2. \end{array}$$

a) Which point dominates all the others?
b) Which point does $x^{(2)}$ and $x^{(3)}$ dominate?
c) Which point is nondominated?
d) Which point is Pareto optimal?

20.5 Consider the points in Problem 20.4. For what values of ϵ do $x^{(2)}$, $x^{(3)}$, and $x^{(4)}$ additively ϵ-dominate $x^{(1)}$? For what values of ϵ do they multiplicatively ϵ-dominate $x^{(1)}$?

20.6 Give an example of two points in a two-dimensional, two-objective minimization problem such that one point does not multiplicatively ϵ-dominate the other point for any value of ϵ.

20.7 Give an example of two Pareto fronts P_1 and P_2, both having the same number of points, for which the union of the P_2 hypervolumes is greater than that of P_1, but the intersection of the P_2 hypervolumes is less than that of P_1.

20.8 Consider the following four-point Pareto front approximation $f(x)$, and three-point Pareto front approximation $f(y)$, to a two-objective minimization problem:

$$\begin{array}{llll} f_1(x^{(1)}) = 3, & f_2(x^{(1)}) = 4, & f_1(y^{(1)}) = 1, & f_2(y^{(1)}) = 3 \\ f_1(x^{(2)}) = 3, & f_2(x^{(2)}) = 3, & f_1(y^{(2)}) = 4, & f_2(y^{(2)}) = 3 \\ f_1(x^{(3)}) = 2, & f_2(x^{(3)}) = 2, & f_1(y^{(3)}) = 4, & f_2(y^{(3)}) = 1 \\ f_1(x^{(4)}) = 5, & f_2(x^{(4)}) = 2. & & \end{array}$$

What is the coverage of x relative to y? What is the coverage of y relative to x? According to the relative coverage values, which Pareto front approximation is better?

20.9 Why do we need to assume that all of the objectives are non-negative in the product aggregation method of Equation (20.26)?

20.10 Explain the difference between elitism and an archive.

20.11 What is the largest number of individuals that can be stored in the archive of an ϵ-MOEA with two objectives, where $f_1 \in [0, f_{i,\max}]$ for $i \in [1, 2]$? What about one with three objectives?

20.12 Figure 20.17 illustrates a rectangle that we can use for the crowding distance calculation in NSGA-II. Suppose we instead use the rectangle that is the largest one that encloses x without including any other points. Sketch that rectangle.

20.13 Consider two points x and y in a two-dimensional objective function space. Suppose the NSGA-II crowding distance $d_1(x)$ is calculated using the nearest neighbors of x in each dimension, and the crowding distance $d_2(x)$ is calculated using the largest rectangle that encloses x without including any other points. Does $d_1(x) < d_1(y)$ imply that $d_2(x) < d_2(y)$?

20.14 Consider Figure 20.29, which illustrates individuals in a population and archive for a two-objective maximization problem using SPEA [Zitzler and Thiele, 1999]. What are the raw cost values of each individual in the population, and the strength values of each individual in the archive?

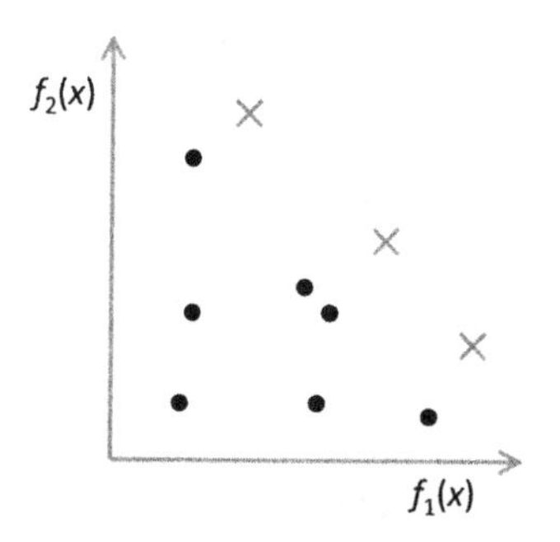

Figure 20.29 Problem 20.14: The circles are the individuals in the population, and the exes are the individuals in the archive.

20.15 Suppose we have three individuals in a two-objective minimization problem:

$$\begin{aligned} f_1(x_1) &= 3, \quad f_2(x_1) = 4, \\ f_1(x_2) &= 4, \quad f_2(x_2) = 3, \\ f_1(x_3) &= 2, \quad f_2(x_3) = 2. \end{aligned}$$

What are the VEBBO ranks r_{ji} in Figure 20.25?

20.16 How many references are listed on Carlos Coello Coello's web site "List of References on Evolutionary Multiobjective Optimization"?

Computer Exercises

20.17 Use exhaustive search with a search domain resolution of 0.01 to find the Pareto set and Pareto front to the problem

$$\min[\cos(x_1 + x_2),\ \sin(x_1 - x_2)]$$

where the search domain for each dimension is $[0, \pi]$.

20.18 Use the weighted sum approach to reduce the two objectives of Problem 20.17 to a single objective. For what values of the weights w_1 and w_2 does the solution of the single objective problem equal the Pareto front of the two-objective problem?

20.19 Run the gender-based EA of Figure 20.11 on a multi-objective problem. Test performance with the following variations: (a) allowing only one parent per subpopulation vs. allowing two parents per subpopulation; (b) allowing each child to be a member of only one subpopulation vs. allowing each child to be a member of more than one subpopulation; (c) replacing duplicate individuals each generation vs. not replacing duplicate individuals.

CHAPTER 21

Expensive, Noisy, and Dynamic Fitness Functions

Evolutionary algorithms often have to solve optimization problems in the presence of a wide range of uncertainties.

—Yaochu Jin and Jürgen Branke [Jin and Branke, 2005]

Anyone who worked on evolutionary algorithms before 1970 was ahead of his time. About a dozen individuals made fundamental contributions to EAs during those early years, and each of them could reasonably be dubbed "the father of evolutionary algorithms," or at least *one* of the fathers.[1] But all of this early work on evolutionary algorithms foundered to some extent because of the lack of computer resources. The EAs in the 1960s had to be very small and simple to run in a reasonable period of time. The computing power in the 1960s was simply not adequate to carry EA research or practice very far.

During the 1970s computing resources began to be more accessible and powerful, and by the 1980s EA research had rebounded from its doldrums to become an active area of investigation. A search of INSPEC, a computerized database of research articles in the area of computers and engineering, shows exactly one publication in the 1970s in the area of GAs, 37 publications in the 1980s, 7924 publications

[1] As far as I know, there were not any females working on EAs during those almost-prehistoric times. The fact that EAs has multiple fathers and no mothers seems appropriate in view of the flexibility of its biological foundations.

in the 1990s, and 35440 publications in the first decade of the 21st century.[2] The computing technology of today is sufficiently powerful so that anyone can write EA software on a desktop PC to solve interesting and challenging problems.

Mainframe computers in the 1960s topped out at clock speeds of 10 MHz. A typical desktop PC in the early 21st century has a clock speed close to 10 GHz, and even faster than that if we consider multiple cores. We have seen a three-order-of-magnitude increase in computational power from 1960 to 2010, but EAs can still take several days to run to completion. This is one reason why there is a strong emphasis in EA research on parallelization. But parallelization entails its own set of problem-specific challenges.

Overview of the Chapter

This chapter discusses how to reduce the computational cost of EAs. In the real world, cost function evaluations can be very computationally expensive. The benchmark functions that we have used so far in this book are simple, and we can evaluate them in a matter of milliseconds. But in the real world a cost function could take several days to evaluate, and in such cases we cannot afford to run an EA that requires thousands of function evaluations. Section 21.1 discusses how to handle expensive cost functions.

Related issues that we encounter in the real world include time-varying cost function evaluations, and noisy cost functions. Cost functions can change with time due to the dynamic and often unpredictable nature of our world, and so Section 21.2 discusses ways to handle dynamic optimization problems. Finally, cost functions are often noisy due to the lack of precision that is available in many problems, or due to the inherent ambiguity in determining the quality of a candidate solution, and so Section 21.3 discusses ways to handle noisy optimization problems.

21.1 EXPENSIVE FITNESS FUNCTIONS

In many real-world problems, a single fitness evaluation can require computation or experiments that take minutes, hours, days, or even longer. Here we discuss how to reduce the time for fitness evaluation to make EAs computationally less demanding.

Anyone who has worked with EAs for real-world applications knows from first-hand experience that fitness function evaluation is the most time-consuming aspect of the algorithm. This is not always the case for benchmarks or academic problems, but it is almost always the case for real-world problems. John Koza goes so far as to say that the computational effort required for fitness function calculation "is usually so great that it will rarely pay to give any consideration at all to any other aspect of the run" [Koza, 1992, Appendix H]. We find a similar statement in [Banzhaf et al., 1998, Section 11.1]: "Almost all of the time consumed by a GP run is spent performing fitness evaluations." Real-world problems often involve fitness functions that include one or more of the following characteristics [Knowles, 2005].

[2] INSPEC is a large, but not exhaustive, research database. Therefore, the numbers here provide a lower bound for the number of publications.

- A single fitness function evaluation requires minutes, hours, or days. This is especially true for fitness functions that must be evaluated experimentally rather than through simulation. Many of the earliest EAs were implemented in experimental systems because of the lack of simulation resources [Rechenberg, 1998], [Rechenberg, 1973].

- Fitness function evaluations cannot be parallelized. This is especially true for fitness functions that must be evaluated experimentally and with limited resources. For example, some optimization problems require experimental setups that are unique, that require extensive human interaction, or that are financially expensive. Sometimes we need human experts to evaluate the fitness of candidate solutions. Some fitness functions cannot be quantified, and instead require subjective evaluations by human experts. This is the case, for example, when generating algorithms that create music or art [Nierhaus, 2010].

- The number of fitness function evaluations is limited by time or some other resource constraints. This is the case for problems that must be solved by a certain deadline, problems for which the EA must run in real time, or fitness functions that must be evaluated experimentally by professionals with unique skills and full schedules.

Some ways to reduce the computational effort required for fitness function evaluations include the following.

- Do not recompute the cost of individuals that have already been evaluated. In many EAs, certain individuals may survive unchanged from one generation to the next. Individuals that are duplicated from generation i to generation $i+1$ do not have to be re-evaluated at generation $i+1$; we already know their cost values, assuming that the problem is not dynamic.

  This idea can be extended to keep track of all candidate solution vectors and their associated cost values as they are encountered during the entire EA. After each generation we store each individual and its cost value in an archive. If we have a fixed population of N individuals, then after T generations we will have an archive with NT individuals (minus duplicates). The archive is not involved in the evolutionary process; we use it only to avoid unnecessary cost evaluations. Every time we need to evaluate a cost, we first look through the archive to see if that specific individual has been evaluated in the past. After many generations, the archive could grow to be quite large, and searching through the archive before each cost evaluation could be expensive. But, depending on the problem, this search process might still be much less expensive than a cost evaluation.

- Fitness function evaluations can be truncated if it is seen that they perform very well or very poorly [Gathercole and Ross, 1997]. If we are halfway through a fitness function evaluation for an individual and we see that the individual is performing very well, then we can prematurely exit the evaluation routine, assign its fitness a high approximate value and save half of the computational effort for that evaluation. Similarly, if we are halfway through a fitness function evaluation and we see that the individual is performing very poorly, then we can prematurely exit the evaluation routine, assign its fitness a low approximate value, and again save half of the computational effort.

- If we need to perform cost function evaluations on a large set of test cases, then we can approximate the cost with a subset of the test cases. For example, suppose that we want to minimize $f(x) = \sum_{i=1}^{M} f_i(x)$ with respect to x. That is, the fitness function is a composite of several sub-fitness-functions. This is often the case when we want to optimize a function across several different operating regions. For example, we might want to optimize robot tracking performance for several different initial conditions and for several different tasks. One way to approach this is to minimize $f_1(x)$ with the first T generations. Then we can minimize $f_1(x)+f_2(x)$ for the next T generations. Then we can minimize $f_1(x) + f_2(x) + f_3(x)$ for the next T generations. We continue this process until we finally minimize $\sum_{i=1}^{M} f_i(x)$ during generations $(M-1)T$ through MT. Another approach is to minimize the combination of a random selection of $f_i(x)$ functions, gradually increasing the number of instances as the generation count increases. This approach is called stochastic sampling [Banzhaf et al., 1998, Section 10.1.5], and is similar to lexicographic ordering (see Section 20.3.3) and the ϵ-constraint method (see Section 20.3.4) in multi-objective optimization.

- If fitness function evaluations are conducted on computers, then standard methods of speeding up software execution can be used. These methods include preallocating arrays, using the optimized features of specific programming languages (for example, array operations rather than loops in MATLAB), reducing computer precision, using lookup tables for complicated functions, and disabling graphics and output operations.

In the remainder of this section we discuss some ways to approximate fitness functions (Section 21.1.1), how the transformation of a fitness function can improve approximation performance (Section 21.1.2), how to use fitness function approximations in an evolutionary algorithm (Section 21.1.3), when to use multiple fitness function approximations in an EA (Section 21.1.4), the danger of overfitting a fitness function approximation (Section 21.1.5), and how to assess the quality of a fitness function approximation (Section 21.1.6).

21.1.1 Fitness Function Approximation

We can create fitness function models to reduce fitness function evaluation effort. We can also use fitness function models to improve EA performance even if computational effort is not a bottleneck. We refer to such models as surrogates, response surfaces, or meta-models [Shi and Rasheed, 2010]. We use the term surrogate because a fitness model can be viewed as a temporary replacement for the more exact fitness evaluation. We use the term meta-model because often the fitness evaluation is itself only an approximation (for example, a simulation that models a physical process), and so a fitness function model is a reduced-order model of a higher-order model.

Suppose that we have a fitness function $f(x)$ that we have evaluated on M individuals $\{x_i\}$. We can use those M fitness function values to estimate the fitness at any point in the search space. We expect that the fitness estimate will have errors; after all, if the estimate were perfect, then we would never need to perform additional evaluations. However, even if the fitness estimate has errors, the errors might be small enough to make the estimate useful. Figure 21.1 illustrates the essential idea of fitness estimation. We generate an estimate $\hat{f}(x)$ on the basis of known fitness function values $f(x_i)$.

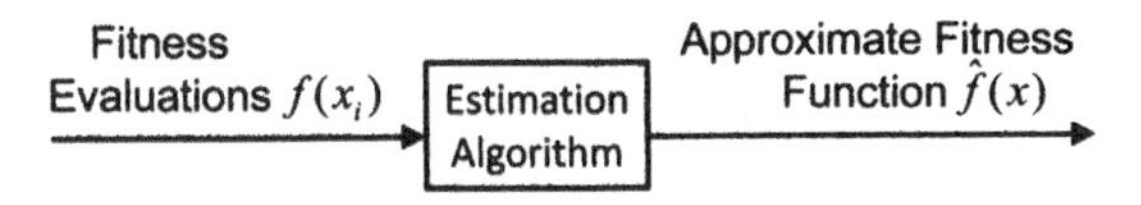

Figure 21.1 Fitness function estimation. We use exact fitness function values $\{f(x_i)\}$ to approximate $f(x)$ for $x \notin \{x_i\}$.

Fitness function approximation to reduce EA computational effort dates back to the 1960s [Dunham et al., 1963], when computational resources were much more scarce than today. Since then, researchers have tried many different algorithms for the estimation algorithm in Figure 21.1. In fact, we could try any approximation or interpolation algorithm. Approaches include the k-nearest neighbor algorithm, radial basis functions, neural networks, fuzzy logic, clustering, decision trees, polynomial models, kriging models, Fourier series, Taylor series, NK models, Gaussian process models, and support vector machines [Jin, 2005], [Shi and Rasheed, 2010]. We do not discuss the details of these approaches in this book, but suffice it to say that virtually any approximation algorithm can be used for EA fitness approximation.

One of the simplest estimation algorithms that we can use in Figure 21.1 is to approximate the fitness of an individual as the fitness of the nearest neighbor that has been evaluated. This approach is called fitness imitation and reduces to a piecewise constant approximation of the fitness landscape. Figure 21.2 illustrates fitness imitation.

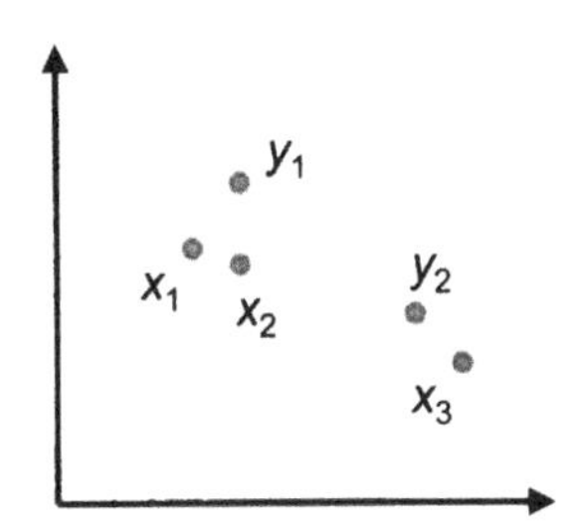

Figure 21.2 Fitness imitation in a two-dimensional search space. Individuals y_1 and y_2 have been evaluated with the fitness function routine or experiment, and so their fitness values are precisely known. Individuals x_1, x_2, and x_3 have not been evaluated. We can approximate their fitness values by assigning each one to the nearest evaluated fitness; that is, $\hat{f}(x_1) = f(y_1)$ and $\hat{f}(x_2) = f(y_2)$ and $\hat{f}(x_3) = f(y_3)$.

One aspect of Figure 21.1 that could be important is how to update the fitness estimate $\hat{f}(\cdot)$ as new data becomes available. We would like the fitness estimation algorithm to be recursive so that we can update $\hat{f}(\cdot)$ with minimal effort each time new fitness information becomes available. However, if we think that the fitness landscape might be dynamic, then we might want the estimation algorithm to discount old values of fitness data in its generation of $\hat{f}(\cdot)$. When we update a fitness approximation using new data, we call it online surrogate updating.

Another approach to fitness approximation involves assigning the fitness of an EA child on the basis of its parents' fitness values. This is called fitness inheritance [Smith et al., 1995]. We can approximate a child's fitness as the average of its parents' fitness values, or as a weighted average that depends on how similar it is to each parent. We can easily generalize this idea for EAs that use any number of parents for each child. We can also extend this idea by approximating a child's fitness as the weighted average of the fitness values of the entire population, again depending on how similar it is to each evaluated individual in the population [Sastry et al., 2001]. We can also use more sophisticated fitness inheritance ideas, such as taking into account the correlations between independent variables and fitness values [Pelikan and Sastry, 2004]. [Ducheyne et al., 2003] concluded that fitness inheritance is effective only for relatively simple problems. In particular, for multi-objective problems, fitness inheritance is effective only if the Pareto front is continuous and convex.

21.1.1.1 Polynomial Models The piecewise constant approximation of fitness imitation is a good starting point because it shows us how to extend fitness approximation to higher order polynomials. For example, we could approximate fitness as the linear function

$$\hat{f}(x) = a(0) + \sum_{k=1}^{n} a(k)x(k) \tag{21.1}$$

where n is the problem dimension, and $x(k)$ is the k-th element of individual x. This is a simple example of a polynomial model, and is also called a response surface. We can calculate the $a(k)$ values by solving the following problem:

$$\min \sum_{i=1}^{M} \left(f(x_i) - \left[a_0 + \sum_{k=1}^{n} a(k)x_i(k) \right] \right)^2 \tag{21.2}$$

where M is the number of individuals for which we have exact fitness values, x_i is the i-th individual for which we have an exact fitness value, and $x_i(k)$ is the k-th element of x_i. The minimization in Equation (21.2) is taken over the $(n+1)$ parameters $a(k)$ for $k \in [0, n]$. We can solve Equation (21.2) with a recursive least squares algorithm [Simon, 2006, Chapter 3]. In this way, as we obtain additional fitness values ($M = 1$, $M = 2$, and so on), only minimal computational effort is required to update the solution of Equation (21.2).

We can write a model that is more accurate than the linear one of Equation (21.1) as follows:

$$\hat{f}(x) = a(0) + \sum_{k=1}^{n} a(k)x(k) + \sum_{j,k=1}^{n} a(j,k)x(j)x(k). \tag{21.3}$$

This is a quadratic model with $(n^2 + n + 1)$ parameters. It can still be solved with recursive least squares because it is linear with respect to the model parameters $a(k)$ and $a(j,k)$. Once we understand the idea of polynomial modeling, we can experiment with various model forms such as

$$\hat{f}(x) = a(0) + \sum_{k=1}^{n} a(k)g(x(k)) + \sum_{j,k=1}^{n} a(j,k)h(x(j),x(k)) \tag{21.4}$$

where $g(\cdot)$ and $h(\cdot)$ can be any functions, linear or nonlinear. For instance, if we have some reason to believe that, for our particular optimization problem, fitness might be nicely represented by trigonometric functions, we could use sine and cosine terms for $g(\cdot)$.

We might want to use methods other than least squares to approximate a fitness model. For instance, instead of finding a model that solves Equation (21.2), we might prefer to find a model that solves

$$\min_{\{a(k)\}} \max_{i} \left| f(x_i) - \left[a_0 + \sum_{k=1}^{n} a(k)x_i(k) \right] \right| \tag{21.5}$$

where the minimization is again taken over the $(n+1)$ parameters $a(k)$. Figure 21.3 shows the difference between minimizing the sum of the squares of the estimation errors and minimizing the maximum estimation error. The least squares criterion of Equation (21.2) is attractive because it is easy to solve analytically, but a min-max criterion might be more robust because it finds the approximation that results in the smallest worst-case error. The min-max approximation will sacrifice approximation errors in easy-to-fit areas of the search space in order to reduce approximation errors in more challenging areas of the search space.

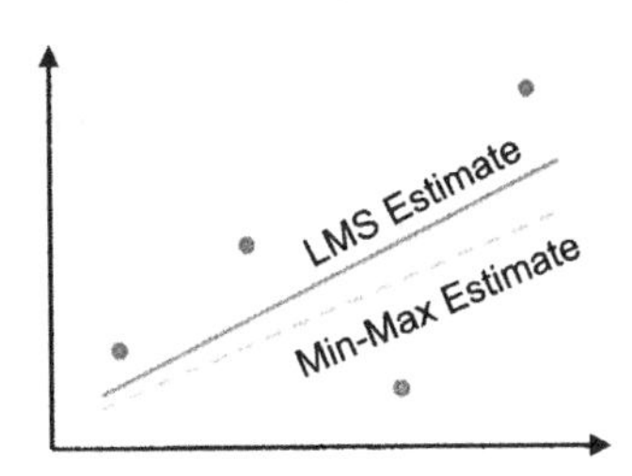

Figure 21.3 Least-mean-square straight-line approximation compared to min-max approximation. The least squares approximation can be solved analytically, but the min-max approximation might be more robust.

21.1.1.2 Design and Analysis of Computer Experiments Design and analysis of computer experiments (DACE) is a stochastic approximation method that includes diagnostic tests to measure the goodness of the approximation [Jones et al., 1998]. Given M fitness function evaluations $f(x_i)$ for n-dimensional vectors x_i, we assume that the fitness function can be approximated as

$$f(x) = \mu + \epsilon(x) \tag{21.6}$$

where μ is a constant (but not necessarily the mean of the evaluated fitness functions $f(x_i)$), and $\epsilon(x)$ is a correction term. DACE assumes that the correction term $\epsilon(x)$

is Gaussian with a mean of μ and a variance of σ^2 for all x; that is, the probability density function (PDF) of $f(x)$ is

$$\text{PDF}(f(x)) = \frac{1}{\sigma\sqrt{2\pi}} \exp\left[\frac{-(f(x)-\mu)^2}{2\sigma^2}\right]. \tag{21.7}$$

However, DACE also makes the important assumption that the $\epsilon(x)$ terms are *not* independent for different values of x; that is, correction terms should be similar for values of x that are similar. DACE assumes that the correlation coefficient ρ_{ij} between $f(x_i)$ and $f(x_j)$ can be expressed as follows:

$$\begin{aligned} d_{ij} &= \sum_{k=1}^{n} \theta_k \left|x_i(k) - x_j(k)\right|^{p_k} \\ \rho_{ij} &= \text{Corr}(f(x_i), f(x_j)) = \exp(-d_{ij}) \end{aligned} \tag{21.8}$$

where $x_i(k)$ is the k-th element of the i-th candidate solution, $p_k \in [1,2]$ and $\theta_k \geq 0$ are model parameters, and $d_{ij} \geq 0$ is a distance metric. We see that for small d_{ij}, x_i and x_j have a correlation that is close to 1. For large d_{ij}, x_i and x_j have a correlation that is close to 0. Given M fitness function evaluations, we collect them in a vector and parameterize them as shown in Equation (21.6):

$$\begin{aligned} f(x) &= \begin{bmatrix} f(x_1) & \cdots & f(x_M) \end{bmatrix}^T \\ &= \mu 1_M + \begin{bmatrix} \epsilon(x_1) & \cdots & \epsilon(x_M) \end{bmatrix}^T \end{aligned} \tag{21.9}$$

where 1_M is the M-element column vector in which each element is equal to 1. Note that we use the notation $f(x)$ to represent the fitness of a single candidate solution x, and also to represent the M-element vector containing the M fitnesses of $\{x_i\}$; the meaning should be clear from the context. The Gaussian PDF of the M fitness functions of Equation (21.9) is then given as

$$\text{PDF}(f(x)) = \frac{1}{(2\pi)^{M/2}|C|^{1/2}} \exp\left[-\frac{(f(x)-\mu 1_M)^T C^{-1}(f(x)-\mu 1_M)}{2}\right] \tag{21.10}$$

where C is the covariance matrix of $f(x)$. Recall that the covariance C_{ij} between two random variables $f(x_i)$ and $f(x_j)$ that have the same variance σ^2 is given as follows [Simon, 2006, Chapter 2]:

$$C_{ij} = \rho_{ij}\sigma^2. \tag{21.11}$$

Therefore, Equation (21.10) can be written as

$$\text{PDF}(f(x)) = \frac{1}{(2\pi)^{M/2}\sigma^M|R|^{1/2}} \exp\left[-\frac{(f(x)-\mu 1_M)^T R^{-1}(f(x)-\mu 1_M)}{2\sigma^2}\right] \tag{21.12}$$

where R is the correlation matrix; the element in the i-th row and j-th column of R is equal to ρ_{ij}.

Given a set of candidate solutions $\{x_i\}$ and a vector of fitness evaluations $f(x)$, we can find the μ and σ values that provide the best fit between the measured $f(x)$ value and the assumed parametric form of $f(x)$. Equation (21.12) gives the PDF of $f(x)$, which is proportional to the likelihood of obtaining a specific $f(x)$ given its assumed random nature. Therefore, to find the best fit between the assumed parametric form of $f(x)$ and the measured values of $f(x)$, we want to find the μ and σ values that maximize PDF($f(x)$) in Equation (21.12). First we consider maximization with respect to μ. We can maximize PDF($f(x)$) with respect to μ by minimizing the negative of the exponential argument with respect to μ. Taking the partial derivative of the negative of the exponential argument with respect to μ and setting it equal to 0 gives

$$\frac{\partial (f(x) - \mu 1_M)^T R^{-1} (f(x) - \mu 1_M)}{\partial \mu} = 0 \tag{21.13}$$

where we are ignoring the $2\sigma^2$ term in the denominator of Equation (21.12) since it is independent of μ. Solving Equation (21.13) gives

$$\begin{aligned} -2f^T(x)R^{-1}1_M + 2\mu 1_M^T R^{-1} 1_M &= 0 \\ \mu &= \frac{f^T(x)R^{-1}1_M}{1_M^T R^{-1} 1_M}. \end{aligned} \tag{21.14}$$

Taking the partial derivative of Equation (21.12) with respect to σ^2 gives

$$\frac{\partial \text{PDF}(f(x))}{\partial \sigma^2} = \frac{1}{2\sigma^2}\left[\frac{(f(x) - \mu 1_M)^T R^{-1}(f(x) - \mu 1_M)}{\sigma^2} - M\right] \text{PDF}(f(x)). \tag{21.15}$$

Setting the above equation equal to 0 gives

$$\sigma^2 = \frac{(f(x) - \mu 1_M)^T R^{-1}(f(x) - \mu 1_M)}{M}. \tag{21.16}$$

Equations (21.14) and (21.16) give the optimal values of μ and σ for fitness function approximation using DACE.

Now consider our M fitness function evaluations $f(x)$ of candidate solutions $\{x_i\}$. Suppose that the fitness functions are correlated as shown in Equation (21.8). Suppose that we obtain another fitness function evalution $f(x^*)$ for another candidate solution x^*. We augment the fitness function vector to obtain the $(M+1)$-element vector

$$\tilde{f}(x) = \begin{bmatrix} f^T(x) & f(x^*) \end{bmatrix}^T. \tag{21.17}$$

We can write the new correlation matrix as

$$\tilde{R} = \begin{bmatrix} R & r \\ r^T & 1 \end{bmatrix} \tag{21.18}$$

where r is the vector of correlations between the M fitness function evaluations $f(x)$ and the additional fitness function evaluation $f(x^*)$. We want to maximize the PDF of Equation (21.12) with respect to $f(x^*)$, which will give us the estimate of the form of Equation (21.6) that best fits the new data $f(x^*)$. This is called a maximum-likelihood estimate. We can maximize the PDF of Equation (21.12) by maximizing

$$(\tilde{f}(x) - \mu 1_M)^T \tilde{R}^{-1} (\tilde{f}(x) - \mu 1_M) = \begin{bmatrix} f(x) - \mu 1_M \\ f(x^*) - \mu \end{bmatrix}^T \begin{bmatrix} R & r \\ r^T & 1 \end{bmatrix}^{-1} \begin{bmatrix} f(x) - \mu 1_M \\ f(x^*) - \mu \end{bmatrix} \tag{21.19}$$

with respect to $f(x^*)$. We can use results from the matrix inversion lemma derivation [Simon, 2006, Chapter 1] to show that

$$\tilde{R}^{-1} = \begin{bmatrix} R & r \\ r^T & 1 \end{bmatrix}^{-1} = \frac{1}{1 - r^T R^{-1} r} \begin{bmatrix} R^{-1} + R^{-1} r r^T R^{-1} & -R^{-1} r \\ -r^T R^{-1} & 1 \end{bmatrix}. \quad (21.20)$$

Substituting this into Equation (21.19) gives

$$\frac{(f(x^*) - \mu)^2 - 2r^T R^{-1}(f(x) - \mu 1_M)(f(x^*) - \mu)}{1 - r^T R^{-1} r} + \text{terms without } f(x^*). \quad (21.21)$$

Since we want to maximize this expression with respect to $f(x^*)$, we take the derivative with respect to $f(x^*)$ and set it equal to zero to obtain

$$\frac{2(f(x^*) - \mu) - 2r^T R^{-1}(f(x) - \mu 1_M)}{1 - r^T R^{-1} r} = 0. \quad (21.22)$$

Solving for $f(x^*)$ gives

$$f(x^*) = \mu + r^T R^{-1}(f(x) - \mu 1_M). \quad (21.23)$$

This equation shows how we can use an existing model to approximate the fitness value of a new point x^*. The mean square error of the approximation is derived in [Jones et al., 1998] as

$$s^2(x^*) = \sigma^2 \left[1 - r^T R^{-1} r + \frac{(1 - 1_M^T R^{-1} r)^2}{1_M^T R^{-1} 1_M} \right]. \quad (21.24)$$

A couple of lines of algebra easily shows that $s(x) = 0$ at the sampled data points (see Problem 21.3) [Jones et al., 1998].

We can use the mean square error to determine suitable sampling points for additional fitness evaluations. There are two areas in the search space where we might be especially interested in obtaining additional fitness evaluations. First, we might be interested in sampling (that is, computing $f(x)$) near the minimum of the fitness approximation in the hope of finding a better solution to the optimization problem. Second, we might be especially interested in sampling in regions where $s(x)$ is large because we have a lot of uncertainty in those regions of the search space. Figure 21.4 illustrates this idea. Sampling additional fitness values near the minimum of the approximation is an exploitation strategy because it involves searching in areas where we already have good results. Sampling additional fitness values in regions of high mean square error is an exploration strategy because it involves searching in areas where we have little information about the fitness function.

Choosing sample points to increase modeling accuracy is called active learning. Active learning usually means we choose the sample points in a learning algorithm to optimize some cost function. In the DACE scenario described above we can choose sample points to reduce the maximum mean square error. Neural network training methods often include active learning [Settles, 2010].

We can obtain additional fitness function accuracy by estimating the optimal values of $\{p_k\}$ and $\{\theta_k\}$ from Equation (21.8). We do this by substituting Equations (21.14) and (21.16) into Equation (21.12), which gives us an expression for

PDF($f(x)$) that depends only on $\{p_k\}$ and $\{\theta_k\}$. We then maximize this expression with respect to $\{p_k\}$ and $\{\theta_k\}$, which gives us an estimate of the optimal correlation coefficients ρ_{ij}, which are the elements of R. We then use this value of R in Equations (21.14) and (21.16) to obtain the best estimates for μ and σ. Any time we get a new candidate solution x^*, we use Equation (21.23) to approximate its fitness.

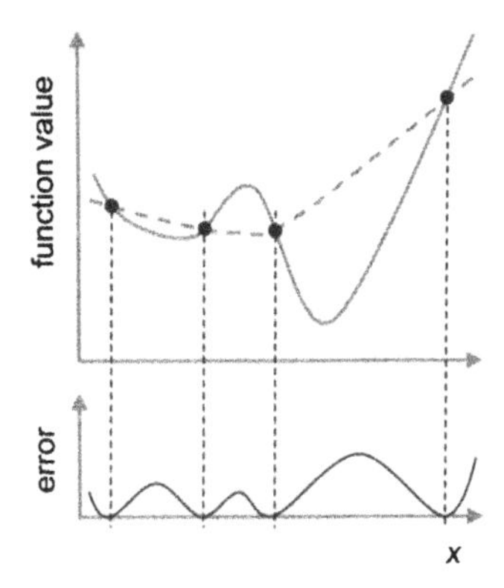

Figure 21.4 The solid line in the top figure represents a function, and the dashed line represents an approximation to it. The curve in the bottom figure represents the mean square error of Equation (21.24). We might be tempted to improve our approximation by sampling additional function values near the minimum of the approximation, but in this case we should sample where the error is highest because that is where the function minimum occurs.

■ EXAMPLE 21.1

We use DACE to estimate the two-dimensional Branin benchmark function

$$f(x) = (x(2) - (5/(4\pi^2))x(1)^2 + 5x(1)/\pi - 6)^2 + 10(1 - 1/(8\pi))\cos(x(1)) + 10 \quad (21.25)$$

where $x(1)$ and $x(2)$ are the two components of a candidate solution ($n = 2$). The domain of the function is $x(1) \in [-5, 10]$ and $x(2) \in [0, 15]$. First we have to decide which sample points to use. Here we arbitrarily decide to use 25 sample points that are evenly distributed in the two-dimensional search domain ($M = 25$). Next we use the `fmincon` function in MATLAB to maximize Equation (21.12) with respect to $\{p_k\}$ and $\{\theta_k\}$, which gives

$$\begin{aligned} p_1 &= 1.6194, & p_2 &= 2 \\ \theta_1 &= 0.020816, & \theta_2 &= 0.00018011. \end{aligned} \quad (21.26)$$

Next we use Equation (21.23) to approximate $f(x)$ on a fine grid. Figure 21.5 shows the results. We see that the approximation captures the essential shape of the Branin function pretty well, and most importantly it captures the function's multimodal characteristic.

□

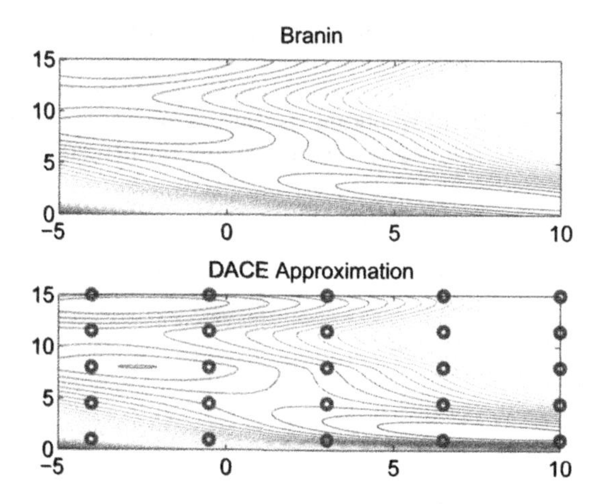

Figure 21.5 Example 21.1 results. The top figure shows the contour plot of the Branin function. The bottom figure shows 25 uniformly-distributed sample points, and the DACE-based approximation of the Branin function.

Example 21.1 used uniform sampling, but other sampling methods might give better approximation results. One popular method is Latin hypercube sampling. This method divides a domain into intervals in each dimension, and then places sample points in such a way that each interval in each dimension contains only one sample point. This approach to sampling can sometimes capture the unpredictable, unknown nature of a function better than uniform sampling. Figure 21.6 illustrates the difference between uniform sampling and Latin hypercube sampling.

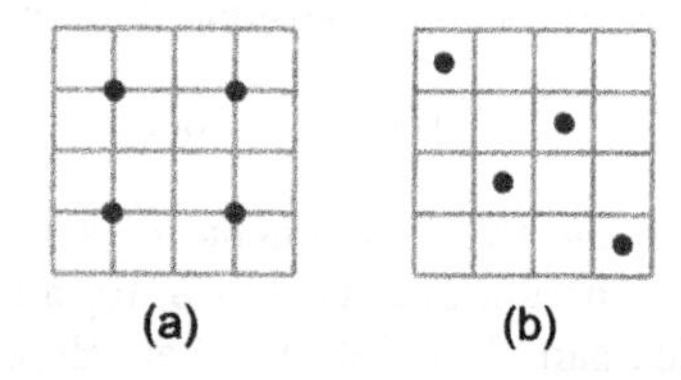

Figure 21.6 The figure on the left shows uniform sampling of four points in a search domain. The figure on the right shows Latin hypercube sampling. Note that there is only one point in each row, and only one point in each column. Several different arrangements of points have this characteristic, and so Latin hypercube sampling is non-unique.

■ EXAMPLE 21.2

We use DACE with Latin hypercube sampling to estimate the two-dimensional Branin function of Example 21.1. We arbitrarily decide to use 21 sample points ($M = 21$). Next we use MATLAB's `fmincon` function to maximize

Equation (21.12) with respect to $\{p_k\}$ and $\{\theta_k\}$, which gives

$$\begin{array}{ll} p_1 = 1, & p_2 = 2 \\ \theta_1 = 0.028227, & \theta_2 = 0.0013912. \end{array} \tag{21.27}$$

Next we use Equation (21.23) to approximate $f(x)$ on a fine grid. Figure 21.7 shows the results. Comparing Figures 21.5 and 21.7, it appears that Latin hypercube sampling gives a better approximation than uniform sampling. In fact, the RMS approximation error of the DACE approximation of Example 21.1, which uses uniform sampling, is 24.9, while the RMS approximation with Latin hypercube sampling is 14.3. Even with fewer sample points Latin hypercube sampling gives an approximation error that is almost 50% better than uniform sampling. We see that the sampling method can have a significant effect on DACE approximation results. Also, the method that we use to maximize Equation (21.12) to find optimal $\{p_k\}$ and $\{\theta_k\}$ values can have significantly affect DACE, although we do not give any examples here.

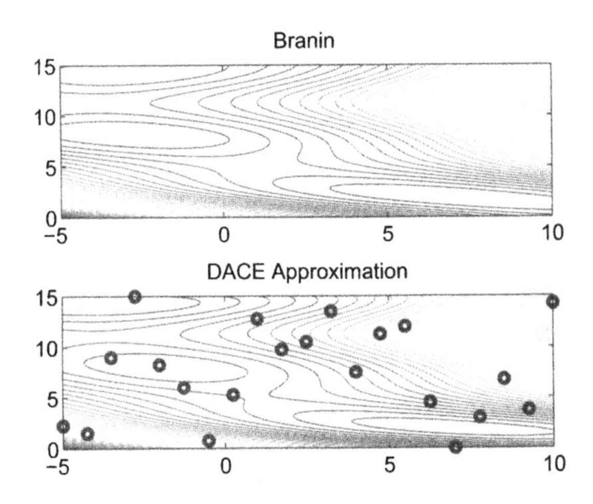

Figure 21.7 Example 21.2 results. The top figure shows the contour plot of the Branin function. The bottom figure shows the 21 sample points, which were obtained using Latin hypercube sampling, and the DACE-based approximation of the Branin function.

□

DACE is a generalization of the kriging algorithm. The kriging algorithm is an approximation method that was named after Daniel Gerhardus Krige, who developed it for geological applications [Krige, 1951]. Although it is named after a person, kriging is usually spelled with a lower-case "k." Kriging is the same as DACE, except that in kriging Equation (21.8) is replaced with

$$\begin{aligned} d_{ij} &= \sum_{k=1}^{n} \theta_k \left| x_i(k) - x_j(k) \right|^2 \\ \rho_{ij} &= \mathrm{Corr}(f(x_i), f(x_j)) = \exp(-d_{ij}). \end{aligned} \tag{21.28}$$

That is, p_k in Equation (21.8) is replaced with the constant 2 [Chung et al., 2003].

21.1.2 Approximating Transformed Functions

Sometimes fitness approximation methods do not perform well. For example, the DACE method of Equation (21.23) requires a matrix inverse, which may not exist. If the inverse does not exist, we can instead use the pseudo-inverse [Golan, 2007]. However, the main point here is that the basis functions that we use to approximate a fitness function may not be suitable for the shape of that fitness function. For example, if we use a Fourier series to approximate a function with irregular behavior and with sharp edges, then we cannot expect good approximation performance at all points in the function domain. In such cases we can transform the original fitness function and then find an approximation to the transformed function. For instance, suppose that we have evaluated a fitness function at M sample points $\{x_i\}$. If the approximation does not perform well, we can try transforming the fitness function samples $f(x_i)$ by taking their natural log:

$$L(x_i) = \log(f(x_i)). \tag{21.29}$$

We then find an approximation to $L(x)$ by using the sample points $L(x_i)$. We denote the approximation as $\hat{L}(x)$. We then invert the transformation to find the approximation for the original function:

$$\hat{f}(x) = \exp(\hat{L}(x)). \tag{21.30}$$

■ **EXAMPLE 21.3**

We use the DACE method of Section 21.1.1.2 on the Goldstein-Price function [Floudas and Pardalos, 1990]:

$$\begin{aligned} a &= 1 + (x(1) + x(2) + 1)^2 \times \\ &\quad (19 - 14x(1) + 3x(1)^2 - 14x(2) + 6x(1)x(2) + 3x(2)^2) \\ b &= 30 + (2x(1) - 3x(2))^2 \times \\ &\quad (18 - 32x(1) + 12x(1)^2 + 48x(2) - 36x(1)x(2) + 27x(2)^2) \\ f(x) &= ab \end{aligned} \tag{21.31}$$

where $x(1)$ and $x(2)$ are in the domain $[-2, 2]$. This function is very flat near its minimum, which occurs at $x^*(1) = 0$ and $x^*(2) = 1$. The function minimum is $f^* = 3$. The flat region defeats the DACE approximation method because the sample points are highly correlated in the flat region, which means some of the columns in R are comprised almost entirely of ones, which means that R is nearly singular. We might be able to overcome this problem by use a pseudo-inverse of R instead of the regular inverse of R. Instead we choose in this example to take the natural log of the sample points as shown in Equation (21.29). This greatly changes the shape of the function; it spreads apart values of $f(x)$ that are small but close together, and brings together values of $f(x)$ that are large, thus compressing the total range of the function while separating function values that are similar. We then use DACE to approximate $L(x)$, and then we calculate the approximation of the original function as shown in Equation (21.30). This simple modification to the approximation process gives us a decent approximation, as shown in Figure 21.8. The approximation does not look great, but at least it results in an invertible R matrix, and it also captures the large flat region in the middle of the domain.

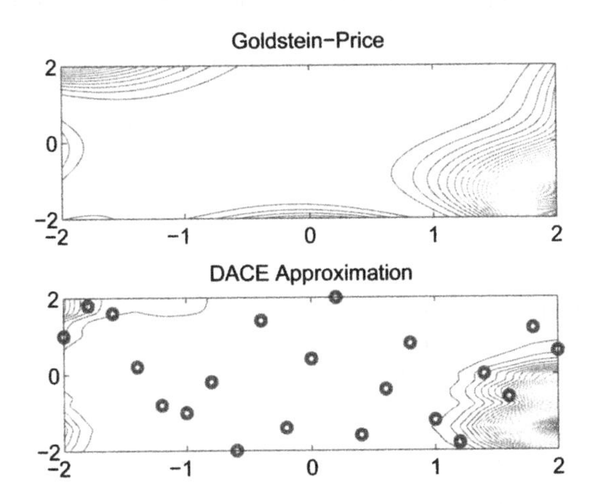

Figure 21.8 Example 21.3 results. The top figure shows the contour plot of the Goldstein-Price function. The bottom figure shows 21 sample points that were obtained using Latin hypercube sampling, and the DACE-based approximation of the function.

□

21.1.3 How to Use Fitness Approximations in Evolutionary Algorithms

Once we have a fitness approximation algorithm, we have several options for how to use it in an EA [Jin, 2005]. First, we could simply replace a fixed fraction r of fitness evaluations with fitness approximations. Assuming that fitness function evaluation is the dominant computational effort in the EA, this would reduce the computational effort from E to $(1-r)E$. However, this idea should not be taken too far. If we replace too many fitness evaluations with approximations, then the EA will take longer to converge, and our attempt at computational savings might be counterproductive. In the extreme case, if we replace all fitness evaluations with approximations, then $r = 1$. The computational effort could indeed be approximately zero in that case, but the EA would never converge to a useful result.

Another option is to create extra children each generation, and use their approximate fitness values to decide which ones to keep for the next generation. We call this idea evolution control, or model management. If we evaluate some individuals with the exact fitness function and other individuals with the approximate fitness function, we call it individual evolution control [Shi and Rasheed, 2010]. We can use various methods to decide which individuals to evaluate exactly and which ones to evaluate approximately. For instance, we could randomly decide which type of evaluation to use on each individual. We could also choose to use exact fitness evaluations only on individuals with good approximate fitness each generation. Individuals whose fitness we evaluate with the exact fitness function are called controlled individuals.

If we evaluate all individuals in certain generations with the exact fitness function, and all individuals in certain generations with the approximate fitness function, we have generation-based evolution control. We can use various methods to decide which generations to evaluate exactly, and which ones to evaluate approximately. For instance, we could deterministically decide to evaluate only every k-th generation with the exact fitness function, where k is a user-defined control parameter. Alternatively, we could randomly decide which type of evaluation to use at each generation. We could also use approximate fitness evaluations until we detect convergence (for example, the best individual has not improved for a certain number of generations, or the standard deviation of the population has fallen below some threshold), and then evaluate the next generation with the exact fitness function. After that, we would go back to approximate fitness evaluations. Generations in which we evaluate all individuals with the exact fitness function are called controlled generations.

Researchers have proposed several varieties of evolution control, including the dynamic approximate fitness based hybrid EA (DAFHEA) [Bhattacharya, 2008], which is illustrated in Figure 21.9.

N = population size
$N_c \leftarrow 5N$
Create N_c random individuals
Evaluate the fitness of the N_c individuals
Use the N_c fitness values to create a fitness approximation $\hat{f}(\cdot)$
Keep the best N individuals for the initial population
While not(termination criterion)
    Use an EA to create N_c children
    Use $\hat{f}(\cdot)$ to approximate the fitness of the children
    Save the N best children (according to $\hat{f}(\cdot)$) for the next generation
    If it is time for a new approximation then
        Compute the fitness $f(x_i)$ for each individual x_i
        Use the fitness values to update $\hat{f}(\cdot)$
    End if
Next generation

Figure 21.9 Outline of the dynamic approximate fitness based hybrid EA (DAFHEA).

DAFHEA typically implements elitism although this is not depicted in Figure 21.9. We could try several variations with Figure 21.9. For instance, we could try values other than $5N$ for N_c. We could try various algorithms for fitness approximation, although the original DAFHEA used support vector machines. We could try various methods to decide when it is time for a new approximation. Some common criteria for this decision are a fixed number of generations, or generating a new approximation when the EA meets some convergence criteria.

Also, we could use trust regions [Betts, 2009] to decide when to generate a new approximation. Trust region methods are based on comparisons of approximate fitness values to actual fitness values. If the approximate fitness values are close to the actual values, then the approximation is good so we can increase the time between generating new approximations. However, if the approximate fitness values are not close to the actual values, then the approximation is poor, so we need to decrease the time between generating new approximations. Suppose that G is the number of generations between computations of a new fitness approximation. Every generation we have N_c children in Figure 21.9. We calculate exact fitness values of N_e of the children, and compare their approximated fitness values with their exact fitness values. If the RMS difference exceeds a given threshold T^+, then we decrease G. If it falls below a given threshold T^-, we increase G. We use $T^+ > T^-$ (strict inequality) to prevent chattering in G. We need to tune the values of T^+ and T^- to obtain a good tradeoff between reduced computational effort (large G) and reasonably accurate fitness approximations (small G). We also need to tune N_e in this approach.

Figure 21.9 is fairly general. To be more specific, we can use fitness approximation to decide which N of the N_c initial individuals to keep in the initial population. This is called informed initialization, in which case we might want to base the fitness approximation on something other than the exact fitness evaluations of the N_c initial individuals. For example, we can construct a fitness approximation algorithm off-line before the EA begins execution.

We can also use fitness approximation in Figure 21.9 only for crossover (if our underlying EA is a GA), or only for migration (if our underlying EA is BBO), or only for a recombination algorithm that depends on the EA we are using. In this case we perform recombination (crossover, or migration, or some other EA-specific recombination method) to create many children (more than N), but we only keep the best N children, based on their approximate fitness values. This would be called, for example, informed crossover, or informed migration.

We can also use fitness approximation in Figure 21.9 only for mutation in the particular EA that we are using. In this case we create many (more than N) mutated versions of the N children, but we only keep the best N versions, based on their approximate fitness values. This is called informed mutation and is similar to the idea in Section 16.4 of implementing opposition-based learning only for the least fit individuals in a population.

We could try various algorithms for updating $\hat{f}(\cdot)$ at the end of DAFHEA depending on how much information we want to retain from the previous approximation. This could depend on how dynamic we believe the fitness function is. We could use multiple fitness approximation models and switch back and forth between the models depending on how accurate they are. This is similar to multiple model approximation as discussed in the next section. However, in the next section we focus on purposely combining low-accuracy models with high-accuracy models in our EA. Here we could try to use multiple models in an attempt to find the highest-accuracy model.

Finally, we can slightly modify Figure 21.9 so that we use the fitness approximation only to decide which children to keep for the next generation. This is called the informed operator approach [Rasheed and Hirsh, 2000] and is illustrated in Figure 21.10.

N = population size
$N_c \leftarrow 5N$
Create N_c random individuals
Evaluate the fitness of the N_c individuals
Use the N_c fitness values to create a fitness approximation $\hat{f}(\cdot)$
Keep the best N individuals for the initial population
While not(termination criterion)
    Use an EA to create N_c children
    Use $\hat{f}(\cdot)$ to approximate the fitness of the children
    Save the N best children (according to $\hat{f}(\cdot)$) for the next generation
    Compute the fitness $f(x_i)$ for each child x_i
    Use the fitness values to update $\hat{f}(\cdot)$
Next generation

Figure 21.10 Outline of the informed operator algorithm.

21.1.4 Multiple Models

We can perform approximate cost function evaluations during the early generations, and perform more accurate evaluations during later generations. This is similar to the idea of using a subset of test cases for cost function evaluation, which we discussed above, but here we use the same number of test cases while performing less accurate evaluations during the early EA generations. As a specific example, suppose that the cost function evaluation involves solving a Riccati equation:

$$P = FPF^T - FPH^T(HPH^T + R)^{-1}HPF^T + Q. \tag{21.32}$$

This type of equation often arises in control and estimation problems, and so is likely to appear in EAs that are trying to optimize a controller or estimator [Simon, 2006]. Given known square matrices F, Q, and R, and the possibly non-square matrix H, we need to solve for the square matrix P. The performance of a control or estimation algorithm is often proportional to the trace of P. A Riccati equation solver can be computationally expensive, but we can use approximate methods to obtain estimates of the solution [Emre and Knowles, 1987]. During the early EA generations we can use rough approximations for the solution of Equation (21.32), and during later generations we can use more accurate approximations.

Figure 21.11 illustrates this process. An EA with a low-accuracy fitness model runs for T_1 generations. After T_1 generations, the EA population is used to initialize the next EA, which uses a medium-accuracy model and runs for T_2 generations. After that EA concludes, its final population is used to initialize the final EA, which uses a high-accuracy model and runs for T_3 generations. This can be extended to as many model accuracy levels as desired. With this approach we need to take special care that each EA is initialized with a diverse population. We can accomplish this by taking measures to ensure that each EA population is sufficiently diverse before it migrates to the next highest level of fitness function approximation. We can also accomplish this by migrating only a few individuals from the lower-accuracy EA to the higher-accuracy EA, and initializing the rest of the higher-accuracy EA population in a way that achieves diversity.

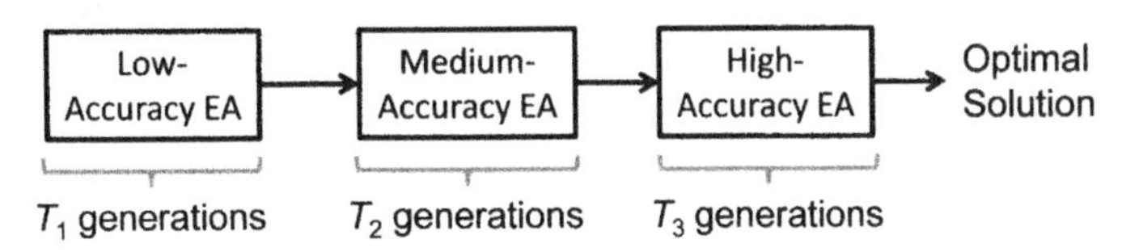

Figure 21.11 Multiple model fitness approximation. EAs with different levels of fitness function approximation run sequentially.

A more tightly integrated approach to multiple model optimization runs EAs with different levels of fitness function approximations in parallel. In this approach, individuals migrate between parallel EAs at specified frequencies [Sefrioui and Périaux, 2000]. This approach, called hierarchical evolutionary computation, includes a couple of different options. First, we could migrate individuals from EAs with higher-accuracy approximations, to EAs with lower-accuracy approximations, as shown in Figure 21.12. Second, we could migrate individuals back and forth between EAs with similar levels of fitness approximations as shown in Figure 21.13. Hierarchical EAs can be used with as many model accuracy levels as desired.

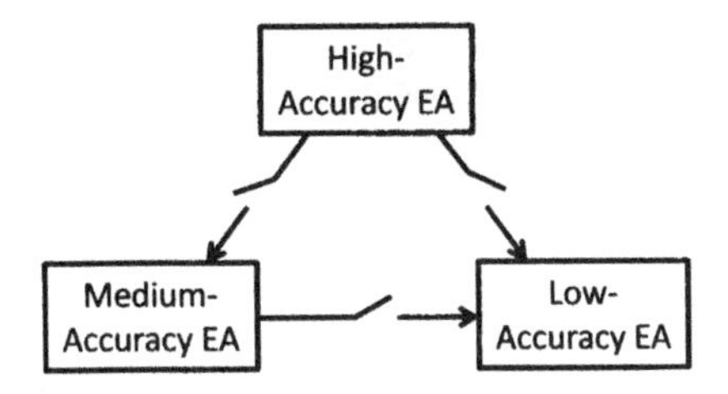

Figure 21.12 Hierarchical EA. This model migrates individuals from EAs with higher-accuracy fitness function approximations to EAs with lower-accuracy approximations. Migration occurs at user-defined frequencies as indicated by the switches in the figure.

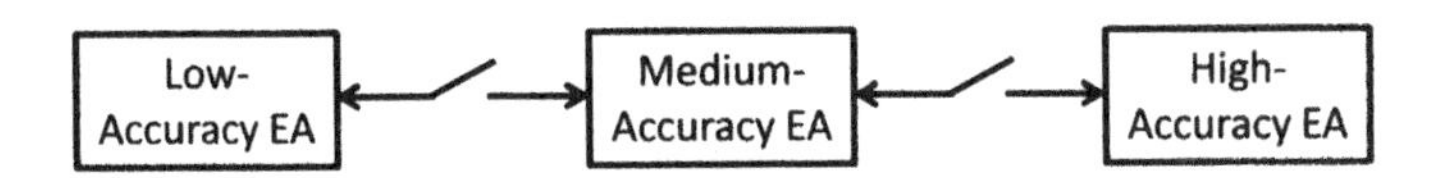

Figure 21.13 Hierarchical EA. This model migrates individuals between EAs with similar levels of fitness function approximations. Migration occurs at user-defined frequencies as indicated by the switches in the figure.

Another approach to using multiple models is to generate multiple models that can be used in various combinations for any individual x. For example, suppose that we have evaluated the fitness of M individuals $\{x_i\}$. We could use a clustering algorithm to divide $\{x_i\}$ in C clusters. Then we could create a fitness approximation model for each of the C clusters. This gives us $\hat{f}_k(x)$ for $k \in [1, C]$. Now when we want to approximate the fitness of an individual x, we can take one of several approaches. For instance we could approximate $f(x)$ as $\hat{f}_k(x)$, where index k is the cluster that is closest to x [Chung and Alonso, 2004]. Alternatively we could approximate $f(x)$ as a weighted combination of $\hat{f}_k(x)$, where the weights sum to 1 and are functions of how far x is from each cluster.

21.1.5 Overfitting

Overfitting can be be a problem in some fitness approximation approaches. The EA designer always needs to be leery of overfitting when using fitness approximation methods. Overfitting is often a problem in neural networks unless the engineer makes an intentional effort to avoid it [Krogh, 2008]. Figure 21.14 shows an example of overfitting for the simple problem of fitting a curve to a set of data points. Even though the higher-order polynomial in the figure matches the data better than the lower-order polynomials, the higher-order polynomial does not generalize very well. We could say that it memorizes the data points while not providing good performance in between the data points. Often we need to accept a higher fitting error at the data points in order to obtain better generalization performance.

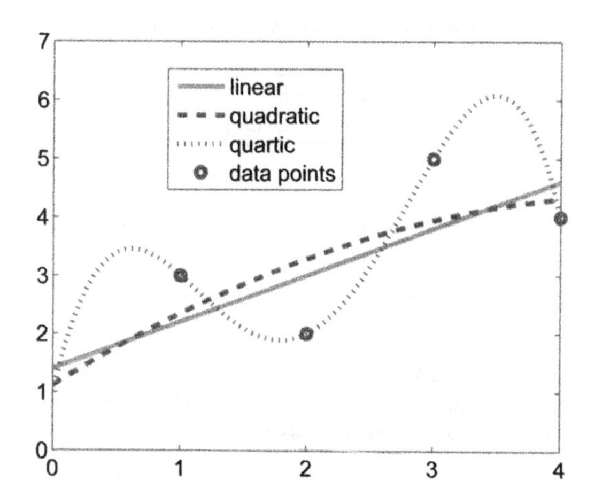

Figure 21.14 This figure shows an example of overfitting. A linear function and a quadratic function seem to fit the data pretty well. A quadratic function matches the data perfectly but includes large oscillations, which indicates that it does not generalize well.

Overfitting can be alleviated with ensemble techniques. An ensemble is a set of individually trained fitness approximations whose predictions are combined when estimating fitness values for previously unencountered points in the search space [Opitz and Maclin, 1999], [Lim et al., 2010].

21.1.6 Evaluating Approximation Methods

Once we have a function approximation, we need to verify that it gives good results before we rely on it in an EA. We should always start by checking the approximation values at the sample points (that is, the points we used to create the approximation). Many approximation methods automatically output functions $\hat{f}(x)$ that exactly match the true function $f(x)$ at the M sample points $\{x_i\}$; that is, $\hat{f}(x) = f(x)$ for $x \in \{x_i\}$. However, we should still check this to verify that we implemented the approximation algorithm correctly.

One method for assessing the accuracy of an approximation method is to choose a few additional sample points beyond those that we used for building our approximation. Say we choose Q additional sample points $\{x_i\}$, called test points, where $i \in [M+1, M+Q]$. We then evaluate the function and the approximation at the test points to see how well the approximation works. We measure RMS approximation error as

$$E_{\text{RMS}}^2 = \frac{1}{Q} \sum_{i=M+1}^{M+Q} (f(x_i) - \hat{f}(x_i))^2 \tag{21.33}$$

and we measure worst-case approximation error as

$$E_{\max} = \max_{i \in [M+1, M+Q]} |f(x_i) - \hat{f}(x_i)|. \tag{21.34}$$

We can use either metric to assess the quality of an approximation method depending on our priorities and our specific problem.

Another method, which is called cross validation or rotation estimation [Geisser, 1993], allows us to assess approximation quality without using additional sample points. As above, suppose that we have M sample points and M function values $f(x_i)$. In cross validation, we compute an approximation using all sample points except for the k-th one, and we call this approximation $\hat{f}_k(x)$. We thus compute M approximations, with each one leaving out one sample point. This gives us M approximations $\hat{f}_k(x)$ for $k \in [1, M]$, where each approximation uses a unique set of $(M - 1)$ sample points. We then evaluate each approximation at the sample point that we did not use when we constructed it. That is, we evaluate $\hat{f}_k(x_k)$ for $k \in [1, M]$. As above, we measure RMS or worst-case approximation error as

$$\begin{aligned} E_{\text{RMS}}^2 &= \frac{1}{M} \sum_{i=1}^{M} (f(x_i) - \hat{f}_i(x_i))^2 \\ E_{\max} &= \max_{i \in [1, M]} |f(x_i) - \hat{f}_i(x_i)|. \end{aligned} \tag{21.35}$$

After we use cross validation to convince ourselves that our approximation approach is correct, we use all M data points to find the function approximation $\hat{f}(x)$ that we will use in our EA.

We can use other metrics for evaluating the quality of a fitness approximation method if we have the resources to compare the approximate fitness values with the exact fitness values. These metrics include comparing the number of individuals that are selected for recombination when using the true fitness values versus the approximate fitness values; the ranks of the individuals that are incorrectly selected when using the approximate fitness values; correlations between true and approximate fitness values; and correlations between true and approximate fitness ranks [Jin et al., 2003].

The preceding sections discussed several different ways to approximate fitness functions. There are also many other methods that we have not discussed, and each method includes several variations and tuning parameters. It is not easy to define the "best" fitness function approximation method. In addition to the RMS and maximum error metrics of Equation (21.35), there are several criteria that we need to consider when evaluating fitness function approximation methods.

- How accurate is the approximation method for the given problem?
- Regardless of the accuracy of the approximation method, how well does an EA perform when using the approximation method? Note that the relative performance of different EAs might vary with different methods. For example, EA #1 might perform the best with approximation method A, while EA #2 might perform the best with approximation method B.
- How much does an approximation method reduce computational effort?
- How complex is the approximation method? This affects the maintainability, extendability, and portability of the code. How easy is the code to modify (maintainability)? How easy is it to add new functions or features (extendability)? How easy is it to port to other computing platforms or other optimization problems (portability)?

Elitism

Finally, we note that expensive fitness functions may be an exception to the rule of always using elitism in our EAs. Up to this point we have generally recommended elitism to save the best individual(s) each generation. However, EAs with few function evaluations may perform better without elitism than with elitism [Torregosa and Kanok-Nukulchai, 2002]. This is because when the population size is small, or the number of evaluations is small, exploration becomes relatively more important than exploitation. Non-elitist EAs can allow for increased exploration.

21.2 DYNAMIC FITNESS FUNCTIONS

Fitness functions often change with time; that is, they are nonstationary. Sometimes they change because the environment of a fitness evaluation experiment changes with time. For example, if we are trying to tune a robot controller, the environment or task of the robot could change with time. The robot parameters itself could also change with time as the sensors and electromechanical components age. Sometimes the requirements of the customer or client change with time – that is, people change their minds about what they want. Sometimes constraints change with time because resources are consumed or replenished. This section discusses how to use EAs to track a dynamic optimum.

■ EXAMPLE 21.4

This example does *not* illustrate a dynamic EA, but simply illustrates the effect on EA performance of dynamics in an optimization benchmark function. We use the simplified dynamic benchmark function generator of Figure C.24, and we use the Ackley function as the basis function. We insert dynamics into the function every 100 generations (that is, $E_{\text{update}} = 100$ generations in Figure C.24), and we use a problem dimension of 10. We run BBO (Chapter 14) with a population size of 50 and an elitism parameter of 2, and we replace duplicate individuals each generation with randomly-generated individuals. Figure 21.15 shows the performance BBO on the stationary Ackley function and the dynamic Ackley function, averaged over 20 Monte Carlo simulations. Performance is identical for the first 100 generations. But we see that for the dynamic function, BBO essentially has to start over every 100 generations because the function changes so drastically. This example illustrates the need for methods in EAs that can intelligently handle dynamic changes in the cost function.

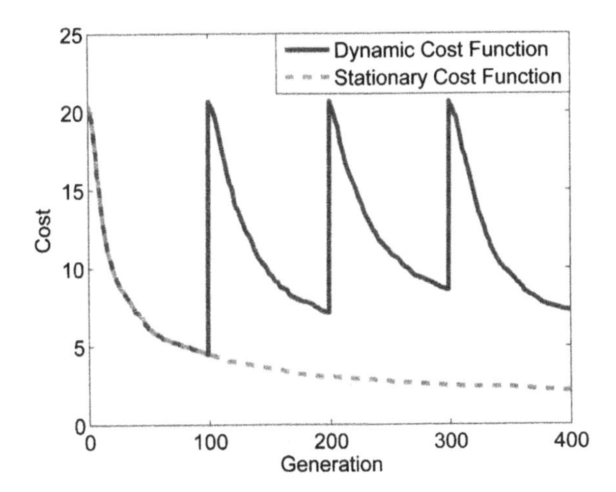

Figure 21.15 Example 21.4: This figure shows BBO performance on the 10-dimensional stationary and dynamic Ackley functions, averaged over 20 Monte Carlo simulations. Dynamics are inserted into the dynamic version of the benchmark every 100 generations, which causes BBO to lose all of the evolutionary progress that it made up to that point.

□

The first challenge in dynamic optimization is detecting a change in the fitness function landscape. We can detect such a change by using some marker individuals and evaluating their fitness values each generation. If their fitness values change significantly (beyond what would be expected from noise) from one generation to the next, then we can infer that the fitness landscape has changed; that is, the optimization problem has changed. But the use of markers is not a foolproof method. Detection of a landscape change is not an either-or proposition. The landscape may change at the marker locations while remaining fixed at the optimal point. Conversely, the landscape could change at the optimal point while remaining fixed at the marker locations. These difficulties are illustrated in Figure 21.16.

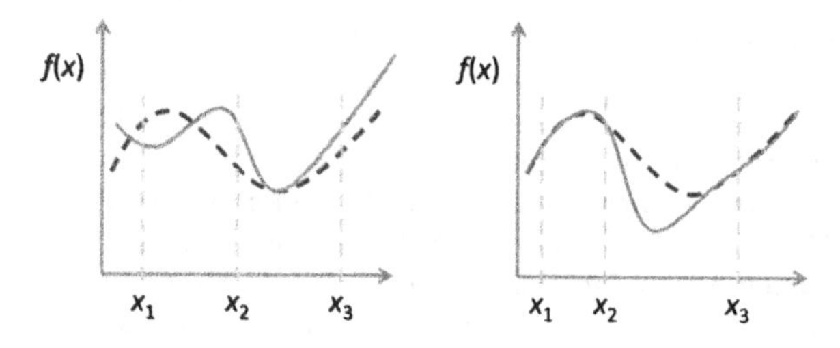

Figure 21.16 Detecting the change in a dynamic fitness function. The dashed line in each graph is the original fitness function, and the solid line is the new fitness function. The three points x_1, x_2, and x_3 are markers that we use to detect a fitness function change. The figure on the left shows that the even though the fitness of the markers has drastically changed, the optimum has not changed. The figure on the right shows the converse: even though the fitness of the markers has not changed, the optimum has changed dramatically.

After we detect a change in the fitness landscape, we can take one of several approaches to track the changing optimum. One possibility is to completely replace the old population with a new random population and restart the evolutionary optimization process. This is an extreme measure that does not reuse any of the information from the previous optimization process. This might be appropriate if the fitness landscape has changed drastically enough that the previous population does not contain any useful information about the new landscape. This essentially entails running a new EA on a new optimization problem.

However, for most practical problems, there is some resemblance between the old fitness landscape and the new landscape. That is, the fitness landscape changes gradually rather than drastically. In this case, we want to explore the new landscape while also exploiting the results of the EA's progress on the old landscape. For instance, we could retain most of the old population, but seed it with some new individuals in our attempt to explore the new landscape. We could also temporarily increase the mutation level to temporarily increase exploration. This approach is called hypermutation [Cobb and Grefenstette, 1993]. The number of new individuals that we introduce in the population, the amount by which we increase mutation, and the number of generations for which we increase mutation, all control the balance between exploration and exploitation. This balance determines how adaptable the EA is to new landscapes, and how much it relies on its past results.

In the following sections we discuss several dynamic EA approaches, including the predictive EA (Section 21.2.1), immigrant-based EAs (Section 21.2.2), and memory-based EAs (Section 21.2.3). We also discuss the challenge of evaluating the performance of EAs on dynamic optimization problems (Section 21.2.4).

21.2.1 The Predictive Evolutionary Algorithm

One approach to dynamic optimization is to combine a forecasting technique with the EA. This gives an algorithm called the predictive EA [Hatzakis and Wallace, 2006]. If an EA is running and its optimum is changing with time in a predictable way, then we can presumably create a model for how it is changing with time. Then when a change in the landscape is detected, we can use that model to seed new population members. The key condition here is that the optimum must be changing "in a predictable way." If this condition is not satisfied, then we may as well re-seed the entire population with a randomly initialized population and simply restart the EA. Figure 21.17 illustrates the basic idea of the predictive EA, and Figure 21.18 shows an example of the dynamic evolution of an EA optimum.

Initialize the EA population
$X^* \leftarrow \emptyset$
While not(termination criterion)
    Run an EA for T generations, or
        until a change in the fitness landscape is detected
    Denote the best individual in the EA as x^*
    $X^* \leftarrow \{X^*, x^*\}$
    Extrapolate the sequence X^* to estimate the new optimum $\hat{x}^*$
    Create a subpopulation S of individuals that are near $\hat{x}^*$
    Replace some of the individuals in the EA population with S
Next generation

Figure 21.17 Outline of a predictive evolutionary algorithm for dynamic optimization.

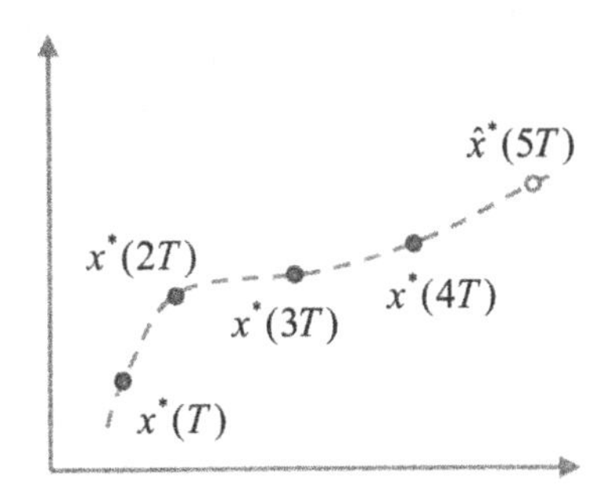

Figure 21.18 Example of the progression of an EA optimum in a two-dimensional space, and its predicted value. $x^*(T)$, $x^*(2T)$, $x^*(3T)$, and $x^*(4T)$ are the EA optima after T, $2T$, $3T$, and $4T$ generations. $\hat{x}^*(5T)$ is the predicted optimum that is used to seed the next EA.

Several implementation details in Figure 21.17 are left to the EA designer. For instance, if we run for a constant number of generations T between extrapolations, how should we decide the value of T? How can we detect a change in the fitness landscape? If we use marker individuals, how many should we use? If we use too few markers, we may miss a change. But if we use too many markers, we may waste time on unnecessary fitness function evaluations.

How can we use the set X^* in Figure 21.17 to estimate the new optimum? That is, what kind of extrapolation algorithm should we use? How can we create the subpopulation S that is near $\hat{x}^*$? We could simply set S equal to the one-element set $\{\hat{x}^*\}$. Another possibility is to set S equal to a set of M mutated versions of $\hat{x}^*$, where M is a user-defined constant. Yet another option is to set S equal to a deterministic set of M individuals in a hypercube or hypersphere surrounding $\hat{x}^*$. If we keep track of the accuracy of our $\hat{x}^*$ prediction from one EA execution to the next, we could use it to determine the size of the hypervolume. If we find that $\hat{x}^*$ is consistently accurate, then we can use an S with a small cardinality and with a small hypervolume. If we find that $\hat{x}^*$ is consistently inaccurate, then we should increase the cardinality of S and its hypervolume. In fact, we could use this idea to adaptively adjust the cardinality and hypervolume of S. Finally, we have to decide which individuals in the old population to replace with S in Figure 21.17. Common options are to either replace the worst individuals in the old population, or to replace randomly-selected individuals in the population.

21.2.2 Immigrant Schemes

There may be situations when we cannot detect changes in the fitness function, or when the fitness function is changing almost continuously. In this case we can continually introduce new individuals into the population in an attempt to make the EA robust to fitness landscape changes. Such algorithms are called immigrant schemes [Yu et al., 2009]. There are two basic approaches to immigrant schemes. A direct immigrant scheme uses the individuals in the population to create new individuals. This is similar to standard recombination and mutation algorithms, which use a parent population to create a child population. For example, a direct immigrant scheme could mutate or recombine elite individuals from the current or past generation to create new individuals. An indirect immigrant scheme creates new individuals based on a model of the population. For example, we could use a PBIL-type algorithm to model the population (see Section 13.2.3) and then create new individuals based on the model.

After we decide to use either a direct or indirect immigrant scheme, we need to decide the following.

1. How should we generate new individuals? We could generate a set of individuals X_e based on elites, as mentioned above. We could also generate a set of random individuals X_r. If we think the fitness landscape might be changing drastically, we could generate a set of dual individuals X_d, which are the same as the opposite individuals of Chapter 16. Finally, we could use a combination of these three options. If we keep track throughout the EA of how well each of these types of individuals perform, we could adapt the number of X_e, X_r, and X_d individuals that we introduce each generation [Yu et al., 2009].

2. How many new individuals should we introduce to the population? Most researchers introduce about $0.2N$ or $0.3N$ new individuals, where N is the population size. If we can detect the amount or frequency of change in the fitness landscape, we could adjust the replacement rate accordingly. For in-

stance, if we detect a large change in the fitness landscape, we might want to introduce more new individuals.

3. Which individuals in the population should we replace with the new individuals? One common answer to this question is to replace randomly-selected individuals. Another answer is to replace the worst individuals. One point to remember here is that new individuals might not be very fit in the population, but their fitness might improve with time as the landscape changes. Therefore, we might want to keep track of an age factor that prevents individuals from being replaced until after they exceed a certain age [Tinós and Yang, 2005].

N = population size
r_r = proportion of random individuals to create each generation
r_e = proportion of elite-based individuals to create each generation
r_d = proportion of dual individuals to create each generation
Create N initial individuals $\{x_i\}$
Evaluate the fitness of the N individuals
While not(termination criterion)
    Use a recombination/mutation method to evolve $\{x_i\}$ for the next generation
    Evaluate the fitness of the individuals $\{x_i\}$
    $X \leftarrow \{Nr_r$ randomly generated individuals$\}$
    $X \leftarrow X \cup \{Nr_e$ mutations of the elites$\}$
    $X \leftarrow X \cup \{Nr_d$ dual individuals$\}$
    Evaluate the fitness of the individuals in X
    Replace individuals in $\{x_i\}$ with individuals from X
    Adapt r_r, r_e, and r_d based on the performance of the new individuals
Next generation

Figure 21.19 Outline of an immigrant-based EA for dynamic optimization.

Figure 21.19 gives an overview of an immigrant-based EA for dynamic optimization. However, it leaves a lot of room for decisions by the EA designer.

1. What values should we use for r_r, r_d, and r_e? As implied earlier, a commonly used value for the total replacement proportion r_T is around 0.2 or 0.3. We usually start with the same number of random, dual, and elite replacement individuals so that $r_r \approx r_d \approx r_e \approx r_T/3$.

2. Should we adapt r_r, r_d, and r_e? We can adapt them to try to improve EA performance, as shown in Figure 21.19, but this makes our algorithm more complicated. [Yu et al., 2009] suggests an adaptation scheme like the following. Each generation we evaluate the fitness of the new individuals. If the group of random individuals performs better than the group of dual individuals and elite individuals, then we make the following assignments:

$$\begin{aligned} r_d &\leftarrow \max(r_{\min}, r_d - \alpha) \\ r_e &\leftarrow \max(r_{\min}, r_e - \alpha) \\ r_r &\leftarrow r_T - r_d - r_e \end{aligned} \tag{21.36}$$

where α controls the speed of adaptation, $r_{\min}$ defines the minimum proportion of each type of new individual to create each generation, and the constant r_T defines the total proportion of new individuals to create each generation. [Yu et al., 2009] uses $\alpha \approx 0.02$ and $r_{\min} = 0.04$. If the group of dual individuals or elite individuals performs best, then we rewrite Equation (21.36) accordingly to increase the number of high-performing individual types that are created during the next generation, and to decrease the number of the other types. This adaptation approach raises the question, How do we decide which type of new individuals perform the "best"? We could decide on the basis of the best random individual, the best dual individual, and the best elite-based individual; or we could decide on the basis of the average performance of the entire group of random individuals, dual individuals, and elite-based individuals.

3. Which individuals in the population $\{x_i\}$ should we replace with the new individuals? We already mentioned this issue briefly. [Yu et al., 2009] replaces the worst individuals, but we could also replace randomly-selected individuals, or a combination of the worst and random individuals. Also, we could group the original population $\{x_i\}$ with the replacement population X and use a stochastic selection mechanism to choose the best N individuals. We could use any of the selection mechanisms discussed in Section 8.7.1.

4. Last but not least, we mention that we need to choose a recombination and mutation method to evolve $\{x_i\}$ for the next generation in Figure 21.19. We could use any EA for recombination, and we could use any of the mutation methods discussed in Section 8.9.

■ EXAMPLE 21.5

In this example we evaluate BBO performance on the same dynamic Ackley function of Example 21.4. We can detect a change in the fitness function because we save two elite individuals each generation, so the population's best cost should decrease each generation. If the best cost increases, then we infer that the cost function has changed.[3] We try two different ways of adapting to a change in the cost function: first, we reinitialize the population to randomly-generated individuals; second, we use the direct immigrant scheme of Figure 21.19. Figure 21.20 shows BBO performance averaged over 20 Monte Carlo simulations for these two restart options. We see that a random restart performs better than the immigrant scheme. The dynamic change in the cost function is so random (a rotation of the bias vector) that replacing only 30% of the population is not drastic enough to outperform a random restart.

[3]This is not a foolproof scheme. A cost function change could result in a *decrease* of the best cost in the population. However, if the best cost decreases from one generation to the next, then we should not complain too much, even if our fitness-function-change detection algorithm fails.

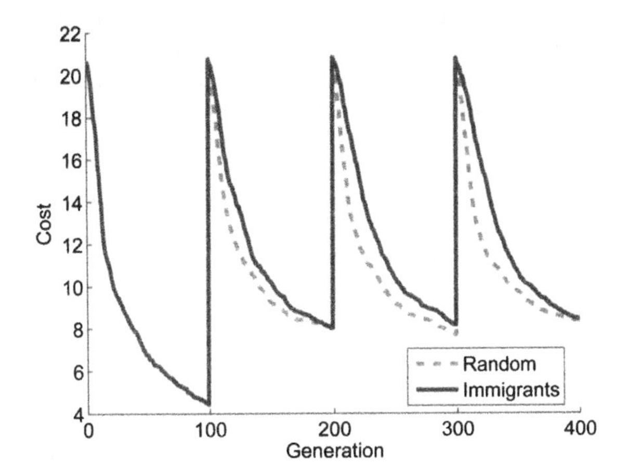

Figure 21.20 Example 21.5 results. This figure shows BBO performance on the 10-dimensional dynamic Ackley function, averaged over 20 Monte Carlo simulations. When the cost function changes via a rotation of the bias vector every 100 generations, we either randomly reinitialize the entire population, or replace 30% of the population with the immigrant scheme of Figure 21.20. Since the cost function dynamics are so random, random restart outperforms the immigrant scheme.

□

■ EXAMPLE 21.6

In this example we again evaluate BBO performance on the dynamic Ackley function. However, in this example a dynamic change in the cost function does not consist of a rotation of the bias vector; instead it consists of a perturbation of the bias vector by a random amount:

$$\begin{aligned} \theta(t) &\leftarrow \theta(t-1) + 0.1(x_{\max} - x_{\min})\rho(t-1) \\ \theta(t) &\leftarrow \min(\theta(t), x_{\max} - x^*) \\ \theta(t) &\leftarrow \max(\theta(t), x_{\min} - x^*) \end{aligned} \tag{21.37}$$

where $[x_{\min}, x_{\max}]$ defines the search space, x^* is the optimizing value of the unbiased cost function, and $\rho(t-1)$ is a random number taken from a zero-mean, unity-variance Gaussian distribution. See Appendix C.4 for additional discussion of the above parameters. The above sequence of assignments ensures that the optimum of $f(x - \theta(t))$ is within the search domain. For the Ackley function, $x^* = 0$, but Equation (21.37) applies equally well to any other benchmark. Equation (21.37) shows that the bias vector $\theta(t)$ changes according to a Gaussian distribution with a standard deviation that is equal to 10% of the range of the search space. This moderate change in the cost function might be more realistic than the less structured bias vector change of Example 21.5. In this example we try three different options for adapting to a change in the cost function: (1) we reinitialize the population to randomly-generated individuals; (2) we use the direct immigrant scheme of Figure 21.19 with r_r, r_e, and r_d each equal to 10% of the population size, and we use new individuals to replace the worst individuals each generation; (3) we ignore the cost function dynamics and do not make any change to the population. Figure 21.21 shows BBO performance averaged over 20 Monte Carlo simulations for these three options. We see that a random restart performs the worst. This is because the dynamic change in the cost function has some structure, so replacing the entire population results in a loss of the information that evolved over the previous 100 generations. Figure 21.21 shows that the immigrant scheme and the "ignore" option both perform about the same.

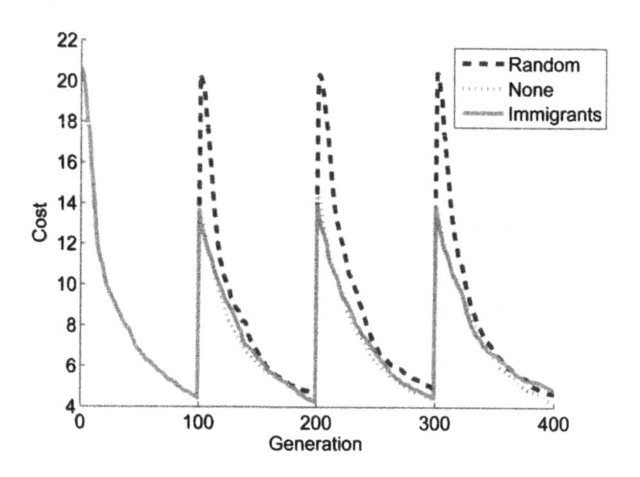

Figure 21.21 Example 21.6 results. This figure shows BBO performance on the 10-dimensional dynamic Ackley function, averaged over 20 Monte Carlo simulations. When the cost function changes via a moderate perturbation of the bias vector every 100 generations, we either randomly reinitialize the entire population, replace 30% of the population with the immigrant scheme of Figure 21.19, or simply do nothing. Since the cost function change is relatively structured, random restart performs the worst.

□

We would generally expect the immigrant approach to perform better than the "ignore" approach. We might be able to improve the performance of the immigrant scheme by adapting the r_r, r_e, and r_d parameters as illustrated in Equation (21.36), which we did not do in the above example. We also note that the comparison shown in Figure 21.21 is not completely fair because the comparison between the three algorithms is made for the same number of generations. The random restart algorithm requires N fitness function evaluations each time the function changes, the immigrate algorithm requires $0.3N$ fitness function evaluations each time the function changes, and the "ignore" option does not require any fitness function evaluations when the function changes. However, the function changes only once each 100 generations. So even though a rigorous comparison between the three approaches should be made on the basis of function evaluations rather than generations, the difference in the number of function evaluations is close enough in this example that we ignore it.

21.2.3 Memory-Based Approaches

Sometimes a cost function changes among a finite set of functions. For example, suppose that we need to optimize a control process in a manufacturing plant. The parameters of the manufacturing process may change periodically as the plant manager changes the task, or replaces machine parts. These changes result in a change of the cost function, but the changes are not random. The changes are deterministic, but the details of these changes may not be available to the controller and the optimization process. In cases like these, we might want to keep track of previous optimal solutions and insert them into the population when we detect a change in the cost function.

Explicit memory-based approaches store good individuals in an archive [Woldesenbet and Yen, 2009]. Whenever a cost function change is detected, individuals are retrieved from the archive and inserted into the population. If the cost function changes to a previously-encountered function, the individuals from the archive will be good solutions and the EA will converge very quickly to the new optimum. Figure 21.22 illustrates this approach. The figure provides only a basic outline and does not address some important details. The following questions provide rich opportunities for additional research.

1. How many individuals should we store in the archive when a change is detected?

2. How large should we allow the archive to grow? Figure 21.22 does not include an upper limit on its size. In practice, we might want to try to detect the "operating point" of the problem. If the operating point is the same as a previously-encountered operating point, then elites will have already been saved to the archive for that operating point, and we will not want to save current elites to the archive unless they are better than the previously-archived values. For example, suppose that after a change is detected in $f(\cdot)$, we search the archive and find that the current elite set is similar to previously-stored individuals. We might then infer that the problem the EA just solved is the same as a previously-solved problem, so there would be no need to save current elites in the archive unless they are better than the archive.

3. How can we detect a change in $f(\cdot)$? We discussed this at the beginning of Section 21.2.

4. When we detect a change in $f(\cdot)$, which individuals in the population should be replaced with archived individuals? Which archived individuals should be inserted into the population?

21.2.4 Evaluating Dynamic Optimization Performance

Evaluating dynamic optimization performance is different than evaluating stationary optimization performance. When evaluating stationary optimization performance, we usually look at the population of the last generation to see how well an EA performed. However, when evaluating dynamic optimization performance, the fitness function changes from one generation to the next. Therefore, looking at only the last generation does not give a good overall indication of EA performance. Instead we need to look at the performance across all generations [Yu et al., 2009]. Two common metrics for dynamic optimization problems are mean best performance $\bar{f}_b$, and mean average performance $\bar{f}_a$:

$$\begin{aligned} \bar{f}_b &= \frac{1}{G}\sum_{i=1}^{G} f_{i,b} \\ \bar{f}_a &= \frac{1}{G}\sum_{i=1}^{G} f_{i,a} \end{aligned} \tag{21.38}$$

where G is the number of generations, $f_{i,b}$ is the best fitness during the i-th generation, and $f_{i,a}$ is the average fitness during the i-th generation. These quantities give us good metrics with which to compare dynamic optimization performance between different EAs. If we run several Monte Carlo simulations, then we will have an additional level of averaging: we will average $\bar{f}_b$ and $\bar{f}_a$ over the Monte Carlo simulations. We discuss performance evaluation in more detail in Appendix B.2.2.

Create the initial population $\{x_i\}$
Archive $A \leftarrow \emptyset$
While not(termination criterion)
    Use a recombination/mutation method to evolve $\{x_i\}$ for the next generation
    Evaluate the fitness of the individuals $\{x_i\}$
    Store the best individuals from $\{x_i\}$ in the elite set E
    If a change is detected in $f(\cdot)$ then
        Replace some of the individuals in the population $\{x_i\}$
            with individuals from the archive A
        Store the elite set E in the archive A
    End if
Next generation

Figure 21.22 Outline of an explicit memory-based approach for dynamic optimization. This algorithm is especially suitable for fitness functions that are periodic with time, or that are equal to one of a finite set of fitness functions.

21.3 NOISY FITNESS FUNCTIONS

Fitness function evaluations in EAs are often accompanied by noise. For example, sensor inaccuracies can cause noise in experimental fitness function evaluations. Also, if we measure fitness function values with simulation software, then approximation errors in our software could cause noise in fitness function evaluations. Ingo Rechenberg, the inventor of the evolution strategy, was probably the first to investigate the effects of noise on EAs [Rechenberg, 1973].

A noisy fitness function evaluation could result in a high fitness being mistakenly assigned to a low-fitness individual. Conversely, it could result in a low fitness being mistakenly assigned to a high-fitness individual. Figure 21.23 illustrates the PDF of two noisy but unbiased fitness functions $f(x_1)$ and $f(x_2)$. We see that the true value of $f(x_1)$ is 0 and the true value of $f(x_2)$ is 4, but the evaluations are noisy. Therefore, x_1 might have an evaluated fitness that is actually greater than that of x_2. This situation would result in an inaccurate assessment of the relative fitness values of x_1 and x_2, which could result in an EA selecting the wrong individual for recombination. That is, noise can deceive an EA.

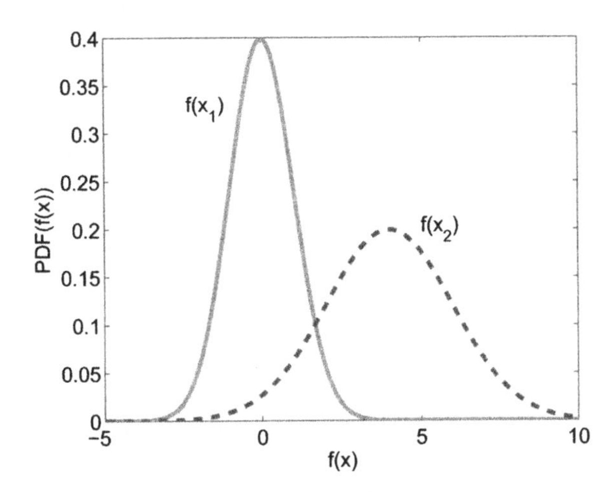

Figure 21.23 This figure depicts the PDFs of two fitness functions. x_2, which has a true value of 4, is more fit than x_1, which has a true value of 0. But depending on the noise that is realized during fitness function evaluation, the EA might think that x_1 is more fit than x_2. This could result in incorrect selection for the next generation.

When we have noisy fitness function evaluations, we cannot be sure which individual is best. Suppose that we have two individuals x_1 and x_2, two true fitness values $f_t(x_1)$ and $f_t(x_2)$ (depicted, for example, as 0 and 4 in Figure 21.23), and two noisy fitness function evaluations $f(x_1)$ and $f(x_2)$. Because of the noise, $f(x_1) > f(x_2)$ does not necessarily imply that $f_t(x_1) > f_t(x_2)$. However, if we know the PDFs of $f(x_1)$ and $f(x_2)$, then we can calculate the probability that $f_t(x_1) > f_t(x_2)$ given specific values of $f(x_1)$ and $f(x_2)$. We will not go through the mathematics here, but we can perform the calculations using standard methods from probability theory [Grinstead and Snell, 1997], [Mitzenmacher and Upfal, 2005]. Note that during EA execution we will not have the PDF of the noisy fitness function since we will not know the true fitness function (the presumed mean of the noisy fitness function). However, we might know the PDF of the true fitness function. This situation is analogous to that shown in Figure 21.23 except that instead of treating the noisy fitness function as a random variable with a mean equal to the true fitness function, we can treat the true fitness function as a random variable with a mean equal to the noisy evaluated fitness function value.

This section discusses three methods for dealing with noisy fitness functions. Section 21.3.1 discusses the resampling approach, Section 21.3.2 discusses the fitness estimation approach, and Section 21.3.3 discusses the Kalman EA, which uses a Kalman filter to estimate fitness function values.

21.3.1 Resampling

resampling One simple approach to reduce noise is to resample the fitness function. If we evaluate a fitness function for a given individual N times, and the noise values of those N samples are independent, then the variance of the average fitness function decreases by a factor of N [Grinstead and Snell, 1997], [Mitzenmacher and Upfal, 2005]. Suppose that the evaluated fitness $g(x)$ of a candidate solution x is given by

$$g(x) = f(x) + w \tag{21.39}$$

where $f(x)$ is the true fitness, and w is zero-mean noise with a variance of σ^2. This means that the measured fitness value $g(x)$ has a mean of $f(x)$ and a variance of σ^2. If we take N independent measurements $\{g_i(x)\}$, then each measurement $g_i(x)$ has a variance of σ^2, the best estimate of the true fitness is

$$\hat{f}(x) = \frac{1}{N}\sum_{i=1}^{N} g_i(x) \tag{21.40}$$

and the variance of $\hat{f}(x)$ is σ^2/N. Figure 21.24 illustrates this idea. The average of a set of N noisy fitness function evaluations is N times as accurate as a single evaluation.

However, the resampling strategy is completely valid only if the fitness function evaluation noise is independent from one sample to the next. For instance, suppose that we measure the fitness of candidate solutions with noisy instrumentation. If the instrumentation noise is time-correlated with itself from one sample time to the next, then averaging N samples does *not* reduce the variance by a factor of N. In this case, the amount by which the variance is reduced depends on the noise correlation from one sample to the next.

If we have N fitness evaluations $\{g_i(x)\}$ of a candidate solution x, then we can find an estimate $\hat{\sigma}^2$ of the variance σ^2 of the fitness estimate as follows:

$$\begin{aligned} \hat{f}(x) &= \frac{1}{N}\sum_{i=1}^{N} g_i(x) \\ \hat{\sigma}^2 &= \frac{1}{N-1}\sum_{i=1}^{N}\left(\hat{f}(x) - g_i(x)\right)^2 . \end{aligned} \tag{21.41}$$

Intuitively, it seems that the equation for $\hat{\sigma}^2$ should have N instead of $(N-1)$ in the denominator, but the $(N-1)$ term is preferred because it gives an unbiased estimate of the variance [Simon, 2006, Problem 3.6]. We can use Equation (21.41) to see how many times we have to sample a noisy fitness function to achieve a desired variance in our fitness value estimate (see Problem 21.7). The desired variance is user-defined and depends on the particular problem. As $N \to \infty$, the variance goes to 0 and our fitness value estimate becomes error-free.

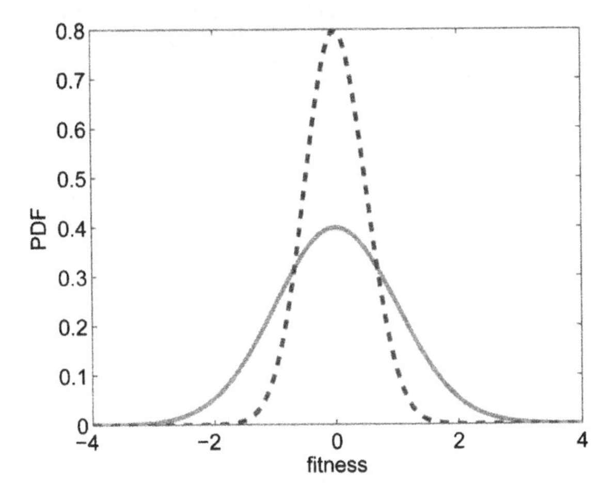

Figure 21.24 This figure illustrates the resampling strategy for noisy fitness function evaluations. The solid line shows the PDF of a noisy fitness function. The dashed line shows the PDF of the average of four fitness function evaluations. Both have a mean of zero, but the averaged evaluation has a variance that is 1/4 that of a single evaluation. The averaged evaluation is likely to be much closer to its mean than a single evaluation.

Some researchers suggest resampling a fixed number of times for each candidate solution. But this ignores the possibility that the fitness evaluation noise may not be the same for all candidate solutions [Di Pietro et al., 2004]. This could be the case, for example, if fitness evaluation is performed with sensors whose noise is proportional to the signal that they measure [Arnold, 2002]. In this case, Equation (21.39) is replaced with

$$g(x) = f(x) + w(x). \tag{21.42}$$

That is, the fitness evaluation noise depends on the specific candidate solution that is being evaluated. In that case, sampling a fixed number of times for each x would be inefficient because it would result in different accuracies for different candidate solutions. But we can still use Equation (21.41) to decide how many times to sample each candidate solution.

The resampling strategy can require many samples to achieve a desired variance. This might not be feasible for expensive fitness functions. Therefore, we might need to combine resampling with one of the fitness function approximation methods discussed in Section 21.1.1. Resampling can also be reduced by performing it only for the best individuals in the population. We might not know which individuals are best on the basis of a single fitness function evaluation, but we can at least get an idea, and we can reserve the computational effort that is required for resampling for the best individuals in the population [Branke, 1998].

Another approach for deciding how many times to resample is based on the fitness inheritance idea that we discussed in Section 21.1.1 [Bui et al., 2005]. Suppose that several parents $\{p_i\}$ produce a child x. Suppose further that the i-th parent has estimated fitness value $g(p_i)$ and standard deviation $\sigma(p_i)$. Before we evaluate the fitness of the child, we can use fitness inheritance to obtain estimates of its fitness $\hat{f}(x)$ and its standard deviation $\hat{\sigma}(x)$. We then evaluate the fitness of the child once. If its fitness evaluation $g(x)$ falls within $\pm 3\hat{\sigma}(x)$ of $\hat{f}(x)$, then we accept $g(x)$ as valid. Otherwise, we infer that $g(x)$ is very noisy, and so we pursue a resampling strategy to reduce the noise. This approach is counterintuitive in some ways. It indicates that the larger the value of the parents' noise, the more likely we are to accept the evaluation of the child. Intuitively, we expect the converse to be true; that is, the noisier the parents' fitness values, the more likely we should be to resample the child's fitness [Syberfeldt et al., 2010].

One approach for reducing the number of fitness evaluations is to resample only a few times at the beginning of the EA, and more often as the EA progresses [Syberfeldt et al., 2010]. This makes sense because the EA search is usually relatively coarse at the beginning of the optimization process. During the early generations the EA is trying to find the general neighborhood of the optimal solution; during later generations the EA is trying to converge to more accurate solutions. We need precise fitness function values only when the EA begins to converge toward the end of the optimization process. That is, precision is relevant only if accuracy is good (see Problem 21.8) [Taylor, 1997].

Many resampling strategies assume that the fitness noise is normally distributed. This is true for many noise phenomena in measurement and computing, but not for all. Many resampling strategies that have been published in the research literature need to be rederived if the noise is not Gaussian.

21.3.2 Fitness Estimation

Equation (21.41) shows a simple averaging method to estimate fitness on the basis of noisy samples. However, we can also use more involved probabilistic methods. For example, [Sano and Kita, 2002] assumes that individuals x_1 and x_2 that are close to each other in the search space have similar fitness values:

$$f(x_1) \sim N(f(x_2), kd). \tag{21.43}$$

That is, the fitness of x_1 is a Gaussian random variable with a mean of $f(x_2)$ and a variance of kd, where k is an unknown parameter and d is the distance between x_1 and x_2. If the fitness evaluation $g(x_1)$ has a variance of σ^2, then

$$g(x_1) \sim N(f(x_2), kd + \sigma^2). \tag{21.44}$$

With these assumptions, we can use maximum-likelihood calculations to estimate the fitness of x_1 and x_2 given noisy evaluations of x_1 and x_2. That is, we use not only the noisy evaluations of an individual to estimate its fitness, but we also use the noisy evaluations of its neighbors [Branke et al., 2001].

21.3.3 The Kalman Evolutionary Algorithm

The Kalman EA is designed for problems with noisy and expensive fitness function evaluations. The Kalman EA, which was originally proposed in the context of genetic algorithms, is an approach to keep track of the uncertainty in fitness values and to allocate fitness evaluations accordingly [Stroud, 2001].

A Kalman filter is an optimal estimator for the states of a linear dynamic system [Simon, 2006]. The Kalman EA assumes that the fitness of a given individual x is constant. We further assume that fitness evaluation noise is not a function of x. With these assumptions, we can use a reduced and simplified scalar form of the Kalman filter to keep track of the uncertainty in each fitness estimate. We denote the variance of a single fitness function evaluation as R. We denote the variance of the fitness estimate of an individual x after k fitness function evaluations as $P_k(x)$. We denote the value of the k-th fitness function evaluation of x as $g_k(x)$. Finally, we denote our estimate of the fitness of x after k fitness function evaluations as $\hat{f}_k(x)$. With this notation we can use Kalman filter theory to write

$$\begin{aligned} \hat{f}_{k+1}(x) &= \hat{f}_k(x) + \frac{P_k(x)(g_{k+1}(x) - \hat{f}_k(x))}{P_k(x) + R} \\ P_{k+1}(x) &= \frac{P_k(x)R}{P_k(x) + R} \end{aligned} \tag{21.45}$$

for $k = 0, 1, 2, \cdots$. We initialize $P_0(x) = \infty$ for all x, which gives

$$\begin{aligned} \hat{f}_1(x) &= g_1(x) \\ P_1(x) &= R. \end{aligned} \tag{21.46}$$

That is, our estimate $\hat{f}_1(x)$ after the first fitness function evaluation $g_1(x)$ is simply equal to that first evaluation. Also, the uncertainty $P_1(x)$ in our fitness estimate after the first evaluation is simply equal to the uncertainty in the evaluation.

Equation (21.45) shows that each time we evaluate the fitness of x, we modify our estimate $\hat{f}(x)$ based on the the previous estimate, its uncertainty, and the most recent fitness function evaluation result $g(x)$. Equation (21.45) shows that

$$\begin{aligned} \lim_{P_k(x)\to 0} \hat{f}_{k+1}(x) &= \hat{f}_k(x) \\ \lim_{P_k(x)\to \infty} \hat{f}_{k+1}(x) &= g_{k+1}(x). \end{aligned} \tag{21.47}$$

In other words, if we are completely certain of the fitness of x (that is, $P_k(x) = 0$), then further evaluations of the fitness of x will not change our estimate of its fitness. On the other hand, if we are completely uncertain of the fitness of x (that is, $P_k(x) \to \infty$), then we will set our estimate of its fitness equal to the next fitness function evaluation result.

Equation (21.45) also shows that each time we evaluate the fitness of x, our uncertainty $P(x)$ in its value decreases (that is, our confidence in its estimated value increases). Equation (21.45) shows that

$$\begin{aligned} \lim_{R\to 0} \hat{f}_{k+1}(x) &= g_{k+1}(x) \\ \lim_{R\to 0} P_{k+1}(x) &= 0 \\ \lim_{R\to \infty} \hat{f}_{k+1}(x) &= \hat{f}_k(x) \\ \lim_{R\to \infty} P_{k+1}(x) &= P_k(x). \end{aligned} \tag{21.48}$$

These results agree with intuition. If the fitness function noise variance R is 0, then the fitness function evaluation is perfect so our estimate is simply equal to the fitness function evaluation result, and the uncertainty in our fitness function estimate is 0. On the other hand, if the fitness function noise variance R is infinite, then the noise is so large that fitness function evaluations do not provide us with any information. In this case, additional fitness function evaluations do not change our estimate of the fitness function value, and neither do they reduce our uncertainty in its value.

The Kalman EA keeps track of the fitness function estimate $\hat{f}_k(x)$ and variance $P_k(x)$ for each individual x from one fitness function evaluation to the next $(k = 1, 2, \cdots)$. We allocate a user-defined fraction F of the available evaluations to generate and evaluate new individuals. We initialize our fitness estimate and variance for each new individual as described in Equation (21.46). We use the fraction $(1 - F)$ of the available evaluations to re-evaluate existing individuals. In this case, we update our fitness estimate and variance as described in Equation (21.45). Each time we have enough resources for a fitness function evaluation, we generate a random number r that is uniformly distributed on $[0, 1]$. If $r < F$ then we perform EA recombination and mutation to generate a new individual, and then we evaluate its fitness; otherwise, we re-evaluate an existing individual.

When it is time to re-evaluate an existing individual, we consider two guiding principles. First, we can generate more information by re-evaluating individuals whose fitness estimate variance is high. Second, we can generate more useful information by re-evaluating individuals whose estimated fitness is high. That is, we do not care too much about obtaining a high precision in the estimate of low-fitness individuals, because we are probably not interested in recombining them for future EA generations. [Stroud, 2001] therefore suggests the following strategy to selecting an individual x_s for re-evaluation:

$$\begin{aligned} \bar{f} &\leftarrow \text{mean of the population's estimated fitness values} \\ \sigma &\leftarrow \text{standard deviation of the population's estimated fitness values} \\ x_s &\leftarrow \arg\max\{P(x) : \hat{f}(x) > \bar{f} - \sigma\} \end{aligned} \tag{21.49}$$

where we have omitted the subscript k on $\hat{f}(x)$ and $P(x)$; we use the most recently-updated values of $\hat{f}(x)$ and $P(x)$ for each individual x in Equation (21.49). The equation shows that among all individuals whose estimated fitness is greater than one standard deviation below the mean, we select the one with the largest uncertainty for re-evaluation. This strategy assumes that $f(x)$ is fitness, so that larger $f(x)$ is better.

We see a lot of room for additional development of the Kalman EA. For instance, how can we extend it to dynamic optimization problems? How can we decide the optimal new-individual fraction F? Is there some way to adapt F based on performance? How can we extend the Kalman EA to more general filtering paradigms, such as H_∞ filtering [Simon, 2006]?

21.4 CONCLUSION

Surveys of EA fitness approximation are given in [Ong et al., 2004], [Jin, 2005], [Knowles and Nakayama, 2008], and [Shi and Rasheed, 2010]. A collection of papers related to EAs for problems with expensive fitness function evaluations is available in [Tenne and Goh, 2010]. Books about dynamic EAs include [Branke, 2002],

[Morrison, 2004], [Yang et al., 2010], and [Simões, 2011]. Jürgen Branke maintains a web site devoted to EAs for dynamic optimization [Branke, 2012].

Fitness approximation is not the only way to handle expensive fitness functions. We can also use grid computing, which entails the use of distributed computing resources that combine to solve a single problem [Melab et al., 2006], [Lim et al., 2007]. Other time-saving approaches for expensive fitness functions include multiple cores for parallelization using a single computer, cluster computing, cloud computing, or other forms of distributed processing [Tomassini and Vanneschi, 2009], [Tomassini and Vanneschi, 2010].

Surveys of EAs in uncertain environments are given in [Jin and Branke, 2005] and [Nguyen et al., 2012]. "Uncertain environments" in this context includes expensive fitness functions, dynamic fitness functions, and noisy fitness functions, which are the main topics of this chapter. However, [Jin and Branke, 2005] also discusses robustness, which is measure of EA solution quality in the presence of variations in either the decision vector or the problem parameters [Eiben and Smit, 2011].

Consider robustness with respect to parameter variations. Many fitness functions can be written as $f(x,p)$, where x is the decision vector and p is a parameter vector. For example, if we are trying to optimize a robot control algorithm, p might refer to some of the physical design parameters of the robot. When we optimize $f(x,p)$, "robustness" refers to the quality of $f(x, p + \Delta p)$, where Δp represents parameter variations. In general, a good solution is often not a robust solution [Keel and Bhattacharyya, 1997]. Figure 21.25 illustrates this situation. The minimum cost is attained at $x = x_1$, but at this value of x the cost function is very sensitive to parameter variations. We can obtain a suboptimal cost at $x = x_2$, which gives a worse cost but much better robustness with respect to parameter variations. Depending on the expected amount of parameter variation, x_2 might be preferable to x_1.

We can also consider robustness with respect to decision vector variations. In this case, when we optimize $f(x)$, "robustness" refers to the quality of $f(x + \Delta x)$, where Δx represents decision vector variations. When we find an optimal decision vector x, the decision vector that we implement in our solution may vary because of implementation issues, manufacturing uncertainties, and other issues. Figure 21.26 illustrates this situation. The minimum cost is attained at $x = x_1$, but at this value of x the cost function is very sensitive to variations in x. We can obtain a suboptimal cost at $x = x_2$, which gives a worse cost but much better robustness with respect to variations in x. Depending on the expected amount of variation, x_2 might be preferable to x_1. We have not discussed robustness in this chapter, but it is an important issue in real-world problems, and [Jin and Branke, 2005] and [Branke, 2002, Chapter 8] give a good overview.

As EAs are applied to more and more real-world problems, EA research efforts may begin to shift toward more of the topics discussed in this chapter, including expensive fitness functions, dynamic fitness functions, and noisy fitness functions. Some interesting areas of future research include EAs for problems that have more than one of these features. For example, we have discussed algorithms for handling dynamic fitness functions and algorithms for handling noisy fitness functions, but how could we combine those algorithms to handle fitness functions that are both dynamic and noisy? Another interesting question is how to allocate a budget of K fitness evaluations to obtain the best results from an EA. Note that "best" in this context could be interpreted in several ways. For example, we could search for the best strategy in terms of optimizing the expected EA result, or in terms of optimizing the worst-case EA result, or in terms of optimizing the best-case-minus-three-sigma result, or some combination.

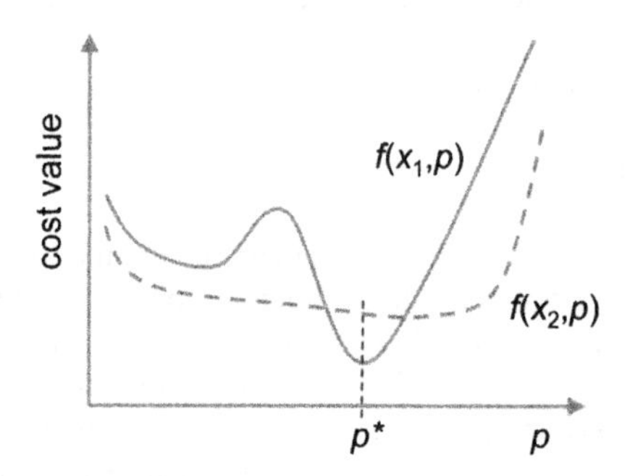

Figure 21.25 The above figure shows cost as a function of parameter value p, where p^* is the nominal parameter value. $x = x_1$ gives the best cost at $p = p^*$, but is not robust. $x = x_2$ gives a worse cost but provides much better robustness with respect to parameter variations.

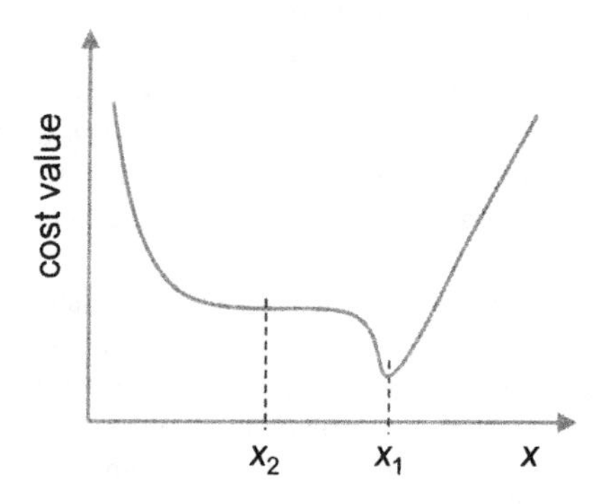

Figure 21.26 The above figure shows cost as a function of independent variable x. We see that $x = x_1$ gives the best cost but is not robust. $x = x_2$ gives a worse cost but provides much better robustness with respect to variations in x.

PROBLEMS

Written Exercises

21.1 Consider a one-dimensional function comprised of the following three points (x, y): $(1, 3)$, $(2, 1)$, and $(3, 4)$.

a) What is the linear least-mean-square fit to this function? What is the RMS error and min-max error of this approximation?

b) What is the linear min-max fit to this function? What is the RMS error and min-max error of this approximation?

21.2 Derive Equation (21.16).

21.3 The text states that $s(x)$ in Equation (21.24) is zero at the sampled data points. Prove it. (Hint: If x^* is one of the sampled data points, then r is one of the columns of R.)

21.4 How many possible Latin hypercube sampling arrangements exist for an $M \times M$ two-dimensional grid?

21.5 Suppose the noisy fitness evaluation of x_1 is uniform between -2 and 2. Suppose the noisy fitness evaluation of x_2 is uniform between -1 and 3. What is the probability that the noisy fitness evaluation of x_1 is greater than that of x_2?

21.6 Prove the following statement from Section 21.3.1: "If we evaluate a fitness function for a given individual N times, and the noise values of those N samples are independent, then the variance of the average fitness function decreases by a factor of N."

21.7 Suppose you have the following 10 noisy fitness function evaluations of a given individual:

$$[101, 102, 98, 97, 99, 103, 104, 101, 97, 98].$$

How many fitness evaluations will you need to average to obtain a variance of 5?

21.8 What is the difference between precision and accuracy? Give both an explanation and an example.

21.9 Suppose we use a Kalman EA to estimate the fitness of individual x. Suppose that our fitness estimate $\hat{f}(x) = 10$ with uncertainty variance $= 1$. Suppose that we obtain a new fitness measurement $g(x) = 11$ with uncertainty variance $=$ 3. What is the new fitness estimate and its uncertainty variance?

Computer Exercises

21.10 Repeat Example 21.1 and 21.2 for the two-dimensional Ackley function with each independent variable in the domain $[-3, +3]$.

21.11 Use Equation (21.38) to evaluate the performance of the three dynamic EA approaches of Example 21.6.

21.12 Write a computer program to experimentally confirm that the average of N noisy terms has a variance that is $1/N$ times the variance of each noisy term.

21.13 Simulate Examples 21.5 and 21.6 while keeping track of the following proportions:

$$
\begin{aligned}
p_r &= \text{how often the random individuals } R \text{ are the best of the new individuals} \\
p_e &= \text{how often the mutated elites } E \text{ are the best of the new individuals} \\
p_d &= \text{how often the dual individuals } D \text{ are the best of the new individuals}
\end{aligned}
$$

where *best* is defined in terms of average cost. That is, after you create new random individuals R, new mutated elite individuals E, and new dual individuals D, how often is the average cost of R better than both the average cost of E and the average cost of D, and so on? Use a high enough generation count so that you can collect enough data to make reasonably certain conclusions.

PART V

APPENDICES

APPENDIX A

Some Practical Advice

Good advice is always certain to be ignored, but that's no reason not to give it.
—Agatha Christie

This appendix provides some practical advice for EA researchers and practitioners. What should we do if our EA does not work? How can we make our EA work better? How can we learn to be successful EA researchers? What should we focus on in our research? The book *A Field Guide to Genetic Programming* has a lot of good practical advice [Poli et al., 2008, Chapter 13]. That advice is specific for genetic programming, but much of it is general enough to apply to any other type of EA also, and so we borrow some of their ideas in this appendix.

A.1 CHECK FOR BUGS

Many students and beginning researchers expect their code, and others' code, to run the first time. There is no bug-free code. All code has bugs. We need to understand our software well enough so that we can find problems in the code, or problems with how we are using the code. That means that we need to step through our code, one line at a time, and inspect variables with a debugger. We

*

need to understand what values are being loaded into what variables, and why. We need to do this even if the code seems to run the first time. Just because the code runs does not mean that it works.

This is not to say that off-the-shelf programs do not work. It is not to say that we need to understand everything about every computer program that we use. But production software is qualitatively different than research and engineering software. If production software does not work then we can usually see the results of the malfunction, and we can take creative measures to work around those problems (for example, reboot, or experiment with a different sequence of button clicks to obtain desired results).

However, if research software does not work, then we usually need to fix the software ourselves. To do that, we need to understand the software. More importantly, even if the software appears to work, it might not be working correctly. If we do not understand the software, then we will never be sure that it is working correctly and we will never be able to really trust our results.

There are no shortcuts in EA research (or in any other engineering research). It is always better to write our own software. If we want to use someone else's software, then we must study the code and understand the code. Otherwise we will probably get what we pay for, especially if the software is free.

A.2 EVOLUTIONARY ALGORITHMS ARE STOCHASTIC

Evolutionary algorithms are stochastic. This means that they will not give the same results every time they run. If we run an EA 10 different times to try to solve a problem, and it does not work any of those 10 times, we might naively conclude that the EA does not work or that the problem is not solvable. However, if the EA has a 20% chance of solving the problem, then there is a 10% chance that it will fail 10 times in a row.

Conversely, if we run an 10 different times and it succeeds 10 times in a row, we might naively conclude that the EA works every time. However, if the EA has 20% chance of failing, then there is a 10% chance that it will succeed 10 times in a row. A good understanding of probability theory is needed to thoroughly understand the stochastic nature of EAs, all that it implies.

A.3 SMALL CHANGES CAN HAVE BIG EFFECTS

Something as innocuous as changing the way that duplicate individuals are handled, or small changes in the mutation rate, or changing the selection method, can drastically change the operation of an EA. We must never assume that a small change is insignificant. That means that when we obtain good results, we must be careful to save the EA settings that gave us those results. We should even save the random number seed so that we can reproduce our results (see Appendix B.2.3). If we get good results and we start fiddling with parameter settings to improve performance, we might lose our good results forever if we forget the parameter settings that we used to get the good results.

A.4 BIG CHANGES CAN HAVE SMALL EFFECTS

In contrast to the above point, sometimes an EA is insensitive to large changes in parameter settings. This is usually good if the EA is performing well because it means the EA is robust to that parameter. But insensitivity to parameter changes is bad if the EA is performing poorly. If the EA is performing poorly, it might be because the problem is too difficult for the EA, or because the problem has not been formulated in a way that is amenable to the EA. We are not guaranteed that an EA can give us good results, so we must not assume that the EA will be successful if we can just find the right parameter settings. Here is where we must balance our perseverance with the possibility that we might be wasting our time. However, most of us err on the side of too little perseverance rather than too much.

A.5 POPULATIONS HAVE LOTS OF INFORMATION

If our EA is not working well, then we should study the population at various generations. The makeup of the population can give us a lot of information. If we see that the population is converging to a single candidate solution, then we know that we need to deal with duplicate individuals more carefully. If we see that individuals that recombine are not improving, then we know that we need to modify our recombination strategy. If we keep track of mutations, then we know if we are mutating too much or too little, or too frequently or too infrequently. If we see that the population is not improving after the first few generations, then we know that we need more exploration and less exploitation. The number of details about EA behavior that we can glean from studying our EA population is limited only by our creativity and perseverance.

A.6 ENCOURAGE DIVERSITY

This is related to the above point. We need to be aware of the diversity (or lack thereof) in our population. If our population does not have enough diversity, then it will probably not perform well. We need to do whatever is required to encourage diversity while not preventing the exploitation of good candidate solutions. We need to remember that we only need to find one, or a few, good solutions to have a successful EA. Greater diversity increases our chances for success.

A.7 USE PROBLEM-SPECIFIC INFORMATION

The better we understand our problem, the better solution we will be able to find. An EA is typically a model-free optimizer, which means that we do not need to incorporate any problem-specific information into the EA. However, if we do incorporate problem-specific information, then we can almost certainly find a better solution than we could otherwise. An EA is a good global optimizer, but a local search method like hill climbing or gradient descent can improve the results of an EA to such an extent that it proves the difference between failure and success.

A.8 SAVE YOUR RESULTS OFTEN

Computer disk space is cheap. We should save our problem settings, old versions of our software, results, and intermediate results. This means that we need to be organized so that we can efficiently wade through all of our stored programs and data to find what we need without wasting too much time.

EA runs are often computationally expensive. This means that we might need to run for days or weeks to get good results. An efficient and organized way of doing this is to run the EA for a day, save our results, and then begin the EA again with the new population seeded with the previous results. This allows the EA to take advantage of previously discovered good candidate solutions while still running for long periods of time.

A.9 UNDERSTAND STATISTICAL SIGNIFICANCE

We need to understand the statistical significance of our experimental results. That means that we need to understand statistics and the no free lunch theorem (see Appendix B). It also means that we need to test our EA results on a validation set of data. An EA can be used to find a good solution to an optimization problem, but it may be just as important to test the performance of the solution on data that has not been used during training. For example, we might use an EA to find parameters that optimize a classifier, and we have some test data that we use for our fitness function evaluation. However, if we stop there, then we have done nothing more than show that the classifier can memorize the training data. The real test of the classifier is to see how well it performs on data that it has not yet seen. This data is called the validation set. There are several ways to divide data into training data and validation data. We do not discuss those ideas in this book except for a short discussion in Section 21.1.6, but it is important that we understand the basic ideas of validation [Hastie et al., 2009].

A.10 WRITE WELL

If we do the best EA research in the world but we do not communicate it to others, then the research is worthless. Research is performed to be communicated. The great English scientist Michael Faraday said, "Work, finish, publish" [Beveridge, 2004, page 121], implying that the research process is not complete until the results are published. Technical writing is a skill that is sorely lacking in today's students, engineers, and researchers. We will be better professionals if we can write better. Learning to write better is not a mysterious process. We learn to write better just like we learn to do anything better: we study good writing and we practice writing.

A.11 EMPHASIZE THEORY

Too much EA research today involves tweaking parameters and hybridizing EAs with other optimization algorithms. Given all of the EAs that are available today, all of their tuning parameters, and all of the other non-evolutionary optimization algorithms that are available, there is a virtually infinite number of ways that EAs can be modified, combined, and tuned to get better performance on benchmarks. But that type of research does not really advance the state-of-the-art. That type of research is extremely incremental and short-sighted in nature, and does not provide any insight beyond its immediate results. Most EAs are sorely lacking in theoretical support and mathematical analysis. If we could emphasize theory and mathematics a little bit more in our research, we could start to make a difference in our fundamental understanding of EAs. A single good theory paper is worth a dozen parameter-tuning papers.

A.12 EMPHASIZE PRACTICE

Too much EA research today focuses on benchmarks. But if we want our research to make a difference in the world, then we need to collaborate with industry and solve problems that are important to practicing engineers and other professionals. Our EAs could spend the remainder of their existence optimizing benchmarks while never seeing the outside of a university. But that is not why EAs exist, or why EAs were invented in the first place [Fogel et al., 1966], [Fogel, 1999]. They were invented to solve real-world problems and to make a contribution to society. Benchmarks are important but they are a means to an end, with the end being the development of EAs for eventual application in the real world. Let us not confuse the means for the end, and let us not forget our ultimate goal in EA research.

APPENDIX B
The No Free Lunch Theorem and Performance Testing

One might expect that there are pairs of search algorithms A and B such that A performs better than B on average ... One of the main results of this paper is that such expectations are incorrect.
—David Wolpert and William Macready [Wolpert and Macready, 1997, page 67]

There are three kinds of lies: lies, damned lies, and statistics.
—Mark Twain [Twain, 2010, page 228]

This appendix discusses two distinct but related issues. This is a critically important chapter for EA students and researchers to read and understand. It is an appendix because it is not directly related to EAs, and because we need to understand EAs and simulations before reading this appendix, but its relegation to the appendix does not lessen its importance.

Section B.1 discusses the no free lunch (NFL) theorem from a intuitive, non-mathematical aspect. The NFL theorem tells us that all algorithms perform equally

*

well, under certain conditions. Section B.2 discusses how EA simulation results are often presented in a misleading way, and how they can instead be correctly presented. Section B.3 contains a note on the relationship between the methodologies used in this book and the research guidelines discussed in this appendix.

B.1 THE NO FREE LUNCH THEOREM

The NFL theorem, which was first formalized by David Wolpert and William McReady [Wolpert and Macready, 1997], is surprising:

> All optimization algorithms perform equally well when averaged over all possible problems.

This means that, in general, no optimization algorithm is better than any other, and none is worse than any other. Note that the NFL theorem is more than a conjecture, or general statement, or rule of thumb; it is a mathematical theorem. For specific types of problems, certain algorithms will work better than others. But the NFL theorem should prevent us from making unsubstantiated claims about our favorite algorithm; it should encourage us to be more modest in our claims.

Technically, the NFL theorem only applies to discrete optimization problems. However, all real-world fitness functions are eventually discretized. After all, we define candidate solutions and measure fitness with digital computers. Therefore, for all practical purposes, the word "discrete" can be removed from the NFL theorem.

Intuitively, we expect that some optimization algorithms are better, on average, than other algorithms. For example, consider algorithm A, which is a hill descending algorithm similar to the algorithms described in Section 2.6; and algorithm B, which is a random number generator. Suppose that both algorithms are trying to find the minimum of some function $f(x)$. If $f(x) = x^2$, then hill descending will usually do much better than random search. However, random search will sometimes, just by chance, generate a number that is very close to the minimum, and it will do better than A on those occasions.

We could argue that there are an infinite number of smooth functions that are similar to $f(x) = x^2$, and hill descending will perform quite well on such functions, while random search will usually not perform very well. However, there are also an infinite number of irregular functions for which random search will actually perform better than hill descending, as illustrated by the following example.

■ EXAMPLE 2.1

Consider the discrete function in Figure B.1. This particular function has a search space size of 10 and was generated randomly. We can implement a hill descending algorithm to find the minimum of this function. We initialize the search at a random point in the search space, and the algorithm examines neighboring values to decide which direction to proceed with its search. If the algorithm gets stuck in a local, non-global minimum $x \in \{4, 7, 10\}$, then it restarts at a random initial value. It takes the hill descending algorithm an average of about 19 function evaluations to find the minimum, while it takes random search an average of exactly 10 function evaluations. The reason hill descending takes so long is that it has to start at $x \in [1, 3]$ to find the global minimum; otherwise it will converge to a local minimum, and it will have to reinitialize with a new random starting point. Any starting point other than $x \in [1, 3]$, will result in the hill descending algorithm wasting time in a local neighborhood that will not lead to the global minimum.

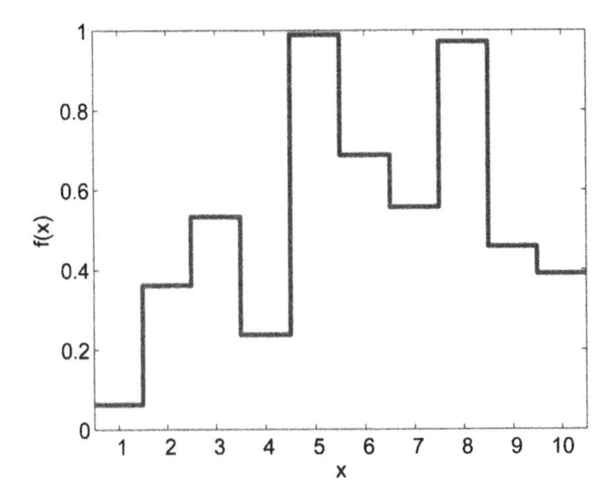

Figure B.1 Example 2.1: For this minimization problem, a hill descending algorithm requires an average of 19 function evaluations to find the global minimum at $x = 1$, while random search requires an average of only 10 function evaluations.

□

There are an infinite number of irregular functions like the one shown in Figure B.1 for which random search will perform better than hill descending. For every regular function for which hill descending performs better, we can find a corresponding irregular function for which random search performs better. We see that, averaged over all possible functions, hill descending and random search both perform equally well.

It may not be surprising that an irregular function like the one in Figure B.1 can be minimized more effectively with random search than with a hill descending algorithm. What is more surprising is that the NFL theorem asserts that a hill *climbing* algorithm performs just as well as a hill *descending* algorithm, on average, for function minimization. If we want to find the minimum of a function, should we use a hill descending algorithm or a hill climbing algorithm? Common sense tells us to use a hill descending algorithm if we want to find a minimum. However, the NFL theorem tells us that a hill climbing algorithm will perform just as well, averaged over all possible functions, as illustrated by the following example.

■ EXAMPLE 2.2

Consider the discrete function shown in Figure B.2. This can be considered a deceptive function because local changes in the cost function lead us to expect that the minimum should be at one of the extreme values of x. A hill descending algorithm uses local information in its search, and therefore will usually get stuck at $x = 1$ or $x = 11$, and will then have to start over. The only time the hill descending algorithm will succeed is if it is initialized at $x \in [5, 7]$. On the other hand, a hill climbing algorithm will always climb to $x = 5$ if starts at $x < 5$, and it will always climb to $x = 7$ if it starts at $x > 7$. After it reaches $x = 5$ or $x = 7$, it will stumble into the global minimum at the next iteration as it searches for the ascending direction. The hill climbing algorithm will never require more than 6 function evaluations to find the global minimum. The hill descending algorithm requires an average of about 16 function evaluations, random search requires an average of 11 function evaluations, and hill climbing requires an average of only 5 function evaluations to find the global minimum. We can concoct an infinite number of functions with the same topology as the function shown in Figure B.2. Even though it goes against our common sense to use a hill climbing algorithm to minimize a function, we see that hill climbing performs just as well as hill descending, averaged over all possible functions.

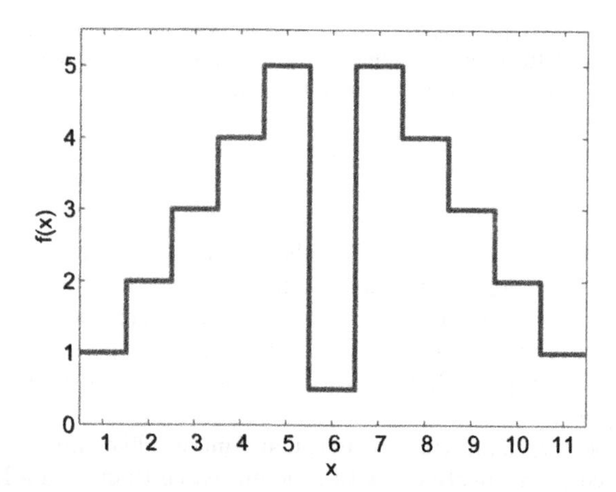

Figure B.2 Example 2.2: For this minimization problem, a hill descending algorithm requires an average of 16 function evaluations to find the global minimum at $x = 6$, random search requires an average of 11 function evaluations, and hill climbing requires an average of only 5 function evaluations. Hill climbing is better at minimization than hill descending for this minimization problem.

□

■ EXAMPLE 2.3

Consider a finely-tuned GA that is programmed to solve some difficult real-world optimization problem. We would think that the GA should do better than a random person randomly guessing solutions. But suppose that you walk out on the street and ask the first person that you meet to pick a number. The number that the person picks is the solution to *some* optimization problem. In fact, that number is the solution to an infinite number of optimization problems. The specific problem that we are trying solve is hopefully better attacked with a GA than with a random number generator. But averaged over all possible problems, random guessing is just as good as a GA.

□

The NFL theorem can be clarified by asking very simple questions about the set of all possible functions, the set of all possible optimization algorithms, and the set of all possible behaviors of optimization algorithms [Whitley and Watson, 2005]. This approach to understanding the NFL theorem is illustrated by the following example.

■ EXAMPLE 2.4

Consider a simple minimization problem which has the domain $x \in [1,3]$. Suppose that we initialize all search algorithms to $x = 1$. There are an infinite number of functions such that $f(1) > f(2) > f(3)$, and there are an equally infinite number of functions such that $f(1) > f(3) > f(2)$. Therefore, there is a one-to-one correspondence between the number of functions that are minimized at $x = 3$ and the number of functions that are minimized at $x = 2$.

Now consider the set of all optimization algorithms A that examine $x = 2$ as their first step (after initialization) in the search process. Denote by $\bar{A}$ the set of all optimization algorithms that examine $x = 3$ as their first step. Since there is a one-to-one correspondence between the number of functions that are minimized at $x = 3$ and the number of functions that are minimized at $x = 2$, we see that $\bar{A}$ will find the minimum before A for exactly half of all possible functions, while A will find the minimum before $\bar{A}$ for the other half of all possible functions.

□

The above discussions do not comprise a proof of the NFL theorem; they are more like explanations or illustrations, and they give us an intuitive sense for why the NFL theorem is true. Rigorous mathematical proofs are presented in [Radcliffe and Surry, 1995] and [Wolpert and Macready, 1997]. The NFL theorem can be stated in various ways, many of which are equivalent. Among other essentially equivalent statements, the NFL theorem asserts the following [Schumacher et al., 2001].

- All optimization algorithms perform equally well when averaged over all possible problems (as stated at the beginning of this appendix).

- For any two optimization algorithms A and B, if the performance of A on problem f is given by $V(A, f)$, then there exists a problem g such that $V(A, f) = V(B, g)$. This statement tells us that no matter how well algorithm A performs on some problem, there is another algorithm B that performs equally well on some other problem. Conversely, no matter how poorly algorithm B performs on some problem, there is some other problem for which algorithm A performs just as poorly.

- An algorithm that achieves performance that is better than random search is like a perpetual motion machine [Schaffer, 1994]. This is called the Law of Conservation of Generalization Performance.

- It is futile to attempt to design an optimization algorithm that is better than random search, unless you can incorporate problem-specific information in the algorithm [English, 1999]. This is called the Law of Conservation of Information.

- In the absence of problem-specific information, we must assume that all possible solutions are equally likely [Dembski and Marks, 2009b]. This is called Bernoulli's Principle of Insufficient Reason.

The NFL theorem should give us some balance in our claims. For example, suppose that we develop a new algorithm, or develop an improved version of some algorithm, and we want to show that it is better than other algorithms so that we can publish our results. Suppose that we have a set of benchmark functions F. We use $\bar{F}$ to denote the complement of F; that is, $\bar{F}$ is the set of all functions other than F. If our favorite algorithm, algorithm A, performs better than algorithm B on a set of benchmark functions F, then the NFL theorem assures us that algorithm B performs better on $\bar{F}$.

Up until now, we have not talked about how to quantify *performance*. There are many ways to measure the performance of an algorithm. For example:

- Performance can be measured by the best solution obtained after a certain number of generations and a certain number of simulations. This can be called *the best of the best*.

- Performance can be measured by the average of the best solutions obtained by a certain number of simulations, each of which run for a certain number of generations. This can be called *the average of the best*.

- Performance can be measured by computing the average fitness of all individuals in the population after a certain number of generations, and finding the best of these averages after a certain number of simulations. This can be called *the best of the average*.

- Performance can be measured by computing the average fitness of all individuals in the population after a certain number of generations, and finding the average of these averages after a certain number of simulations. This can be called *the average of the average*.

- Performance can be measured by the standard deviation of the best solutions found after a certain number of generations and a certain number of simulations.

- Any of the above performance measures can be combined in any way to form hybrid performance measures.

- Any of the above performance measures can be averaged over several problems.

It seems that there are as many ways to quantify performance as there are optimization problems. In fact, there are an infinite number of both.

This brings us to another important feature of the NFL theorem: the NFL theorem applies regardless of how we measure performance. Sometimes an algorithm is touted as being more robust than other algorithms. The term *robustness* might mean that the algorithm has good performance over a wide variety of functions, or that the algorithm is relatively insensitive to fitness function evaluation noise, fitness function parameter variations, or algorithm parameter variations. The NFL theorem assures us that all algorithms are equally robust. Conversely, it also tells us that all algorithms are equally specialized [Schumacher et al., 2001]. If algorithm A is more robust than algorithm B on function set F, then algorithm B is more robust than algorithm A on function set $\bar{F}$.

We see that in some ways the NFL theorem is nonintuitive. But if we look at it in other ways, it is highly intuitive. Given this second, intuitive, perspective of the NFL theorem, it is not surprising that it shows up in the literature long before its formal proofs in the mid-1990s. For example, Gregory Rawlins wrote [Rawlins, 1991, page 7]:

> ... it is sometimes suggested that GAs are universal in that they can be used to optimize any function. These statements are true in only a very limited sense; any algorithm satisfying one of these claims can expect to do no better than random search over the space of all functions.

Reactions to the NFL theorem from optimization researchers has varied. Some say that the NFL theorem is not of any practical interest because of its condition, "averaged over *all* possible problems." In the real world (as opposed to the world of mathematical theory), we are not interested in all possible problems, but we are rather interested in problems that arise in practical applications. Real-world problems are not usually random or deceptive, but rather have some structure. This means that a hill descending algorithm will perform better than random search or a hill climbing algorithm on real-world minimization problems; and, in fact, that is what we usually observe. This gives the yin to the NFL theorem's yang:

> *Not* all optimization algorithms perform equally well when averaged over all problems of interest.

But this does not mean that the NFL theorem is irrelevant to practicing engineers. The NFL theorem provides a solid foundation for what we think we know intuitively. That is, to find a good solution to an optimization problem, we need to incorporate problem-specific knowledge into our search algorithm. This is how we "pay for lunch" and obtain better performance than, say, random search. If we know that good solutions to a given problem tend to cluster together (that is, the search space has some regularity), then we know that recombination will tend to perform well. If we have a problem that has a less regular search space, then we know that mutation will be more important.

For example, suppose that we are trying to find proportional-integral-derivative (PID) control parameters to optimize the performance of a control system. We could design a search that calculates gradient information (that is, the sensitivity of the performance to the PID parameters), and that uses that information in its search. This means that we are incorporating problem-specific information in our search. This PID tuning algorithm avoids the NFL theorem because gradient information is not available for almost all functions. We are paying for lunch with our gradient information, so we have moved outside of the realm of the NFL theorem and our EA will probably do better than random search. Another way of stating this is to say that the NFL theorem views the optimization process as a blind search without any problem-specific information incorporated into the search. But effective real-world search algorithms are not blind; that is, they *do* incorporate problem-specific information [Culberson, 1998].

Another example is inversion in the traveling salesman problem (see Section 18.4.1). Multiple applications of inversion guarantee a tour without crossed edges. Therefore, inversion spends most of its time on tours that are better than average.

The NFL theorem also tells us, indirectly, why problem representation is important (see Section 8.3). If a problem's search space has a regular structure, then we should represent that problem in a way that preserve's the regularity of the search space; then we can obtain good results with a structured search. If we represent a problem in such a way that the representation does not have any structure, then we might as well use a random search.

Whitley and Watson give the following practical implications of the NFL theorem [Whitley and Watson, 2005].

- The design of an optimization algorithm entails a tradeoff between generality, and effectiveness for a specific problem. An algorithm that works well on a wide range of benchmarks may not work well for a particular real-world problem. Conversely, an algorithm that works poorly on standard benchmarks may give good results for a real-world problem. Simple algorithms often give good results; complicated algorithms can be designed to give better results for a given problem. We have to decide how much time and effort we want to expend to tune our algorithm for a given problem, and how important it is to achieve a marginally better optimization result.

- If we incorporate problem-specific information in our optimization algorithm, we should be able to get better results than we would get with a more general algorithm (as long as we use the problem-specific information correctly). This is an example of a general principle that engineers and scientists have observed for a long time. Performance and generality trade off with each other. Tools and algorithms that are designed to work on a wide variety of problems end up not performing well on any problems.

- The representation of the optimization problem can have a significant effect on the performance of an optimization algorithm. There are an infinite number of ways to represent any optimization problem. Choosing an appropriate representation can require a lot of work before the algorithm is even implemented, but it can pay high dividends in the long run (see Section 8.3).

- Do not assume that an optimization algorithm that works well on benchmarks will work well on real-world problems; likewise, do not assume that an algorithm that works poorly on benchmarks will work poorly on real-world problems. This implication of the NFL theorem casts doubt on the practical importance of most EA papers due to their emphasis on benchmark problems, and their lack of emphasis on real-world problems.

Current research in the area of NFL theorems includes finding function sets over which the NFL theorem does not hold. Recall that the NFL theorem holds over all possible optimization problems; it does not hold over all sets of optimization problems. For example, a hill descending algorithm will clearly do better than random search at solving problems with a single minimum. Less intuitive is the result that the NFL theorem does not hold over the set of functions that can be described by polynomials of a single variable with bounded complexity [Christensen and Oppacher, 2001]. Also, the NFL theorem does not hold over certain types of co-evolutionary problems [Wolpert and Macready, 2005].

Finding other functional sets over which the NFL theorem does not hold, and finding algorithms that provide a free lunch over those sets, could have important implications for the design of practical optimization algorithms. These questions are related to issues such as function compressibility, problem description length, and the distinction between infinite sets of functions and finite sets of functions (that is, permutation closures) [Schumacher et al., 2001], [Lattimore and Hutter, 2011].

B.2 PERFORMANCE TESTING

This section discusses some issues that are related to statistics and the presentation of EA simulation results in theses, dissertations, and technical papers. Anyone who publishes or studies EA research needs to have a clear grasp of the ideas of this section. This section shows how we can reduce the bias in the presentation of our results. It also shows us how to recognize the bias in the results of others. Perhaps most importantly, it emphasizes that we need to have a healthy skepticism of others when we read their research papers, and a healthy skepticism of ourselves when we document our own research results.

Section B.2.1 gives an overview of some of the problems that we commonly see in EA research papers. Section B.2.2 shows how authors often present simulation results to make any point that they want; that is, how they present results in a misleading way (hopefully unintentionally). It also shows how we can instead present simulation results in clear and honest way. Section B.2.3 makes a few important points about the random number generators that we use in our simulations. Section B.2.4 reviews the t-test, which can tell us if differences between two sets of simulation results are statistically significant. Section B.2.5 reviews the F-test, which can tell us if differences between more than two sets of simulation results are statistically significant.

B.2.1 Overstatements Based on Simulation Results

Mark Twain's quote at the beginning of this chapter, the book *How to Lie with Statistics* [Huff and Geis, 1993], and this section, all tell us essentially the same things about EA research. When we see EA results in a paper or a book, there is always some bias. The bias may be intentional or it may be unintentional; it may be explicit or it may be implicit; it may be obvious or it may be subtle; it may be insignificant or it may lead to entirely wrong conclusions; but there is always some bias. When we read a paper and draw conclusions from EA results, we must remember that we would draw different conclusions if the author had chosen to present a different set of results, or even if he had chosen to present the same results in a different way.

Given the NFL theorem, which states that all optimization algorithms perform equally well when averaged over all possible optimization problems, it seems futile to test our algorithms on benchmark problems. As Darrell Whitley wrote, "From a theoretical point of view, comparative evaluation of search algorithms is a dangerous, if not dubious, enterprise" [Whitley and Watson, 2005, page 333]. If we write a paper about optimization algorithm A and show that it performs better than algorithm B on a set of benchmark functions F, then the NFL theorem assures us that B performs better than A on the set of functions $\bar{F}$.

However, this should not discourage us too much. If the point of our paper is to show the superiority of A over B on the set F, then the paper has succeeded. Remember that the NFL theorem does *not* say that all algorithms perform equally well over all problem sets F; it says that all algorithms perform equally well when averaged over all possible problems. This means that algorithm A might indeed perform better than B for specific problems, or for specific types of problems.

This indicates that benchmarking is a worthwhile effort, but that we have to keep the NFL theorem in mind to modulate the conclusions that we draw. A typical EA paper demonstrates that A performs better than B on the set of benchmarks F, and then concludes something like:

> We therefore see that A is better than B for function optimization.

Claims like this are clearly and fundamentally incorrect in view of the NFL theorem. But typical EA papers do include statements like that in the abstract, introduction, and conclusion. A more modest claim would be:

> We therefore see that A is better than B for optimizing functions that have the characteristics of the problems that we examined in this paper.

However, typical EA papers that present benchmark performance do not make any effort to examine the characteristics of the benchmarks. What is it about the benchmark set F that makes A perform better than B? Does A perform better than B because the functions in F are differentiable, or because they are multimodal, or because they have continuous second derivatives, or because they are constrained, or for some other of a million possible reasons? Questions like this are difficult to answer and so they are usually ignored. Therefore, an even better (that is, more modest) claim in an EA paper would be:

> We therefore see that A is better than B for optimizing the functions that we examined in this paper.

This claim is better because it does not make any attempt to generalize beyond the empirical results presented in the paper. In other words, it does not try to claim too much. However, even this claim is probably too much because of the many ways that algorithms can be tuned and implemented.

Another NFL-related danger of presenting benchmark results is that the benchmark functions may not be in the same class as interesting, real-world problems. We assume that most EA researchers are interested in the eventual application of their research to real-world problems. If an EA paper demonstrates good performance for algorithm A on the set of benchmarks F, what does that have to do with the real world? Actually, in view of the NFL theorem, it may indicate that A performs *poorly* on real-world problems. After all, if A performs better on the set F, which includes only benchmark problems but no real-world problems, then B will perform better on the set $\bar{F}$, which includes (among others) all real-world problems. The constant race to obtain better performance on benchmark problems is leading us to a set of algorithms that have been fine-tuned for good benchmark performance but that may be useless in engineering applications. As John Hooker writes, "The tail wags the dog as problems begin to design algorithms" [Hooker, 1995].

In view of this discussion, it seems that EA papers should place more emphasis on applications and less emphasis on benchmarks. Sometimes it is difficult to get a paper accepted for publication if it does not include results from the currently-fashionable set of benchmark functions. However, this over-reliance on standard benchmarks gives us a false confidence. It is rare to see a real application in an EA journal today unless it is in a special "applications" issue of the journal. But in view of the NFL theorem, applications in EA papers should be the rule rather than the exception. It is only by testing EA performance on real-world problems that we can have any confidence that the EA will be useful.

B.2.2 How to Report (and How Not to Report) Simulation Results

This section shows how authors often present simulation results in a misleading way (hopefully unintentionally), and how they can present results to communicate whatever message the author desires. It also shows, using very little statistical background, how simulation results can be presented in a more clear and honest way. We defer a more rigorous discussion of statistics to Section B.2.4.

Suppose that we want to compare the performance of evolutionary algorithms A and B on some benchmark. Suppose that we run algorithms A and B on some benchmark problem. Each algorithm runs for T generations. At each generation we measure the cost f_{min} of the best individual in the entire population. This gives us a set of data that looks like the following:

$$\begin{aligned} \text{Algorithm } A\text{: } f_{A,\text{min}} &= \{f_{A0}, f_{A1}, f_{A2}, \cdots, f_{AT}\} \\ \text{Algorithm } B\text{: } f_{B,\text{min}} &= \{f_{B0}, f_{B1}, f_{B2}, \cdots, f_{BT}\} \end{aligned} \tag{B.1}$$

where f_{A0} and f_{B0} are the cost values of the best individual after initialization, and f_{Ai} and f_{Bi} are the cost values of the best individual after the i-th generation. The performance of algorithm A can be quantified by the cost of the best individual that it found during the T generations:

$$\text{Algorithm } A \text{ metric: } \min_{i\in[0,T]} f_{Ai}. \tag{B.2}$$

Assuming that we are using elitism, f_{Ai} is monotonically nonincreasing, so

$$\text{Algorithm } A \text{ metric: } \min_{i\in[0,T]} f_{Ai} = f_{AT}. \tag{B.3}$$

If we run algorithms A and B on a given benchmark and we want to see which algorithm performs best, we can simply compare f_{AT} and f_{BT}. We might conclude

$$\left.\begin{array}{l} A \text{ is better than } B \text{ if } f_{AT} < f_{BT} \\ B \text{ is better than } A \text{ if } f_{BT} < f_{AT} \end{array}\right\} \text{ Invalid Reasoning.} \tag{B.4}$$

As noted above, this reasoning is not valid. We must not use flawed logic like that shown in Equation (B.4). Of primary importance is the fact that EAs are stochastic; that is, they depend on the output of a random number generator. Therefore, in general, a given EA will produce different results each time it runs. If we run the experiment once we might conclude that algorithm A is best, but if we run it again we might conclude that algorithm B is best. This is why it is critically important to run multiple simulations when comparing EAs, and to use the results from all of the simulations when comparing algorithms. Using multiple simulations, each seeded with a different random number seed, is called a Monte Carlo simulation, Monte Carlo experiment, or Monte Carlo method [Robert and Casella, 2010].

So now suppose that we have grasped the concept of Monte Carlo simulation. We therefore run algorithms A and B on some problem, and we run the algorithms M times each, where M is the number of Monte Carlo simulations. Each time we run a simulation we get a different result for f_{AT} and f_{BT}. We use the notation f_{ATk} and f_{BTk} to denote the cost of algorithms A and B at the end of the T-th generation of the k-th Monte Carlo simulation. We can write our simulation data as follows:

$$\begin{aligned} \text{Algorithm } A \text{ results: } & \{f_{AT1}, f_{AT2}, \cdots, f_{ATM}\} \\ \text{Algorithm } B \text{ results: } & \{f_{BT1}, f_{BT2}, \cdots, f_{BTM}\}. \end{aligned} \tag{B.5}$$

Now we can do a few different things with these results. First, we can compare the average performance of the two algorithms:

$$\begin{aligned} \bar{f}_A &= \frac{1}{M}\sum_{k=1}^{M} f_{ATk} \\ \bar{f}_B &= \frac{1}{M}\sum_{k=1}^{M} f_{BTk}. \end{aligned} \tag{B.6}$$

Second, we can compare the variance of the performance of the two algorithms:[1]

$$\begin{aligned} \sigma_A^2 &= \frac{1}{M-1}\sum_{k=1}^{M} (f_{ATk} - \bar{f}_A)^2 \\ \sigma_B^2 &= \frac{1}{M-1}\sum_{k=1}^{M} (f_{BTk} - \bar{f}_B)^2. \end{aligned} \tag{B.7}$$

Third, we can compare the best-case performance of the two algorithms:

$$\begin{aligned} \bar{f}_{A,\text{best}} &= \min_{k\in[1,M]} f_{ATk} \\ \bar{f}_{B,\text{best}} &= \min_{k\in[1,M]} f_{BTk}. \end{aligned} \tag{B.8}$$

Fourth, we can compare the worst-case performance of the two algorithms:

$$\begin{aligned} \bar{f}_{A,\text{worst}} &= \max_{k\in[1,M]} f_{ATk} \\ \bar{f}_{B,\text{worst}} &= \max_{k\in[1,M]} f_{BTk}. \end{aligned} \tag{B.9}$$

[1]Intuitively we expect that the denominator in Equation (B.7) should be M instead of $M-1$. However, using $M-1$ in the denominator gives a better estimate of the variance [Simon, 2006, Problem 3.6].

Each of these metrics quantifies a different type of performance. The average cost values of Equation (B.6) tell us how good the algorithms perform on average. The variances of Equation (B.7) tell us the consistency of the performance of the algorithms. The best-case cost values of Equation (B.8) tell us which algorithm we can expect to give the best results if we run it multiple times. The worst-case cost values of Equation (B.9) tell us which algorithm we can expect to give the best results if we can only run the algorithm once and we happen to seed the random number generator in such a way that we get unusually bad performance. [Eiben and Smit, 2011] presents some additional discussion of EA performance metrics.

Suppose that we run algorithms A and B on a benchmark problem. We run the algorithms M times each, and we calculate the following metrics:

$$\begin{array}{llll} \bar{f}_A = 14, & \sigma_A^2 = 4, & \bar{f}_{A,\text{best}} = 7, & \bar{f}_{A,\text{worst}} = 23 \\ \bar{f}_B = 16, & \sigma_B^2 = 3, & \bar{f}_{B,\text{best}} = 6, & \bar{f}_{B,\text{worst}} = 25. \end{array} \tag{B.10}$$

Any of the following statements could appear in a paper that we write about this experiment.

1. Algorithm A obtains an average minimum cost of 14, while algorithm B obtains an average minimum cost of 16. This shows that algorithm A performs better.

2. Algorithm A has a variance of 4, while algorithm B has a variance of 3. This shows that algorithm B is more robust.

3. Algorithm A obtained a best cost of 7, while algorithm B obtained a best cost of 6. This shows that algorithm B performs better.

4. Algorithm A obtained a cost that was never worse than 23, while algorithm B obtained a cost that was never worse than 25. This shows that algorithm A performs better.

None of these statements are exactly false, but they show how we interpret statistics according to preconceived ideas. That is, we are not objective. We tend to present results in a way that favors whichever algorithm agrees with our bias.

Consider the first statement above. It is indeed true that, on average, algorithm A performs better that algorithm B. However, why did we choose to include that statement in our paper instead of statements 2 or 3? Furthermore, how significant is the improved performance that is noted in statement 1? Algorithm A performs only 2 units better than algorithm B, which is within the standard deviations of the algorithms. Figure B.3 shows a plot of the mean and the standard deviations of the algorithms. The figure on the left implies that algorithm A is significantly better than algorithm B, but the figure on the right shows that the improved performance is not as impressive as we might otherwise think. Any given simulation of the algorithms could easily result in algorithm B outperforming algorithm A if they performed one standard deviation differently than their averages.

Box plots are a nice way to plot EA performance results. Box plots include five pieces of data for each algorithm: the smallest result, the lower quartile, the median, the upper quartile, and the largest result of the M Monte Carlo simulations [McGill et al., 1978]. The lower quartile is the value that is greater than 25% of all results, the median is the value that is greater than 50% of all results, and the upper quartile is the value that is greater than 75% of all results.[2] The MATLAB® Statistics Toolbox™ has a `boxplot` function to create box plots. Figure B.4 shows an example of a box plot. Box plots show a lot of relevant information in a single graph. They show median results, they show typical results as the middle 50%, and they show extreme results. This makes box plots a concise and informative way to present results and to compare algorithms.

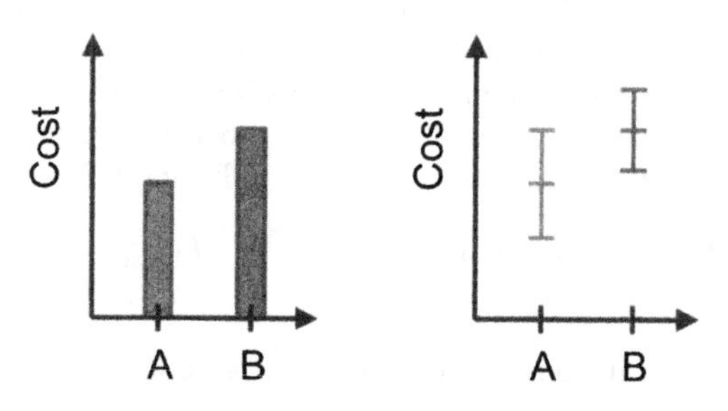

Figure B.3 The means (left figure), and the means and standard deviations (right figure), of the performance of two hypothetical algorithms. The figure on the left makes algorithm A look more impressive than the figure on the right.

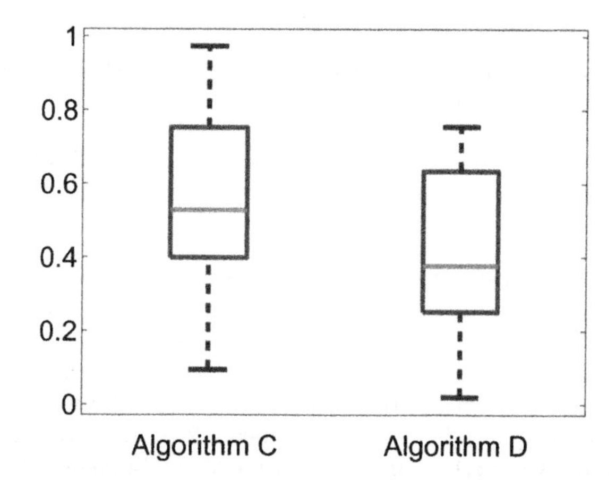

Figure B.4 Typical box plot showing performance results of two hypothetical algorithms. Each box shows the middle 50% of the set of results for an algorithm. The line inside each box shows the median result. The lines above and below each box (connected by dashed lines to each box) show the maximum and minimum results.

However, there are not any foolproof ways to prevent bias in the presentation of simulation results. For example, suppose that we run some simulations for algo-

[2]Note that box plots contain five quartiles: the 0%, the 25%, the 50%, the 75%, and the 100% quartiles. The 0% quartile is the minimum value, the 50% quartile is the median value, and the 100% quartile is the maximum value.

rithms A and B, and that we collect five quartile values for each set of simulations. Suppose that we do this for six benchmark functions. Our results might look something like Table B.1. We see that algorithm A (the current state of the art) performs better on benchmark functions 1, 2, and 3, while algorithm B (our newly proposed algorithm) performs better on benchmarks 4, 5, and 6. What should we do in this situation? We have spent many months developing, refining, coding, debugging, and testing our new algorithm. Furthermore, we need publications to get our graduate degree, or to get tenure and funding. Many researchers in this situation are tempted to write a paper that includes benchmarks 4, 5, and 6, and that omits the results from benchmarks 1, 2, and 3.

	F_1	F_2	F_3	F_4	F_5	F_6
Algorithm A	13	21	16	45	24	33
Algorithm B	16	30	22	36	17	28

Table B.1 Sample simulation results for two algorithms on six benchmarks. The hypothetical numbers are the mean values of a set of Monte Carlo simulations. Algorithm A is the state of the art, and algorithm B is the researcher's newly developed algorithm. Given these results, many researchers will choose to publish data from benchmarks 4, 5, and 6, and will omit the results from benchmarks 1, 2, and 3.

These types of practices are unethical. Of all the widespread practices in research and publishing today, this is not nearly the worst, but most of us would agree that it is dishonest. However, ethics is not always black-and-white. This situation occurs in modulated forms far more frequently than the obvious selective presentation illustrated in Table B.1. For instance, if we obtain preliminary results that do not speak well of our research, it is easy to convince ourselves that those results are suspect for some reason,[3] and to then obtain more extensive results using only the benchmarks that speak well of our research.

Peer review standards encourage this subtle form of dishonesty. If a newly proposed algorithm, or variation of an existing algorithm, does not give better results than the state of the art, it is often dismissed and rejected from being published. This occurs even though the state of the art has, by definition, evolved over many decades, while new algorithms are, by definition, in their infancy and have not yet had a chance to mature. Proposals and papers during the 1960s that were related to genetic algorithm research were routinely rejected because GAs were not yet as good as established optimization approaches. A criticism that was levied in 1966 is typical: "Everything you have suggested [for genetic algorithm research] can be put much more clearly and sharply in terms of tree searches" [Fogel, 1999, page xi]. The research establishment is ostensibly open to new ideas, but like any establishment, we tend to draw in on ourselves and discourage those who think outside the box.

Instead of encouraging incremental results and insisting on continuous improvements in benchmark results, peer reviewers should instead encourage novelty and

[3]Perhaps we are unsure of the mathematical formulation of the benchmark. Perhaps the benchmark takes too much CPU time and we want to focus on faster benchmarks. Perhaps the benchmark is not as popular as others. Perhaps the benchmark has some characteristics that convince us that it is atypical. We can find many reasons to exclude any benchmark, if we look hard enough.

creativity, because we cannot predict which new algorithms or technologies will make an impact in the future. Also, in view of the NFL theorem, peer reviewers should not just encourage, but should insist on authors discussing the shortcomings and drawbacks of their proposed algorithms. Authors should not feel pressured to hide negative results, but should be required to publicize negative results in the interest of openness and transparency.

Success Rate

One common performance metric in EAs is success rate [Suganthan et al., 2005], [Liang et al., 2006], [Mallipeddi and Suganthan, 2010]. That is, given some benchmark cost function $f(x)$ with known minimum f^*, we conduct M simulations of our EA, each for a specific number of function evaluations F_{max}. We consider a given simulation a success if it finds some individual x such that $f(x) < f^* + \epsilon$, where ϵ is a positive success threshold. The success rate S is the percentage of simulations that are successful.

This metric has become so popular that many editors and reviewers will not publish a paper without it. However, there are a few problems with this performance metric. First, the choice of ϵ is arbitrary and can have a significant effect on the apparent merits of two competing algorithms. If $\epsilon = \epsilon_1$ then algorithm A might outperform algorithm B, but if $\epsilon = \epsilon_2 \neq \epsilon_1$ then algorithm B might outperform algorithm A. Second, the number of function evaluations F_{max} is arbitrary and can also have an effect on the apparent merits of two competing algorithms. If $F_{\text{max}} = F_1$ then algorithm A might outperform algorithm B, but if $F_{\text{max}} = F_2 \neq F_1$ then algorithm B might outperform algorithm A. Third, the number of simulations M needs to be unrealistically large to obtain a confident estimate of the success rate S. We might use $M = 20$ and find $S = 30\%$, and then come back the next day and run 20 more simulations and find $S = 40\%$. We might need to use a very large value of M to obtain a reasonably small standard deviation in S, but such large values of M might require many days or weeks for data collection [Clerc, 2012b]. When success rate is reported in publications, it is never accompanied by the standard deviation of success rate, and this negates its usefulness as a metric.

Population Size

For a given problem, EA #1 might perform best with population size N_1, while EA #2 might perform best with population size N_2. Therefore, when we compare two EAs, we might first need to tune each EA separately to find their optimal population sizes. On the other hand, the purpose of our comparison might be to see which EA performs best with a given population size. In this case we should use the same population size for both EAs. For instance, EA performance with small population sizes might be an important research topic. So the question of what population sizes to use when comparing EAs is not clear cut, but depends on the purpose of the comparison.

B.2.3 Random Numbers

To obtain valid EA simulation results, it is essential to clearly understand random number generators. The random number generator can have a significant influence on EA performance. The quality of an EA solution is not necessarily correlated with the quality of the random number generator in the EA [Clerc, 2012b]. For some problems and some EAs better random number generation leads to better EA performance, while for other problems and other EAs worse random number generation leads to better EA performance.

Most random numbers are not really random. In fact, the definition and existence of randomness are subject to some dispute. Random number generators in computers do not generate truly random numbers; they generate *pseudo*-random numbers. Pseudo-random numbers appear to be random, but they are not really random because they are generated with deterministic algorithms.

The rest of this section focuses on MATLAB for the sake of illustration, but it can easily be generalized to any other programming environment. Consider MATLAB's `rand` function, which returns a random number that is uniformly distributed on the open interval $(0, 1)$. Suppose we begin a MATLAB session using MATLAB Version 7.13.0.564 (R2011b) on a 32-bit Windows XP computer. If we use `rand` to generate five random numbers, we obtain the numbers

$$0.8147,\ 0.9058,\ 0.1270,\ 0.9134,\ 0.6324. \tag{B.11}$$

This looks pretty random. But if we log off the computer, come back the next day, start MATLAB again, and use `rand` to again generate five random numbers, we obtain the exact same numbers as those shown above! Suddenly the numbers do not appear to be very random. This is because MATLAB uses a deterministic algorithm to generate pseudo-random numbers, and that algorithm has the same initial state each time MATLAB begins.

This fact has important implications for EA testing. Suppose we turn on our computer, run an EA, and obtain the performance result f_1. Now suppose we log off, come back the next day, log on, start MATLAB, and run the EA again. We will obtain exactly the same result because the random numbers that we generate in our EA for selection, mutation, and other purposes, will be exactly the same as the previous day. Although an EA is stochastic and hence should theoretically give a different result each time it runs, in practice it will give the same result from one day to the next because it relies on a pseudo-random number generator that returns numbers that are not truly random, and that is initialized to the same state each time MATLAB begins.

Now suppose that we turn on the computer, run the EA, and obtain the result f_1. We leave MATLAB running, come back the next day, and run the EA again. This time we will, in general, obtain a different result. This is because we did not reinitialize the random number generator by restarting MATLAB. So the second time we run the EA, MATLAB will generate a different string of random numbers than the first time we run the EA.

If we want to generate the same string of random numbers, we can use the MATLAB `rng(seed)` function to reinitialize the random number generator to the given integer seed.[4] This initializes the random number generator so that it begins in the given state and thus generates the same sequence of random numbers from one execution to the next. For instance, the MATLAB commands

[4]Function `rng` was introduced around the year 2011. Earlier versions of MATLAB provide other functions to initialize the random number generator.

```
rng(489)
rand, rand, rand, rand, rand                    (B.12)
```

generate the random numbers 0.9780, 0.8578, 0.0599, 0.0894, 0.4774, no matter when the commands are performed, and no matter how long MATLAB has been running beforehand. With this approach we can create identical EA results from one simulation to the next even though EA results are intrinsically stochastic. Sometimes an EA bug or result shows up only under special circumstances. To re-create the bug or result, we need to initialize the random number generator with the same seed. That is why it is always a good idea to keep track of the initial random number generator seed each time we run an EA. Keeping track of the seed allows us to precisely reproduce results, which greatly expedites debugging.

If we want to generate a different string of random numbers each time we run our EA, we can again use the MATLAB `rng` function, but instead provide it with a random seed. For example, an integer that represents the current date and time is pretty random, and so using the date and time as a seed results in a different sequence of random numbers each time we run MATLAB. The MATLAB command `clock` returns a six-element vector containing the current year, month, day, hours, minutes, and seconds. So the MATLAB commands

```
Seed = sum(100*clock);
rng(Seed);
rand, rand, rand, rand, rand                    (B.13)
```

generate a different sequence of random numbers each time we execute them. With this approach, we can create randomly-varying EA results from one simulation to the next, even though we run our simulation the first thing in the morning after we start MATLAB. This approach also allows us to keep track of the random number seed so that we can record it and use it later in case we need to debug our code.

Now suppose that we want to perform a simple comparison of two EAs. The first random process in our EAs is the generation of the initial population. So we run the first EA and obtain some results. We then run the second EA and obtain different results. However, if we do not reinitialize the random number generator between simulations, the two EAs will start with entirely different initial populations. This may give an unfair advantage to one EA. If we want to more fairly compare different EAs, they should start with the same initial population. We could do this with the following command sequence:

```
Seed = sum(clock);
rng(Seed);
EA1;
rng(Seed);
EA2;                                            (B.14)
```

This sequence ensures that the random number generator is initialized identically for both EAs, so both EAs will begin with the same random population, assuming that the random population is created the same way in EA1 and EA2.

B.2.4 T-Tests

The t-test was invented in 1908 by William Sealy Gosset, an Irish chemist who used it to monitor the quality of beer for the brewery where he worked [Box, 1987]. He published the method under the pen name "Student" because his employer did not want its competitors to know that they were using statistical methods. The t-test has many applications, but in this section we restrict our discussion to a very specific application related to the interpretation of EA experiments, and we present its use without any derivations. Standard statistics books include additional details and derivations of the t-test [Salkind, 2007].

The basic question that the t-test answers is, How can we tell if two sets of experimental results are significantly different? When we say *significantly different* here, we do not mean how large the difference is; we instead refer to the question of whether or not the difference is fundamental, or if it is instead due to random fluctuations. For instance, suppose that we measure the performance of two students on two separate quizzes during a course:

$$\begin{aligned} &\text{Student } A\text{: } 69\%,\ 84\% \\ &\text{Student } B\text{: } 66\%,\ 83\%. \end{aligned} \tag{B.15}$$

Student A had a higher score than student B on both quizzes. But we would probably not conclude that there was a difference between the two students; we would attribute the superior performance of student A to chance, and we would hypothesize that the two students have essentially identical performance and ability. But now suppose that we have 10 quizzes on which to base our judgment:

$$\begin{aligned} &\text{Student } A\text{: } 69,\ 84,\ 75,\ 93,\ 92,\ 88,\ 68,\ 74,\ 89,\ 81\% \\ &\text{Student } B\text{: } 66,\ 83,\ 72,\ 88,\ 95,\ 83,\ 71,\ 71,\ 84,\ 80\%. \end{aligned} \tag{B.16}$$

Student A had a higher score eight times out of 10. Now we would start to suspect that student A really is a better performer than student B, and that their differences in performance are *not* simply due to random variations. The differences are not large but it appears that they are statistically significant. But how can we quantify the probability that this hypothesis is true or false? In particular, what is the probability that we would obtain the results of Equation (B.16) if students A and B were equal performers? That is, what is the probability that the differences in the results of Equation (B.16) are simply due to random variations? This is exactly the question that the t-test answers. The hypothesis that the differences in the results of students A and B are simply due to random variations is called the null hypothesis. Note that if students A and B are equal performers and we give each student 10 quizzes, then it is equally likely that either student A or B performs better most of the time.

Now we return to the problem of analyzing EA simulation results. Suppose that we run algorithms A and B on some optimization problem. We know the importance of Monte Carlo simulations, so we run each algorithm M times and compute the average performance and variance of each algorithm as shown in Equations (B.6) and (B.7):

$$\begin{aligned} &\text{Algorithm } A\text{: Mean } = \bar{f}_A,\ \text{Variance } = \sigma_A^2 \\ &\text{Algorithm } B\text{: Mean } = \bar{f}_B,\ \text{Variance } = \sigma_B^2. \end{aligned} \tag{B.17}$$

What is the probability that we obtain these results if the performance of algorithms A and B is fundamentally identical? This is the question that the t-test answers. The t-test statistic is defined as

$$t = \frac{|\bar{f}_A - \bar{f}_B|}{\sqrt{(\sigma_A^2 + \sigma_B^2)/M}}. \tag{B.18}$$

We see that t measures the difference between the outcomes of algorithms A and B. If there is a big difference between $\bar{f}_A$ and $\bar{f}_B$, then t will be large. However, if σ_A^2 or σ_B^2 are large, then that indicates that there is a large variation in the performance of algorithm A or B, which dilutes the effect of a large difference between $\bar{f}_A$ and $\bar{f}_B$, and which makes t small. Large M tends to make t large since a large number of experiments is a more reliable indicator of performance than a small number of experiments.

After we compute the t-test statistic, we compute a quantity called the degrees of freedom:

$$d = \frac{(M-1)(\sigma_A^2 + \sigma_B^2)^2}{\sigma_A^4 + \sigma_B^4}. \tag{B.19}$$

The degrees of freedom indirectly tells us how large t needs to be to give us a specified level of confidence that there is a statistically significant difference in the performance of algorithms A and B.

After we compute t and d, we look in a t-test table to find the probability that the difference in the performance of algorithms A and B is due to random variation rather than a fundamental difference between the two algorithms. These probabilities are called p-values. Table B.2 shows some t-test values. More complete tables can be found in statistics books or on the internet. To approximate values between the coordinates in the table, we can use any reasonable interpolation method. We can also use t-test functions in statistical software, including MATLAB, Microsoft Excel®, and many other software packages.

■ EXAMPLE 2.5

Suppose we run algorithms A and B on some optimization problem. We run each algorithm six times ($M = 6$) and obtain the following results, which are written in the format of Equation (B.5):

$$\begin{aligned} &\text{Algorithm } A\text{: } \{f_{ATk}\} = \{30.02, 29.99, 30.11, 29.97, 30.01, 29.99\} \\ &\text{Algorithm } B\text{: } \{f_{BTk}\} = \{29.89, 29.93, 29.72, 29.98, 30.02, 29.98\}. \end{aligned} \tag{B.20}$$

To estimate the probability of obtaining these differences simply due to random variation and not due to any fundamental difference between the performance of the two algorithms, we first calculate the mean and variance of the results as shown in Equations (B.6), (B.7), and (B.17):

$$\begin{aligned} \bar{f}_A &= 30.015, \qquad \sigma_A^2 = 0.0497 \\ \bar{f}_B &= 29.920, \qquad \sigma_B^2 = 0.1079. \end{aligned} \tag{B.21}$$

Next we use Equations (B.18) and (B.19) to compute the t-test statistic and the degrees of freedom:

$$t = 1.959, \quad d = 7.0306. \tag{B.22}$$

Finally we look in Table B.2 and see that for $d = 7.0306$, the t-test statistic is equal to 1.959 at a p-value of about 9%. This means that if the performance of algorithms A and B is fundamentally equivalent, there is a 9% probability that we would get the results shown in Equation (B.20). Equivalently, we can say that if the performance of algorithms A and B is equivalent, there is a 91% probability that we would see smaller variations that those shown in Equation (B.5). Whether these probabilities are significant enough for us to conclude that the algorithms are fundamentally different is a qualitative judgment.

□

There are several assumptions that underlie the t-test discussion of this section.

1. First, we assume that algorithm A could be either better or worse than algorithm B, so we use the two-sided t-test in this section. That is why the caption of Table B.2 indicates that it is for two-sided t-tests. The column headings (that is, the p-values) of Table B.2 would change if we wanted to run one-sided t-tests, but one-sided tests are generally not relevant for the analysis of EA experiments.

2. The t-test assumes that each experiment's results follows a Gaussian distribution. That is, if we run many simulations for algorithm A and plot the results in a histogram, we will see a Gaussian curve; the same assumption applies to algorithm B also. Since the limiting values in a Gaussian distribution are $\pm\infty$, there are no computer simulations or physical experiments that are truly Gaussian. But many processes can be approximated quite well with Gaussian distributions. The central limit theorem assures us that many experiments and simulations have results that are nearly Gaussian [Grinstead and Snell, 1997], so Gaussianity is a reasonable assumption. However, verifying that our process really is Gaussian is a separate problem. In general, it is safe to assume Gaussianity in the absence of evidence to the contrary.

3. The t-test assumes that we have only two sets of data. If we have more than two sets of data – for instance, results from algorithms A, B, and C – then we cannot use pair-wise t-tests on the pairs A/B, A/C, and B/C to test for statistically significant differences. We discuss this in more detail below in Section B.2.5.

4. This section assumes that the two sample sizes are the same; that is, we run M experiments for both algorithms A and B. If we have a different number of experiments for the two algorithms, then we need to slightly modify the equations of this section.

d	50%	40%	30%	20%	10%	5%	2%	1%	0.5%	0.1%
1	1.000	1.376	1.963	3.078	6.314	12.71	31.82	63.66	127.3	636.6
2	0.816	1.061	1.386	1.886	2.920	4.303	6.965	9.925	14.09	31.60
3	0.765	0.978	1.250	1.638	2.353	3.182	4.541	5.841	7.453	12.92
4	0.741	0.941	1.190	1.533	2.132	2.776	3.747	4.604	5.598	8.610
5	0.727	0.920	1.156	1.476	2.015	2.571	3.365	4.032	4.773	6.869
6	0.718	0.906	1.134	1.440	1.943	2.447	3.143	3.707	4.317	5.959
7	0.711	0.896	1.119	1.415	1.895	2.365	2.998	3.499	4.029	5.408
8	0.706	0.889	1.108	1.397	1.860	2.306	2.896	3.355	3.833	5.041
9	0.703	0.883	1.100	1.383	1.833	2.262	2.821	3.250	3.690	4.781
10	0.700	0.879	1.093	1.372	1.812	2.228	2.764	3.169	3.581	4.587
11	0.697	0.876	1.088	1.363	1.796	2.201	2.718	3.106	3.497	4.437
12	0.695	0.873	1.083	1.356	1.782	2.179	2.681	3.055	3.428	4.318
13	0.694	0.870	1.079	1.350	1.771	2.160	2.650	3.012	3.372	4.221
14	0.692	0.868	1.076	1.345	1.761	2.145	2.624	2.977	3.326	4.140
15	0.691	0.866	1.074	1.341	1.753	2.131	2.602	2.947	3.286	4.073
16	0.690	0.865	1.071	1.337	1.746	2.120	2.583	2.921	3.252	4.015
17	0.689	0.863	1.069	1.333	1.740	2.110	2.567	2.898	3.222	3.965
18	0.688	0.862	1.067	1.330	1.734	2.101	2.552	2.878	3.197	3.922
19	0.688	0.861	1.066	1.328	1.729	2.093	2.539	2.861	3.174	3.883
20	0.687	0.860	1.064	1.325	1.725	2.086	2.528	2.845	3.153	3.850
21	0.686	0.859	1.063	1.323	1.721	2.080	2.518	2.831	3.135	3.819
22	0.686	0.858	1.061	1.321	1.717	2.074	2.508	2.819	3.119	3.792
23	0.685	0.858	1.060	1.319	1.714	2.069	2.500	2.807	3.104	3.767
24	0.685	0.857	1.059	1.318	1.711	2.064	2.492	2.797	3.091	3.745
25	0.684	0.856	1.058	1.316	1.708	2.060	2.485	2.787	3.078	3.725
26	0.684	0.856	1.058	1.315	1.706	2.056	2.479	2.779	3.067	3.707
27	0.684	0.855	1.057	1.314	1.703	2.052	2.473	2.771	3.057	3.690
28	0.683	0.855	1.056	1.313	1.701	2.048	2.467	2.763	3.047	3.674
29	0.683	0.854	1.055	1.311	1.699	2.045	2.462	2.756	3.038	3.659
30	0.683	0.854	1.055	1.310	1.697	2.042	2.457	2.750	3.030	3.646
40	0.681	0.851	1.050	1.303	1.684	2.021	2.423	2.704	2.971	3.551
50	0.679	0.849	1.047	1.299	1.676	2.009	2.403	2.678	2.937	3.496
60	0.679	0.848	1.045	1.296	1.671	2.000	2.390	2.660	2.915	3.460
80	0.678	0.846	1.043	1.292	1.664	1.990	2.374	2.639	2.887	3.416
100	0.677	0.845	1.042	1.290	1.660	1.984	2.364	2.626	2.871	3.390

Table B.2 Two-sided t-test table. Each row corresponds to a degree of freedom d, and each column corresponds to a p-value probability. The numbers in the table show the t values for the given degrees of freedom d and the given p-value probability that differences between two sets of experiments are entirely due to random variation.

Many people misinterpret t-test results. As we noted earlier in this section, the t-test gives the probability that our results would be obtained if the performance of algorithms A and B were fundamentally equivalent. We can write this as

$$p = \Pr(R = r | A = B). \tag{B.23}$$

In words, the p-value is equal to the probability that the results R are equal to the obtained values r, given that the performance of algorithms A and B is equivalent.[5] Here we correct some common misunderstandings about the p-value.

1. $p \neq \Pr(A = B)$. That is, the p-value is *not* equal to the probability that the performance of the algorithms is equivalent.

2. $1 - p \neq \Pr(A \neq B)$. This is very similar to the above statement. That is, we can *not* use the p-value to derive the probability that the performance of the algorithms is different.

3. $p \neq \Pr(A = B | R = r)$. That is, the p-value is *not* equal to the probability that the performance of the algorithms is equivalent, given the results that we obtained. From Bayes' theorem [Grinstead and Snell, 1997] we know that

$$\begin{aligned} \Pr(A = B | R = r) &= \frac{\Pr(R = r | A = B) \Pr(A = B)}{\Pr(R = r)} \\ &= \frac{p \Pr(A = B)}{\Pr(R = r)}. \end{aligned} \tag{B.24}$$

   So $\Pr(A = B | R = r)$, which is the probability that the performance of algorithms A and B is equivalent given the results that we observed, is directly proportional to the p-value. But if we want to compute a numerical value for $\Pr(A = B | R = r)$, we need to know not only p, but we also need to know $\Pr(A = B)$ (that is, the a-priori probability that the performance of the two algorithms is equivalent), and $\Pr(R = r)$ (that is, the a-priori probability that we obtained the results that we obtained).

4. $p \neq \Pr(R = r)$. That is, the p-value is *not* equal to the a-priori probability of obtaining the results that we obtained. As Equation (B.23) shows, p is equal to the a-posterior probability of obtaining the results that we observed, *given* that the performance of the algorithms is equivalent.

5. Suppose that p is a small number, so we conclude from Equation (B.24) that $A \neq B$. Then $(1 - p)$ is not equal to the probability that a second experiment would give the same conclusion.

6. The p-value does not tell us quantitatively *how* different the two algorithms perform. However, the p-value does have a positive correlation with the magnitude of the difference, so a larger p-value indicates a larger difference.

It may seem like nitpicking to obsess about the details of the meaning of the p-value, but there can be large differences between the true meaning of the p-value and the common but false interpretations noted above [Johnson, 1999], [Schervish, 1996], [Sterne and Smith, 2001].

[5]Note that the notation $A = B$ does *not* mean that the two algorithms are equivalent, but rather that their *performances* are equivalent.

B.2.5 F-Tests

We can use the F-test, like the t-test, for a variety of tasks. This section discusses one simple application of the F-test that is particularly useful for analyzing EA results. As with the t-test, we restrict our discussion of the F-test to a very specific application related to the interpretation of EA experiments, and we present its use without any derivations.

Suppose that we test algorithms 1, 2, and 3 on an optimization problem. We want to use the results to judge whether there is a statistically significant difference between the three algorithms. As mentioned in the previous section, we can *not* simply perform pair-wise t-tests on algorithms 1 and 2, 1 and 3, and then 2 and 3. As a simple intuitive explanation for why pair-wise t-testing does not work, suppose that we roll a fair, six-sided die. We know that we have a $1/6 \approx 17\%$ chance of rolling a 1. Now suppose that we roll the die three times. We have a probability of $1-(5/6)^3 \approx 42\%$ of rolling a 1 on at least one of those throws. The probability that an event happens after one trial, is not the same as the probability that it happens after more than one trial. The t-test gives us the probability of observing certain differences between two algorithms given that the performance of two algorithms is identical. However, if we perform the test more than once, then the probability of obtaining those differences increases.

Now that we know we cannot use the t-test, let us discuss how to use the F-test. Suppose that we run G separate algorithms on some optimization problem. These separate algorithms could actually be the same algorithm but with G different parameter settings (for example, G different mutation rates). We run each algorithm M times, and compute the average performance and the variance of each algorithm as shown in Equations (B.6) and (B.7):

$$\text{Mean } = \bar{f}_g, \text{ Variance } = \sigma_g^2 \text{ for } g \in [1, G]. \tag{B.25}$$

What is the probability that we obtain these results if the performance of the G algorithms is fundamentally identical? That is, what is the probability that these results are *not* due to any fundamental difference between the G algorithms, but are simply due to random experimental variations? This is the question that the F-test answers. The F-statistic is calculated as

$$\begin{aligned}
\bar{f} &= \frac{1}{G}\sum_{g=1}^{G} \bar{f}_g \\
S_w &= \frac{1}{G}\sum_{g=1}^{G} \sigma_g^2 \\
S_b &= \frac{M}{G-1}\sum_{g=1}^{G} (\bar{f}_g - \bar{f})^2 \\
F &= S_b / S_w
\end{aligned} \tag{B.26}$$

where $\bar{f}$ is the average performance metric of all algorithms, S_w is the within-group variance and measures the average variance of the algorithms, S_b is the between-group variance and measures the variance of the performance of all algorithms, and F is the F-statistic. We see that a large difference beteen the performance of the algorithms corresponds to a large F.

After we compute the F-statistic, we compute quantities that are called the numerator degree of freedom D_n and the denominator degree of freedom D_d:

$$\begin{aligned} D_n &= G - 1 \\ D_d &= G(M - 1). \end{aligned} \tag{B.27}$$

The degrees of freedom indirectly tell us how large F needs to be to give us a specified level of confidence that there is a statistically significant difference in the performance of the algorithms.

After we compute F, D_n, and D_d, we look in an F-test table to find the probability that the difference in the performance of the G algorithms is due to random variation rather than a fundamental difference between two or more of the algorithms. Since we have two degree-of-freedom parameters (D_n and D_d), we need a separate table for each probability level. Tables B.3 and B.4 show some F-test thresholds for probability values of 5% and 1%. More complete tables can be found in statistics books or on the internet. To approximate values between the coordinates in the tables, we can use any reasonable interpolation method. We can also use F-test functions in statistical software, including MATLAB, Microsoft Excel®, and many other software packages.

Note that the F-statistics in Tables B.3 and B.4 decrease with D_n and D_d. That is, as the number of groups G and the number of Monte Carlo experiments M increase, we require smaller differences between the algorithms' performance metrics to conclude that those differences are *not* simply due to random variations. For example, consider Table B.3 with $D_n = D_d = 3$. If $F = 9.27$ there is a 5% probability that the observed differences are due to random variation. If $F > 9.27$ there is a less than 5% probability that the observed differences are due to random variation, so there is a greater than 95% probability that the observed differences are not due random variation. Compare this with $D_n = D_d = 5$. In this case, if $F = 5.05$ there is a 5% probability that the observed differences are due to random variation. If $F > 5.05$ there is a less than 5% probability that the observed differences are due to random variation, so there is a greater than 95% probability that the observed differences are not due random variation.

	$D_n = 1$	2	3	4	5	6	7	8
$D_d = 1$	161.47	199.49	215.74	224.50	230.07	234.00	236.77	238.95
2	18.51	18.99	19.16	19.24	19.29	19.32	19.35	19.37
3	10.12	9.55	9.27	9.11	9.01	8.94	8.88	8.84
4	7.70	6.94	6.59	6.38	6.25	6.16	6.09	6.04
5	6.60	5.78	5.40	5.19	5.05	4.95	4.87	4.81
6	5.98	5.14	4.75	4.53	4.38	4.28	4.20	4.14
7	5.59	4.73	4.34	4.12	3.97	3.86	3.78	3.72
8	5.31	4.45	4.06	3.83	3.68	3.58	3.50	3.43
9	5.11	4.25	3.86	3.63	3.48	3.37	3.29	3.22
10	4.96	4.10	3.70	3.47	3.32	3.21	3.13	3.07

Table B.3 F-test table for 5% probability. Each row corresponds to a denominator degree of freedom D_d, and each column corresponds to a numerator degree of freedom D_n. The numbers in the table show the F values that allow us to conclude that there is a 5% or less probability that differences between multiple sets of experiments are entirely due to random variation.

	$D_n = 1$	2	3	4	5	6	7
$D_d = 1$	4063.25	4992.22	5404.03	5636.51	5760.41	5889.88	5889.88
2	98.50	98.99	99.15	99.26	99.30	99.34	99.34
3	34.11	30.81	29.45	28.70	28.23	27.91	27.67
4	21.19	17.99	16.69	15.97	15.52	15.20	14.97
5	16.25	13.27	12.05	11.39	10.96	10.67	10.45
6	13.74	10.92	9.77	9.14	8.74	8.46	8.25
7	12.24	9.54	8.45	7.84	7.46	7.19	6.99
8	11.25	8.64	7.59	7.00	6.63	6.37	6.17
9	10.56	8.02	6.99	6.42	6.05	5.80	5.61
10	10.04	7.55	6.55	5.99	5.63	5.38	5.20

Table B.4 F-test table for 1% probability. Each row corresponds to a denominator degree of freedom D_d, and each column corresponds to a numerator degree of freedom D_n. The numbers in the table show the F values that allow us to conclude that there is a 1% or less probability that differences between multiple sets of experiments are entirely due to random variation.

■ EXAMPLE 2.6

Suppose we run algorithms A, B, and C on some optimization problem. We run each algorithm four times ($M = 4$) and obtain the following results, which are written in the format of Equation (B.5):

$$\begin{aligned} &\text{Algorithm } A\text{: } \{f_{ATk}\} = \{4,\ 5,\ 3,\ 2\} \\ &\text{Algorithm } B\text{: } \{f_{BTk}\} = \{6,\ 4,\ 4,\ 5\} \\ &\text{Algorithm } C\text{: } \{f_{CTk}\} = \{5,\ 7,\ 6,\ 6\}. \end{aligned} \tag{B.28}$$

To estimate the probability of obtaining these differences simply due to random variation and not due to any fundamental difference between the performance of the three algorithms, we first calculate the mean and variance of the results as shown in Equations (B.6), (B.7), and (B.25):

$$\begin{aligned} \bar{f}_A &= 3.50, & \sigma_A^2 &= 1.67 \\ \bar{f}_B &= 4.75, & \sigma_B^2 &= 0.92 \\ \bar{f}_C &= 6.00, & \sigma_C^2 &= 0.67. \end{aligned} \tag{B.29}$$

Next we use Equation (B.26) to compute the F-statistic:

$$\begin{aligned} \bar{f} &= 4.75 \\ S_w &= 1.08 \\ S_b &= 6.25 \\ F &= 5.77. \end{aligned} \tag{B.30}$$

Next we use Equation (B.27) to compute the degrees of freedom:

$$\begin{aligned} D_n &= 2 \\ D_d &= 9. \end{aligned} \tag{B.31}$$

Now we look in Table B.3, which is the 5% F-test table, and we see that for these degree-of-freedom values the F-statistic is 4.25. Our F-statistic is 5.77, so if the performance of algorithms A, B, and C is fundamentally equivalent, the probability is less than 5% that we would get the results shown in Equation (B.28). Equivalently, we can say that if the performance of algorithms A, B, and C is equivalent, the probability is more than 95% that we would see smaller variations that those shown in Equation (B.28). We can also look in Table B.4, which is the 1% F-test table, to see that for $D_n = 2$ and $D_d = 9$, the F-statistic is equal to 8.02. Our F-statistic is 5.77, so if the performance of algorithms A, B, and C is fundamentally equivalent, the probability is greater than 1% that we would see the differences shown in Equation (B.28). Equivalently, we can say that if the performance of algorithms A, B, and C is equivalent, the probability is less than 99% that we would see smaller variations that those shown in Equation (B.28). Combining these results with simple linear interpolation, we conclude

$$\begin{aligned} F = 4.25 &\quad \text{gives} \quad p = 5\% \\ F = 8.02 &\quad \text{gives} \quad p = 1\% \\ \text{therefore, } F = 5.77 &\quad \text{gives} \quad p \approx 3.4\%. \end{aligned} \tag{B.32}$$

That is, if the performance of algorithms A, B, and C is fundamentally equivalent, the probability is about 3.4% that we would see the differences shown in Equation (B.28). At this point we can perform pair-wise t-tests, or simpler tests, to see where the algorithm differences lie. Equation (B.29) shows that algorithm C appears to be significantly different than algorithms A and B, so we can conclude that the 3.4% F-test probability is mostly due to algorithm C.

□

Like the t-test, the F-test assumes that each experiment's results follows a Gaussian distribution. Unlike the t-test, F-test results can be greatly affected by non-normal distributions, so F-test results may not be valid if the Gaussian assumption is violated. In these cases we can use non-parametric tests, which do not assume that the data follows any particular probability distribution function. Many non-parametric statistical tests are available to the EA researcher [Good and Hardin, 2009], [Kanji, 2006], including the commonly-used Wilcoxon test [Corder and Foreman, 2009].

B.3 CONCLUSION

The astute reader will notice that this book violates many of the principles discussed in this appendix. For instance, when comparing various algorithms and EA variations in this book, we did not perform any statistical testing. This inconsistency is purposeful, and is not due to laziness, hypocrisy, or time pressure. We could have easily included statistical tests in this book. But the purpose of this book is primarily instruction rather than research. Therefore, the chapters are *not* laid out or presented in a way that is suitable for a journal or research monograph. If we had adhered to peer-review standards in this book, then it would have been overly full of technical and numerical details that would have obscured the simplicity of the algorithms and results.

The overall goal of this book is to provide a simple, broad, down-to-earth, basic education in the area of EAs. The specific goal of this appendix is to encourage researchers and peer reviewers to think more carefully about standards in EA research. Some additional resources that provide excellent guidelines for conducting EA experiments and reporting results include [Barr et al.,] and [Črepinšek et al., 2013].

APPENDIX C

Benchmark Optimization Functions

> ...algorithms can become customized for a particular set of test problems; this is troubling if the test problems do not represent the types of problems that evolutionary algorithms are best suited for in practice.
>
> —Darrell Whitley [Whitley et al., 1996]

This appendix presents some standard benchmark optimization problems that can be used to compare optimization algorithms. We use $x = [x_1, \ldots, x_n]$ to represent the n-dimensional domain of the function, and $f(x)$ to represent the scalar function value.

Appendix C.1 presents unconstrained optimization benchmarks, Appendix C.2 presents constrained optimization benchmarks, and Appendix C.3 presents multi-objective optimization benchmarks, for which $f(x)$ is a vector. Appendix C.4 presents dynamic optimization benchmarks, Appendix C.5 presents noisy optimization benchmarks, and Appendix C.6 presents traveling salesman benchmarks.

Some optimization algorithms are naturally biased toward certain types of search spaces. Because of this, it is important to modify the benchmarks in this appendix by incorporating offsets and rotation matrices in the problem. We discuss this important point in Appendix C.7.

*

Benchmarks are useful and important for obtaining comparative results between different EAs. But in the final analysis, it is more interesting and useful to test optimization algorithms on problems that have applications in the real world.

C.1 UNCONSTRAINED BENCHMARKS

The problem is to minimize $f(x)$ over all x. We use x^* to represent the optimizing value of x, and $f(x^*)$ is the minimum value of $f(x)$:

$$x^* = \arg\min_x f(x). \tag{C.1}$$

Many of the benchmarks that we present in this section are from [Bäck, 1996], [Cai and Wang, 2006], and [Yao et al., 1999]. Detailed information about the unconstrained benchmarks and evaluation metrics for EA competitions at the 2005 IEEE Congress on Evolutionary Computation can be found in [Suganthan et al., 2005], and [Ali et al., 2005] also includes many unconstrained benchmarks. [Floudas et al., 2010] is an entire book that is devoted to the definition of unconstrained optimization benchmarks. We restrict the benchmarks presented here to those functions that can be defined for any number of dimensions n. Many other benchmarks have been proposed, including some with a fixed number of dimensions. But we think that it is more interesting to test optimization algorithms on benchmarks whose dimensionality can be varied so that performance can be explored as a function of the number of dimensions. The domains that we specify in the following subsections are common, but researchers have also used many other domains.

C.1.1 The Sphere Function

The sphere function is given as

$$\begin{aligned} f(x) &= \sum_{i=1}^{n} x_i^2 \\ x^* &= 0 \\ f(x^*) &= 0 \end{aligned} \tag{C.2}$$

where $x_i \in [-5.12, +5.12]$. This is called function 1 in Ken De Jong's thesis [De Jong, 1975], and it is Problem 1.1 and Problem 2.17 in [Schwefel, 1995]. Figure C.1 shows a plot of $f(x)$ in two dimensions. This is a very simple optimization problem, and almost any reasonable algorithm should be able to find its minimum accurately, but it provides a good preliminary test for optimization algorithms. It also provides a good benchmark for comparison between algorithms, because many optimization problems are approximately quadratic near their minimum.

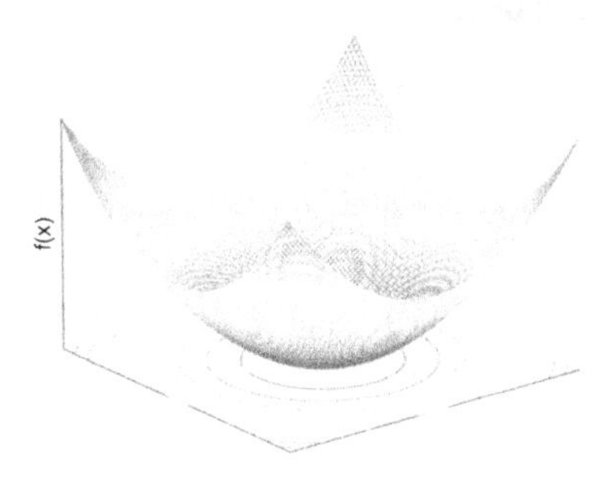

Figure C.1 The two-dimensional sphere function.

C.1.2 The Ackley Function

The Ackley function is given as

$$
\begin{aligned}
f(x) &= 20 + e - 20\exp\left(-0.2\sum_{i=1}^{n} x_i^2/n\right) - \exp\left(\sum_{i=1}^{n}(\cos 2\pi x_i)/n\right) \\
x^* &= 0 \\
f(x^*) &= 0
\end{aligned}
\tag{C.3}
$$

where $x_i \in [-30, +30]$. This benchmark was proposed in [Ackley, 1987b]. Figure C.2 shows a plot of $f(x)$ in two dimensions. Its many local minima make it a challenge for optimization algorithms.

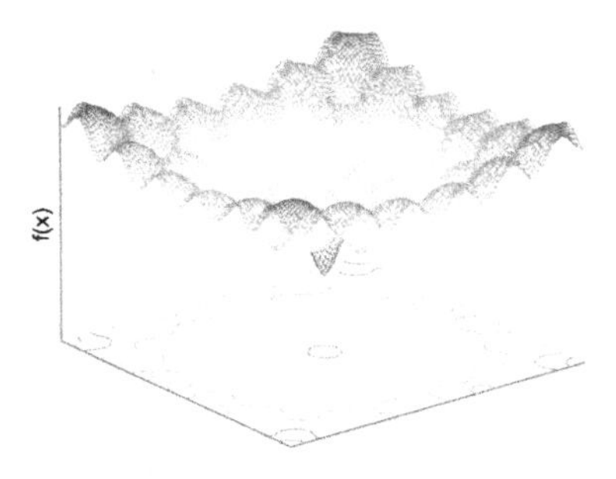

Figure C.2 The two-dimensional Ackley function.

C.1.3 The Ackley Test Function

The Ackley test function is given as

$$f(x) = \sum_{i=1}^{n-1} 3(\cos(2x_i) + \sin(2x_{i+1})) + \exp(-0.2)\sqrt{x_i^2 + x_{i+1}^2} \tag{C.4}$$

where $x_i \in [-30, +30]$. Note that x^* and $f(x^*)$ are not known for this problem. This benchmark is similar to the Ackley function with its many hills and valleys, as shown in Figure C.3.

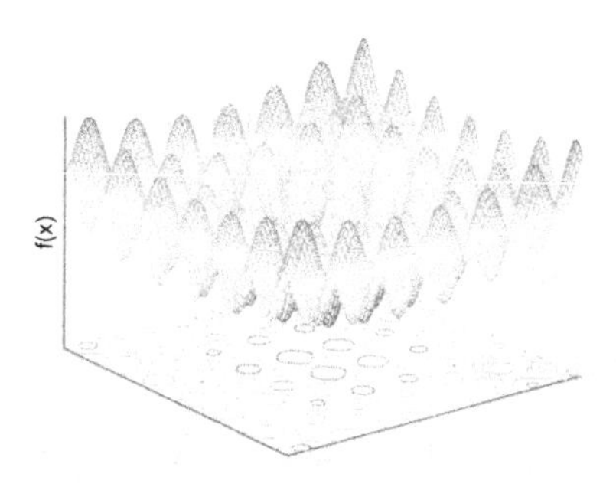

Figure C.3 The two-dimensional Ackley test function.

C.1.4 The Rosenbrock Function

The Rosenbrock function is given as

$$\begin{aligned} f(x) &= \sum_{i=1}^{n-1} \left[100(x_{i+1} - x_i^2)^2 + (x_i - 1)^2\right] \\ x^* &= [1, \cdots, 1] \\ f(x^*) &= 0 \end{aligned} \tag{C.5}$$

where $x_i \in [-2.048, +2.048]$. This benchmark was proposed in [Rosenbrock, 1960], it is called Function 2 in De Jong's thesis [De Jong, 1975], and it is Problem 2.4, 2.24, and 2.25 in [Schwefel, 1995]. Figure C.4 shows a plot of $f(x)$ in two dimensions. It has a long, narrow, banana-shaped valley that makes it a challenge for optimization algorithms.

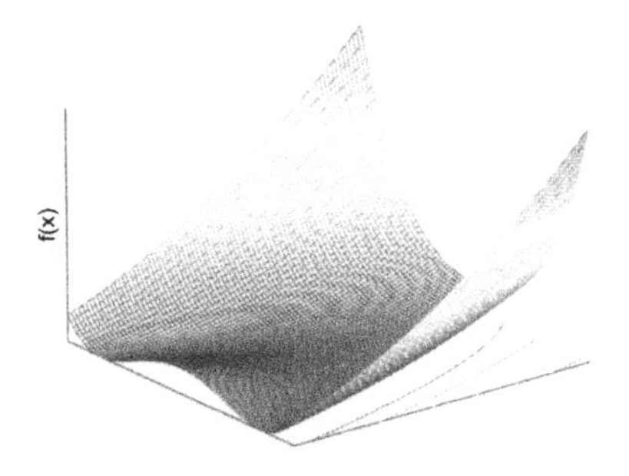

Figure C.4 The two-dimensional Rosenbrock function.

C.1.5 The Fletcher Function

The Fletcher function, also called the Fletcher-Powell function, is given as

$$\begin{aligned}
f(x) &= \sum_{i=1}^{n}(A_i - B_i)^2 \\
A_i &= \sum_{i=1}^{n}(a_{ij}\sin\alpha_j + b_{ij}\cos\alpha_j) \\
B_i &= \sum_{i=1}^{n}(a_{ij}\sin x_j + b_{ij}\cos x_j) \\
\alpha_i &\in [-\pi, \pi], \quad i \in \{1, \cdots, n\} \\
a_{ij}, b_{ij} &\in [-100, 100], \quad i, j \in \{1, \cdots, n\} \\
x^* &= \alpha \\
f(x^*) &= 0
\end{aligned} \tag{C.6}$$

where $x_i \in [-\pi, +\pi]$. This benchmark was proposed in [Fletcher and Powell, 1963] and is called Problem 2.13 in [Schwefel, 1995]. Figure C.5 shows a plot of $f(x)$ in two dimensions for specific values of a_{ij}, b_{ij}, and α_i. This function is interesting because it changes with each realization of a_{ij}, b_{ij}, and α_i. These parameters are often set with a uniform random number generator.

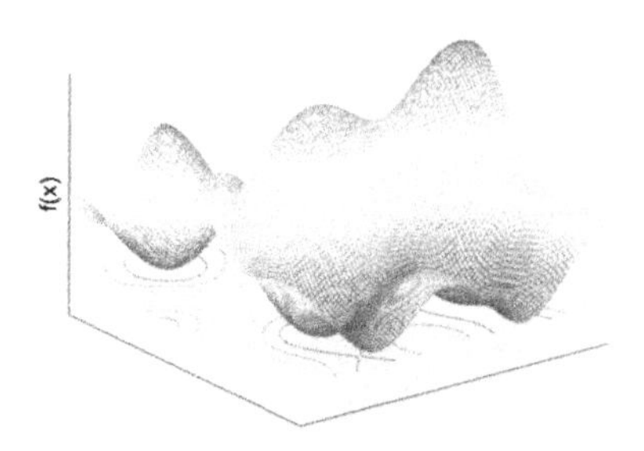

Figure C.5 The two-dimensional Fletcher function.

C.1.6 The Griewank Function

The Griewank function, which is sometimes spelled Griewangk, is given as

$$\begin{aligned} f(x) &= 1 + \sum_{i=1}^{n} x_i^2/4000 - \prod_{i=1}^{n} \cos\left(x_i/\sqrt{i}\right) \\ x^* &= 0 \\ f(x^*) &= 0 \end{aligned} \quad \text{(C.7)}$$

where $x_i \in [-600, +600]$. This benchmark is discussed in [Bäck et al., 1997a, Section B2.7]. Figure C.6 shows a plot of $f(x)$ in two dimensions. This function has many local optima, and the product term in $f(x)$ causes a lot of interdependence among the components of x.

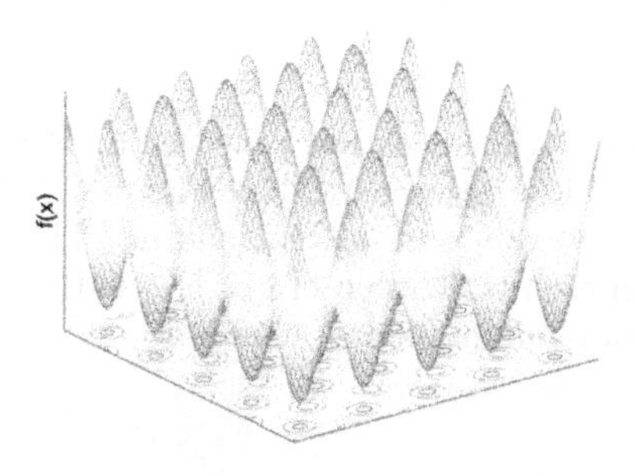

Figure C.6 The two-dimensional Griewank function.

C.1.7 The Penalty #1 Function

The penalty #1 function is given as

$$\begin{aligned}
f(x) &= \frac{\pi}{n}\left\{10\sin^2(\pi x_1) + \sum_{i=1}^{n-1}(x_i - 1)^2[1 + 10\sin^2(\pi x_{i+1})] + (x_n - 1)^2\right\} + \sum_{i=1}^{n} u_i \\
u_i &= \begin{cases} k(x_i - a)^m & x_i > a \\ 0 & -a \le x_i \le a \\ k(-x_i - a)^m & x_i < -a \end{cases} \\
y_i &= 1 + (x_i + 1)/4 \\
x^* &= [1, \cdots, 1] \\
f(x^*) &= 0
\end{aligned} \tag{C.8}$$

where $x_i \in [-50, +50]$. This benchmark is given in [Yao et al., 1999]. Values for k, a, and m are not given but we usually use $k = 100$, $a = 10$, and $m = 4$. Figure C.7 shows a plot of $f(x)$ in two dimensions. This function only has one minimum, but the function is very shallow at the minimum, so it is a challenge to find the minimum with high accuracy.

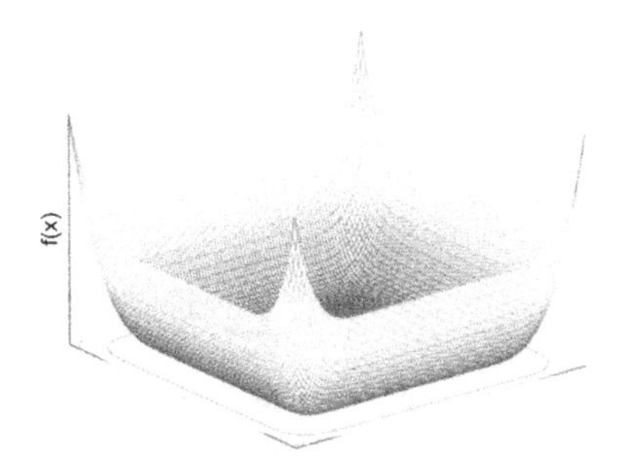

Figure C.7 The two-dimensional penalty #1 function.

C.1.8 The Penalty #2 Function

The penalty #2 function is given as

$$\begin{aligned}
f(x) &= \sum_{i=1}^{n} u_i + 0.1\left\{10\sin^2(3\pi x_1) + \right. \\
&\qquad \left. \sum_{i=1}^{n-1}(x_i - 1)^2[1 + \sin^2(3\pi x_{i+1})] + (x_n - 1)^2[1 + \sin^2(2\pi x_n)]\right\} \\
x^* &= [1, \cdots, 1] \\
f(x^*) &= 0
\end{aligned} \tag{C.9}$$

where $x_i \in [-50, +50]$, and u_i is given in Equation (C.8). This benchmark is given in [Yao et al., 1999]. Like the penalty #1 function, values for k, a, and m are not given but we usually use $k = 100$, $a = 5$, and $m = 4$. Figure C.8 shows a plot of $f(x)$ in two dimensions. Like the penalty #1 function, the penalty #2 function only has one minimum, but the function is very shallow at the minimum, so it is a challenge to find the minimum with high accuracy.

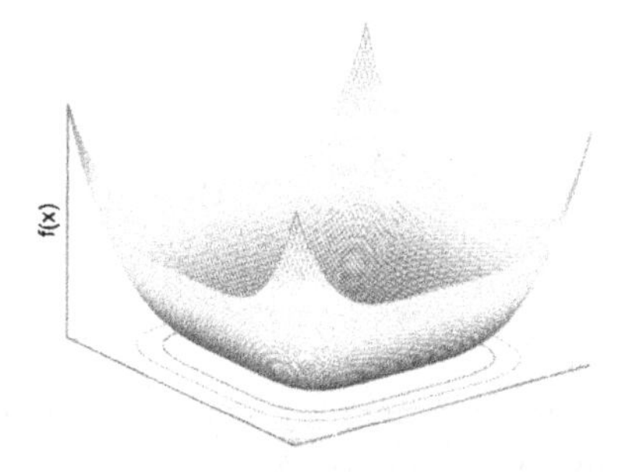

Figure C.8 The two-dimensional penalty #2 function.

C.1.9 The Quartic Function

The quartic function is given as

$$\begin{aligned} f(x) &= \sum_{i=1}^{n} i x_i^4 \\ x^* &= 0 \\ f(x^*) &= 0 \end{aligned} \tag{C.10}$$

where $x_i \in [-1.28, +1.28]$. Noise is often added to $f(x)$, but that does not change the argument of the minimum. This benchmark is called function 4 in De Jong's thesis [De Jong, 1975], and it is also given in [Yao et al., 1999]. The quartic function can also be written with x_i raised to the second instead of the fourth power, in which case it is called the hyper-ellipsoid function, or the weighted sphere function [Ros and Hansen, 2008]. However, sometimes the hyper-ellipsoid function is written as follows [Yao and Liu, 1997]:

$$f(x) = \sum_{i=1}^{n} 2^i x_i^2. \tag{C.11}$$

Figure C.9 shows a plot of $f(x)$ in two dimensions. Like the penalty functions, the quartic function only has only one minimum, but the function is very shallow at the minimum, so it is a challenge to find the minimum with high accuracy.

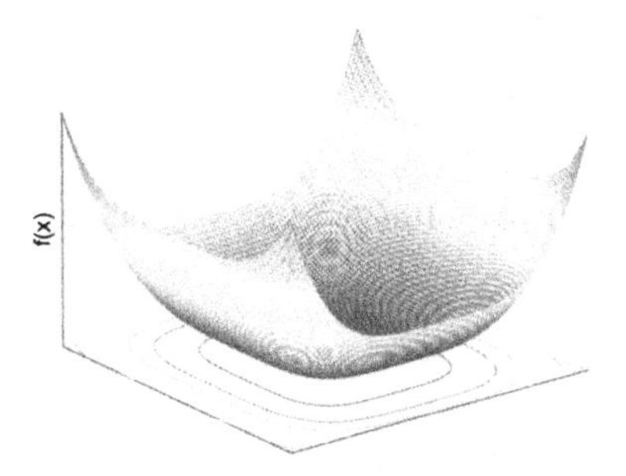

Figure C.9 The two-dimensional quartic function.

C.1.10 The Tenth Power Function

The tenth power function is given as

$$\begin{aligned} f(x) &= \sum_{i=1}^{n} x_i^{10} \\ x^* &= 0 \\ f(x^*) &= 0 \end{aligned} \tag{C.12}$$

where $x_i \in [-5.12, +5.12]$. This benchmark was proposed in [Schwefel, 1995] as Problem 2.23, and is also given in [Yao et al., 1999]. Figure C.10 shows a plot of $f(x)$ in two dimensions. Like the quartic and penalty functions, the the tenth power function only has only one minimum, but the function is very shallow at the minimum, so it is a challenge to find the minimum with high accuracy.

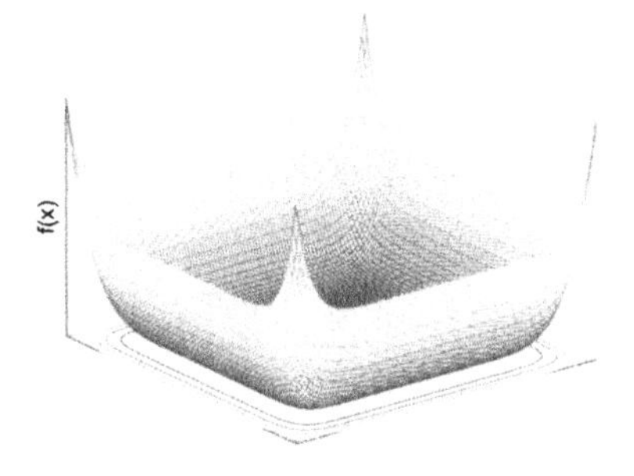

Figure C.10 The two-dimensional tenth power function.

C.1.11 The Rastrigin Function

The Rastrigin function is given as

$$\begin{aligned} f(x) &= 10n + \sum_{i=1}^{n} x_i^2 - 10\cos(2\pi x_i) \\ x^* &= 0 \\ f(x^*) &= 0 \end{aligned} \tag{C.13}$$

where $x_i \in [-5.12, +5.12]$. This benchmark was proposed in [Rastrigin, 1974], and is also given in [Yao et al., 1999]. Figure C.11 shows a plot of $f(x)$ in two dimensions. The Rastrigin function looks similar to the Griewank function. The number of local minima in the Rastrigin function increases exponentially with n [Beyer and Schwefel, 2002].

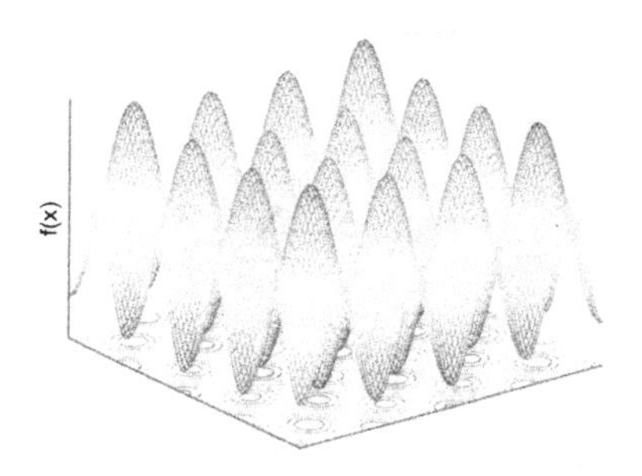

Figure C.11 The two-dimensional Rastrigin function.

C.1.12 The Schwefel Double Sum Function

The Schwefel double sum function, also called Schwefel's ridge function [Price et al., 2005], Schwefel 1.2, and the quadric function, is given as

$$\begin{aligned} f(x) &= \sum_{i=1}^{n} \left(\sum_{j=1}^{i} x_j \right)^2 \\ x^* &= 0 \\ f(x^*) &= 0 \end{aligned} \tag{C.14}$$

where $x_i \in [-65.536, +65.536]$. This benchmark is also called the rotated hyper-ellipsoid function [Ros and Hansen, 2008]. It was proposed in [Schwefel, 1995] as Problem 1.2 and Problem 2.9, and is also given in [Yao et al., 1999]. It is a quadratic function whose condition number is proportional to n^2. Figure C.12 shows a plot of $f(x)$ in two dimensions.

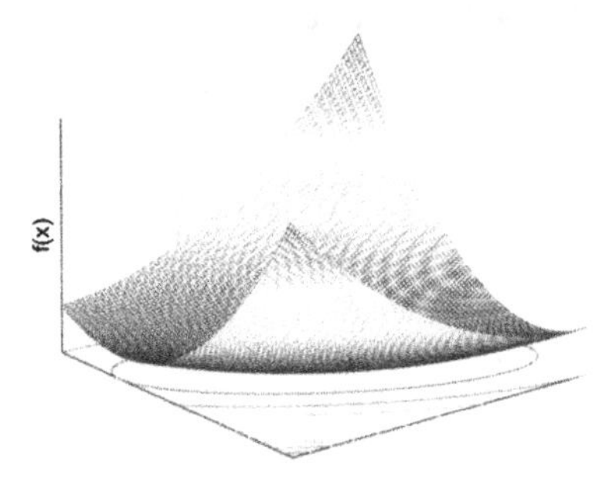

Figure C.12 The two-dimensional Schwefel double sum function.

C.1.13 The Schwefel Max Function

The Schwefel max function, also called the Schwefel 2.21 function, is given as

$$\begin{aligned} f(x) &= \max_i (|x_i| : i \in \{1, \cdots, n\}) \\ x^* &= 0 \\ f(x^*) &= 0 \end{aligned} \tag{C.15}$$

where $x_i \in [-100, +100]$. This benchmark was proposed in [Schwefel, 1995], and is also given in [Yao et al., 1999]. It is nondifferentiable. Figure C.13 shows a plot of $f(x)$ in two dimensions.

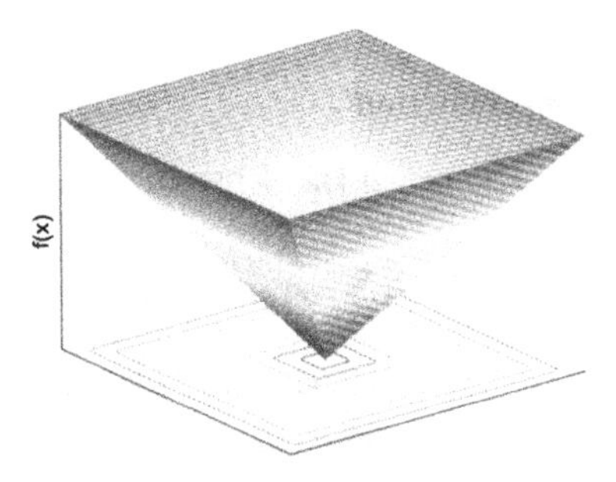

Figure C.13 The two-dimensional Schwefel max function.

C.1.14 The Schwefel Absolute Function

The Schwefel absolute function, also called the Schwefel 2.22 function, is given as

$$\begin{aligned} f(x) &= \sum_{i=1}^{n} |x_i| + \prod_{i=1}^{n} |x_i| \\ x^* &= 0 \\ f(x^*) &= 0 \end{aligned} \tag{C.16}$$

where $x_i \in [-10, +10]$. This benchmark was proposed as Problem 2.22 in [Schwefel, 1995], and is also given in [Yao et al., 1999]. It is nondifferentiable. Figure C.14 shows a plot of $f(x)$ in two dimensions.

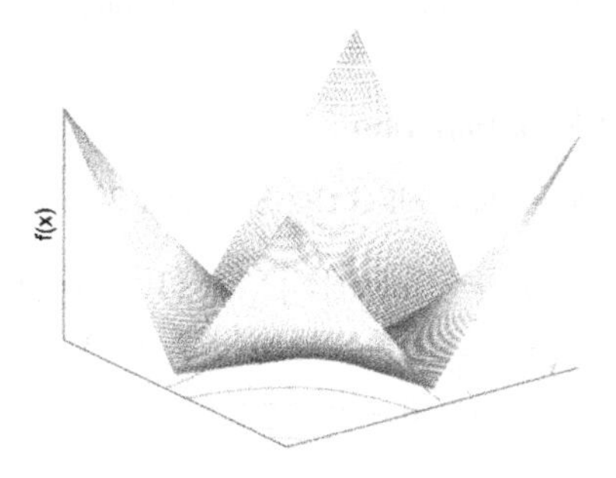

Figure C.14 The two-dimensional Schwefel absolute function.

C.1.15 The Schwefel Sine Function

The Schwefel sine function, also called the Schwefel 2.26 function, is given as

$$\begin{aligned} f(x) &= -\sum_{i=1}^{n} x_i \sin \sqrt{|x_i|} \\ x^* &= [420.9687, \cdots, 420.9867] \\ f(x^*) &= -12965.5 \end{aligned} \tag{C.17}$$

where $x_i \in [-500, +500]$. This benchmark was proposed in [Schwefel, 1995] as Problems 2.3 and 2.26, and is also given in [Yao et al., 1999]. It has many local minima. Figure C.15 shows a plot of $f(x)$ in two dimensions.

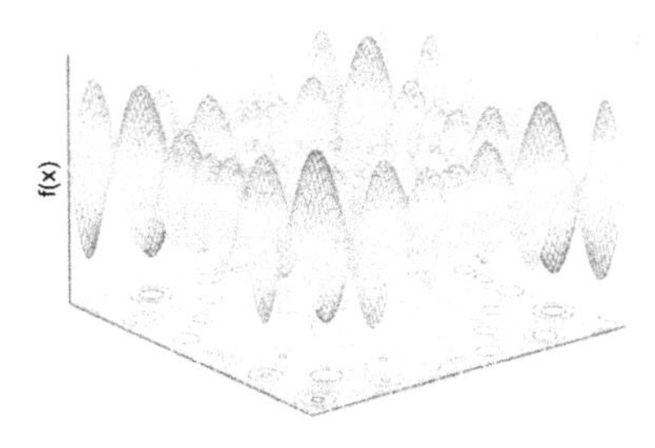

Figure C.15 The two-dimensional Schwefel sine function.

C.1.16 The Step Function

The step function is given as

$$\begin{aligned} f(x) &= \sum_{i=1}^{n} (\text{floor}(x_i + 0.5))^2 \\ x^* &= 0 \\ f(x^*) &= 0 \end{aligned} \tag{C.18}$$

where $x_i \in [-100, +100]$, and where the *floor* function returns the smallest integer less than or equal to its argument. This benchmark is called function 3 in De Jong's thesis [De Jong, 1975], and it is also given in [Yao et al., 1999]. It is not differentiable, and it has many plateaus. Figure C.16 shows a plot of $f(x)$ in two dimensions.

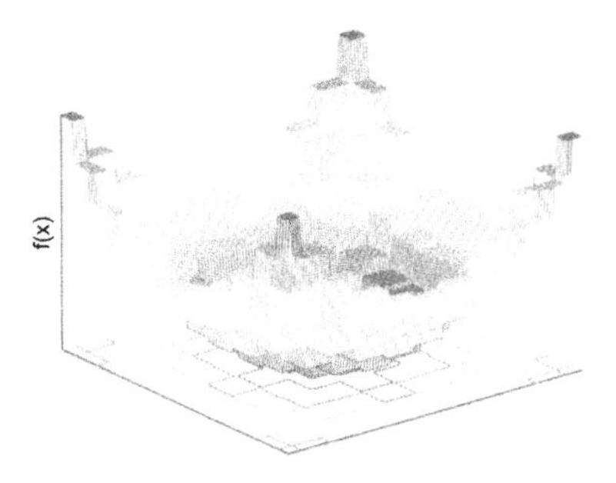

Figure C.16 The two-dimensional step function.

C.1.17 The Absolute Function

The absolute function is given as

$$\begin{aligned} f(x) &= \sum_{i=1}^{n} |x_i| \\ x^* &= 0 \\ f(x^*) &= 0 \end{aligned} \tag{C.19}$$

where $x_i \in [-10, +10]$. This benchmark is Problem 2.20 in [Schwefel, 1995]. It is not differentiable. Figure C.17 shows a plot of $f(x)$ in two dimensions.

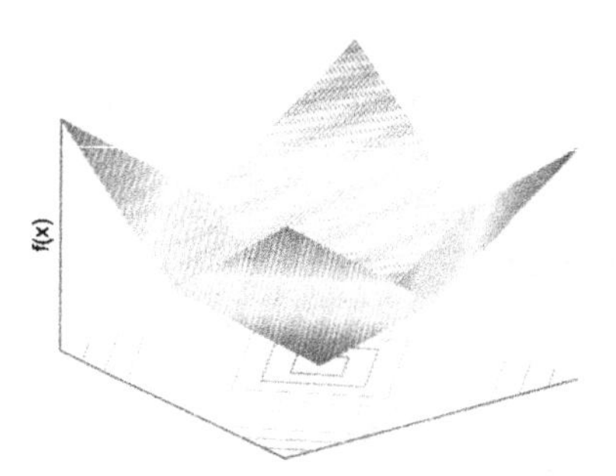

Figure C.17 The two-dimensional absolute function.

C.1.18 Shekel's Foxhole Function

Shekel's foxhole function is given as

$$\begin{aligned} f(x) &= \left[\frac{1}{500} + \sum_{j=1}^{2} 5 \frac{1}{j + \sum_{i=1}^{n} (x_i - a_{ij})^6} \right]^{-1} \\ x^* &= [-32, \cdots, -32] \\ f(x^*) &\approx 1 \end{aligned} \tag{C.20}$$

where $x_i \in [-65.536, +65.536]$, and where a_{ij} is the element in the ith row and jth column of a. For two dimensions ($n = 2$), a is given as

$$\begin{aligned} a &= \begin{bmatrix} b_0 & \cdots & b_0 \\ b_1 & \cdots & b_5 \end{bmatrix} \\ b_0 &= \begin{bmatrix} -32 & -16 & 0 & 16 & 32 \end{bmatrix} \\ b_i &= (16(i-1) - 32) \begin{bmatrix} 1 & 1 & 1 & 1 & 1 \end{bmatrix}. \end{aligned} \tag{C.21}$$

This benchmark is called function 5 in De Jong's thesis [De Jong, 1975], and it is also given in [Yao et al., 1999]. It has multiple local minima, not all of which are the same value, and it has a steep drop to its minimum, as shown by its plot in Figure C.18. Additional rows can be augmented to a if $n > 2$ citeBersini.

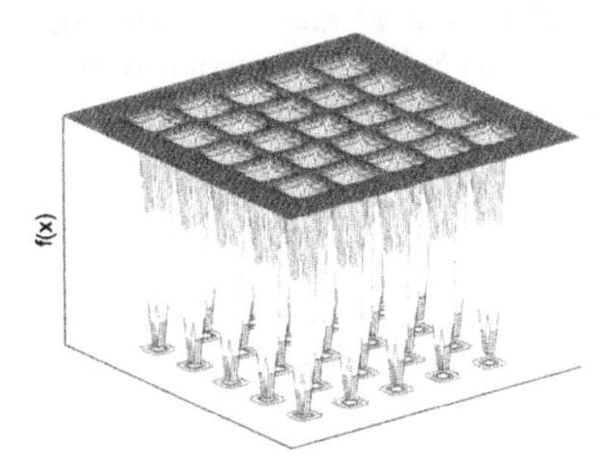

Figure C.18 The two-dimensional Shekel foxhole function.

C.1.19 The Michalewicz Function

The Michalewicz function is given as

$$f(x) = -\sum_{i=1}^{n} \sin x_i \sin^{2m}(i x_i^2/\pi) \tag{C.22}$$

where $x_i \in [0, \pi]$, and m is a parameter that controls the difficulty of the search. Note that x^* and $f(x^*)$ are not known for this problem. This benchmark is given in [Michalewicz, 1996]. It has long narrow valleys with a sudden drop-off to the minimum, as shown by its plot in Figure C.19.

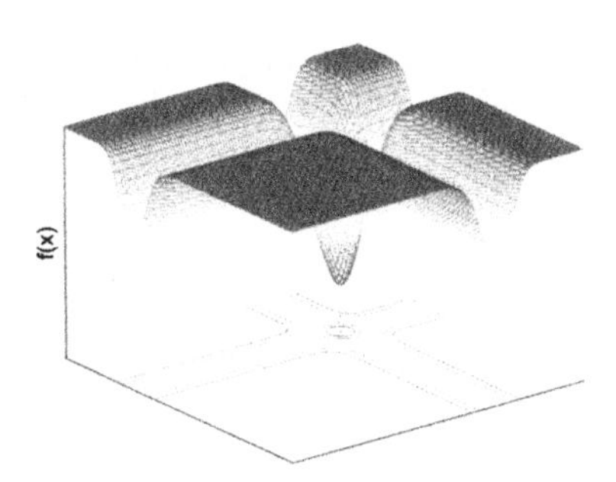

Figure C.19 The two-dimensional Michalewicz function with $m = 10$.

C.1.20 The Sine Envelope Function

The sine envelope function is given as

$$f(x) = -\sum_{i=1}^{n-1} \frac{\sin^2 \sqrt{x_i + x_{i+1} - 0.5}}{(0.001(x_i^2 + x_{i+1}^2) + 1)^2} \tag{C.23}$$

where $x_i \in [-100, +100]$. Note that x^* and $f(x^*)$ are not known for this problem. This benchmark, also called the Schaeffer function [Cheng et al., 2008], has many valleys and local minima, as shown by its plot in Figure C.20.

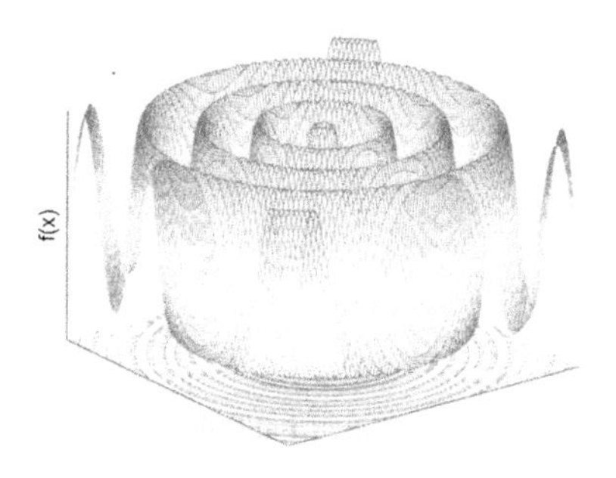

Figure C.20 The two-dimensional sine envelope function.

C.1.21 The Eggholder Function

The eggholder function is given as

$$f(x) = -\sum_{i=1}^{n-1} (x_{i+1} + 47) \sin \sqrt{|x_{i+1} + x_i/2 + 47|} + x_i \sin \sqrt{|x_i - x_{i+1} - 47|} \quad \text{(C.24)}$$

where $x_i \in [-512, +512]$. Note that x^* and $f(x^*)$ are not known for this problem. This benchmark is given in [Wu and Chow, 2007]. Its two-dimensional plot is shown in Figure C.21.

Figure C.21 The two-dimensional eggholder function.

C.1.22 The Weierstrass Function

The Weierstrass function is given as

$$\begin{aligned} f(x) &= \sum_{i=1}^{n} \left\{ \sum_{k=0}^{k_{\max}} \left[a^k \cos\left(2\pi b^k (x_i + 0.5)\right)\right] \right\} - n \sum_{k=0}^{k_{\max}} \left[a^k \cos(\pi b^k)\right] \\ x^* &= 0 \\ f(x^*) &= 0 \end{aligned} \tag{C.25}$$

where $x_i \in [-5, +5]$, $a = 0.5$, $b = 3$, and $k_{\max} = 20$. This benchmark is given in [Liang et al., 2005]. It has the interesting property that as $n \to \infty$, it is continuous everywhere but differentiable nowhere, and it is nonmonotonic everywhere. A plot of the two-dimensional Weierstrass function is shown in Figure C.22.

Figure C.22 The two-dimensional Weierstrass function.

C.2 CONSTRAINED BENCHMARKS

A constrained optimization problem involves the minimization of $f(x)$ over all x such that $x \in \mathcal{F} \in \mathcal{R}^n$, where $\mathcal{F}$ is the feasible set and n is the problem dimension. We use x^* to represent the optimizing value of x, and $f(x^*)$ is the constrained minimum of $f(x)$:

$$\begin{aligned} &x^* = \arg\min_x f(x) \\ \text{such that} \quad &g_i(x) \leq 0 \text{ for } i \in [1, m] \text{ and } h_j(x) = 0 \text{ for } j \in [1, p]. \end{aligned} \tag{C.26}$$

This problem includes $(m + p)$ constraints, m of which are inequality constraints and p of which are equality constraints. Many problems of this form have long and involved forms for $f(x)$, $g_i(x)$, and $h_j(x)$, and it thus requires a lot of space simply to write the problem. Therefore, we only show simple constrained benchmarks in this section, while giving references where some longer and more complicated benchmarks can be found.

Constrained benchmark functions are given in [Araujo et al., 2009], [Coello Coello, 2000a], [Coello Coello, 2002], [Deb, 2000], [Mezura-Montes and Coello Coello, 2005], and [Runarsson and Yao, 2000]. Detailed information about the constrained benchmarks and evaluation metrics for EA competitions at the 2006 and 2010 IEEE Congress on Evolutionary Computation can be found in [Liang et al., 2006] and [Mallipeddi and Suganthan, 2010]. Note that [Floudas and Pardalos, 1990] is an entire book that is devoted to the definition of constrained optimization benchmarks. Constrained multi-objective benchmarks can be found in [Deb et al., 2001].

The constrained benchmarks in this section are all taken from [Mallipeddi and Suganthan, 2010] and were used in a constrained EA competition at the 2010 Congress on Evolutionary Computation (CEC). In the problem statements below we use o_i to refer to a random offset and we use M to refer to a random rotation matrix (see Section C.7).

C.2.1 The C01 Function

The C01 function is given as

$$\begin{aligned} f(x) &= -\left|\frac{\sum_{i=1}^n \cos^4 z_i - 2\prod_{i=1}^n \cos^2 z_i}{\sum_{i=1}^n i z_i^2}\right| \\ g_1(x) &= 0.75 - \prod_{i=1}^n z_i \le 0 \\ g_2(x) &= \sum_{i=1}^n z_i - 7.5n \le 0 \\ x_i &\in [0, 10] \end{aligned} \tag{C.27}$$

where $z_i = x_i - o_i$ for $i \in [1, n]$.

C.2.2 The C02 Function

The C02 function is given as

$$\begin{aligned} f(x) &= \max_i z_i \\ g_1(x) &= 10 - \frac{1}{n}\sum_{i=1}^n \left(z_i^2 - 10\cos(2\pi z_i) + 10\right) \le 0 \\ g_2(x) &= \frac{1}{n}\sum_{i=1}^n \left(z_i^2 - 10\cos(2\pi z_i) + 10\right) - 15 \le 0 \\ h(x) &= \frac{1}{n}\sum_{i=1}^n \left(y_i^2 - 10\cos(2\pi y_i) + 10\right) - 20 = 0 \\ x_i &\in [-5.12, +5.12] \end{aligned} \tag{C.28}$$

where $z_i = x_i - o_i$ and $y_i = z_i - 0.5$ for $i \in [1, n]$.

C.2.3 The C03 Function

The C03 function is given as

$$\begin{aligned} f(x) &= \sum_{i=1}^{n-1} \left[100(z_i^2 - z_{i+1})^2 + (z_i - 1)^2\right] \\ h(x) &= \sum_{i=1}^{n-1} (z_i - z_{i+1})^2 = 0 \\ x_i &\in [-1000, 1000] \end{aligned} \tag{C.29}$$

where $z_i = x_i - o_i$ for $i \in [1, n]$.

C.2.4 The C04 Function

The C04 function is given as

$$\begin{aligned} f(x) &= \max_i z_i \\ h_1(x) &= \frac{1}{n}\sum_{i=1}^{n} z_i \cos\sqrt{|z_i|} = 0 \\ h_2(x) &= \sum_{i=1}^{n/2-1} (z_i - z_{i+1})^2 = 0 \\ h_3(x) &= \sum_{i=n/2+1}^{n} (z_i^2 - z_{i+1}) = 0 \\ h_4(x) &= \sum_{i=1}^{n} z_i = 0 \\ x_i &\in [-50, 50] \end{aligned} \tag{C.30}$$

where $z_i = x_i - o_i$ for $i \in [1, n]$.

C.2.5 The C05 Function

The C05 function is given as

$$\begin{aligned} f(x) &= \max_i z_i \\ h_1(x) &= \frac{1}{n}\sum_{i=1}^{n} \left[-z_i \sin\left(\sqrt{|z_i|}\right)\right] = 0 \\ h_2(x) &= \frac{1}{n}\sum_{i=1}^{n} \left[-z_i \cos\left(0.5\sqrt{|z_i|}\right)\right] = 0 \\ x_i &\in [-600, 600] \end{aligned} \tag{C.31}$$

where $z_i = x_i - o_i$ for $i \in [1, n]$.

C.2.6 The C06 Function

The C06 function is given as

$$\begin{aligned}
f(x) &= \max_i z_i \\
y_i &= (z_i + 483.6106156535)M - 483.6106156535 \\
h_1(x) &= \frac{1}{n}\sum_{i=1}^{n}\left[-y_i \sin\left(\sqrt{|y_i|}\right)\right] = 0 \\
h_2(x) &= \frac{1}{n}\sum_{i=1}^{n}\left[-y_i \cos\left(0.5\sqrt{|y_i|}\right)\right] = 0 \\
x_i &\in [-600, 600]
\end{aligned} \tag{C.32}$$

where $z_i = x_i - o_i$ for $i \in [1, n]$.

C.2.7 The C07 Function

The C07 function is given as

$$\begin{aligned}
f(x) &= \sum_{i=1}^{n-1}\left[100(z_i^2 - z_{i+1})^2 + (z_i - 1)^2\right] \\
g(x) &= 0.5 - \exp\left[\frac{0.1}{n}\sum_{i=1}^{n} y_i^2\right] - 3\exp\left[\frac{1}{n}\sum_{i=1}^{n}\cos(0.1y_i)\right] + \exp(1) \le 0 \\
x_i &\in [-140, 140]
\end{aligned} \tag{C.33}$$

where $y_i = x_i - o_i$ and $z_i = x_i - o_i + 1$ for $i \in [1, n]$.

C.2.8 The C08 Function

The C08 function is given as

$$\begin{aligned}
f(x) &= \sum_{i=1}^{n-1}\left[100(z_i^2 - z_{i+1})^2 + (z_i - 1)^2\right] \\
g(x) &= 0.5 - \exp\left[\frac{0.1}{n}\sum_{i=1}^{n} y_i^2\right] - 3\exp\left[\frac{1}{n}\sum_{i=1}^{n}\cos(0.1y_i)\right] + \exp(1) \le 0 \\
x_i &\in [-140, 140]
\end{aligned} \tag{C.34}$$

where $y_i = (x_i - o_i)M$ and $z_i = x_i - o_i + 1$ for $i \in [1, n]$.

C.2.9 The C09 Function

The C09 function is given as

$$\begin{aligned} f(x) &= \sum_{i=1}^{n-1} \left[100(z_i^2 - z_{i+1})^2 + (z_i - 1)^2\right] \\ h(x) &= \sum_{i=1}^{n} y \sin\sqrt{|y_i|} = 0 \\ x_i &\in [-500, 500] \end{aligned} \tag{C.35}$$

where $y_i = x_i - o_i$ and $z_i = x_i + 1 - o_i$ for $i \in [1, n]$.

C.2.10 The C10 Function

The C10 function is given as

$$\begin{aligned} f(x) &= \sum_{i=1}^{n-1} \left[100(z_i^2 - z_{i+1})^2 + (z_i - 1)^2\right] \\ h(x) &= \sum_{i=1}^{n} y_i \sin\sqrt{|y_i|} = 0 \\ x &\in [-500, 500] \end{aligned} \tag{C.36}$$

where $y_i = (x_i - o_i)M$ and $z_i = x_i + 1 - o_i$ for $i \in [1, n]$.

C.2.11 The C11 Function

The C11 function is given as

$$\begin{aligned} f(x) &= \frac{1}{n} \sum_{i=1}^{n} \left[-z_i \cos\left(2\sqrt{|z_i|}\right)\right] \\ h(x) &= \sum_{i=1}^{n-1} \left[100(y_i^2 - y_{i+1})^2 + (y_i - 1)^2\right] = 0 \\ x_i &\in [-100, 100] \end{aligned} \tag{C.37}$$

where $y_i = x_i + 1 - o_i$ and $z_i = (x_i - o_i)M$ for $i \in [1, n]$.

C.2.12 The C12 Function

The C12 function is given as

$$\begin{aligned} f(x) &= \sum_{i=1}^{n} z_i \sin\sqrt{|z_i|} \\ h(x) &= \sum_{i=1}^{n} (z_i^2 - z_{i+1})^2 = 0 \\ g(x) &= \sum_{i=1}^{n-1} [z_i - 100\cos(0.1 z_i) + 10] \leq 0 \\ x_i &\in [-1000, 1000] \end{aligned} \tag{C.38}$$

where $z_i = x_i - o_i$ for $i \in [1, n]$.

C.2.13 The C13 Function

The C13 function is given as

$$\begin{aligned} f(x) &= \frac{1}{n}\sum_{i=1}^{n} \left[-z_i \sin\sqrt{|z_i|}\right] \\ g_1(x) &= -50 + \frac{1}{100n}\sum_{i=1}^{n} z_i^2 \leq 0 \\ g_2(x) &= \frac{50}{n}\sum_{i=1}^{n} \sin\left(\frac{\pi z_i}{50}\right) \leq 0 \\ g_3(x) &= 75 - 50\left[\sum_{i=1}^{n} \frac{z_i^2}{4000} - \prod_{i=1}^{n} \cos\left(\frac{z_i}{\sqrt{i}}\right) + 1\right] \leq 0 \\ x_i &\in [-500, 500] \end{aligned} \tag{C.39}$$

where $z_i = x_i - o_i$ for $i \in [1, n]$.

C.2.14 The C14 Function

The C14 function is given as

$$\begin{aligned} f(x) &= \sum_{i=1}^{n-1} \left[100(z_i^2 - z_{i+1})^2 + (z_i - 1)^2\right] \\ g_1(x) &= \sum_{i=1}^{n} \left[-y_i \cos\sqrt{|y_i|}\right] - n \leq 0 \\ g_2(x) &= \sum_{i=1}^{n} \left[y_i \cos\sqrt{|y_i|}\right] - n \leq 0 \\ g_3(x) &= \sum_{i=1}^{n} \left[y_i \sin\sqrt{|y_i|}\right] - 10n \leq 0 \\ x_i &\in [-1000, 1000] \end{aligned} \tag{C.40}$$

where $y_i = x_i - o_i$ and $z_i = x_i - o_i + 1$ for $i \in [1, n]$.

C.2.15 The C15 Function

The C15 function is given as

$$\begin{aligned}
f(x) &= \sum_{i=1}^{n-1} \left[100(z_i^2 - z_{i+1})^2 + (z_i - 1)^2\right] \\
g_1(x) &= \sum_{i=1}^{n} \left[-y_i \cos\sqrt{|y_i|}\right] - n \leq 0 \\
g_2(x) &= \sum_{i=1}^{n} \left[y_i \cos\sqrt{|y_i|}\right] - n \leq 0 \\
g_3(x) &= \sum_{i=1}^{n} \left[y_i \sin\sqrt{|y_i|}\right] - 10n \leq 0 \\
x_i &\in [-1000, 1000]
\end{aligned} \tag{C.41}$$

where $y_i = (x_i - o_i)M$ and $z_i = x_i - o_i + 1$ for $i \in [1, n]$.

C.2.16 The C16 Function

The C16 function is given as

$$\begin{aligned}
f(x) &= \sum_{i=1}^{n} \frac{z_i^2}{4000} - \prod_{i=1}^{D} \cos\left(\frac{z_i}{\sqrt{i}}\right) + 1 \\
g_1(x) &= \sum_{i=1}^{n} \left[z_i^2 - 100\cos(\pi z_i) + 10\right] \leq 0 \\
g_2(x) &= \prod_{i=1}^{n} z_i \leq 0 \\
h_1(x) &= \sum_{i=1}^{n} \left[z_i \sin\sqrt{|z_i|}\right] = 0 \\
h_2(x) &= \sum_{i=1}^{n} \left[-z \sin\sqrt{|z_i|}\right] = 0 \\
x_i &\in [-10, 10]
\end{aligned} \tag{C.42}$$

where $z_i = x_i - o_i$ for $i \in [1, n]$.

C.2.17 The C17 Function

The C17 function is given as

$$\begin{aligned} f(x) &= \sum_{i=1}^{n-1} (z_i - z_{i+1})^2 \\ g_1(x) &= \prod_{i=1}^{n} z_i \leq 0 \\ g_2(x) &= \sum_{i=1}^{n} z_i \leq 0 \\ h(x) &= \sum_{i=i}^{n} z_i \sin\left(4\sqrt{|z_i|}\right) = 0 \\ x_i &\in [-10, 10] \end{aligned} \tag{C.43}$$

where $z_i = x_i - o_i$ for $i \in [1, n]$.

C.2.18 The C18 Function

The C18 function is given as

$$\begin{aligned} f(x) &= \sum_{i=1}^{n-1} (z_i - z_{i+1})^2 \\ g(x) &= \frac{1}{n}\sum_{i=1}^{n} \left[-z_i \sin\sqrt{|z_i|}\right] = 0 \\ h(x) &= \frac{1}{n}\sum_{i=1}^{n} \left[z_i \sin\sqrt{|z_i|}\right] = 0 \\ x_i &\in [-50, 50] \end{aligned} \tag{C.44}$$

where $z_i = x_i - o_i$ for $i \in [1, n]$.

C.2.19 Summary of Constrained Benchmarks

Here we give a summary of the 18 CEC 2010 benchmarks presented above. The estimated ratio ρ between the size of the feasible set and the size of the search space indicates how difficult it is to satisfy the constraints (see Equation (19.53)). Table C.1 summarizes the 18 constrained benchmarks.

Function	N_e	N_i	ρ $(n = 10)$	ρ $(n = 30)$
C01	0	2	0.997689	1.000000
C02	1	2	0.000000	0.000000
C03	1	0	0.000000	0.000000
C04	4	0	0.000000	0.000000
C05	2	0	0.000000	0.000000
C06	2	0	0.000000	0.000000
C07	0	1	0.505123	0.503725
C08	0	1	0.379512	0.375278
C09	1	0	0.000000	0.000000
C10	1	0	0.000000	0.000000
C11	1	0	0.000000	0.000000
C12	1	1	0.000000	0.000000
C13	0	3	0.000000	0.000000
C14	0	3	0.003112	0.006123
C15	0	3	0.003210	0.006023
C16	2	2	0.000000	0.000000
C17	1	2	0.000000	0.000000
C18	1	1	0.000000	0.000000

Table C.1 Summary of 18 CEC 2010 constrained optimization benchmarks. N_e is the number of equality constraints, N_i is the number of inequality constraints, and ρ is the ratio of the size of the feasible set to the size of the search space for the 10-dimensional and 30-dimensional versions of each problem.

C.3 MULTI-OBJECTIVE BENCHMARKS

A multi-objective optimization problem (MOP) involves the minimization of $f(x)$ over all x, where $f(x)$ is a vector, and x is the n-dimensional decision vector. Vector minimization is undefined in the normal sense of the word, and so we define the Pareto set P_s and the Pareto front P_f in Section 20.1. We can then pose an MOP as the problem of finding the "best" possible P_s and P_f. We can define "best" in a number of different ways as we discuss in Section 20.2.

Detailed information about the multi-objective benchmarks and evaluation metrics for EA competitions at the 2007 and 2009 IEEE Congress on Evolutionary Computation can be found in [Huang et al., 2007] and [Zhang et al., 2009]. Additional multi-objective benchmark problems can be found in [Zitzler et al., 2000]. Constrained multi-objective benchmarks can be found in [Deb et al., 2001]. Approaches for designing new multi-objective test problems can be found in [Deb et al., 2002b] and [Zhang et al., 2009]. The literature proposes many multi-objective benchmarks and new ones continually appear in the literature. In this section we only show the unconstrained MOPs from the CEC 2009 competition [Zhang et al., 2009]. The reader can find additional multi-objective benchmarks (both constrained and unconstrained) in the references above. The dimension of the independent variable in the benchmarks below is variable, but the CEC 2009 competition used $n = 30$.

C.3.1 Unconstrained Multi-Objective Optimization Problem 1

This two-objective problem is defined as

$$\begin{aligned} f_1(x) &= x_1 + \frac{2}{|J_1|} \sum_{j \in J_1} \left[x_j - \sin(6\pi x_1 + j\pi/n)\right]^2 \\ f_2(x) &= 1 - \sqrt{x_1} + \frac{2}{|J_2|} \sum_{j \in J_2} \left[x_j - \sin(6\pi x_1 + j\pi/n)\right]^2 \end{aligned} \tag{C.45}$$

where the sets J_1 and J_2 are defined as

$$\begin{aligned} J_1 &= \{j \in [2, n] : j \text{ is odd}\} \\ J_2 &= \{j \in [2, n] : j \text{ is even}\}. \end{aligned} \tag{C.46}$$

The search space is

$$\begin{aligned} x_1 &\in [0, 1] \\ x_j &\in [-1, 1] \text{ for } j \in [2, n]. \end{aligned} \tag{C.47}$$

The Pareto front is

$$\begin{aligned} f_1^* &\in [0, 1] \\ f_2^* &= 1 - \sqrt{f_1^*}. \end{aligned} \tag{C.48}$$

The Pareto set is

$$\begin{aligned} x_1^* &\in [0, 1] \\ x_j^* &= \sin(6\pi x_1 + j\pi/n) \;\text{ for } j \in [2, n]. \end{aligned} \tag{C.49}$$

C.3.2 Unconstrained Multi-Objective Optimization Problem 2

This two-objective problem is defined as

$$\begin{aligned} f_1 &= x_1 + \frac{2}{|J_1|} \sum_{j \in J_1} y_j^2 \\ f_2 &= 1 - \sqrt{x_1} + \frac{2}{|J_2|} \sum_{j \in J_2} y_j^2 \end{aligned} \tag{C.50}$$

where J_1 and J_2 are the same as in Unconstrained MOP 1, and y_j is defined as

$$y_j = \begin{cases} x_j - \left[0.3x_1^2 \cos(24\pi x_1 + 4j\pi/n) + 0.6x_1\right] \cos(6\pi x_1 + j\pi/n) & \text{if } j \in J_1 \\ x_j - \left[0.3x_1^2 \cos(24\pi x_1 + 4j\pi/n) + 0.6x_1\right] \sin(6\pi x_1 + j\pi/n) & \text{if } j \in J_2. \end{cases} \tag{C.51}$$

The search space is

$$\begin{aligned} x_1 &\in [0, 1] \\ x_j &\in [-1, 1] \text{ for } j \in [2, n]. \end{aligned} \tag{C.52}$$

The Pareto front is

$$\begin{aligned} f_1^* &\in [0, 1] \\ f_2^* &= 1 - \sqrt{f_1^*}. \end{aligned} \tag{C.53}$$

The Pareto set is

$$\begin{aligned} x_1^* &\in [0,1] \\ x_j^* &= \begin{cases} \left[0.3(x_1^*)^2 \cos(24\pi x_1^* + 4j\pi/n) + 0.6x_1^*\right] \cos(6\pi x_1^* + j\pi/n) & \text{if } j \in J_1 \\ \left[0.3(x_1^*)^2 \cos(24\pi x_1^* + 4j\pi/n) + 0.6x_1^*\right] \sin(6\pi x_1^* + j\pi/n) & \text{if } j \in J_2. \end{cases} \end{aligned} \tag{C.54}$$

C.3.3 Unconstrained Multi-Objective Optimization Problem 3

This two-objective problem is defined as

$$\begin{aligned} f_1 &= x_1 + \frac{2}{|J_1|}\left[4\sum_{j\in J_1} y_j^2 - 2\prod_{j\in J_1}\cos\left(20y_j\pi/\sqrt{j}\right) + 2\right] \\ f_2 &= 1 - \sqrt{x_1} + \frac{2}{|J_2|}\left[4\sum_{j\in J_2} y_j^2 - 2\prod_{j\in J_2}\cos\left(20y_j\pi/\sqrt{j}\right) + 2\right] \end{aligned} \tag{C.55}$$

where J_1 and J_2 are the same as in Unconstrained MOP 1, and y_j is defined as

$$y_j = x_j - x_1^{0.5[1+3(j-2)/(n-2)]} \text{ for } j \in [2,n]. \tag{C.56}$$

The search space is

$$x_j \in [-1,1] \text{ for } j \in [1,n]. \tag{C.57}$$

The Pareto front is

$$\begin{aligned} f_1^* &\in [0,1] \\ f_2^* &= 1 - \sqrt{f_1^*}. \end{aligned} \tag{C.58}$$

The Pareto set is

$$\begin{aligned} x_1^* &\in [0,1] \\ x_j^* &= (x_1^*)^{0.5[1+3(j-2)/(n-2)]} \text{ for } j \in [2,n]. \end{aligned} \tag{C.59}$$

C.3.4 Unconstrained Multi-Objective Optimization Problem 4

This two-objective problem is defined as

$$f_1 = x_1 + \frac{2}{|J_1|}\sum_{j\in J_1} h(y_j) \tag{C.60}$$

$$f_2 = 1 - \sqrt{x_1} + \frac{2}{|J_2|}\sum_{j\in J_2} h(y_j) \tag{C.61}$$

where J_1 and J_2 are the same as in Unconstrained MOP 1, y_j is defined as

$$y_j = x_j - \sin(6\pi x_1 + j\pi/n) \text{ for } j \in [2,n] \tag{C.62}$$

and $h(\cdot)$ is defined as

$$h(t) = \frac{|t|}{1 + e^{2|t|}}. \tag{C.63}$$

The search space is

$$\begin{aligned} x_1 &\in [0,1] \\ x_j &\in [-2,2] \text{ for } j \in [2,n]. \end{aligned} \tag{C.64}$$

The Pareto front is

$$\begin{aligned} f_1^* &\in [0,1] \\ f_2^* &= 1-(f_1^*)^2. \end{aligned} \tag{C.65}$$

The Pareto set is

$$\begin{aligned} x_1^* &\in [0,1] \\ x_j^* &= \sin(6\pi x_1^* + j\pi/n) \text{ for } j \in [2,n]. \end{aligned} \tag{C.66}$$

C.3.5 Unconstrained Multi-Objective Optimization Problem 5

This two-objective problem is defined as

$$f_1 = x_1 + \left(\frac{1}{2N}+\epsilon\right)|\sin(2N\pi x_1)| + \frac{2}{|J_1|}\sum_{j\in J_1} h(y_j) \tag{C.67}$$

$$f_2 = 1 - x_1 + \left(\frac{1}{2N}+\epsilon\right)|\sin(2N\pi x_1)| + \frac{2}{|J_2|}\sum_{j\in J_2} h(y_j) \tag{C.68}$$

where J_1 and J_2 are the same as in Unconstrained MOP 1, N is an integer ($N = 10$ in the CEC 2009 competition), ϵ is a positive real number ($\epsilon = 0.5$ in the CEC 2009 competition), y_j is defined as

$$y_j = x_j - \sin(6\pi x_1 + j\pi/n) \text{ for } j \in [2,n] \tag{C.69}$$

and $h(\cdot)$ is defined as

$$h(t) = 2t^2 - \cos(4\pi t) + 1. \tag{C.70}$$

The search space is

$$\begin{aligned} x_1 &\in [0,1] \\ x_j &\in [-1,1] \text{ for } j \in [2,n]. \end{aligned} \tag{C.71}$$

The Pareto front contains $(2N+1)$ discrete points:

$$(f_{1i}^*, f_{2i}^*) = (i/(2N), 1 - i/(2N)) \text{ for } i \in [1, 2N+1]. \tag{C.72}$$

The Pareto set also contains $(2N+1)$ discrete points, but they cannot be expressed analytically and so we do not show them here.

C.3.6 Unconstrained Multi-Objective Optimization Problem 6

This two-objective problem is defined as

$$\begin{aligned} f_1 &= x_1 + \max\left\{0, 2\left(\frac{1}{2N}+\epsilon\right)\sin(2N\pi x_1)\right\} + z_1 \\ f_2 &= 1 - x_1 + \max\left\{0, 2\left(\frac{1}{2N}+\epsilon\right)\sin(2N\pi x_1)\right\} + z_2 \end{aligned} \tag{C.73}$$

where N is an integer ($N = 2$ in the CEC 2009 competition), ϵ is a positive real number ($\epsilon = 0.1$ in the CEC 2009 competition), z_i is defined as

$$z_i = \frac{2}{|J_i|}\left(4\sum_{j\in J_i} y_j^2 - 2\prod_{j\in J_i}\cos\left(20y_j\pi/\sqrt{j}\right) + 2\right) \quad \text{for } i \in [1,2] \tag{C.74}$$

the sets J_1 and J_2 are the same as in Unconstrained MOP 1, and y_j is defined as

$$y_j = x_j - \sin(6\pi x_1 + j\pi/n) \quad \text{for } j \in [2,n]. \tag{C.75}$$

The search space is

$$\begin{aligned} x_1 &\in [0,1] \\ x_j &\in [-1,1] \text{ for } j \in [2,n]. \end{aligned} \tag{C.76}$$

The Pareto front contains one discrete point $(0, 1)$, and the following N disconnected segments:

$$\begin{aligned} f_1^* &= \bigcup_{i=1}^{N}\left[\frac{2i-1}{2N}, \frac{2i}{2N}\right] \\ f_2^* &= 1 - f_1^*. \end{aligned} \tag{C.77}$$

The Pareto set consists of discrete points, but they cannot be expressed analytically and so we do not show them here.

C.3.7 Unconstrained Multi-Objective Optimization Problem 7

This two-objective problem is defined as

$$\begin{aligned} f_1 &= x_1^{1/5} + \frac{2}{|J_1|}\sum_{j\in J_1} y_j^2 \\ f_2 &= 1 - x_1^{1/5} + \frac{2}{|J_2|}\sum_{j\in J_2} y_j^2 \end{aligned} \tag{C.78}$$

where J_1 and J_2 are the same as in Unconstrained MOP 1, and y_j is defined as

$$y_j = x_j - \sin(6\pi + j\pi/n) \quad \text{for } j \in [2,n]. \tag{C.79}$$

The search space is

$$\begin{aligned} x_1 &\in [0,1] \\ x_j &\in [-1,1] \text{ for } j \in [2,n]. \end{aligned} \tag{C.80}$$

The Pareto front is

$$\begin{aligned} f_1^* &\in [0,1] \\ f_2^* &= 1 - f_1^*. \end{aligned} \tag{C.81}$$

The Pareto set is

$$\begin{aligned} x_1^* &\in [0,1] \\ x_j^* &= \sin(6\pi x_1 + j\pi/n) \text{ for } j \in [2,n]. \end{aligned} \tag{C.82}$$

C.3.8 Unconstrained Multi-Objective Optimization Problem 8

This three-objective problem is defined as

$$\begin{aligned} f_1 &= \cos(0.5x_1\pi)\cos(0.5x_2\pi) + \frac{2}{|J_1|}\sum_{j\in J_1}\left[x_j - 2x_2\sin(2\pi x_1 + j\pi/n)\right]^2 \\ f_2 &= \cos(0.5x_1\pi)\sin(0.5x_2\pi) + \frac{2}{|J_2|}\sum_{j\in J_2}\left[x_j - 2x_2\sin(2\pi x_1 + j\pi/n)\right]^2 \\ f_3 &= \sin(0.5x_1\pi) + \frac{2}{|J_3|}\sum_{j\in J_3}\left(x_j - 2x_2\sin(2\pi x_1 + j\pi/n)\right]^2 \end{aligned} \tag{C.83}$$

where the sets J_1, J_2, and J_3 are defined as

$$\begin{aligned} J_1 &= \{j \in [3,n] : j-1 \text{ is a multiple of } 3\} \\ J_2 &= \{j \in [3,n] : j-2 \text{ is a multiple of } 3\} \\ J_3 &= \{j \in [3,n] : j \text{ is a multiple of } 3\}. \end{aligned} \tag{C.84}$$

The search space is

$$\begin{aligned} x_1 &\in [0,1] \\ x_2 &\in [0,1] \\ x_j &\in [-2,2] \text{ for } j \in [3,n]. \end{aligned} \tag{C.85}$$

The Pareto front is

$$(f_1^*, f_2^*, f_3^*) \quad \text{such that} \quad \begin{aligned} & f_1^* \in [0,1], f_2^* \in [0,1], f_3^* \in [0,1], \text{ and} \\ & (f_1^*)^2 + (f_2^*)^2 + (f_3^*)^2 = 1. \end{aligned} \tag{C.86}$$

The Pareto set is

$$\begin{aligned} x_1^* &\in [0,1] \\ x_2^* &\in [0,1] \\ x_j^* &= 2x_2^*\sin(2\pi x_1^* + j\pi/n) \text{ for } j \in [3,n]. \end{aligned} \tag{C.87}$$

C.3.9 Unconstrained Multi-Objective Optimization Problem 9

This three-objective problem is defined as

$$\begin{aligned} f_1 &= 0.5\left[\max\left\{0, \left(1+\epsilon\right)\left(1 - 4(2x_1-1)^2\right)\right\} + 2x_1\right]x_2 + z_1 \\ f_2 &= 0.5\left[\max\left\{0, \left(1+\epsilon\right)\left(1 - 4(2x_1-1)^2\right)\right\} - 2x_1 + 2\right]x_2 + z_2 \\ f_3 &= 1 - x_2 + \frac{2}{|J_3|}\sum_{j\in J_3}\left[x_j - 2x_2\sin(2\pi x_1 + j\pi/n)\right]^2 \end{aligned} \tag{C.88}$$

where ϵ is a positive real number ($\epsilon = 0.1$ in the CEC 2009 competition), z_i is defined as

$$z_i = \frac{2}{|J_1|}\sum_{j\in J_1}\left[x_j - 2x_2\sin(2\pi x_1 + j\pi/n)\right]^2 \text{ for } i \in [1,2] \tag{C.89}$$

and the sets J_1, J_2, and J_3 are the same as in Unconstrained MOP 8. The search space is

$$\begin{aligned} x_1 &\in [0,1] \\ x_2 &\in [0,1] \\ x_j &\in [-2,2] \text{ for } j \in [3,n]. \end{aligned} \tag{C.90}$$

The Pareto front has two sections: the first section is

$$\begin{aligned} f_3^* &\in [0,1] \\ f_1^* &\in [0,(1-f_3)/4] \\ f_2^* &= 1 - f_1^* - f_3^* \end{aligned} \tag{C.91}$$

and the second section is

$$\begin{aligned} f_3^* &\in [0,1] \\ f_1^* &\in [3(1-f_3)/4, 1] \\ f_2^* &= 1 - f_1^* - f_3^*. \end{aligned} \tag{C.92}$$

The Pareto set is

$$\begin{aligned} x_1^* &\in [0, 0.25] \cup [0.75, 1] \\ x_2^* &\in [0,1] \\ x_j^* &= 2x_2 \sin(2\pi x_1 + j\pi/n) \text{ for } j \in [3,n]. \end{aligned} \tag{C.93}$$

C.3.10 Unconstrained Multi-Objective Optimization Problem 10

This three-objective problem is defined as

$$\begin{aligned} f_1 &= \cos(0.5x_1\pi)\cos(0.5x_2\pi) + \frac{2}{|J_1|}\sum_{j\in J_1}\left[4y_j^2 - \cos(8\pi y_j) + 1\right] \\ f_2 &= \cos(0.5x_1\pi)\sin(0.5x_2\pi) + \frac{2}{|J_2|}\sum_{j\in J_2}\left[4y_j^2 - \cos(8\pi y_j) + 1\right] \\ f_3 &= \sin(0.5x_1\pi) + \frac{2}{|J_3|}\sum_{j\in J_3}\left[4y_j^2 - \cos(8\pi y_j) + 1\right] \end{aligned} \tag{C.94}$$

where the sets J_1, J_2, and J_3 are the same as in Unconstrained MOP 8, and y_j is defined as

$$y_j = x_j - 2x_2 \sin(2\pi x_1 + j\pi/n) \text{ for } j \in [3,n]. \tag{C.95}$$

The search space, Pareto front, and Pareto set are the same as in Unconstrained MOP 8.

C.4 DYNAMIC BENCHMARKS

Researchers have suggested various dynamic benchmark problems over the years [Branke, 1999]. Some constrained dynamic problems are given in [Nguyen and Yao, 2009], and some multi-objective dynamic problems are given in [Ray et al., 2009a]. Several combinatorial dynamic benchmarks have been suggested [Yang, 2008a], including dynamic knapsack and traveling salesman problems [Branke et al., 2006], [Mavrovouniotis and Yang, 2011]. But we restrict our discussion here to continuous dynamic benchmarks.

This section summarizes the optimization problems in [Li et al., 2008], which contains the continuous benchmarks and evaluation metrics that were used for the dynamic optimization competition at the 2009 IEEE Congress on Evolutionary Computation (CEC 2009). The dynamic benchmarks are based on some of the unconstrained problems of Appendix C.1. The dynamic benchmarks include offsets and rotation matrices (see Appendix C.7), the incorporation of time-varying functions, and the summing (or "composition") of several such functions. We give a complete description of the CEC 2009 dynamic benchmarks in Section C.4.1, and then we suggest a highly simplified version of the benchmarks in Section C.4.2.

C.4.1 The Complete Dynamic Benchmark Description

Consider one of the the n-dimensional functions $f(x)$ of Section C.1. First we normalize the magnitude of the benchmark. We do this to make sure that the time-varying function that we add later has the desired relative effect. We normalize the magnitude of the benchmark by scaling it as follows:

$$f'(x) = \frac{Cf(x)}{f_{\max}}, \text{ where } C = 2000. \tag{C.96}$$

The constant C is chosen to give the same magnitude to all scaled benchmarks so that the effect of the time-varying component that we add later will have the same impact on all scaled benchmarks.

Now we discuss the determination of $f_{\max}$ in Equation (C.96). For the dynamic benchmarks we typically use a baseline function $f(x)$ that generally increases as x increases. Although many of the functions of Section C.1 have a lot of local peaks and valleys, many of the functions are close to their maximum when each element of x is at its maximum value. Therefore, the quantity $f_{\max}$ in Equation (C.96) is estimated as

$$f_{\max} \approx f(x_{\max}Q) \tag{C.97}$$

where Q is the rotation matrix that we discuss below, and $x_{\max}$ is defined element-by-element:

$$\begin{aligned} x &= \begin{bmatrix} x_1 & \cdots & x_n \end{bmatrix} \text{ where } x_i \in [x_{i,\min}, x_{i,\max}] \\ \Longrightarrow x_{\max} &= \begin{bmatrix} x_{1,\max} & \cdots & x_{n,\max} \end{bmatrix}. \end{aligned} \tag{C.98}$$

Next we shift $f'(x)$ to obtain $f'(x-\theta)$, where θ is a random n-element bias vector. Each element of the bias vector is uniformly distributed in such a way that the optimum of $f'(x-\theta)$ is uniformly distributed on the domain of x. For instance, suppose that we are using the Ackley function as the baseline function.

The domain of the Ackley function is $x_i \in [-30, 30]$ for $i \in [1, n]$. The optimum of the unbiased Ackley function is located at $x_i^* = 0$ for $i \in [1, n]$. Therefore, each element of θ should be uniformly distributed on $[-30, 30]$ for all i. That way, each element of the optimizing value of $f'(x - \theta)$ is uniformly distributed on $[-30, 30]$. This helps ensure an even playing field when comparing different EAs, as discussed in Appendix C.7.1.

Next we rotate the scaled and shifted benchmark to obtain $f'((x-\theta)Q)$, where Q is a random orthogonal rotation matrix. This is an additional step to help ensure an even playing field when comparing different EAs, as discussed in Appendix C.7.2. Note that Q is also used in Equation (C.97) to approximate $f_{\max}$.

Next we add a time-varying function $\phi(t)$ to obtain $f'((x - \theta)Q) + \phi(t)$. The function $\phi(t)$ can be modified from one generation to the next as follows:

$$\begin{aligned} \phi(t) &\leftarrow \phi(t-1) + \Delta\phi \\ \phi(t) &\leftarrow \min(\phi(t), \phi_{\max}) \\ \phi(t) &\leftarrow \max(\phi(t), \phi_{\min}) \end{aligned} \tag{C.99}$$

where t is the function update iteration number (not necessarily the EA generation number), and $\phi_{\min}$ and $\phi_{\max}$ define the minimum and maximum allowable values of $\phi(t)$. The variation $\Delta\phi$ can take several forms. We first discuss dynamics that are referred to in [Li et al., 2008], [Li and Yang, 2008] as small-step dynamics:

$$\text{small step: } \Delta\phi = \alpha\phi_{\text{range}} r(t-1)\phi_s \tag{C.100}$$

where α is a constant, ϕ_{range} is the allowable range of $\phi(t)$, ϕ_s is a constant that defines the severity of the $\phi(t)$ change, and $r(t-1)$ is a random number uniformly distributed on $[-1, 1]$. [Li et al., 2008] uses

$$\begin{aligned} \alpha &= 0.04 \\ \phi_s &= 5 \\ \phi_{\min} &= 10 \\ \phi_{\max} &= 100 \\ \phi_{\text{range}} &= \phi_{\max} - \phi_{\min}. \end{aligned} \tag{C.101}$$

The initial value of $\phi(t)$ at $t = 0$ is a random number taken from a uniform distribution between $\phi_{\min}$ and $\phi_{\max}$. We see from Equations (C.99)–(C.101) that with small-step dynamics, $\phi(t)$ changes by no more than 18 each generation. We see from Equation (C.96) that $f'(x) \in [-2000, 2000]$ (approximately). Therefore, the maximum change in $\phi(t)$ in one generation relative to the maximum value of $f'(x)$ is $18/2000 = 0.9\%$ for the small-step change.

Note that Equation (C.99) does not apply at each generation; it only applies once in a while. [Li et al., 2008] suggests that Equation (C.99) be implemented once every 10,000 function evaluations, and that the EA run for a total of 600,000 function evaluations. In addition, we use the rotation matrix Q to rotate θ (the bias vector) every 10,000 function evaluations:

$$\theta(t) \leftarrow \theta(t-1)Q. \tag{C.102}$$

Finally, we generate m of these scaled, shifted, rotated, time-varying functions, and add them together to obtain a dynamic composition function:

$$F(x,t) = \sum_{i=1}^{m} w_i \left[f'((x - \theta_i(t))Q_i) + \phi_i(t) \right] \tag{C.103}$$

where each w_i is a weighting value defined by performing the following four statements in order:

$$\begin{aligned} w_i &\leftarrow \exp\left[-\left(\frac{\sum_{k=1}^{n}(x_k - \theta_{ik}(t))^2)}{2n} \right)^{1/2} \right] \\ w_{\max} &\leftarrow \max\{w_i\} \\ w_i &\leftarrow \begin{cases} w_i & \text{if } w_i = w_{\max} \\ w_i(1 - w_{\max}^{10}) & \text{if } w_i \neq w_{\max} \end{cases} \\ w_i &\leftarrow \frac{w_i}{\sum_{j=1}^{m} w_j} \end{aligned} \tag{C.104}$$

for $i \in [1, m]$. Note that $w_i \in [0, 1]$, and as x gets farther from θ_i (the optimum of the i-th shifted function), w_i decreases. [Li et al., 2008] uses $m = 10$. Each $\theta_i(t)$ vector in Equation (C.103) is a random n-element vector that is rotated every 10,000 function evaluations, and $\theta_{ik}(t)$ in Equation (C.104) is the k-th element of $\theta_i(t)$. Each Q_i matrix is a random but time-invariant $n \times n$ rotation matrix, and each $\phi_i(t)$ function is a random scalar function defined by Equation (C.100) and updated every 10,000 function evaluations. Each of the m functions that are summed in Equation (C.103) has a different time-varying component. Therefore, when we add these m functions together, we obtain a composite function whose minimizing value might change from one generation to the next.

Summing up the results in the above paragraphs, we obtain the algorithm of Figure C.23 for generating a dynamic benchmark function.

Figure C.23 describes the definition of dynamic benchmark functions for small-step dynamics. [Li et al., 2008] and [Li and Yang, 2008] suggest a total of six types of dynamics.

1. Small-step dynamics are summarized above in Equations (C.99)–(C.101).

2. Large-step dynamics are described as follows:

$$\text{large step: } \Delta\phi = \phi_{\text{range}} \left[\alpha \, \text{sign}(r(t-1)) + (\alpha_{\max} - \alpha) r(t-1) \right] \phi_s \tag{C.105}$$

where $r(t-1)$ is a random number uniformly distributed on $[-1, 1]$. The only new constant in the above equation is $\alpha_{\max}$, which [Li et al., 2008] sets as

$$\alpha_{\max} = 0.1. \tag{C.106}$$

We see from Equations (C.101), (C.105), and (C.106) that with large-step dynamics, $\phi(t)$ changes by no more than 45 in a single generation. We see from Equation (C.96) that $f'(x) \in [-2000, 2000]$ (approximately). Therefore, the maximum change in $\phi(t)$ in one generation relative to the maximum value of $f'(x)$ is $45/2000 = 2.25\%$ for the large-step change.

Begin initialization
  $f(\cdot)$ = baseline function from Section C.1
  $[x_{\min}, x_{\max}]$ = n-dimensional search domain
  x^* = n-dimensional optimizing value of $f(x)$
  E_{update} = number of function evaluations between dynamic updates
    (Typically $E_{\text{update}} = 10,000$)
  m = number of functions to combine in benchmark (typically $m = 10$)
  For $i = 1$ to m
    Generate random rotation matrix Q_i (see Section C.7.2)
    Generate random bias vector θ_i such that $x^* + \theta_i \in [x_{\min}, x_{\max}]$
  Next i
  $f_{\max} = f(x_{\max}Q)$
  $C = 2000$
  Function definition: $f'(x) = Cf(x)/f_{\max}$
  $\phi(0) \leftarrow U[\phi_{\min}, \phi_{\max}]$
  $E \leftarrow 0$ = number of function evaluations
End initialization
When we're ready to evaluate the benchmark function for a candidate solution x
  Use Equation (C.104) to calculate w_i for $i \in [1, m]$
  $E \leftarrow E + 1$
  If (E mod E_{update}) = 0 then
    Use Equations (C.99)–(C.101) to update $\phi_i(t)$ for $i \in [1, m]$
    Use Equation (C.102) to update $\theta_i(t)$ for $i \in [1, m]$
  End if
  Use Equation (C.103) to evaluate the candidate solution x
Next benchmark evaluation

Figure C.23 Function definition for an n-dimensional dynamic function based on the standard benchmark $f(\cdot)$ with small-step dynamics. (E mod E_{update}) is the remainder after the integer division E/E_{update}.

3. Random dynamics are described as follows:

$$\text{random: } \Delta\phi = \phi_s \rho(t-1) \tag{C.107}$$

where $\rho(t-1)$ is a random number taken from a zero-mean, unity-variance Gaussian distribution. Since a Gaussian random number is unbounded, $\phi(t)$ can change from its minimum to its maximum value (or vice versa) in a single generation. However, 99.7% of the time, the change in $\phi(t)$ will be within 3σ, which is $3\phi_s = 15$. With random dynamics the 3σ change in $\phi(t)$ in one generation relative to the maximum value of $f'(x)$ is $15/2000 = 0.75\%$.

4. Chaotic dynamics are described as follows:

$$\text{chaotic: } \phi(t) = A\left[\phi(t-1) - \phi_{\min}\right]\left[1 - \frac{\phi(t-1) - \phi_{\min}}{\phi_{\text{range}}}\right] \tag{C.108}$$

where $r(t-1)$ is a random number uniformly distributed on $[-1, 1]$. The only new constant in the above equation is A, which [Li et al., 2008] defines as

$$A = 3.67. \tag{C.109}$$

5. Recurrent dynamics are described as follows:

$$\text{recurrent: } \phi(t) = \phi_{\text{min}} + \frac{\phi_{\text{range}}\left[\sin(2\pi(t-1)/P + \zeta) + 1\right]}{2}. \tag{C.110}$$

These are the only deterministic dynamics defined in [Li et al., 2008]. The only new constants in the above equation are P (the period) and ζ (the initial phase), which [Li et al., 2008] defines as

$$\begin{aligned} P &= 12E_{\text{update}} \\ \zeta &= U[0, 2\pi] \end{aligned} \tag{C.111}$$

where E_{update} is the number of function evaluations between dynamic updates, and $U[0, 2\pi]$ is a random number uniformly distributed between 0 and 2π.

6. Noisy recurrent dynamics are described as follows:

$$\text{noisy recurrent: } \phi(t) = \phi_{\text{min}} + \frac{\phi_{\text{range}}\left[\sin(2\pi(t-1)/P + \zeta) + 1\right]}{2} + \rho_s \rho(t-1). \tag{C.112}$$

where $\rho(t-1)$ is a random number taken from a zero-mean, unity-variance Gaussian distribution. The only new constant in the above equation is ρ_s, the severity of the noisy dynamics, which [Li et al., 2008] defines as

$$\rho_s = 0.8. \tag{C.113}$$

We can modify Figure C.23 to implement any of the above types of dynamics. We only need to change one line in Figure C.23 to change the type of dynamics. We modify the line that says, "Use Equations (C.99)–(C.101) to update $\phi_i(t)$ for $i \in [1, m]$."

1. If we want small-step dynamics, we implement Figure C.23 as written.
2. If we want large-step dynamics, we use Equation (C.105) to update $\phi_i(t)$.
3. If we want random dynamics, we use Equation (C.107) to update $\phi_i(t)$.
4. If we want chaotic dynamics, we use Equation (C.108) to update $\phi_i(t)$.
5. If we want recurrent dynamics, we use Equation (C.110) to update $\phi_i(t)$.
6. If we want noisy recurrent dynamics, we use Equation (C.112) to update $\phi_i(t)$.

[Li et al., 2008] suggests five different functions to use as basis functions $f(\cdot)$ in Figure C.23: the sphere function (Section C.1.1), the Rastrigin function (Section C.1.11), the Weierstrass function (Section C.1.22), the Griewank function (Section C.1.6), and the Ackley function (Section C.1.2). Note that each of these functions in their original, unshifted versions has the optimizing solution $x^* = 0$.

C.4.2 A Simplified Dynamic Benchmark Description

Figure C.23 shows that there are several interacting dynamics in the benchmark functions, including the weights $\{w_i\}$ which are themselves functions of the candidate solution x, the dynamic variables $\phi_i(t)$, and the dynamic bias variables $\theta_i(t)$. However, it seems that the essence of the dynamics can be captured by the bias variables; the other variables provide only second-order effects. Furthermore, there is no need to add multiple functions together to obtain a function with a lot of dynamics; in other words, we can use $m = 1$ in Figure C.23 and still obtain good dynamic benchmarks. This results in Figure C.24, which is a simple but effective dynamic benchmark function generator.

Finally, we mention that we could use methods other than Equation (C.102) to update the bias $\theta(t)$. Equation (C.102) consists of a rotation of $\theta(t)$ around the origin of the search space. However, there are many other reasonable ways to update $\theta(t)$. We could change $\theta(t)$ in some other predictable way (linearly or periodically, for example), or we could generate a random $\theta(t)$ at each dynamic change. We could use different methods for changing $\theta(t)$ to represent dynamics in specific real-world problems.

Begin initialization
    $f(\cdot)$ = baseline function from Section C.1
    $[x_{\min}, x_{\max}]$ = n-dimensional search domain
    x^* = n-dimensional optimizing value of $f(x)$
    E_{update} = number of function evaluations between dynamic updates
    Generate random rotation matrix Q (see Section C.7.2)
    Generate random bias θ such that $x^* + \theta \in [x_{\min}, x_{\max}]$
End initialization
When we're ready to evaluate the benchmark function for a candidate solution x
    $E \leftarrow E + 1$
    If (E mod E_{update}) = 0 then
        Use Equation (C.102) to update $\theta(t)$
    End if
    Use $F(x, t) = f((x - \theta(t))Q)$ to evaluate the candidate solution x
Next benchmark evaluation

Figure C.24 Simplified function definition for an n-dimensional dynamic function based on the standard benchmark $f(\cdot)$.

C.5 NOISY BENCHMARKS

Noisy benchmark problems for EAs are easy to generate. We simply take a standard, non-noisy benchmark function and add noise. We can add various types of noise: noise with statistics that are independent of the given candidate solution x, as shown in Equation (21.39); noise with statistics that somehow vary with x, as shown in Equation (21.42); Gaussian noise; uniform noise; or any other type of noise that we want to use with our EA.

C.6 TRAVELING SALESMAN PROBLEMS

The TSPLIB web site has a collection of over 100 TSP benchmarks [Reinelt, 2008]. The simplest TSP in the collection is the Ulysses16 benchmark, which is based on 16 cities that the legendary Greek king Ulysses visited in the Mediterranean during his journeys [Grötschel and Padberg, 2001]. The largest TSP on the web site is a programmable logic array problem with 85,900 nodes. The Center for Discrete Mathematics & Theoretical Computer Science maintains a web site with large-scale TSP benchmarks, the largest of which contains over 20 million nodes [Demetrescu, 2012].

Each TSP is defined by a file with the extension `TSP` – for example, `ULYSSES.TSP`. A TSP file includes the coordinates of each city's latitude and longitude in the format `DDD.MM`, where `DDD` specifies degrees and `MM` specifies minutes. The TSP file also specifies the "edge weight type" of the problem, either `EUC_2D` or `GEO`, which indicates how to calculate distances between cities.

For `EUC_2D` problems, we need to calculate the Euclidean distance $D(i, k)$ between cities i and k as follows:

$$\begin{aligned} \Delta B &= B_i - B_k \\ \Delta L &= L_i - L_k \\ D(i,k) &= \text{round}\sqrt{\Delta B^2 + \Delta L^2} \end{aligned} \tag{C.114}$$

where B_i and L_i are the latitude and longitude of city i, and the `round` function rounds to the nearest integer. Rounding is not strictly necessary, but it is traditionally performed for TSPLIB problems, and so we may want to use rounding for fair comparisons between different TSP algorithms and previously published results.

For `GEO` problems, we need to calculate the geographical distance between cities i and k assuming that the Earth is a perfect sphere:

$$\begin{aligned} q_1 &= \cos(L_i - L_k) \\ q_2 &= \cos(B_i - B_k) \\ q_3 &= \cos(B_i + B_k) \\ D(i,k) &= \lfloor R \arccos\{[(1+q_1)q_2 - (1-q_1)q_3]/2\} + 1 \rfloor \end{aligned} \tag{C.115}$$

where $R = 6378.388$ km is the radius of the Earth, and the floor function $\lfloor \cdot \rfloor$ returns the largest integer that is less than or equal to its argument. The floor function and the addition of 1 at the end of the $D(i, k)$ calculation are not necessary in general, but are used in the standard geographical distance calculation in the TSPLIB benchmarks to round up to the nearest integer.

Derivation of Geographical Distance

We can derive Equation (C.115) by first converting latitude and longitude to Cartesian coordinates, which gives the coordinates of city i as

$$\begin{aligned} x_i &= R \cos B_i \cos L_i \\ y_i &= R \cos B_i \sin L_i \\ z_i &= R \sin B_i. \end{aligned} \tag{C.116}$$

We obtain similar equations for the Cartesian coordinates x_k, y_k, and z_k of the k-th city. Now remember that we can write the dot product between two vectors A and B as

$$A \cdot B = |A| \cdot |B| \cos\theta \tag{C.117}$$

where θ is the angle between the vectors. Therefore, we can write the dot product between the vectors that define city i and city k as

$$\begin{bmatrix} R\cos B_i \cos L_i \\ R\cos B_i \sin L_i \\ R\sin B_i \end{bmatrix} \cdot \begin{bmatrix} R\cos B_k \cos L_k \\ R\cos B_k \sin L_k \\ R\sin B_k \end{bmatrix} = R^2 \cos\theta \tag{C.118}$$

where θ is the angle between cities i and k. Dividing both sides by R^2 and expanding the above equation gives

$$\cos B_i \cos L_i \cos B_k \cos L_k + \cos B_i \sin L_i \cos B_k \sin L_k + \sin B_i \sin B_k = \cos\theta \tag{C.119}$$

which can be simplified to

$$\cos B_i \cos B_k (\cos L_i \cos L_k + \sin L_i \sin L_k) + \sin B_i \sin B_k = \cos\theta. \tag{C.120}$$

We can use standard trigonometric identities to write the above equation as

$$\begin{aligned} \frac{1}{2}\left[\cos(B_i + B_k) + \cos(B_i - B_k)\right]\cos(L_i - L_k) + & \\ \frac{1}{2}\left[\cos(B_i - B_k) - \cos(B_i + B_k)\right] &= \cos\theta. \end{aligned} \tag{C.121}$$

Solving for θ gives

$$\theta = \arccos\left\{\frac{1}{2}\left[q_1(q_2 + q_3) + q_2 - q_3\right]\right\}. \tag{C.122}$$

Two points on a sphere of radius R that are separated by an angle of θ have a distance between them on the surface of the sphere of $R\theta$, which gives Equation (C.115).[1]

Other Distance Metrics

Other distance metrics can be also found at [Reinelt, 2008], including three-dimensional Euclidean distance; Manhattan distance, which assumes that a route follows roads that are laid out on an orthogonal grid; maximum distance, which measures distance along the coordinate that requires the farthest travel distance; pseudo-Euclidean distance, which is the same as Equation (C.114) except that instead of rounding to the nearest integer we round to the next highest integer; and finally, a special distance function that is related to x-ray crystallography.

[1]This calculation is based on Jasper Spaans's web site at http://jsp.vs19.net/lr/sphere-distance.php.

Other Combinatorial Problems

Both symmetric and asymmetric TSPs can be found at [Reinelt, 2008]. The web site also includes related types of problems, including the following.

1. The sequential ordering problem is an asymmetric TSP that has precedence constraints [Dorigo and Stützle, 2004]. That is, given n cities, find the tour that results in the shortest distance while requiring that city i_m be visited before city k_m for $m \in [1, M]$, where M is the number of constraints.

2. The capacitated vehicle routing problem includes $(n - 1)$ nodes and one depot [Toth and Vigo, 2002]. The problem is to use trucks to make required deliveries from the depot to the nodes, assuming that each node has a specific delivery demand and that all trucks have identical capacities. Each tour begins at the depot, makes deliveries to a certain number of nodes, and then returns to the depot. The cost function could be the total distance traveled by all trucks or the total time required to make the deliveries.

3. The Hamiltonian path problem is the problem of discovering a path that visits each node of a graph exactly once [Balakrishnan, 1997]. The Hamiltonian cycle problem includes the additional requirement that the path returns to its starting point. Figure C.25 shows an example of two Hamiltonian path problems. The connected graph on the left has a Hamiltonian path: the path $1 \to 3 \to 2 \to 5 \to 4$ is a solution. However, the graph on the right does not have a Hamiltonian path. In the figure on the right we can find a path that visits all nodes once, but the path will also visit some nodes more than once (for example, $5 \to 2 \to 3 \to 1 \to 3 \to 4$).

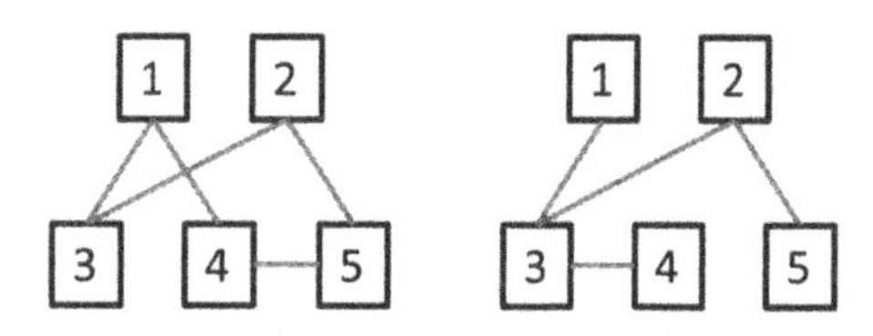

Figure C.25 Two connected graphs. The one on the left has a Hamiltonian path but the one on the right does not.

C.7 UNBIASING THE SEARCH SPACE

In this section we return to the discussion of continuous optimization problems. Some EAs naturally perform well on certain benchmarks simply because of the coincidental alignment of the quirks of the benchmarks with the quirks of the EAs. This does not speak well of the performance of the EAs; it only indicates that the features of many artificial benchmarks are not representative of real-world problems. This section discusses the use of offsets and rotation matrices in optimization benchmarks to make them more challenging and more realistic.

C.7.1 Offsets

Some EAs are naturally biased toward certain types of search spaces. For example, we saw in Section 16.2 that certain types of opposition-based learning (OBL) tend to move candidate solutions closer to the center of the search domain. Therefore, OBL naturally performs well on problems whose solution is near the center of the search space. However, this good performance by OBL is misleading; it is an artificial side effect of the fact that many benchmarks have solutions near the center of the search space. Another example is differential evolution (DE), which modifies candidate solutions on the basis of difference vectors. DE will therefore perform well on constrained problems whose feasible regions all lie parallel to each other. Again, this good performance by DE may be misleading; it is an artificial side effect of the fact that many benchmarks have parallel feasible regions. Many of the benchmarks in this appendix have solutions that are precisely at the center of the search space. It is not fair to use such benchmarks to evaluate EA performance.

Sometimes, although not always, we can directly see that an algorithm is biased and more easily finds a solution point when it lies at the center of the search space [Clerc, 2012b]. For example, we can run our algorithm on the two-dimensional problem with the search space $x_1 \in [-1, +1]$ and $x_2 \in [-1, +1]$ and with the cost function $f(x) = 1$ for all x, and plot the population in the two-dimensional search domain after many generations. If the algorithm is unbiased then the distribution will be uniform. However, for many algorithms the distribution will be more dense around the point $x = (0, 0)$. In that case we can conclude that the algorithm is biased. The density difference may be not visible, though, so the only safe way to evaluate optimization algorithms is to never use cost functions whose solution is at the center of the search domain (or even on a diagonal, in fact).

We can modify biased benchmarks and make them unbiased by adding offsets to the independent variables of the problem [Liang et al., 2005], [Suganthan et al., 2005]. Consider the sphere function of Equation (C.2), repeated here:

$$f(x) = \sum_{i=1}^{n} x_i^2 \tag{C.123}$$

with the optimum $x^* = 0$. If the search space is $x_i \in [-C, C]$ for all i, then the optimum is at the center of the search space. We therefore modify Equation (C.123) as follows:

$$f(x) = \sum_{i=1}^{n} (x_i - o_i)^2 \quad \text{where } o_i \sim U[-C, C] \text{ for } i \in [1, n]. \tag{C.124}$$

We could also use a distribution other than the uniform distribution to generate o_i, but we generally limit o_i to the domain $[-C, C]$ to ensure that the global optimum of Equation (C.124) lies in the search space. The shifted sphere function of Equation (C.124) has the same shape as the original sphere function, but its solution is located at a random point in the search space. This helps ensure that no particular EA has an unfair advantage during benchmark evaluation. Figure C.26 shows a shifted version of the sphere function.

When comparing EAs on the shifted sphere function, we should run several Monte Carlo simulations, each with a different shift. This approach, outlined in Figure C.27, allows us to determine the best EA for sphere-like functions while avoiding bias that is due to the location of the optimum. After completing the loop in Figure C.27 we will have M results for the first EA, M results for the second EA, and so on. We can compare the performances of the EAs by taking the average of each set of M results, or by taking the best, or the worst, or any other measure, depending on what quantity is of most interest (see Appendix B.2).

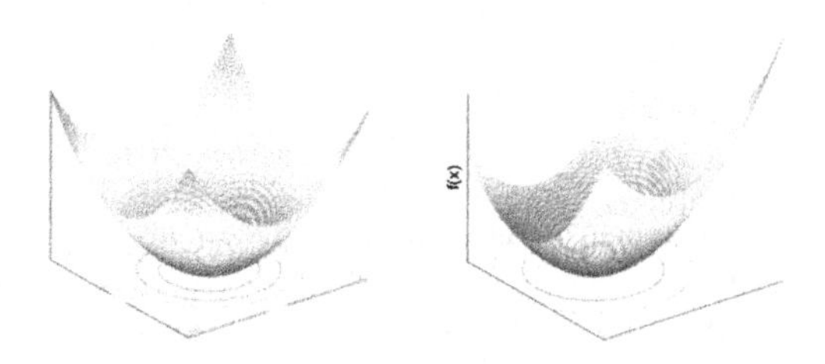

Figure C.26 The figure on the left is the original, unshifted two-dimensional sphere function. The figure on the right is the same function shifted along both independent variables.

P = number of EAs to evaluate
M = number of Monte Carlo simulations
For $j = 1$ to M
    Generate a random o_i for $i \in [1, n]$
    For $p = 1$ to P
        Evaluate the p-th EA's performance on $f(x - o)$
    Next EA
Next Monte Carlo simulation

Figure C.27 Outline of a Monte Carlo simulation for the evaluation of EA performance on n-dimensional shifted problems. The bias that is due to the location of the optimum has been removed.

C.7.2 Rotation Matrices

The search process of some EAs is naturally biased towards searches along a single independent variable. For example, a mutation or hill climbing strategy that changes one independent variable at a time searches along a single dimension of the problem each iteration. These types of optimization algorithms perform well on problems whose gradients are parallel to the independent variables. However, this good performance may be misleading; it is an artificial side effect of the fact that many benchmarks have gradients that are parallel to the unit vectors of the search space. Many of the benchmarks in this appendix have such gradients. It is not fair to use such benchmarks to evaluate EA performance.

Because of this, it is important to modify benchmarks by incorporating rotation matrices in the problem [Salomon, 1996], [Suganthan et al., 2005]. Consider the

Schwefel max function of Equation (C.15), repeated here for ease of reference:

$$f(x) = \max_i (|x_i| : i \in \{1, \cdots, n\}) \tag{C.125}$$

where our objective is to minimize $f(x)$. A search process that simply decreases one element of x at a time will perform quite well on this problem. Such a simple search process might also perform well on real-world problems, but there are many real-world problems that require a more sophisticated search strategy. We therefore modify Equation (C.125) as follows:

$$f(x) = \max_i (|y_i| : i \in \{1, \ldots, n\}) \text{ where } y = xQ. \tag{C.126}$$

The n-element vectors x and y are row vectors, and Q is an $n \times n$ rotation matrix. A rotation matrix is a matrix which, when multiplied by a vector, rotates that vector in its n-dimensional domain [Golan, 2007]. A rotation matrix is equivalent to an orthogonal matrix, and an orthogonal matrix is defined as a matrix whose transpose is equal to its determinant, and whose determinant is equal to 1:

$$Q^{-1} = Q^T, \text{ and } |Q| = 1. \tag{C.127}$$

A random rotation matrix can be generated with QR decomposition [Golan, 2007]. QR decomposition entails finding an orthogonal matrix Q and an upper triangular matrix R such that $QR = D$ for a given matrix D. Any real square matrix D has a QR decomposition. If we generate an $n \times n$ matrix D with random entries and find its QR decomposition, then the Q matrix is equal to a random rotation matrix. Therefore, we can generate a random $n \times n$ rotation matrix Q in MATLAB as follows:

$$D = \text{randn}(n);$$
$$[Q, R] = \text{QR}(D);$$

where `randn(n)` is a MATLAB function that generations an $n \times n$ matrix with each entry taken from a Gaussian distribution with a mean of zero and variance of one, and `QR` is MATLAB's QR decomposition function. Other linear algebra libraries and software packages have similar functions. The shifted Schwefel max function of Equation (C.126) has the same shape as the original Schwefel max function but is rotated with respect to the origin of the search space. Therefore, the gradient of the objective function is no longer parallel to the independent variable dimensions. This helps ensure that no particular EA has an unfair advantage during benchmark evaluation. Figure C.28 shows the Schwefel max function rotated by a few degrees in the counterclockwise direction (looking down on the plot).

When comparing EAs on the rotated Schwefel max function, we should run several Monte Carlo simulations, each with a different rotation matrix. This approach is similar to that shown in Figure C.27 and is outlined in Figure C.29. This approach allows us to determine the best EA performance while avoiding the bias that may be present due to the parallel nature of the gradient of the original function. After completing the loop in Figure C.29 we will have M results for the first EA, M results for the second EA, and so on. We can compare the performances of the EAs by taking the average of each set of M results, or by taking the best, or the worst, or any other measure, depending on what quantity is of most interest (see Appendix B.2). We can combine the logic of Figures C.27 and C.29 to obtain benchmark comparisons on functions $f((x-o)Q)$ that are both shifted and rotated.

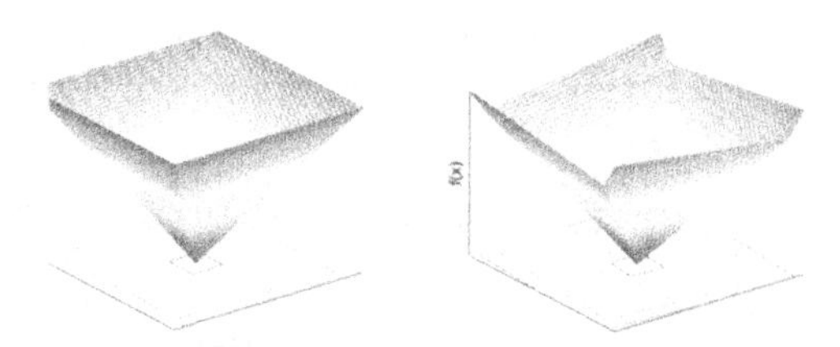

Figure C.28 The figure on the left is the two-dimensional Schwefel max function. The figure on the right is the same function rotated a few degrees in the counterclockwise direction.

P = number of EAs to evaluate
M = number of Monte Carlo simulations
For $j = 1$ to M
    Generate a random rotation matrix Q
    For $p = 1$ to P
        Evaluate the p-th EA's performance on $f(xQ)$
    Next EA
Next Monte Carlo simulation

Figure C.29 Outline of a Monte Carlo simulation for the evaluation of EA performance on rotated problems. The bias that is due to parallel alignment of the gradient with the coordinate system has been removed.

REFERENCES

Aarts, E. and Korst, J. (1989). *Simulated Annealing and Boltzmann Machines: A Stochastic Approach to Combinatorial Optimization and Neural Computing.* John Wiley & Sons.

Aarts, E., Lenstra, J., and van Laarhoven, P. (2003). Simulated annealing. In Aarts, E. and Lenstra, J., editors, *Local Search in Combinatorial Optimization*, pages 91–120. Princeton University Press.

Ackley, D. (1987a). *A Connectionist Machine for Genetic Hillclimbing.* Kluwer Academic Publishers.

Ackley, D. (1987b). An empirical study of bit vector function optimization. In Davis, L., editor, *Genetic Algorithms and Simulated Annealing*, pages 170–215. Pitman Publishing.

Adami, C. (1997). *Introduction to Artificial Life.* Springer.

Adler, F. and Nuernberger, B. (1994). Persistence in patchy irregular landscapes. *Theoretical Population Biology*, 45(1):41–75.

Aguirre, A., Rionda, S., Coello Coello, C., Lizárraga, G., and Mezura-Montes, E. (2004). Handling constraints using multiobjective optimization concepts. *International Journal for Numerical Methods in Engineering*, 59(15):1989–2017.

Ahn, C. and Ramakrishna, R. (2007). Multiobjective real-coded Bayesian optimization algorithm revisited: Diversity preservation. *Genetic and Evolutionary Computation Conference*, London, England, pages 593–600.

Akat, S. and Gazi, V. (2008). Particle swarm optimization with dynamic neighborhood topology: Three neighborhood strategies and preliminary results. *IEEE Swarm Intelligence Symposium*, St. Louis, Missouri, pages 1–8.

Alami, J. and El Imrani, A. (2008). Using cultural algorithm for the fixed-spectrum frequency assignment problem. *Journal of Mobile Communication*, 2(1):1–9.

Alami, J., El Imrani, A., and Bouroumi, A. (2007). A multipopulation cultural algorithm using fuzzy clustering. *Applied Soft Computing*, 7(2):506–519.

Alexander, R. (1996). *Optima for Animals.* Princeton University Press.

Ali, M., Khompatraporn, C., and Zabinsky, Z. (2005). A numerical evaluation of several stochastic algorithms on selected continuous global optimization test problems. *Journal of Global Optimization*, 31(4):635–672.

Allenson, R. (1992). Genetic algorithms with gender for multi-function optimisation. Technical report, Edinburgh Parallel Computing Centre. EPCC-SS92-01.

Altenberg, L. (1994). Emergent phenomena in genetic programming. *Conference on Evolutionary Programming*, San Diego, California, pages 233–241.

Anderson, M. and Oates, T. (2007). A review of recent research in metareasoning and metalearning. *AI Magazine*, 28(1):7–16.

Andre, D., Bennett, F., and Koza, J. (1996). Discovery by genetic programming of a cellular automata rule that is better than any known rule for the majority classification problem. *Genetic Programming Conference*, Palo Alto, California, pages 28–31.

Angeline, P. (1996a). An investigation into the sensitivity of genetic programming to the frequency of leaf selection during subtree crossover. *Genetic Programming Conference*, Palo Alto, California, pages 21–29.

Angeline, P. (1996b). Two self-adaptive crossover operators for genetic programming. In Angeline, P. and Kinnear, K., editors, *Advances in Genetic Programming: Volume 2*, pages 89–110. The MIT Press.

Angeline, P. (1997). Subtree crossover: Building block engine or macromutation? *Genetic Programming Conference*, Palo Alto, California, pages 9–17.

Applegate, D., Bixby, R., Chvatal, V., and Cook, W. (2007). *The Traveling Salesman Problem.* Princeton University Press.

Araujo, M., Wanner, E., Guimarães, F., and Takahashi, R. (2009). Constrained optimization based on quadratic approximations in genetic algorithms. In Mezura-Montes, E., editor, *Constraint-Handling in Evolutionary Optimization*, pages 193–217. Springer.

Arnold, D. (2002). *Noisy Optimization with Evolution Strategies.* Kluwer Academic Publishers.

Ashlock, D. (2009). *Evolutionary Computation for Modeling and Optimization.* Springer.

Åström, K. and Wittenmark, B. (2008). *Adaptive Control.* Dover Publications.

Atashpaz-Gargari, E. and Lucas, C. (2007). Imperialist competitive algorithm: An algorithm for optimization inspired by imperialistic competition. *IEEE Congress on Evolutionary Computation*, Singapore, pages 4661–4667.

Auger, A., Bader, J., Brockhoff, D., and Zitzler, E. (2012). Hypervolume-based multiobjective optimization: Theoretical foundations and practical implications. *Theoretical Computer Science*, 425:75–103.

Axelrod, R. (1997). The dissemination of culture: A model with local convergence and global polarization. *Journal of Conflict Resolution*, 41(2):203–226.

Axelrod, R. (2006). *The Evolution of Cooperation: Revised Edition.* Basic Books. First published in 1984.

Bäck, T. (1996). *Evolutionary Algorithms in Theory and Practice.* Oxford University Press.

Bäck, T., Fogel, D., and Michalewicz, Z. (1997a). *Handbook of Evolutionary Computation.* Taylor and Francis.

Bäck, T., Hammel, U., and Schwefel, H. (1997b). Evolutionary computation: Comments on the history and current state. *IEEE Transactions on Evolutionary Computation*, 1(1):3–17.

Bäck, T. and Schwefel, H.-P. (1993). An overview of evolutionary algorithms for parameter optimization. *Evolutionary Computation*, 1(1):1–23.

Baker, J. (1987). Reducing bias and inefficiency in the selection algorithm. *International Conference on Genetic Algorithms and Their Application*, Cambridge, Massachusetts, pages 14–21.

Balakrishnan, V. (1997). *Schaum's Outline of Graph Theory*. McGraw-Hill, 13th edition.

Balasubramaniam, P. and Kumar, A. (2009). Solution of matrix Riccati differential equation for nonlinear singular system using genetic programming. *Genetic Programming and Evolvable Machines*, 10(1):71–89.

Ball, W. and Coxeter, H. (2010). *Mathematical Recreations and Essays*. Dover, 13th edition.

Baluja, S. (1994). Population-based incremental learning. Technical report, Carnegie Mellon University. CMU-CS-94-163.

Baluja, S. and Caruana, R. (1995). Removing the genetics from the standard genetic algorithm. *12th International Conference on Machine Learning*, Tahoe City, California, pages 38–46.

Baluja, S. and Davies, S. (1998). Fast probabilistic modeling for combinatorial optimization. *Conference on Artificial Intelligence/Innovative Applications of Artificial Intelligence*, pages 469–476.

Bandyopadhyay, S., Saha, S., Maulik, U., and Deb, K. (2008). A simulated annealing-based multiobjective optimization algorithm: AMOSA. *IEEE Transactions on Evolutionary Computation*, 12(3):269–283.

Banks, A., Vincent, J., and Anyakoha, C. (2007). A review of particle swarm optimization. Part I: Background and development. *Natural Computing*, 6(4):467–484.

Banks, A., Vincent, J., and Anyakoha, C. (2008). A review of particle swarm optimization. Part II: Hybridisation, combinatorial, multicriteria and constrained optimization, and indicative applications. *Natural Computing*, 7(1):109–124.

Bankston, J. (2005). *Gregor Mendel and the Discovery of the Gene*. Mitchell Lane Publishers.

Banzhaf, W. (1990). The "molecular" traveling salesman. *Biological Cybernetics*, 64(1):7–14.

Banzhaf, W., Nordin, P., Keller, R., and Francone, F. (1998). *Genetic Programming*. Morgan Kauffman Publishers.

Barr, R., Golden, B., Kelly, J., Resende, M., and Stewart, W. Designing and reporting on computational experiments with heuristic methods. *Journal of Metaheuristics*, 1(1).

Barricelli, N. (1954). Esempi numerici di processi di evoluzione. *Methodos*, 6:45–68. The English translation of the title is *Numerical models of evolutionary processes*.

Basturk, B. and Karaboga, D. (2006). An artificial bee colony (ABC) algorithm for numeric function optimization. *IEEE Swarm Intelligence Symposium*, Indianapolis, Indiana.

Becerra, R. and Coello Coello, C. (2004). A cultural algorithm with differential evolution to solve constrained optimization problems. In Lemaitre, C., Reyes, C., and Gonzalez, J., editors, *Advances in Artificial Intelligence – IBERAMIA 2004: 9th Ibero-American Conference on AI, Puebla, Mexico, November 22-26, 2004*, pages 881–890. Springer.

Bellman, R. (1961). *Adaptive Control Processes: A Guided Tour.* Princeton University Press.

Benatchba, K., Admane, L., and Koudil, M. (2005). Using bees to solve a data-mining problem expressed as a max-sat one. In Mira, J. and Álvarez, J., editors, *Artificial Intelligence and Knowledge Engineering Applications: A Bioinspired Approach*, pages 212–220. Springer.

Bernstein, D. (2006). Optimization r us. *IEEE Control Systems Magazine*, 26(5):6–7.

Betts, J. (2009). *Practical Methods for Optimal Control and Estimation Using Nonlinear Programming.* Society for Industrial & Applied Mathematics, 2nd edition.

Beveridge, W. (2004). *The Art of Scientific Investigation.* Blackburn Press.

Beyer, H.-G. (1998). On the dynamics of EAs without selection. *Foundations of Genetic Algorithms*, Amsterdam, The Netherlands, pages 5–26.

Beyer, H.-G. (2010). *The Theory of Evolution Strategies.* Springer.

Beyer, H.-G. and Deb, K. (2001). On self-adaptive features in real-parameter evolutionary algorithms. *IEEE Transactions on Evolutionary Computation*, 5(3):250–269.

Beyer, H.-G. and Schwefel, H.-P. (2002). Evolution strategies: A comprehensive introduction. *Natural Computing*, 1(1):3–52.

Beyer, H.-G. and Sendhoff, B. (2008). Covariance matrix adaptation revisited: The CMSA evolution strategy. In Rudolph, G., Jansen, T., Lucas, S., Poloni, C., and Beume, N., editors, *Parallel Problem Solving from Nature - PPSN X*, pages 123–132. Springer.

Bhattacharya, M. (2008). Reduced computation for evolutionary optimization in noisy environment. *Genetic and Evolutionary Computation Conference*, Atlanta, Georgia, pages 2117–2122.

Bishop, C. (2006). *Pattern Recognition and Machine Learning.* Springer.

Bishop, J. (1989). Stochastic searching networks. *First IEE Conference on Artificial Neural Networks*, London, England, pages 329–331.

Biswas, A., Dasgupta, S., Das, S., and Abraham, A. (2007a). A synergy of differential evolution and bacterial foraging optimization for global optimization. *Neural Network World*, 17(6):607–626.

Biswas, A., Dasgupta, S., Das, S., and Abraham, A. (2007b). Synergy of PSO and bacterial foraging optimization – A comparative study on numerical benchmarks. In Corchado, E., Corchado, J., and Abraham, A., editors, *Innovations in Hybrid Intelligent Systems*, pages 255–263. Springer.

Blum, C. (2005a). Ant colony optimization: Introduction and recent trends. *Physics of Life Reviews*, 2(4):353–373.

Blum, C. (2005b). Beam-ACO – Hybridizing ant colony optimization with beam search: An application to open shop scheduling. *Computers & Operations Research*, 32(6):1565–1591.

Blum, C. (2007). Ant colony optimization: Introduction and hybridizations. *Seventh International Conference on Hybrid Intelligent Systems*, Kaiserlautern, Germany, pages 24–29.

Blum, C. and Dorigo, M. (2004). The hypercube framework for ant colony optimization. *IEEE Transactions on Systems, Man, and Cybernetics - Part B: Cybernetics*, 34(2):1161–1172.

Bonabeau, E., Theraulaz, G., and Dorigo, M. (1999). *Swarm Intelligence: From Natural to Artificial Systems.* Oxford University Press.

Bonacich, P., Shure, G., Kahan, J., and Meeker, R. (1976). Cooperation and group size in the n-person prisoners' dilemma. *The Journal of Conflict Resolution*, 20(4):687–706.

Boslaugh, S. and Watters, P. (2008). *Statistics in a Nutshell.* O'Reilly Media.

Bosman, P. and Thierens, D. (2003). The balance between proximity and diversity in multiobjective evolutionary algorithms. *IEEE Transactions on Evolutionary Computation*, 7(2):174–188.

Box, G. (1957). Evolutionary operation: A method for increasing industrial productivity. *Journal of the Royal Statistical Society, Series C (Applied Statistics)*, 6(2):81–101.

Box, J. (1987). Guinness, Gosset, Fisher, and small samples. *Statistical Science*, 2(1):45–52.

Branke, J. (1998). Creating robust solutions by means of evolutionary algorithms. In Eiben, A., Bäck, T., Schoenauer, M., and Schwefel, H.-P., editors, *Parallel Problem Solving from Nature - PPSN V*, pages 119–128. Springer.

Branke, J. (1999). Efficient fitness estimation in noisy environments. *Memory enhanced evolutionary algorithms for changing optimization problems*, Washington, District of Columbia, pages 1875–1882.

Branke, J. (2002). *Evolutionary Optimization in Dynamic Environments.* Kluwer Academic Publishers.

Branke, J. (2012). Evolutionary Algorithms for Dynamic Optimization Problems (EvoDOP). `http://people.aifb.kit.edu/jbr/EvoDOP`.

Branke, J., Orbayi, M., and Uyar, S. (2006). The role of representations in dynamic knapsack problems. In Rothlauf, F., editor, *Applications of Evolutionary Computing*, pages 764–775. Springer.

Branke, J., Schmidt, C., and Schmec, H. (2001). Efficient fitness estimation in noisy environments. *Genetic and Evolutionary Computation Conference*, San Francisco, California, pages 243–250.

Bratton, D. and Kennedy, J. (2007). Defining a standard for particle swarm optimization. *IEEE Swarm Intelligence Symposium*, Honolulu, Hawaii, pages 120–127.

Bremermann, H., Rogson, M., and Salaff, S. (1966). Global properties of evolution processes. In Pattee, H., Edlsack, E., Fein, L., and Callahan, A., editors, *Natural Automata and Useful Simulations*, pages 3–41. Spartan Books.

Brest, J. (2009). Constrained real-parameter optimization with ϵ-self-adaptive differential evolution. In Mezura-Montes, E., editor, *Constraint-Handling in Evolutionary Optimization*, pages 73–93. Springer.

Brest, J., Zamuda, A., Boskovic, B., Maucec, M., and Zumer, V. (2009). Dynamic optimization using self-adaptive differential evolution. *IEEE Congress on Evolutionary Computation*, Trondheim, Norway, pages 415–422.

Bringmann, K. and Friedrich, T. (2010). An efficient algorithm for computing hypervolume contributions. *Evolutionary Computation*, 18(3):383–402.

Bui, L., Abbass, H., and Essam, D. (2005). Fitness inheritance for noisy evolutionary multi-objective optmization. *Genetic and Evolutionary Computation Conference*, Washington, District of Columbia, pages 779–785.

Bureerat, S. and Sriworamas, K. (2007). Population-based incremental learning for multiobjective optimisation. In Saad, A., Dahal, K., Sarfraz, M., and Roy, R., editors, *Soft Computing in Industrial Applications*, pages 223–232. Springer.

Burke, E. (2003). *High-Tech Cycling.* Human Kinetics, 2nd edition.

Cai, C. and Wang, Y. (2006). A multiobjective optimization-based evolutionary algorithm for constrained optimization. *IEEE Transactions on Evolutionary Computation*, 10(6):658–675.

Cakir, B., Altiparmak, F., and Dengiz, B. (2011). Multi-objective optimization of a stochastic assembly line balancing: A hybrid simulated annealing algorithm. *Computers & Industrial Engineering*, 60(3):376–384.

Carlisle, A. and Dozier, G. (2001). An off-the-shelf PSO. *Particle Swarm Optimization Workshop*, Indianapolis, Indiana, pages 1–6.

Carlson, S. and Shonkwiler, R. (1998). Annealing a genetic algorith over constraints. *IEEE International Conference on Systems, Man, and Cybernetics*, San Diego, California, pages 3931–3936.

Černý, V. (1985). Thermodynamical approach to the travelling salesman problem: An efficient simulation algorithm. *Journal of Optimization Theory and Applications*, 45(1):41–51.

Chafekar, D., Shi, L., Rasheed, K., and Xuan, J. (2005). Multiobjective GA optimization using reduced models. *IEEE Transactions on Systems, Man, and Cybernetics – Part C: Applications and Reviews*, 35(2):261–265.

Chen, S. and Montgomery, J. (2011). Selection strategies for initial positions and initial velocities in multi-optima particle swarms. *Genetic and Evolutionary Computation Conference*, Dublin, Ireland, pages 53–60.

Chen, Y.-L. and Liu, C.-C. (1994). Multiobjective VAr planning using the goal-attainment method. *IEE Proceedings on Generation, Transmission and Distribution*, 141(3):227–232.

Cheng, C., Wang, W., Xu, D., and Chau, K. (2008). Optimizing hydropower reservoir operation using hybrid genetic algorithm and chaos. *Water Resources Management*, 22(7):895–909.

Choi, S. and Moon, B. (2003). Normalization in genetic algorithms. *Genetic and Evolutionary Computation Conference*, Chicago, Illinois, pages 862–873.

Chow, C. and Liu, C. (1968). Approximating discrete probability distributions with dependence trees. *IEEE Transactions on Information Theory*, IT-14(3):462–467.

Christensen, S. and Oppacher, F. (2001). What can we learn from no free lunch? A first attempt to characterize the concept of a searchable function. *Genetic and Evolutionary Computation Conference*, San Francisco, California, pages 1219–1226.

Chuan-Chong, C. and Khee-Meng, K. (1992). *Principles and Techniques in Combinatorics*. World Scientific.

Chuang, C.-L. and Jiang, J.-A. (2007). Integrated radiation optimization: Inspired by the gravitational radiation in the curvature of space-time. *IEEE Congress on Evolutionary Computation*, Singapore, pages 3157–3164.

Chung, H.-S. and Alonso, J. (2004). Multiobjective optimization using approximation model-based genetic algorithms. *10th AIAA/ISSMO Symposium on Multidisciplinary Analysis and Optimization*, Albany, New York.

Chung, H.-S., Choi, S., and Alonso, J. (2003). Supersonic business jet design using a knowledge-based genetic algorithm with an adaptive, unstructured grid methodology. *21st AIAA Applied Aerodynamics Conference*, Orlando, Florida.

Clement, P. (1959). A class of triple-diagonal matrices for test purposes. *SIAM Review*, 1(1):50–52.

Clerc, M. (1999). The swarm and the queen: Towards a deterministic and adaptive particle swarm optimization. In *IEEE Congress on Evolutionary Computing*, pages 1951–1957.

Clerc, M. (2004). Discrete particle swarm optimization, illustrated by the traveling salesman problem. In Onwubolu, G. and Babu, R., editors, *New Optimization Techniques in Engineering*, pages 219–239. Springer.

Clerc, M. (2006). *Particle Swarm Optimization.* John Wiley & Sons.

Clerc, M. (2012a). Particle Swarm Optimization. `http://clerc.maurice.free.fr/pso`.

Clerc, M. (2012b). Randomness matters. Technical report. `http://clerc.maurice.free.fr/pso`.

Clerc, M. and Kennedy, J. (2002). The particle swarm – Explosion, stability, and convergence in a multidimensional complex space. *IEEE Transactions on Evolutionary Computation*, 6(1):58–73.

Clerc, M. and Poli, R. (2006). Stagnation analysis in particle swarm optimisation or what happens when nothing happens. Technical report, University of Essex. `http://clerc.maurice.free.fr/pso`.

Cobb, H. and Grefenstette, J. (1993). Genetic algorithms for tracking changing environments. *International Conference on Genetic Algorithms*, Urbana-Champaign, Illinois, pages 523–530.

Coello Coello, C. (1999). A comprehensive survey of evolutionary-based multiobjective optimization techniques. *Knowledge and Information Systems*, 1(3):269–308.

Coello Coello, C. (2000a). Constraint-handling using an evolutionary multiobjective optimization technique. *Civil Engineering and Environmental Systems*, 17(4):319–346.

Coello Coello, C. (2000b). Use of a self-adaptive penalty approach for engineering optimization problems. *Computers in Industry*, 41(2):113–127.

Coello Coello, C. (2002). Theoretical and numerical constraint-handling techniques used with evolutionary algorithms: A survey of the state of the art. *Computer Methods in Applied Mechanics and Engineering*, 191(11–12):1245–1287.

Coello Coello, C. (2006). Evolutionary multi-objective optimization: A historical view of the field. *IEEE Computational Intelligence Magazine*, 1(1):28–36.

Coello Coello, C. (2009). Evolutionary multi-objective optimization: Some current research trends and topics that remain to be explored. *Frontiers of Computer Science in China*, 3(1):18–30.

Coello Coello, C. (2012a). List of references on constraint-handling techniques used with evolutionary algorithms. `www.cs.cinvestav.mx/~constraint`.

Coello Coello, C. (2012b). List of references on evolutionary multiobjective optimization. `www.lania.mx/~ccoello/EMOO/EMOObib.html`.

Coello Coello, C. and Becerra, R. (2002). Constrained optimization using an evolutionary programming-based cultural algorithm. In Parmee, I., editor, *Adaptive Computing in Design and Manufacture V*, pages 317–328. Springer.

Coello Coello, C. and Becerra, R. (2003). Evolutionary multiobjective optimization using a cultural algorithm. *Swarm Intelligence Symposium*, Indianapolis, Indiana, pages 6–13.

Coello Coello, C., Lamont, G., and Van Veldhuizen, D. (2007). *Evolutionary Algorithms for Solving Multi-Objective Problems.* Springer.

Coello Coello, C. and Mezura-Montes, E. (2011). Constraint-handling in nature-inspired numerical optimization: Past, present and future. *Swarm and Evolutionary Computation*, 1(4):173–194.

Coit, D. and Smith, A. (1996). Penalty guided genetic search for reliability design optimization. *Computers and Industrial Engineering*, 30(4):895–904.

Coit, D., Smith, A., and Tate, D. (1996). Adaptive penalty methods for genetic optimization of constrained combinatorial problems. *INFORMS Journal on Computing*, 8(2):173–182.

Collard, P. and Aurand, J. (1994). DGA: An efficient genetic algorithm. *11th European Conference on Artificial Intelligence*, Amsterdam, The Netherlands, pages 487–492.

Collard, P. and Gaspar, A. (1996). "Royal-road" landscapes for a dual genetic algorithm. *12th European Conference on Artificial Intelligence*, Budapest, Hungary, pages 213–217.

Collette, Y. and Siarry, P. (2004). *Multiobjective Optimization: Principles and Case Studies.* Springer.

Colorni, A., Dorigo, M., and Maniezzo, V. (1991). Distributed optimization by ant colonies. *European Conference on Artificial Life*, Paris, France, pages 134–142.

Corder, G. and Foreman, D. (2009). *Nonparametric Statistics for Non-Statisticians.* John Wiley & Sons.

Cordon, O., Herrera, F., de Viana, F., and Moreno, L. (2000). A new ACO model integrating evolutionary computation concepts: The best-worst ant system. *From Ant Colonies to Artificial Ants: Second International Workshop on Ant Algorithms*, Brussels, Belgium, pages 22–29.

Corfman, K. and Lehmann, D. (1994). The prisoner's dilemma and the role of information in setting advertising budgets. *Journal of Advertising*, 23(2):35–48.

Courant, R. (1943). Variational methods for the solution of problems of equilibrium and vibrations. *Bulletin of the American Mathematical Society*, 49(1):1–23.

Cover, T. and Thomas, J. (1991). *Elements of Information Theory.* Wiley-Interscience.

Cramer, N. (1985). A representation for the adaptive generation of simple sequential programs. *International Conference on Genetic Algorithms and Their Application*, Pittsburgh, Pennsylvania, pages 183–187.

Črepinšek, M., Liu, S.-H., and Mernik, L. (2012). A note on teaching-learning-based optimization algorithm. *Information Sciences*, 212:79–93.

Črepinšek, M., Liu, S.-H., and Mernik, M. (2013). Replication and comparison of computational experiments in applied evolutionary computing: Common pitfalls and guidelines to avoid them. *Information Sciences*, submitted for publication.

Culberson, J. (1998). On the futility of blind search. *Evolutionary Computation*, 6(2):109–127.

Darwin, C. (1859). *On the Origin of Species by Means of Natural Selection, or The Preservation of Favoured Races in the Struggle for Life.* John Murray, Albemarle Street.

Darwin, C., Neve, M., and Messenger, S. (2002). *Autobiographies.* Penguin Classics.

Das, S., Biswas, A., Dasgupta, S., and Abraham, A. (2009). Bacterial foraging optimization algorithm: Theoretical foundations, analysis, and applications. In Abraham, A., Hassanien, A.-E., Siarry, P., and Engelbrecht, A., editors, *Foundations of Computational Intelligence – Volume 3: Global Optimization*, pages 23–56. Springer.

Das, S. and Suganthan, P. (2011). Differential evolution: A survey of the state-of-the-art. *IEEE Transactions on Evolutionary Computation*, 15(1):4–31.

Das, S., Suganthan, P., and Coello Coello, C. (2011). Guest editorial: Special issue on differential evolution. *IEEE Transactions on Evolutionary Computation*, 15(1):1–3.

Dasgupta, S., Das, S., Abraham, A., and Biswas, A. (2009). Adaptive computational chemotaxis in bacterial foraging optimization: An analysis. *IEEE Transactions on Evolutionary Computation*, 13(4):919–941.

Davis, L. (1985). Job shop scheduling with genetic algorithms. *International Conference on Genetic Algorithms and Their Application*, Pittsburgh, Pennsylvania, pages 136–140.

Davis, L. and Steenstrup, M. (1987). Genetic algorithms and simulated annealing: An overview. In Davis, L., editor, *Genetic Algorithms and Simulated Annealing*, pages 1–11. Pitman Publishing.

Davis, T. and Principe, J. (1991). A simulated annealing like convergence theory for the simple genetic algorithm. *International Conference on Genetic Algorithms*, San Diego, California, pages 174–181.

Davis, T. and Principe, J. (1993). A Markov chain framework for the simple genetic algorithm. *Evolutionary Computation*, 1(3):269–288.

De Bonet, J., Isbell, C., and Viola, P. (1997). MMIC: Finding optima by estimating probability densities. In Mozer, M., Jordan, M., and Petsche, T., editors, *Advances in Neural Information Processing Systems 9*, pages 424–430. MIT Press.

de Franca, F., Coelho, G., Von Zuben, F., and Attux, R. (2008). Multivariate ant colony optimization in continuous search spaces. In *Genetic and Evolutionary Computation Conference*, pages 9–16.

de Garis, H. (1990). Genetic programming: Building artificial nervous systems with genetically programmed neural network modules. *Seventh International Conference on Machine Learning*, Austin, Texas, pages 132–139.

De Jong, K. (1975). *An Analysis of the Behaviour of a Class of Genetic Adaptive Systems.* PhD thesis, University of Michigan.

De Jong, K. (1992). Genetic algorithms are NOT function optimizers. *Second Workshop on Foundations of Genetic Algorithms*, Vail, Colorado, pages 5–17.

De Jong, K. (2002). *Evolutionary Computation.* The MIT Press.

De Jong, K., Fogel, D., and Schwefel, H.-P. (1997). A history of evolutionary computation. In Bäck, T., Fogel, D., and Michalewicz, Z., editors, *Handbook of Evolutionary Computation*, pages A2.3:1–12. Oxford University Press.

de Oca, M. and Stützle, T. (2008). Convergence behavior of the fully informed particle swarm optimization algorihtm. *Genetic and Evolutionary Computation Conference*, Atlanta, Georgia, pages 71–78.

Deb, K. (2000). An efficient constraint handling method for genetic algorithms. *Computer Methods in Applied Mechanics and Engineering*, 186(2–4):311–338.

Deb, K. (2009). *Multi-Objective Optimization using Evolutionary Algorithms.* John Wiley & Sons.

Deb, K. and Agrawal, R. (1995). Simulated binary crossover for continuous search space. *Complex Systems*, 9(2):115–148.

Deb, K. and Agrawal, S. (1999). A niched-penalty approach for constraint handling in genetic algorithms. *International Conference on Artificial Neural Nets and Genetic Algorithms*, Portoroz, Slovenia, pages 235–242.

Deb, K., Agrawal, S., Pratap, A., and Meyarivan, T. (2000). A fast elitist non-dominated sorting genetic algorithm for multi-objective optimization: NSGA-II. In Schoenauer, M., Deb, K., Rudolph, G., Yao, X., Lutton, E., Merelo, J., and Schwefel, H.-P., editors, *Parallel Problem Solving from Nature - PPSN VI*, pages 849–858. Springer.

Deb, K. and Goldberg, D. (1989). An investigation of niche and species formation in genetic function optimization. *International Conference on Genetic Algorithms*, Fairfax, Virginia, pages 42–50.

Deb, K., Mohan, M., and Mishra, S. (2005). Evaluating the ϵ-domination based multi-objective evolutionary algorithm for a quick computation of Pareto-optimal solutions. *Evolutionary Computation*, 13(4):501–525.

Deb, K., Pratap, A., Agarwal, S., and Meyarivan, T. (2002a). A fast and elitist multiobjective genetic algorithm: NSGA-II. *IEEE Transactions on Evolutionary Computation*, 6(2):182–197.

Deb, K., Pratap, A., and Meyarivan, T. (2001). Constrained test problems for multi-objective evolutionary optimization. In Zitzler, E., Deb, K., Thiele, L., Coello Coello, C., and Corne, D., editors, *Evolutionary Multi-Criterion Optimization: First International Conference, EMO 2001*, pages 284–298. Springer.

Deb, K., Thiele, L., Laumanns, M., and Zitzler, E. (2002b). Scalable multi-objective optimization test problems. *World Congress on Computational Intelligence*, Honolulu, Hawaii, pages 825–830.

Dechter, R. (2003). *Constraint Processing.* Morgan Kaufmann.

Deep, K. and Thakur, M. (2007). A new crossover operator for real coded genetic algorithms. *Applied Mathematics and Computation*, 188(1):895–911.

del Valle, Y., Venayagamoorthy, G., Mohagheghi, S., Hernandez, J.-C., and Harley, R. (2008). Particle swarm optimization: Basic concepts, variants and applications in power systems. *IEEE Transactions on Evolutionary Computation*, 12(2):171–195.

Delahaye, J.-P. and Mathieu, P. (1995). Complex strategies in the iterated prisoner's dilemma. In Albert, A., editor, *Chaos and Society*, pages 283–292. IOS Press.

Delsuc, F. (2003). Army ants trapped by their evolutionary history. *Public Library of Science Biology*, 1(2):e37.

Dembski, W. and Marks, R. (2009a). Bernoulli's principle of insufficient reason and conservation of information in computer search. *IEEE Conference on Systems, Man and Cybernetics*, San Antonio, Texas, pages 2647–2652.

Dembski, W. and Marks, R. (2009b). Conservation of information in search: Measuring the cost of success. *IEEE Transactions on Systems, Man, and Cybernetics – Part A: Systems and Humans*, 39(5):1051–1060.

Dembski, W. and Marks, R. (2010). The search for a search: Measuring the information cost of higher level search. *Journal of Advanced Computational Intelligence and Intelligent Informatics*, 14(5):475–486.

Demetrescu, C. (2012). 9th DIMACS Implementation Challenge – Shortest Paths. `www.dis.uniroma1.it/challenge9`.

Deneubourg, J.-L., Aron, S., Goss, S., and Pasteels, J. (1990). The self-organizing exploratory pattern of the Argentine ant. *Journal of Insect Behavior*, 3(2):159–168.

DePaulo, B., Kashy, D., Kirkendol, S., Wyer, M., and Epstein, J. (1996). Lying in everyday life. *Journal of Personality and Social Psychology*, 70(5):979–995.

Devroye, L. (1978). Progressive global random search of continuous functions. *Mathematical Programming*, 15(1):330–342.

Di Pietro, A., While, L., and Barone, L. (2004). Applying evolutionary algorithms to problems with noisy, time-consuming fitness functions. *IEEE Congress on Evolutionary Computation*, Portland, Oregon, pages 1254–1261.

Dominguez, J. and Pulido, G. (2011). A comparison on the search of particle swarm optimization and differential evolution on multi-objective optimization. *IEEE Congress on Evolutionary Computation*, New Orleans, Louisiana, pages 1978–1985.

Doran, R. (2007). The gray code. *Journal of Universal Computer Science*, 13(11):1573–1597.

Dorigo, M., Birattari, M., and Stützle, T. (2006). Ant colony optimization: Artificial ants as a computational intelligence technique. *IEEE Computational Intelligence Magazine*, 1(4):28–39.

Dorigo, M. and Gambardella, L. (1997a). Ant colonies for the traveling salesman problem. *BioSystems*, 43(2):73–81.

Dorigo, M. and Gambardella, L. (1997b). Ant colony system: A cooperative learning approach to the traveling salesman problem. *IEEE Transactions on Evolutionary Computation*, 1(1):53–66.

Dorigo, M., Maniezzo, V., and Colorni, A. (1996). Ant system: Optimization by a colony of cooperating agents. *IEEE Transactions on Systems, Man, and Cybernetics – Part B: Cybernetics*, 26(1):29–41.

Dorigo, M. and Stützle, T. (2004). *Ant Colony Optimization*. The MIT Press.

Dorigo, M. and Stützle, T. (2010). Ant colony optimization: Overview and recent advances. In Gendreau, M. and Potvin, J.-Y., editors, *Handbook of Metaheuristics*, pages 227–263. Springer.

Droste, S., Jansen, T., and Wegener, I. (2002). On the analysis of the (1+1) evolutionary algorithm. *Theoretical Computer Science*, 276(1–2):51–81.

Du, D., Simon, D., and Ergezer, M. (2009). Biogeography-based optimization combined with evolutionary strategy and immigration refusal. *IEEE Conference on Systems, Man, and Cybernetics*, San Antonio, Texas, pages 1023–1028.

Duan, Q., Gupta, V., and Sorooshian, S. (1993). Shuffled complex evolution approach for effective and efficient global minimization. *Journal of Optimization Theory and Applications*, 76(3):501–521.

Duan, Q., Sorooshian, S., and Gupta, V. (1992). Effective and efficient global optimization for conceptual rainfall-runoff models. *Water Resources Research*, 28(4):1015–1031.

Ducheyne, E., De Baets, B., and De Wulf, R. (2003). Is fitness inheritance useful for real-world applications? *Second International Conference on Evolutionary Multi-Criterion Optimization*, Faro, Portugal, pages 31–42.

Dueck, G. (1993). New optimisation heuristics: The great deluge algorithm and the record-to-record travel. *Journal of Computational Physics*, 104(86):86–92.

Dueck, G. and Scheuer, T. (1990). Threshold accepting: A general purpose optimization algorithm appearing superior to simulated annealing. *Journal of Computational Physics*, 90(1):161–175.

Dunham, B., Fridshal, D., Fridshal, R., and North, J. (1963). Design by natural selection. *Synthese*, 15(2):254–259.

Durham, W. (1992). *Coevolution: Genes, Culture, and Human Diversity*. Stanford University Press.

Dyson, G. (1998). *Darwin Among the Machines*. Basic Books.

Eberhart, R. and Kennedy, J. (1995). A new optimizer using particle swarm theory. *International Symposium on Micro Machine and Human Science*, Nagoya, Japan, pages 39–43.

Eberhart, R. and Shi, Y. (2000). Comparing inertia weights and constriction factors in particle swarm optimization. *IEEE Congress on Evolutionary Computation*, San Diego, California, pages 84–88.

Eberhart, R. and Shi, Y. (2001). Particle swarm optimization: Developments, applications and resources. *IEEE Congress on Evolutionary Computation*, Seoul, Korea, pages 81–86.

Edgeworth, F. (1881). *Mathematical Physics.* Kegan Paul.

Ehrgott, M. (2005). *Multicriteria Optimization.* Springer.

Ehrnborg, C. and Rosén, T. (2009). The psychology behind doping in sport. *Growth Hormone & IGF Research*, 19(4):285–287.

Eiben, A. (2000). Multiparent recombination. In Bäck, T., Fogel, D., and Michalewicz, Z., editors, *Evolutionary Computation 1: Basic Algorithms and Operators*, pages 289–307. Institute of Physics Publishing.

Eiben, A. (2001). Evolutionary algorithms and cnstraint satisfaction: Definitions, survey, methodology, and research directions. In Kallel, L., Naudts, B., and Rogers, A., editors, *Theoretical Aspects of Evolutionary Computating*, pages 13–30. Springer.

Eiben, A. (2003). Multiparent recombination in evolutionary computing. In Ghosh, A. and Tsutsui, S., editors, *Advances in Evolutionary Computing*, pages 175–192. Springer-Verlag.

Eiben, A. and Bäck, T. (1998). Empirical investigation of multiparent recombination operators in evolution strategies. *Evolutionary Computation*, 5(3):347–365.

Eiben, A. and Schippers, C. (1996). Multi-parent's niche: n-ary crossovers on nk-landscapes. In Ebeling, W., Rechenberg, I., Schwefel, H.-P., and Voigt, H.-M., editors, *Parallel Problem Solving from Nature – PPSN IV*, pages 319–328. Springer.

Eiben, A. and Smit, S. (2011). Parameter tuning for configuring and analyzing evolutionary algorithms. *Swarm and Evolutionary Computation*, 1(1):19–31.

Eiben, A. and Smith, J. (2010). *Introduction to Evolutionary Computing.* Springer.

Elbeltagi, E., Hegazy, T., and Grierson, D. (2005). Comparison among five evolutionary-based optimization algorithms. *Advanced Engineering Informatics*, 19(1):43–53.

Ellis, T. and Yao, X. (2007). Evolving cooperation in the non-iterated prisoner's dilemma: A social network inspired approach. *IEEE Congress on Evolutionary Computation*, Singapore, pages 736–743.

Elton, C. (1958). *Ecology of Invasions by Animals and Plants.* Chapman & Hall.

Emre, E. and Knowles, G. (1987). A Newton-like approximation algorithm for the steady-state solution of the Riccati equation for time-varying systems. *Optimal Control Applications and Methods*, 8(2):191–197.

Engelbrecht, A. (2003). *Computational Intelligence.* John Wiley & Sons.

English, T. (1999). Some information theoretic results on evolutionary optimization. *IEEE Congress on Evolutionary Computation*, Washington, District of Columbia, pages 788–795.

Ergezer, M. (2011). Oppositional biogeography-based optimization. Technical report, Cleveland State University. Doctoral dissertation proposal, unpublished.

Ergezer, M. and Simon, D. (2011). Oppositional biogeography-based optimization for combinatorial problems. *IEEE Congress on Evolutionary Computation*, New Orleans, Louisiana, pages 1496–1503.

Ergezer, M., Simon, D., and Du, D. (2009). Oppositional biogeography-based optimization. *IEEE Conference on Systems, Man, and Cybernetics*, San Antonio, Texas, pages 1035–1040.

Erol, O. and Eksin, I. (2006). New optimization method: Big bang-big crunch. *Advances in Engineering Software*, 37(2):106–111.

Eshelman, L., Caruana, R., and Schaffer, J. (1989). Biases in the crossover landscape. *International Conference on Genetic Algorithms*, Fairfax, Virginia, pages 10–19.

Eshelman, L. and Schaffer, J. (1993). Real-coded genetic algorithms and interval schemata. In Whitley, D., editor, *Foundations of Genetic Algorithms 2*, pages 187–202. Morgan Kaufmann.

Eskandari, H. and Geiger, C. (2008). A fast Pareto genetic algorithm approach for solving expensive multiobjective optimization problems. *Journal of Heuristics*, 14(3):203–241.

Eusuff, M. and Lansey, K. (2003). Optimization of water distribution network design using the shuffled frog leaping algorithm (SFLA). *Journal of Water Resources Planning and Management*, 129(3):210–225.

Eusuff, M., Lansey, K., and Pasha, F. (2006). Shuffled frog-leaping algorithm: A memetic meta-heuristic for discrete optimization. *Engineering Optimization*, 38(2):129–154.

Evans, M., Hastings, N., and Peacock, B. (2000). *Statistical Distributions.* Wiley-Interscience.

Farmani, R. and Wright, J. (2003). Self-adaptive fitness formulation for constrained optimization. *IEEE Transactions on Evolutionary Computation*, 7(5):445–455.

Fausett, L. (1994). *Fundamentals of Neural Networks.* Prentice Hall.

Fealy, M. (2006). *The Great Pawn Hunter Chess Tutorial.* AuthorHouse.

Feoktistov, V. (2006). *Differential Evolution: In Search of Solutions.* Springer.

Fernandes, M., Martins, T., and Rocha, A. (2009). Fish swarm intelligent algorithm for bound constrained global optimization. *International Conference on Computational and Mathematical Methods in Science and Engineering*, Gijón, Spain.

Fish, F. (1995). Kinematics of ducklings swimming in formation: Consequences of position. *Journal of Experimental Zoology*, 273(1):1–11.

Fleming, P., Purshouse, R., and Lygoe, R. (2005). Many-objective optimization: An engineering design perspective. In Coello Coello, C., Hernández Aguirre, A., and Zitzler, E., editors, *Evolutionary Multi-Criterion Optimization*, pages 14–32. Springer.

Fletcher, R. and Powell, M. (1963). A rapidly convergent descent method for minimization. *The Computer Journal*, 6(2):163–168.

Floudas, C. and Pardalos, P. (1990). *A Collection of Test Problems for Constrained Global Optimization Algorithms.* Springer.

Floudas, C., Pardalos, P., Adjiman, C., Esposito, W., Gümüs, Z., Harding, S., Klepeis, J., Meyer, C., and Schweiger, C. (2010). *Handbook of Test Problems in Local and Global Optimization.* Springer.

Fogel, D. (1988). An evolutionary aproach to the traveling salesman problem. *Biological Cybernetics*, 60(2):139–144.

Fogel, D. (1990). A parallel pocessing approach to a multiple traveling salesman problem using evolutionary programming. *Fourth Annual Parallel Processing Symposium*, Fullerton, California, pages 318–326.

Fogel, D., editor (1998). *Evolutionary Computation: The Fossil Record.* Wiley-IEEE Press.

Fogel, D. (2000). What is evolutionary computation? *IEEE Spectrum*, 37(2):26–32.

Fogel, D. (2006). George Friedman – Evolving circuits for robots. *IEEE Computational Intelligence Magazine*, 1(4):52–54.

Fogel, D. and Anderson, R. (2000). Revisiting Bremermann's genetic algorithm: I. Simultaneous mutation of all parameters. *IEEE Congress on Evolutionary Computation*, San Diego, California, pages 1204–1209.

Fogel, L. (1999). *Intelligence through Simulated Evolution: Forty Years of Evolutionary Programming.* John Wiley & Sons.

Fogel, L., Owens, A., and Walsh, M. (1966). *Artificial Intelligence through Simulated Evolution*. John Wiley & Sons.

Fonseca, C. and Fleming, P. (1993). Genetic algorithms for multiobjective optimization: Formulation, discussion and generalization. *International Conference on Genetic Algorithms*, Urbana-Champaign, Illinois, pages 416–423.

Fonseca, C. and Fleming, P. (1995). An overview of evolutionary algorithms in multiobjective optimization. *Evolutionary Computation*, 3(1):1–16.

Formato, R. (2007). Central force optimization: A new metaheuristic with applications in applied electromagnetics. *Progress in Electromagnetics Research*, 77:425–491.

Formato, R. (2008). Central force optimization: A new nature inspired computational framework for multidimensional search and optimization. In Krasnogor, N., Nicosia, G., Pavone, M., and Pelta, D., editors, *Nature Inspired Cooperative Strategies for Optimization (NICSO 2007)*, pages 221–238. Springer.

Forsyth, R. (1981). BEAGLE – A Darwinian approach to pattern recognition. *Kybernetes*, 10(3):159–166.

Fourman, M. (1985). Compaction of symbolic layout using genetic algorithms. *International Conference on Genetic Algorithms*, Pittsburgh, Pennsylvania, pages 141–153.

Fox, B. and McMahon, M. (1991). Genetic operators for sequencing problems. In Rawlins, G., editor, *Foundations of Genetic Algorithms*, pages 284–300. Morgan Kaufmann Publishers.

Francqis, O. (1998). An evolutionary strategy for global minimization and its Markov chain analysis. *IEEE Transactions on Evolutionary Computation*, 2(3):77–90.

Fraser, A. (1957). Simulation of genetic systems by automatic digital computers: I. Introduction. *Australian Journal of Biological Sciences*, 10(3):484–491.

Friedberg, R. (1958). A learning machine: Part I. *IBM Journal of Research and Development*, 2(1):2–13.

Friedberg, R., Dunham, B., and North, J. (1958). A learning machine: Part II. *IBM Journal of Research and Development*, 3(3):282–287.

Friedman, G. (1998). Selective feedback computers for engineering synthesis and nervous system analogy. In Fogel, D., editor, *Evolutionary Computation: The Fossil Record*, pages 30–84. Wiley-IEEE Press.

Furuta, H., Maeda, K., and Watanabe, E. (1995). Application of genetic algorithm to aesthetic eesign of bridge structures. *Computer-Aided Civil and Infrastructure Engineering*, 10(6):415–421.

Galinier, P., Hamiez, J.-P., Hao, J.-K., and Porumbel, D. (2013). Recent advances in graph vertex coloring. In Zelinka, I., Snášel, V., and Abraham, A., editors, *Handbook of Optimization*. `ebooks.com`.

Gallagher, M., Wood, I., Keith, J., and Sofronov, G. (2007). Bayesian inference in estimation of distribution algorithms. *IEEE Congress on Evolutionary Computation*, Singapore, pages 127–133.

Gambardella, L. and Dorigo, M. (1995). Ant-Q: A reinforcement learning approach to the traveling salesman problem. *Twelfth International Conference on Machine Learning*, Tahoe City, California, pages 252–260.

Gandomi, A. and Alavi, A. (2012). Krill herd: A new bio-inspired optimization algorithm. *Communications in Nonlinear Science and Numerical Simulation*, 17(12):4831–4845.

Gathercole, C. and Ross, P. (1994). Dynamic training subset selection for supervised learning in genetic programming. In Davidor, Y., Schwefel, H.-P., and Männer, R., editors, *Parallel Problem Solving from Nature – PPSN III*, pages 312–321. Springer.

Gathercole, C. and Ross, P. (1997). Small populations over many generations can beat large populations over few generations in genetic programming. *Second Annual Conference on Genetic Programming*, Palo Alto, California, pages 111–118.

Geem, Z., editor (2010a). *Harmony Search Algorithms for Structural Design Optimization*. Springer.

Geem, Z., editor (2010b). *Music-Inspired Harmony Search Algorithm*. Springer.

Geem, Z. (2010c). *Recent Advances in Harmony Search Algorithm*. Springer.

Geem, Z., Kim, J.-H., and Loganathan, G. (2001). A new heuristic optimization algorithm: Harmony search. *Simulation*, 76(2):60–68.

Geisser, S. (1993). *Predictive Inference*. Chapman & Hall.

Geman, S. and Geman, D. (1984). Stochastic relaxation, gibbs distributions, and the bayesian restoration of images. *IEEE Transactions on Pattern Analysis and Machine Intelligence*, 6(6):721–741.

Gendreau, M. (2003). An introduction to tabu search. In Glover, F. and Kochenberger, G., editors, *Handbook of Metaheuristics*, pages 37–54. Springer.

Gendreau, M. and Potvin, J.-Y. (2010). Tabu search. In Gendreau, M. and Potvin, J.-Y., editors, *Handbook of Metaheuristics*, pages 41–59. Springer.

Giraldeau, L.-A. and Caraco, T. (2000). *Social Foraging Theory*. Princeton University Press.

Glover, F. and Laguna, M. (1998). *Tabu Search*. Springer.

Glover, F. and McMillan, C. (1986). The general employee scheduling problem: An integration of MS and AI. *Computers and Operations Research*, 13(5):563–573.

Goh, C. and Tan, K. (2007). An investigation on noisy environments in evolutionary multiobjective optimization. *IEEE Transactions on Evolutionary Computation*, 11(3):354–381.

Golan, J. (2007). *The Linear Algebra a Beginning Graduate Student Ought to Know*. Springer.

Goldberg, D. (1989a). *Genetic Algorithms in Search, Optimization, and Machine Learning*. Addison Wesley.

Goldberg, D. (1989b). Messy genetic algorithms: Motivation, analysis, and first results. *Complex Systems*, 3(5):493–530.

Goldberg, D. (1991). Real-coded genetic algorithms, virtual alphabets, and blocking. *Complex Systems*, 5(2):139–167.

Goldberg, D. and Lingle, R. (1985). Alleles, loci, and the traveling salesman problem. *International Conference on Genetic Algorithms and Their Application*, Pittsburgh, Pennsylvania, pages 154–159.

Gómez, J., Barrera, J., Rojas, J., Macias-Samano, J., Liedo, J., Cruz-Lopez, L., and Badii, M. (2005). Volatile compounds released by disturbed females of Cephalonomia stephanoderis (Hymenoptera: Bethylidae): A parasitoid of the coffee berry borer Hypothenemus hampei (Coleoptera: Scolytidae). *Florida Entomologist*, 88(2):180–187.

González, C., Lozano, J., and Larrañaga, P. (2000). Analyzing the PBIL algorithm by means of discrete dynamical systems. *Complex Systems*, 12(4):465–479.

González, C., Lozano, J., and Larrañaga, P. (2001). The convergence behavior of the PBIL algorithm: A preliminary approach. In Kůrková, V., Steele, N., Neruda, R., and Kárný, M., editors, *Artificial Neural Nets and Genetic Algorithms*, pages 228–231. Springer-Verlag.

González, C., Lozano, J., and Larrañaga, P. (2002). Mathematical modeling of discrete estimation of distribution algorithms. In Larrañaga, P. and Lozano, J., editors, *Estimation of Distribution Algorithms*, pages 147–163. Kluwer Academic Publishers.

Good, P. and Hardin, J. (2009). *Common Errors in Statistics.* John Wiley & Sons, 3rd edition.

Goss, S., Aron, S., Deneubourg, J., and Pasteels, J. (1989). Self-organized shortcuts in the Argentine ant. *Naturwissenschaften*, 76(12):579–581.

Gotelli, N. (2008). *A Primer of Ecology.* Sinauer Associates.

Gray, R. (2011). *Entropy and Information Theory.* Springer.

Greene, M. and Gordon, D. (2007). Structural complexity of chemical recognition cues affects the perception of group membership in the ants Linephithema humile and Aphaenogaster cockerelli. *Journal of Experimental Biology*, 210(5):897–905.

Grefenstette, J., Gopal, R., Rosmaita, B., and Van Gucht, D. (1985). Genetic algorithms for the TSP. *International Conference on Genetic Algorithms and Their Application*, Cambridge, Massachusetts, pages 160–165.

Gregory, R. and Karney, D. (1969). *A Collection of Matrices for Testing Computational Algorithms.* John Wiley & Sons.

Grieco, J. (1988). Realist theory and the problem of international cooperation: Analysis with an amended prisoner's dilemma model. *The Journal of Politics*, 50(3):600–624.

Grinstead, C. and Snell, J. (1997). *Introduction to Probability.* American Mathematical Society.

Grötschel, M. and Padberg, M. (2001). The optimized odyssey. *AIROnews*, 6(2):1–7.

Guntsch, M. and Middendorf, M. (2002). Applying population based ACO to dynamic optimization problems. *Third International Workshop on Ant Algorithms*, Brussels, Belgium, pages 111–122.

Gustafson, S. and Burke, E. (2006). Speciating island model: An alternative parallel evolutionary algorithm. *Parallel and Distributed Computing*, 66(8):1025–1036.

Gutierrez, A., Lanza, M., Barriuso, I., Valle, L., Domingo, M., Perez, J., and Basterrechea, J. (2002). Comparison of different PSO initialization techniques for high dimensional search space problems: A test with FSS and antenna arrays. *5th European Conference on Antennas and Propagation*, Rome, Italy, pages 965–969.

Gutin, G. and Punnen, A., editors (2007). *The Traveling Salesman Problem and Its Variations.* Springer.

Gutjahr, W. (2000). A graph-based ant system and its convergence. *Future Generation Computer Systems*, 16(9):873–888.

Gutjahr, W. (2008). First steps to the runtime complexity analysis of ant colony optimization. *Computers & Operations Research*, 35(9):2711–2727.

Hadj-Alouane, A. and Bean, J. (1993). A genetic algorithm for the multiple choice integer program. Technical report, Department of Industrial & Operations Engineering, University of Michigan. `http://ioe.engin.umich.edu/techrprt/pdf/TR92-50.pdf`.

Hadj-Alouane, A. and Bean, J. (1997). A genetic algorithm for the multiple choice integer program. *Operations Research*, 45(1):92–101.

Hajela, P. and Lin, C.-Y. (1997). Genetic search strategies in multicriterion optimal design. *Structural and Multidisciplinary Optimization*, 4(2):99–107.

Hamida, S. and Schoenauer, M. (2000). An adaptive algorithm for constrained optimization problems. In Schoenauer, M., Deb, K., Rudolph, G., Yao, X., Lutton, E., Merelo,

J., and Schwefel, H.-P., editors, *Parallel Problem Solving from Nature – PPSN VI*, pages 529–538. Springer.

Hamida, S. and Schoenauer, M. (2002). ASCHEA: New results using adaptive segregational constraint handling. *IEEE Congress on Evolutionary Computation*, Honolulu, Hawaii, pages 884–889.

Hamilton, W. (1971). Geometry for the selfish herd. *Journal of Theoretical Biology*, 31(2):295–311.

Hansen, N. (2010). The CMA evolution strategy: A comparing review. In Lozano, J., Larrañga, P., Inza, I., and Bengoetxea, E., editors, *Towards a New Evolutionary Computation: Advances on Estimation of Distribution Algorithms*, pages 75–102. Springer.

Hansen, N., Müller, S., and Koumoutsakos, P. (2003). Reducing the time complexity of the derandomized evolution strategy with covariance matrix adaptation (CMA-ES). *Evolutionary Computation*, 11(1):1–18.

Hansen, N. and Ostermeier, A. (2001). Completely derandomized self-adaptation in evolution strategies. *Evolutionary Computation*, 9(2):159–195.

Hanski, I. (1999). Habitat connectivity, habitat continuity, and metapopulations in dynamic landscapes. *Oikos*, 87(2):209–219.

Hanski, I. and Gilpin, M. (1997). *Metapopulation Biology*. Academic Press.

Hao, J.-K. and Middendorf, M., editors (2012). *Evolutionary Computation in Combinatorial Optimization*. Springer.

Harding, S. (2006). *Animate Earth*. Chelsea Green Publishing Company.

Harik, G. (1995). Finding multimodal solutions using restricted tournament selection. *International Conference on Genetic Algorithms*, Pittsburgh, Pennsylvania, pages 24–31.

Harik, G. (1999). Linkage learning via probabilistic modeling in the ECGA. Technical report, Illinois Genetic Algorithms Laboratory, University of Illinois. IlliGAL Report No. 99010.

Harik, G., Lobo, F., and Goldberg, D. (1999). The compact genetic algorithm. *IEEE Transactions on Evolutionary Computation*, 3(4):287–297.

Harik, G., Lobo, F., and Sastry, K. (2010). Linkage learning via probabilistic modeling in the extended compact genetic algorithm (ecga). In Pelikan, M., Sastry, K., and Cantú-Paz, E., editors, *Scalable Optimization via Probabilistic Modeling*, pages 39–62. Springer.

Harrald, P. and Fogel, D. (1996). Evolving continuous behaviors in the iterated prisoner's dilemma. *Biosystems*, 37(1–2):135–145.

Hastie, T., Tibshirani, R., and Friedman, J. (2009). *The Elements of Statistical Learning*. Springer, 2nd edition.

Hastings, A. and Higgins, K. (1994). Persistence of transients in spatially structured models. *Science*, 263(5150):1133–1136.

Hastings, W. (1970). Monte Carlo sampling methods using Markov chains and their applications. *Biometrika*, 57(1):97–109.

Hatzakis, I. and Wallace, D. (2006). Dynamic multi-objective optimization with evolutionary algorithms: A forward-looking approach. *Genetic and Evolutionary Computation Conference*, Seattle, Washington, pages 1201–1208.

Haupt, R. and Haupt, S. (2004). *Practical Genetic Algorithms*. John Wiley & Sons, 2nd edition.

Hauptman, A., Elyasaf, A., Sipper, M., and Karmon, A. (2009). GP-Rush: Using genetic programming to evolve solvers for the rush hour puzzle. *Genetic and Evolutionary Computation Conference*, Montreal, Canada, pages 955–962.

Hauptman, A. and Sipper, M. (2007). Evolution of an efficient search algorithm for the mate-in-n problem in chess. *European Conference on Genetic Programming*, Valencia, Spain, pages 78–89.

He, S., Wu, Q., and Saunders, J. (2009). Group search optimizer: An optimization algorithm inspired by animal searching behavior. *IEEE Transactions on Evolutionary Computation*, 13(5):973–990.

Heinrich, B. (2002). *Why We Run*. Harper Perennial.

Helwig, S. and Wanka, R. (2008). Theoretical analysis of initial particle swarm behavior. In Rudolph, G., Jansen, T., Lucas, S., Poloni, C., and Beume, N., editors, *Parallel Problem Solving from Nature - PPSN X*, pages 889–898. Springer.

Henderson, D., Jacobson, S., and Johnson, A. (2003). The theory and practice of simulated annealing. In Glover, F. and Kochenberger, G., editors, *Handbook of Metaheuristics*, pages 287–320. Springer.

Herrera, F., Lozano, M., and Verdegay, J. (1998). Tackling real-coded genetic algorithms: Operators and tools for behavioural analysis. *Artificial Intelligence Review*, 12(4):265–319.

Hofmeyr, S. and Forrest, S. (2000). Architecture for an artificial immune system. *Evolutionary Computation*, 8(4):443–473.

Holland, J. (1975). *Adaptation in Natural and Artificial Systems*. The University of Michigan Press.

Hölldobler, B. and Wilson, E. (1990). *The Ants*. The Belknap Press of Harvard University Press.

Hölldobler, B. and Wilson, E. (1994). *Journey to the Ants*. The Belknap Press of Harvard University Press.

Hölldobler, B. and Wilson, E. (2008). *The Superorganism: The Beauty, Elegance, and Strangeness of Insect Societies*. W. W. Norton & Company.

Homaifar, A., Qi, C., and Lai, S. (1994). Constrained optimization via genetic algorithms. *Simulation*, 62(4):242–253.

Hooker, J. (1995). Testing heuristics: We have it all wrong. *Journal of Heuristics*, 1(1):33–42.

Horn, J., Nafpliotis, N., and Goldberg, D. (1994). A niched Pareto genetic algorithm for multiobjective optimization. *IEEE Conference on Evolutionary Computation*, Orlando, Florida, pages 82–87.

Horne, E. and Jaeger, R. (1988). Territorial pheromones of female red-backed salamanders. *Ethology*, 78(2):143–152.

Horoba, C. and Neumann, F. (2010). Approximating Pareto-optimal sets using diversity strategies in evolutionary multi-objective optimization. In Coello Coello, C., Dhaenens, C., and Jourdan, L., editors, *Advances in Multi-Objective Nature Inspired Computing*, pages 23–44. Springer.

Houck, C., Joines, J., and Kay, M. (1995). A genetic algorithm for function optimization: A Matlab implementation. Technical report, North Carolina State University.

Hsiao, Y.-T., Chuang, C.-L., Jiang, J.-A., and Chien, C.-C. (2005). A novel optimization algorithm: Space gravitational optimization. *IEEE International Conference on Systems, Man and Cybernetics*, Waikoloa, Hawaii, pages 2323–2328.

Hu, T., Harding, S., and Banzhaf, W. (2010). Variable population size and evolution acceleration: A case study with a parallel evolutionary algorithm. *Genetic Programming and Evolvable Machines*, 11(2):205–225.

Huang, H. and Wang, F. (2002). Fuzzy decision-making design of chemical plant using mixed-integer hybrid differential evolution. *Computers and Chemical Engineering*, 26(12):1649–1660.

Huang, V., Qin, A., Deb, K., Zitzler, E., Suganthan, P., Liang, J., Preuss, M., and Huband, S. (2007). Problem definitions for performance assessment on multi-objective optimization algorithms. Technical report. `www.ntu.edu.sg/home/EPNSugan/index_files/cec-benchmarking.htm`.

Huff, D. and Geis, I. (1993). *How to Lie with Statistics*. W. W. Norton & Company.

Iba, H. and de Garis, H. (1996). Extending genetic programming with recombinative guidance. In Angeline, P. and Kinnear, K., editors, *Advances in Genetic Programming: Volume 2*, pages 69–88. The MIT Press.

Igelnik, B. and Simon, D. (2011). The eigenvalues of a tridiagonal matrix in biogeography. *Applied Mathematics and Computation*, 218(1):195–201.

Ingber, L. (1996). Adaptive simulated annealing: Lessons learned. *Control and Cybernetics*, 25(1):33–54.

Ito, K., Akagi, S., and Nishikawa, M. (1983). A multiobjective optimization approach to a design problem of heat insulation for thermal distribution piping network systems. *Journal of Mechanisms, Transmissions, and Automation in Design*, 105(2):206–213.

Jaszkiewicz, A. and Zielniewicz, P. (2006). Pareto memetic algorithm with path relinking for bi-objective traveling salesperson problem. *European Journal of Operational Research*, 193(3):885–890.

Jayalakshmi, G., Sathiamoorthy, S., and Rajaram, R. (2001). A hybrid genetic algorithm – A new approach to solve traveling salesman problem. *International Journal of Computational Engineering Science*, 2(2):339–355.

Jefferson, D., Collins, R., Cooper, C., Dyer, M., Flowers, M., Korf, R., Taylor, C., and Wang, A. (2003). Evolution as a theme in artificial life: The genesys/tracker system. In Langton, C., Taylor, C., Farmer, J., and Rasmussen, S., editors, *Artificial Life II*, pages 549–578. Westview Press.

Jensen, T. and Toft, B. (1994). *Graph Coloring Problems*. John Wiley & Sons.

Jin, Y. (2005). A comprehensive survey of fitness approximation in evolutionary computation. *Soft Computing*, 9(1):3–12.

Jin, Y. and Branke, J. (2005). Evolutionary optmization in uncertain environments – A survey. *IEEE Transactions on Evolutionary Computation*, 9(3):303–317.

Jin, Y., Hüskin, M., and Sendhoff, B. (2003). Quality measures for approximate models in evolutionary computation. *Genetic and Evolutionary Computation Conference*, Chicago, Illinois, pages 170–173.

Jofré, P., Reisenegger, A., and Fernández, R. (2006). Constraining a possible time variation of the gravitational constant through "gravitochemical heating" of neutron stars. *Physical Review Letters*, 97(13):131102.

Johnson, D. (1999). The insignificance of statistical significance testing. *Journal of Wildlife Management*, 63(3):763–772.

Joines, J. and Houck, C. (1994). On the use of non-stationary penalty functions to solve nonlinear constrained optimization problems with GA's. *IEEE World Congress on Computational Intelligence*, Orlando, Florida, pages 579–584.

Jones, D., Schonlau, M., and Welch, W. (1998). Efficient global optimization of expensive black-box functions. *Journal of Global Optimization*, 13(4):455–492.

Joslin, D. and Clements, D. (1999). Squeaky wheel optimization. *Journal of Artificial Intelligence Research*, 10:353–373.

Kanji, G. (2006). *100 Statistical Tests.* Sage Publications.

Karaboga, D. and Akay, B. (2009). A comparative study of artificial bee colony algorithm. *Applied Mathematics and Computation*, 214(1):108–132.

Karaboga, D. and Basturk, B. (2007). A powerful and efficient agorithm for numerical function optimization: Artificial bee colony (ABC) algorithm. *Journal of Global Optimization*, 39(3):459–471.

Karaboga, D. and Basturk, B. (2008). On the performance of artificial bee colony (ABC) algorithm. *Applied Soft Computing*, 8(1):687–697.

Karaboga, D., Gorkemli, B., Ozturk, C., and Karaboga, N. (2013). A comprehensive survey: Artificial bee colony (ABC) algorithm and applications. *Artificial Intelligence Review*, in print.

Kaveh, A. and Talatahari, S. (2010). A novel heuristic optimization method: Charged system search. *Acta Mechanica*, 213(3–4):267–289.

Kazarlis, S. and Petridis, V. (1998). Varying fitness functions in genetic algorithms: Studying the rate of increase of the dynamic penalty terms. In Eiben, A., Bäck, T., Schoenauer, M., and Schwefel, H.-P., editors, *Parallel Problem Solving from Nature – PPSN V*, pages 211–220. Springer.

Keel, L. and Bhattacharyya, S. (1997). Robust, fragile, or optimal? *IEEE Transactions on Automatic Control*, 42(8):1098–1105.

Kemeny, J., Snell, J., and Thompson, G. (1974). *Introduction to Finite Mathematics.* Prentice-Hall.

Kennedy, J. (1998). Thinking is social: Experiments with the adaptive culture model. *Journal of Conflict Resolution*, 42(1):56–76.

Kennedy, J. and Eberhart, R. (1997). A discrete binary version of the particle swarm algorithm. *IEEE Conference on Systems, Man, and Cybernetics*, Orlando, Florida, pages 4104–4109.

Kennedy, J. and Eberhart, R., editors (2001). *Swarm Intelligence.* Morgan Kaufmann.

Kern, S., Müller, S., Hansen, N., Büche, D., Ocenasek, J., and Koumoutsakos, P. (2004). Learning probability distributions in continuous evolutionary algorithms – A comparative review. *Natural Computing*, 3(1):77–112.

Keynes, R., editor (2001). *Charles Darwin's Beagle diary.* Cambridge University Press.

Khare, V., Yao, X., and Deb, K. (2003). Performance scaling of multi-objective evolutionary algorithms. In Fonseca, C., Fleming, P., Zitzler, E., Thiele, L., and Deb, K., editors, *Evolutionary Multi-Criterion Optimization: Second International Conference, EMO 2003*, pages 376–390. Springer.

Khatib, W. and Fleming, P. (1998). The stud GA: A mini revolution? In Eiben, A., Bäck, T., Schoenauer, M., and Schwefel, H.-P., editors, *Parallel Problem Solving from Nature – PPSN V*, pages 683–691. Springer.

Kim, D. (2006). Memory analysis and significance test for agent behaviours. *Genetic and Evolutionary Computation Conference*, Seattle, Washington, pages 151–158.

Kim, H.-S. and Cho, S.-B. (2000). Application of interactive genetic algorithm to fashion design. *Engineering Applications of Artificial Intelligence*, 13(6):635–644.

Kinnear, K. (1993). Evolving a sort: Lessons in genetic programming. *International Conference on Neural Networks*, San Francisco, California, pages 881–888.

Kinnear, K. (1994). Alternatives in automatic function definition: A comparison of performance. In Kinnear, K., editor, *Advances in Genetic Programming*, pages 119–141. MIT Press.

Kirk, D., editor (2004). *Optimal Control Theory*. Dover.

Kirkpatrick, S., Gelatt, C., and Vecchi, M. (1983). Optimization by simmulated annealing. *Science*, 220(4598):671–680.

Kjellström, G. (1969). Network optimization by random variation of component values. *Ericsson Technics*, 25(3):133–151.

Kleidon, A. (2004). Amazonian biogeography as a test for Gaia. In Schneider, S., Miller, J., Crist, E., and Boston, P., editors, *Scientists Debate Gaia*, pages 291–296. MIT Press.

Knowles, J. (2005). ParEGO: A hybrid algorithm with on-line landscape approximation for expensive multiobjective optimization problems. *IEEE Transactions on Evolutionary Computation*, 10(1):50–66.

Knowles, J. and Corne, D. (2001). Approximating the nondominated front using the Pareto archived evolution strategy. *Evolutionary Computation*, 8(2):149–172.

Knowles, J. and Nakayama, H. (2008). Meta-modeling in multiobjective optimization. In Branke, J., Deb, K., Miettinen, K., and Slowinski, R., editors, *Multiobjective Optimization*, pages 245–284. Springer.

Konak, A., Coit, D., and Smith, A. (2006). Multi-objective optimization using genetic algorithms: A tutorial. *Reliability Engineering and System Safety*, 91(9):992–1007.

Kondoh, M. (2006). Does foraging adaptation create the positive complexity-stability relationship in realistic food-web structure? *Journal of Theoretical Biology*, 238(3):646–651.

Koza, J., editor (1992). *Genetic Programming: On the Programming of Computers by Means of Natural Selection*. The MIT Press.

Koza, J., editor (1994). *Genetic Programming II: Automatic Discovery of Reusable Programs*. The MIT Press.

Koza, J. (1997). Classifying protein segments as transmembrane domains using genetic programming and architecture-altering operations. In Bäck, T., Fogel, D., and Michalewicz, Z., editors, *Handbook of Evolutionary Computation*, pages G6.1:1–5. Oxford University Press.

Koza, J. (2010). Human-competitive results produced by genetic programming. *Genetic Programming and Evolvable Machines*, 11(3–4):251–284.

Koza, J., Al-Sakran, L., and Jones, L. (2008). Automated ab initio synthesis of complete designs of four patented optical lens systems by means of genetic programming. *Artificial Intelligence for Engineering Design, Analysis and Manufacturing*, 22(3):249–273.

Koza, J., Bennett, F., Andre, D., and Keane, M., editors (1999). *Genetic Programming III: Darwinian Invention and Problem Solving*. Morgan Kaufmann.

Koza, J., Keane, M., Streeter, M., Mydlowec, W., Yu, J., and Lanza, G., editors (2005). *Genetic Programming IV: Routine Human-Competitive Machine Intelligence*. The MIT Press.

Koziel, S. and Michalewicz, Z. (1998). A decoder-based evolutionary algorithm for constrained parameter optimization problems. In Eiben, A., Bäck, T., Schoenauer, M., and Schwefel, H.-P., editors, *Parallel Problem Solving from Nature – PPSN V*, pages 231–240. Springer.

Koziel, S. and Michalewicz, Z. (1999). Evolutionary algorithms, homomorphous mappings, and constrained parameter optimization. *Evolutionary Computation*, 7(1):19–44.

Krause, J. and Ruxton, G., editors (2002). *Living in Groups*. Oxford University Press.

Krige, D. (1951). A statistical approach to some basic mine valuation problems on the Witwatersrand. *Journal of the Chemical, Metallurgical and Mining Society of South Africa*, 52(6):119–139.

Krishnanand, K. and Ghose, D. (2009). Glowworm swarm optimization for simultaneous capture of multiple local optima of multimodal functions. *Swarm Intelligence*, 3(2):87–124.

Krogh, A. (2008). What are artificial neural networks? *Nature Biotechnology*, 6(2):195–197.

Kursawe, F. (1991). A variant of evolution strategies for vector optimization. In Schwefel, H.-P. and Männer, R., editors, *Parallel Problem Solving from Nature - PPSN I*, pages 193–197. Springer.

Kvasnicka, V., Pelikan, M., and Pospichal, J. (1996). Hill climbing with learning (an abstraction of genetic algorithm). *Neural Network World*, 6(5):773–796.

Lam, A. and Li, V. (2010). Chemical-reaction-inspired metaheuristic for optimization. *IEEE Transactions on Evolutionary Computation*, 14(3):381–399.

Lampinen, J. (2002). A constraint handling approach for the differential evolution algorithm. *IEEE Congress on Evolutionary Computation*, Honolulu, Hawaii, pages 1468–1473.

Langdon, W. (2000). Size fair and homologous tree genetic programming crossovers. *Genetic Programming and Evolvable Machines*, 1(1–2):95–119.

Langdon, W. and Poli, R., editors (2002). *Foundations of Genetic Programming*. Springer.

Larrañaga, P. (2002). A review on estimation of distribution algorithms. In Larrañaga, P. and Lozano, J., editors, *Estimation of Distribution Algorithms: A New Tool for Evolutionary Computation*, pages 57–100. Kluwer Academic Publishers.

Larrañaga, P., Etxeberria, R., Lozano, J., and Peña, J. (1999a). Optimization by learning and simulation of Bayesian and Gaussian networks. Technical report, University of the Basque Country. `http://citeseerx.ist.psu.edu/viewdoc/summary?doi=10.1.1.41.1895`.

Larrañaga, P., Etxeberria, R., Lozano, J., and Peña, J. (2000). Combinatorial optimization by learning and simulation of Bayesian networks. *Sixteenth Conference on Uncertainty in Artificial Intelligence*, Stanford, California, pages 343–352.

Larrañaga, P., Karshenas, H., Bielza, C., and Santana, R. (2012). A review on probabilistic graphical models in evolutionary computation. *Journal of Heuristics*, 18(5):795–819.

Larrañaga, P., Kuijpers, C., Murga, R., Inza, I., and Dizdarevic, S. (1999b). Genetic algorithms for the travelling salesman problem: A review of representations and operators. *Artificial Intelligence Review*, 13(2):129–170.

Larrañaga, P. and Lozano, J., editors (2002). *Estimation of Distribution Algorithms: A New Tool for Evolutionary Computation*. Kluwer Academic Publishers.

Latané, B., Nowak, A., and Liu, J. (1994). Measuring emergent social phenomena: Dynamism, polarization, and clustering as order parameters of social systems. *Behavioral Science*, 39(1):1–24.

Lattimore, T. and Hutter, M. (2011). No free lunch versus Occam's razor in supervised learning. *Solomonoff 85th Memorial Conference*, Melbourne, Australia.

Laumanns, M., Thiele, L., and Zitzler, E. (2003). Running time analysis of evolutionary agorithms on vector-valued pseudo-Boolean functions. *IEEE Transactions on Evolutionary Computation*, 8(2):170–182.

Lawler, E., Lenstra, J., Rinnooy Kan, A., and Shmoys, D., editors (1985). *The Traveling Salesman Problem*. John Wiley & Sons.

Le Riche, R., Knopf-Lenoir, C., and Haftka, R. (1995). A segregated genetic algorithm for constrained structural optimization. *International Conference on Genetic Algorithms*, Pittsburgh, Pennsylvania, pages 558–565.

Lee, K. and Geem, Z. (2006). A new meta-heuristic algorithm for continuous engineering optimization: Harmony search theory and practice. *Computer Methods in Applied Mechanics and Engineering*, 194(36–38):3902–3933.

Leguizamón, G. and Coello Coello, C. (2009). Boundary search for constrained numerical optimization problems. In Mezura-Montes, E., editor, *Constraint-Handling in Evolutionary Optimization*, pages 25–49. Springer.

Lehman, J. and Stanley, K. (2011). Abandoning objectives: Evolution through the search for novelty alone. *Evolutionary Computation*, 19(2):189–223.

Lenton, T. (1998). Gaia and natural selection. *Nature*, 394(6692):439–447.

Li, C. and Yang, S. (2008). A generalized approach to construct benchmark problems for dynamic optimization. In Li, X., editor, *Simulated Evolution and Learning*, pages 391–400. Springer.

Li, C., Yang, S., Nguyen, T., Yu, E., Yao, X., Jin, Y., Beyer, H.-G., and Suganthan, P. (2008). Benchmark generator for CEC'2009 competition on dynamic optimization. Technical report. `www.ntu.edu.sg/home/EPNSugan/index_files/cec-benchmarking.htm`.

Li, X., Shao, Z., and Qian, J. (2003). An optimizing method based on autonomous animats: Fish-swarm algorithm. *Systems Engineering – Theory & Practice*, 22(11):32–38.

Li, Y., Zhang, S., and Zeng, X. (2009). Research of multi-population agent genetic algorithm for feature selection. *Expert Systems with Applications*, 36(9):11570–11581.

Liang, J., Runarsson, T., Mezura-Montes, E., Clerc, M., Suganthan, P., Coello Coello, C., and Deb, K. (2006). Problem definitions and evaluation criteria for the CEC 2006 special session on constrained real-parameter optimization. Technical report. `www.ntu.edu.sg/home/EPNSugan/index_files/cec-benchmarking.htm`.

Liang, J., Suganthan, P., and Deb, K. (2005). Novel composition test functions for numerical global optimization. *Swarm Intelligence Symposium*, Pasadena, California, pages 68–75.

Lim, D., Jin, Y., Ong, Y.-S., and Sendhoff, B. (2010). Generalizing surrogate-assisted evolutionary computation. *IEEE Transactions on Evolutionary Computation*, 14(3):329–355.

Lim, D., Ong, Y.-S., Jin, Y., Sendhoff, B., and Lee, B.-S. (2007). Efficient hierarchical parallel genetic algorithms using grid computing. *Future Generation Computer Systems*, 23(4):658–670.

Lima, C. and Lobo, F. (2004). Parameter-less optimization with the extended compact genetic algorithm and iterated local search. *Genetic and Evolutionary Computation Conference*, Seattle, Washington, pages 1328–1339.

Lima, S. (1995). Back to the basics of anti-predatory vigilance: The group-size effect. *Animal Behaviour*, 49(1):11–20.

Lis, J. and Eiben, A. (1997). A multi-sexual genetic algorithm for multiobjective optimization. *IEEE International Conference on Evolutionary Computation*, Indianapolis, Indiana, pages 59–64.

Löbbing, M. and Wegener, I. (1995). The number of knight's tours equals 33,439,123,484,294 – counting with binary decision diagrams. *Electronic Journal of Combinatorics*, 3(1):5.

Lohn, J., Hornby, G., and Linden, D. (2004). An evolved antenna for deployment on NASA's space technology 5 mission. In O'Reilly, U.-M., Riolo, R., Yu, G., and Worzel, W., editors, *Genetic Programming Theory and Practice II*, pages 301–315. Kluwer Academic Publishers.

Lomolino, M. (2000a). A call for a new paradigm of island biogeography. *Global Ecology and Biogeography*, 9(1):1–6.

Lomolino, M. (2000b). A species-based theory of insular zoogeography. *Global Ecology and Biogeography*, 9(1):39–58.

López Jaimes, A., Coello Coello, C., and Urías Barrientos, J. (2009). Online objective reduction to deal with many-objective problems. *5th International Conference on Evolutionary Multi-Criterion Optimization*, Nantes, France, pages 423–437.

Lovelock, J. (1990). Hands up for the Gaia hypothesis. *Nature*, 344(6262):100–102.

Lovelock, J., editor (1995). *Gaia.* Oxford University Press.

Lozano, J., Larrañaga, P., Inza, I., and Bengoetxea, E., editors (2006). *Towards a New Evolutionary Computation: Advances on Estimation of Distribution Algorithms.* Springer.

Łukasik, S. and Żak, S. (2009). Firefly algorithm for continuous constrained optimization tasks. *1st International Conference on Computational Collective Intelligence*, Wrocław, Poland, pages 97–106.

Lundy, M. and Mees, A. (1986). Convergence of an annealing algorithm. *Mathematical Programming*, 34(1):111–124.

Ma, H. (2010). An analysis of the equilibrium of migration models for biogeography-based optimization. *Information Sciences*, 180(18):3444–3464.

Ma, H., Ni, S., and Sun, M. (2009). Equilibrium species counts and migration model tradeoffs for biogeography-based optimization. *IEEE Conference on Decision and Control*, Shanghai, China, pages 3306–3310.

Ma, H. and Simon, D. (2010). Biogeography-based optimization with blended migration for constrained optimization problems. *Genetic and Evolutionary Computation Conference*, Portland, Oregon, pages 417–418.

Ma, H. and Simon, D. (2011a). Analysis of migration models of biogeography-based optimization using markov theory. *Engineering Applications of Artificial Intelligence*, 24(6):1052–1060.

Ma, H. and Simon, D. (2011b). Blended biogeography-based optimization for constrained optimization. *Engineering Applications of Artificial Intelligence*, 24(3):517–525.

Ma, H. and Simon, D. (2013). Variations of biogeography-based optimization and markov analysis. *Information Sciences*, 220:492–506.

Ma, H., Simon, D., and Fei, M. (2013). On the statistical mechanics approximation of biogeography-based optimization. *Submitted for publication.*

MacArthur, R. (1955). Fluctuations of animal populations and a measure of community stability. *Ecology*, 36(3):533–536.

MacArthur, R. and Wilson, E. (1963). An equilibrium theory of insular zoogeography. *Evolution*, 17(4):373–387.

MacArthur, R. and Wilson, E. (1967). *The Theory of Island Biogeography.* Princeton University Press.

Mahfoud, S. (1992). Crowding and preselection revisited. Technical report, Illinois Genetic Algorithms Laboratory, University of Illinois. IlliGAL Report No. 92004.

Mahfoud, S. (1995a). A comparison of parallel and sequential niching methods. *International Conference on Genetic Algorithms*, Pittsburgh, Pennsylvania, pages 136–143.

Mahfoud, S. (1995b). Niching methods for genetic algorithms. Technical report, Illinois Genetic Algorithms Laboratory, University of Illinois. IlliGAL Report No. 95001.

Mahnig, T. and Mühlenbein, H. (2000). Mathematical analysis of optimization methods using search distributions. *Genetic and Evolutionary Computation Conference*, Las Vegas, Nevada, pages 205–208.

Malisia, A. (2008). Improving the exploration ability of ant-based algorithms. In Tizhoosh, H. and Ventresca, M., editors, *Oppositional Concepts in Computational Intelligence*, pages 121–142. Springer.

Mallipeddi, R. and Suganthan, P. (2010). Problem definitions and evaluation criteria for the CEC 2010 competition on constrained real-parameter optimization. Technical report, Nanyang Technological University. `www.ntu.edu.sg/home/EPNSugan/index_files/cec-benchmarking.htm`.

Maniezzo, V., Gambardella, L., and de Luigi, F. (2004). Ant colony optimization. In Onwubolu, G. and Babu, R., editors, *New Optimization Techniques in Engineering*, pages 101–122. Springer.

Margulis, L. (1996). Gaia is a tough bitch. In Brockman, J., editor, *The Third Culture: Beyond the Scientific Revolution*, pages 129–151. Touchstone.

Marriott, K. and Stuckey, P. (1998). *Programming with Constraints: An Introduction*. The MIT Press.

Mavrovouniotis, M. and Yang, S. (2011). Ant colony optimization with immigrants schemes in dynamic environments. In Schaefer, R., Cotta, C., Kolodziej, J., and Rudolph, G., editors, *Parallel Problem Solving from Nature - PPSN XI*, pages 371–380. Springer.

May, R. (1973). *Stability and Complexity in Model Ecosystems*. Princeton University Press.

McCann, K. (2000). The diversity-stability debate. *Nature*, 405(6783):228–233.

McConaghy, T., Palmers, P., Gielen, G., and Steyaert, M. (2008). Genetic programming with reuse of known designs for industrially scalable, novel circuit design. In Riolo, R., Soule, T., and Worzel, B., editors, *Genetic Programming Theory and Practice V*, pages 159–184. Springer.

McGill, R., Tukey, J., and Larsen, W. (1978). Variations of box plots. *The American Statistician*, 32(1):12–16.

McNab, B. (2002). *The Physiological Ecology of Vertebrates*. Cornell University.

McTavish, T. and Restrepo, D. (2008). Evolving solutions: The genetic algorithm and evolution strategies for finding optimal parameters. In Smolinski, T., Milanova, M., and Hassanien, A., editors, *Applications of Computational Intelligence in Biology*, pages 55–78. Springer.

Mehrabian, R. and Lucas, C. (2006). A novel numerical optimization algorithm inspired from weed colonization. *Ecological Informatics*, 1(4):355–366.

Melab, N., Cahon, S., and Talbi, E.-G. (2006). Grid computing for parallel bioinspired algorithms. *Journal of Parallel and Distributed Computing*, 66(8):1052–1061.

Mendes, R., Kennedy, J., and Neves, J. (2004). The fully informed particle swarm: Simpler, maybe better. *IEEE Transactions on Evolutionary Computation*, 8(3):204–210.

Metropolis, N. (1987). The beginning of the Monte Carlo method. *Los Alamos Science*, 15:125–130.

Metropolis, N., Rosenbluth, A., Rosenbluth, M., Teller, A., and Teller, E. (1953). Equations of state calculations by fast computing machines. *The Journal of Chemical Physics*, 21(6):1087–1092.

Meuleau, N., Peshkin, L., Kim, K.-E., and Kaelbling, L. (1999). Learning finite-state controllers for partially observable environments. *Conference on Uncertainty in Artificial Intelligence*, Stockholm, Sweden, pages 427–436.

Meuleau02, N. and Dorigo, M. (2002). Ant colony optimization and stochastic gradient descent. *Artificial Life*, 8(2):103–121.

Mezura-Montes, E. and Coello Coello, C. (2005). A simple multimembered evolution strategy to solve constrained optimization problems. *IEEE Transactions on Evolutionary Computation*, 9(1):1–17.

Mezura-Montes, E. and Coello Coello, C. (2008). Constrained optimization via multiobjective evolutionary algorithms. In Knowles, J., Corne, D., and Deb, K., editors, *Multiobjective Problem Solving from Nature*, pages 53–75. Springer.

Mezura-Montes, E. and Palomeque-Oritiz, A. (2009). Parameter control in differential evolution for constrained optimization. *IEEE Congress on Evolutionary Computation*, Trondheim, Norway, pages 1375–1382.

Mezura-Montes, E., Reyes-Sierra, M., and Coello Coello, C. (2008). Multi-objective optimization using differential evolution: A survey of the state-of-the-art. In Chakraborty, U., editor, *Advances in Differential Evolution*, pages 173–196. Springer.

Michalewicz, Z. (1996). *Genetic Algorithms + Data Structures = Evolution Programs*. Springer.

Michalewicz, Z. and Attia, N. (1994). Evolutionary optimization of constrained problems. *Third Annual Conference on Evolutionary Programming*, San Diego, California, pages 98–108.

Michalewicz, Z., Dasgupta, D., Riche, R. L., and Schoenauer, M. (1996). Evolutionary algorithms for constrained engineering problems. *Computers & Industrial Engineering*, 30(4):851–870.

Michalewicz, Z. and Janikow, C. (1991). Handling constraints in genetic algorithms. *International Conference on Genetic Algorithms*, Breckenridge, Colorado, pages 151–157.

Michalewicz, Z. and Nazhiyath, G. (1995). Genocop III: A co-evolutionary algorithm for numerical optimization problems with nonlinear constraints. *IEEE Conference on Evolutionary Computation*, Perth, Western Australia, pages 647–651.

Michalewicz, Z. and Schoenauer, M. (1996). Evolutionary algorithms for constrained parameter optimization problems. *Evolutionary Computation*, 4(1):1–32.

Michiels, W., Aarts, E., and Korst, J. (2007). *Theoretical Aspects of Local Search*. Springer.

Milinski, H. and Heller, R. (1978). Influence of a predator on the optimal foraging behavior of sticklebacks. *Nature*, 275(5681):642–644.

Miller, J. and Smith, S. (2006). Redundancy and computational efficiency in Cartesian genetic programming. *IEEE Transactions on Evolutionary Computation*, 10(2):167–174. 2006.

Mitchell, M. (1998). *An Introduction to Genetic Algorithms*. The MIT Press.

Mitzenmacher, M. and Upfal, E. (2005). *Probability and Computing: Randomized Algorithms and Probabilistic Analysis*. Cambridge University Press.

Morales, A. and Quezada, C. (1998). A univeral eclectic genetic algorithm for constrained optimization. *Sixth European Congress on Intelligent Techniques and Soft Computing*, Aachen, Germany, pages 518–522.

Morrison, R. (2004). *Designing Evolutionary Algorithms for Dynamic Environments*. Springer.

Mühlenbein, H., Mahnig, T., and Ochoa, A. (1999). Schemata, distributions and graphical models in evolutionary optimization. *Journal of Heuristics*, 5(2):215–247.

Mühlenbein, H. and Paaβ, G. (1996). From recombination of genes to the estimation of distributions: I. Binary parameters. In Voigt, H.-M., Ebeling, W., Rechenberg, I., and Schwefel, H.-P., editors, *Parallel Problem Solving from Nature – PPSN IV*, pages 178–187. Springer.

Mühlenbein, H. and Schlierkamp-Voosen, D. (1993). Predictive models for the breeder genetic algorithm: I. Continuous parameter optimization. *Evolutionary Computation*, 1(1):25–49.

Mühlenbein, H. and Schlierkamp-Voosen, D. (1997). The equation for response to selection and its use for prediction. *Evolutionary Computation*, 5(3):303–346.

Mühlenbein, H. and Voigt, H.-M. (1995). Gene pool recombination for the breeder genetic algorithm. *First Metaheuristics International Conference*, Breckenridge, Colorado, pages 19–25.

Muller, S., Marchetto, J., Airaghi, S., and Kournoutsakos, P. (2002). Optimization based on bacterial chemotaxis. *IEEE Transactions on Evolutionary Computation*, 6(1):16–29.

Munroe, E. (1948). *The Geographical Distribution of Butterflies in the West Indies*. PhD thesis, Cornell University.

Nemhauser, G. and Wolsey, L. (1999). *Integer and Combinatorial Optimization*. John Wiley & Sons.

Neri, F. and Tirronen, V. (2010). Recent advances in differential evolution: A survey and experimental analysis. *Artificial Intelligence Review*, 33(1):61–106.

Neshat, M., Adeli, A., Sepidnam, G., Sargolzaei, M., and Toosi, A. (2012). A review of artificial fish swarm optimization methods and applications. *International Journal on Smart Sensing and Intelligent Systems*, 5(1):107–148.

Neumann, F. and Witt, C. (2009). Runtime analysis of a simple ant colony optimization algorithm. *Algorithmica*, 54(2):243–255.

Newton, M. (2004). *Savage Girls and Wild Boys*. Picador.

Nguyen, T., Yang, S., and Branke, J. (2012). Evolutionary dynamic optimization: A survey of the state of the art. *Swarm and Evolutionary Computation*, 6:1–24.

Nguyen, T. and Yao, X. (2009). Benchmarking and solving dynamic constrained problems. *IEEE Congress on Evolutionary Computation*, Trondheim, Norway, pages 690–697.

Nierhaus, G. (2010). *Algorithmic Composition: Paradigms of Automated Music Generation*. Springer.

Niknam, T. and Amiri, B. (2010). An efficient hybrid approach based on PSO, ACO and k-means for cluster analysis. *Applied Soft Computing*, 10(1):183–197.

Nix, A. and Vose, M. (1992). Modeling genetic algorithms with Markov chains. *Annals of Mathematics and Artificial Intelligence*, 5(1):79–88.

Noel, M. and Jannett, T. (2005). A new continuous optimization algorithm based on sociological models. *American Control Conference*, Portland, Oregon, pages 237–242.

Noman, N. and Iba, H. (2008). Accelerating differential evolution using an adaptive local search. *IEEE Transactions on Evolutionary Computation*, 12(1):107–125.

Nordin, P., Francone, F., and Banzhaf, W. (1996). Explicitly defined introns and destructive crossover in genetic programming. In Angeline, P. and Kinnear, K., editors, *Advances in Genetic Programming: Volume 2*, pages 111–134. The MIT Press.

Noren, S., Biedenbach, G., Redfern, J., and Edwards, E. (2008). Hitching a ride: The formation locomotion strategy of dolphin calves. *Functional Ecology*, 22(2):278–283.

Nourani, Y. and Andresen, B. (1998). A comparison of simulated annealing cooling strategies. *Journal of Physics A: Mathematical and General*, 31(41):8373–8385.

Okubo, A. and Levin, S. (2001). *Diffusion and Ecological Problems*. Springer.

Oliver, I., Smith, D., and Holland, J. (1987). A study of permutation crossover operators on the traveling salesman problem. *International Conference on Genetic Algorithms*, Cambridge, Massachusetts, pages 224–230.

Omran, M. (2008). Using opposition-based learning with particle swarm optimization and barebones differential evolution. In Lazinica, A., editor, *Particle Swarm Optimization*, pages 373–384. InTech.

Omran, M., Engelbrecht, A., and Salman, A. (2009). Bare bones differential evolution. *European Journal of Operational Research*, 196(1):128–139.

Omran, M. and Mahdavi, M. (2008). Global-best harmony search. *Applied Mathematics and Computation*, 198(2):643–656.

Omran, M., Simon, D., and Clerc, M. (2013). Linearized biogeography-based optimization. *Submitted for publication.*

O'Neill, M. and Ryan, C. (2003). *Grammatical Evolution*. Springer.

Ong, Y., Nair, P., Keane, A., and Wong, K. (2004). Surrogate-assisted evolutionary optimization frameworks for high-fidelity engineering design problems. In Jin, Y., editor, *Knowledge Incorporation in Evolutionary Computation*, pages 307–332. Springer.

Ong, Y.-S., Krasnogor, N., and Ishibuchi, H. (2007). Special issue on memetic algorithms. *IEEE Transactions on Systems, Man, and Cybernetics – Part B: Cybernetics*, 37(1):2–5.

Onwubolu, G. and Davendra, D., editors (2009). *Differential Evolution: A Handbook for Global Permutation-Based Combinatorial Optimization*. Springer.

Opitz, D. and Maclin, R. (1999). Popular ensemble methods: An empirical study. *Journal of Artificial Intelligence Research*, 11:169–198.

O'Reilly, U. and Oppacher, F. (1995). The troubling aspects of a building block hypothesis for genetic programming. In Whitley, L. and Vose, M., editors, *Foundations of Genetic Algorithms, Volume 3*, pages 73–88. Morgan Kaufmann.

Orvosh, D. and Davis, L. (1993). Shall we repair? Genetic algorithms, combinatorial optimization, and feasibility constraints. *International Conference on Genetic Algorithms*, Urbana-Champaign, Illinois, page 650.

Otten, R. and van Ginneken, L. (1989). *The Annealing Algorithm*. Kluwer Academic Publishers.

Palmer, C. and Kershenbaum, A. (1994). Representing trees in genetic algorithms. *IEEE Conference on Evolutionary Computation*, Orlando, Florida, pages 379–384.

Pan, Q., Tasgetiren, M., and Liang, Y. (2008). A discrete differential evolution algorithm for the permutation flowshop scheduling problem. *Computers & Industrial Engineering*, 55(4):795–816.

Paquet, U. and Engelbrecht, A. (2003). A new particle swarm optimiser for linearly constrained optimisation. *IEEE Congress on Evolutionary Computation*, Canberra, Australia, pages 227–233.

Pardalos, P. and Mavridou, T. (1998). The graph coloring problem: A bibliographic survey. In Zhu, D.-Z. and Pardalos, P., editors, *Handbook of Combinatorial Optimization*, pages 331–395. Kluwer Academic Publishers.

Paredis, J. (2000). Coevolutionary algorithms. In Bäck, T., Fogel, D., and Michalewicz, Z., editors, *Evolutionary Computation 2*, pages 224–238. Institute of Physics.

Pareto, V. (1896). *Cours d'Economie Politique*. Guillaumin.

Parks, G. and Miller, I. (1998). Selective breeding in a multiobjective genetic algorithm. In Eiben, A., Bäck, T., Schoenauer, M., and Schwefel, H.-P., editors, *Parallel Problem Solving from Nature - PPSN V*, pages 250–259. Springer.

Passino, K. (2002). Biomimicry of bacterial foraging. *IEEE Control Systems Magazine*, 22(3):52–67.

Paul, T. and Iba, H. (2003). Optimization in continuous domain by real-coded estimation of distribution algorithm. In Abraham, A., Köppen, M., and Franke, K., editors, *Design and Application of Hybrid Intelligent Systems*, pages 262–271. IOS Press.

Peña, J., Robles, V., Larrañaga, P., Herves, V., Rosales, F., and Pérez, M. (2004). GA-EDA: Hybrid evolutionary algorithm using genetic and estimation of distribution algorithms. In Orchard, B., Yang, C., and Ali, M., editors, *Innovations in Applied Artificial Intelligence*, pages 361–371. Springer.

Pedersen, M. (2010). Good parameters for particle swarm optimization. Technical report, Hvass Laboratories. www.hvass-labs.org.

Pedersen, M. and Chipperfield, A. (2010). Simplifying particle swarm optimization. *Applied Soft Computing*, 10(2):618–628.

Pelikan, M. (2005). *Hierarchical Bayesian Optimization Algorithm*. Springer.

Pelikan, M., Goldberg, D., and Cantú-Paz, E. (1999). BOA: The Bayesian optimization algorithm. *Genetic and Evolutionary Computation Conference*, Orlando, Florida, pages 525–532.

Pelikan, M., Goldberg, D., and Lobo, F. (2002). A survey of optimization by building and using probabilistic models. *Computational Optimization and Applications*, 21(1):5–20.

Pelikan, M. and Mühlenbein, H. (1998). The bivariate marginal distribution algorithm. In Benitez, J., Cordon, O., Hoffmann, F., and Roy, R., editors, *Advances in Soft Computing: Engineering Design and Manufacturing*, pages 521–535. Springer.

Pelikan, M. and Sastry, K. (2004). Fitness inheritance in the Bayesian optimization algorithm. *Genetic and Evolutionary Computation Conference*, Seattle, Washington, pages 48–59.

Petroski, H. (1992). *To Engineer Is Human: The Role of Failure in Successful Design*. Vintage.

Pétrowski, A. (1996). A clearing procedure as a niching method for genetic algorithms. *IEEE Conference on Evolutionary Computation*, Nagoya, Japan, pages 798–803.

Pham, D., Ghanbarzadeh, A., Koç, E., Otri, S., Rahim, S., and Zaidi, M. (2006). The bees algorithm – A novel tool for complex optimisation problems. *2nd International Virtual Conference on Intelligent Production Machines and Systems*, pages 454–459.

Pincus, M. (1968a). A closed form solution of certain programming problems. *Operations Research*, 16(3):690–694.

Pincus, M. (1968b). A Monte Carlo method for the approximate solution of certain types of constrained optimization problems. *Operations Research*, 18(6):1225–1228.

Pitcher, T. and Parrish, J. (1993). Functions of shoaling behaviour in teleosts. In Pitcher, T., editor, *Behaviour of Teleost Fishes*, pages 363–439. Chapman & Hall.

Poli, R. (2003). A simple but theoretically-motivated method to control bloat in genetic programming. *Sixth European Conference on Genetic Programming*, Essex, England, pages 211–223.

Poli, R. (2008). Dynamics and stability of the sampling distribution of particle swarm optimisers via moment analysis. *Journal of Artificial Evolution and Applications*, 2008. Article ID 761459, 10 pages, doi:10.1155/2008/761459.

Poli, R., Kennedy, J., and Blackwell, T. (2007). Particle swarm optimization: An overview. *Swarm Intelligence*, 1(1):33–57.

Poli, R., Langdon, W., and McPhee, N. (2008). *A Field Guide to Genetic Programming.* Published via `http://lulu.com` and freely available at `http://www.gp-field-guide.org.uk`.

Poli, R., McPhee, N., and Rowe, J. (2004). Exact schema theory and Markov chain models for genetic programming and variable-length genetic algorithms with homologous crossover. *Genetic Programming and Evolvable Machines*, 5(1):31–70.

Poli, R., Rowe, J., and McPhee, N. (2001). Markov chain models for GP and variable-length GAs with homologous crossover. *Genetic and Evolutionary Computation Conference*, San Francisco, California, pages 112–119.

Poundstone, W. (1993). *Prisoner's Dilemma.* Anchor.

Powell, D. and Skolnick, M. (1993). Using genetic algorithms in engineering design optimization with non-linear constraints. *International Conference on Genetic Algorithms*, Urbana-Champaign, Illinois, pages 424–431.

Preble, S., Lipson, M., and Lipson, H. (2005). Two-dimensional photonic crystals designed by evolutionary algorithms. *Applied Physics Letters*, 86(6):061111.

Price, K. (1997). Differential evolution versus the functions of the 2nd ICEO. *IEEE Conference on Evolutionary Computation*, Indianapolis, Indiana, pages 153–157.

Price, K. (2013). Differential evolution. In Zelinka, I., Snášel, V., and Abraham, A., editors, *Handbook of Optimization.* `ebooks.com`.

Price, K. and Storn, R. (1997). Differential evolution: Numerical optimization made easy. *Dr. Dobb's Journal*, pages 18–24.

Price, K., Storn, R., and Lampinen, J. (2005). *Differential Evolution.* Springer.

PSC (2012). Particle Swarm Central. `www.particleswarm.info`.

Păun, G. (2003). Membrane computing. In Lingas, A. and Nilsson, B., editors, *Fundamentals of Computation Theory*, pages 177–220. Springer.

Pyke, G. (1978). Optimal foraging in bumblebees and coevolution with their plants. *Oecologia*, 36(3):281–293.

Qin, A., Huang, V., and Suganthan, P. (2009). Differential evolution algorithm with strategy adaptation for global numerical optimization. *IEEE Transactions on Evolutionary Computation*, 13(2):39–417.

Qing, A. (2009). *Differential Evolution: Fundamentals and Applications in Electrical Engineering.* John Wiley & Sons.

Quammen, D. (1997). *The Song of the Dodo: Island Biogeography in an Age of Extinction.* Scribner.

Quijano, N., Passino, K., and Andrews, B. (2006). Foraging theory for multizone temperature control. *IEEE Computational Intelligence Magazine*, 1(4):18–27.

Rabanal, P., Rodríguez, I., and Rubio, F. (2007). Using river formation dynamics to design heuristic algorithms. In Akl, S., Calude, C., Dinneen, M., Rozenberg, G., and Wareham, H., editors, *Unconventional Computation*, pages 163–177. Springer.

Radcliffe, N. and Surry, P. (1995). Fundamental limitations on search algorithms: Evolutionary computing in perspective. In Van Leeuwen, J., editor, *Computer Science Today: Recent Trends and Developments (Lecture Notes in Computer Science, No. 1000)*, pages 275–291. Springer-Verlag.

Rahnamayan, S., Tizhoosh, H., and Salama, M. (2008). Opposition-based differential evolution. *IEEE Transactions on Evolutionary Computation*, 12(1):64–79.

Rao, R. and Patel, V. (2012). An elitist teaching-learning-based optimization algorithm for solving complex constrained optimization problems. *International Journal of Industrial Engineering Computations*, 3(4):535–560.

Rao, R. and Savsani, V. (2012). *Mechanical Design Optimization Using Advanced Optimization Techniques*. Springer.

Rao, R., Savsani, V., and Vakharia, D. (2011). Teaching-learning-based optimization: A novel method for constrained mechanical design optimization problems. *Computer-Aided Design*, 43(3):303–315.

Rao, R., Savsani, V., and Vakharia, D. (2012). Teaching-learning-based optimization: A novel optimization method for continuous non-linear large scale problems. *Information Sciences*, 183(1):1–15.

Rashedi, E., Nezamabadi-pour, H., and Saryazdi, S. (2009). GSA: A gravitational search algorithm. *Information Sciences*, 179(13):2232–2248.

Rashedi, E., Nezamabadi-pour, H., and Saryazdi, S. (2010). BGSA: Binary gravitational search algorithm. *Natural Computing*, 9(3):727–745.

Rasheed, K. and Hirsh, H. (2000). Informed operators: Speeding up genetic-algorithm-based design optimization using reduced models. *Genetic and Evolutionary Computation Conference*, Las Vegas, Nevada, pages 628–635.

Rashid, M. and Baig, A. (2010). Improved opposition-based PSO for feedforward neural network training. *International Conference on Information Science and Applications*, Seoul, Korea, pages 1–6.

Rastrigin, L. (1974). *Extremal Control Systems*. Nauka. In Russian.

Rawlins, G., editor (1991). *Foundations of Genetic Algorithms*. Morgan Kaufmann Publishers.

Ray, T., Isaacs, A., and Smith, W. (2009a). A memetic algorithm for dynamic multi-objective optimization. In Goh, C.-K., Ong, Y.-S., and Tan, K., editors, *Multi-Objective Memetic Algorithms*, pages 353–367. Springer.

Ray, T. and Liew, K. (2003). Society and civilization: An optimization algorithm based on the simulation of social behavior. *IEEE Transactions on Evolutionary Computation*, 7(4):386–396.

Ray, T., Singh, H., Isaacs, A., and Smith, W. (2009b). Infeasibility driven evolutionary algorithm for constrained optimization. In Mezura-Montes, E., editor, *Constraint-Handling in Evolutionary Optimization*, pages 145–165. Springer.

Rechenberg, I. (1973). *Evolutionsstrategie – Optimierung Technischer Systeme nach Prinzipien der Biologischen Evolution*. Frommann-Holzboog. The English translation of the title is *Evolution strategy – Optimization of Technical Systems according to Principles of Biological Evolution*.

Rechenberg, I. (1998). Cybernetic solution path of an experimental problem. In Fogel, D., editor, *Evolutionary Computation: The Fossil Record*, pages 301–310. Wiley-IEEE Press. First published in 1964.

Reeves, C. (1993). *Modern Heuristic Techniques for Combinatorial Problems*. John Wiley & Sons.

Reeves, C. and Rowe, J. (2003). *Genetic Algorithms: Principles and Perspectives.* Kluwer Academic Publishers.

Reinelt, G. (2008). TSPLIB. `http://comopt.ifi.uni-heidelberg.de/software/TSPLIB95`.

Reynolds, R. (1994). An introduction to cultural algorithms. *Third Annual Conference on Evolutionary Computing*, Madison, Wisconsin, pages 131–139.

Reynolds, R. (1999). Cultural algorithms: Theory and applications. In Corne, D., Dorigo, M., Glover, F., Dasgupta, D., Moscato, P., Poli, R., and Price, K., editors, *New Ideas in Optimization*, pages 367–378. McGraw-Hill.

Reynolds, R., Ashlock, D., Yannakakis, G., Togelius, J., and Preuss, M. (2011). Tutorials: Cultural algorithms: Incorporating social intelligence into virtual worlds. *IEEE Conference on Computational Intelligence and Games*, Seoul, South Korea, pages J1–J5.

Reynolds, R. and Chung, C. (1997). Knowledge-based self-adaptation in evolutionary programming using cultural algorithms. *IEEE International Conference on Evolutionary Computation*, Indianapolis, Indiana, pages 71–76.

Ritscher, T., Helwig, S., and Wanka, R. (2010). Design and experimental evaluation of multiple adaptation layers in self-optimizing particle swarm optimization. *IEEE Congress on Evolutionary Computation*, Barcelona, Spain, pages 1–8.

Ritzel, B., Eheart, J., and Ranjithan, S. (1995). Using genetic algorithms to solve a multiple objective groundwater pollution containment problem. *Water Resources Research*, 30(5):1589–1603.

Robert, C. and Casella, G. (2010). *Monte Carlo Statistical Methods.* Springer.

Ronald, S. (1998). Duplicate genotypes in a genetic algorithm. *IEEE World Congress on Computational Intelligence*, Anchorage, Alaska, pages 793–798.

Ros, R. and Hansen, N. (2008). A simple modification in CMA-ES achieving linear time and space complexity. In Rudolp, G., Jansen, T., Lucas, S., Poloni, C., and Beume, N., editors, *Parallel Problem Solving from Nature - PPSN X*, pages 296–305. Springer.

Rosca, J. (1997). Analysis of complexity drift in genetic programming. *Second Annual Conference on Genetic Programming*, Palo Alto, California, pages 286–294.

Rosenberg, R. (1967). *Simulation of genetic populations with biochemical properties.* PhD thesis, University of Michigan.

Rosenbrock, H. (1960). An automatic method for finding the greatest or least value of a function. *The Computer Journal*, 3(3):175–184.

Rosenkrantz, D., Stearns, R., and Lewis, P. (1977). An analysis of several heuristics for the traveling salesman problem. *SIAM Journal on Computing*, 6(3):563–581.

Ross, T. (2010). *Fuzzy Logic with Engineering Applications.* John Wiley & Sons, 3rd edition.

Rossi, F., van Beek, P., and Walsh, T. (2006). *Handbook of Constraint Programming.* Elsevier.

Rothlauf, F. and Goldberg, D. (2003). Redundant representations in evolutionary computation. *Evolutionary Computation*, 11(4):381–415.

Rubinstein, A. (1986). Finite automata play the repeated prisoner's dilemma. *Journal of Economic Theory*, 39(1):83–96.

Rudlof, S. and Köppen, M. (1996). Stochastic hill climbing by vectors of normal distributions. *First Online Workshop on Soft Computing*, Nagoya, Japan, pages 60–70.

Rudolph, G. (1992). Parallel approaches to stochastic global optimization. In Joosen, W. and Milgrom, E., editors, *Parallel Computing: From Theory to Sound Practice*, pages 256–267. IOS Press.

Rudolph, G. and Agapie, A. (2000). Convergence properties of some multi-objective evolutionary algorithms. *IEEE Congress on Evolutionary Computation*, San Diego, California, pages 1010–1016.

Rudolph, G. and Schwefel, H.-P. (2008). Simulated evolution under multiple criteria conditions revisited. *IEEE World Congress on Computational Intelligence*, Hong Kong, pages 249–261.

Runarsson, T. and Yao, X. (2000). Stochastic ranking for constrained evolutionary optimization. *IEEE Transactions on Evolutionary Computation*, 4(3):284–294.

Sakawa, M. (2002). *Genetic Algorithms and Fuzzy Multiobjective Optimization*. Springer.

Salkind, N. (2007). *Statistics for People Who (Think They) Hate Statistics*. Sage Publications.

Salomon, R. (1996). Reevaluating genetic algorithm performance under coordinate rotation of benchmark functions. *BioSystems*, 39(3):263–278.

Salustowicz, R. and Schmidhuber, J. (1997). Probabilistic incremental program evolution. *Evolutionary Computation*, 5(2):123–141.

Sano, Y. and Kita, H. (2002). Optimization of noisy fitness functions by means of genetic algorithms using history of search with test of estimation. *IEEE Congress on Evolutionary Computation*, Honolulu, Hawaii, pages 360–365.

Santana, R. (1998). Estimation of distribution algorithms with Kikuchi approximations. *Evolutionary Computation*, 13(1):67–97.

Santana, R. (2003). A Markov network based factorized distribution algorithm for optimization. *14th European Conference on Machine Learning*, Cavtat, Croatia, pages 337–348.

Santana, R. and Echegoyen, C. (2012). Matlab Toolbox for Estimation of Distribution Algorithms (MATEDA-2.0). `www.sc.ehu.es/ccwbayes/members/rsantana/software/matlab/MATEDA.html`.

Santana, R., Larrañaga, P., and Lozano, J. (2008). Adaptive estimation of distribution algorithms. In Cotta, C., Sevaux, M., and Sörensen, K., editors, *Adaptive and Multilevel Metaheuristics*, pages 177–197. Springer.

Santana-Quintero, L., Montaño, A., and Coello Coello, C. (2010). A review of techniques for handling expensive functions in evolutionary multi-objective optimization. In Tenne, Y. and Goh, C.-K., editors, *Computational Intelligence in Expensive Optimization Problems*, pages 29–60. Springer.

Sareni, B. and Krähenbühl, L. (1998). Fitness sharing and niching methods revisited. *IEEE Transactions on Evolutionary Computation*, 2(3):97–106.

Sastry, K. and Goldberg, D. (2000). On extended compact genetic algorithm. *Genetic and Evolutionary Computation Conference*, Las Vegas, Nevada, pages 352–359.

Sastry, K., Goldberg, D., and Pelikan, M. (2001). Don't evaluate, inherit. *Genetic and Evolutionary Computation Conference*, San Francisco, California, pages 551–558.

Savicky, P. and Robnik-Sikonja, M. (2008). Learning random numbers: A Matlab anomaly. *Applied Artificial Intelligence*, 22(3):254–265.

Savla, K., Frazzoli, E., and Bullo, F. (2008). Traveling salesperson problems for the Dubins vehicle. *IEEE Transactions on Automatic Control*, 53(6):1378–1391.

Sayadi, M., Ramezanian, R., and Ghaffari-Nasab, N. (2010). A discrete firefly metaheuristic with local search for makespan minimization in permutation flow shop scheduling problems. *International Journal of Industrial Engineering Computations*, 1(1):1–10.

Schaffer, C. (1994). A conservation law for generalization performance. *11th International Conference on Machine Learning*, Boca Raton, Florida, pages 259–265.

Schaffer, J. (1985). Multiple objective optimization with vector evaluated genetic algorithms. *International Conference on Genetic Algorithms and Their Application*, Pittsburgh, Pennsylvania, pages 93–100.

Schervish, M. (1996). P values: What they are and what they are not. *The American Statistician*, 50(3):203–206.

Schmidhuber, J. (1987). *Evolutionary principles in self-referential learning, or on learning how to learn: The meta-meta-... hook.* PhD thesis, Technische Universität München.

Schoenauer, M. and Michalewicz, Z. (1996). Evolutionary computation at the edge of feasibility. In Ebeling, W., Rechenberg, I., Schwefel, H.-P., and Voigt, H.-M., editors, *Parallel Problem Solving from Nature - PPSN IV*, pages 245–254. Springer.

Schoenauer, M., Sebag, M., Jouve, F., Lamy, B., and Maitournam, H. (1996). Evolutionary identification of macro-mechanical models. In Angeline, P. and Kinnear, K., editors, *Advances in Genetic Programming*, volume 2, pages 467–488. MIT Press.

Schoenauer, M. and Xanthakis, S. (1993). Constrained GA optimization. *International Conference on Genetic Algorithms*, Urbana-Champaign, Illinois, pages 573–580.

Schrijver, A. (2005). On the history of combinatorial optimization (till 1960). In Aardal, K., Nemhauser, G., and Weismantel, R., editors, *Discrete Optimization*, volume 12 of *Handbooks in Operations Research and Management Science*, pages 1–68. Elsevier.

Schultz, T. (1999). Ants, plants and antibiotics. *Nature*, 398(6730):747–748.

Schultz, T. (2000). In search of ant ancestors. *Proceedings of the National Academy of Sciences*, 97(26):14028–14029.

Schumacher, C., Vose, M., and Whitley, L. (2001). The no free lunch and problem description length. *Genetic and Evolutionary Computation Conference*, San Francisco, California, pages 565–570.

Schütze, O., Lara, A., and Coello Coello, C. (2011). On the influence of the number of objectives on the hardness of a multiobjective optimization problem. *IEEE Transactions on Evolutionary Computation*, 15(4):444–454.

Schwefel, H.-P. (1977). *Numerische Optimierung von Computer-Modellen.* Birkhauser. The English translation of the title is *Evolutionary Strategy and Numerical Optimization.*

Schwefel, H.-P. (1981). *Numerical Optimization of Computer Models.* John Wiley & Sons. Translation of [Schwefel, 1977] along with some additional material.

Schwefel, H.-P. (1995). *Evolution and Optimum Seeking.* John Wiley & Sons. Expanded version of [Schwefel, 1981].

Schwefel, H.-P. and Mendes, M. (2010). 45 years of evolution strategies. *SIGEVOlution*, 4(2):2–8.

Sebag, M. and Ducoulombier, A. (1998). Extending population-based incremental learning to continuous search spaces. In Eiben, A., Bäck, T., Schoenauer, M., and Schwefel, H.-P., editors, *Parallel Problem Solving from Nature - PPSN V*, pages 418–427. Springer.

Sefrioui, M. and Périaux, J. (2000). A hierarchical genetic algorithm using multiple models for optimization. In Schoenauer, M., Deb, K., Rudolph, G., Yao, X., Lutton, E., Merelo, J., and Schwefel, H.-P., editors, *Parallel Problem Solving from Nature - PPSN VI*, pages 879–888. Springer.

Selvakumar, A. and Thanushkodi, K. (2007). A new particle swarm optimization solution to nonconvex economic dispatch problems. *IEEE Transactions on Power Systems*, 22(1):42–51.

Seneta, E. (1966). Markov and the birth of chain dependence theory. *International Statistical Review*, 64(3):255–263.

Settles, B. (2010). Active learning literature survey. Technical report, University of Wisconsin-Madison. `www.cs.cmu.edu/~bsettles/pub/settles.activelearning.pdf`.

Shah-Hosseini, H. (2007). Problem solving by intelligent water drops. *IEEE Congress on Evolutionary Computation*, Singapore, pages 3226–3231.

Shakya, S. and Santana, R., editors (2012). *Markov Networks in Evolutionary Computation*. Springer.

Shi, L. and Rasheed, K. (2010). A survey of fitness approximation methods applied in evolutionary algorithms. In Tenne, Y. and Goh, C.-K., editors, *Computational Intelligence in Expensive Optimization Problems*, pages 3–28. Springer.

Shi, Y. and Eberhart, R. (1999). Empirical study of particle swarm optimization. *IEEE Congress on Evolutionary Computation*, Washington, District of Columbia, pages 1945–1950.

Shibani, Y., Yasuno, S., and Ishiguro, I. (2001). Effects of global information feedback on diversity. *Advances in Genetic Programming*, 45(1):80–96.

Simões, A. (2011). *Evolutionary Algorithms in Dynamic Optimization Problems*. Lambert Academic Publishing.

Simon, D. (2005). Research in the balance. *IEEE Potentials*, 24(2):17–21.

Simon, D. (2006). *Optimal State Estimation*. John Wiley & Sons.

Simon, D. (2008). Biogeography-based optimization. *IEEE Transactions on Evolutionary Computation*, 12(6):702–713.

Simon, D. (2011a). A dynamic system model of biogeography-based optimization. *Applied Soft Computing*, 11(8):5652–5661.

Simon, D. (2011b). A probabilistic analysis of a simplified biogeography-based optimization algorithm. *Evolutionary Computation*, 19(2):167–188.

Simon, D. (2012). Biogeography-Based Optimization Web Site. `http://embeddedlab.csuohio.edu/BBO`.

Simon, D., Ergezer, M., and Du, D. (2009). Population distributions in biogeography-based optimization algorithms with elitism. *IEEE Conference on Systems, Man, and Cybernetics*, San Antonio, Texas, pages 1017–1022.

Simon, D., Ergezer, M., Du, D., and Rarick, R. (2011a). Markov models for biogeography-based optimization. *IEEE Transactions on Systems, Man and Cybernetics – Part B: Cybernetics*, 41(1):299–306.

Simon, D., Rarick, R., Ergezer, M., and Du, D. (2011b). Analytical and numerical comparisons of biogeography-based optimization and genetic algorithms. *Information Sciences*, 181(7):1224–1248.

Singh, H., Ray, T., and Smith, W. (2010). Surrogate assisted simulated annealing (SASA) for constrained multi-objective optimization. *IEEE Congress on Evolutionary Computation*, Barcelona, Spain, pages 1–8.

Smith, A. and Tate, D. (1993). Genetic optimization using a penalty function. *International Conference on Genetic Algorithms*, Urbana-Champaign, Illinois, pages 499–505.

Smith, R., Dike, B., and Stegmann, S. (1995). Fitness inheritance in genetic algorithms. *Symposium on Applied Computing*, Nashville, Tennessee, pages 345–350.

Smith, S. (1980). *A Learning System Based on Genetic Adaptive Algorithms*. PhD thesis, University of Pittsburgh.

Šobotník, J., Hanus, R., Kalinová, B., Piskorski, R., Cvačka, J., Bourguignon, T., and Roisin, Y. (2008). (E,E)-α-Farnesene, an alarm pheromone of the termite Prorhinotermes canalifrons. *Journal of Chemical Ecology*, 34(4):478–486.

Socha, K. and Dorigo, M. (2008). Ant colony optimization for continuous domains. *European Journal of Operational Research*, 185(3):1155–1173.

Solnon, C. (2010). *Ant Colony Optimization and Constraint Programming*. John Wiley & Sons.

Spears, W. and De Jong, K. (1997). Analyzing GAs using Markov models with semantically ordered and lumped states. In Belew, R. and Vose, M., editors, *Foundations of Genetic Algorithms*, volume 4, pages 85–100. Morgan Kaufmann.

Srinivas, N. and Deb, K. (1994). Multiobjective optimization using nondominated sorting in genetic algorithms. *Evolutionary Computation*, 2(3):221–248.

Stanley, K. and Miikkulainen, R. (2002). Evolving neural networks through augmenting topologies. *Evolutionary Computation*, 10(2):99–127.

Stephens, D. and Krebs, J. (1986). *Foraging Theory*. Princeton University Press.

Sterne, J. and Smith, G. (2001). Sifting the evidence – what's wrong with significance tests? *Physical Therapy*, 81(8):1464–1469.

Stone, L. (2009). *Zebras*. Lerner Publications Company.

Storn, R. (1996a). Differential evolution design of an IIR-filter. *IEEE Conference on Evolutionary Computation*, Nagoya, Japan, pages 268–273.

Storn, R. (1996b). On the usage of differential evolution for function optimization. *Conference of the North American Fuzzy Information Processing Society*, Berkeley, California, pages 519–523.

Storn, R. and Price, K. (1996). Minimizing the real functions of the ICEC'96 contest by differential evolution. *IEEE Conference on Evolutionary Computation*, Nagoya, Japan, pages 842–844.

Storn, R. and Price, K. (1997). Differential evolution – A simple and efficient heuristic for global optimization over continuous spaces. *Journal of Global Optimization*, 11(4):341–359.

Stroud, P. (2001). Kalman-extended genetic algorithm for search in nonstationary environments with noisy fitness evaluations. *IEEE Transactions on Evolutionary Computation*, 5(1):66–77.

Stützle, T. and Hoos, H. (2000). MAX-MIN ant system. *Future Generation Computer Systems*, 16(8):889–914.

Su, C. and Lee, C. (2003). Network reconfiguration of distribution systems using improved mixed-integer hybrid differential evolution. *IEEE Transactions on Power Delivery*, 18(3):1022–1027.

Suganthan, P., Hansen, N., Liang, J., Deb, K., Chen, Y.-P., Auger, A., and Tiwari, S. (2005). Problem definitions and evaluation criteria for the CEC 2005 special session on real-parameter optimization. Technical report. `www.iitk.ac.in/kangal/papers/k2005005.pdf`, `www.ntu.edu.sg/home/EPNSugan/index_files/cec-benchmarking.htm`.

Sun, J., Lai, C.-H., and Wu, X.-J. (2011). *Particle Swarm Optimisation: Classical and Quantum Perspectives*. CRC Press.

Sverdlik, W. and Reynolds, R. (1993). Incorporating domain specific knowledge into version space search. *Fifth International Conference on Tools with Artificial Intelligence*, Boston, Massachusetts, pages 216–223.

Syberfeldt, A., Ng, A., John, R., and Moore, P. (2010). Evolutionary optimisation of noisy multi-objective problems using confidence-based dynamic resampling. *European Journal of Operational Research*, 204(3):533–544.

Syswerda, G. (1991). Schedule optimization using genetic algorithms. In Davis, L., editor, *Handbook of Genetic Algorithms*, pages 332–349. Van Nostrand Reinhold.

Syswerda, G. (2010). Differential evolution research – trends and open questions. In Chakraborty, U., editor, *Advances in Differential Evolution*, pages 1–32. Springer.

Szu, H. and Hartley, R. (1987). Fast simulated annealing. *Physics Letters A*, 122(3–4):157–162.

Takahama, T. and Sakai, S. (2009). Solving difficult constrained optimziation problems by the ϵ constrained differential evolution with gradient-based mutation. In Mezura-Montes, E., editor, *Constraint-Handling in Evolutionary Optimization*, pages 51–72. Springer.

Tan, K., Khor, E., and Lee, T. (2010). *Multiobjective Evolutionary Algorithms and Applications*. Springer.

Tanaka, M. and Tanino, T. (1992). Global optimization by the genetic algorithm in a multiobjective decision support system. *International Conference on Multiple Criteria Decision Making*, Taipei, Taiwan, pages 261–270.

Tao, G. and Michalewicz, Z. (1998). Inver-over operator for the TSP. In Eiben, A., Bäck, T., Schoenauer, M., and Schwefel, H.-P., editors, *Parallel Problem Solving from Nature - PPSN V*, pages 803–812. Springer.

Taylor, J. (1997). *An Introduction to Error Analysis: The Study of Uncertainties in Physical Measurements*. University Science Books, 2nd edition.

Tenne, Y. and Goh, C.-K., editors (2010). *Computational Intelligence in Expensive Optimization Problems*. Springer.

Teodorović, D. (2003). Transport modeling by multi-agent systems: A swarm intelligence approach. *Transportation Planning and Technology*, 26(4):289–312.

Tereshko, V. (2000). Reaction-diffusion model of a honeybee colony's foraging behaviour. In Schoenauer, M., Deb, K., Rudolph, G., Yao, X., Lutton, E., Merelo, J., and Schwefel, H.-P., editors, *Parallel Problem Solving from Nature - PPSN VI*, pages 807–816. Springer.

Tessema, B. and Yen, G. (2006). A self adaptive penalty function based algorithm for constrained optimization. *IEEE Congress on Evolutionary Computation*, Vancouver, Canada, pages 246–253.

Thiele, L., Miettinen, K., Korhonen, P., and Molina, J. (2009). A preference-based evolutionary algorithm for multi-objective optimization. *Evolutionary Computation*, 17(3):411–436.

Thomson, I. (2010). *Culture Wars and Enduring American Dilemmas*. The University of Michigan Press.

Tilman, D., May, R., Lehman, C., and Nowak, M. (1994). Habitat destruction and the extinction debt. *Nature*, 371(3):65–66.

Tinoco, J. and Coello Coello, C. (2013). hypDE: A hyper-heuristic based on differential evolution for solving constrained optimization problems. In Schütze, O., Coello Coello, C., Tantar, A.-A., Tantar, E., Bouvry, P., Mora, P. D., and Legrand, P., editors, *EVOLVE - A Bridge between Probability, Set Oriented Numerics, and Evolutionary Computation II*, pages 267–282. Springer.

Tinós, R. and Yang, S. (2005). Genetic algorithms with self-organized criticality for dynamic optimziation problems. *IEEE Congress on Evolutionary Computation*, Edinburgh, United Kingdom, pages 2816–2823.

Tizhoosh, H. (2005). Opposition-based learning: A new scheme for machine intelligence. *International Conference on Computational Intelligence for Modelling, Control and Automation*, Vienna, Austria, pages 695–701.

Tizhoosh, H., Ventresca, M., and Rahnamayan, S. (2008). Opposition-based computing. In Tizhoosh, H. and Ventresca, M., editors, *Oppositional Concepts in Computational Intelligence*, pages 11–28. Springer.

Tomassini, M. and Vanneschi, L. (2009). Introduction: Special issue on parallel and distributed evolutionary algorithms, Part I. *Genetic Programming and Evolvable Machines*, 10(4):339–341.

Tomassini, M. and Vanneschi, L. (2010). Guest editorial: Special issue on parallel and distributed evolutionary algorithms, Part II. *Genetic Programming and Evolvable Machines*, 11(2):129–130.

Torregosa, R. and Kanok-Nukulchai, W. (2002). Weight optimization of steel frames using genetic algorithm. *Advances in Structural Engineering*, 5(2):99–111.

Toth, P. and Vigo, D., editors (2002). *The Vehicle Routing Problem.* The Society for Industrial and Applied Mathematics.

Tripathi, P., Bandyopadhyay, S., and Pal, S. (2007). Multi-objective particle swarm optimization with time variant inertia and acceleration coefficients. *Information Sciences*, 177(22):5033–5049.

Tsutsui, S. (2004). Ant colony optimisation for continuous domains with aggregation pheromones metaphor. *Fifth International Conference on Recent Advances in Soft Computing (RASC-04)*, Nottingham, United Kingdom, pages 207–212.

Turing, A. (1950). Computing machinery and intelligence. *Mind*, 59(236):433–460.

Turner, G. and Pitcher, T. (1986). Attack abatement: A model for group protection by combined avoidance and dilution. *The American Naturalist*, 128(2):228–240.

Twain, M. (2010). *Autobiography of Mark Twain, Volume 1.* University of California Press.

Tylor, E. (2009). *Primitive Culture: Researches into the Development of Mythology, Philosophy, Religion, Art and Custom*, volume 1. Cornell University Library. Originally published in 1871.

Tylor, E. (2011). *Researches into the Early History of Mankind and the Development of Civilization.* University of California Libraries. Originally published in 1878.

Ufuktepe, U. and Bacak, G. (2005). Applications of graph coloring. *International Conference on Computational Science and Applications*, Singapore, pages 465–477.

van Laarhoven, P. and Aarts, E. (2010). *Simulated Annealing: Theory and Applications.* Springer.

Van Veldhuizen, D. and Lamont, G. (2000). Multiobjective evolutionary algorithms: Analyzing the state-of-the-art. *Evolutionary Computation*, 8(2):125–147.

Vavak, F. and Fogarty, T. (1996). Comparison of steady state and generational genetic algorithms for use in nonstationary environments. *IEEE Conference on Evolutionary Computation*, Nagoya, Japan, pages 192–195.

Ventresca, M. and Tizhoosh, H. (2007). Simulated annealing with opposite neighbors. *IEEE Symposium on Foundations of Computational Intelligence*, Honolulu, Hawaii, pages 186–192.

Volk, T. (1997). *Gaia's Body: Toward a Physiology of Earth.* Springer.

Vorwerk, K., Kennings, A., and Greene, J. (2009). Improving simulated annealing-based FPGA placement with directed moves. *IEEE Transactions on Computer-Aided Design of Integrated Circuits and Systems*, 28(2):179–192.

Vose, M. (1990). Formalizing genetic algorithms. *IEEE Workshop on Genetic Algorithms, Neural Networks, and Simulated Annealing Applied to Signal and Image Processing*, Glasgow, Scotland.

Vose, M. (1999). *The Simple Genetic Algorithm: Foundations and Theory.* MIT Press.

Vose, M. and Liepins, G. (1991). Punctuated equilibria in genetic search. *Complex Systems*, 5(1):31–44.

Vose, M. and Wright, A. (1998a). The simple genetic algorithm and the Walsh transform: Part I, Theory. *Evolutionary Computation*, 6(3):253–273.

Vose, M. and Wright, A. (1998b). The simple genetic algorithm and the Walsh transform: Part II, The inverse. *Evolutionary Computation*, 6(3):275–289.

Waghmare, G. (2013). Comments on "A note on teaching-learning-based optimization algorithm". *Information Sciences*, in print.

Wallace, A. (2006). *The Geographical Distribution of Animals (two volumes).* Adamant Media Corporation. First published in 1876.

Wang, H., Yang, S., Ip, W., and Wang, D. (2009). Adaptive primal-dual genetic algorithms in dynamic environments. *IEEE Transactions on Systems, Man, and Cybernetics – Part B: Cybernetics*, 39(6):1348–1361.

Wedde, H., Farooq, M., and Zhang, Y. (2004). Beehive: An efficient fault-tolerant routing algorithm inspired by honey bee behavior. In Dorigo, M., Birattari, M., Blum, C., Gambardella, L., Mondada, F., and Stützle, T., editors, *Ant Colony Optimization and Swarm Intelligence: 4th International Workshop, ANTS 2004*, pages 83–94. Springer.

Welland, M. (2009). *Sand: The Never-Ending Story.* University of California Press.

Welsch, R. and Endicott, K. (2005). *Taking Sides: Clashing Views in Cultural Anthropology.* McGraw-Hill, 2nd edition.

Wesche, T., Goertler, G., and Hubert, W. (1987). Modified habitat suitability index model for brown trout in southeastern Wyoming. *North American Journal of Fisheries Management*, 7(2):232–237.

Wetzel, C. and Insko, C. (1982). The similarity-attraction relationship: Is there an ideal one? *Journal of Experimental Social Psychology*, 18(3):253–76.

Whigham, P. (1995). A schema theorem for context-free grammars. *IEEE Conference on Evolutionary Computation*, Perth, Western Australia, pages 178–181.

Whitley, D. (1989). The GENITOR algorithm and selection pressure: Why rank-based allocation of reproductive trials is best. *International Conference on Genetic Algorithms*, Fairfax, Virginia, pages 116–121.

Whitley, D. (1994). A genetic algorithm tutorial. *Statistics and Computing*, 4(2):65–85.

Whitley, D. (1999). A free lunch proof for gray versus binary encodings. *Genetic and Evolutionary Computation Conference*, Orlando, Florida, pages 726–733.

Whitley, D. (2001). An overview of evolutionary algorithms: Practical issues and common pitfalls. *Information and Software Technology*, 43(14):817–831.

Whitley, D., Rana, S., Dzubera, J., and Mathias, K. (1996). Evaluating evolutionary algorithms. *Artificial Intelligence*, 85(1–2):245–276.

Whitley, D., Rana, S., and Heckendorn, R. (1998). The island model genetic algorithm: On separability, population size and convergence. *Journal of Computing and Information Technology*, 7(1):33–47.

Whitley, D. and Watson, J. (2005). Complexity theory and the no free lunch theorem. In Burke, E. and Kendall, G., editors, *Search Methodologies: Introductory Tutorials in Optimization and Decision Support Techniques*, pages Chapter 11, 317–339. Springer.

Whittaker, R. and Bush, M. (1993). Dispersal and establishment of tropical forst assemblages, Krakatoa, Indonesia. In Miles, J. and Walton, D., editors, *Primary Succession on Land*, pages 147–160. Blackwell Science.

Wienke, D., Lucasius, C., and Kateman, G. (1992). Multicriteria target vector optimization of analytical procedures using a genetic algorithm: Part I. Theory, numerical simulations and application to atomic emission spectroscopy. *Analytica Chimica Act*, 265(2):211–225.

Wilson, P. and Macleod, M. (1993). Low implementation cost IIR digital filter design using genetic algorithms. *IEE/IEEE Workshop on Natural Algorithms in Signal Processing*, Chelmsford, England, pages 4/1–4/8.

Winchester, S. (2008). *The Day the World Exploded*. Collins.

Winston, P. and Horn, B. (1989). *Lisp*. Addison Wesley, 3rd edition.

Woldesenbet, Y. and Yen, G. (2009). Dynamic evolutionary algorithm with variable relocation. *IEEE Transactions on Evolutionary Computation*, 13(3):500–513.

Wolpert, D. and Macready, W. (1997). No free lunch theorems for optimization. *IEEE Transactions on Evolutionary Computation*, 1(1):67–82.

Wolpert, D. and Macready, W. (2005). Coevolutionary free lunches. *IEEE Transactions on Evolutionary Computation*, 9(6):721–735.

Worden, L. and Levin, S. (2007). Evolutionary escape from the prisoner's dilemma. *Journal of Theoretical Biology*, 245(3):411–422.

Wright, A., Poli, R., Stephens, C., Langdon, W., and Pulavarty, W. (2004). An estimation of distribution algorithm based on maximum entropy. *Genetic and Evolutionary Computation Conference*, Seattle, Washington, pages 343–354.

Wright, S. (1987). *Primal-Dual Interior-Point Methods*. Society for Industrial Mathematics.

Wu, J. and Vankat, J. (1995). Island biogeography theory and applications. In Nierenberg, W., editor, *Encyclopedia of Environmental Biology*, pages 317–379. Academic Press.

Wu, S. and Chow, T. (2007). Self-organizing and self-evolving neurons: A new neural network for optimization. *IEEE Transactions on Neural Networks*, 18(2):385–396.

Wyatt, T. (2003). *Pheromones and Animal Behaviour: Communication by Smell and Taste*. Cambridge University Press.

Xinchao, Z. (2010). A perturbed particle swarm algorithm for numerical optimization. *Applied Soft Computing*, 10(1):119–124.

Yang, C. and Simon, D. (2005). A new particle swarm optimization technique. *International Conference on Systems Engineering*, Las Vegas, Nevada, pages 164–169.

Yang, C.-H., Tsai, S.-W., Chuang, L.-Y., and Yang, C.-H. (2011). A modified particle swarm optimization for global optimization. *International Journal of Advancements in Computing Technology*, 3(7):169–189.

Yang, S. (2003a). Non-stationary problem optimization using the primal-dual genetic algorithm. *IEEE Congress on Evolutionary Computation*, Canberra, Australia, pages 2246–2253.

Yang, S. (2003b). PDGA: The primal-dual genetic algorithm. *International Conference on Hybrid Intelligent Systems*, Melbourne, Australia, pages 214–223.

Yang, S. (2008a). Genetic algorithms with memory- and elitism-based immigrants in dynamic environments. *Evolutionary Computation*, 16(3):385–416.

Yang, S., Ong, Y.-S., and Jin, Y., editors (2010). *Evolutionary Computation in Dynamic and Uncertain Environments*. Springer.

Yang, S. and Yao, X. (2005). Experimental study on population-based incremental learning algorithms for dynamic optimization problems. *Soft Computing*, 9(11):815–834.

Yang, S. and Yao, X. (2008a). Population-based incremental learning with associative memory for dynamic environments. *IEEE Transactions on Evolutionary Computation*, 12(5):542–561.

Yang, S. and Yao, X. (2008b). Population-based incremental learning with associative memory for dynamic environments. *IEEE Transactions on Evolutionary Computation*, 12(5):542–561.

Yang, X.-S., editor (2008b). *Nature-Inspired Metaheuristic Algorithms*. Luniver Press.

Yang, X.-S. (2009a). Cuckoo search via Lévy flights. *World Congress on Nature & Biologically Inspired Computing*, Coimbatore, India, pages 210–214.

Yang, X.-S. (2009b). Firefly algorithm, Lévy flights and global optimization. In Ellis, R. and Petridis, M., editors, *Research and Development in Intelligent Systems XXVI*, pages 209–218. Springer.

Yang, X.-S. (2010a). Firefly algorithm, stochastic test functions and design optimisation. *International Journal of Bio-Inspired Computation*, 2(2):78–84.

Yang, X.-S. (2010b). Firefly algorithms for multimodal optimization. In Watanabe, O. and Zeugmann, T., editors, *Stochastic Algorithms: Foundations and Applications*, pages 169–178. Springer.

Yang, X.-S. (2010c). A new metaheuristic bat-inspired algorithm. In González, J., Pelta, D., Cruz, C., Terrazas, G., and Krasnogor, N., editors, *Nature Inspired Cooperative Strategies for Optimization*, pages 65–74. Springer.

Yang, Z., Tang, K., and Yao, X. (2008). Large scale evolutionary optimization using cooperative coevolution. *Information Sciences*, 178(15):2985–2999.

Yao, X. and Liu, Y. (1997). Fast evolution strategies. In Angeline, P., Reynolds, R., McDonnell, J., and Eberhart, R., editors, *Evolutionary Programming VI*, pages 151–161. Springer.

Yao, X., Liu, Y., and Lin, G. (1999). Evolutionary programming made faster. *IEEE Transactions on Evolutionary Programming*, 3(2):82–102.

Yen, G. (2009). An adaptive penalty function for handling constraint in multi-objective evolutionary optimization. In Mezura-Montes, E., editor, *Constraint-Handling in Evolutionary Optimization*, pages 121–143. Springer.

Yoshida, H., Kawata, K., and Fukuyama, Y. (2001). A particle swarm optimization for reactive power and voltage control considering voltage security assessment. *IEEE Transactions on Power Systems*, 15(4):1232–1239.

Yu, X., Tang, K., Chen, T., and Yao, X. (2009). Empirical analysis of evolutionary algorithms with immigrants schemes for dynamic optimization. *Memetic Computing*, 1(1):3–24.

Yuan, B., Orlowska, M., and Sadiz, S. (2007). On the optimal robot routing problem in wireless sensor networks. *IEEE Transactions on Knowledge and Data Engineering*, 19(9):1251–1261.

Zaharie, D. (2002). Critical values for the control parameters of differential evolution algorithms. *International Conference on Soft Computing*, Brno, Czech Republic, pages 62–67.

Zavala, A., Aguirre, A., and Diharce, E. (2009). Continuous constrained optimization with dynamic tolerance using the COPSO algorithm. In Mezura-Montes, E., editor, *Constraint-Handling in Evolutionary Optimization*, pages 1–23. Springer.

Zhan, Z.-H., Zhang, J., Li, Y., and Chung, H. (2009). Adaptive particle swarm optimization. *IEEE Transactions on Systems, Man, and Cybernetics – Part B: Cybernetics*, 39(6):1362–1381.

Zhang, J. and Sanderson, A., editors (2009). *Adaptive Differential Evolution*. Springer.

Zhang, Q., Sun, J., and Tsang, E. (2005). An evolutionary algorithm with guided mutation for the maximum clique problem. *IEEE Transactions on Evolutionary Computation*, 9(2):192–200.

Zhang, Q., Zhou, A., Zhao, S., Suganthan, P., Liu, W., and Tiwari, S. (2009). Multiobjective optimization test instances for the CEC 2009 special session and competition. Technical report. `www.ntu.edu.sg/home/EPNSugan/index_files/cec-benchmarking.htm`.

Zhu, Y., Yang, Z., and Song, J. (2006). A genetic algorithm with age and sexual features. *International Conference on Intelligent Computing*, Kunming, China, pages 634–640.

Zimmerman, A. and Lynch, J. (2009). A parallel simulated annealing architecture for model updating in wireless sensor networks. *IEEE Sensors Journal*, 9(11):1503–1510.

Zitzler, E., Deb, K., and Thiele, L. (2000). Comparison of multiobjective evolutionary algorithms: Empirical results. *Evolutionary Computation*, 8(2):173–195.

Zitzler, E., Laumanns, M., and Bleuler, S. (2004). A tutorial on evolutionary multiobjective optimization. In Gandibleux, X., Sevaux, M., Sörensen, K., and T'Kindt, V., editors, *Metaheuristics for Multiobjective Optimisation*, pages 3–38. Springer.

Zitzler, E., Laumanns, M., and Thiele, L. (2001). SPEA2: Improving the strength Pareto evolutionary algorithm. *EUROGEN 2001: Evolutionary Methods for Design, Optimisation and Control with Applications to Industrial Problems*, Athens, Greece, pages 95–100.

Zitzler, E. and Thiele, L. (1999). Multiobjective evolutionary algorithms: A comparative case study and the strength Pareto approach. *IEEE Transactions on Evolutionary Computation*, 3(4):257–271.

Zitzler, E., Thiele, L., and Bader, J. (2010). On set-based multiobjective optimization. *IEEE Transactions on Evolutionary Computation*, 14(1):58–79.

Zitzler, E., Thiele, L., Laumanns, M., Fonseca, C., and da Fonseca, V. (2003). Performance assessment of multiobjective optimizers: An analysis and review. *IEEE Transactions on Evolutionary Computation*, 7(2):117–132.

Zlochin, M., Birattari, M., Meuleau, N., and Dorigo, M. (2004). Model-based search for combinatorial optimization: A critical survey. *Annals of Operations Research*, 131(1–4):373–395.

INDEX

B

C

E

F

G

H

I

J

K

L

M

N

Q

R

T

U

V

W

X

Printed and bound by CPI Group (UK) Ltd, Croydon, CR0 4YY

16/04/2025

14658366-0003